Molecular and Cellular Mechanisms of Anesthetics

Dr. Carl Koller, the first to employ cocaine as a local anesthetic, 1884.
The photograph, taken in Vienna in 1885, was kindly provided by
his daughter, Mrs. Hortense Becker.

Molecular and Cellular Mechanisms of Anesthetics

Edited by

Sheldon H. Roth

Faculty of Medicine
The University of Calgary
Calgary, Alberta, Canada

and

Keith W. Miller

Massachusetts General Hospital
Harvard Medical School
Boston, Massachusetts

Plenum Medical Book Company
New York and London

Library of Congress Cataloging in Publication Data

Main entry under title:

Molecular and cellular mechanisms of anesthetics.

Based on the Third International Conference on Molecular and Cellular Mechanisms of Anesthesia, University of Calgary, June 13–15, 1984.
Includes bibliographies and index.
1. Anesthetics—Physiological effect—Congresses. 2. Anesthetics—Metabolism—Congresses. 3. Anesthetics—Side effects—Congresses. 4. Nerve tissue—Effect of drugs on—Congresses. I. Roth, Sheldon H. II. Miller, Keith W. III. International Conference on Molecular and Cellular Mechanisms of Anesthesia (3rd: 1984: University of Calgary) [DNLM: 1. Anesthetics—adverse effects—congresses. 2. Anesthetics—metabolism—congresses. 3. Cells—physiology—congresses. QV 81 M7175 1984]
RD82.M63 1986 615′.781 85-28274
ISBN-13: 978-1-4684-5035-4 e-ISBN-13: 978-1-4684-5033-0
DOI: 10.1007/978-1-4684-5033-0

Softcover reprint of the hardcover 1st edition 1986

233 Spring Street, New York, N.Y. 10013

Plenum Medical Book Company is an imprint of Plenum Publishing Corporation

Foreword

B. Raymond Fink

Sheldon Roth and Keith Miller have asked me to record that the Third Conference on Molecular and Cellular Mechanisms of Anesthesia was held in Calgary last May "in my honor." Such was my dear friends' gracious way of continuing a series that began at the University of Washington, where I hosted two, four, or five previous ones, [1, 3–6] depending on how far back one wishes to count. At that, Seattle took up where Paris left off in 1951.[2]

These occasions create their own unforgettable memories. This book captures the fine, invigorating ambience of the University of Calgary and the exciting explorations and companionship of a gathering in a frontier territory of neuroscience. So, floreant symposia. They have progressively refined the quarry, from pathway to synapse to lipoprotein membrane to receptor and single channel, in heuristic convergences of neuronal physiology, biochemistry, and pharmacology. Nevertheless, the anesthesiologist in me senses a certain disquiet, a certain claustrophobia provoked by the narrow confines of micropipettes. How much more tubular must tunnel vision become before the desired broad view emerges? At present, the advances in molecular neurobiology seem continually to increase the apparent complexity of the total problem and the conceptual distance between the reductionists in the laboratories and the holists in the operating rooms. Happily, what is also growing is the excitement in trying to bridge the gap.

Perhaps it would be timely to regard general anesthesia not as a state but as a syndrome. It is worth recalling yet again that the extraordinary diversity of molecules that can induce general anesthesia is matched by the diversity in detail of the anesthetic syndromes they produce, and more than matched by the diversity of proteins and lipoproteins in neuronal membranes. Thus it may be an oversimplification to regard general anesthesia, or even the loss of consciousness, as a single "state" of depression of the nervous system, and closer to reality to regard each as a pharmacological syndrome in a pharmacological spectrum of syndromes. If this is so, then pressure reversal of anesthesia presumably also possesses, or at least traverses, an underlying hidden multiplicity. In the old days of diethyl ether we used to divide anesthesia into a number of stages and planes distinguished on the basis of motor responses, and made guesses or experimental inferences about the related sites of central nervous depression. It was a time of unified theories, focusing on which phase—aqueous,

B. Raymond Fink • Department of Anesthesia, University of Washington, Seattle, Washington 98195.

lipid, or lipoprotein—was the seat of what one might call the biophysical field effect of anesthetics. A needle-eyed reappraisal is now proceeding, centering on receptor–ligand interactions, where the diversity alluded to previously can be easily accommodated.

One of the most striking features of induction of anesthesia with diethyl ether was the progressive impairment of speech function. Slurred speech was the earliest sign of the assault on the central computer, soon merging into incoherence and utter inability to heed the physician. The computer was still able to output sounds, but they didn't make sense. On a cold day, when vapor pressure under the open drop mask stayed low, more minutes might go by before verbal activity completely ceased and the patient lapsed into unconsciousness, although not yet ready for the knife. At which level are we to seek the molecular mechanisms of anesthesia? Anesthesia takes the patient down a long staircase, where whole flights can be jumped or negotiated via the balusters without prejudice to a happy landing. On just which floor is our micropipette installed?

In seeking a molecular understanding of anesthesia we are in effect seeking a molecular knowledge of our innermost selves. The resolution of the problem is not around the corner. One may be confident that conferences on molecular mechanisms of anesthesia will continue far into the future. The mystery of anesthesia, like that of consciousness, has not yet yielded to molecularity. Humiliating, godlike ignorance is wielded daily by the anesthesiologist who puts the patient's person wholly or partly (reversibly) in limbo. Was Alcmaeon of Croton (c. 599 B.C.) so wide of the mark in considering the brain to be the central fireplace of the human microcosm? Or Plato (427–347 B.C.), when he made it the seat of the eternal, rational soul? I hear you mutter, "Measure me a soul." I agree. Well then, measure me a consciousness. Do I sense laughter? I hope so. Laughter, were it measurable deep down in the membranes, might give us a clue to consciousness. Anesthetized patients don't laugh. Full recovery has not occurred until there is laughter, or at least a smile.

Let's try another argument. What is the nervous system, any nervous system, but an energy computer? Feel a surface, prick the skin, sense a ray, smell a rat. Every disturbance transduces energy to information, a bioenergetic hybrid amenable to calculating operations by the nervous system. The lability of bioenergetic systems is surely why a living computer can be anesthetized, but an IBM one cannot. I take a certain comfort in the thought that electronic computers of the *n*th generation remain doomed to stupidity if, like our present ones, they continue to enjoy a free lunch. No electronic computer has to forage or generate its energy. Yet foraging is of the very essence of animate intelligence and a central element in the operation of natural selection. Anesthesia puts a stop to foraging. So, I suggest, an acid test of true intelligence in a computer will be: can it be anesthetized? Can its operations be suspended by permeating its circuits with a drug? If technology ever achieves that, it will have produced an artificial living computer, operating with a built-in supply of energy whose maintenance is part of the program, so that everything the computer calculates has to take into account its self-interested renewal and storage of energy. Such a computer will possess feelings. In my book, true AI spells the end of computer free lunch and the beginning of computer angor animi: the quasihuman intelligent computer will be afraid to die.

REFERENCES

1. *Cellular Biology and Toxocity of Anesthetics.* Baltimore, Williams & Wilkins, 1972.
2. *Mécanismes de la Narcose.* Paris, Coll. Internat. C.N.R.S., 1950.
3. *Molecular Mechanisms of Anesthesia,* 1. New York, Raven Press, 1975.
4. *Molecular Mechanisms of Anesthesia,* 2. New York, Raven Press, 1980.
5. Neurophysiology in relation to anesthesiology. *Anesthesiology* 1967; 28:1–200.
6. *Toxicity of Anesthetics.* Baltimore, Williams & Wilkins, 1968.

Preface

One hundred and thirty-eight years after the first demonstration of the use of ether at the Massachusetts General Hospital, a distinguished group of anesthesiologists and scientists representing diverse disciplines met on the campus of the University of Calgary for the Third International Conference on Molecular and Cellular Mechanisms of Anesthesia. In his Foreword in this volume, B. Raymond Fink has outlined the history of this series of meetings, which he himself originated. He has now passed the torch to us, and we will attempt to preserve the tradition he has so ably created. This volume aims to provide an overview of the progress achieved in the last 5 years, to summarize the present situation, and to indicate future trends. We hope that it will prove a useful source to workers in the field, and to this end we have also included reference tables in an appendix.

The mechanisms of general anesthesia still remain a mystery because of our rudimentary understanding of how the brain functions as a whole and of what regulates consciousness. We thus lack a clear definition of general anesthesia. Physiological approaches to the problem are dealt with in the first section of the book and lead to some general conclusions about actions at the cellular level on both inhibitory and excitatory synapses. Whether all the effects reported are responsible for general anesthesia remains to be ascertained.

The second and third sections deal with actions at relatively simple levels of organization, namely, postsynaptic membranes and excitable channels, where the targets are better defined and mechanisms of action can be approached by postulating models or schemes that describe the interactions between anesthetics and their targets. At some of these targets, notably the acetylcholine receptor and the sodium channel, good progress is being made, and furthermore, studies in molecular biology and in techniques for reconstituting purified proteins into lipid bilayers are now underway, which will enable the underlying mechanisms to be elucidated in even more detail in the future.

In the fourth section detailed studies using both structural and functional methods aim at achieving molecular descriptions in even simpler model systems. It is to be hoped that the power of such methods will eventually become directly applicable to more physiologically relevant structures and lead to more complete descriptions of the underlying mechanisms.

Studies of the pressure reversal of general anesthesia have provided impetus to the field for over a decade. In spite of the experimental difficulties, work described in the fifth section continues to exploit this tool to refine our knowledge of the physiological and cellular site of general anesthetic action.

Although there is a definite requirement for anesthetics in surgical procedures, they provide no therapeutic role, and the risk/benefit tradeoff to the patient has to be approached almost entirely from the level of reducing risk. The last two sections thus focus on the toxicities of local and general anesthetics and the processes underlying them.

Finally, in the Appendix we have gathered together some physicochemical and pharmacological data for gaseous, volatile, alcohol, and barbiturate anesthetics. Many of these data are often difficult to find, and we hope this compilation will fulfill the needs expressed by many in the field. If this is so, please support our effort and help improve this feature of future volumes by offering your suggestions and contributions.

Funds for the Third International Conference on Molecular and Cellular Mechanisms of Anesthesia, University of Calgary, June 13–15, 1984, were provided by the Alberta Heritage Foundation for Medical Research, Medical Research Council of Canada, Travel Alberta, University of Calgary, Burroughs–Wellcome Inc. of Canada, Janssen Pharmaceutica, Ohio Medical Anesthetics, Organon Canada Ltd., Perdue Frederick Inc., the Upjohn Company of Canada, and Winthrop Laboratories. We extend our thanks to these donors whose contributions enhanced the quality of the meeting.

We are also grateful to the following departments of anesthesia, which helped to defray some of the costs incurred in publishing this book:

Harvard Medical School, Boston, and affiliated hospitals:
- Beth Israel Medical Center, Boston
- Brigham and Women's Hospital, Boston
- Children's Hospital Medical Center, Boston
- Massachusetts General Hospital, Boston

Beth Israel Medical Center, New York
Columbia University, New York
Johns Hopkins Hospital, Baltimore
Stanford University School of Medicine, Stanford
University of Alabama Medical Center, Birmingham
University of Calgary, Foothills Provincial Hospital, Calgary
University of California, San Francisco
University of Iowa College of Medicine, Iowa City
University of Massachusetts Medical Center, Worcester
University of Virginia Medical Center, Charlottesville
University of Washington, Seattle

In spite of all the work reported in this book, the central mystery of general anesthesia remains to tantalize and challenge us in future years, but on the periphery encouraging progress is being made. Understanding gained here will both increase our understanding of the clinical pharmacology of these agents and point to new approaches for solving the central problem.

Sheldon H. Roth
Keith W. Miller

Calgary and Boston

Contents

I. Cellular Actions of Anesthetics

II. Postsynaptic Actions of Anesthetics

I

CELLULAR ACTIONS OF ANESTHETICS

1

Cellular and Synaptic Effects of General Anesthetics

Krešimir Krnjević

ABSTRACT. Ideas about the effects produced by anesthetics have ranged widely, from very general depressions of either metabolism or excitability by a ubiquitous physicochemical change in membrane properties, to highly specific mechanisms, such as a presynaptic reduction in transmitter release or a selective enhancement of inhibitory transmission.

The latter has gained much prominence in recent years, through the demonstration that barbiturates (and possibly some other narcotics) enhance both the action of GABA and inhibitory postsynaptic potentials (IPSPs) at a variety of sites. The net result is a striking potentiation and especially prolongation of IPSPs. The stabilizing influence of stronger inhibition is therefore a major feature of anesthesia.

Another way of stabilizing cell potentials is through increased K conductance (G_K). Several kinds of K^+ currents have been described all of which tend to prevent depolarization and firing. Thus neurons can be considered to have two essential limiting states: in one, characterized by very low K^+ conductances, cells are highly responsive to any input. In the second, large outward currents of K^+ prevent responses to any but the most powerful inputs. A predominance of the first state in the brain (?cortex) could be a requisite for conscious awareness, whereas a shift to the second state could provide a cellular basis for loss of awareness. Because the G_K of most cells is to an important extent controlled by internal Ca^{2+}, a relatively simple explanation for a shift to the second state would be diminished sequestration, outward transport, or effectiveness of intracellular Ca^{2+}, as a result of anesthetic action. These and other relevant mechanisms are discussed.

For the study of anesthesia we need to have some idea of what we are looking for. Like many other drugs, anesthetics if applied in large quantity will inactivate all neurons. But most drugs are not useful general anesthetics. Two essential questions therefore are (1) What features of neural activity (and where) are especially characteristic of the conscious state? and (2) What particular aspects of anesthetic action would interfere especially with such neural activity?

ANESTHESIA

By definition, anesthesia is the loss of sensation and, when general, usually includes a loss of awareness of surroundings, as well as of consciousness. The latter is entirely subjec-

Krešimir Krnjević • Departments of Anaesthesia Research and Physiology, McGill University, Montréal, Québec H3G 1Y6, Canada.

tive and therefore can only be reported on or inferred indirectly (for some useful recent discussion see Refs. 40, 42, 86). The conscious state is not all-or-none; it appears to vary both quantitatively and qualitatively. For example, in sleep, there may be substantial loss of awareness of surroundings but without loss of self-consciousness (especially during dreams). Marked reversible changes in level of conscious state typically occur (a) physiologically, at the onset of sleep or awakening; (b) when the brain is deprived of an adequate supply of O_2 and other metabolic substrates; and (c) when certain drugs are given.

Mechanism of Consciousness

That the brain is somehow essential for the conscious state is largely beyond dispute. It is less clear though that the conscious state depends on the function of a certain part of the brain and, if so, where this is situated. There is wide agreement that the cerebral cortex plays a vital role, but there is also reason to believe that this cannot reach its full expression without an essential contribution of some deeper or lower region(s) of the brain.[83]

Cortical Activity Is Involved in Consciousness

Is the critical feature simply a general enhancement of neural activity? The answer is clearly no, on the basis of much recording of ongoing unit activity.[98] Yet there is a tendency for unit firing to be reduced in slow-wave sleep and even more during anesthesia[79]; particularly clear and consistent changes are observed if output units (cortifugal cells, such as cells of origin of the pyramidal tract) and nonoutput units are analyzed separately.[98] On behavioral arousal, output units show a characteristic series of changes, including an increase in the probability of firing, of antidromic invasion, and of synaptic activation. But nonoutput cells tend to show opposite changes, as might be expected if many of them are inhibitory interneurons (as argued very plausibly by Steriade[98]).

Since Bremer's pioneer observations,[17] numerous authors have confirmed that in sleep or anesthesia, the primary (shortest-latency) sensory-evoked responses are either unchanged or enhanced, but the subsequent components of the field response (or repetitive unitary discharge) are substantially reduced or may even vanish.[1,16,18,29,35,41,49,67] Two examples of such changes are illustrated in Fig. 1A and B.

Significance of Repetitive Discharge

It has been a long-held belief that relatively prolonged, repetitive neuronal activity is essential for both consciousness and the laying down of memory,[39,47,49] presumably because it takes time to produce whatever "structural" changes embody memory traces. In keeping with this is the finding by Libet[66] that direct cortical stimulation (in man) by relatively strong *single* shocks fails to reach consciousness, whereas much weaker repetitive stimulation does if it lasts a minimum of about one tenth of a second. Judging by Libet's data, the field responses evoked in the cortex by the effective tetani do not show either marked progressive facilitation or prolonged afterdischarges. On the other hand, the excited cells appear to be situated deep in the cortex, and therefore, conscious experience may well require the activation of corticofugal cells.[83,98]

According to recent investigations on neuronal properties, repetitive activity, especially in "bursts," can be correlated with an increased probability of generating Ca^{2+}-mediated action potentials[105] and, therefore, substantial Ca^{2+} influx[61,62,73] (Fig. 2). In view of the manifold intracellular messenger functions of Ca^{2+},[88] one can expect that Ca^{2+} can change the functional properties of a variety of membrane or other cellular proteins, either directly,

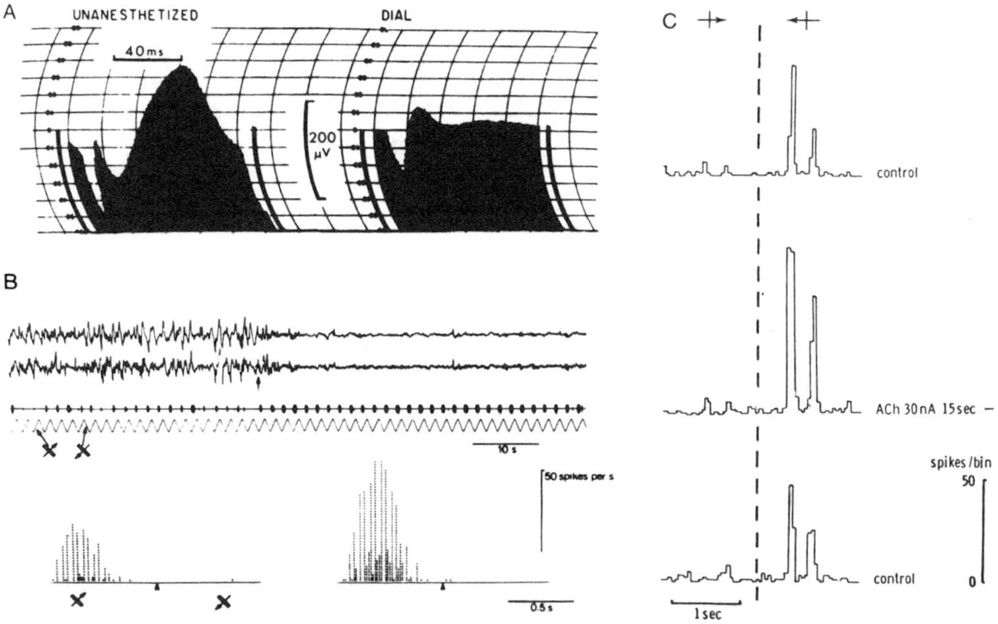

Figure 1. Cortical sensory responses, arousal, and acetylcholine (ACh). (A) Averaged responses in auditory cortex of cat to repeated clicks, before and after injection of anesthetic (Dial, 75 mg/kg i.p.) (from Fig. 6, Goldstein *et al.*[41]). Anesthetic eliminates particularly the late component of response. (B) Responses of unit in layer 6 of striate cortex are enhanced during arousal. Unit firing is illustrated in line 3 from above. Regularly reversing movement of stimulus in visual field is illustrated in line 4, and histograms below show average of 40 responses before and after arousal (cf. EEG traces above) (from Fig 2, Livingstone and Hubel[67]). (C) In a comparable experiment, histograms of firing of directionally sensitive unit in visual cortex of cat show marked enhancement during iontophoretic release of ACh, without loss of field specificity (Fig. 2 from Sillito and Kemp[96]). Thus, both arousal (B) and local ACh (C) produce a similar facilitation of cortical response, without changing its functional characteristics.

or through the activation of calmodulin and related protein kinases. This would provide a mechanism for shorter- or longer-term modifications in neural behavior. For example, a detailed scheme of the well-known phenomenon of long-term potentiation in the hippocampus, which is widely seen as a paradigm of learning, postulates the following main steps: Ca^{2+} influx, activation of a Ca^{2+}-sensitive protease, and consequent increase in the density of glutamate receptors on hippocampal cells.[68] Another thoroughly investigated mechanism of learning in the mollusk *Hermissenda* appears to be mediated by Ca^{2+} influx and a resulting long-term *depression* of a fast-adapting K^+ current.[2]

Conditions Fostering Repetitive Activity

In most cells (and particularly neurons), the membrane potential is normally stabilized by a high conductance for K^+ (G_K), which maintains a relatively negative resting level and ensures that any depolarizing trend is countered by an efflux of K^+.[48] Any tendency toward repetitive firing is further opposed by the fact that most K^+ channels appear to be voltage-dependent; that is, the probability of their being open increases with depolarization. One can therefore state axiomatically that a high responsiveness and ability to discharge repetitively require a diminished availability of K^+ channels. Indeed, the excitability (the capacity to generate an all-or-none action potential) of an extraordinarily wide range of cells throughout the animal kingdom can be much enhanced (or revealed, when normally not evident) by agents

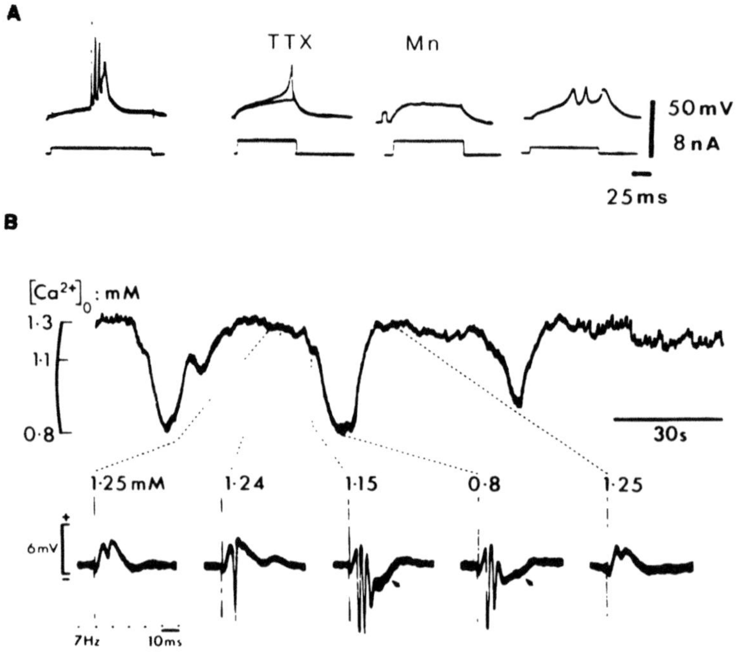

Figure 2. Major Ca currents are generated during hippocampal burst discharges. (A) Calcium component in CA3 pyramidal cell spike burst evoked by depolarizing pulse persists in presence of tetrodotoxin (TTX), but it is abolished by Mn. Final trace shows recovery of Ca spikes in Mn-free solution (from Fig. 1 in Wong and Prince[105]). (B) Large reductions in extracellular free Ca^{2+} concentration, recorded with Ca^{2+}-sensitive microelectrode in CA1 region of the hippocampus (*in situ*), occurred periodically during fimbrial tetanic stimulation (at 7 Hz, maintained throughout). As illustrated by oscilloscope traces below, large drops in extracellular Ca^{2+} coincided with bursts of population spikes (from Fig. 10, Krnjević *et al.*[62]).

such as Ba^{2+} and tetraethylammonium (TEA), which selectively suppress G_K.[3,33,36,59,80,85,95,101,104]

In the cerebral cortex, Ba^{2+} and TEA not only facilitate action potentials or enhance repetitive firing,[38,59,60] but they both tend to promote Ca^{2+}-mediated action potentials[23,30] (Fig. 3). Of even greater physiological interest, however, is the well-established fact that the endogenous neurotransmitter acetylcholine (ACh) acts very much like Ba^{2+} in reducing both resting and voltage dependent G_K in neocortex[60] and archicortex.[10,11,32,44] ACh thus greatly facilitates firing, after discharges, and presumably Ca currents of many cortical cells, including identified corticofugal neurons,[55] evoked by various kinds of excitatory inputs[55,63,96,106] (Fig. 1C).

From a variety of investigations, there is strong evidence that most regions of the cortex receive a specific cholinergic innervation.[28,51] This has its origin in the basal region of the forebrain and forms an extensive network of fibers throughout all regions of the cortex.[37,57,58,70] By releasing ACh, this cholinergic innervation facilitates the responses of cortical neurons to various inputs[56] (Fig. 4). Its essential role for higher mental function has been revealed by the severe loss of memory and other higher cortical function that characterizes Alzheimer's (and related) disease, which can be ascribed mainly to degeneration of these cholinergic fibers.[25] In keeping with this is the drastic and permanent loss of consciousness that follows selective interruptions of vascular supply to the deep region of the forebrain.[27]

In summary, there is good reason to believe that heightened responsiveness of (some)

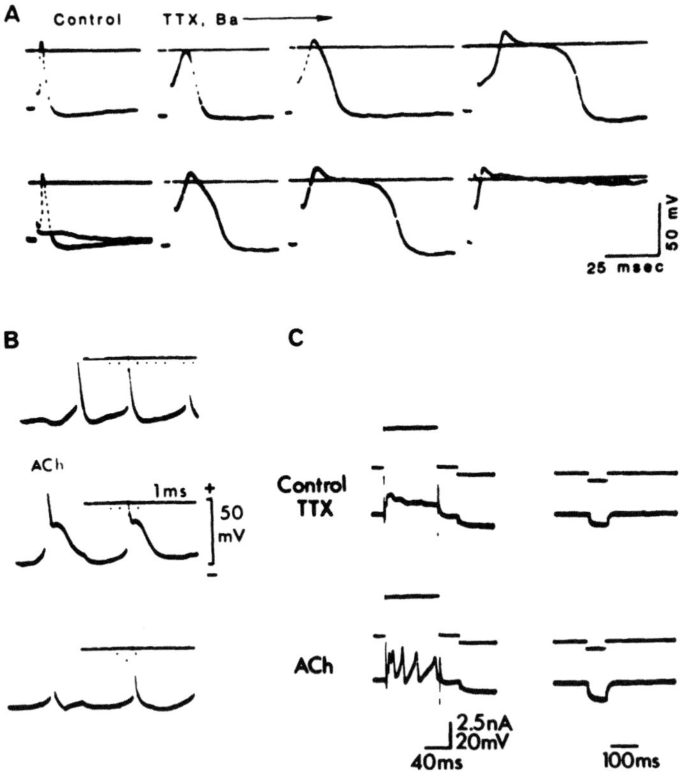

Figure 3. Barium and ACh both inactivate K currents and so facilitate Ca currents in cortical neurons. (A) Action potentials from two cortical cells in culture showed marked increase in action potential duration in saline containing Ba (5–15 mM) and TTX (q0^{-7}–10^{-6} g/ml) (from Fig. 5 in Dichter *et al.*[30]). (B) Action potentials of pericruciate neuron in cortex of cat show delayed repolarization during application of ACh (from Fig. 6 in Krnjević *et al.*[60]). (C) After treatment of hippocampal slice with TTX, ACh restores ability of CA1 neuron to fire spikes, presumably mediated by calcium. Note, in traces at right, increased membrane resistance in presence of ACh (Fig. 5 from Benardo and Prince[10]).

corticofugal neurons is an essential feature of awareness—perhaps because memory traces can then be laid down by intracellular processes triggered by Ca^{2+} influx. The increase in responsiveness is made possible by a lowered K^+ permeability of cortical neurons, under the influence particularly of acetylcholine released by cholinergic fibers from basal forebrain nuclei. One may further postulate that these nuclei (basalis, of medial septum, and of diagonal band) are the principal elements of the centrencephalic system postulated by Penfield[83] as essential for consciousness. Any mechanical or functional interference with this region of the forebrain (or interruption of the main inputs from below that maintain its tonic activity) can thus be expected to disturb the cholinergic activation of the cortex and therefore the conscious state.

ANESTHETIC ACTION

If, as argued previously, a condition of low G_K of some or many cortical neurons is an essential feature of the conscious state, the converse should be true; a lowering of consciousness or anesthesia should be induced by agents that facilitate outward currents, notably of K^+ (but also of Cl^-, see below). What factors control the opening of K^+ channels?

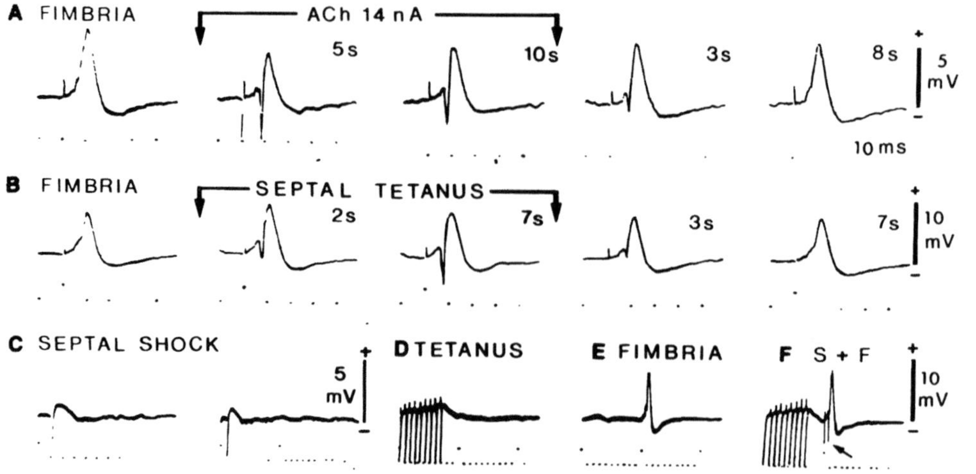

Figure 4. Acetylcholine, applied in hippocampal CA1 region, and septal tetanic stimulation cause similar strong facilitation of population spikes (in rat under urethane). Neither single nor tetanic septal stimulation alone evoked more than minimal field response in hippocampal CA1 area (C and D), but preceding septal tetanus markedly facilitated response to single fimbrial shocks as illustrated in E and F, as well as A and B at faster speed (Fig. 1 from Krnjević and Ropert[56]).

Membrane K^+ Channels

Various types of K^+ channels have been described, including the delayed rectifier[48] and the Ca^{2+}-dependent channel.[54,69] In view of the recent demonstration that the Ca^{2+}-dependent K^+ channel is also strongly voltage-dependent[9,64,92] (Fig. 5), it is no longer so evident that there is an essential difference between these two types of K^+ channels (or indeed most other types that have been described). At any rate, the Ca^{2+}-dependent channels may functionally predominate, owing to their widespread (almost ubiquitous) distribution in both excitable and nonexcitable cells and their exceptionally high unitary conductance.[64]

Changes in intracellular free Ca^{2+} concentration ($[Ca]_i$) can therefore be expected to lead to rapid corresponding changes in membrane G_K and therefore excitability. The maintenance of the very low normal $[Ca]_i$—through Ca^{2+} outward transport, as well as sequestration in mitochondria and other intracellular structures[4,5,14,19,88]—requires much energy; hence, any conditions that reduce the supply of vital substrates (such as O_2 and glucose), or agents that somehow interfere with metabolism and Ca^{2+} removal, should promote a rise in $[Ca]_i$ and so raise G_K.

Evidence That General Anesthetics Lower Neuronal Excitability by Raising G_K

In many iontophoretic studies of cortical neurons[21,55] and neurons in other brain areas,[13,15,26,34,72,84] ACh-evoked excitatory responses have proved to be particularly susceptible to general anesthetics (Fig. 6). This has been interpreted as in keeping with opposite selective effects of ACh and anesthetics on neuronal G_K.[52]

More direct evidence that anesthetics tend to hyperpolarize cells by enhancing G_K has come from intracellular recordings: in several types of invertebrate neurons,[22,24,87,93,84] in the particularly relevant mammalian cortical cells, by Carlen *et al.*[20] and by Nicoll and Madison.[78] The latter were performed on hippocampal cells, using doses of general anesthetics in, or close to, the range effective for clinical anesthesia. In Nicoll and Madison's

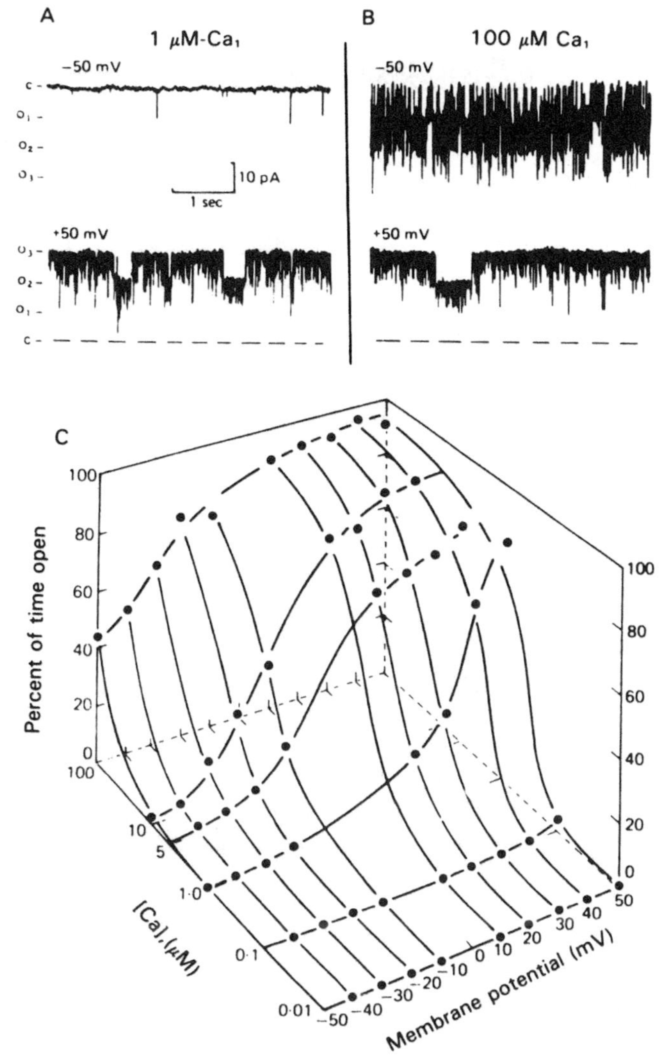

Figure 5. Intracellular free Ca^{2+} activates K conductance. In patch-clamp experiments on cultured muscle cells: (A) 1 μM [Ca] activates all-or-none currents when membrane is held at +50 mV but not at −50 mV. (B) Higher [Ca] (100 μM) activates K channels even at −50 mV. (C) Composite graph illustrates clear voltage-dependence of the Ca-sensitive K channels. Note in A and B that K currents are downward at −50 and upward at +50 mV (from Fig. 8 in Barrett *et al.*[9]).

experiments (Fig. 7), a wide variety of anesthetics (including barbiturates, ether, halothane, alphaxolone, and chloralose) had a consistent hyperpolarizing effect—clearly ascribable to enhanced G_K—in both the frog spinal cord and rat hippocampal slices. Moreover, there was a clear correlation between the anesthetic potency and the hyperpolarizing potency (Fig. 7D).

These results thus strongly indicate that general anesthetics do activate G_K of cortical neurons, and they are consistent with the possibility that this effect is caused by an increase in $[Ca]_i$.

Anesthetics and Intracellular Free Ca^{2+}

Direct evidence that anesthetics, in clinically effective doses, significantly interfere with Ca^{2+} sequestration and/or outward transport is both limited and controversial. Impressive reductions in mitochondrial Ca^{2+} uptake by hexobarbitone, halothane, ether, chlo-

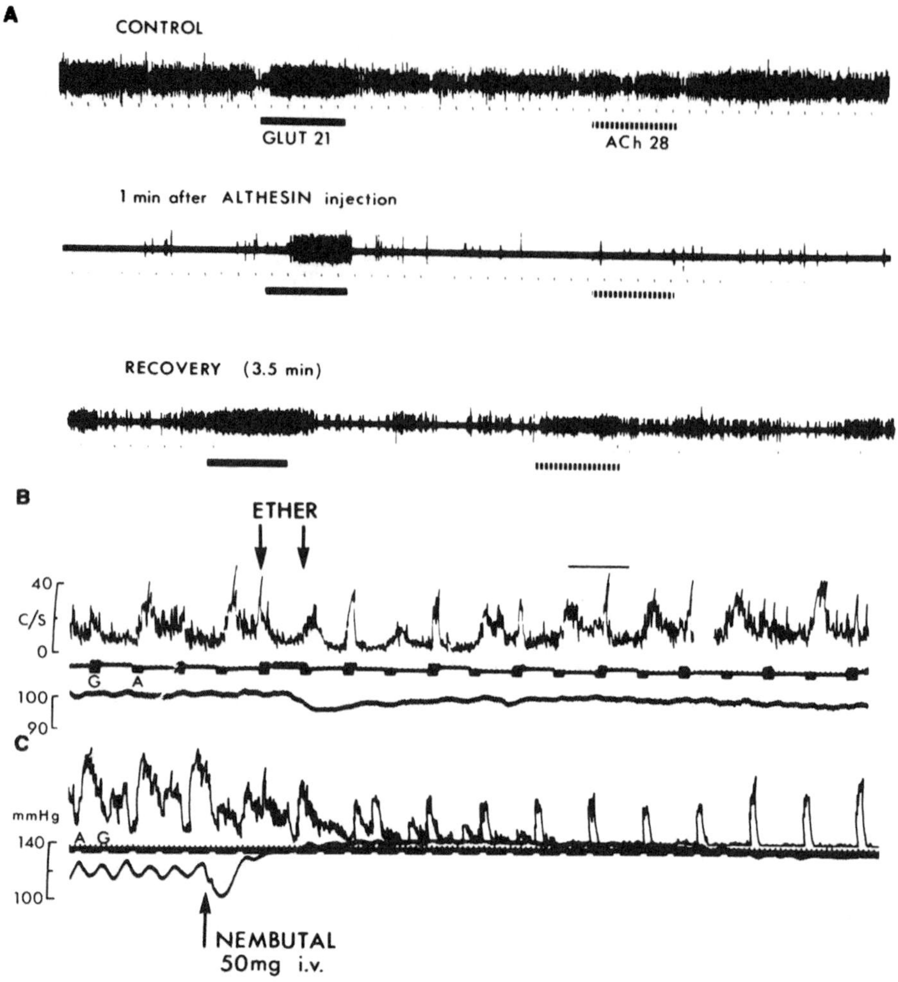

Figure 6. General anesthetics selectively block ACh-evoked excitation of cortical neurons (in cats). (A) Ongoing unit firing and excitation evoked by glutamate (GLUT) or ACh (iontophoretic doses are indicated in nA). Note striking reduction in ongoing activity and loss of response to ACh after Althesin injection (0.1 mg/kg i.v.). (B) Ratemeter traces of unit-firing in cat cortex evoked by alternating applications of glutamate (G) and ACh (A). The inhalation of ether for about 1 minute caused a temporary, selective reduction of responses to ACh and a relative increase in responses to glutamate. (C) Similar unit responses were evoked in cat cortex by alternating applications of ACh (A) and glutamate (G); in this case, the signal for applications is a gap in the signal trace below. Intravenous injection of Nembutal was followed by a prolonged disappearance of responses to ACh, and only sharp responses to glutamate remained (unpublished observations by R. Catchlove, K. Krnjević, H. Maretić, and E. Puil).

roform, alphaxalone, and propanidid have been observed *in vitro*.[65,91,99,100] Various anesthetics also markedly reduced brain mitochondrial Ca^{2+} uptake in experiments on rats *in vivo*[99] (Table I). The experiments on brain mitochondria are particularly relevant; contrary to some earlier opinion,[12,74] it is now clear that brain mitochondria probably regulate $[Ca]_i$ at levels very close to the normal physiological value.[75,76] On the other hand, some authors (Willow and Bygrave[102]) have seen the opposite effect, that is, increased mitochondrial uptake in the presence of barbiturates. One has to keep in mind that other nonmitochondrial intracellular structures also play a significant role in binding Ca^{2+}: such as endoplasmic

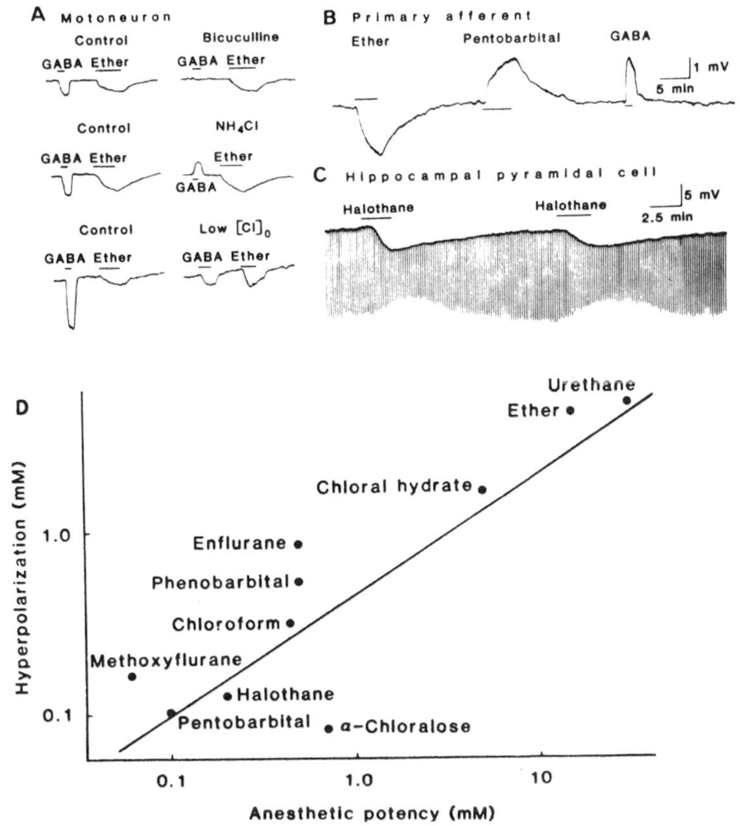

Figure 7. General anesthetics cause hyperpolarization of vertebrate neurons by increasing K conductance. (A) Sucrose gap records from frog ventral root illustrate hyperpolarizing action of GABA (0.1 mM) and ether (20 mM). In contrast to effect of GABA, which was blocked, reduced, or reversed by bicuculline, NH_4Cl, or removal of extracellular Cl (respectively), the effect of ether was virtually unchanged. (B) Sucrose gap records from frog dorsal root again show a hyperpolarizing effect of ether (40 mM) in contrast to GABA and pentobarbital (both 0.1 mM), which depolarize by increasing Cl conductance. (C) Rat hippocampal pyramidal cell *in vitro* shows repeatable hyperpolarizing effect of halothane (2 mM) and accompanying fall in membrane resistance. (D) Graph demonstrates linear relationship between minimum dose required to produce hyperpolarization of spinal motoneurons and clinically effective dose of various agents (from Figs. 1 and 2, in Nicoll and Madison[78]).

reticulum, other microsomes, and soluble Ca^{2+}-binding proteins. And, of course, outward transport by Ca^{2+}-ATPase and Ca^{2+}/Na^{+} exchange are also important regulators of $[Ca]_i$.[5,14,19,31,88] Therefore, the overall effect of anesthetics on *free* Ca^{2+} is the most critical feature. Indeed, there is evidence that nonmitochondrial Ca^{2+} uptake is depressed by anesthetics.[46] Even small changes in $[Ca]_i$, possibly at or very near the limit of resolution, may be sufficient to alter a cell's firing characteristics, because of the marked voltage-dependence of the Ca^{2+}-sensitive K^+ channel. According to Barrett *et al.*'s[9] patch-clamp observations on single K^+ channels in muscle cells, near the resting potential a large change in probability of channel opening requires $[Ca]_i$ to rise well above 10^{-6} M; but when the potential is at +50 mV, changes in $[Ca]_i$ well below 10^{-7} M would greatly increase G_K (Fig. 5). Careful measurements of $[Ca]_i$ are needed, preferably by a technique, such as the Ca^{2+} microelectrode, that is not itself sensitive to the anesthetics. An alternative approach is the use of an appropriate Ca^{2+} sensitive photoprobe like QUIN 2. (See Morris, this volume, Chapter 6.)

Table I. Effect of Althesin, Hexobarbitone, and Halothane on the Ratio for the Distribution of $^{45}Ca^{2+}$ between Mitochondrial and Postmitochondrial Fractions in Rat Brain in Vivo[a,d]

Treatment	Relative content of $^{45}Ca^{2+}$ (mitochondrial fraction/ postmitochondrial fraction)
Control	6.78 ± 0.76
Althesin	2.79 ± 0.87[b]
Hexobarbitone	2.29 ± 0.74[b]
Halothane	3.95 ± 0.58[c]

[a]Three general anesthetics reduce Ca^{2+} transport into mitochondria. Values in table are means and standard errors, the number of animals being 10 for Althesin and hexobarbitone, and six for halothane.
[b]Significantly different from controls at $p < 0.005$.
[c]Significantly different from controls at $p < 0.01$. Doses of anesthetics were: Althesin, 12 mg/Kg i.v.; hexobarbitone, 90 mg/Kg i.p.; halothane, 3% in oxygen.
[d] From Sweetman and Esmail.[99]

Other Mechanisms of Enhancement by Anesthetics

One such is clearly indicated by Baker and Schapira's[6] finding that aequorin fluorescence is consistently enhanced by a wide variety of anesthetics, even when $[Ca]_i$ is kept constant by appropriate buffering. The authors therefore argue that since aequorin is probably typical of other proteins that bind Ca^{2+} with a high affinity, a disturbance of membrane proteins that regulate G_K could explain an enhancement of the Ca^{2+} affinity of K^+ channels by anesthetics (independent of any changes in $[Ca]_i$). A comparable sensitization of red blood cell K^+ channels to Ca^{2+} seems to be caused by the drug propranolol.[97]

A further possibility is that anesthetics can increase the membrane permeability to H^+ and K^+ by directly altering the physical characteristics of phospholipid bilayers, without acting on specific K^+ channels.[7,50] The latter scheme, on the other hand, goes against abundant evidence that membrane K^+ channels are discrete protein units, with all-or-none opening characteristics,[71,92] which can be isolated and reinserted functionally into artificial lipid membrane.[64]

Other Hyperpolarizing Mechanisms of Anesthetics

So far the emphasis in this paper has been mainly on modulations of neuronal responsiveness through changes in G_K, in other words, a mechanism intrinsic to the cell. But a state of low membrane conductance (resulting from low G_K) and therefore high responsiveness (here equated with "consciousness") can also be very effectively reversed by a high Cl conductance (G_{Cl}). In most neurons that have been studied, resting G_{Cl} shows little voltage-dependence and therefore plays no major role in *intrinsic* regulation of cell firing, but it is often strikingly and specifically enhanced by synaptic inhibitory action. Thus, in the cortex, powerful and widespread Cl^--mediated IPSPs are generated by numerous GABA-releasing inhibitory cells.[53]

A remarkable property of GABA receptors is that they can be readily modulated by a variety of agents, notably benzodiazepine tranquilizers acting on a specific receptor that appears to be closely associated with the GABA receptor site,[43] as well as by barbiturates at a different but also related site.[81,90,103] The evidence for a functionally significant interaction between GABA receptor and barbiturates came originally from the observation that pentobarbitone potentiated greatly not only IPSPs but also the action of iontophoretically applied GABA.[8,77] A particularly striking aspect of this action was a pronounced increase in duration of both IPSPs and GABA action. More detailed studies of pentobarbitone, particularly by Barker and his colleagues, have shown that when applied in a very high dose the anesthetic can even directly enhance G_{Cl}. Chapters 2, 3, and 4 in this volume provide further much more detailed descriptions of the barbiturate-mediated potentiation of GABA and IPSPs. The possibility that enhanced inhibition in the cortex contributes significantly to loss of awareness thus has strong experimental evidence in its support (as well as a more speculative behavioral basis[82]).

There is evidence that a comparable kind of potentiation is also produced by several other anesthetic agents, including gases like halothane and steroids such as alphalaxone.[45,94] It is of much interest that the widely used general anesthetics should greatly facilitate both intrinsically and extrinsically generated outward currents (through G_K and G_{Cl}) and thus counteract the low G_K state of high responsiveness that may be an essential feature of the conscious state. On the face of it, this common property of most anesthetics could account for their usefulness in clinical (and experimental) practice. No doubt, other actions, for example on receptors for excitatory transmitters, the release of transmitters from nerve endings, or membrane currents mediated by Na^+ or Ca^{2+} channels—all of which have been emphasized by various authors[89]—do in various degrees also contribute to the overall syndrome, giving a special character to the anesthetic state produced by each agent.

ACKNOWLEDGMENTS. The author is grateful to the Medical Research Council of Canada for its financial support.

REFERENCES

1. Adrian ED: Afferent discharges to the cerebral cortex from peripheral sense organs. *J Physiol* 1941; 100:159–191.
2. Alkon DL, Lederhendler L, Shoukimas JJ: Primary changes of membrane currents during retention of associative learning. *Science* 1982; 215:693–695.
3. Atwater I, Ribalet B, Rojas E: Mouse pancreatic B-cells: Tetraethylammonium blockage of the potassium permeability increase induced by depolarization. *J Physiol* 1979; 288:561–574.
4. Baker, PF: Transport and metabolism of calcium ions in nerve. *Progr Biophys* 1972;24:177–223.
5. Baker PF: Regulation of intracellular Ca and Mg in squid axons. *Fed Proc* 1976;35:2589–2595.
6. Baker PF, Schapira AHV: Anaesthetics increase light emission from aequorin at constant-ionized calcium. *Nature* 1980;284:168–169.
7. Bangham AD , Standish MM, Miller N: Cation permeability of phospholipid model membranes: Effect of narcotics. *Nature* 1965;208:1295–1297.
8. Barker JL, and Ransom BR: Pentobarbitone pharmacology of mammalian central neurones grown in tissue culture. *J Physiol* 1978;280:355–372.
9. Barrett JN, Magleby KL, Pallotta BS: Properties of single calcium-activated potassium channels in cultured rat muscle. *J Physiol* 1982;331:211–230.
10. Benardo LS, Prince DA: Ionic mechanisms of cholinergic excitation in mammalian hippocampal pyramidal cells. *Brain Res* 1982;249:333–344.
11. Ben-Ari Y, Krnjević, K, Reinhardt W, Ropert N: Intracellular observations on disinhibitory action of acetylcholine in hippocampus. *Neuroscience* 1981;6:2475–2484.
12. Blaustein MP, Ratzlaff RW, Kendrick NC, Schweitzer ES: Calcium buffering in presynaptic nerve terminals. 1. Evidence for involvement of a non-mitochondrial ATP-dependent sequestration mechanism. *J Gen Physiol* 1978;72:15–41.
13. Bloom FE, Costa E, Salmoiraghi GC: Anaesthesia and the responsiveness of individual neurons of the caudate nucleus of the cat to acetylcholine, norepinephrine and dopamine administered by microelectrophoresis. *J Pharmacol Exp Ther* 1965;150:244–252.

14. Borle A: Control, modulation and regulation of cell calcium. *Rev Physiol Biochem Pharmacol* 1981;90:13–153.
15. Bradley PB, Dray A: Modification of the responses of brain stem neurones to transmitter substances by anaesthetic agents. *Br J Pharmacol* 1973;48:212–224.
16. Brazier MAB: The action of anaesthetics on the nervous system, with special reference to the brain stem reticular system, in Adrian ED, Bremer F, Jasper HH (eds): *Brain Mechanisms and Consciousness*. Toronto, Ryerson Press, 1954,pp 163–199.
17. Bremer F: Etude oscillographique des activités sensorielles du cortex cérébral. *C R Soc Biol Paris* 1937;124:842–847.
18. Bremer F, Bonnet V: Interprétation des réactions rythmiques prolongées des aires sensorielles de l'écorce cérébrale. *Electroenceph Clin Neurophysiol* 1950;2:389–400.
19. Carafoli E: The transport of calcium across the inner membrane of mitochondria, in Carafoli E (ed): *Membrane Transport of Calcium*. London, Academic Press, 1982,pp 109–139.
20. Carlen L, Gurevich N, Durand D: Ethanol in low doses augments calcium-mediated mechanisms measured intracellularly in hippocampal neurons. *Science* 1982;215:306–309.
21. Catchlove RFH, Krnjević K, Maretić H: Similarity between effects of general anesthetics and dinitrophenol on cortical neurones. *Can J Physiol Pharmacol* 1972;50:1111–1114.
22. Chalazonitis N: Selective actions of volatile anesthetics on synaptic transmission and autorhythmicity in single identifiable neurons. *Anesthesiology* 1967;28:111–122.
23. Connors BW, Gutnick MJ, Prince DA: Electrophysiological properties of neocortical neurons in vitro. *J Neurophysiol* 1982;48:1302–1320.
24. Cote IL, Zbicz KL, Wilson WA : Barbiturate-induced slow outward currents in Aplysia neurones. *Nature* 1978;274:594–596.
25. Coyle JT, Price DL, DeLong MR: Alzheimer's disease: A disorder of cortical cholinergic innervation. *Science* 1983;219:1184–1190.
26. Crawford JM: Anaesthetic agents and the chemical sensitivity of cortical neurones. *Neuropharmacology* 1970;9:31–46.
27. Dandy WE: The location of the conscious center in the brain—the corpus striatum. *Bull Johns Hopkins Hosp* 1946;79:34–58.
28. DeFeudis FV: *Central Cholinergic Systems and Behaviour*. London, Academic Press, 1974.
29. Dempsey EW, Morison RS: The electrical activity of a thalamocortical relay system. *Am J Physiol* 1943;138:283–296.
30. Dichter MA, Lisak J, Biales B: Action potential mechanism of mammalian cortical neurons in cell culture. *Brain Res* 1983;289:99–107.
31. DiPolo R, Beaugé L: The calcium pump and sodium-calcium exchange in squid axons. *Ann Rev Physiol* 1983;45:313–324.
32. Dodd J, Dingledine R, Kelly JS: The excitatory action of acetylcholine on hippocampal neurones of the guinea pig and rat maintained *in vitro*. *Brain Res* 1981;207:109–127.
33. Douglas WW, Taraskevich PS: Calcium component to action potentials in rat pars intermedia cells. *J Physiol* 1980;309:623–630.
34. Duggan AW, Headley PM, Lodge D: Acetylcholine-sensitive cells in the caudal medulla of the rat: Distribution, pharmacology and effects of pentobarbitone. *Br J Pharmacol* 1974;54:23–31.
35. Evarts EV: Photically evoked responses in visual cortex units during sleep and waking. *J Neurophysiol* 1963;26:229–248.
36. Fatt P, Ginsborg BL: The ionic requirements for the production of action potentials in crustacean muscle fibres. *J Physiol* 1958;142:516–543.
37. Fibiger HC: The organization and some projections of cholinergic neurons of the mammalian forebrain. *Brain Res Rev* 1982;4:327–388.
38. Fry JP, Herz A, Zieglgänsberger W: Lack of barium-induced excitation in the cerebral cortex and hippocampus of morphine tolerant/dependent rats. *J Physiol* 1979;291:48–49P.
39. Gerard RW: Biological roots of psychiatry. *Science* 1955;122:225–230.
40. Globus GG, Maxwell G, Savodnick I (ed): *Consciousness and the Brain, A Scientific and Philosophical Inquiry*. New York, Plenum Press, 1976.
41. Goldstein MH Jr., Kiang NY-S, Brown RM: Responses of the auditory cortex to repetitive acoustic stimuli. *J Acoust Soc Am* 1959;31:356–364.
42. Gregory RL: *Mind in Science*. Cambridge, Cambridge University Press, 1981.
43. Haefely W, Pieri L, Polc P, Schaffner R: General pharmacology and neuropharmacology of benzodiazepine derivatives. *Handbook Exp Pharmacol* 1981;55:9–262.

44. Halliwell JV, Adams PR: Voltage-clamp analysis of muscarinic excitation in hippocampal neurons. *Brain Res* 1982;250:71–92.
45. Harrison NL, Simmonds MA: Alphaxalone selectively potentiates responses to GABA and muscimol in rat cuneate nucleus *in vitro*. *J Physiol* 1984;346:42P.
46. Harris RA: Ethanol and pentobarbital inhibition of intrasynaptosomal sequestration of calcium. *Biochem Pharmacol* 1981;30:3209–3215.
47. Hebb DO: *The Organization of Behavior: A Neuropsychological Theory*. New York, Wiley, 1949.
48. Hodgkin AL: *The conduction of the nervous impulse*, Liverpool, Liverpool University Press, 1964.
49. Jasper HH: Pathophysiological studies in brain mechanisms in different states of consciousness, in Eccles J C (ed): *Brain and Conscious Experience*, New York, Springer-Verlag, 1966,pp 256–282.
50. Johnson SM, Bangham AD: The action of anesthetics on phospholipid membranes. *Biochim Biophys Acta* 1969;193:92–104.
51. Krnjević K: Central cholinergic pathways. *Fed Proc* 1969;28:113–120.
52. Krnjević K: Central actions of general anaesthetics, in Halsey MJ, Millar R, Sutton JA (eds): *Molecular Mechanisms in General Anaesthesia*. Churchill Livingston, 1974, New York, pp65–89.
53. Krnjević K: Inhibitory action of GABA and GABA-mimetics on vertebrate neurons. in: Roberts E, Chase TN, Tower DB (eds): *GABA in Nervous System Function*. New York, Raven Press, 1976,pp269–281.
54. KrnjevićK, Lisiewicz A: Injections of calcium ions into spinal motoneurones. *J Physiol* 1972;225:363–390.
55. Krnjević K, Phillis JW: Acetylcholine-sensitive cells in the cerebral cortex. *J Physiol* 1963;166:296–327.
56. Krnjević K, Ropert N: Septo-hippocampal pathway modulates hippocampal activity by a cholinergic mechanism. *Can J Physiol Pharmacol* 1981;59:911–914.
57. Krnjević K, Silver A: A histochemical study of cholinergic fibres in the cerebral cortex. *J Anat (London)* 1965;99:711–759.
58. Krnjević K, Silver A: Acetylcholinesterase in the developing forebrain. *J Anat (London)* 1966;100:63–89.
59. Krnjević K, Pumain R, Renaud L: Effects of Ba^{2+} and tetraethylammonium on cortical neurones. *J Physiol* 1971a;215:223–245.
60. Krnjević K, Pumain R, Renaud L: The mechanism of excitation by acetylcholine in the cerebral cortex. *J Physiol* 1971b;215:247–268.
61. Krnjević K, Morris ME, Reiffenstein RJ: Changes in extracellular Ca^{2+} and K^{+} activity accompanying hippocampal discharges. *Can J Physiol Pharmacol* 1980;58:579–583.
62. Krnjević, K, Morris ME, Reiffenstein RJ: Stimulation-evoked changes in extracellular K^{+} and Ca^{2+} in pyramidal layers of the rat's hippocampus. *Can J Physiol Pharmacol* 1982;60:1643–1657.
63. Lamour Y, Dutar P, Jobert A: Excitatory effect of acetylcholine on different types of neurons in the first somatosensory neocortex of the rat: Laminar distribution and pharmacological characteristics. *Neuroscience* 1982;7:1483–1494.
64. Latorre R, Miller C: Conduction and selectivity in potassium channels. *J Membrane Biol* 1983;71:11–30.
65. Lee SL, Alto LE, Dhalla NS: Subcellular effects of some anesthetic agents on rat myocardium. *Can J Physiol Pharmacol* 1979;57:65–70.
66. Libet B: Electrical stimulation of cortex in human subjects and conscious sensory aspects, in Iggo A (ed): *Handbook of Sensory Physiology*, Vol. II, *Somatosensory System*. Berlin, Springer-Verlag, 1973,pp 744–790.
67. Livingstone, MS, Hubel DH: Effects of sleep and arousal on the processing of visual information in the cat. *Nature* 1981;291:554–561.
68. Lynch G, Baudry M: The biochemistry of memory: A new and specific hypothesis. *Science* 1984;224:1057–1063.
69. Meech R W: Intracellular calcium injection causes increased potassium conductance in *Aplysia* nerve cells. *Comp Biochem Physiol* 1972;42A:493–499.
70. Mesulam M-M, Mufson E J, Wainer B H, Levey A I: Central cholinergic pathways in the rat: An overview based on an alternative nomenclature (Ch1-Ch6). *Neuroscience* 1983;10:1185–1201.
71. Miller C: Integral membrane channels: Studies in model membranes. *Physiol Rev* 1983;63:1209–1242.
72. Morin-Surun MP, Champagnat J, Denavit-Saubié M, Moyanova S: The effects of acetylcholine on bulbar respiratory related neurones. *Naunyn-Schmiedeberg's Arch Pharmacol* 1984;325:205–208.
73. Morris ME, Krnjević K, MacDonald JF: Changes in intracellular free Ca ion concentration evoked by electrical activity in cat spinal neurons *in situ*. *Neuroscience* 1985; 14:563–580.
74. Mullins LJ: Steady-state calcium fluxes: Membrane versus mitochondrial control of ionized calcium in axoplasm. *Fed Proc* 1976;35:2583–2588.
75. Nicholls DG: Calcium transport and proton electrochemical potential gradient in mitochondria from guinea-pig cerebral cortex and rat heart. *Biochem J* 1978;170:511–522.

76. Nicholls DG, Scott ID: The regulation of brain mitochondrial calcium-ion transport. *Biochem J* 1980;186:833–839.
77. Nicoll RA: The effects of anaesthetics on synaptic excitation and inhibition in the olfactory bulb. *J Physiol* 1972;223:803–814.
78. Nicoll RA, Madison DV: General anesthetics hyperpolarize neurons in the vertebrate central nervous system. *Science* 1982;217:1055–1057.
79. Noda H, Adey WR: Neuronal activity in the association cortex of the cat during sleep, wakefulness and anesthesia. *Brain Res* 1973;54:243–259.
80. Ohmori H: Studies of ionic currents in the isolated vestibular hair cell of the chick. *J Physiol* 1984;350:561–581.
81. Olsen RW: Drug interactions at the GABA receptor-ionophore complex. *Ann Rev Pharmacol Toxicol* 1982;22:247–277.
82. Pawlow I: Innere Hemmung der bedingten Reflexe and der Schlafein und derselbe Prozeß. *Skand Arch Physiol* 1923;44:42–58.
83. Penfield W: *The Excitable Cortex in Conscious Man.* Liverpool, Liverpool University Press, 1958.
84. Zbicz KL, Wilson WA: Barbiturate enhancement of spike frequency adaptation in *Aplysia* giant neurons. *J Pharmacol Exp Ther* 1981;217:222–227.
85. Pitman RM: Intracellular citrate or externally applied tetraethylammonium ions produce calcium-dependent action potentials in an insect motoneurone cell body. *J Physiol* 1979;291:327–337.
86. Popper KR, Eccles JC: *The Self and Its Brain, An Argument for Interactionism.* London, Springer-Verlag, 1977.
87. Prichard J W: Effect of phenobarbital on a leech neuron. *Neuropharmacology* 1972;11:585–590.
88. Rasmussen H: *Calcium and cAMP as synarchic messengers.* New York, Wiley, 1981.
89. Richards CD: Actions of general anaesthetics on synaptic transmission in the CNS. *Br J Anaesth* 1983;55:201–207.
90. Richter JA, Holtman JR: Barbiturates: Their *in vivo* effects and potential biochemical mechanisms. *Prog Neurobiol* 1982;18:275–319.
91. Rosenberg H, Haugaard N: The effects of halothane on metabolism and calcium uptake in mitochondria of the rat liver and brain. *Anesthesiology* 1973;39:44–53.
92. Sakmann B, Neher E: Patch clamp techniques for studying ionic channels in excitable membranes. *Ann Rev Physiol* 1984;46:455–472.
93. Sato M, Austin, GM Yai H: Increase in permeability of the postsynaptic membrane to potassium produced by "Nembutal." *Nature* 1967;215:1506–1508.
94. Scholfield CN: Potentiation of inhibition by general anesthetics in neurones of the olfactory cortex in vitro. *Pflugers Arch* 1980;383:249–255.
95. Schwindt PC, Crill WE: Properties of a persistent inward current in normal and TEA-injected motoneurons. *J Neurophysiol* 1980;43:1700–1724.
96. Sillito AS, Kemp JA: Cholinergic modulation of the functional organization of the cat visual cortex. *Brain Res* 1983;289:143–155.
97. Simons TJB: Effect of propranolol on calcium-dependent potassium permeability of red cell membranes. *J Physiol* 1980;300:35–36P.
98. Steriade M: Cortical long-axoned cells and putative interneurons during the sleep-waking cycle. *Behav Brain Sci* 1978;3:465–514.
99. Sweetman AJ, Esmail AF: Evidence for the role of calcium ions and mitochondria in the maintenance of anaesthesia in the rat. *Biochem Biophys Res Comm* 1975;64:885–890.
100. Sweetman AJ, Lovett GS, Markham A, Cameron I: Effect of the general anaesthetic, propanidid on oxidative phosphorylation and calcium transport in rat liver mitochondria. *Biochem Pharmacol* 1981;30:1921–1924.
101. Werman R, Grundfest H: Graded and all-or-none electrogenesis in arthropod muscle II. The effects of alkali-earth and onium ions on lobster muscle fibres. *J Gen Physiol* 1961;44:997–1027.
102. Willow M, Bygrave FL: Effects of pentobarbitone on $^{45}Ca^{2+}$ transport by rat brain mitochondria. *J Neurochem* 1982;39:557–562.
103. Willow M, Johnston GAR: Pharmacology of barbiturates: Electrophysiological and neurochemical studies. *Int Rev Neurobiol* 1983;24:15–49.
104. Wong BS, Binstock L: Inhibition of potassium conductance with external tetraethylammonium in Myxicola giant axons. *Biophys J* 1980;32:1037–1042.
105. Wong RKS, Prince DA: Participation of calcium spikes during intrinsic burst firing in hippocampal neurons. *Brain Res* 1978;159:385–390.
106. Woody CD, Swartz BE, Gruen E: Effects of acetylcholine and cyclic GMP on input resistance of cortical neurons in awake cats. *Brain Res* 1978;158:373–395.

2

Barbiturate and Benzodiazepine Actions on Mouse Neurons in Cell Culture

Robert L. Macdonald, John H. Skerritt, and Mary Ann Werz

ABSTRACT. Barbiturates and benzodiazepines have been shown to have multiple actions on neuronal membrane properties and on synaptic transmission. We have used mouse neurons in cell culture to study two effects of these compounds: modification of GABA responses and reduction of voltage-dependent calcium conductance. We have used both intracellular recording and voltage clamp techniques. The barbiturates, pentobarbital and phenobarbital, and the benzodiazepines, diazepam and clonazepam, enhanced GABA responses at subanesthetic concentrations. Beta carbolines, which are benzodiazepine receptor ligands, were also studied. The convulsant β-carboline DMCM antagonized GABA responses whereas the "proconvulsant" β-carbolines, βCCME and βCCET, did not modify GABA responses. The anticonvulsant β-carboline CCPR enhanced GABA whereas βCCME, βCCET, and βCCPR all antagonized diazepam enhancement of GABA responses. These results are consistent with the proposal that DMCM is an inverse agonist, βCCME and βCCET are partial inverse agonists or antagonists, and βCCPR is a partial agonist at the benzodiazepine receptor.

Barbiturates and benzodiazepines also reduced the duration and amplitude of calcium-dependent action potentials recorded from mouse spinal cord and dorsal ganglion neurons at concentrations that blocked release of neurotransmitter and uptake of calcium by synaptosomes. Voltage clamp analysis of the effect on cesium-loaded neurons with cesium replacement of extracellular potassium ions demonstrated that the effect of the barbiturates was to block calcium conductance and not to enhance potassium conductance.

The potential role enhancement of GABAergic synaptic transmission and blockade of presynaptic calcium entry and neurotransmitter release in anesthesia is discussed.

INTRODUCTION

Barbiturates and benzodiazepines have similar but not identical clinical actions. Both drugs are anxiolytic, anticonvulsant, and sedative. Barbiturates are also used as anesthetics. In addition to having multiple behavioral effects, both compounds have been shown to have numerous actions in the central nervous system.[12, 21] Barbiturates and benzodiazepines have been demonstrated to enhance GABAergic synaptic transmission and to block release of neurotransmitter by reducing presynaptic calcium entry. In addition, barbiturates have been shown to block excitatory synaptic transmission.

Robert L. Macdonald, John H. Skerritt, and Mary Ann Werz • Department of Neurology, University of Michigan, Ann Arbor, Michigan 48104.

We have used mouse neurons in cell culture to study the multiple actions of barbiturates and benzodiazepines. [22,21] In this paper we will review their actions on postsynaptic GABA responses and on calcium entry.

POSTSYNAPTIC GABA RECEPTORS AND COUPLED CHLORIDE CHANNELS

GABA is the most widespread inhibitory neurotransmitter in the central nervous system. When applied by iontophoresis or pressure application, GABA and GABA analogues such as THIP (4, 5, 6, 7-tetrahydroisoxazolo-[5, 4-(c)-3-ol][19]) produce membrane hyperpolarization of mouse neurons in cell culture[3,34,36] and most vertebrate neurons *in vivo* [9,18] (Fig. 1A1). Since GABA responses reverse polarity at the chloride ion equilibrium potential, the hyperpolarization is due to activation of membrane chloride ion conductance. Following injection of chloride ion into neurons to reduce chloride ion equilibrium potential from about -65 mV to about -30mV, GABA responses become depolarizing at membrane potentials more negative than -30mV (Fig. 1A2). GABA and THIP responses are blocked by the GABA antagonists picrotoxin (Fig. 1B), bicuculline, penicillin, and pentylenetetrazol.[23]

The postsynaptic GABA receptor has been shown to be an oligomeric complex composed of at least three interconnected elements.[31] The receptor contains distinct binding sites for GABA, picrotoxin/barbiturates, and benzodiazepines and is coupled to a chloride ion channel. Dose–response curves for GABA responses on neurons in cell culture are generally sigmoidal with Hill coefficients greater than one,[2,7,15] suggesting that either more than one GABA molecule binds to the receptor complex to produce chloride channel openings or chloride channels are not activated independently. The specific antagonist bicuculline competitively antagonizes GABA responses[16] and binding while picrotoxin noncompetitively antagonizes GABA responses, suggesting that the convulsants bind to different sites on the GABA receptor complex.

GABA-activated single chloride ion channels have been studied recently[4,17] using the

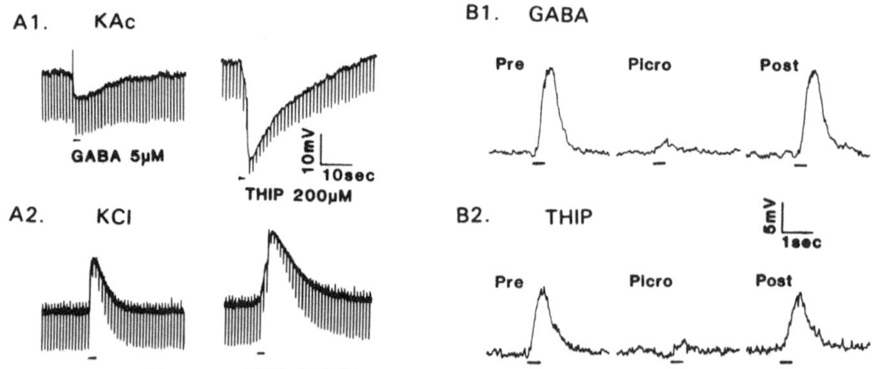

Figure 1. GABA and THIP responses recorded from spinal cord neurons in cell culture. (A) GABA was applied to neurons from a miniperfusion pipette containing 5 µM GABA or 200 µM THIP. (A1) When intracellular recordings were made with 4 M KAc-filled intracellular micropipettes, membrane hyperpolarization and increased conductance were produced. (A2) When 3 M KCl-filled intracellular micropipettes were used and chloride was injected into the neurons, GABA and THIP produced membrane depolarization with increased conductance. Resting membrane potentials were −57 mV (A1) and −70 mV (A2). (B) GABA and THIP were applied by iontophoresis to spinal cord neurons hyperpolarized to −90 mV; 3 M KCl-filled intracellular micropipettes were used. Picrotoxin applied by pressure ejection from a pipette containing 100 µM reversibly blocked the GABA and THIP responses. Reproduced with permission from Skerritt and Macdonald.[36]

patch clamp technique.[3] Patches were taken from mouse spinal cord neurons in cell culture. GABA activates a chloride ion channel with a main conductance state of about 30 picosiemens and a subconductance state of about 19 picosiemens. Channel openings occur in bursts with two populations of open times. The openings can be described as exponentially distributed with decay constants of 2.4 msec and 29 msec. Thus, GABA activates chloride ion channels with at least two conducting states and two distributions of open time. The basis for and significance of the multiple conductance states and open time time distributions are uncertain at the present time.

GABA RESPONSES AND BARBITURATES

Barbiturates have been shown to have actions on postsynaptic neurotransmitter responses,[25,26,30] enhancing GABAergic synaptic transmission and reducing excitatory synaptic transmission. The interactions of barbiturates with postsynaptic GABA and glutamate responses have been studied using mouse neurons in cell culture.[21,22] Neurons were impaled with either potassium acetate- or potassium chloride-filled micropipettes, and GABA was applied either by pressure application or by iontophoresis. For pharmacological investigation of GABA responses, KCl-filled micropipettes were routinely used since a large separation of membrane potential and E_{Cl} is obtained after chloride injection. Thus, pharmacological modulation of GABA responses is not limited by a lack of electrochemical driving force for chloride ions. Barbiturates were applied either by superfusion or by local pressure ejection. Both pentobarbital and phenobarbital enhance the amplitude of iontophoretic GABA responses.[3,25,26,35] Pentobarbital enhancement, however, is characterized by both an increase in the amplitude of the response and a prolongation of the response. Pentobarbital is more potent than phenobarbital, producing threshold enhancement at about 2 μM, whereas phenobarbital acts above 20 μM. Pentobarbital and phenobarbital produce 50% enhancements at 25 μM and 200 μM, respectively. Both barbiturates produce a leftward shift in the GABA dose–response curve without altering the apparent maximal response. Thus, there appears to be a decrease in the apparent K_d. We have proposed that barbiturates enhance GABA responses by increasing the affinity of GABA for its receptor.[25,26] This proposal has received support in binding studies where it has been demonstrated that low-affinity GABA binding can be enhanced by pentobarbital.[1,32,45] The enhancement of GABA binding is chloride dependent. It is not clear whether the enhancement is due to an increased number of GABA binding sites or to an increase in receptor affinity. The mechanism of barbiturate enhancement of GABA responses has also been studied using noise analysis of GABA current fluctuations produced on mouse spinal cord neurons in cell culture.[42] In these studies, mean channel conductance was not altered, but pentobarbital prolonged the open lifetime of chloride channels. Thus, the available data suggest that barbiturates enhance postsynaptic GABA action by increasing the affinity of GABA receptors for GABA and consequently prolonging mean channel open time for the chloride channels. The mechanism of blockade of glutamate and acetylcholine responses remains uncertain.

GABA RESPONSES AND BENZODIAZEPINE LIGANDS

Benzodiazepines were initially demonstrated to enhance GABAergic synaptic transmission and the response to exogenously applied GABA.[8,24,33] Subsequently, specific binding sites for radiolabeled benzodiazepines in brain membrane preparations were described,[28,41] and benzodiazepines were demonstrated to enhance the low-affinity binding of GABA to GABA receptors.[11,39] Thus, it has been assumed that benzodiazepines bind to a site on the

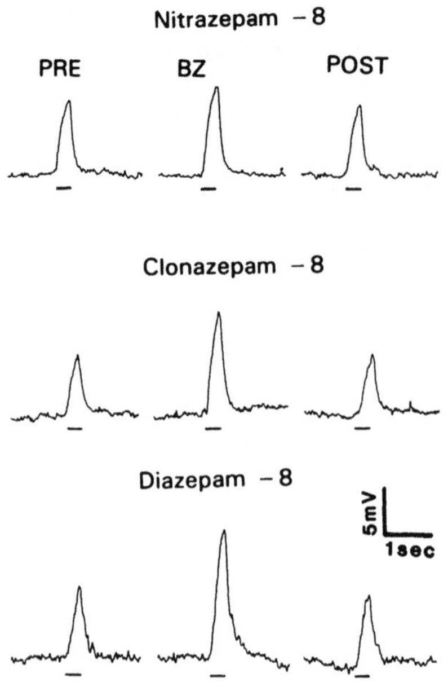

Figure 2. Benzodiazepine enhanced GABA responses evoked on spinal cord neurons. Nitrazepam, clonazepam, and diazepam (10 nM) enhanced iontophoretically produced GABA responses. Recordings were made using KCl-containing recording micropipettes, and membrane potentials were held at −79, −80, and −76 mV respectively. The post responses were obtained 2 min after removal of the benzodiazepine-containing pipettes.

GABA receptor to enhance the affinity of GABA to bind to its receptor. The benzodiazepines diazepam, clonazepam, and nitrazepam produced dose-dependent enhancement of iontophoretic GABA responses on spinal cord and cortical neurons in cell culture[8,24,25,37,44] (Fig. 2). On spinal cord neurons, the threshold for clonazepam and diazepam action was about 1 nM with peak effects being seen at 10 to 100 nM, whereas nitrazepam was somewhat less potent with threshold effects around 10 nM.[37] Diazepam did not modify iontophoretic responses to either glycine or glutamic acid.

Recently, there has been considerable interest in identifying endogenous ligands for the benzodiazepine binding sites.[5] During these investigations, β-carboline-3-carboxylate esters have been described as being benzodiazepine receptor ligands.[5] Although it is unlikely that these compounds are the endogenous ligand for benzodiazepine receptors, they have been demonstrated to interact with benzodiazepine receptors in a complex and interesting manner. The alkyl esters of β-carboline-3-carboxylic acid are potent ligands for the benzodiazpine binding sites and have a variety of pharmacological profiles.[6] Some of these compounds act as convulsants (methyl-6,7-dimethoxy-4-ethyl-β-carboline-3-carboxylate, DMCM), whereas others are proconvulsant and anxiogenic (the ethyl and methyl esters, βCCET and βCCME) or anticonvulsant and anxiolytic (the propyl ester, βCCPR). We have investigated the actions of a series of β-carboline derivatives on iontophoretic GABA responses of mouse neurons in cell culture[38] (Fig. 3). DMCM reduced GABA responses in all spinal cord neurons studied. βCCME and βCCET did not alter GABA responses at concentrations up to 10 mM. However, CCPR weakly, but significantly, enhanced GABA responses at concentrations above 10 nM. In addition, βCCME, βCCET, and βCCPR all antagonized the enhancement of GABA responses produced by diazepam. These data are consistent with the proposal that the convulsant DMCM acts at benzodiazepine receptors to reduce the action of GABA,[6] thus acting as "inverse agonists." βCCME and βCCET appear to be either pure antagonists or possibly weak partial inverse agonists. βCCPR appears to be a partial agonist, and the benzodiazepines, diazepam, and clonazepam appear to be full agonists. Thus, this series of benzodiazepine and β-carboline benzodiazepine receptor ligands have the full range of action from full agonist to full inverse agonist on benzodiazepine receptors and coupled

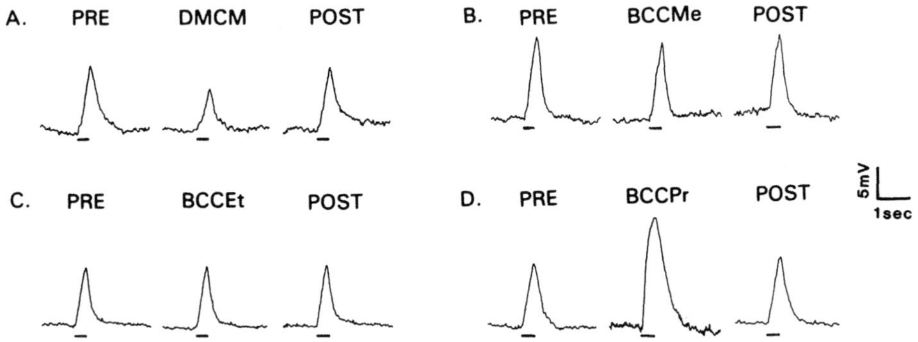

Figure 3. β-Carbolines had differing actions on GABA responses. GABA responses were produced by iontophoresis, and KCl-containing micropipettes were used for intracellular recordings. (A) DMCM (100 nM) decreased GABA responses (membrane potential = −75 mV). (B and C) βCCMe and βCCEt (100 nM) did not alter GABA responses (membrane potential = −72 mV and −78 mV, respectively). (D) βCCPr (10 μM) enhanced GABA responses (membrane potential = −82mV). Reproduced with permission from Skerritt and Macdonald.[38]

GABA receptors. The behavioral correlate of benzodiazepine receptor ligand action appears to be anticonvulsant or anxiolytic action if an enhancement of GABA response is produced or convulsant or anxiogenic action if a reduction in GABA response is produced.

CALCIUM CONDUCTANCE AND BARBITURATES

Barbiturates have been demonstrated to reduce calcium entry into presynaptic terminals[4] and to reduce presynaptic release of neurotransmitter.[15] It has been assumed that the blockade of calcium entry by barbiturates interrupts excitation–secretion coupling, and that reduction of calcium entry is the primary event in reduction of transmitter release. Although the action of barbiturates on presynaptic terminals of vertebrate neurons has not been directly investigated, barbiturates have been shown to reduce the duration of presynaptic calcium-dependent action potentials and thus calcium entry at squid giant synapses at concentrations that reduced release of neurotransmitter at the synapse.[29] Furthermore, in the R2 cell of *Aplysia* abdominal ganglion, barbiturates decreased the maximal rate of rise of somatic calcium-dependent action potentials.[10] Thus, in invertebrates barbiturates appeared to reduce calcium entry in both presynaptic terminals and cell bodies. By analogy, we have demonstrated that barbiturates reduce the duration of somatic calcium-dependent action potentials of mouse neurons in cell culture[15] at concentrations that reduce spontaneous release of neurotransmitter- and potassium-stimulated uptake of $^{45}Ca^{2+}$. Thus, it is likely that actions of barbiturates on voltage-dependent calcium movement in presynaptic terminals and cell bodies are similar. Study of calcium entry into cell bodies may be informative, therefore, concerning regulation of calcium entry into presynaptic terminals. The dose dependency of the action of barbiturates on calcium-dependent action potentials was studied in spinal cord and dorsal root ganglion neurons.[15,43] In both cell types, pentobarbital produced threshold shortening at 25 μM and produced maximal reduction in action potential duration by 500 μM. Phenobarbital was about fivefold less potent and barbiturate acid had no action up to 2 mM.

Barbiturates could modify calcium-dependent action potential duration by directly blocking calcium channels. However, barbiturates could also reduce calcium-dependent action potential duration indirectly by enhancing voltage- or calcium-dependent potassium conductance which would result in less net inward current and thus less net calcium entry. We have studied the actions of barbiturates on calcium conductance by recording from

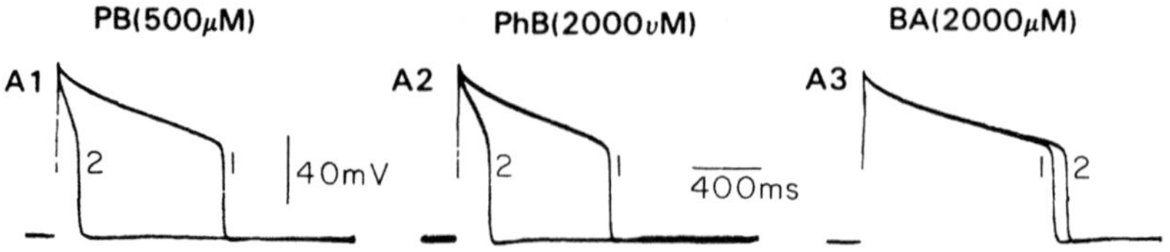

Figure 4. Barbiturates decreased calcium-dependent action potential duration of DRG neurons following blockade of potassium conductance by intracellular cesium injection. Action potentials evoked from a DRG neuron bathed in medium that substituted cesium for potassium. Cesium was also injected intracellularly since the recording micropipette was filled with 3 M CsCl. Substantial blockade of potassium conductance was achieved, suggested by the prolongation of calcium-dependent action potentials to durations of 500 msec to 2 sec and by the loss of the afterhyperpolarization. Pentobarbital (A_1) and phenobarbital (A_2) decreased the calcium component of the action potential under these conditions, whereas the inactive barbiturate, barbituric acid, was without effect (A_3).

mouse dorsal root ganglion neurons in cell culture with intracellular cesium ion injection and replacement of extracellular potassium ion by cesium ion.[27,43] Cesium ion is a potent potassium channel blocker. In the presence of cesium ion substitution for potassium ion, long-duration voltage-dependent action potentials can be produced in both spinal cord and dorsal root ganglion neurons. Barbiturates remain potent in reducing the duration of these action potentials (Fig. 4), suggesting that barbiturates act directly on calcium channels rather than on potassium channels. However, to directly demonstrate this action, we applied the single-electrode voltage clamp technique to single dorsal root ganglion neurons in cell culture. After the replacement of potassium by cesium ion and the use of cesium recording electrodes, impalements of dorsal root ganglion cells and depolarizing voltage step clamps of 15 msec produced net inward currents at large positive potentials. These inward calcium currents were entirely blocked by cadmium. In the presence of cadmium only leak conductance was left. Barbiturates reduced the inward current without altering either the range of activation of the voltage-dependent calcium conductance or the apparent calcium equilibrium potential (Fig. 5). Furthermore, we investigated the membrane chord conductance by mea-

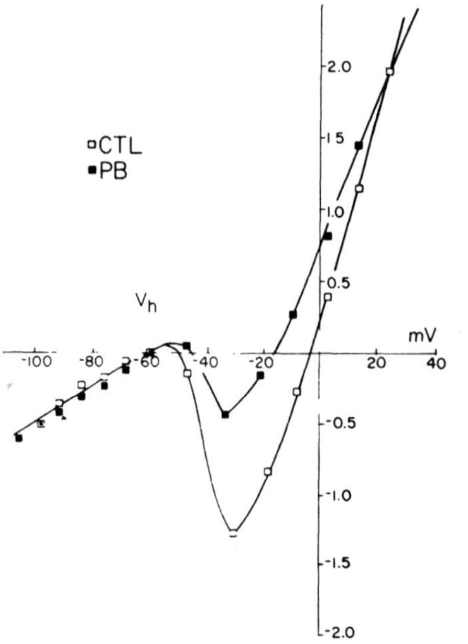

Figure 5. Pentobarbital reduced the amplitude of calcium inward currents. Current voltage relation obtained when command steps were applied from a holding potential of −60 mV to a dorsal root ganglion neuron bathed in potassium-free medium that substituted cesium for potassium and contained 5 mM tetrethylammonium chloride is illustrated. The recording micropipette contained 3 M CsCl. Pentobarbital (500 μM) did not affect the current–voltage relation obtained with hyperpolarizing step commands but did reduce the inward current evoked by depolarizing commands. Pentobarbital did not affect the voltage-dependence or reversal potential of the inward current. CTL = control. PB = pentobarbital.

suring instantaneous current-voltage curves at large positive potentials that are above the activation range of the voltage-dependent calcium conductance. In this range the chord conductance is linear as expected and conductance was reduced by barbiturates. Thus, it is apparent that barbiturates reduce voltage-dependent calcium current by directly reducing calcium conductance and not indirectly by increasing potassium conductance. A comparison of the rate of inactivation of calcium current with and without application of pentobarbital revealed that the barbiturate appeared to be accelerating calcium channel inactivation as the primary action. It would appear that barbiturates increase the rate of calcium channel inactivation as well as directly blocking calcium channels.

CALCIUM CONDUCTANCE AND BENZODIAZEPINES

Benzodiazepines have been demonstrated also to modify presynaptic calcium entry.[20] As with barbiturates, we have investigated the action of benzodiazepines on calcium-dependent action potentials on mouse neurons in cell culture.[40] Diazepam reduced calcium-dependent action potential duration at concentrations above 1 μM in a dose-dependent fashion. In addition, the convulsant benzodiazepine Ro 5-4864 also reduced calcium-dependent action potential duration at similar concentrations. This reduction in calcium-dependent action potential duration was also independent of effects on resting membrane potential or conductance and occurred after intracellular injection of the potassium channel blocker cesium, suggesting that the mechanism of action of benzodiazepines is also to decrease calcium ion conductance.

WHAT IS THE MECHANISM UNDERLYING BARBITURATE AND BENZODIAZEPINE ANESTHETIC ACTIONS?

It is quite clear that barbiturates and benzodiazepine receptor ligands have multiple actions on synaptic transmission and excitability. Which of these many actions may relate to sedative, hypnotic, or anesthetic actions of these compounds?

Although it is not possible to specify a behavioral action of a drug using *in vitro* preparations, certain suggestions can be made. First, although barbiturates and benzodiazepines both enhance postsynaptic GABAergic inhibition, they produce this enhancement at concentrations that are not anesthetic. Phenobarbital produces enhancement at concentrations above 20 μM, and the benzodiazepines, diazepam and clonazepam, enhance GABA responses above 1 nM. Furthermore, anticonvulsant and anxiolytic or convulsant and anxiogenic actions of the benzodiazepine receptor ligards correlate well with enhancement or reduction of GABA responses. Thus, it is likely that threshold enhancement of GABAergic synaptic transmission correlates perhaps with anticonvulsant and anxiolytic actions of barbiturates and benzodiazepines.

Could maximal enhancement of GABAergic synaptic transmission be responsible for anesthesia? This cannot be ruled out in the present experiments, but certain observations suggest that this is not the case. For benzodiazepines there is a large safety factor between concentrations effective in producing anticonvulsant and anxiolytic actions and those producing sedation or anesthesia. However, large enhancements of GABA responses are produced at low concentrations of benzodiazepines, concentrations that are not sedative or anesthetic. These results suggest that enhancement of GABAergic transmission is not the sole cause for sedation and anesthesia.

What is the relationship of blockade of presynaptic calcium entry and transmitter release to the behavioral effects of benzodiazepine and barbiturates? Barbiturates and benzodiazepines both block presynaptic calcium entry and transmitter release at concentrations above those effective in modifying postsynaptic GABAergic responses. This is consistent with a large safety factor for benzodiazepines between anticonvulsant and anxiolytic action and anesthetic action. Furthermore, for barbiturates, the calcium current block occurs at anesthetic concentrations. Nonetheless, the occurrence of transmitter release block and blockade of presynaptic calcium entry at anesthetic concentrations does not prove that this underlies the anesthetic effects of these drugs.

In conclusion, it would appear that small amounts of enhancement of GABAergic synaptic transmission can account perhaps for anxiolytic and anticonvulsant actions of these drugs but are not anesthetic. A reasonable hypothesis is that anesthesia produced by these compounds is due to one of three mechanisms. First, enhancement of GABAergic synaptic transmission may produce a continuum of behavioral effects. At low concentrations, anxiolytic and anticonvulsant effects may be produced, but at higher amounts of enhancement of GABAergic inhibition anesthesia may be produced. Alternatively, it is possible that the combination of the enhancement of GABAergic synaptic transmission with presynaptic blockade of both excitatory and inhibitory synaptic transmitter release may account for anesthesia. Finally, it is possible that enhancement of GABAergic synaptic transmission does not underlie anesthesia, and that the presynaptic action of barbiturates and benzodiazepines to block transmitter release may be the sole mechanism underlying anesthesia. However, definitive evidence that these mechanisms underlie anesthesia awaits demonstration of one of these actions in an intact nervous system preparation.

REFERENCES

1. Asano T, Ogasawara N: Chloride-dependent stimulation of GABA and benzodiazepine receptor binding by pentobarbital. *Brain Res* 1981;225:212–216.
2. Barker JL, Ransom BR: Amino acid pharmacology of mammalian central neurones grown in tissue culture. *J Physiol* 1978a;280:331–354.
3. Barker JL, Ransom BR: Pentobarbitone pharmacology of mammalian central neurones grown in tissue culture. *J Physiol (London)* 1978b;280:355–372.
4. Blaustein MP, Ector AC: Barbiturate inhibition of calcium uptake by depolarized nerve terminals *in vitro*. *Mol Pharmacol* 1975;11:369–378.
5. Braestrup C, Nielsen M, Olsen CE: Urinary and brain beta-carboline-3-carboxylates as potent inhibitiors of brain benzodiazepine receptors. *Proc Nat Acad Sci US-Biol Sci* 1980;77:2288–2288.
6. Braestrup C, Nielsen M, Honore T, Jensen LH, Petersen EN: Benzodiazepine receptor ligands with positive and negative efficiacy. *Neuropharmacology* 1983;22:1451–1457.
7. Choi DW, Fischbach GD: GABA conductance of chick spinal cord and dorsal root ganglion neurons in cell culture. *J Neurophysiol* 1981;45:605–620.
8. Choi DW, Farb, DH, Fischbach GD: Lordiazepoxide selectively augments GABA action in spinal cord cell cultures. *Nature* 1977;269:342–344.
9. Curtis DR, Johnston GAR: Amino acid transmitters in the mammalian central nervous system. *Ergeb Physiol Biol Chem Exp Pharmacol* 1974;69:97–188.
10. Goldring JM, Blaustein MP: Effect of pentobarbital on Na and Ca action potentials in an invertebrate neuron. *Brain Res* 1982;240:273–283.
11. Guidotti A, Toffano G, Costa E: An endogenous protein modulates the affinity of GABA and benzodiazepine receptors in rat brain. *Nature (London)* 1978;275:553–555.
12. Haefely W, Polc P: Electrophysiolgical studies on the interaction of anxiolytic drugs with GABAergic mechanisms, in Malick JB, Enna S, Yamaura HI (eds): *Anxiolytics: Neurochemical, Behavioral and Clinical Perspectives*. New York, Raven Press, 1983; pp113–145.
13. Hamill OP, Marty A, Neher E, Sakmann B, Sigworth FJ: Improved patch-clamp techniques for high-resolution current recording from cells and cell-free membrane patches. *Pflugers Arch* 1981;391:85–100.
14. Hamill OP, Bormann J, Sakmann B: Activation of multiple-conductance state chloride channels in spinal neurones by glycine and GABA. *Nature* 1983;305:805–808.
15. Heyer EJ, Macdonald RL: Barbiturate reduction of calcium-dependent action potentials: Correlation with anesthetic action. *Brain Res* 1982;236:157–171.

16. Heyer EJ, Nowak LM, Macdonald RL: Bicuculline: A convulsant with synaptic and nonsynaptic actions. *Neurology* 1981;31:1381–1390.
17. Jackson MB, Lecar H, Mathers DA, Barker JL: Single channel currents activated by γ-aminobutyric acid, muscimol, and (-)-pentobarbital in cultured mouse spinal neurons. *J Neurosci* 1982;2:889–894.
18. Krnjevic K: Chemical nature of synaptic transmission in vertebrates. *Physiol Rev* 1974;54:418–450.
19. Krongsgaard-Larsen P, Johnston GAR, Lodge D, Curtis DR: A new class of GABA agonist. *Nature (London)* 1977;268:53–55.
20. Leslie SW, Friedman, Mitchell B, Coleman RR: Effects of chlordiazepoxide on depolarization-induced calcium influx into synaptosomes. *Biochem Pharmacol* 1980;29:2439–2443.
21. Macdonald RL: Neuropharmacology of spinal cord and dorsal root ganglion neurons in primary dissociated cell culture, in Davidoff R (ed): *Spinal Cord Pharmacology*. New York, Marcel Dekker Pub Co, 1983a; pp381–408.
22. Macdonald RL: Mechanisms of anticonvulsant drug action, in Meldrum BS, Pedley TA (eds): *Recent Advances in Epilepsy*, Vol 1. New York, Chruchill Livingston, 1983b,pp 1–23.
23. Macdonald RL, Barker JL: Specific antagonism of GABA-mediated postsynaptic inhibition in cultured mammalian spinal cord neurons: A common mode of convulsant action. *Neurology* 1978a,28:325–330.
24. Macdonald RL, Barker JL: Benzodiazepines specifically modulate GABA-mediated postsynaptic inhibition in cultured mammalian neurons. *Nature* 1978b;271:563–564.
25. Macdonald RL, Barker JL: Enhancement of GABA-mediated postsynaptic inhibition in cultured mammalian spinal cord neurons: A common mode of anticonvulsant action. *Brain Res* 1979a;167:323–336.
26. Macdonald RL, Barker JL: Anticonvulsant and anesthetic barbiturates have different postsynaptic actions in cultured mammalian neurons. *Neu'ology* 1979b;29:432–447.
27. Macdonald RL, Werz, MA: Barbiturates decrease voltage-dependent conductance of mouse neurons in dissociated cell culture. *Soc Neurosci Abstr* 1982;*8*:492.
28. Mohler H, Okada T: Benzodiazepine receptor: Demonstration in the central nervous system. *Science* 1977;198:1 849–851.
29. Morgan KG, Bryant SH: Pentobarbital: Presynaptic effect in the squid giant synapse. *Experientia* 1977;33:487–488.
30. Nicoll RA: Pentobarbital: Action on frog motorneurons. *Brain Res* 1975;96:119–123.
31. Olsen RW: Drug interactions at the GABA receptor ionophore complex. *Annu Rev Pharmacol Toxicol* 1982;22:245–277.
32. Olsen RW, Snowman AM: Chloride-dependent enhancement by barbiturates of γ-aminobutyric acid receptor binding. *J Neurosci* 1982;2:1812.
33. Polc P, Haefely W: Effects of two benzodiazepines, phenobarbitone, and baclofen on synaptic transmission in cat cuneate nucleus. *Nauyn-Schmiedeberg's Arch Pharmac* 1976;294:121–131.
34. Ransom, BR, Bullock, PN, Nelson, PG: Mouse spinal cord in cell culture. III. Neuronal chemosensitivity and its relationship to synaptic activity. *J Neurophysiol* 1977;40:1163–1177.
35. Schulz DW, Macdonald RL: Barbiturate enhancement of GABA-mediated inhibition and activation of chloride ion conductance: Correlation with anticonvulsant and anesthetic actions. *Brain Res* 1981;209:177–188.
36. Skerritt J, Macdonald RL: Diazepam enhances the action but not the binding of the GABA analog THIP. *Brain Res* 1984a;297:181–186.
37. Skerritt J, Macdonald RL: Benzodiazepine receptor ligand actions on GABA responses. Benzodiazepines, CL 218872, Zopiclone. *Eur J Pharmacol* 1984b;101:127–134.
38. Skerritt J, Macdonald RL: Benzodiazepine receptor ligand action on GABA responses. Beta-carbolines, purines. *Eur J Pharmacol* 1984c;101:135–141.
39. Skerritt JH, Willow M, Johnston GAR: Diazepam enhancement of low affinity GABA binding to rat brain membranes. *Neurosci Lett* 1982;29:63–66.
40. Skerritt JH, Werz MA, McLean MJ, Macdonald RL: Diazepam and its anomalous *p*-chloro-derivative Ro 5-4864: Comparative effects on mouse neurons in cell culture. *Brain Res* 1984;310:99–105.
41. Squires RF, Braestrup C: Benzodiazepine receptors in rat brain. *Nature* 1977;266:732–734.
42. Study RE, Barker JL: Diazepam and (-)-pentobarbital: Fluctuation analysis reveals different mechanisms for potentiation of γ/aminobutyric acid responses in cultured central neurons. *Proc Natl Acad Sci USA* 1981;78:7180–7184.
43. Werz MA, Macdonald RL: Barbiturates decrease voltage-dependent calcium conductance of mouse neurons in dissociated cell culture. *Mol Pharmacol* 1985;28:269–277.
44. White WF, Dichter MA, Snodgrass SR: Benzodiazepine binding and interactions with GABA receptor complex in living cultures of rat cerebral cortex. *Brain Res* 1981;215:162–176.
45. Willow M, Johnston GAR: Enhancement by anesthetic and convulsant barbiturates of GABA binding to rat brain synaptosomal membranes. *J Neurosci* 1981;1:364–367.

3

Postsynaptic Actions of Pentobarbital in Cultured Mouse Spinal Neurons and Rat Hippocampal Neurons

David G. Owen, Jeffery L. Barker, Menachem Segal, and Robert E. Study

ABSTRACT. Three distinct postsynaptic properties of pentobarbital have been studied in cultured mouse spinal cord and rat hippocampal neurons. At subanesthetic concentrations (10–100 μM), pentobarbital inhibited Ca-dependent action potentials through an unknown mechanism but probably not indirectly through an effect on potassium conductances (G_K). At higher concentrations but still within this range, pentobarbital potentiated Cl^- current responses to exogenously applied GABA by prolonging the average open-channel lifetime. Current fluctuation analysis of responses to applied GABA and analysis of GABA-mediated postsynaptic currents (PSCs) suggest that inhibitory PSC (IPSC) duration is determined by channel lifetime. Pentobarbital lengthened IPSCs presumably by increasing channel lifetime. A direct response to pentobarbital occurred at concentrations in excess of 100 μM and consisted of an increase in Cl^- conductance (G_{Cl}). Evidence has been obtained to suggest that pentobarbital opens the same Cl^- channel as GABA but probably not via the GABA receptor. Rather, pentobarbital may bind a separate site, closely coupled to the GABA receptor and Cl^- channel.

It is suggested that the ability of pentobarbital to modulate calcium conductances (G_{Ca}) and GABA-activated G_{Cl} may contribute significantly to its anticonvulsive effects but not its anesthetic property. Rather, the latter may be a function of the ability of pentobarbital to directly activate G_{Cl}.

INTRODUCTION

The barbiturates have been widely used clinically as anesthetics and in anticonvulsant therapy.[12,13] This group of drugs has consequently inspired considerable research aimed at elucidating the mechanism underlying their clinical properties. It has become apparent that several loci of action probably contribute to a greater or lesser extent to the observed clinical effects. One member of this family of compounds, pentobarbital (PB), has been studied in some depth in this laboratory, using the cultured spinal cord and hippocampal neuron as model systems. It is quite well established now, from a variety of experimental systems, that PB is able to act in a number of distinct ways. These actions include (1) potentiation of

David G. Owen, Jeffery L. Barker, and Robert E. Study • Laboratory of Neurophysiology, NINCDS, NIH, Bethesda, Maryland 20205. *Menachem Segal* • Weizmann Institute, Rehovot, Israel.

GABA function,[2,16,17] (2) depression of excitatory amino acid responsivity,[2,16,17,31] (3) inhibition of calcium conductance,[5,13,21] and (4) direct activation of the GABA-gated Cl^- conductance[16,19,32] (Study and Barker, in preparation). We have observed all these actions in the cultured mouse spinal cord and rat hippocampal neuron. In particular, the preparation has allowed us to examine in detail the mechanisms underlying the actions of PB intimately associated with the GABA-activated Cl^- channel, both in conjunction with and independently of GABA. Furthermore, we have been able to demonstrate a good correlation between the modulatory effects of PB on postsynaptic responses to exogenously applied GABA and GABA released synaptically in cultures of hippocampal neurons.

METHODS

Mouse spinal cord and rat hippocampal neurons were cultured according to previously published methods.[30] Electrophysiological recordings were made after 3–4 weeks in culture, using microelectrodes filled with 3 M KCl, 3 M K_2Ac, or 3 M CsCl. Voltage-clamp was achieved using a conventional two electrode system. The basic recording medium consisted of (mM): 140 NaCl, 5 KCl, 4 $MgCl_2$, 4 $CaCl_2$, 5.6 glucose, and 5 HEPES/NaOH, pH 7.4. In some experiments 1 μM TTX and 20 mM TEA were included and the Ca/Mg ratio altered, in which cases NaCl was adjusted to maintain isoosmolarity. Purified (+) and (−) PB were obtained from Dr. R. Willette. All drugs were applied to cells by pressure ejection from micropipettes.

RESULTS

Pentobarbital Potentiation of GABA Function

Application of GABA to voltage-clamped hippocampal neurones caused an increase in membrane conductance, as indicated by an increase in the current responses to small hyperpolarizing voltage pulses (Fig. 1B). There was no net change in the steady-state current during the response, indicating that the holding potential (−20 mV) corresponded to the reversal potential for GABA (E_{GABA}). A brief application of (−) PB (100 μM) potentiated the response to applied GABA without altering the driving force or directly changing membrane conductance itself (Fig. 1B). The result suggests that PB acts by increasing the macroscopic conductance response to GABA. A similar macroscopic effect of Diazepam

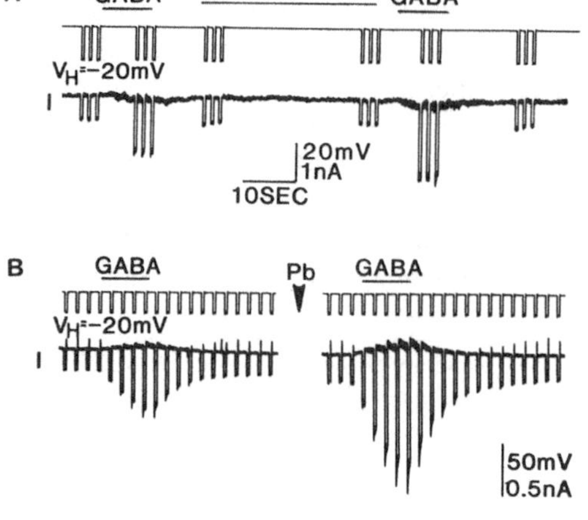

Figure 1. Diazepam and pentobarbital enhance GABA-evoked conductance. (A) The cell was voltage-clamped at −20 mV, and 20 mV hyperpolarizing commands were applied to estimate resting conductance. GABA was applied to the cell at the inversion potential for the current response under control conditions and immediately after pressure application of 10 μM diazepam. Diazepam enhances the GABA-activated conductance by about 50%, while having little effect of its own on resting conductance. (B) In another cell GABA-activated conductance (left) is markedly enhanced by 100 μM pentobarbital. In neither case was there a change in the current response to GABA during drug application. Reproduced with permission from Segal and Barker.[33]

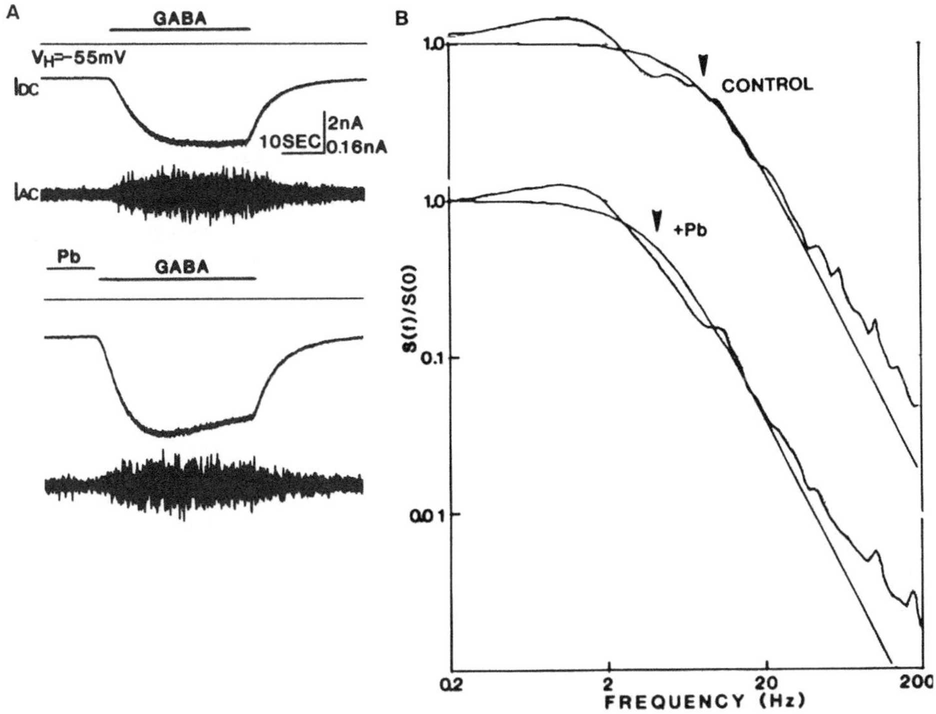

Figure 2. Pentobarbital potentiates GABA-evoked responses and increases channel lifetime. (A) The cell was voltage-clamped at −55 mV, and 20 μM GABA was applied by pressure under control conditions (top) and immediately after application of 100 μM pentobarbital (bottom). The drug increases the current response amplitude from 3.1 nA to 4.7 nA. (B) Spectral analysis of the membrane current variance associated with the responses shows that both spectra are well fitted by a single Lorentzian term and that pentobarbital shifts the corner frequency from 8.2 Hz to 4.1 Hz, corresponding to a shift in estimated channel lifetime from 19.4 msec in control to 38.5 msec in the presence of the drug. Reproduced with permission from Segal and Barker.[33]

(DZ) was also found in these cells (Fig. 1A) although the mechanism underlying this effect differs (see below).

Analysis of current fluctuations, associated with the GABA-evoked current response under voltage-clamp, suggested that this effect of PB could largely be accounted for by an increase in the average lifetime of the Cl^- channel mediating the GABA response (Fig. 2). In contrast, current fluctuation analysis showed that the potentiation of GABA-evoked current by 10 μM DZ (Fig. 1) was primarily due to an increase in the opening frequency of channels with only slight lengthening of the channel lifetime. Neither PB nor DZ altered the single-channel conductance (~18 pS).

Does Pentobarbital Facilitate GABA-Mediated Transmission?

It was clear that PB could modulate the postsynaptic response to exogenously applied GABA in hippocampal neurons as had been previously described for cultured spinal neurons.[2] However, could PB modulate synaptic responses, thought to be mediated by endogenous GABA, in hippocampal cell cultures?

To answer this question, synaptic events were studied in hippocampal cell cultures by two methods. (1) Inhibitory postsynaptic potentials (IPSPs), recorded with a single intracellular electrode, were evoked by triggering action potentials in the presynaptic neuron with

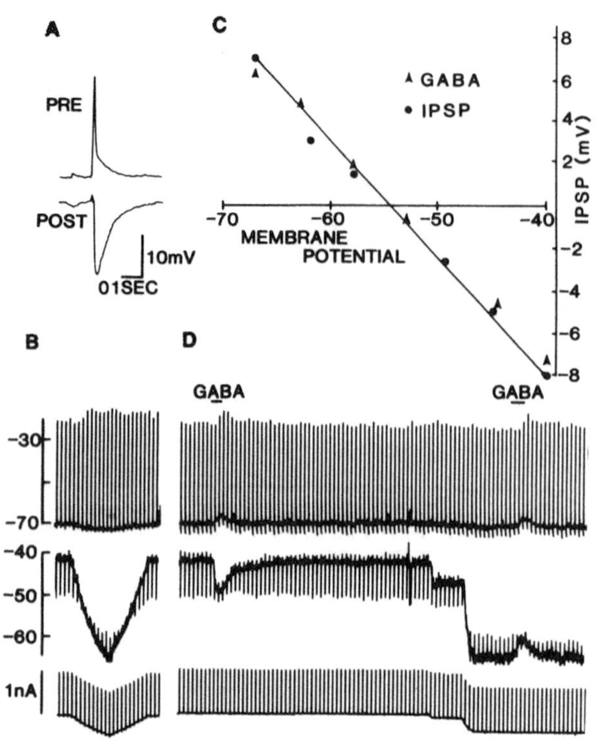

Figure 3. IPSPs evoked in cultured hippocampal neurons invert at the same potential as responses to GABA. Simultaneous intracellular recordings were made from pairs of neurons using a KCl-filled electrode to trigger an action potential (attenuated by the chart recorder) in the presynaptic cell (A), and a Cs_2SO_4-filled electrode (A–D) to record postsynaptic potentials. The membrane potential of the postsynaptic cell was varied by DC current injection (bottom traces) while ether evoking spikes in presynaptic neurons (upward current deflections in bottom traces produce truncated spikes in top traces) at 1 Hz to generate IPSPs (B) or applying GABA to elicit voltage responses (D). Note that the responses to GABA shunt the IPSPs. A plot of IPSP amplitude as a function of membrane potential (C) shows that GABA-evoked responses invert in polarity at the same potential as IPSPs (about −54 mV). In all chart records the top trace is of the presynaptic neuron, the middle trace of the postsynaptic neuron, and the bottom trace of the current pulses applied to the presynaptic neuron. Reproduced with permission from Segal and Barker.[33]

another intracellular electrode. (2) Voltage-clamp was used to record inhibitory postsynaptic currents (IPSCs). In this case synaptic events were triggered by pressure application of glutamate (Glu) to a presynaptic cell (Fig. 4, inset).

GABA-mediated IPSPs were identified on the basis of their inversion potential and sensitivity to applied picrotoxin, a GABA antagonist. Using Cs_2SO_4-filled recording electrodes, picrotoxin-sensitive (not shown) IPSPs reversed at about −55 mV, which coincided with the reversal potential for responses to applied GABA in these cells (Fig. 3). Similarly, flurries of PSCs, evoked by Glu (Fig. 4), inverted at about −60 mV and could summate to cause a long-lasting conductance increase in the postsynaptic cell. Again, IPSCs showing picrotoxin sensitivity were selected for analysis (Fig. 4). Based on these criteria. GABA-mediated IPSCs decayed exponentially with a single time constant of about 24 msec at −40 mV and were slightly prolonged at more depolarized potentials (27–30 msec at +10 mV) (Fig. 5). These values correspond closely to the average open-times of channels activated by applied GABA (Fig. 2), suggesting that the duration of IPSCs evoked under these recording conditions is determined primarily by the open-channel lifetime. The decay of IPSCs was voltage-sensitive, becoming slower at depolarized potentials (Fig. 5). This resembles the known voltage sensitivity of channels activated by applied GABA (Study *et al.*, submitted for publication). PB caused a prolongation of the duration of IPSCs evoked by Glu application without altering the amplitude of the IPSC appreciably (Fig. 4B).

Pentobarbital Inhibition of Ca^{2+} Spikes

In order to examine the effects of PB on Ca^{2+} conductances in spinal cord neurons, the recording conditions were designed to eliminate all other voltage-dependent conductances

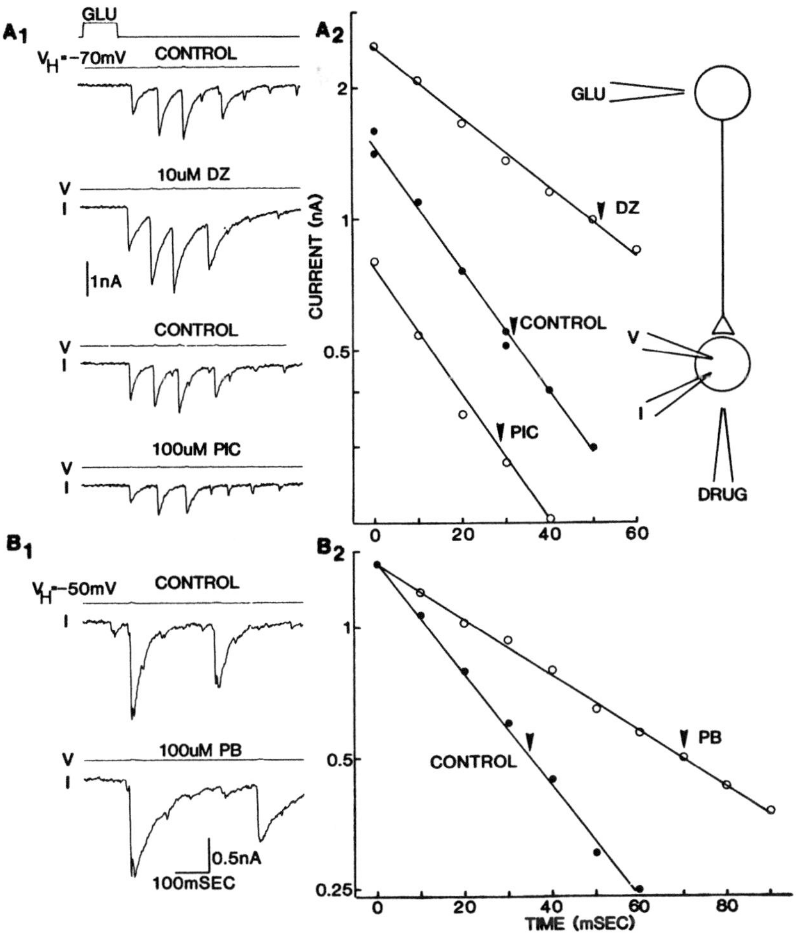

Figure 4. (A1) Effects of diazepam and picrotoxin on IPSC amplitude and τ_{IPSC}. The cell was voltage-clamped at −70 mV, and IPSCs were triggered, under control conditions and in the presence of diazepam or picrotoxin, by application of glutamate onto a remote neuron. In each panel the top tracing is a pen recorder tracing of voltage and the bottom is a pen recorder tracing of current. (A2) Semilogarithmic plots of representative IPSCs under control and drug conditions show that diazepam increases IPSC amplitude from 1.8 nA to 2.7 nA and τ_{IPSC} from 15 msec to 20 msec. Picrotoxin, on the other hand, reduces the amplitude but does not alter the time constant of decay of IPSCs. Insert shows diagrammatic representation of the experimental method used to evoke IPSCs (B1) Pentobarbital increases τ_{IPSC}. The cell was voltage-clamped at −50 mV, and IPSCs were evoked by brief pulses of glutamate applied to the presynaptic cell under control conditions and in the presence of 100 μM pentobarbital. The drug markedly prolongs the time course of the IPSC with little change in its amplitude. (B2) Semilogarithmic plots of IPSC decay in control and during pentobarbital show that in each condition the IPSC decays as a single exponential with the drug shifting τ_{IPSC} from about 40 msec to about 81 msec. Reproduced with permission from *Neuroscience Letters*.[4]

that may be activated over the same range of membrane potentials. Thus, TTX and TEA were included in the medium and electrodes were filled with 3M CsCl.[1] Under these conditions these cells have negligible outward K^+ currents but may exhibit a Ca^{2+}-dependent Cl^- conductance which is often reflected as a regenerative afterdepolarization.[27] Administration of 50 μM (−)PB completely and reversibly prevented the triggering of an all-or-none Ca^{2+}-dependent action potential in response to depolarizing current injection (Fig. 6A). On the other hand, the (+) isomer of PB caused only a shortening in the action potential duration, at

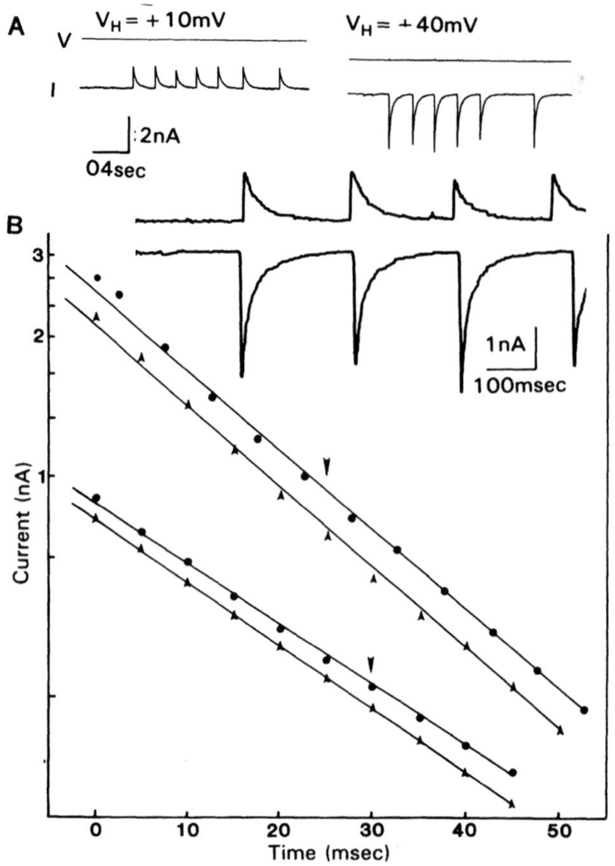

Figure 5. IPSCs decay with a single exponential time constant. (A) Chart recording of IPSCs generated by glutamate application to a remote neuron. Inward-going IPSCs were evoked when the cell was voltage clamped at −40 mV (right), while outward-going IPSCs were triggered at +10 mV (left). (B) When the decay of membrane current is plotted, it is evident that the IPSCs decay with a single time constant at both potentials. τ_{IPSC} is 23 msec and 24 msec for two representative IPSCs measured at −40 mV (top two plots) and 27 msec and 30 msec at +10 mV (bottom two plots). The inset shows IPSCs at +10 mV (top) and −40 mV (bottom) digitized on a Nicolet oscilloscope. Reproduced with permission from Segal and Barker.[33]

this concentration, without significant change in the peak amplitude or rate of rise and decay (Fig. 6B).

The (−)PB effect on Ca^{2+} spikes was dose-dependent and effective at concentrations as low as 10 μM. At this concentration the Ca^{2+} action potential was considerably shortened with little effect on the peak amplitude (Fig. 7A). Significantly, on the same cell there was no detectable potentiation of responses to GABA at this concentration (Fig. 7B). Voltage-clamp analysis is required to elucidate the mechanism underlying this effect of PB on Ca^{2+} spike generation. However, since steps were taken to eliminate those K^+ conductances activated by depolarization, it is reasonable to speculate that PB interacts directly with the Ca^{2+} conductance in some way. This might include promoting inactivation as described in invertebrate neurons.[21]

Although the observation that PB inhibits Ca^{2+} conductances in this and other systems is not a novel one,[5,13,21] the concentration range for this effect is somewhat lower than previously reported. This property of PB may therefore be more significant than previously supposed in its contribution to the subanesthetic, anticonvulsant actions of the drug.

Direct Agonist Activity of Pentobarbital

At concentrations expected to produce clinical anaesthesia (>100 μM) (−)PB directly caused a conductance increase in spinal neurons. The current evoked under voltage-clamp

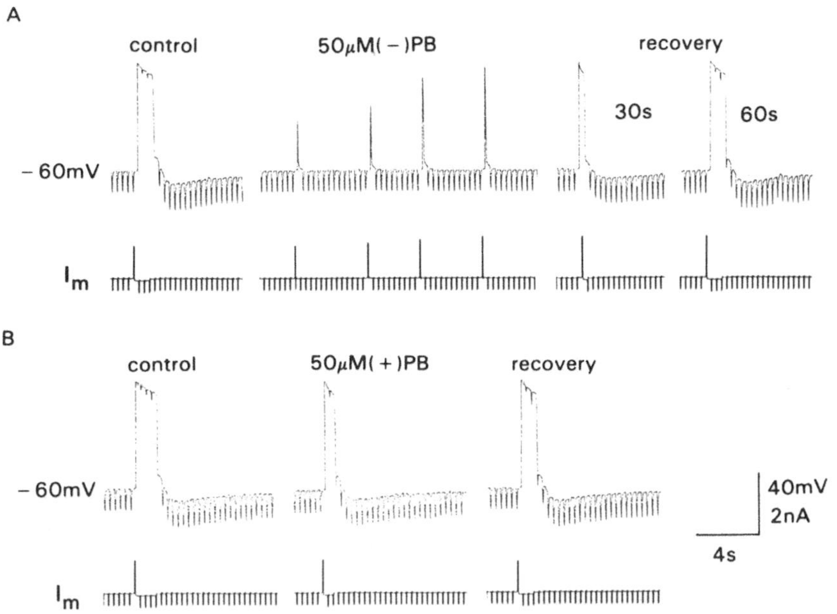

Figure 6. (−) Pentobarbital is a much more potent inhibitor of Ca-dependent action potentials than is (+) pentobarbital. (A) All-or-none spike is triggered in a spinal neuron by a 50 msec depolarizing current pulse (I_m) from the resting potential of −60 mV. Application of 50 μM (−) PB (middle panel) prevents activation even with larger current pulses. This effect was reversible (right panel). (B) In the same cell, 50 μM (+) PB caused a shortening of the spike with little effect on spike amplitude. Recording was made with CsCl-filled electrode and in the presence of TEA_o (20 mM) and TTX (1 μM).

reversed in polarity at the same potential as GABA current in the same cell (Fig. 8), namely, −20 mV (using KCl-filled electrodes). The dose–response curve for the (−)PB response lies to the right of the GABA response curve but is apparently parallel and also has a steep slope (Fig. 9). The (+) isomer had no activity at 200 μM on a cell responding to 100 μM (−)PB. Other structural analogues having anesthetic properties, such as (−)secobarbital. (−)DMBB, and (−)PMBB, did exhibit activity at this concentration, whereas less potent anesthetics such

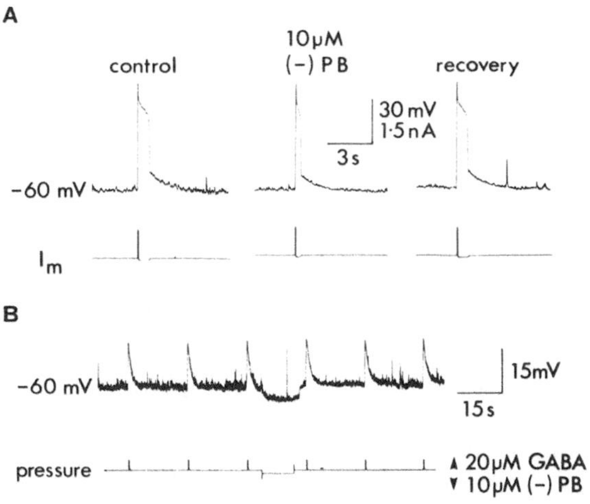

Figure 7. (−) PB blocks Ca spikes without potentiating GABA. (A) A dose of 10 μM (−) PB reduces the duration of a Ca spike evoked with depolarizing current injection (I_m) (left and middle panels). The inhibitory effect was reversible (right panel). (B) At the same concentration and in the same cell. (−) PB failed to potentiate the responses to brief pulses of 20 μM GABA applied by pressure ejection from a micropipette. The timing of drug applications is indicated in the lower tracing (pressure). The application of (−) PB was made at the same location with respect to the cell soma and with the same pressure as in A.

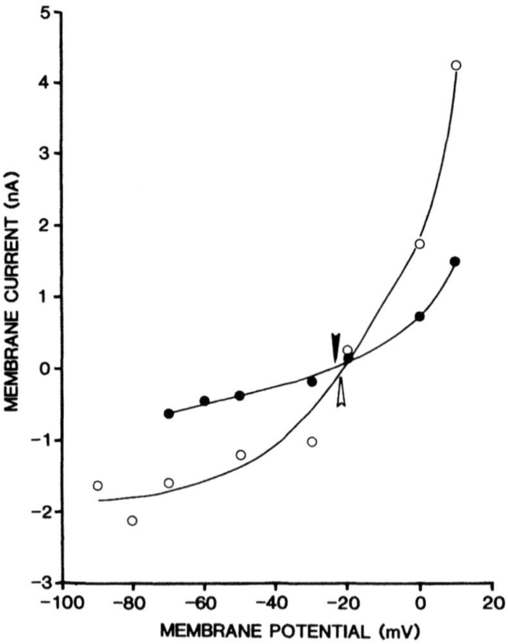

Figure 8. Reversal of GABA- and (−)PB-activated membrane currents on a cultured spinal neuron. Plotted values represent net current produced by GABA (15 μM) (open circles) and (−)PB (100μM) (filled circles) in a voltage-clamped cell at the indicated potentials (abscissa). Arrows indicate potentials at which no net current was observed [−22 mV for GABA, −23 mV for (−)PB].

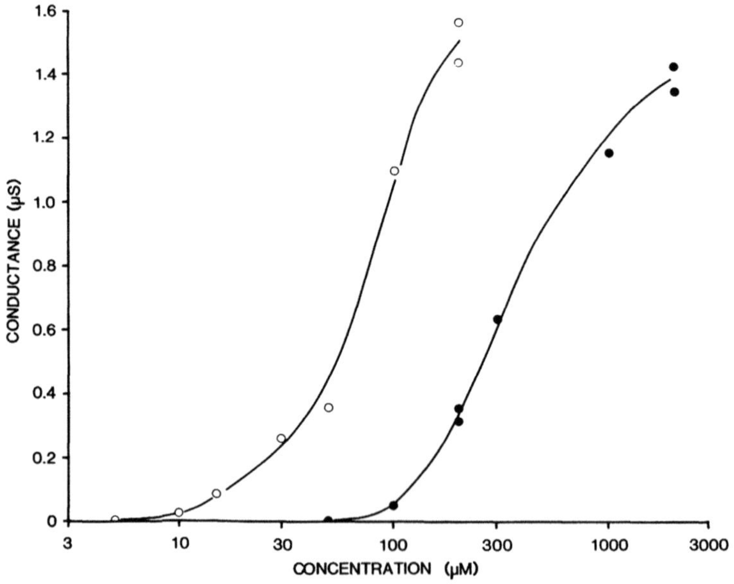

Figure 9. Concentration-response curves for GABA (open circles) and (−)PB (filled circles) on a single cultured mouse spinal neuron. The response was measured as the net conductance change induced by each agent at the reversal potential and was determined using 500-msec voltage command pulses under voltage-clamp. The voltage pulse amplitude was reduced from 20 mV to 2 mV as the concentration of either agonist was raised in order to prevent clamp saturation. Data were obtained from three cells. The maximum values shown are not necessarily the maximum attainable with these agents but rather the maximum that could be measured before recording instability or clamp saturation during voltage jumps prevented reliable measurements.

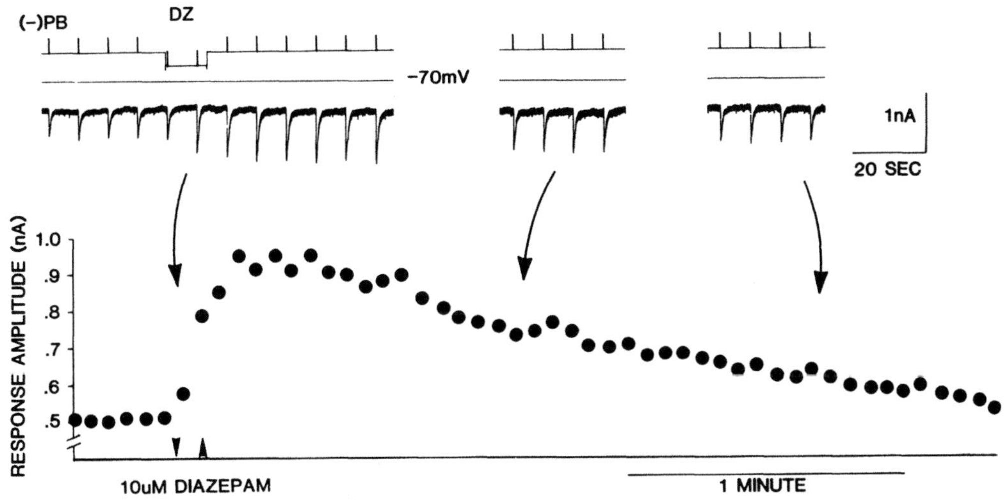

Figure 10. Effect of DZ on (−)PB-induced responses in a voltage-clamped cultured mouse spinal neuron. (−)PB (100 μM) was applied using 200 msec pressure pulses from a pipette positioned 10–20 μ from the cell soma. Diazepam (17 μM) was applied also by pressure for 10 sec from another pipette. Selected chart records of the application pressure (top trace), voltage (clamped at −70 mV, middle trace), and current (bottom trace) are shown at the top. These results are plotted below.

as hexobarbital and phenobarbital were ineffective at up to 200 μM (not shown). Drugs that are known to interact with GABA function also modified responses to (−)PB, bicuculline blocking the response (not shown) and DZ potentiating PB currents (Fig. 10).

The power spectrum of current fluctuations, associated with PB-induced responses under voltage-clamp, was fit by a Lorentzian function consistent with activation of a two-state (open/closed) ion channel having, on average, an open time of about 70 msec, considerably longer than that gated by GABA of about 25 msec (Fig. 11). In a small proportion of cells, however, PB apparently activated channels whose lifetime approximated that of GABA (Fig. 11). Current-variance plots for GABA and PB yielded single linear regressions whose slopes indicated an elementary conductance of about 30 pS in each case. The linearity of these plots confirmed that the probability of opening of both GABA- and PB-activated channels was low, an assumption made in the fluctuation analysis.

DISCUSSION

These results show quite clearly that (−)PB has several distinct actions in the model systems studied. These are (1) potentiation of GABA function, (2) depression of a voltage-activated Ca^{2+} conductance, and (3) direct activation of an inhibitory Cl^- conductance. All three effects are potentially inhibitory to electrical excitibility in the central nervous system (CNS) and together may contribute to the overall clinical properties of the drug.

The clinical actions of PB are, broadly speaking, twofold: first, at lower doses (50–100 μM), an anticonvulsant action and second, at doses in excess of 100 μM, a sedative–hypnotic effect. The observation that PB can potentiate the effects of synaptically released GABA lends support to the already prevalent idea[13,32] (Study and Barker, in preparation) that the anticonvulsive property of this barbiturate may reflect its ability to promote inhibitory influences exerted through GABAergic systems in the CNS. In addition, the results of fluctuation and IPSC analyses have suggested the following: (1) The duration of GABA-

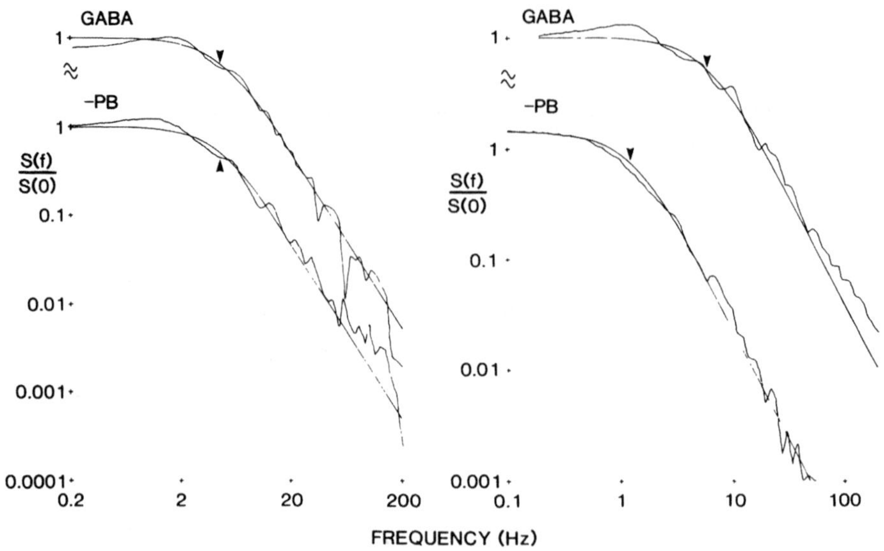

Figure 11. Representative power-density spectra of responses to GABA and (−)PB for two cultured mouse spinal neurons voltage-clamped at −60 mV. The left panel shows normalized spectra (partially overlapped) for (−)PB (200 μM) and GABA (15 μM) on one neuron, with the corner frequency, in each case, indicated by an arrowhead. The corner frequency (f_c) is related to the average open time (τ) of the channel by $\tau = \frac{1}{2}.\pi.f_c$. In this case f_c corresponds to an average channel open time of ~32 msec for both (−)PB and GABA. The right panel shows another cell where the corner frequency was significantly different for the two agents: ca. 1.5 Hz for (−)PB and ~8 Hz for GABA, corresponding to average channel open times of 106 msec and 20 msec, respectively. Note that the scales are different for the two cells.

mediated synapic currents is probably limited by the average lifetime of the channel and not diffusion of transmitter away from receptor sites. (2) PB is able to interact with "synaptic" GABA-receptor-channel complexes, resulting in longer-lasting inhibitory events. (3) This potentiation probably results from an increase in the average lifetime of open channels with little change in the number of channels opening per quantal event. Since the probability of reopening of channels is apparently low [from (1)], this would result in the observed effect, namely, a negligible change in PSC amplitude but significantly longer duration. (4) From the amplitude of evoked IPSCs and the elementary current estimated from noise analysis (e.g., Fig. 3), it follows that on average approximately 1700 channels are open at the peak of the IPSC. Table 1 summarizes the results of noise and IPSC analyses for hippocampal cultures and compares these with data from hippocampal slice.[7] There is reasonable agreement in both the time course of IPSCs and the functional effect of PB in these preparations suggesting a similarity in the basic mechanisms underlying GABA and PB function in the two systems. The mechanism of action of PB is clearly distinguishable from that of other GABA-modulating agents that interact with benzodiazepine binding sites. Thus, whereas PB lengthens the open time of channels gated with GABA, the latter, in general, have only slight effects on the channel open time as estimated from noise analysis experiments.[26] For example, the benzodiazepines, irazepine (IRZ), and lorazepam (LRZ) potentiate the effect of GABA, largely by increasing the effective opening frequency of GABA-activated channels.[26] These agents then resemble DZ in their effects on Cl^- channel kinetics.[26] Members of the β-carboline group of compounds, which also interact with CNS benzodiazepine receptors,[6] can also modulate GABA function. For example, the so-called inverse agonists, methyl β-carboline-3-carboxylate (CCM) and methyl 6,7-dimethoxy-4-ethyl β-carboline-3-carboxylate

Table I. Comparison of τ_{IPSC} and τ_{GABA} in Rat Hippocampal Neurons

	τ_{IPSC} (ms)	τ_{GABA} (ms)	Ratio
Culture[33]			
−60 mV	20	20	1.0
+10 mV	30	30	1.0
Picrotoxin	18	21	0.8
Pentobarbital	70	50	1.4
Diazepam	30	26	1.1
Slice[7]			
−60 mV	12	—	—
+10 mV	20	—	—
Pentobarbital	43	—	—

(DMCM), attenuate GABA function in cultured mouse spinal neurons.[26] These effects result from a decrease in the opening frequency of channels, again with little alteration in the channel open time. Similarly, butyl β-carboline-3-carboxylic acid ester (CCB) potentiates GABA function by increasing channel opening frequency and not open time.[26] There does therefore appear to be a separation in the loci of action of barbiturate and benzodiazepine receptor ligands with respect to the gating process of GABA activated Cl^- channels. This point is illustrated in Fig. 12, which summarizes, diagrammatically, the effects of various GABA-modulating agents on the properties of GABA-activated Cl^- channels in spinal cord neurons.

It appears, by reference to behavioral studies reported in the literature, that compounds which can modulate GABA function *in vitro,* as described previously, may have convulsant or anticonvulsant properties *in vivo* (Table II). For example, in addition to DZ, LRZ, and IRZ, the benzodiazepines, chlorazepam, flunitrazepam, and chlordiazepoxide (CDX), which are depressant *in vivo,* all potentiate GABA function *in vitro.* The β-carboline group of compounds, which include CCB, CCM, and DMCM and also ethyl β-carboline-3-carboxylate and hydroxymethyl β-carboline-3-carboxylate, exhibit a spectrum of activities *in vitro* ranging from inhibition to potentiation of GABA function. This spectrum is paralleled *in vivo* in terms of anticonvulsant, proconvulsant, and convulsant properties. The benzodiazepine receptor antagonist RO15-1788 has no direct effects on GABA function *in vitro* (except at high concentrations, personal observation) but does inhibit DZ-mediated potentiation of GABA.[26] *In vivo,* this compound correspondingly antagonizes convulsant or anticonvulsant actions of benzodiazepines and β-carbolines without itself having effects except at high doses. Finally, *in vitro* GABA antagonists such as penicillin, pentylenetetrazol, and picrotoxin are convulsant *in vivo.* Observations of these kinds have contributed to the notion that modulation of GABA function may underlie the clinical properties of certain drugs used in the control of anxiety and epilepsy. This may also be true of PB, and consistent with this idea, PB does alter GABA-activated channel activity *in vitro* at concentrations within the range (50–100 μM) required for its anticonvulsant effects. It is important to bear in mind, however, that not all anticonvulsants need necessarily act through GABA systems, and some certainly appear not to do so. For example, diphenhydramine, ethosuximide, sodium valproate, and carbamazepine do not modulate GABA function in cultured mouse spinal neurons. So although the suggestion that modulation of intrinsic GABA function may be a significant factor underlying the anticonvulsant property of PB is quite tenable, other actions unrelated to GABA systems may be involved. With this in mind it is noteworthy that inhibition of

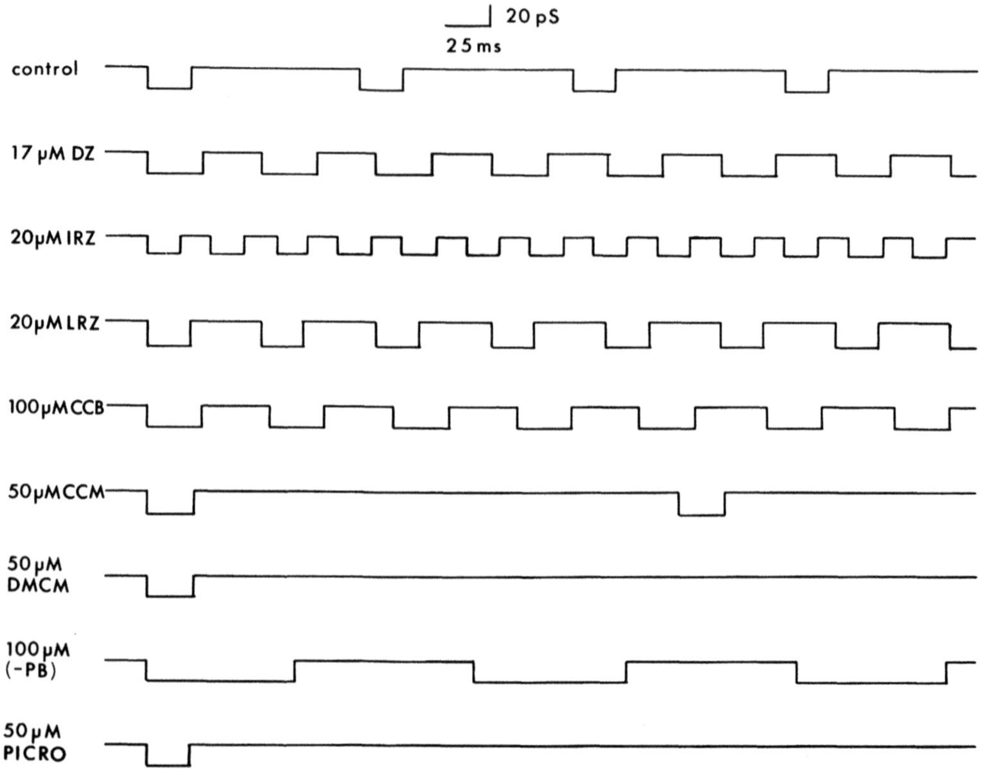

Figure 12. Schematic summary of the way in which GABA-modulating agents alter single-channel parameters. Each line represents the average behavior of GABA-gated Cl^- channels in cultured spinal neurons, either in the presence of GABA (20 μM) alone or together with modulators of GABA function. The downward-going steps represent conductance increases lasting the duration of the step. Intervals between steps do not represent the *actual* average intervals between channel openings, but do indicate the *relative* opening frequencies between the different conditions. The channel behavior described here was determined from current fluctuation experiments at room temperature[3,25,34] and is represented in terms of the average elementary conductance, channel lifetime, and relative opening frequency. Abbreviations of drug names are defined in the text.

Ca^{2+} conductances in spinal neurones can occur at quite low concentrations of PB, certainly at subanesthetic levels. If this effect holds for Ca^{2+} conductances at nerve terminals, it obviously may have important consequences for synaptic transmission. Whether or not PB can select for Ca^{2+} conductances on excitatory versus inhibitory terminals is an intriguing thought.

Based on concentration dependence alone, neither GABA modulation nor reduction in Ca conductance can account for the observed anesthetic effects of PB. The well-documented direct effect of PB[16,19,32] (Study and Barker, in preparation), confirmed in this study, may underlie this clinical property, however. Certainly the concentration range of PB required to elicit the direct inhibitory response in spinal neurons is consistent with clinically effective doses. The results suggest that PB behaves as a GABA mimetic since (1) the dose–response curve is parallel to that of GABA with an equally steep slope, suggestive of a cooperative ligand binding step; (2) PB opens Cl^- ion channels with identical conductance to those operated by GABA; and (3) the PB effect is sensitive to agents known to interact directly with the GABA receptor (bicuculline) or allosterically (DZ). However, the data do not support the idea that PB activates the GABA receptor directly since the estimated channel

Table II. Review of the Behavioral Properties of Neuroactive Agents in Relation to GABA Function in the in Vitro Mouse Spinal Neuron[a]

Agent	GABA[6] ratio	GABA[24,34,36] binding	Behavioral[12,29] effects		GABA[3,17,26,35] function (electrophys.)
DZ	2.30	↑	Anticonflict[28]	hypnotic,	↑
CLZ	2.12	—		sedative,	↑
FNZ	2.45	—		anticonvulsant,	↑
LRZ	1.75	—		muscle relaxant[11]	↑
IRZ	—	—	Anticonflict[28]		↑
CDX	2.23	—			↑
RO15-1788	1.22	—	V. weak BZ-like cpd. Anticonvulsant at high conc.[22] Antagonizes: sed/ hyp/anticonv. effects of BZs.[10,15,22] CCEs proconv. effect,[22] DMCM-evoked convulsions[6]		No effect
CCB	—	—	Anticonvulsant[b]		↑
CCE	0.86	no effect	Antagonizes FLZ-anticonv. effect.[9] Proconvulsant.[9,23] Anxiogenic.[20] ↓ IRZ/PhB-anticonv. effects[28]		↓
CCM	0.61	no effect	Convulsant[6]		↓
HMC	—	no effect	↑ sleep latency.[18] Antagonizes FLZ-induced hypnosis[18]		↓
DMCM	0.46	no effect	Convulsant[6]		↓
PICRO	—	no effect	Convulsant		↓
PCN	—	—	Convulsant		↓
PTZ	—	—	Convulsant		
(−)PB	—	↑	Hypnotic/sedative/anticonvulsant		↑
DPH	—	—	Anticonvulsant	(absence—poorly)	No effect
ETHO.	—	—	Anticonvulsant	(PTZ/absence—good)	No effect
VALP.	—	—	Anticonvulsant	(PTZ/absence/kindled)	No effect
CARBAM.	—	—	Anticonvulsant	(PCN/PTZ/max. and min. electroshock)	No effect

[a]GABA ratio is the relative affinity of test ligand for brain benzodiazepine (BZ) receptor in the presence and absence of GABA or muscimol; e.g., values greater than 1.0 indicate an increased affinity due to GABA and reflect a positive, reciprocal cooperativity in binding between GABA and the BZ receptor ligand. Superscripts denote references. For explanation of drug abbreviations, see text.
[b]Skolnick, personal communication.

lifetime for PB is not [PB]-dependent, as would be expected in that case. In other words, PB ought to be able to potentiate its own GABA-like response. For the same reason, PB is probably not simply potentiating ambient GABA since it is known that the duration of GABA-activated channels in the presence of PB is [PB]-dependent. Binding data confirm that PB does not compete with GABA for its receptor.[25] A more attractive scheme may be envisaged requiring at least two cooperative binding sites for PB, allosterically coupled to the open form of the Cl^- channel in such a way as to stabilize that conformation. At low concentrations of PB, binding only one site, the effect would be to stabilize active channels promoted by GABA binding to its site. At high [PB]s the allosteric interaction between the PB binding sites and the open conformation of the channel might be sufficiently intensified so as to significantly stabilize spontaneous channel openings, resulting in an apparent agonist effect.

It is, perhaps, surprising that this drug has such a diversity of actions all of which can lead to a consistent overall effect, inhibition in the CNS. It might not be so surprising after all if, as in other systems such as the opiate[14] and possibly benzodiazepine systems,[8] there is an endogenous counterpart to the drug (PB) in the CNS. Regulation of the expression of such a substance could prove to be a much more powerful clinical tool.

REFERENCES

1. Adams PR: The calcium current of a vertebrate neurone, in Salankim J (ed): *Advances in Physiological Sciences*. New York, Pergamon, 1981,pp 135–138.
2. Barker JL, Ransom BR: Pentobarbital pharmacology of mammalian central neurones grown in tissue culture. *J Physiol (London)* 1978;280:331–354.
3. Barker JL, McBurney RN, Mathers DA: Convulsant-induced depression of amino acid responses in cultured mouse spinal neurones studied under voltage clamp. *Br J Pharmacol* 1983;80:619–629.
4. Barker JL, Owen DG, Segal M: GABA actions on the excitibility of cultured CNS neurons. *Neurosci Lett* 1984;47:313–318.
5. Blaustein MP, Ector AC: Barbiturate inhibition of calcium uptake by depolarized nerve terminals *in vitro*. *Molec. Pharmacol* 1975;11:369–378.
6. Braestrup C, Schmiechen R, Neef G, Nielsen M, Petersen EN: Interaction of convulsive ligands with benzodiazepine receptors. *Science* 1982;216:1241–1243.
7. Collingridge GL, Gage PW, Robertson B: Inhibitory postsynaptic currents in rat hippocampal neurones. *J Physiol* 1984;356:551–564.
8. Costa E, Corda MG, Guidotti A: On a brain polypeptide functioning as a putative effector for the recognition sites of benzodiazepine and beta-carboline derivatives. *Neuropharmacology* 1983;22:1481–1492.
9. Cowen PJ, Green AR, Nutt DJ, Martin IL: Ethyl-β-carboline carboxylate lowers seizure threshold and antagonizes flurazepam-induced sedation in rats. *Nature* 1981;290:54–55.
10. Darragh A, Scully M, Lambe R, Brick I. O'Boyle C, Downie WW: Investigation in man of the efficacy of a benzodiazepine antagonist, RO 15-1788. *Lancet* 1981;2:8–10.
11. Garattini S, Mussini E, Randall LO, in Garattini S, Mussini E, Randall LO (eds): *The Benzodiazepines*. New York, Raven Press, p 685.
12. Harvey SC: Hypnotics and sedatives, in Gilman AG, Goodman LS, Gilman A (eds): *The Pharmacological Basis of Therapeutics*. New York, Macmillan. 1980.
13. Heyer EJ, Macdonald RL: Barbiturate reduction of calcium-dependent action potentials: Correlation with anaesthetic action. *Brain Res* 1982;236:157–171.
14. Hughes J, Smith TW, Kosterlitz HW, Fothergill LA, Morgan BA, Morris HR: Identification of two related pentapeptides from the brain with potent opiate agonist activity. *Nature* 1975;258:577–579.
15. Hunkeler W, Mohler M, Pieri L, Dole P, Bonetti E, Cumin R, Schaffner R, Haefely W: Selective antagonists of benzodiazepines. *Nature* 1981;290:514–516.
16. Macdonald RL, Barker JL: Different actions of anticonvulsants and anaesthetic barbiturates revealed by use of cultured mammalian neurones. *Science* 1978;200:775–777.
17. Macdonald RL, Barker JL: Enhancement of GABA-mediated inhibition in cultured mammalian spinal cord neurons: a common mode of anticonvulsant action. *Brain Res* 1979;167:323–336.
18. Mendelson WB, Cain M, Cook JM, Paul SM, Skolnick P: A benzodiazepine receptor antagonist decreases sleep and reverses the hypnotic actions of flurazepam. *Science* 1983;219:414–416.
19. Nicoll RA: Presynaptic action of barbiturates in the frog spinal cord. *Proc Nat Acad Sci USA* 1975;72:1460–1463.
20. Ninan PT, Insel TM, Cohen RM, Cook JM, Skolnick P, Paul SM: Benzodiazepine receptor-mediated experimental "anxiety" in primates. *Science* 1982;218:1332–1334.
21. Nishi K, Oyama Y: Accelerating effects of pentobarbitone on the inactivation process of the calcium current in Helix neurones. *Br. J. Pharmacol* 1983;79:645–654.
22. Nutt DJ, Cowen PJ, Little HJ: Unusual interactions of benzodiazepine receptor antagonists. *Nature* 1982;295:436–438.
23. Oakley NR, Jones BJ: The proconvulsant and diazepam-reversing effects of ethyl β-carboline-3-carboxylate. *Eur J Pharmacol* 1980; 68:381.
24. Olsen RW, Ticku MK, Van Ness PC, Greenlee D: Effects of drugs on α-aminobutyric acid receptors, uptake, release and synthesis *in vitro*. *Brain Res* 1978;139:277–294.
25. Olsen RW, L-Lundberg F, Snowman A, Stephenson FA: Barbiturate interactions with the benzodiazepine-GABA receptor complexes in mammalian brain, in Paul SM (ed): *Pharmacology of Benzodiazepines*. New York, Macmillan, 1982.
26. Owen DG, Study RE, Gratz E, Barker JL: Pharmacological modulation of GABA responses in cultured mouse spinal neurones. *Soc Neurosci Abstr* 1982;8:239.5.

27. Owen DG, Segal M, Barker JL: A Ca-dependent Cl^- conductance is present in cultured mammalian spinal neurones. *Nature* 1984;311:567–570.
28. Petersen EN, Paschelke G, Kehr W, Nielsen M, Braestrup C: Does the reversal of the anticonflict effect of phenobarbital by β-CCE and FG 7142 indicate benzodiazepine receptor-mediated anxiogenic properties? *Eur J Pharmacol* 1982;82:217–221.
29. Rall TW, Schlafer LS: Drugs effective in the therapy of the epilepsies, in Gilman AG, Goodman LS, Gilman A (eds): New York, Macmillan. 1980.
30. Ransom BR, Neale E, Henkart M, Bullock PN, Nelson PG: Mouse spinal cord in cell culture. I. Morphology and intrinsic neuronal electrophysiologic properties. *J Neurophysiol* 1977;40:1132–1150.
31. Richards CD, Smaje JC: The actions of halothane and pentobarbitone on the sensitivity of neurones in the guinea-pig prepiriform cortex to ionotophoretically applied L-glutamate. *J Physiol (London)* 1974;239:103–105.
32. Schultz D, Macdonald RL: Barbiturate enhancement of GABA-mediated inhibition and activation of chloride ion conductance: Correlation with anticonvulsant and anaesthetic actions. *Brain Res* 1981;209:177–188.
33. Segal M, Barker JL: Rat hippocampal neurons in culture: Voltage clamp analysis of inhibitory synaptic connections. *J Neurophysiol* 1984;52:469–487.
34. Skerritt JH, Willow M, Johnson GAR: Diazepam enhancement of low affinity GABA binding to rat brain membranes. *Neurosci Lett* 1982;29:63–66.
35. Study RE, Barker JL: Diazepam and (−)-pentobarbital: Fluctuation analysis reveals different mechanisms for potentiation of γ-aminobutyric acid responses in cultured central neurones. *Proc Natl Acad Sci USA* 1981;78:7180–7184.
36. Willow M, Johnson GAR: Enhancement of GABA binding by pentobarbitone. *Neurosci Lett* 1980;18:323–327.

4

Selective and Differential Effects of Barbiturates on Neuronal Activity

Sheldon H. Roth, Kok-Swang Tan, and M. Bruce MacIver

ABSTRACT. The differential and selective actions of barbiturates were demonstrated using two *in vitro* preparations (rat hippocampal brain slice and crayfish single isolated sensory neuron).

Pentobarbital (Pb) and phenobarbital (Ph) both produce a concentration-dependent alteration of the electrically evoked field potential amplitude (FPA) of hippocampal CA1-neurons. However, Pb produced a biphasic response; low concentrations (20–100 μM) enhanced and higher concentrations (greater than 150 μM) depressed the amplitude. At all effective concentrations Ph only depressed the FPA. A comparison of the stereoisomers of Pb revealed that the R(+) isomer was approximately fivefold more effective on a molar basis at enhancing and approximately 1.5 times at depression the FPA. A biphasic response of Pb was also observed on both electrically induced and spontaneous unit discharge activity.

The effects on the senory neuron were more dramatic. Pb depressed the stretch-induced discharge activity; in contrast, Ph enhanced activity. The R(+) isomer of Pb was approximately 2.5 times more effective than the S(−) isomer. Examining the effects of other barbiturates (1, 3-dimethylbutyl barbituric acid, methohexital, barbital, and amobarbital) demonstrated that the depressant activity appeared to be dependent on substitution of a primary methyl group on the C_5 carbon chain.

In general, the effects of Pb and Ph on membrane electrical properties of the sensory neuron were similar; both depolarized the cell, increased membrane resistance, decreased spike threshold, and decreased amplitude and rate of rise of the action potential. A major difference, however, was the effect on the inhibitory postsynaptic potential (IPSP). It was shown that Ph was less effective on the IPSP, whereas the primary effect of Pb was enhancement of the GABAergic inhibitory modulation resulting in depression of activity.

These results support the concept that anesthetic barbiturates have a primary action at GABAergic sites in contrast to the anticonvulsant and convulsant barbiturates. The effects may be dependent on chemical structure suggesting the existence of a recognition site or multiple mechanisms.

INTRODUCTION

Anesthesia is a complex phenomenon that can be produced by a variety of chemical substances.[48] Although the search for a mechanism(s) of action of anesthetics has continued for several decades, how and where these agents act remains to be resolved.

Sheldon H. Roth, Kok-Swang Tan, and M. Bruce MacIver • Department of Pharmacology and Therapeutics, Faculty of Medicine, The University of Calgary, Calgary, Alberta T2N 4N1, Canada.

Historically, the actions of anesthetics have been described as nonspecific in nature.[53] Bernard wrote over a century ago, "An anesthetic is not a special poison for the nervous system," and implied that anesthetics affect most biological cells in a nonspecific manner.[6] Anesthetics are not related by chemical structure or composition and include simple gases, volatile ethers, alcohols, barbiturates, steroids, and opioids. This apparent lack of structural requirement for activity is a feature unique to neuropharmacology.[12] A common feature of all anesthetics is a relatively high lipid solubility, and this physical property, often defined as an oil/water partition coefficient, correlates extremely well with "anesthetic" potency on a variety of organisms, isolated tissues, and model systems.[41,53] It is this simple correlation that provided the basis of the original lipid solubility hypothesis of anesthesia and has maintained the concept of a common physical mechanism of action.[48] The correlation provides only a rule for efficacy, demonstrates the dependence or ability of the anesthetic to adsorb into the membrane phase, and is useful to predict relative potencies. However, it does little to provide the required data to support a theory of mechanism. Since (1) a lack of structural requirement and dependence on lipid solubility are apparent for activity, (2) a chemical antagonist has never been found, (3) a specific receptor has not been identified or characterized, and (4) high pressure can reverse anaesthesia, the concept of a universal nonspecific mechanism of action has been maintained.

A unified or common mechanism of action of anesthetics is compatible with current observations.[24,40,42,49,50] This theme is present throughout this volume.

It may not be possible to elucidate the precise cellular or molecular mechanisms of anesthetics until site/state of anesthesia and the structure/function of excitable membranes are well defined. There is considerable evidence from neurophysiological studies, however, that all anesthetics do not produce a similar state of anesthesia (see Krnjević, Chapter 1, this volume). Each agent can produce a very characteristic pattern of effects,[49,63] and these differences cannot be explained in terms of kinetics.[39] The pharmacological profiles of anesthetics include both activation and depression, which may reflect selective alteration of facilitatory or inhibitory processes.[7,11,57,63] Selectivity has also been demonstrated at the level of the synapse[33,40] and membrane.[32]

In this chapter, the effects of barbiturate anesthetics on two *in vitro* preparations are described. It is to be noted that *in vitro* studies are not directed at the phenomenon of anesthesia *per se,* but provide quantitative data at the molecular and cellular levels which ultimately will explain, at least in part, the mechanisms involved in the state of anesthesia. The advantages and disadvantages of *in vivo* and *in vitro* studies have been documented.[27]

Isolated *in vitro* preparations provide the advantages of controlled environments, known applied drug concentrations, precise placement of stimulating and recording electrodes, and quantitative measurements of neuronal and synaptic activity.

Barbiturates were chosen for this study for a number of reasons; lipid solubility is an important factor for potency yet does not explain all the quantitative or qualitative membrane (cellular) actions[18]; these agents exhibit a multiplicity of actions[4,19,34,55], (see also chapter by Olsen, this volume); relative minor structural changes produce compounds with very selective (anticonvulsant, convulsant) activities[29] (see also chapters by Olsen, Macdonald, and Ticku, this volume); a variety of analogues including stereoisomers are readily available or can be synthesized; and a great deal of information is published on the mechanisms of action.[18] Our objectives were to demonstrate that barbiturate anesthetics can selectively alter the function of neuronal and synaptic activity, that concentration-dependent biphasic effects occur at these levels, that different neurons (i.e., membranes) are affected in a differential manner, that the neuronal membrane is capable of discriminating minor structural modifications, and that in addition to lipid solubility, the effects of barbiturates on excitable membrane properties may be dependent on chemical structure and/or molecular orientation.

METHODS

Hippocampal Brain Slice Preparation

Adult male Sprague-Dawley rats (200–300 g) were anesthetized with ether. The heart was stopped by a blow to the back of the thorax. The brain was quickly removed and placed in ice-cold oxygenated artificial cerebrospinal fluid (CSF) solution. Thin (400-μm) transverse slices of the dissected dentate–hippocampal formation were made on a Campden vibratome and placed at the gas/liquid interface on a nylon mesh screen in a McIlwain tissue chamber.[45] The chamber was maintained at 35°C and constantly perfused with oxygenated prewarmed CSF (1.0 ml/min).

Bipolar metal electrodes were placed in stratum radiatum to stimulate excitatory synaptic inputs to CA1 pyramidal neurons (see Fig. 1). Extracellular recording electrodes (2 M NaCl, 2–10 Mohm) were placed in the CA1 cell body layer to record synaptically evoked field potentials and unit discharge activity. Paired stimulus pulses of 0.01–0.05 msec duration (25–120 μA) were delivered at 0.1 Hz. Recorded signals were amplified (×1000), filtered (1 Hz–10 KHz, bandpass), and digitally stored for later analysis. The artificial CSF physiological solution[26,43] contained (in mM) NaCl 134; KCl 4.5; $CaCl_2$ 1.6; KH_2PO_4 1.25; $MgSO_4$ 2; $NaHCl_3$ 16; and glucose 10. CSF was saturated with 95%/5% oxygen/CO_2, pH 7.4 ± 0.1.

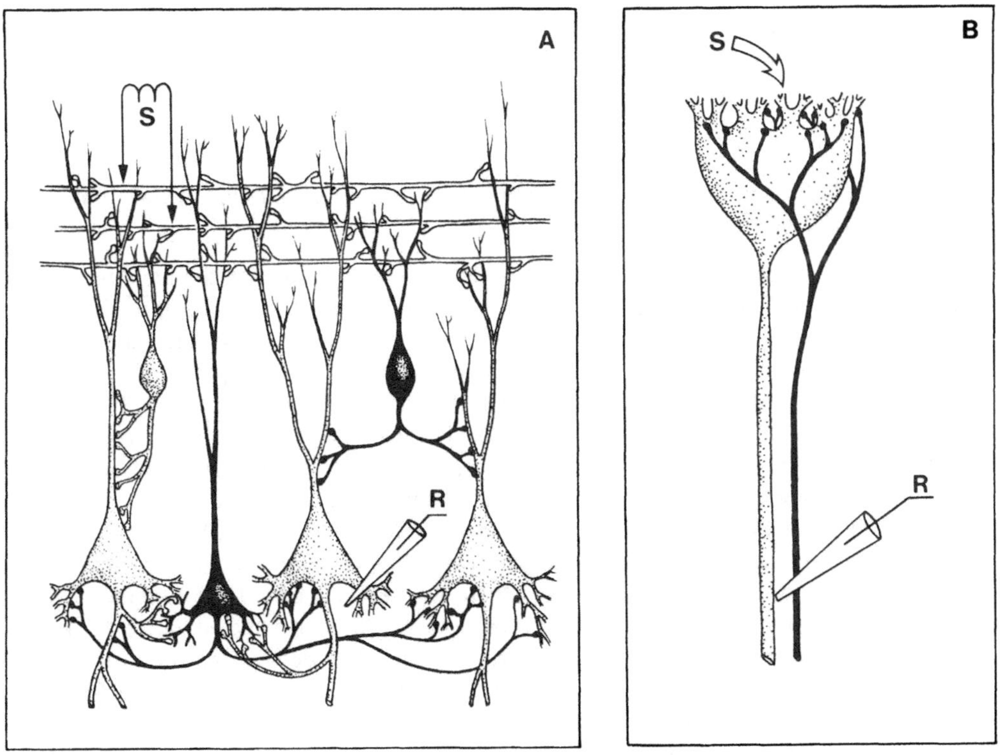

Figure 1. Schematic representations of the rat hippocampal CA1 region (A) and isolated crayfish sensory neuron (B). The integrated circuitry of the hippocampal formation consists of a heterogenous population of afferent input fibers which are synaptically connected to a mixed population of excitatory and inhibitory (darkened) interneurons and to CA1 pyramidal neurons. In contrast, the single sensory neuron is devoid of complicated circuitry consisting of only a primary neuron with inhibitory (GABAergic) fibers terminating onto the dendritic region. S, stimulus; R, recording electrode.

Isolated Crayfish Sensory Neuron

As previously described,[54] isolated abdominal stretch receptor neurons were dissected with a length of dorsal nerve (5–10 mm) from crayfish (*Procambarus clarkii*) cooled in ice water. The sensory neuron with attached tonic muscle fibers was mounted in a glass tissue bath continously perfused with modified Van Harreveld's solution—Van H (flowrate 7.0 ml/min, pH 7.0 ± 0.1, temperature 10.0 ± 0.1°C). Glass suction electrodes were used to record extracellular action potentials from the dorsal nerve of a stimulated neuron at maintained stretch (see Fig. 1). A microprocessor-based counter was used to generate frequency–time histograms. The control rhythmical firing level was approximately 5 Hz.

In separate experiments, a glass microelectrode (20–40 Mohm) was inserted into the soma of unstretched sensory neurons. Current pulses were injected while membrane potentials were simultaneously recorded using a WPI M707 microprobe system. Antidromic action potentials and GABA-mediated inhibitory postsynaptic potentials (IPSP) were generated using extracellular glass suction electrodes on the axons of the sensory neuron and inhibitory fiber.[56] Measurements of membrane resting potential, membrane resistance, spike threshold, peak amplitudes of both orthodromic and antidromic action potentials, duration of action potentials, and rates of depolarization and repolarization were performed as previously reported.[30,54]

The modified control physiological perfusion solution (Van H) contained (in mM): NaCl 205; KCl 5.4; $CaCl_2$ 13.5; $MgCl_2$ 2.6; and Trizma 10.

Drug Solutions. All drug solutions were prepared fresh prior to each experiment in the appropriate physiological solution and pH adjusted.

RESULTS

Selective Effects on Synaptic Field Potentials and Unit Discharge Activity of CA1 Pyramidal Neurons

The effects of pentobarbital and phenobarbital were examined on the amplitude of the electrically evoked field potentials from the CA1 region of the hippocampal slice. Field potential recordings provide a well-characterized and stable response that is suitable to determine accurate drug responses.[2,46] The field potential consists of a positive waveform on which a single negative spike is superimposed. The positive wave represents the dendritic current in response to synaptic activation, and the negative spike represents the synchronized discharge of postsynaptic CA1; pyramidal neurons.[2] Both barbiturates produced a concentration-dependent alteration of the field potential amplitude as shown in Fig. 2. The dose–response curve for pentobarbital was biphasic; low concentrations (20–100 μM) produced an enhancement, whereas higher concentrations (>100 μM) reversed the response toward control values; and concentrations greater than 150 μM depressed the amplitude below control values. Complete reversible depression occurred at approximately 600 μM. In contrast, phenobarbital produced only a concentration-dependent depression of the field potential at concentrations of 20–500 μM (see Fig. 2). Concentrations less than 20 μM did not produce enhancement. The major effect of both barbiturates appeared to be on the population spike; depression resulted at concentrations that slightly depressed the IPSP (see inset of Fig. 2).

To examine further the selectivity of the drug response, the stereoisomers of pentobarbital were examined. The effects of each isomer were qualitatively similar; however, the R(+) isomer was found to be approximately fivefold more efficacious on a molar basis

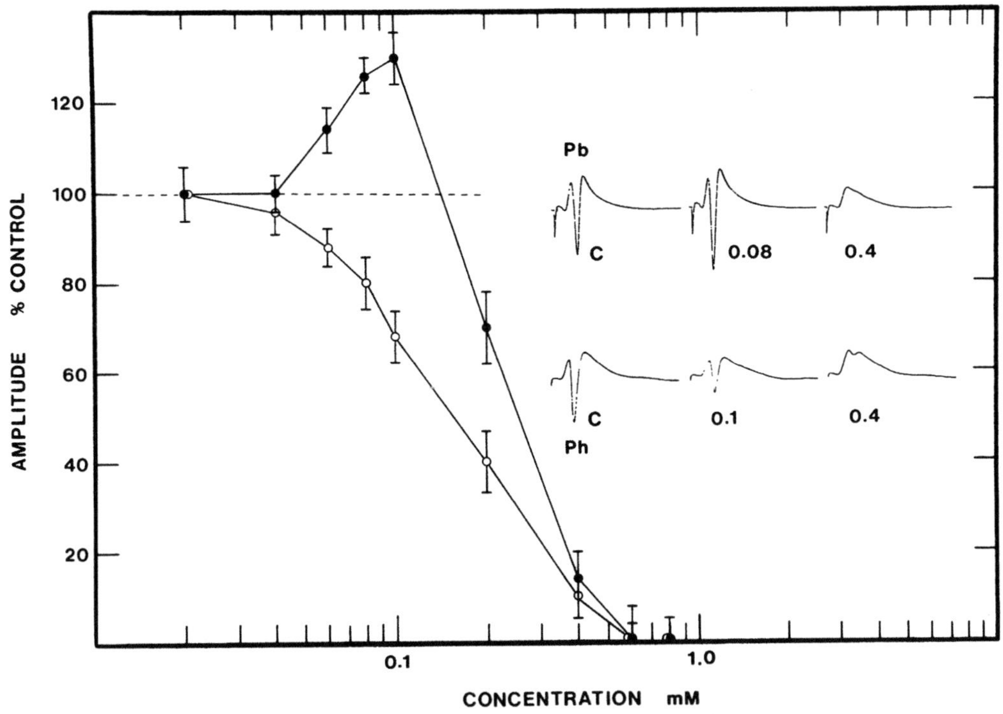

Figure 2. Dose–response curves for pentobarbital (Pb,●) and phenobarbital (Ph,○) on the amplitude (% control) of the electrically evoked field potentials of CA1 pyramidal neurons. Each point represents the mean ± S.E.M. of at least four experiments. (Inset) Examples of field potential recordings in the presence of Pb and Ph (C = control, concentrations are in mM).

enhancing the amplitude, and approximately 1.5 times more potent on the depressant portion of the curve (see Fig. 3). The dose–response curves appeared similar in shape to each other as well as to the racemic compound. Maximal depression of both isomers occured at 500–600 μM, which was comparable to the racemic pentobarbital. It appeared that the R(+) isomer was capable of providing a greater maximal enhancement of field potential amplitude, although this may not be significant.

To determine whether similar effects could be observed on individual cell responses, single unit recordings were made in the presence and absence of selected drug concentrations. A biphasic response of pentobarbital was also observed on unit discharge; low concentrations (80 μM) enhanced both the electrically evoked unit discharge and background activity (Fig. 4A). In submerged slice preparations, spontaneous activity was also enhanced (Fig. 4B). These results were indicative of a drug-induced recruitment of units. At higher concentrations (e.g., 400 μM), all activity was depressed.

In contrast to the biphasic effects of pentobarbital, phenobarbital only depressed the synaptically evoked and spontaneous unit discharge activity of CA1 neurons (Fig. 5), similar to the effects on field potentials.

Differential Effects of Barbiturates on an Isolated Neuron

The isolated sensory neuron of the crayfish was utilized to gain a better appreciation of the effects of the barbiturates at a cellular and membrane level. This preparation is far simpler than the hippocampal slice, well characterized, stable for very prolonged periods,

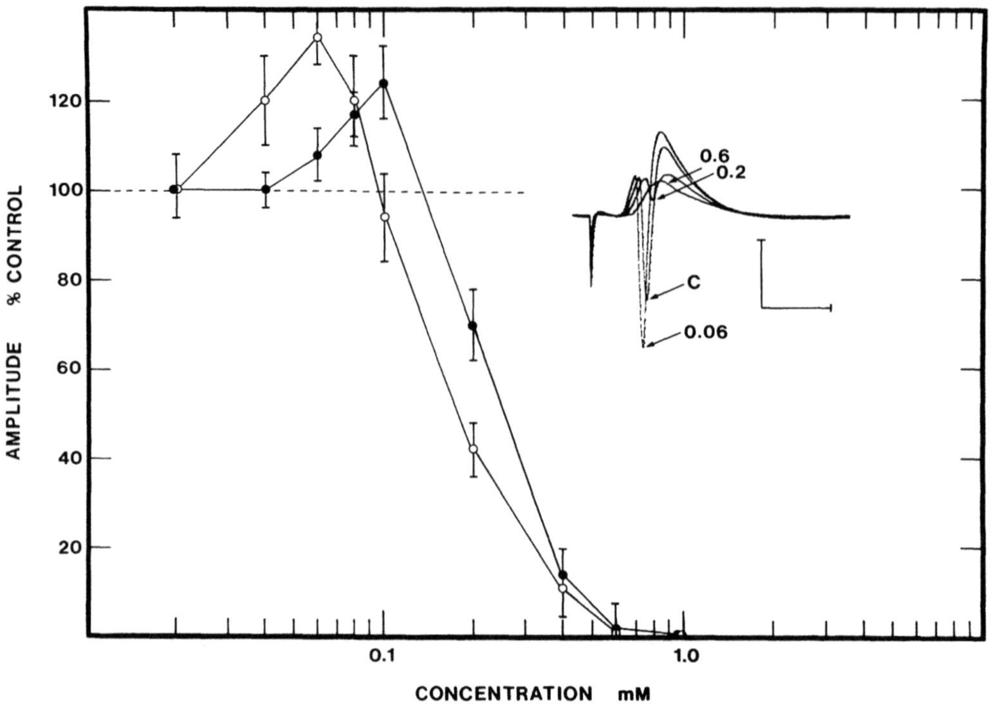

Figure 3. Dose–response curves for (R(+) (○) and S(−) (●) stereoisomers of pentobarbital on the amplitude (% control) of the electrically evoked field potentials of CA1 neurons. Each point represents the mean ± S.E.M. of four experiments (Inset) Superimposed field potential recordings in the presence of three concentrations (in mM) of R(+) pentobarbital compared to control (C). Calibration: 5 mV; 10 msec.

and is essentially an isolated single neuron which can be physiologically stimulated by a stretch input via the muscle fibers. The only synapse is an inhibitory GABAergic pathway which can be selectively activated with electrical stimulation producing an IPSP.[38,56]

Both barbiturates exhibited a concentration-dependent alteration of the stretch-induced discharge activity (Fig. 6). Pentobarbital produced a three-phase effect; initial depression, followed by partial recovery, and enhancement on washout (see inset, Fig. 6). On occasion, a slight, brief enhancement occurred on exposure. Unexpectedly, phenobarbital produced a mirror image of the pentobarbital response: initial excitation, slight recovery, and some degree of depression on washout. At concentrations greater than 300 μM, phenobarbital produced a characteristic burst pattern of activity similar to that observed for other anesthetics[30,49] and TEA (S. H. Roth and K.-S. Tan, unpublished results). As previously reported[52] and shown in Table I, barbiturates can either depress or enhance firing activity of the sensory neuron. The depressant activity appears to be dependent on a primary methyl substitution of the carbon chain at the fifth position of the ring.

The effects of the stereoisomers of pentobarbital on discharge activity (not shown) were similar; the R(+) isomer was approximately 2.5-fold more effective on a molar basis.

It was interesting that the effects of pentobarbital and phenobarbital on membrane electrical properties were qualitatively similar (see Table II). The data are presented relative to control values.[54] Compared to the marked and opposite effects of the two barbiturates on discharge activity, the effects on spike threshold, membrane resistance, action potential peak amplitude, and rate of rise of action potential were similar. At equimolar concentrations of the barbiturates, pentobarbital was more effective in raising membrane resting potential and

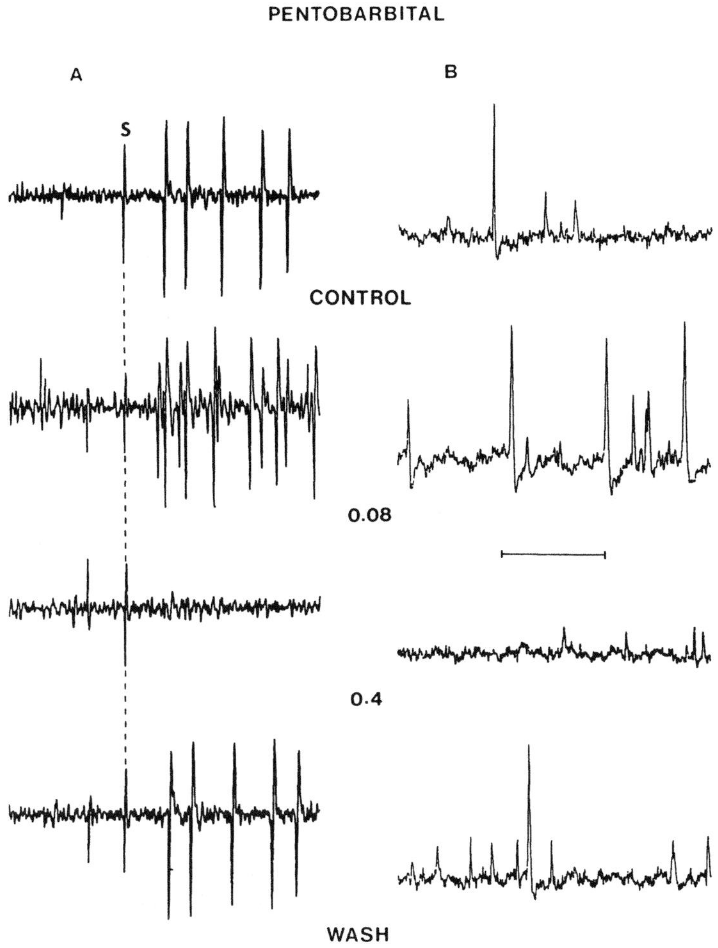

Figure 4. Pentobarbital (0.08 mM) enhanced both the electrically evoked (A) and spontaneous (B) unit discharge activities of CA1 neurons. Higher concentrations (e.g., 0.4 mM) depressed activity. Calibration bar equals 20 msec for A and 10 msec for B. S = stimulus.

depressing the amplitude of the IPSP. To further examine the GABA-mimetic actions of the barbiturates, exogenous GABA (5 μM) was applied and these effects were compared to those of the barbiturates. As shown in Fig. 7, the effects on the IPSP amplitude and membrane resistance (expressed as % change) and membrane resting potential (mV) were similar for GABA and pentobarbital; however, phenobarbital altered these responses by less than 5% (data from Table II).

DISCUSSION

The multiple actions of barbiturates have been well documented[5,18,21,25,37,41,61] (see also chapters by Owen, Olsen, Macdonald, and Ticku, this volume). Differential effects of anesthetic/sedative (e.g., pentobarbital) and anticonvulsant (e.g., phenobarbital) barbiturates have been demonstrated at various levels of complexity.[3,16,28,31,34,37,39,61] These multiple and differential actions are not compatible with a common mode of action as previously suggested for anesthetics.[48,49,53] The differences in biological activity that result

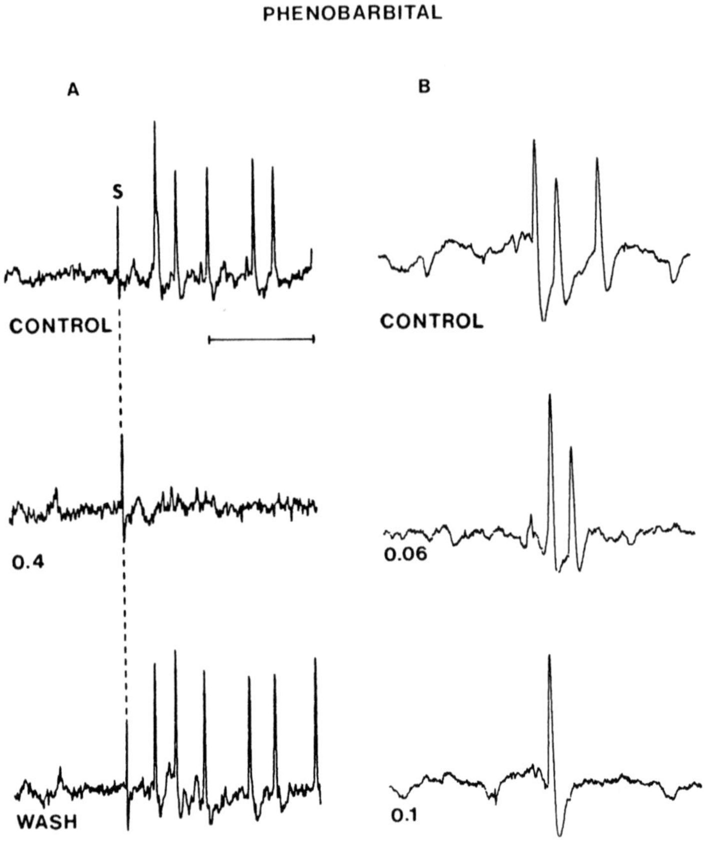

Figure 5. Phenobarbital depressed both the electrically evoked (A) and spontaneous (B) unit discharge activities of CA1 neurons. Concentrations are in mM. Calibration bar equals 20 msec for A and 5 msec for B. S = stimulus.

from minor structure alterations of barbiturates strongly support mechanisms involving specific structure-dependent interactions. It is also possible that the multiplicity of effects is a result of barbiturates interacting with multifunctional site(s) in a selective manner, or very specific interactions with a "barbiturate receptor."[4,37,60]

It is well established that barbiturates enhance GABAergic inhibitory synaptic transmission,[3,22,36,61,64] as well as inhibiting a variety of excitatory synaptic pathways.[13,44] In addition, these agents are lipid soluble, which enables them to penetrate into and thus perturb membrane structure resulting in alterations of excitability.[18,35] This type of interaction is common to all anesthetics and has provided support for a physical nonspecific mechanism of action. A number of barbiturate effects can be correlated with lipid solubility,[14] yet the very different activities observed (convulsant, anticonvulsant, and anesthetic) cannot be explained on the basis of a common mechanism of lipid solubility.[16] The physical property of lipophilicity is a necessary requirement for activity, but is not the major determinant for effect. Moreover, the significant differences in biological activity for barbiturate isomers further support a specific structure-dependent interaction. The difficulty persists as to which of the actions, alone or in combination, contribute to the anesthetic effect.

The data presented in this chapter are consistent with many other studies. In agreement with other reports,[21,64] the effects of pentobarbital differ from those of phenobarbital on mammalian (hippocampal) neurons. The results also confirm previous observations that

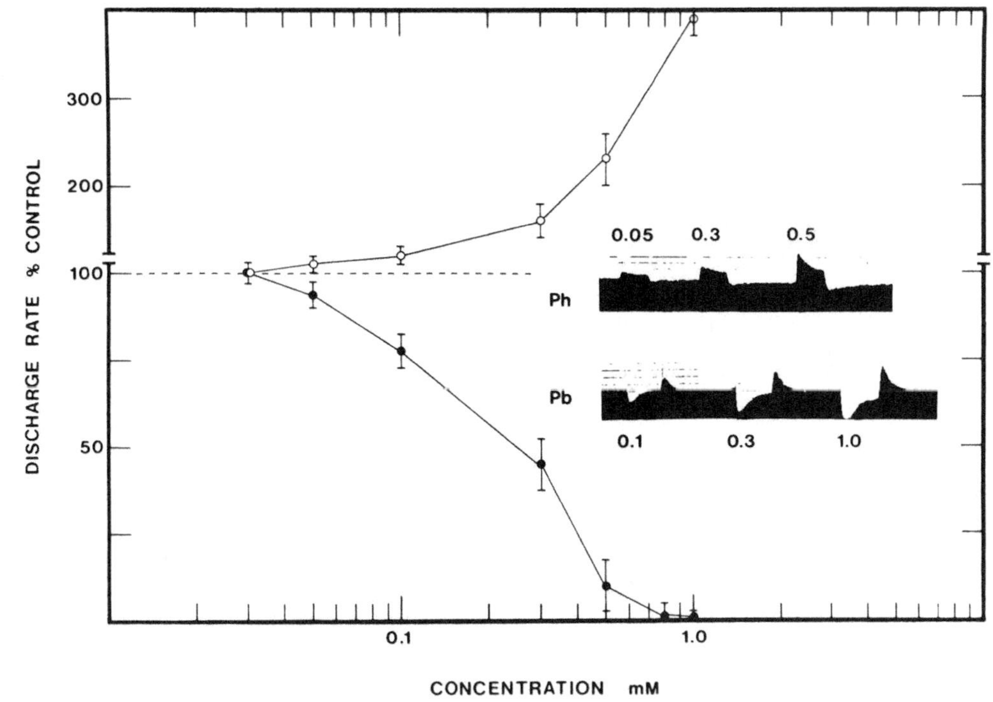

Figure 6. Dose–response curves for phenobarbital (Ph,○) and pentobarbital (Pb,●) on the discharge activity of the sensory neuron. (Inset) Discharge frequency histograms show the concentration-dependent enhancement by Ph and depression by Pb. Control discharge rate was approximately 5 Hz. Each bin equals 10 sec.

Table I. Effect of Barbiturates on the Discharge Activity of the Sensory Neuron

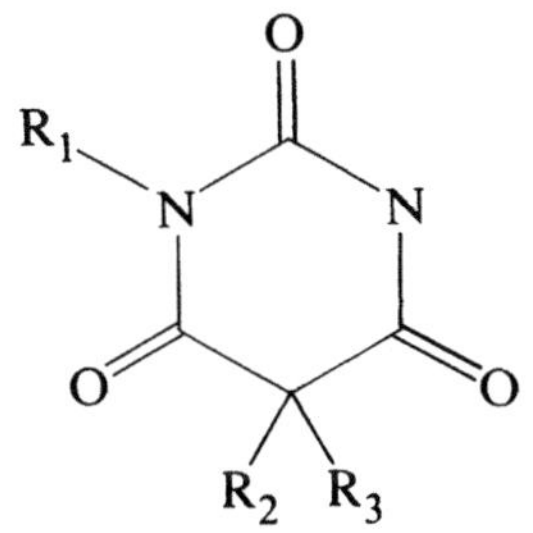

	Ring Substitution			Equieffective concentrations (mM)	
Barbiturate	R_1	R_2	R_3	Depression	Enhancement
Pentobarbital	H	$-CH_2CH_3$	$-CH(CH_3)CH_2CH_2CH_3$	0.5	
1,3-Dimethylbutyl barbituric acid	H	$-CH_2CH_3$	$-CH(CH_3)CH_2CH(CH_3)_2$	0.2	
Methohexital	CH_3	$-CH_2CH{=}CH$	$-CH(CH_3)C{\equiv}C\ CH_2CH_3$	0.1	
Phenobarbital	H	$-CH_2CH_3$			0.3
Barbital	H	$-CH_2CH_3$	$-CH_2CH_3$		0.6
Amobarbital	H	$-CH_2CH_3$	$-CH_2CH_2CH(CH_3)_2$		0.7

Table II. Effects of Phenobarbital and Pentobarbital on Membrane Electrical Properties and Synaptic Transmission

	Pentobarbital (0.5 mM)	Phenobarbital (0.5 mM)
ΔRP (Membrane resting potential)	+2.5 mV	+0.5 mV
Δth (Spike threshold)	−5.0%	−3.6%
ΔR_m (Membrane resistance)	+6.1%	+4.0%
ΔA (Action potential peak amplitude)	−2.9%	−1.0%
ΔdV/dt (First derivative of action potential)	−10.8%	−6.9%
ΔIPSP (Inhibitory postsynaptic potential peak amplitude)	−25.4%	−3.5%

pentobarbital can produce a concentration-dependent biphasic response.[8,13,23,51,52,62,64] This emphasizes the dual action of the barbiturate at cellular and synaptic levels and predicts the progressive spectrum of effects observed *in vivo*.[7,63] A comparison of the stereoisomers of pentobarbital revealed significant differences in intrinsic activities; the R(+) isomer was at least 2.5-fold more effective than the S(−) isomer. Our data are in agreement with studies that have suggested the (+) isomer to be more excitatory[4,10,19] and may in part explain the reported observations that the (+) isomer is excitatory and the (−) isomer is depressant.[19]

The use of the sensory neuron provided a different approach. The isolated single neuron is uncomplicated in morphology, relatively simple in function, and well characterized. It can be stimulated via a physiological input (stretch) to the dendritic region, and the integrated ouput behavior can easily be monitored using conventional extracellular techniques.[49] The results emphasize the dramatic ability of the single neuron to discriminate minor structural differences of barbiturates. The precise location of a methyl substitution on the fifth carbon chain was crucial to the resulting effects.[52] To explain the opposite effects of pentobarbital and phenobarbital, intracellular techniques were employed. The results, shown in Table II, were unexpected. Equimolar concentrations of the two barbiturates produced qualitatively similar effects. Both compounds depolarized the cell, increased membrane resistance, decreased spike threshold, and decreased amplitude and rate of rise of the action potential. Pentobarbital appeared to be slightly more effective in these properties. The most significant difference, however, was the effect on the IPSP, which indicated phenobarbital less effective as a GABAmimetic (see Fig. 7 and Table II), in agreement with other reports.[20,37,60,61] Thus, an explanation for phenobarbital enhancing discharge activity is a primary drug-induced depolarization and increase in membrane resistance that leads to a decrease in spike threshold. In contrast, the primary effect of pentobarbital is enhancement of the GABAergic inhibitory modulation of the stretch input (see Fig. 8). This resulted in depression of the stretch-induced

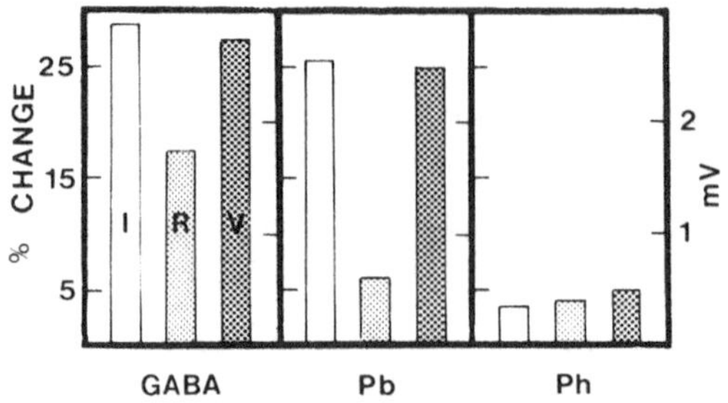

Figure 7. Bar histogram comparing the effects of GABA (5 μM), pentobarbital (Pb, 500 μM), and phenobarbital (Ph, 500 μM) on the evoked IPSP (I), membrane resistance (R) expressed as % change from control, and resting membrane potential (V; mV) of the sensory neuron.

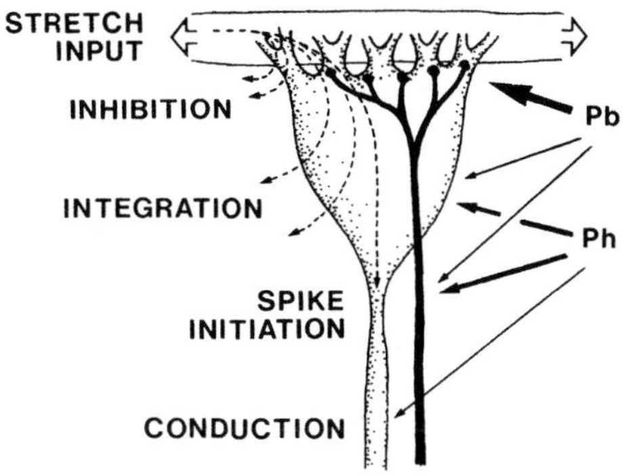

Figure 8. A schematic diagram indicating (arrows) the possible sites of action for pentobarbital (Pb) and phenobarbital (Ph) on the crayfish sensory neuron. Stretch-induced depolarizing currents (generator potential) are modulated by GABAergic inhibition on basal dendrites and cell body. Leakage across the somal membrane (integration) results in further attentuation. Spike initiation occurs at the axon hillock.

currents (generator potential) and thus a decrease in firing frequency. The intact neuron therefore provided a better assay system, which emphasizes that drug effects on the integrated system may not be reflected at the level of individual membrane effects. In other studies, the excitatory effects of phenobarbital have been associated in part with the depression of a potassium conductance. Excess Ca^{2+} was found to antagonize the enhancement, which also provided some support for the role of Ca^{2+} in the excitatory mechanisms.[9,17,58,59]

The results and interpretations are consistent with other studies and provide further support that anesthetic barbiturates may have a primary action at GABAergic sites,[1,3,37] in contrast to anticonvulsant and convulsant barbiturates, which appear to act via different primary mechanisms. We have demonstrated that barbiturates can interact in a selective and differential manner at the cellular level, and that a multisite mechanism should be considered.

It is not only the nature of the anesthetic agent (e.g., lipid solubility, structure, steric configuration) but also the composition of the membrane that must be taken into consideration to explain differences of effect.[15] Even at the level of the single neuron, differences in effect on separate membrane regions can markedly alter overall drug responses (see Fig. 8). It still remains controversial which components or subunits of the excitable membrane are directly involved in the interactions, and since it is not within the context of this chapter to address this problem, the reader is referred to other sections of this volume. Given the complexity, dynamics, and heterogenicity of the excitable membrane,[48] it is highly probable that multiple sites are involved,[47] and that selectivity occurs as a result of differential membrane and intramembrane localization. This concept supports the existence of "recognition sites" for anesthetics, which are not to be interpreted as receptors, and may be characterized with future investigations.

Although the precise mechanisms of action still remain unresolved, the emergence of new ideas will eventually provide a solution to the mystery of anesthesia.

ACKNOWLEDGMENTS. This study was supported by the Medical Research Council of Canada and the Alberta Heritage Foundation for Medical Research. We wish to thank Dr. M. H. Benn, J. Tooth, and Dr. B. H. Bland for their contributions and assistance and Ms. Christy Von Niessen for secretarial assistance.

REFERENCES

1. Alger BE, Nicoll RA: Feed-forward dendritic inhibition in rat hippocampal pyramidal cells studied in vitro. *J Physiol* 1982;328:105–123.

2. Andersen P, Bliss TVP, Škrede KK: Unit analysis of hippocampal population spikes. *Exp Brain Res* 1971;13:208–221.
3. Azanza MJ: Benzodiazepines, barbiturates and baclofen interaction within the GABA-receptors. *Gen Pharmacol* 1981;12:123–128.
4. Barker JL, Mathers DA: GABA receptors and the depressant action of pentobarbital. *Trends in Neurosciences* 1981;4:10–13.
5. Barker JL, Ransom BR: Pentobarbitone pharmacology of mammalian central neurones grown in tissue culture. *J Physiol* 1978;280:355–372.
6. Bernard C: Lecons sur les anesthesiques et sur l'asphyxie, Paris, Bailliere, 1875.
7. Bowyer JF, Winters WD: The effects of various anesthetics on amygdaloid kindled seizures. *Neuropharmacology* 1981;29:199–209.
8. Caldwell KK, Harris RA: Effects of anesthetic and anticonvulsant drugs on calcium-dependent efflux of potassium from human erythrocytes. *Eur J Pharmacol* 1985;107:119–125.
9. Carlen PL, Gurevich N, O'Beirne M: Electrophysiological evidence for increased calcium mediated potassium conductance by low-dose sedative hypnotic drugs, in Rubin EP, Weiss D, Putney JW (eds): *Calcium in Biological Systems; Ca^{+2}* as a Regulator of Membrane Permeability. New York, Plenum Press, 1984,pp 193–200.
10. Downes H, Perry RS, Osthind RE, Karler R: A study of the excitatory effects of barbiturates. *J Pharmacol Exp Ther* 1970;175:672–679.
11. Frank GB, Sanders HD: A proposed common mechanism of action for general & local anaesthetics in the CNS. *Br J Pharm Chemother* 1963;21:1–9.
12. Franks NP, Lieb WR: Molecular mechanisms of general anaesthesia. *Nature* 1982;300:487–493.
13. Fung S-C, Fillenz M: The actions of barbiturates on release of noradrenaline from rat hippocampal synaptosomes. *Neuropharmacology* 1984;23:1113–1116.
14. Hansch C, Anderson SM: The structure–activity relationship in barbiturates and its similarity to that in other narcotics. *J Med Chem* 1967;10:745–753.
15. Harris RA, Groh GI: Membrane disordering effects of anesthetics are enhanced by gangliosides. *Anesthesiology* 1985;62:115–119.
16. Harrison NL, Simmonds MA: Two distinct interactions of barbiturates and dhlormethiazole with the $GABA_A$ receptor complex in rat cuneate nucleus *in vitro*. *Br J Pharmacol* 1983;80:387–394.
17. Heyer EJ, Macdonald RL: Barbiturate reduction of calcium-dependent action potentials: correlation with anesthetic action. *Brain Res* 1982;236:157–171.
18. Ho IK, Harris RA: Mechanism of action of barbiturates. *Ann Rev Pharmacol Toxicol* 1981;21:83–111.
19. Huang L-YM, Barker JL: Pentobarbital: Stereospecific actions of (+) and (−) isomers revealed on cultured mammalian neurons. *Science* 1980;207:195–197.
20. Huguenard JR, Wilson WA: Suppression of repetitive firing of neurons by dephenylbarbituric acid. *J Pharmacol Exp Ther* 1985;232:228–231.
21. Iadarola MJ, Fanelli RJ, McNamara JO, Wilson WA: Comparison of the effects of diphenylbarbituric acid, phenobarbital, pentobarbital and secobarbital on GABA-mediated inhibition and benzo diazepine binding. *J Pharmacol Exp Ther* 1985;232:127–133.
22. Jackson MB, Lecar H, Mathers DA, Barker JL: Single channel currents activated by γ-aminobutyric acid, muscimol and (−) pentobarbital in cultured mouse spinal neurons. *J Neurosci* 1982;2:889–894.
23. Johnston D: Phenobarbital: concentration-dependent biphasic effect on aplysia burst-firing neurons. *Neurosci Lett* 1978;10:175–180.
24. Judge SE: Effect of general anaesthetics on synaptic ion channels. *Br J Anaesth* 1983;55:191–200.
25. Kleinhaus AL, Prichard JW: Interaction of divalent cations and barbiturates on four identified leech neurons. *Comp Biochem Physiol* 1979;63C:351–357.
26. Krnjevic K, Morris ME, Reiffenstein RJ: Stimulation-evoked changes in extracellular K and Ca in pyramidal layers of the rat's hippocampus. *Can J Physiol Pharmacol* 1982;60:1643–1657.
27. Leung L-WS: Differential effects of pentobarbital and ether on the synaptic transmission of the hippocampal CA 1 region in the rat. *Electroenceph Clin Neurophysiol* 1981;51:291–305.
28. Macdonald RL, Barker JL: Different actions of anticonvulsant and anaesthetic barbiturates revealed by use of cultured mammalian neurons. *Science* 1978;200:775–777.
29. Macdonald RL, Barker JL: Anticonvulsant and anaesthetic barbiturates: Different postsynaptic actions in cultured mammalian neurons. *Neurology* 1979;29:432–447.
30. MacIver MB, Roth, SH: The effects of halothane on the neuronal output, membrane properties and synaptic transmission of an isolated neuron. *Proc West Pharmacol Soc* 1980;23:405–411.
31. Mathers DA, Barker JL:, (−) Pentobarbital opens ion channels of long duration in cultured mouse spinal neurons. *Science* 1980;209:507–509.

32. Miller KW, Pang K-YY: General anaesthetics can selectively perturb lipid bilayer membranes. *Nature (London)* 1976;263:253–255.
33. Nicoll R A: Selective actions of barbiturates on synaptic transmission, in Killam KF (ed): *Psychopharmacology: A Generation of Progress*. New York, Raven, pp 1337–1348.
34. Nicoll RA: Differential postsynaptic effects of barbiturates on chemical transmission. in Otsuka M, Hall Z (eds): *Neurobiology of Chemical Transmission*. New York, Wiley, pp 267–278.
35. Nicoll RA, Madison DV: General anaesthetics hyperpolarize neurons in the vertebrate central nervous system. *Science* 1982;217:1055–1057.
36. Nicoll RA, Eccles JC, Oshima T, Rubia F: Prolongation of hippocampal inhibitory postsynaptic potentials by barbiturates. *Nature* 1975;258:625.
37. Olsen RW: Drug interactions at the GABA receptor–ionophore complex. *Ann Rev Pharmacol Toxicol* 1982;22:245–277.
38. Ozawa S, Tsuda K: Membrane permeability change during inhibitory transmitter action in crayfish stretch receptor cell. *J Neurophysiol* 1973;36:805–816.
39. Paton WDM: Unconventional anesthetic molecules, in Halsey MJ, Millar RA, Sutton JA (eds): *Molecular Mechanisms in General Anesthesia*. Edinburgh, Churchill-Livingstone, 1974;pp 48–60.
40. Richards CD: The action of anesthetics on synaptic transmission. *Gen Pharmacol* 1978;9:287–293.
41. Richards CD: In search of the mechanisms of anaesthesia. *Trends Neurosci* 1980;3:9–13.
42. Richards CD: Actions of general anaesthetics on synaptic transmission in the CNS. *Br J Anaesth* 1983;55:201–207.
43. Richards CD, Sercombe R: Calcium, magnesium and the electrical activity of guinea-pig olfactory cortex *in vitro*. *J Physiol (London)* 1970;211:571–584.
44. Richards CD, Smaje JC: The actions of halothane and pentobarbitone on the sensitivity of neurones in the guinea-pig preperiform cortex to iontophoretically applied L-glutamate. *J Physiol* 1974;239:103–105.
45. Richards CD, Tegg WJG: A superfusion chamber suitable for maintaining mammalian brain tissue slices for electrical recording. *Br J Pharmacol* 1977;59:526P.
46. Richards CD, White AE: The actions of volatile anaesthetics on synaptic transmission in the dentate gyrus. *J Physiol* 1975;252:241–247.
47. Richards CD, Martin K, Gregory S, Keightley CA, Hesketh TR, Smith GA, Warren GB, Metcalfe JC: Degenerate perturbations of protein structure as the mechanism of anaesthetic action. *Nature (London)* 1978;276:775–779.
48. Roth SH: Physical mechanisms of anaesthesia. *Annu Rev Pharmacol Toxicol* 1979;19:159–178.
49. Roth SH: Membrane and cellular actions of anaesthetic agents. *Fed Proc* 1980a;39:1595–1599.
50. Roth SH: Differential effects of anaesthetics on neuronal activity, in Fink BR (ed): *Molecular Mechanisms of Anaesthesia, Progress in Anaesthesiology* Vol. II. New York, Raven Press, 1980b,pp 119–128.
51. Roth SH, Bland BH, MacIver MB: Biphasic actions of pentobarbital on synaptic transmission. *Prog Neuro-Psychopharmacol Biol Psychiatry* 1983a;7:821–825.
52. Roth SH, Tan K-S, Tooth J: Neuronal membranes can distinguish between structural differences of barbiturates. *Proc West Pharmacol Soc* 1983b;26:243–246.
53. Seeman P: The membrane actions of anesthetics and tranquillizers. *Pharmacol Rev* 1972;24:583–655.
54. Sewell RDE, Tan K-S, Roth SH: Evidence for excitatory and depressant non-receptor-mediated membrane effects of benzodiazepines in the crayfish. *Neurosci Lett* 1984;45:59–63.
55. Skerritt JH, Macdonald RL: Multiple actions of convulsant barbiturates on mouse neurons in cell culture. *J Pharmacol Exp Ther* 1984;230:82–88.
56. Sokolove PG, Roth SH: Effect of harmaline on the crayfish stretch receptor: Blockade of a GABA-mediated inhibitory synapse. *Neuropharmacology* 1978;17:729–735.
57. Stockard J, Bickford R: The neurophysiology of anesthesia. (ed. Gordon, E.) A basis and practice of neuroanesthesia. Amsterdam, Excepta Med., 1975,pp 3–46.
58. Tan K-S, Roth, SH: Interaction of calcium ions and barbiturates on the firing activity of a single neuron. *Biophys J* 1982;37:63a.
59. Tan K-S, Roth SH: Comparative effects of extracellular ions on the discharge firing activity of an isolated neuron. *IRCS Med Sci* 1983;11:296–297.
60. Ticku MK: Interaction of depressant, convulsant and anticonvulsant barbiturates with the [^{3}H]diazepam binding site of the benzodiazepine-GABA-receptor-ionophore complex. *Biochem Pharmacol* 1981;30:1573–1579.
61. Whittle SR, Turner AJ: Differential effects of sedative and anticonvulsant barbiturates on specific [^{3}H] GABA binding to membrane preparations from rat brain cortex. *Biochem Pharmacol* 1982;31:2891–2895.
62. Willow M, Johnston GAR: Dual action of pentobarbitone on GABA binding: role of binding site integrity. *J Neurochem* 1981;37:1291–1294.
63. Winters WD, Kott KS: Continuum of sedation, activation and hypnosis or hallucinosis: A comparison of low

dose effects of pentobarbital, diazepam or gamma-hydroxybutyrate in the cat. *Neuropharmacology* 1979;18:877–884.

64. Wong EHF, Leeb-Lundberg LMF, Teichberg VI, Olsen RW: γ-Aminobytyric acid activation of ^{36}Cl-flux in rat hippocampal slices and its potentiation by barbiturates. *Brain Res* 1984;303:267–275.

5

Putative Capacitance Increase in Hippocampal CA1 Cells following Ethanol Application

Peter L. Carlen, Timothy J. Blaxter, Evan B. Friedman, and Dominique Durand

ABSTRACT. There is some suggestion in the literature that anesthetics, including ethanol, can increase neuronal membrane capacitance (C_m). We examined this possibility in central vertebrate neurons using intracellular recordings from rat hippocampal $[CA]_i$ cells *in vitro*. Ethanol was perfused or focally pressure-ejected. The voltage decays resulting from short, hyperpolarizing constant-current pulses (0.5 msec, 5–11 nA) were averaged and analyzed by computer-assisted techniques. The following electrotonic parameters were extracted from the data: τ_o (or τ_m), the final neuron time constant; τ_1, the first peeled time constant; R_N, the input resistance.

Previous modeling has shown that τ_m is very sensitive to distal shunts not well demonstrated by changes in R_N. Since $\tau_m = R_m \cdot C_m$, and $R_N \propto R_m$, if R_N remains unchanged or decreases following ethanol exposure and τ_m relatively increases, an increased C_m can be deduced as the simplest explanation, although a distal conductance decrease cannot be ruled out. In 6/11 cells with 20 mM ethanol, 8/11 cells with 100 mM ethanol, and 3/3 cells with 200 mM ethanol, a relatively increased τ_m (up to 75%) resulted with either little change or a decrease in R_N, suggesting that ethanol, in these cells, increased membrane capacitance. How this putative ethanol-induced increase in C_m relates to other ethanol-induced membrane perturbations, such as increased fluidity, or to anesthesia remains to be elucidated.

INTRODUCTION

There are several possible mechanisms of anesthesia including altered ionic mechanisms such as raised potassium conductance[15,16,19] or depressed sodium conductance, increased membrane fluidity and membrane expansion,[23–25] and altered responses to neurotransmitters.[1] Cellular electrophysiologists using intracellular recordings have routinely estimated anesthetic-induced effects from alterations in input resistance (R_N) by measuring voltage deflections from long intracellular constant current pulse injections, and by changes in resting membrane potential (RMP). However, from intracellular current injection techniques it is also possible to estimate anesthetic-induced changes in membrane capacitance in central

Peter L. Carlen, Timothy J. Blaxter, and Evan B. Friedman • Addiction Research Foundation Clinical Institute and Playfair Neuroscience Unit, Toronto Western Hospital, Departments of Medicine and Physiology, University of Toronto, Toronto, Ontario M5S 2S1, Canada. *Dominique Durand* • Applied Neural Control Laboratory, Department of Biomedical Engineering, Case Western Reserve University, Cleveland, Ohio 44106.

vertebrate neurons, a concept introduced by Puil and co-workers.[9,20] Ethanol is known to enhance membrane fluidity,[10,17] and in lipid bilayers, another alcohol anesthetic, benzyl alcohol, was shown to raise membrane capacitance.[22]

This chapter will discuss techniques that have been developed to measure drug-induced changes (increase) in the neuronal time constant (τ_m), which, when coupled with decreased or unchanged R_N, can be used to deduce, under certain circumstances, changes in membrane capacitance. Preliminary data, using sedative to anaesthetic doses of ethanol, will be presented.

METHODS

τ_m Estimation

The technique is simply to inject short hyperpolarizing constant current pulses (0.5–1.0 msec, 5–15 nA) through the intracellular recording electrode (see Fig. 3B). This method was suggested by Jack and Redman for modeling the passive electrical parameters of motoneurons[14] and used by Iansek and Redman[13] and Carlen, Werman and Yaari[3] in motoneurons, by Durand *et al.*[8] in rat hippocampal cells, and by Durand and Carlen[7] in rat CA1 cells. This same technique was used for modeling focal conductance changes in a passive resistance–capacitance (RC) ladder model of a neuron.[2] In this model, it was found that, as predicted theoretically, the voltage decays are multiexponential [equation (1)].

$$V = C_0 e^{-t/\tau_0} + C_1 e^{-t/\tau_1} + C_2 e^{-t/\tau_2} \tag{1}$$

Although the final membrane time constant (τ_o or τ_m) measured using either the long (e.g., 100 msec) or short (0.5–1 msec) constant current pulses is identical if R_N is linear,[2] the short pulse offers some advantages, as illustrated in dentate granule cells[6]:

1. The total amount of charge injected is less with the short pulse.
2. The short current duration is less likely to trigger time-dependent nonlinearities generated by long pulses.
3. The linear dynamic range is larger with short pulses.
4. The coefficients of the exponential terms with fast time constants (i.e., τ_1, τ_2 . . .) are larger and allow a better resolution for the measurement of these time constants.[2]
5. The adjustment of the bridge balance does not affect the voltage decay from the short pulse.

However, it is crucial to adjust the electrode frequency response of the recording system.[8] The input resistance, R_N, can be measured by integrating the area under the curve, or, through a simple equation using the coefficients of the short pulse decay [equation (1)], it is also possible to estimate R_N directly [equation (2)].[8]

$$R_N = \Sigma_i \frac{C_i}{I(1 - e^{-w/\tau_i})} \tag{2}$$

Changes in τ_m, R_N, and C_m

For a simple single RC network without an attached cable, the estimation of changes in C_m based on changes in τ_m and R_N is straightforward. Since $\tau_m = R_m \cdot C_m$, where C_m is the specific membrane capacitance, R_m is the specific membrane resistance, and $R_m \propto R_N$ [see

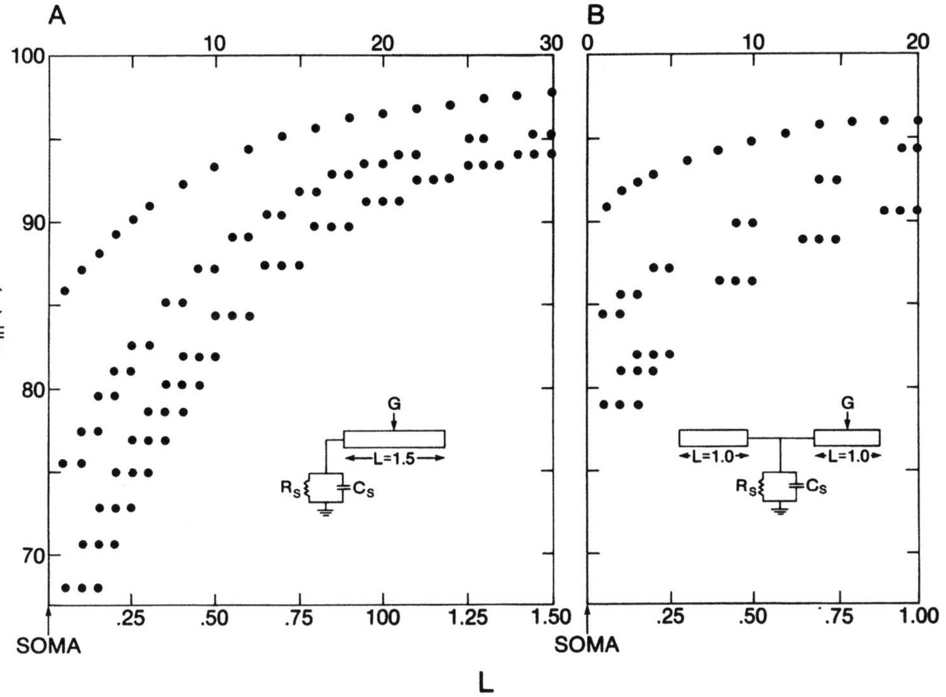

Figure 1. The effect of shunt magnitude and location on input resistance measured in the soma from short or long somatic constant current pulses. R_N estimations using short pulse (integrating the area under the curve) and long pulse (measuring the peak saturating voltage) were superimposable. Each point represents a shunt, and its location is indicated on the abscissa. The top curve points represents the decrease in R_N (as a % of the maximum R_N without any shunt) from a single shunt on the cable, the middle curve from two shunts, each in adjacent dendritic segment. (two points), and the lowest curve from three adjacent shunts (three points). (A) The model with L = 1.5 is used. (B) The model with two dendritic branches, each with an L = 1.0 is used. The shunts are placed in one dendritic tree. Reprinted with permission from Carlen and Durand.[2]

equation (3)]. If τ_m increases and R_N decreases, one can conclude that C_m has increased. This is simply deduced in cells without an attached "dendritic cable,"[9] but not in the typical mammalian central neuron with its large and often asymmetric dendritic trees. Furthermore, in these neurons, drug-induced changes in membrane capacitance or resistance may be focal rather than diffuse. The passive electrical modeling done in this laboratory[2] showed that when a focal conductance increase was placed at increasing distances in the dendritic tree away from the "somatic" recording electrode, the estimated R_N decrease was proportionally smaller (Fig. 1). However, the decrease in τ_m resulting from the focal conductance increase was relatively constant no matter where on the cable model the conductance increase was placed (Fig. 2). In a "cable model" neuron, even distal shunts will be accurately reflected by a decreased τ_m. Therefore, in the case where a drug such as ethanol can cause a decreased R_N and an increased τ_m, even though τ_m is more sensitive than R_N to distal shunts (Figs. 1 and 2), one can assume an increased C_m to explain the increased τ_m.

As previously discussed, one can estimate changes in C_m if one knows the values of τ_m and R_m. τ_m is easily estimated from the multiexponential voltage decay following the short pulse or from the initial phase of a long pulse. Estimating R_m is more difficult in these neurons involving several assumptions and detailed morphological data.[6,8] For a neuron modeled by a simple equivalent cylinder,[21] the R_N is related to the specific membrane and morphological parameters by the following equation:

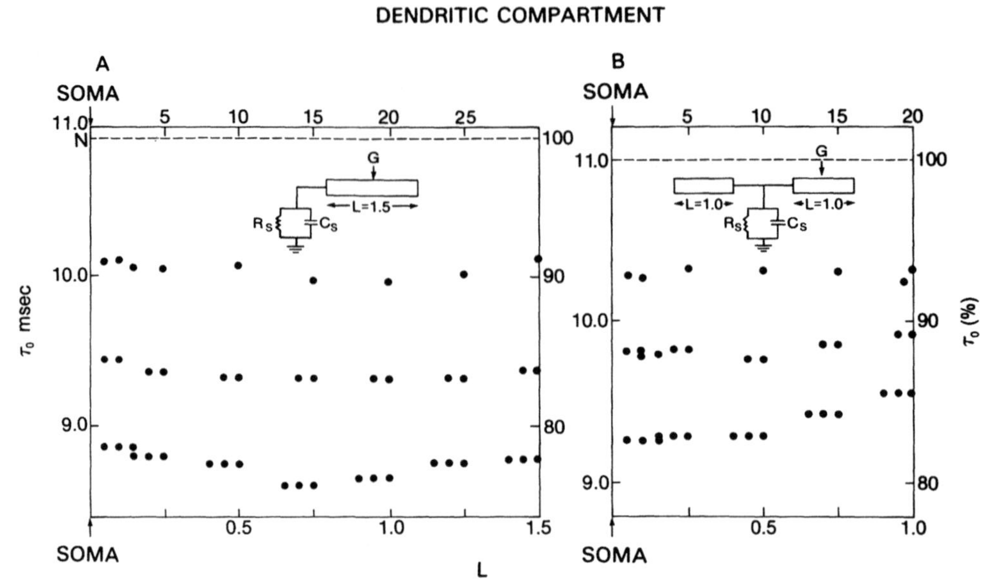

Figure 2. Effects of shunt magnitude and location on estimated τ_o's. The top row of points are the τ_o's from one shunt, the middle row from two shunts, each in adjacent dendritic segments (two points), and the lowest row from three adjacent shunts (three points). (A) The model with L = 1.5 was used. Shunts were placed in various compartments along the cable, and the corresponding τ_o's were estimated. (B) The model with L = 1.0 was used. Shunts were placed at various locations in one of the two equivalent dendritic cables, and the corresponding τ_o was estimated. Reprinted with permission from Carlen and Durand.[2]

$$R_N = \frac{2}{\pi}\sqrt{\frac{RmRi}{D^3}} \times \text{Coth } 2l\sqrt{\frac{Ri}{RmD}} \tag{3}$$

where l = length of equivalent cable (cm), d = diameter of cable (cm), and R_i = axoplasmic resistance (Ohm.cm). Changes in dendritic R_m are not accurately reflected by the measurement of R_N as estimated from the somatic recording electrode, although, as mentioned previously, changes in τ_m closely reflect focal changes in R_m wherever this may occur on the neuronal cable model (i.e., in the dendritic tree). From equation (3) it is clear that if R_i, l, and D are not affected by ethanol, then an increased C_m can be deduced in the following circumstances: (1) R_N remains constant and τ_m increases; (2) R_N decreases and τ_m remains the same or increases.

In Vitro Recording

Rat hippocampal slices 300–400 μM thick were prepared by standard techniques using a tissue chopper or an Oxford vibratome. The slices were maintained on a mesh and perfused by oxygenated (95% O_2, 5% CO_2) artificial CSF at 34–36°C with the following composition (in mM): NaCl (126), $NaHCO_3$ (26),KCl (2), KH_2 PO_4 (1.25), Ca Cl_2 (2), $MgSO_4$ (2), and dextrose (10).

Intracellular recordings from 25 CA1 pyramidal cells to date were obtained with 3 M K acetate (90–150 M ohms) or 3 M KCl (60–120 M ohms) glass microelectrodes (Frederick Haer, o.d. 1.2 mm). Cells had action potentials greater than 75 mV and input resistances (R_N) greater than 20 M ohms. The capacity compensation was carefully adjusted to set the frequency response of the recording system at 4 kHz. The bridge circuit was balanced after cell penetration. The voltage responses from long hyperpolarizing constant current pulses

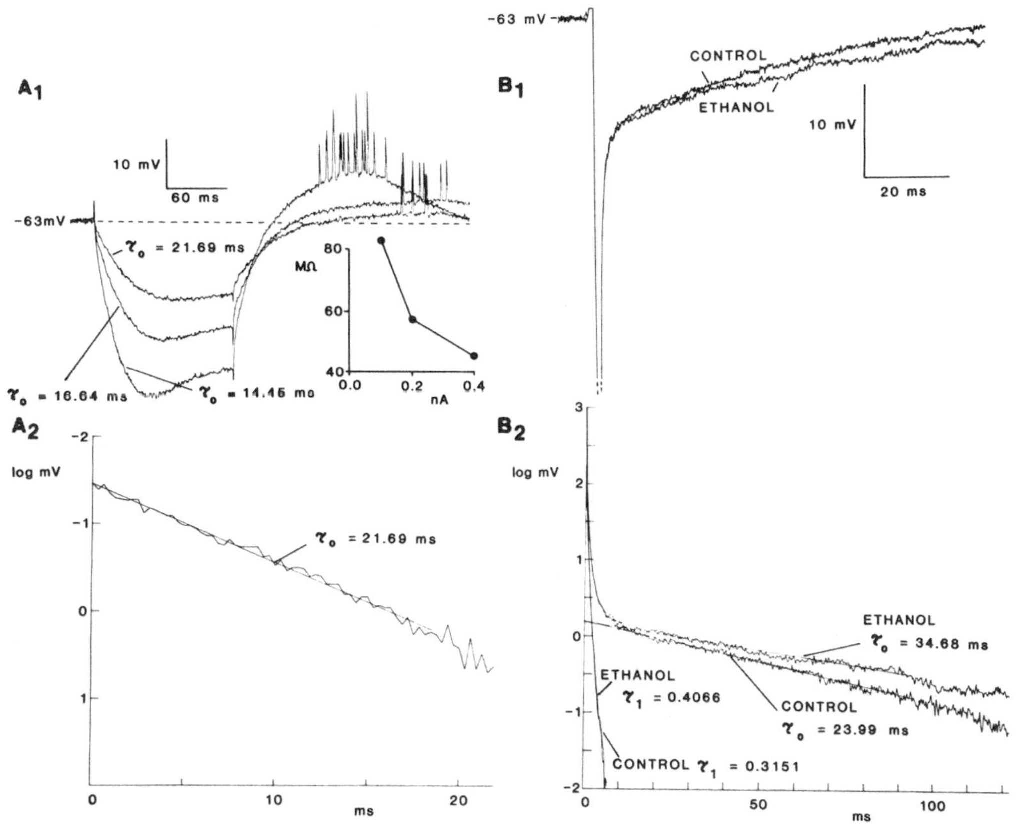

Figure 3. Analysis of short and long current pulses in CA1 neurons. (A) Long pulses. (1) Current–voltage relationship. Each curve is an average of 20 sweeps. Currents were 0.1, 0.2, and 0.4 nA and 100 msec. Note that the largest long pulse produced the largest "off" response and that the relationship was nonlinear as shown in the inset plot of R_N. τ_0's are shown for each sample as calculated from the semilog decay. Note the decreasing τ_0 with increasing long-pulse current. (2) Semilog plot of part of 0.1 nA sample in 1. Region line shows the calculated τ_0 for the 0.1 nA example. Note that τ_1 cannot be estimated in this undifferentiated decay. (B) Short pulses. (1) Each curve is the average of 20 responses to the injection of 11 nA, 0.5 msec pulses. Upper curve is the control response, lower curve the response after exposure of the neuron to ethanol (20 mM). (2) Semilog plot of part of the curves in 1. Regression lines show the calculated τ_0 and τ_1. Lower curve is control and upper curve is after exposure of the neuron to ethanol. $V_m = -63$ mV. R_N calculated from 0.1 nA long pulse increased nonsignificantly by 4% following ethanol application.

(100 msec, 0.05–0.60 nA) at 1–2 Hz were used to generate an IV curve (Fig. 3A$_1$). Short hyperpolarizing constant current pulses (0.5 msec, 3–15 nA) were injected at 1–2 Hz (Fig. 3B$_1$). Both long and short pulses were sampled at 10 KHz, stored on a floppy disk and/or on magnetic tape and were averaged 10–20 times.[3] Long and short pulses were also sometimes generated during DC hyperpolarizations of the neurons prior to drug application and during DC depolarization to control RMP postdrug application if the drug caused a hyperpolarization. Ethanol (20–200 mM) was perfused or applied by several pressure-ejected drops over the somatic and dendritic regions.

RESULTS

The CA1 cells frequently showed marked nonlinearity in the voltage response to relatively small hyperpolarizing currents (Fig. 3, insert). A small "sag" was often seen in the response to the larger currents, probably reflecting the Q current described by Halliwell and

Table I. R_N *and* τ_M *Changes Following Ethanol*

Dose	$\downarrow R_N \pm \uparrow \tau_m$	No changes or slight $\uparrow R_N$ (< 10%) & $\uparrow \tau_m$	Equivocal changes	Relatively $\downarrow \tau_m$
20-mM perfusion	3	3	3	2
100-mM perfusion	6	2	2	1
200-mm drops	2	1	0	0
Totals	11	6	5	3

Adams.[11] Usually τ_m, as estimated from the short-pulse voltage decay, was higher than that estimated from even the smallest long pulse. For example, in Fig. 3, τ_o estimated from the long pulse was 21.69 msec and from the short pulse was 23.99 msec. Often the difference was much greater, evidently because the long pulse is more likely to activate a time-dependent conductance increase, even though the long-pulse voltage response may look strictly passive. This difference in long- and short-pulse τ_o's was also noted in dentate granule cells by Durand *et al.*[8]

Ethanol, at all doses, enhanced postspike long-lasting afterhyperpolarizations, decreased spontaneous spiking when present, and often hyperpolarized the cell 1–4 mV with a moderate conductance increase as has been reported with doses of 20 mM or less.[4] These responses were all insensitive to intracellular chloride injection and are probably due to increased calcium-mediated potassium conductance.[5]

The changes in τ_m following ethanol application did not, in most cases, reflect the changes in R_N: i.e., in 17/25 cases following ethanol application, τ_m was relatively increased compared to changes in R_N (Table I and Fig. 3). For all the low-dose cases, the mean control τ_o was 26.3 ± 10.8 msec and postethanol τ_o was 25.3 ± 7.5 msec, a nonsignificant difference using the student t test by the individual difference method. Mean R_N control was 52 ± 27 M ohms, and postethanol it was 52 ± 28 M ohms. The mean individual difference was −2.3 M ohms, significant at $p < 0.01$. These changes were more consistently seen with higher doses: i.e., with 100- and 200-mM doses, 11/14 cells showed changes consistent with a relatively increased C_m with an increased τ_m (up to 75%) compared to the concomitant change in R_N. For the high doses, the mean control τ_o was 31.5 ± 17.2 msec, and following ethanol τ_o was 43.0 ± 24.6 msec, a significant difference ($p < 0.05$). Control mean R_N was 52 ± 18 M ohms, and postethanol it was 53 ± 21 M ohms; the mean individual differences were not significantly different. Note in Fig. $3B_2$ that τ_1 also increased following ethanol. As per Rall,[21] the normalized electronic length

$$L = \frac{\pi}{\sqrt{\frac{\tau_0}{\tau_1} - l}} \tag{4}$$

is where L is the ratio of the physical length of the equivalent cylinder, l, divided by the space constant of the cylinder, λ. If the neuron can be properly fitted by this cable model, the parallel increases in τ_o and τ_1 following ethanol would leave L relatively unaffected. If a significant (> 1 mV) ethanol-induced hyperpolarization occurred (in about 50% of cells), repolarizing DC current was passed to rule out R_N changes associated with the hyperpolarizations *per se*. Control experiments showed that predrug passive hyperpolarization was often associated with moderately decreased R_N.

DISCUSSION

These preliminary data show that in 6/11 cells following the mildly intoxicating dose of 20 mM ethanol and in 11/14 cells following the anesthetic to toxic doses of 100–200 mM ethanol, the CA1 neuronal τ_m relatively increased compared to the decreased or unchanged R_N. The simplest explanation of the changes in these 17 cells is an increase in neuronal membrane capacitance, provided ethanol does not affect the cellular morphology or R_i. However, one cannot rule out a perisomatic decrease in R_m with a concomitant greater increase in more electrotonically distal dendritic membrane R_m thereby causing an overall increase in total neuronal resistance reflected by an increased τ_m. Ethanol is known to be freely and evenly diffusable throughout the brain in both the extra- and intracellular compartments.[27] Therefore, unless one postulates specific membrane-bound "receptors" to ethanol with a differential spatial distribution over the neuronal somatic and dendritic surfaces, the simplest explanation is a diffusely increased membrane capacitance. To our knowledge, there is no literature describing modeling of localized or diffuse changes of specific neuronal C_m in central mammalian neurons. An independent analysis of possible capacitance changes following ethanol exposure by using phase angle shift or frequency analysis of the short-pulse decay would be helpful.[9,20] However, the only system that can unequivocally separate changes in capacitance from resistance is a single RC network such as obtained with a spherical cell without an attached dendritic or axonal tree.

What is the relevance of this putative increase in neuronal membrane capacitance following anesthetic doses of ethanol? Increased capacitance with an increased τ_m would decrease the effectiveness of a relatively short and abrupt depolarizing current pulse to reach spike threshold because of the greater time necessary to charge up fully the relatively increased membrane capacitance. This phenomenon could contribute significantly to general anesthesia. Puil and Gimbarzevsky[20] have also presented data compatible with relatively increased membrane capacitance in trigeminal sensory neurons following 2–4% isoflurane anesthesia. However, anesthetics also cause decreased R_N[5,19,20] usually attributed to increased K^+ conductance, which would also tend to diminish neuronal excitability. Possibly this putative increase in C_m is actually a concomitant but not a cause of the general anesthesic state in neurons. It might be related to increased membrane fluidity, one of the most popular mechanisms of anesthesia.[24] When ethanol was used at a concentration that reduced the inward Na^+ current in squid giant axon by 50%, the membrane capacity measured at 100 KHz was not greatly affected.[12] Both increased C_m and increased membrane ionic permeability were measured in lipid bilayers following anesthetic doses of benzyl alcohol and chloroform.[22] Anesthetic doses of ethanol expanded and raised the "specific volume" of erythrocyte membrane,[25,26] findings that could be interpreted as a decrease in membrane thickness. Decreased membrane thickness which would increase C_m has been suggested with anesthetic agents.[18] Further work *in vivo* and *in vitro* using relevant concentrations of anesthetics is necessary to assess the importance of membrane capacitance changes as part of the mechanism of anesthesia.

ACKNOWLEDGMENTS. We thank Mary Cairoli for secretarial assistance. This work was supported by grants from the Medical Research Council of Canada, the Alcoholic Beverages Medical Research Foundation, the Ontario Mental Health Foundation, the Canadian Geriatrics Research Society, and the National Institutes of Health.

REFERENCES

1. Barker JL: Selective depression of postsynaptic excitation by general anesthetics, in Fink BR (ed): *Molecular Mechanisms of Anaesthesia*. Progress in Anesthesiology, Vol. 1. New York, Raven Press, 1975,pp 135–153.

2. Carlen PL, Durand D: Modelling the postsynaptic location and magnitude of tonic conductance changes resulting from neurotransmitter or drugs. *Neuroscience* 1981;6:839–846.
3. Carlen PL, Werman R, Yaari Y: Post synaptic conductance associated with presynaptic inhibition in cat lumbar motoneurons. *J Physiol Lond* 1980;298:539–556.
4. Carlen PL, Gurevich N, Durand D: Ethanol in low doses augments calcium-mediated mechanisms measured intracellularly in hippocampal neurons. *Science* 1982;215:306–309.
5. Carlen PL, Gurevich N, O'Beirne M: Electrophysiological evidence for increased calcium-mediated potassium conductance by low-dose sedative-hypnotic drugs, in Rubin EP, Weiss D, Putney JW (eds): *Calcium in Biological Systems; Ca^{+2}* as a Regulator of Membrane Permeability. New York, Plenum Press, pp 193–200.
6. Durand D: The somatic shunt cable model for neurons. *Biophysic J* 1984;46:645–653.
7. Durand D, Carlen PL: Decreased neuronal inhibition *in-vitro* after long-term administration of ethanol. *Science* 1984;224:1359–1361.
8. Durand D, Carlen PL, Gurevich N, Ho A, Kunov H: Electrotonic parameters of rat dentate granule cells measured using short current pulses and HRP staining. *J Neurophysiol* 1983;50:1080–1097.
9. Gimbarzevsky B, Muira RM, Puil E: Impedance profiles of peripheral and central neurons. *Can J Physiol Pharm* 1984;62:460–462.
10. Goldstein DB, Chin JH: Interaction of ethanol with biological membranes. *Fed Proc* 1981;40:2073–2076.
11. Halliwell JV, Adams PR: Voltage clamp analysis of muscarinic excitation in hippocampal neurons. *Brain Res* 1982;250:71–92.
12. Haydon DA, Urban BW: The action of alcohol and other nonionic active substances on the sodium current of the squid giant axon. *J Physiol Lond* 1983;341:411–427.
13. Iansek R, Redman SJ: An analysis of the cable properties of spinal motoneurones using a brief intracellular current pulse. *J Physiol Lond* 1973;234:613–636.
14. Jack JJB, Redman SJ: An electrical description of the motoneurone, and its application to the analysis of synaptic potentials. *J Physiol Lond* 1971;215:321–352.
15. Krnjevic K: Excitable membrane and anesthetics, in Fink BR (ed): *Cellular Biology and Toxicity of Anesthetics.* Baltimore, William & Wilkins, 1972, pp 3–9.
16. Krnjevic K: Is general anesthesia induced by neuronal asphyxia, in Fink BR (ed): *Molecular Mechanisms of Anesthesia,* Vol. 1. New York, Raven Press, pp 93–98.
17. Michaelis EK, Michaelis ML: Physico-chemical interactions between alcohol and biological membranes, in Smart K, Glaser F, Kalant H, Popham R, Schmidt W (eds): *Research Advances in Alcohol and Drug Problems,* Vol. 7. New York, Plenum Press, 1983,pp 127–173.
18. Miller KW: Intermolecular forces. *Br J Anaesth* 1974;46:190–201.
19. Nicoll RA, Madison DV: General anesthetics hyperpolarize neurons in the vertebrate central nerous system. *Science* 1982;217:1055–1057.
20. Puil E, Gimbarzevsky B: Alterations in membrane electrical properties of trigeminal ganglion neurons during general anaesthesia. *Soc Neurosci Abstr* 1984;10:561(1984).
21. Rall W: Core conductor theory and cable properties of neurons, in: *Handbook of Physiology. The Nervous System.* Bethesda, American Physiological Society, Sect. 1, Vol. V, Part 1, 1977, Chap 3, pp 39–37.
22. Reyes J, Latorre R: Effect of the anesthetics benzyl alcohol and chloroform on bilayers made from monolayers. *Biophys J* 1979;28:259–280.
23. Roth SH: Physical mechanisms of anesthesia. *Annu Rev Pharmacol Toxicol* 1979;19:159–178.
24. Seeman P: The membrane actions of anesthetics and tranquilizers. *Pharmacol Rev* 1972;24:583–655.
25. Seeman P: The membrane expansion theory of anesthesia: Direct evidence using ethanol and a high-precision density meter. *Experientia* 1973;30:759–760.
26. Seeman P: The membrane expansion theory of anesthesia, in Fink B R (ed); *Molecular Mechanisms of Anaesthesia, Progress in Anesthesiology,* Vol 1, New York, Raven Press, 1975, pp 243–251.
27. Sunahara GI, Kalant H, Schofield M, Grupp L: Regional distribution of ethanol in the rat brain. *Can J Physiol Pharmacol* 1978;56:988–992.

6

General Anesthetics and Intracellular Free Calcium Ions

Mary E. Morris

ABSTRACT. Increasing evidence suggests that the depression of neuronal excitability by general anesthetics may be attributable to neuronal hyperpolarization, secondary to an increase in membrane K^+ conductance (G_k). Although it has been hypothesized that the ΔG_k is activated by an intraneuronal accumulation of Ca^{2+}, direct proof is lacking. One way this might be provided is by measurement of changes in ion levels. Suggestive but indirect evidence is given by the small, reversible increases in $[K^+]_o$ (and depression of the K^+ accumulation resulting from stimulation) which can be recorded with ion-selective microelectrodes in cortical slices in the presence of pentobarbital. Although anesthetic-induced changes in intraneuronal Ca^{2+} have not yet been measured, success in (1) recording activity-related increases in $[Ca^{2+}]_i$ in mammalian central neurons with ion-selective microelectrodes, and (2) obtaining preliminary results with the fluorescent indicator, QUIN 2 and thymocytes, which show that dose-dependent increases in $[Ca^{2+}]_i$ are evoked by 10^{-3}–10^{-4} M pentobarbital, encourages further assessment.

INTRODUCTION

An important and particularly sensitive way in which anesthetics depress central synaptic transmission may be their ability to directly mimic and potentiate mechanisms that normally operate to damp down neuronal excitability. The most relevant of these would be that which would selectively antagonize events involved in the generation of awareness and the conscious state. The slow excitation of cortical neurons by activation of ascending cholinergic pathways may play an important role in evoking the repetitive discharges that have been implicated in memory and states of arousal.[12,14,15,22,40] Indeed, the sensitivity of these neurons to acetylcholine (ACh) has been found to show a selectively greater vulnerability to the effects of

Abbreviations used in this chapter: ACh, acetylcholine; Ca, calcium; Ca^{2+}, calcium ion; $[Ca^{2+}]$, calcium ion concentration; $[Ca^{2+}]_i$, intracellular calcium ion concentration; CNS, central nervous system; DMSO, dimethylsulfoxide; DNP, dinitrophenol; GABA, γ-aminobutyric acid; G_k, potassium conductance; K, potassium; K^+, potassium ion; [K], potassium concentration; $[K^+]_o$, extracellular potassium ion concentration; HEPES, *N*-2-hydroxyethylpiperazine-*N*′-2-ethanesulfonic acid; V_m, membrane potential; QUIN 2, methoxyquinoline derivative of bis (*o*-aminophenoxy)ethane-N,N,N′,N′-tetracetic acid; Δ, change; [ ], concentration.

Mary E. Morris • Departments of Anaesthesia and Pharmacology and the Playfair Neuroscience Unit, University of Toronto, Toronto, Ontario M5S 1A1, Canada.

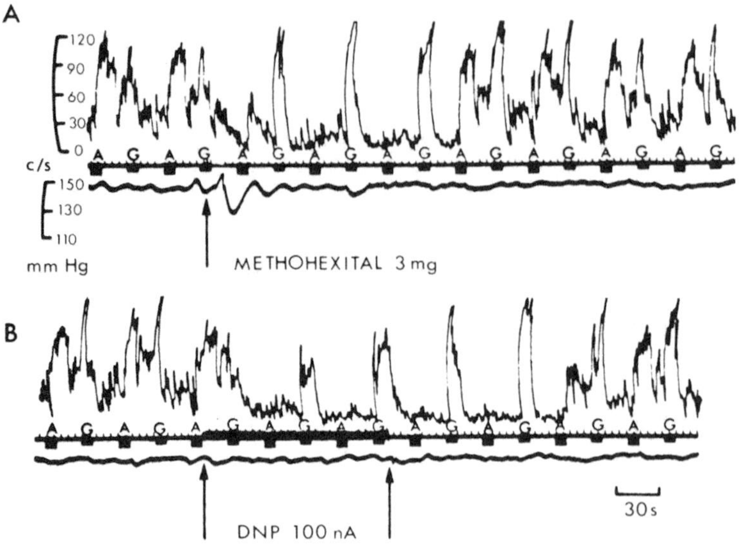

Figure 1. Selective susceptibility of acetylcholine (ACh)-evoked responses of central neuron. Polygraph traces of single cell firing frequency (counts/s), recorded with extracellular microelectrode in cat cerebral cortex. Regularly alternating applications of ACh (A) and L-glutamate (G) by microiontophoresis are indicated by black signals on time trace. Lower traces show simultaneous recording of arterial pressure. (A) IV injection of methohexital eliminates response to ACh, while that to glutamate is spared. (B) Similar action of dinitrophenol (DNP), applied with 100 nA of iontophoretic current. Reproduced with permission from Catchlove *et al.*[7]

anesthetics than that of their response to the excitatory amino acid glutamate,[7] as can be seen in the example of Fig. 1. The results of this and related studies[10,19] led to a hypothesis that anesthetics may act [in ways resembling metabolic inhibition by 2-4-dinitrophenol (DNP) and anoxia] by causing an increase in membrane conductance to K^+ (G_K), which would hyperpolarize and depress the excitability of neurons in general, but specifically counteract the decrease in G_K effected by the muscarinic action of ACh.[12–14] Furthermore, since the injection of Ca^{2+} into central neurons was found to decrease excitability (see intracellular recordings of Fig. 2) by evoking hyperpolarization and an increase in G_K,[16,24,25] Krnjević also proposed that this action of anesthetics might arise secondary to an intracellular accumulation of Ca^{2+}.

Studies in invertebrates[8,39] and with phospholipid artificial membranes[3] have indeed supported the idea that anesthetics hyperpolarize excitable membranes by increasing their permeability to K^+. In addition, there have been observations in the vertebrate central nervous system (CNS) showing that anesthetics depress the presynaptic excitability of primary afferent terminals, at the same time that the efficiency of synaptic transmission is facilitated.[27,29] These have provided indirect evidence for a K^+-mediated hyperpolarization of presynaptic terminals. Recently, however, more direct and convincing data for hyperpolarizing effects of general anesthetics on vertebrate neurons in the hippocampus and spinal cord have been described[6,36] and attributed to an increase in K^+ permeability. Even though some anesthetics (barbiturates, α-chloralose) have GABA(γ-aminobutyric acid)-mimetic actions,[4,35,36] it would appear that all the agents may increase G_K, and more specifically there is suggestive evidence that at least ethanol and barbiturates can enhance Ca^{2+}-activated G_K.[5,6]

Recent advances in the direct measurement of ion levels at the cellular level[21,37,42,44] provide the potential for further support and assessment of the hypothesis linking anesthetics

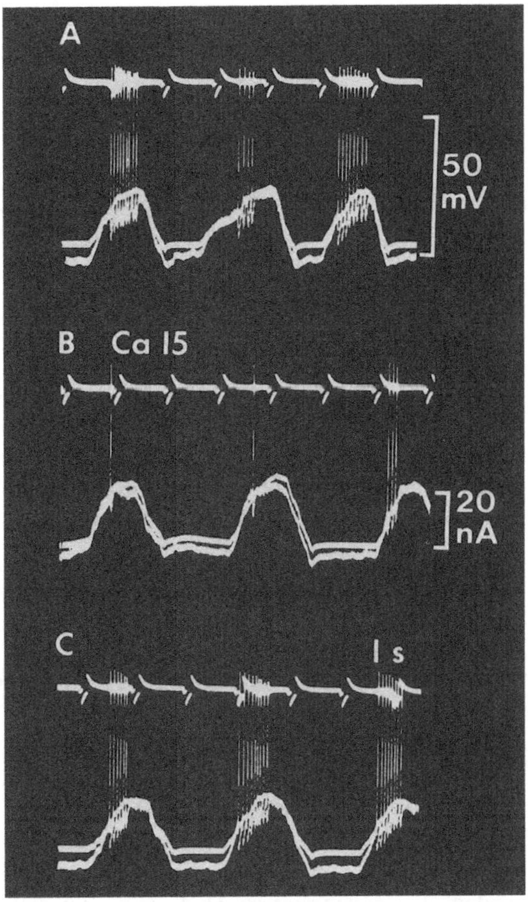

Figure 2. An increase in intraneuronal $Ca^{2+}{}_i$ depresses excitability of cat spinal cord motoneuron. Middle record (B) shows depression of repetitive firing evoked by intracellular depolarization during injection of Ca^{2+} by current of 15 nA (control in A, recovery in C; current monitor superimposed on membrane potential). Reproduced with permission from Krnjević and Lisiewicz.[16]

and Ca^{2+}-activated G_K. The results of preliminary experiments using (1) K^+- and Ca^{2+}-selective microelectrodes and (2) a fluorescent quinoline Ca^{2+} indicator, QUIN 2, are described.

METHODS

Ion-Selective Microelectrode Studies

Ion-selective microelectrodes were prepared from single (Corning 233009) and theta-type double-barrel (R & D Scientific Glass Co.) borosilicate glass capillary tubing, using a variety of techniques.[17,20,21,26,30,34,42,44] They contained either the K^+ liquid ion exchanger (Corning 477317) or Simon's neutral Ca^{2+} carrier, ETH1001[37] and were used to record resting and stimulus- and drug-evoked changes in intracellular and extracellular ion levels in the mammalian CNS. The voltage signal of the reference channel or electrode (filled with 1 M NaCl or 3 M KCl) was subtracted with a Keithley Model 604 Differential Electrometer from that of the ion-sensing electrode to give the pure ion signal for K or Ca (V_K or V_{Ca}) and also used, where possible, for separate recording of either the DC focal (field) potential (V_F) or membrane potential (V_m).

In vivo recordings were made in cats in both the cuneate nucleus of a decerebrate preparation[18,27,28] and the spinal cord of barbiturate-anesthetized animals.[34] *In vitro* studies

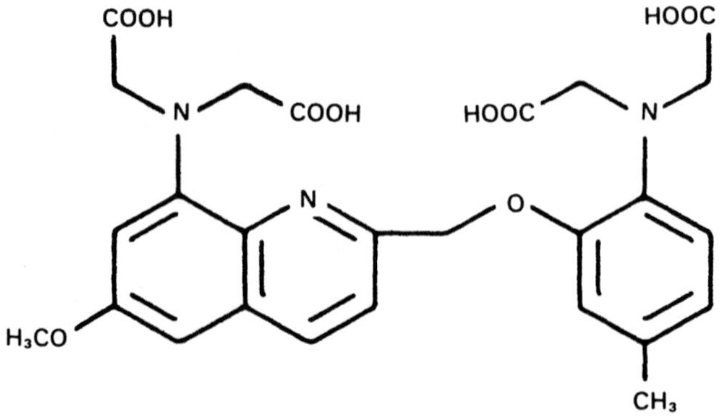

Figure 3. Structure of fluorescent quinoline Ca^{2+} indicator, QUIN 2.

using 400 μm slices of guinea pig cerebral cortex (with constant bath superfusion at 33–35°C with a Ringer solution of composition—122 mM NaCl, 3.1 mM KCl, 1.2 mM $MgSO_4$, 0.4 mM KH_2PO_4, 25 mM $MaHCO_3$, and 10 mM D-glucose, pH 7.4)—were also carried out.[32]

Fluorescence Measurement of $[Ca^{2+}]_i$

Suspensions of thymocytes (40–50 × 10^7 cells/ml) were prepared from minced, homogenized rat thymus for fluorescence measurement of cytosolic $[Ca^{2+}]$, according to the recently developed technique of Tsien *et al.*[43] They were incubated in a HEPES-buffered RPMI-1640 medium at 37°C for 45 min with 15 μM of the acetoxymethyl ester of the fluorescent quinoline derivative, QUIN 2 (Lancaster Synthesis Ltd.), whose structure is shown in Fig. 3. The concentration of the dimethylsulfoxide solvent in the incubating mixture was ≤0.1%, and the final intracellular concentration of QUIN 2 was <1 mM. Resting and drug-induced changes in the relative peak fluorescence intensity of cells (resuspended at a concentration of 4–5 × 10^6 cells/ml in Ringer's solution of the following composition: 140 mM NaCl, 3.0 mM $CaCl_2$, 1.0 mM $MgCl_2$, 10 mM D-glucose, 10 mM HEPES, and 0.2 mM EGTA, pH 7.4) were measured with an SPF-500 spectrofluorometer (American Instrument Co.) at 339 nm excitation and 495 nm emission wavelengths, using slits of 5 and 15–20 nm, respectively. Calibration of Ca^{2+} levels was carried out with 1 μM of the ionophore, 4-Br-A23187,[9] to obtain maximal Ca^{2+} saturation of the probe, which was then quenched with 4 mM $MnCl_2$ to provide the residual autofluorescence of the cells. Substitutions were then made in the following equations, using a value of 115 for K_d, the dissociation constant for Ca^{2+},[43] for calculation of the equivalent $[Ca^{2+}]$ of any selected fluorescence, F:

$$F_{min} = [1/6\ (F_{max} - \text{autofluorescence}) + \text{autofluorescence}] \quad (1)$$

$$[Ca^{2+}] = K_d\ (F - F_{min})/(F_{max} - F) \quad (2)$$

RESULTS AND DISCUSSION

Changes in Extracellular $[K^+]$

Initial studies, in which general anesthetics were shown to produce differential depression of pre- and postsynaptic neuronal excitability and facilitate the efficiency of synaptic

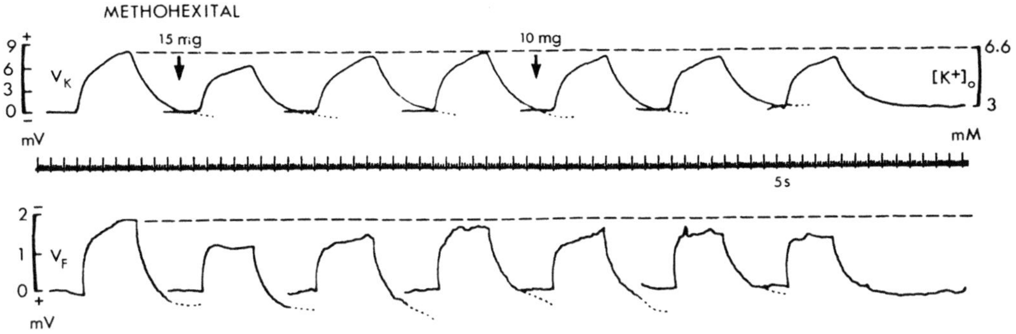

Figure 4. Depressant effect of methohexital on $[K^+]_o$ accumulations, evoked in synaptic regions of cat cuneate nucleus by tetanic stimulation of forelimb nerves (50 Hz, with voltage supramaximal for A fiber activation, ≤ 10 V). V_K = differential recording of K^+ signal between ion-sensitive and reference electrodes. V_F = focal potential recorded by reference electrode alone. Overlap of traces shows control records at left; responses at 2, 4, and 6 min after IV injection of 15 mg (4.7 mg/kg) methohexital, and responses at 3, 8, and 13 min after a further injection of 10 mg (3.1 mg/kg). Dotted horizontal lines mark control peak levels. From Morris.[28]

transmission in the cuneate nucleus of the decerebrate, unanesthetized cat, provided indirect evidence for membrane hyperpolarization.[27,29] Experiments were therefore subsequently carried out to attempt to test the hypothesis that general anesthetics hyperpolarize neurons by increasing G_K[12,14] and to determine whether this was a Ca^{2+}-mediated phenomenon.

In the first experiments, $[K^+]_o$ levels were measured in the synaptic regions of the cuneate nucleus with ion-selective microelectrodes, during administrations of a number of structurally unrelated intravenous and inhalational anesthetics.[28] These agents depressed the release of K^+ evoked by stimulation of peripheral nerves and believed to arise mainly from presynaptic terminals, as shown in the upper traces of Fig. 4, following two separate injections of the short-acting barbiturate, methohexital. Such an effect is consistent with hyperpolarization, the failure of impulse invasion into terminal regions, and a decreased transmembrane driving force for K^+. The apparent inability to detect any or even a small change in the resting $[K^+]_o$ might be attributable to (1) a high signal-to-noise ratio, (2) underestimation, as a result of dead space at the tip of the electrode, (3) efficient regional uptake, and (4) the relatively low doses of anesthetics agents, which were used to avoid hypotension.

In a further series of experiments, $[K^+]_o$ levels in cortical slices were recorded during direct bath applications of pentobarbital (0.1–1 mM) for periods of 5–30 min. A similar depression of K^+ accumulations, evoked by electrical transtissue field stimulation, was observed; however, resting $[K^+]_o$ was elevated (see examples of Fig. 5A and B). The observed postsuperfusion undershoot of $[K^+]_o$ further suggests that the augmented levels of $[K^+]_o$ are not due to interference with Na^+-K^+-ATPase-dependent membrane pumping (which instead appears to be activated), but rather to an increase in transmembrane outward movement of K^+. Differences between the results of these *in vitro* and the earlier *in vivo* experiments may be due to differences in (1) anesthetic dose, (2) geometry, synaptic organization, and other characteristics of brain regions, and (3) voltage drift and rejection capacity of recording electrodes.

Changes in Intracellular $[Ca^{2+}]$

Although the extracellular studies that measured change in $[K^+]_o$ supported the idea that anesthetics may alter G_K, they did not provide direct evidence, correlate with neuronal hyperpolarization, or demonstrate dependence on Ca^{2+}. The development of a highly selec-

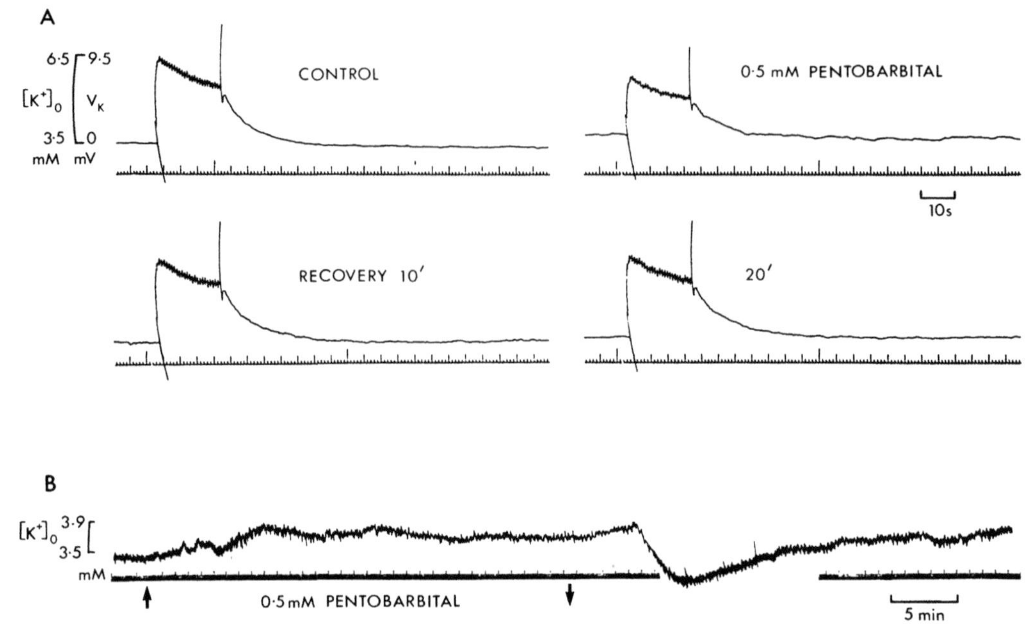

Figure 5. Effects of pentobarbital on resting and evoked levels of $[K^+]_o$ of cerebral cortex. Recordings made with ion-selective microelectrodes at depths of ≤200 μm in guinea pig transverse cortical slices. Records show voltage signal (V_K) and calibrated $[K^+]$ values. (A) An application of 0.5 mM pentobarbital for 8 min raises basal level of $[K^+]_o$ and depresses K^+ accumulation evoked by tetanic field stimulation (100 Hz, 20 mA, 20 sec); resting $[K^+]_o$ declines below control level during initial recovery. (B) Effect of prolonged superfusion of 0.5 mM pentobarbital for 32 min to raise $[K^+]_o$ by greater than 0.5 mM. Undershoot during early recovery period suggests stimulation of Na^+-K^+ pump activity.

tive neutral carrier for Ca^{2+}[37] did, however, promise that direct measurement of changes in intracellular Ca^{2+} could be made. Technical limitations have been encountered in the manufacture of the fine-tipped electrodes needed for successful recording of $[Ca^{2+}]_i$ in small cells.[41] In some cells, where two separate electrodes can be inserted, it has been possible to overcome these by the incorporation of polyvinylchloride in the membrane of the ion-sensing electrode.[1,23,42] However, in spite of our efforts,[31,34] this improvement has not been successfully applied to double-barreled electrodes, which are necessary for differential recordings in very small, nonsyncytial cells.

In spite of the difficulties in the fabrication of sufficiently responsive and stable double-barreled electrodes, it has been possible to demonstrate in studies of *in vivo* mammalian central neurons that motoneurons with resting and action potentials greater than 60 mV have a basal $[Ca^{2+}]_i$ of 1.6 μM (S.D. ± 1.64, $n = 36$), and graded activity-dependent increases of $[Ca^{2+}]_i$[34] can be evoked and measured. The example of Fig. 6A shows such an increase, evoked by antidromic activation of spikes in a neuron of the cat spinal cord by 20 Hz stimulation of a ventral root and associated with a posttetanic hyperpolarization. Promisingly, we have also been able to record $[Ca^{2+}]_i$ from pyramidal cells of the rat hippocampus *in vivo*.[33] Although resting levels appear to be high (8.5 μM; S.D. ± 13.9, $n = 22$; for cells with mean resting potential greater than 50 mV), extremely large stimulus-evoked increases in $Ca^{2+}{}_i$ (109 μM; S.D. ± 17.9; $n = 62$) have been recorded in these small cells of the brain.

Although we have not yet attempted to measure possible anesthetic-induced changes in $[Ca^{2+}]_i$ in these *in vivo* preparations, both membrane hyperpolarization and an increase in

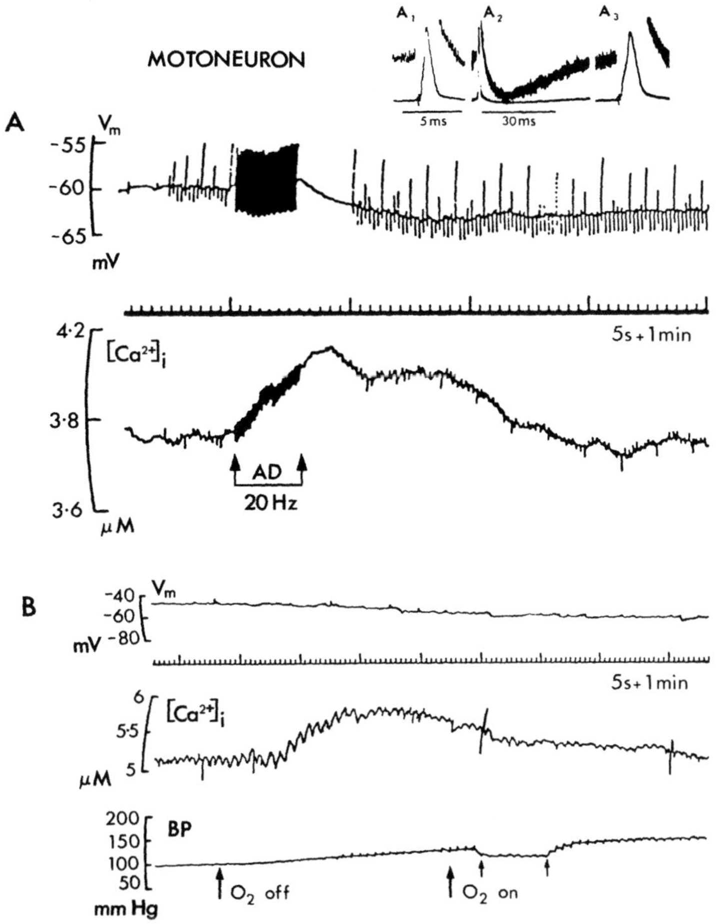

Figure 6. Increases of intracellular free Ca^{2+} in motoneuron are evoked by stimulation and hypoxia. Recordings with ion-selective microelectrode in cat spinal cord. (A) Antidromic activation by ventral root stimulation (20 Hz for 40 sec) causes a large rise in $[Ca^{2+}]_i$, accompanied by hyperpolarizing shift of membrane potential (V_m). Inset shows fast- and slow-sweep oscilloscope traces of spike activity: A_1 and A_2 before stimulation, A_3 during tetanus. (B) Effect of interruption of 0_2 supply to cause progressive hyperpolarization (V_m) and transient increase in $[Ca^{2+}]_i$. (Small arrows in bottom trace mark period of cardiac arrhythmia, which produced a fall in arterial pressure.) Modified from Morris *et al.*[31]; reproduced with permission.

$[Ca^{2+}]_i$ were apparent with anoxia (Fig. 6B). This observation is consistent with the demonstration in hippocampal neurons that anoxia induces hyperpolarization, a decrease in membrane resistance, and an increase in $[K^+]_o$,[11] and supports the proposal of an increase in G_K due to intracellular Ca^{2+} accumulation.[12,13] Further assessment of the effects of both anoxia and anesthetics will require improved fabrication of electrodes and more complete and controlled studies on neurons both *in vivo* and in dissociated culture and tissue slices.

In the meantime, the development of a new fluorescent indicator for Ca^{2+}, QUIN 2,[43] which can be easily incorporated into small cells, has encouraged an alternate approach. Preliminary pilot studies using suspensions of rat thymocytes (which, like neurons, possess Ca^{2+}-activated G_K,[38] but few mitochondria) have shown dose-dependent increases of $[Ca^{2+}]_i$ by up to 40% from resting values of 82.5 (S.D. ± 48.2) nM (see Fig. 7A and B), in the presence of greater than 0.1 mM pentobarbital. In spite of a direct depressant interaction

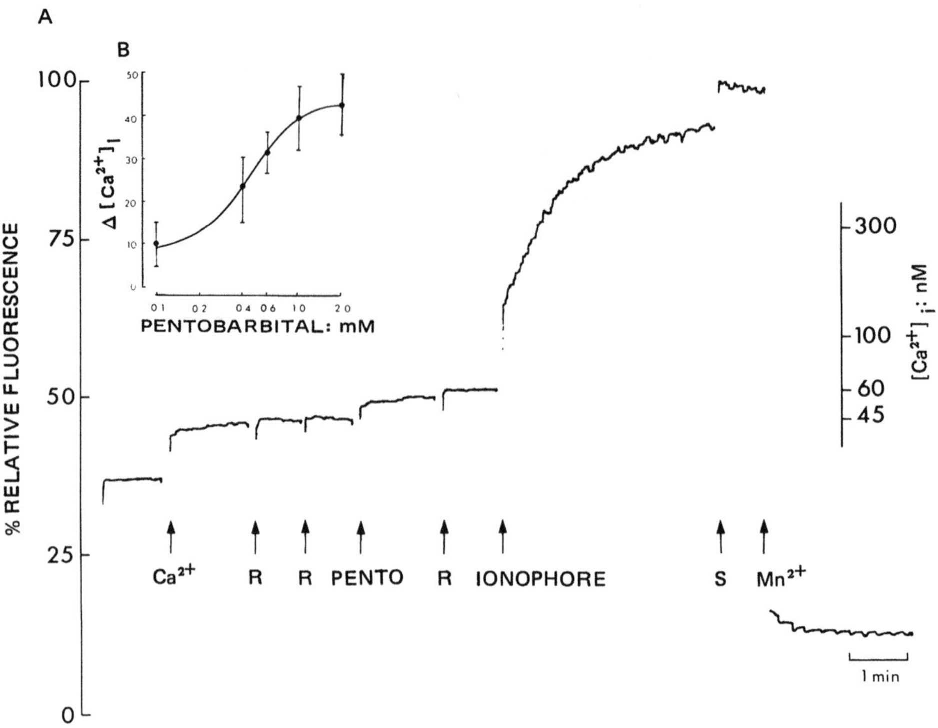

Figure 7. Effect of pentobarbital on $Ca^{2+}{}_i$. (A) Spectrofluorometric recording of changes in relative fluorescence intensity of rat thymocyte cell suspension measured with QUIN 2 Ca^{2+} indicator. Trace during control period at left shows basal level of $Ca^{2+}{}_i$ in cells and response to 1.8 mM $CaCl_2$. Addition of sodium pentobarbital (final concentration of 1 mM) was preceded and followed by stirring with control additions of identical volumes of Ringer solution (R). Final calibration was made with the ionophore, 4-Br-A23187 and stirring (S) to obtain maximal fluorescence and the quenching action of 4 mM $MnCl_2$. (B) Dose–response curve of % increase in $Ca^{2+}{}_i$ in 12 different cell preparations (n = 2–7 at each drug concentration; bars mark S.E.).

with the probe in the absence of cells (in contrast to a potentiating action of anesthetics on aequorin signals[2]), the addition of pentobarbital produced a consistently greater (≃ twofold) increase in fluorescence of suspensions of indicator-loaded cells than of unloaded cells. These preliminary observations therefore suggest that there may be an effect of this anesthetic agent to elevate $[Ca^{2+}]_i$ and have encouraged our further efforts to verify and more accurately estimate drug-induced changes in the calcium levels of suspensions of acutely dissociated and cultured neurons.

CONCLUSIONS

There have been few experiments to directly test the hypothesis that anesthetics alter membrane excitability of neurons by producing a Ca^{2+}-activated increase in G_K.[12,14] In studies with the squid axon, Baker and Schapira[2] observed that several anesthetic agents elevated the intracellular calcium signal. They were forced to conclude that this was due to a direct action on the aequorin indicator, although it was suggested that there could still be a sensitizing action on calcium-mediated events. Supportive evidence is given by the more recent report of the enhancement by several sedative drugs of the calcium dependence of after hyperpolarizations of hippocampal pyramidal cells.[5,6] However, the hyperpolarizing actions that Nicoll and Madison have demonstrated for a number of anesthetics on this same preparation have been attributed mainly to an increase in membrane G_K, which does not appear to be Ca^{2+}-activated.[36]

Changes induced by barbiturate and other anesthetics in stimulus-evoked and resting levels of $[K^+]_o$ have been recorded with ion-selective microelectrodes in synaptic regions in the medulla and cerebral cortex. These provide indirect evidence that anesthetic agents are able to increase membrane G_K. Preliminary data from recent experiments with thymocytes and the fluorescent Ca^{2+} indicator QUIN 2 show an effect of pentobarbital to increase apparent free cytosolic Ca^{2+} levels. These results, as well as our initial success with electrode measurements of intraneuronal Ca^{2+}, encourage our further experimental efforts to develop and use techniques to search for possible anesthetic-induced changes in $Ca^{2+}{}_i$ in mammalian central neurons of *in vitro* and *in vivo* preparations.

ACKNOWLEDGMENTS. The support of the Medical Research Council of Canada and the technical assistance of Edna Leech, David Conner, and Peta Allard are acknowledged with appreciation. Donations of Ca^{2+} exchanger by Professor W. Simon and of 4-Br-A23187 by Dr. C. Deber, as well as the advice of Dr. S. Grinstein, are gratefully recognized.

REFERENCES

1. Alvarez-Leefmans FJ, Rink TJ, Tsien RY: Free calcium ions in neurones of *Helix aspersa* measured with ion-selective micro-electrodes. *J Physiol* 1981;315:531–548.
2. Baker PF, Schapira AHV: Anaesthetics increase light emission from aequorin at constant ionised calcium. *Nature* 1980;284:168–169.
3. Bangham AD, Standish MM, Miller N: Cation permeability of phospholipid model membranes: Effect of narcotics. *Nature* 1965;208:1295–1297.
4. Barker JL, Ransom BR: Pentobarbital pharmacology of mammalian central neurons grown in tissue culture. *J Physiol* 1978;280:355–372.
5. Carlen PL: Enhanced neuronal K^+ conductance; possible common mechanisms for sedative-hypnotic drug action. *Can J Physiol Pharmacol* 1985;63:831–837.
6. Carlen PL, Gurevich N, Durand D: Ethanol in low doses augments calcium-mediated mechanisms measured intracellularly in hippocampal neurons. *Science* 1982;215:306–309.
7. Catchlove RFH, Krnjević K, Maretić H: Similarity between effects of general anesthetics and dinitrophenol on cortical neurones. *Can J Physiol Pharmacol* 1972;50:1111–1114.
8. Chalazonitis N: Selective actions of volatile anesthetics on synaptic transmission and autorhythmicity in single identifiable neurons. *Anesthesiology* 1967;28:111–123.
9. Debono M, Molloy RM, Dorman DE, Paschal JW, Babcock DF, Deber C M, Pfeiffer D R: Synthesis and characterization of halogenated derivatives of the ionophore A23187: Enhanced calcium ion transport specificity by the 4-bromo derivative. *Biochemistry* 1981;20:6865–6872.
10. Godfraind JM, Kawamura H, Krnjević K, Pumain, R: Actions of dinitrophenol and some other metabolic inhibitors on cortical neurones. *J Physiol* 1971;215:199–222.
11. Hansen AJ, Hounsgaard J, Jahnsen H: Anoxia increases potassium conductance in hippocampal nerve cells. *Acta Physiol Scand* 1982;115:301–310.
12. Krnjević K: Central actions of general anaesthetics, in Halsey MJ, Millar R, Sutton JA (ed): *Molecular Mechanisms in General Anaesthesia* New York, Churchill Livingston, 1974;pp 65–89.
13. Krnjević K: Coupling of neuronal metabolism and electrical activity, in Ingvar DH, Lassen NA (eds): *Brain Work, Alfred Benzon Symposium,* Vol VIII. Copenhagen, Munksgaard, 1975a, pp 65–78.
14. Krnjević K: Is general anesthesia induced by neuronal asphyxia? in Fink BR (ed): *Molecular Mechanisms of Anesthesia. Progress in Anesthesiology,* Vol 1. New York, Raven Press, 1975b, pp 93–98.
15. Krnjević K: Acetylcholine as modulator of amino acid-mediated synaptic transmission, in Lombardini JB, Kenny AD (eds): *The Role of Peptides and Amino Acids as Neurotransmitters.* New York, Alan R. Liss, Inc., 1981, pp 127–141.
16. Krnjević K, Lisiewicz A: Injections of calcium ions into spinal motoneurones. *J Physiol* 1972;225:363–390.
17. Krnjević K, Morris ME: Extracellular accumulation of K^+ evoked by activity of primary afferent fibers in the cuneate nucleus and dorsal horn of cats. *Can J Physiol Pharmacol* 1974;52:852–871.

18. Krnjević K, Morris ME: Input–output relation of transmission through cuneate nucleus. *J Physiol* 1976;257:791–815.
19. Krnjević K, Pumain R, Renaud L: The mechanism of excitation by acetylcholine in the cerebral cortex. *J Physiol* 1971;215:247–268.
20. Krnjević K, Morris ME, Reiffenstein RJ: Stimulation-evoked changes in extracellular K^+ and Ca^{2+} in pyramidal layers of the rat's hippocampus. *Can J Physiol Pharmacol* 1982;60:1643–1657.
21. Lux HD: Fast recording ion specific microelectrodes: Their use in pharmacological studies in the CNS *Neuropharmacology* 1974;13:509–517.
22. Magoun HW: *The Waking Brain.* Springfield, IL, Charles C Thomas, 1958, pp 1–138.
23. Marban E, Rink TJ, Tsien RW, Tsien RY: Free calcium in heart muscle at rest and during contraction measured with Ca^{2+}-selective microelectrodes. *Nature* 1983;80:845–850.
24. Meech RW: Intracellular calcium injection causes increased potassium conductance in *Aplysia* nerve cells. *Comp Biochem Physiol* 1972;42A:493–499.
25. Meech RW: The sensitivity of *Helix aspersa* neurones to injected calcium ions. *J Physiol* 1974;237:259–277.
26. Morris ME: Hypoxia and extracellular potassium activity in the guinea-pig cortex. *Can J Physiol Pharmacol* 1974;52:872–882.
27. Morris ME: Facilitation of synaptic transmission by general anaesthetics. *J Physiol.* 1978a;284:307–325.
28. Morris ME: Brain extracellular potassium and general anaesthetics. *Can J Physiol Pharmacol* 1978b;56:863–872.
29. Morris ME: Synaptic facilitation by general anesthetics, in Fink B R (ed): *Progress in Anesthesiology,* Vol. 2. New York, Raven Press, 1979, pp 463–468.
30. Morris ME, Krnjević K: Some measurements of extracellular potassium activity in the mammalian central nervous system, in Hebert NC, Berman PR (eds): *Proceedings of Workshop on Ion-selective Microelectrodes, Advances in Experimental Medicine and Biology,* Vol. 50. New York, Plenum Press, 1974, pp 129–143.
31. Morris ME, Macdonald JF: Measurements of intracellular [Ca^{2+}] in cultured mammalian neurones. *Neurosci Abstr* 1982;8:909.
32. Morris ME, Potashner SJ: Pentobarbital and extracellular K^+ activity of the cerebral cortex. *Proc Can Fed Biol Soc* 1980;23:186.
33. Morris ME, Krnjević K, Ropert N: Changes in free Ca^{2+} recorded inside hippocampal pyramidal neurons in response to fimbrial stimulation. *Soc Neurosci Abs* 1983;9:395.
34. Morris ME, Krnjević K, Macdonald JF: Changes in intracellular free Ca ion concentration evoked by electrical activity in cat spinal neurons in situ. *Neuroscience* 1985;14:563–580.
35. Nicoll RA: Pentobarbital action on frog motoneurons. *Brain Res* 1975;96:119–123.
36. Nicoll RA, Madison DV: General anesthetics hyperpolarize neurons in the vertebrate central nervous system. *Science* 1982;217:1055–1057.
37. Oehme M, Kessler M, Simon W: Neutral carrier Ca^{2+}-microelectrodes. *Chimia* 1976;30:204–206.
38. Rink TJ, Deutsch C: Calcium-activated potassium channels in lymphocytes. *Cell Calcium* 1983;4:463–473.
39. Sato M, Austin GM, Yai H: The actions of volatile anaesthetics on synaptic transmission. *Nature* 1967;215:1506–1508.
40. Spehlmann R, Smathers CC Jr: The effects of acetylcholine and of synaptic stimulation on the sensorimotor cortex of cats. II. Comparison of the neuronal responses to reticular and other stimuli. *Brain Res* 1974;74:243–253.
41. Tsien RY: Intracellular measurement of ion activities. *Annu Rev Bioeng* 1983;12:91–116.
42. Tsien RY, Rink TJ: Neutral carrier ion-selective microelectrodes for measurement of intracellular free calcium. *Biochim Biophys Acta* 1980;559:623–638.
43. Tsien RY, Pozzan T, Rink TJ: Calcium homeostasis in intact lymphocytes: Cytoplasmic free calcium monitored with a new, intracellularly trapped fluorescent indicator. *J Cell Biol* 1982;94:325–334.
44. Walker JL: Ion specific liquid ion exchanger microelectrodes. *Anal Chem* 1971;43:89A–93A.

7

Actions of Local Anesthetics on Identified Leech Neurons

Anna L. Kleinhaus, Jay Yang, and Jørgen Johansen

ABSTRACT. The actions of tetrodotoxin and procaine on identified (Retzius and sensory) neurons in leech segmental ganglia were compared.

Tetrodotoxin (15 μM) depressed the amplitude of the Na^+-dependent action potential of sensory neurons, but 100 μM had no effect on the Na^+-dependent action potential of the Retzius cell.

Procaine (0.1–10 mM) decreased the amplitude and maximum rates of de- and repolarization in both types of cells, but was more effective in the nociceptive (N) cell than in the R cell at physiological pH. At alkaline (8.3) pH, the sensitivity of the R cell approached that of the N cell at pH 7.4, suggesting that the drug acted intracellularly after passage through membrane lipids in its uncharged form.

Calcium antagonized the procaine-induced slowing of repolarization, suggesting that it reduced Ca^{2+} entry necessary for this process. Two additional findings support this concept: procaine reduced the Ca^{2+}-dependent afterhyperpolarization produced in leech neurons following repetitive activity and inhibited the Ca^{2+}-dependent action potentials recorded from R and N cells in Na^+-free solutions in a dose-dependent manner.

Simultaneous administration of procaine and tetrodotoxin in N cells reduced the maximum rate of depolarization of the action potential by an amount intermediate between that expected from two models assuming identical and independent "receptor" sites, respectively, indicating that the drugs act on separate but not entirely independent sites.

These findings imply the existence of two types of Na channels in mature leech neurons. Whereas the differential sensitivity to tetrodotoxin may be structural, that to procaine might be a consequence of variable access to an otherwise identical site. Correlation of such differences among identified neurons should further our understanding of pharmacological responses of intact nervous systems.

INTRODUCTION

The complexity of the mammalian central nervous system greatly complicates the elucidation of the mechanism of action of drugs that act on it. A clear understanding of the pharmacological responses of this system can be arrived at only by careful analysis of drug actions on its components. For example, although local anesthetics can produce excitatory as well as inhibitory effects in the central nervous system, they are mainly known for their ability to block the increase in Na^+ permeability associated with excitation.[36]

Anna L. Kleinhaus, Jay Yang, and Jørgen Johansen • Department of Neurology, Yale University School of Medicine, New Haven, Connecticut 06510. Present address of J. Y.: Department of Anatomy and Neurobiology, Washington University School of Medicine, St. Louis, Missouri 63110

For the last several years, this laboratory has been involved in the study of the ionic mechanisms underlying excitation in identified neurons in the central nervous system of the leech.[21–24,26,28,51] In these studies, we reported that differential drug sensitivities to agents known to block Na^+ and K^+ conductances in other excitable membranes segregated functionally identified neurons in an orderly fashion. Whereas the rising phase of the action potential in these neurons was largely dependent on the extracellular Na^+ concentration, they differed in their sensitivity to tetrodotoxin (TTX).[15,24,28] Some neurons such as the Retzius (R cell) were completely resistant to the drug, whereas the sensory neurons (N, P, T cells) were sensitive to it.

In this chapter, we review some of our work on leech neurons showing that in R and N cells (1) differential TTX sensitivity correlated with differential procaine sensitivity, (2) as in other preparations, TTX and procaine acted at different sites on the Na^+ channel,[51] and (3) in both types of cells, procaine blocked Ca^{2+}-dependent potentials in a dose-dependent fashion without distinction for the cell type.

METHODS

Segmental ganglia dissected from *Macrobdella decora* were fixed with magnet-stabilized pins to the bottom of a 1.5-ml chamber on a microscope slide. The work described here was done on the R cells and the cells sensitive to nociceptive stimuli N cells.

R Cells

The R cells are two large neurons easily visible in all ganglia, named after Retzius the anatomist, who first described them. They contain serotonin,[30,33,34,39] and evidence exists that they are involved in dermal mucus secretion,[29] inhibitory to body wall muscle,[32,41] and participating in the swimming and feeding networks.[31,47]

N Cells

These are two pair of bilateral neurons extensively studied by Nicholls and colleagues.[2–4,14,35] These neurons respond selectively to noxious stimuli to the skin and can be recognized easily by the shape of their action potential and by their anatomical location. Recently they have been subdivided into neurons of the medial type (N_{med}) and those of the lateral type (N_{lat}) by physiological[4,16] and pharmacological[16,40] means. In this study all data are from N_{med} cells because procaine produced a drastic depolarization in lateral N cells.[16,48] In order to facilitate access of the drugs to the neuronal membrane, the capsule overlying the neurons was removed. The standard Ringer's solution (at room temperature) used in these experiments contained (mM): NaCl 120, KCl 4, $CaCl_2$ 2, glucose 10, and either tris(hydroxymethyl)amino methane (Tris) 10 or *N*-2-hydroxyethyl-piperazine-*N'*-2-ethane sulfonic acid (Hepes) 10 with sufficient 10 N HCl or NaOH to achieve a pH of 7.4. Sodium-free solutions contained choline chloride or sucrose in equimolar concentration and 25 mM tetraethylammonium chloride (TEA) (Eastman). Procaine-HCl (Sigma) and TTX (Sanyo) were dissolved in this solution and any resulting pH change corrected as needed. Recording microelectrodes were pulled from 1 mm ultradot glass (Haer); when filled with 4 M potassium acetate or 3 M potassium chloride their d.c. resistance was between 20 and 40 megohms. Details of the procedure and experimental stimulating and recording apparatus were previously described.[22,28,51] Continuous monitoring of the entire experiment was done on a Hewlett-Packard 7402A pen recorder. Permanent recording was done on a PDP-8 computer system which digitalized the traces at 23 kHz and stored them for later reproduction on an incremental plotter and off-line analysis. Among the action potential parameters calculated are the maximum rates of depolarization (dV/dT_{dep}) and repolarization (dV/dT_{rep}). Because technical limitations prohibit uniform voltage clamping of these highly branched neurons,[28] we used dV/dT_{dep} as a measure of the inward Na^+ current occurring during excitation.[13] In our earlier work we established that Na^+ conductance is much greater than the nonsodium conductance of the membrane in these neurons.[24,28] We can therefore assume that dV/dT_{dep} is proportional to the maximum sodium conductance.[7] Although dV/dT_{rep} does not provide information regarding the ionic mechanisms responsible for repolarization, it reflects the rate at which this process occurs, and previous work has shown that potassium permeability is at least in part involved in it[22,23,26]; it correlated well with the duration of the action potential measured at half-amplitude. Experiments where both procaine and TTX were administered simultaneously were analyzed using dose–response relationships based on a one-receptor and a two-independent-receptor models.[18,51] Because dV/dT_{dep} rather than a direct measurement of Na^+ conductance was used, ED_{50}/K was assumed to be 4.[18,37]

RESULTS

Effect of Procaine on the Depolarization of the Na^+-Dependent Action Potential

Figure 1 reviews the evidence that the cells referred to in this chapter can indeed be separated into TTX-sensitive (N, P, T cells) and TTX-insensitive (R cell) types. The Na^+-dependent action potentials of the R cell (upper pair of traces) were largely unaffected by 100

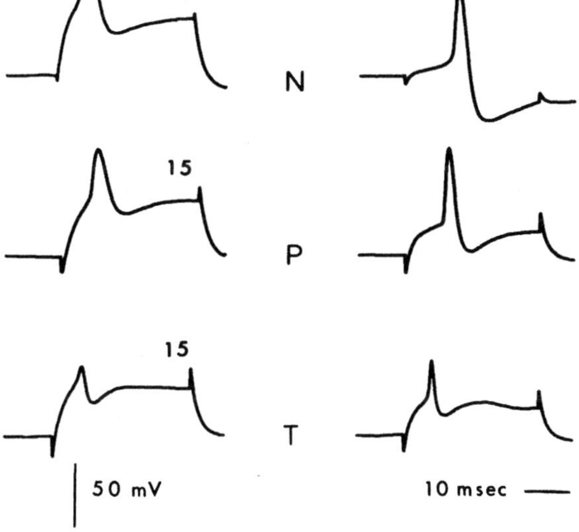

Figure 1. Differential sensitivity of R and sensory neurons to TTX. Traces are intracellular action potentials recorded with a single microelectrode. Cells were stimulated by current injection with bridge compensation. Ganglia were soaked in 100 μM TTX (R cell) and 15 μM TTX (N, P, and T cells) for 2 hr before being studied; left and right traces in each row are from the same cell. The amplitude of the R cell action potential in TTX is slightly depressed, but the threshold for its elicitation is slightly less than after washout. In the N, P and T neurons, the amplitude of the action potentials is depressed in TTX. Traces made by incremental plotter from digitized data.

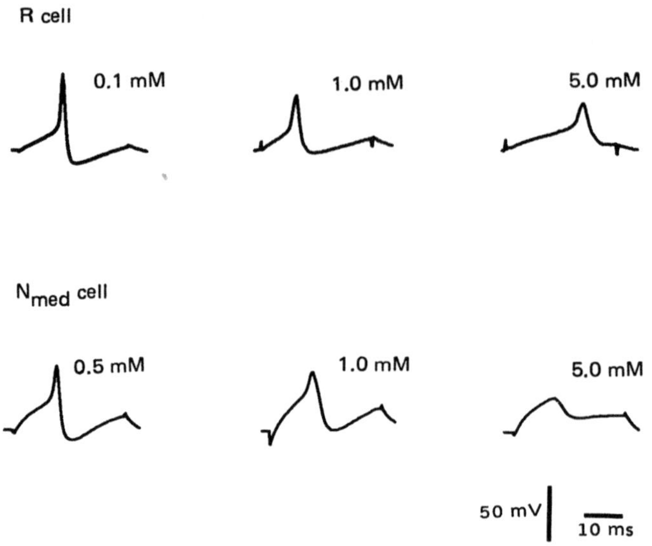

Figure 2. Effect of procaine on R and N cell action potential. Action potentials recorded in R (top row) and N_{med} cells from two ganglia. Increasing drug concentrations decreased amplitude of action potentials in both cell types. Reprinted with permission from Yang *et al.*[51]

μM TTX, whereas those of the sensory cells were reversibly depressed by a much lower (15 μM) concentration of the drug.[24,28] A similar variation in sensitivity to procaine (0.1–5.0 mM) in the same cell types is shown in Fig. 2.[51] The dose-dependent decrease in dV/dT_{dep} produced by the local anesthetic is illustrated in Fig. 3A for R and N_{med} cells. The difference in sensitivity of dV/dT_{dep} between the R and N_{med} cells was significant at the $p < 0.001$ level at all concentrations above 0.5 mM and could not be overcome by prolonged (>30 min) exposure or surgical removal of the connective capsule overlying the neurons. Two salient points can be noted on the graph: (1) the N_{med} was three times as sensitive to procaine as the R cell at concentrations producing 50% inhibition of dV/dT_{dep}, and (2) the curve for the R cell appeared to reach a saturation level for the procaine-insensitive Na^+ channels. However, if we assume that the depression of dV/dT_{dep} observed at 10 mM procaine in the R cell represents the maximum depression of the procaine-sensitive Na^+ channels, then the ED_{50} for both cell types was almost identical [0.72 mM (R) and 0.75 mM (N_{med})].

The relative insensitivity of the R cell could, however, be overcome by raising the pH of the perfusing solution to 8.3. Under these conditions (Fig. 4), 3 mM procaine decreased dV/dT_{dep} of the R cell to 10% of control, which was about the same amount of depression produced by 3 mM procaine at physiological pH in the N_{med} cell. It therefore seems that in leech neurons as in other preparations, local anesthetics act intracellularly[11,12,38,44]; it is also likely that access to the ''active site'' after extracellular application is more restricted in the R cell than in the N_{med} cell.

One series of experiments was done to determine the consequences of the interaction of procaine and TTX on dV/dT_{dep} in the N_{med} cell. Experiments in which the two drugs were given together, at concentrations that produced 50% depression of dV/dT_{dep} when administered separately, produced 77% depression of dV/dT_{dep}. Given the values of $ED_{50} = 0.75$ mM for procaine and $ED_{50} = 2.55$ μM for TTX determined by fitting equation (2) in ref. 51 to Fig. 3A and the results with TTX, if the drugs were to act at the same site, the amount of depression of dV/dT_{dep} would be expected to be 72%; a two-site model predicts a 89% depression given our experimental results.[18,51]

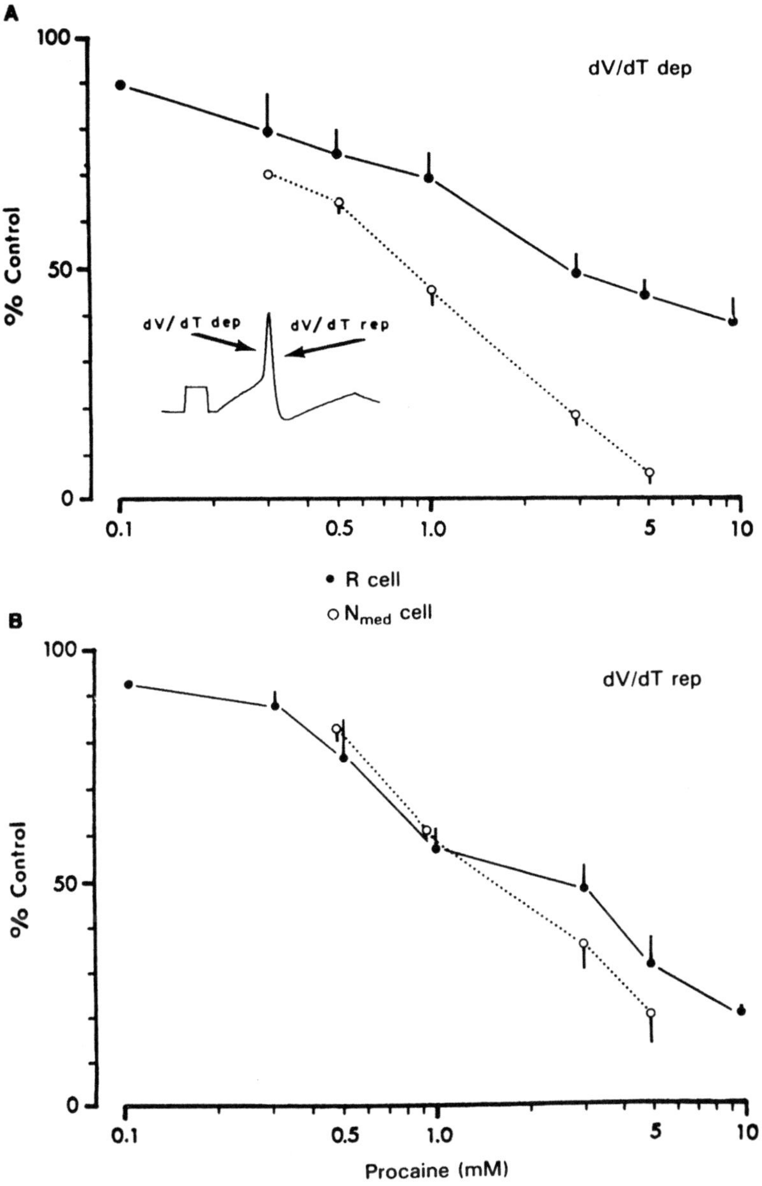

Figure 3. Dose–response curves for procaine action potential parameters in R and N neurons. Points on graphs (A and B) are means ± S.E. expressed as percentage of control. Each (filled circles, R cell; open circles, N cell) is average for four to eight cells except 0.1 mM point (R) and 0.3 mM (N) which is from two values. Action potentials parameters were calculated by PDP-8 computer from digitized samples, examples of which are shown in Figs. 1, 2 and 7 and inset. Arrows show points on action potential at which parameters used in graph were determined. Note that dose–response is shifted to the left for the N_{med} cell. Reprinted with permission from Yang *et al.*[51]

Effect of Procaine on the Repolarization of Na^+-Dependent Action Potentials

Procaine also decreased the dV/dT_{rep} and prolonged the duration of the action potential in both cell types in a dose-dependent manner. This effect could be due to a direct effect on the repolarizing K^+ conductance or to an indirect effect resulting from the reduced rate and

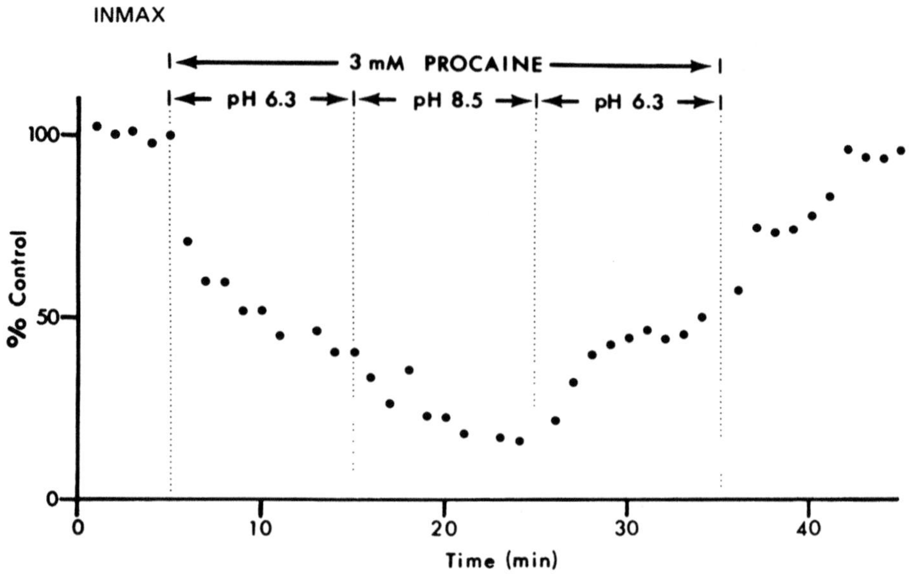

Figure 4. Alkaline pH increases sensitivity of R cell action potential to procaine. An R cell was exposed to procaine 3 mM (indicated by horizontal line) at two pH values . dV/dT_{dep} (INMAX) was reduced to about 50% of control by procaine at pH 6.3. Switching to a solution at pH 8.5 containing the same concentration of procaine reversibly reduced the maximum rate of depolarization of the action potential to about 10% of its control value.

amplitude of the spike. Alternatively, the drug's effect on repolarization could be a consequence of its effect on Ca^{2+} entry. The repolarization process in leech neurons is dependent on Ca^{2+}[22,50]; several drugs (barbiturates,[25,27] physostigmine[20]) slow this process presumably by interfering with Ca^{2+} entry. To test whether the procaine-induced prolongation was Ca^{2+} sensitive, we performed a series of experiments during which we exposed the cells to procaine in the presence of a constant concentration of divalent cations with varying ratios of Ca^{2+} to Mg^{2+}. In solutions where Ca^{2+} was the dominant divalent cation, the action potential was significantly shorter ($p < 0.02$) than in those where Mg^{2+} was at a higher concentration (Fig. 5). These findings are compatible with the idea that the effect of Ca^{2+} on

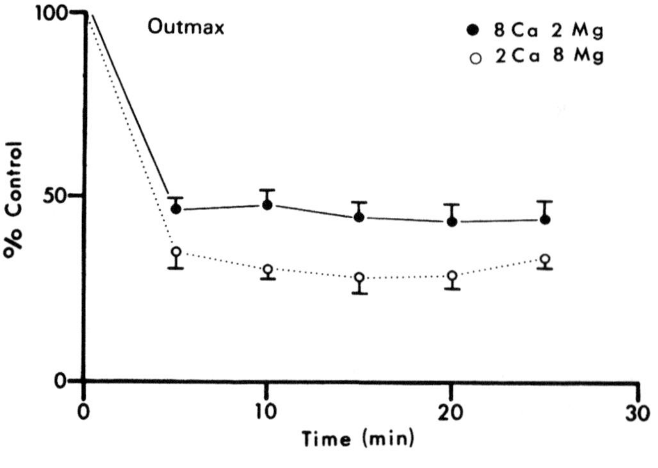

Figure 5. Calcium antagonism of procaine effect on repolarization mechanisms in R cell. Outmax (dV/dT_{rep}) was depressed by procaine in Ringer containing 8 mM Mg^{2+} and 2 mM Ca^{2+} (open circles). When the Ca/ Mg ratio was reversed in favor of Ca^{2+}, partial reversal of procaine's effect occurred (filled circles). Data points are means ± S.E. for six cells from different ganglia. Differences in values under the two conditions were significant at the $P < 0.02$ level.

repolarization may be specific rather than due to a shift in the voltage dependence of the repolarizing K^+ conductance produced by divalent cations.[9]

Effect of Procaine on Ca^{2+}-Dependent Phenomena

Procaine's effects on repolarization were compatible with the view that the drug interfered with Ca^{2+} entry through Ca^{2+} conductance mechanisms. We therefore examined the effects of procaine on processes directly dependent on Ca^{2+}, i.e., on the afterhyperpolarization (PAH) that follows repetitive activity in R and N cells[3,14,49] and on Ca^{2+}-dependent action potentials observed in the same cells in Na^+-free, TEA-containing Ringer.[21,23,26] Figure 6 shows that procaine nearly abolished the postactivity hyperpolarization produced in the R cell by a train of action potentials (second record). Following removal of the drug (third record), the PAH evoked by the same number of action potentials was fully restored. Part of the reduction in PAH could be due to the decrease in amplitude of the Na^+-dependent action potential. However, at the concentration of procaine used for these experiments, the dV/dT_{dep} in the R cell was depressed by less than 40%.

Direct proof of procaine's action on the Ca^{2+} conductance mechanism was obtained by the experiments depicted in Fig. 7. It is clear from the records that the dV/dT_{dep} (upper row of traces in both sets of traces) and the amplitude of the Ca^{2+}-dependent action potentials in R and N_{med} cells were decreased in a dose-dependent manner by procaine. The decrease in dV/dT_{dep} of the Ca^{2+}-dependent potential, like that of the Na^+-dependent one, was enhanced in alkaline solutions in both cell types. It appears that no differential sensitivity to procaine exists among the Ca^{2+}-dependent conductance mechanisms of the R and the N_{med} cells.

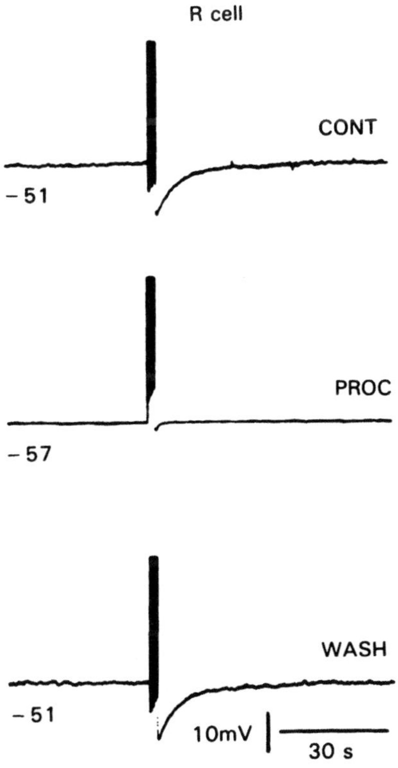

Figure 6. Procaine reduced hyperpolarization following activity in R cells. Ink traces are intracellular records taken from an R cell in 8 mM Ca^{2+}, 2 mM Mg^{2+}. Train of impulses in R cell evoked hyperpolarization of about 15 mV. Five minutes' exposure to 3 mM procaine greatly reduced amplitude of afterhyperpolarization (middle trace) produced by same stimulus train. Complete recovery occurred 12 min after washout.

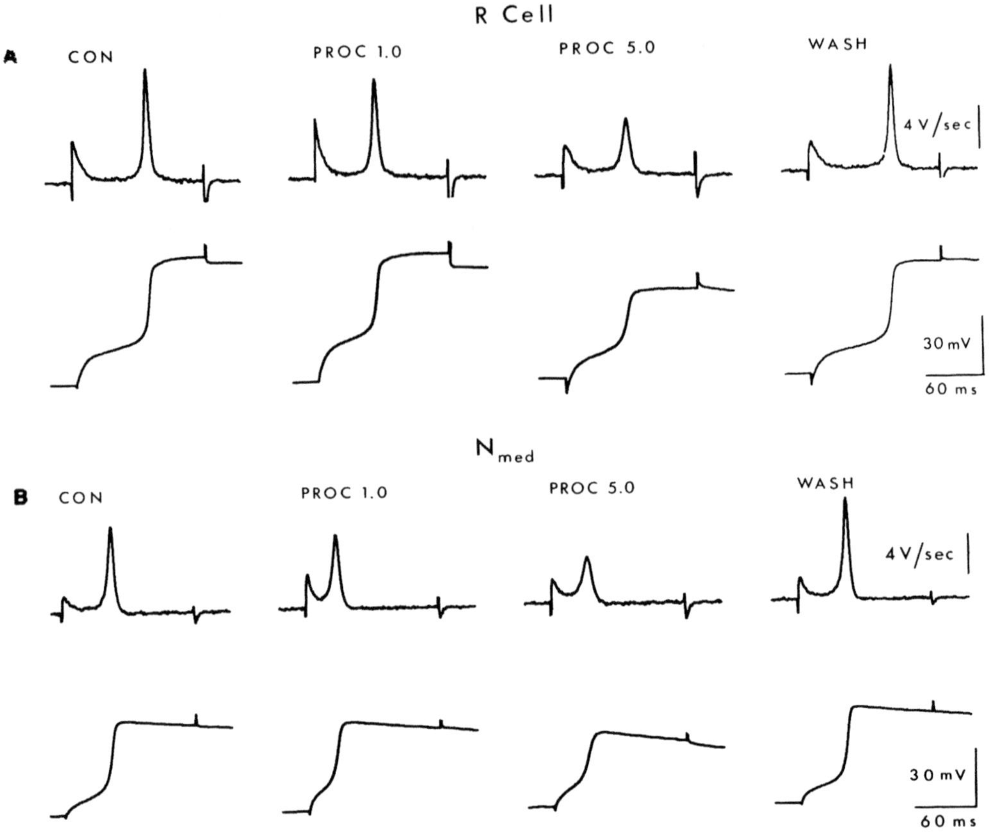

Figure 7. Procaine blocks Ca^{2+}-dependent action potentials in dose-dependent manner in R and N_{med} cells. Action potentials (lower traces) were recorded in Na^+-free, TEA-containing solutions; Ca^{2+} concentration 3 mM. Upper traces in A and B represent output of differentiating circuit. Procaine decreased dV/dT_{dep} and amplitude of Ca^{2+} action potentials in dose-dependent manner. Effect was fully reversible upon drug removal.

DISCUSSION

The results described in this chapter demonstrate a substantial difference in sensitivity to TTX and procaine among Na^+-dependent excitation mechanisms in mature leech neurons juxtaposed in the same ganglion. The observed difference between the R and N cell sensitivities to TTX were not abolished by experimental manipulations which should have maximized access of the drug to its active site.[24,28,51] This suggests that the "receptor sites" for TTX vary among leech neurons. In contrast, differential sensitivity to procaine was overcome by alkalinization, suggesting that in leech neurons, as in other preparations, the drug acts intracellularly[11,12,44] and, further, that the difference between the two cell types may be due to differential access to an otherwise identical site.[51] Further support for the idea that the Na^+ conductance mechanisms of the TTX- and procaine-sensitive neurons may be very similar to those of other excitable membranes comes from the experiments in which TTX and procaine were administered simultaneously. Our data produced a result intermediate between that predicted from one site and two independent site models, respectively[18] suggesting that the TTX and procaine sites are different but not entirely independent. These findings are in agreement with other studies in which TTX and saxitoxin were found to bind at sites distinct from local anesthetic binding sites.[8,10,43,45]

Processes that reflect conductances involved in repolarization mechanisms were affected in both cell types equally, indicating that in this preparation, as in others, procaine was not selective for the Na^+ channel.[44] In the absence of a uniform spatial clamp, it is impossible to determine whether the drug's effect on repolarization was a consequence of procaine's direct action on K^+ conductance, on Na^+ inactivation,[6,19,42] or resulted from the drug's interference with Ca^{2+}-dependent repolarization mechanisms. Our experiments with divalent cations showed that increased Ca^{2+} concentration antagonized procaine's action on repolarization specifically and that the reversal of procaine's effect was not due to nonspecific screening of membrane charges.[5] These findings are also consistent with earlier work suggesting a role for Ca^{2+} in repolarization of R cells[22,25,27,50] based on the sensitivity of that process to Ca^{2+} concentration. This view was further substantiated by procaine's dose-dependent decrease of Ca^{2+}-dependent potentials in Na^+-free Ringer (Fig. 7). A direct effect of local anaesthetics on Ca^{2+} currents has been reported in other preparations as well.[1,17,46]

At present it is technically more difficult to provide correlations between physiologically identified neurons and drug sensitivities in the mammalian central nervous system. Because of this, explanation of paradoxical effects of pharmacological agents in mammals may proceed at a rate slower than that of the invertebrate. The differences in sensitivity to procaine and TTX that exist among mature leech neurons are unlikely to be limited to the leech, and mapping of such variations might provide further insight into the organization of nervous systems and their pharmacological responses.

ACKNOWLEDGMENTS. This work was supported by USPHS Grants NS-06208-18 and 5-R01-NS 18054-02 and a grant from the Swebilius Trust Fund. We are indebted to Mr. Hector Goico for technical assistance and to Mr. Frank Esposito for his enthusiastic help in the maintenance and design of electronic equipment.

REFERENCES

1. Akaike N, Ito H, Nishi K, Oyama Y: Further analysis of inhibitory effects of propalanol and local anesthetics on the calcium current in Helix neurons. *Br J Pharmacol* 1982;76:37–43.
2. Baylor DA, Nicholls JG: Chemical and synaptic connexions between cutaneous mechanoreceptor neurones in the central nervous system of the leech. *J Physiol* 1969a;203:571–589.
3. Baylor DA, Nicholls JG: After-effects of nerve impulses on signalling in the central nervous system of the leech. *J Physiol* 1969b;203:591–609.
4. Blackshaw SE, Nicholls JG, Parnas I: Physiological responses, receptive fields and terminal arborizations of nociceptive cells in the leech. *J Physiol* 1982;323:251–260.
5. Blaustein MP, Goldman DE: Competitive action of calcium and procaine on lobster axon. *J Gen Physiol* 1966;49:1043–1063.
6. Cahalan MD: Local anesthetic block of sodium channels in normal and pronase treated squid giant axon. *Biophys J* 1978;23:285–312.
7. Cohen I, Attwell D, Strichartz G: The dependence of the maximum rate of rise of the action potential upstroke on membrane properties. *Proc Roy Soc B* 1981;214:85–98.
8. Colquhoun D, Henderson R, Ritchie J M: The binding of labeled tetrodotoxin to nonmyelinated nerve fibres. *J Physiol* 1972;227:95–126.
9. Frankenhaeuser B, Hodgkin AL: The action of calcium on the electrical properties of squid axons. *J Physiol* 1957;137:218–244.
10. Henderson R, Ritchie JM, Strichartz GR: The binding of labeled saxitoxin to the channels in nerve membranes. *J Physiol* 1973;235:783–804.
11. Hille B: The pH-dependent rate of action of local anesthetics on the node of Ranvier. *J Gen Physiol* 1977a;69:475–496.
12. Hille B: Local anesthetics: Hydrophilic and hydrophobic pathways for the drug-receptor reaction. *J Gen Physiol* 1977b;69:497–515.

13. Hodgkin AL, Katz B: The effect of sodium ions on the electrical activity of the giant axon of the squid. *J Physiol* 1949;108:37–77.
14. Jansen JKS, Nicholls JG: Conductance changes, an electrogenic pump and the hyperpolarization of leech neurones following impulses. *J Physiol* 1973;229:635–655.
15. Johansen J, Kleinhaus AL: Differential sensitivity to TTX of Na-potentials in nociceptive neurons from four species of leeches. *Neurosci Abst* 1985;11:314.
16. Johansen J, Yang J, Kleinhaus AL: Actions of procaine on specific nociceptive cells in leech central nervous system. *J Neurosci* 1984;4:1253–1261.
17. Josephson I, Sperelakis N: Local anesthetic blockade of Ca-mediated action potentials in cardiac muscle. *Eur J Pharmacol* 1976;40:201–208.
18. Kao CY, Walker SE: Active groups of saxitoxin and tetrodotoxin as deduced from actions of saxitoxin analogues on frog muscle and squid axon *J Physiol* 1982;323:619–637.
19. Khodorov B: Some aspects of the pharmacology of sodium channels in nerve membrane. *Biochem Pharmacol* 1979;28:1451–1459.
20. King WM, Yang J, Lent CM: Action potential prolongation: an effect of physostigmine (eserine) upon Retzius cells in the leech C.N.S. *Comp Biochem Physiol* 1984;79C:71–76.
21. Kleinhaus AL: Divalent cations and the action potential of the leech Retzius cell. *Pflueger's Arch* 1976;363:97–104.
22. Kleinhaus AL: Segregation of leech neurons by the effect of sparteine on action potential duration. *J Physiol* 1980;299:309–321.
23. Kleinhaus AL, Prichard JW: Calcium dependent action potentials produced in leech Retzius cells by tetraethylammonium chloride. *J Physiol* 1975;246:351–361.
24. Kleinhaus AL, Prichard, JW: Sodium-dependent tetrodotoxin-resistant action potentials in a leech neuron. *Brain Res* 1976;102:363–373.
25. Kleinhaus AL, Prichard JW: A calcium reversible action of barbiturates on the leech Retzius cell. *J Pharmacol Exp Ther* 1977a;201:332–339.
26. Kleinhaus AL, Prichard JW: Close relation between TEA responses and Ca-dependent membrane phenomena of four identified leech neurons. *J Physiol* 1977b;270:181–194.
27. Kleinhaus AL, Prichard JW: Interaction of divalent cations and barbiturates on four identified leech neurons. *Comp Biochem Physiol* 1979;63C:351–357.
28. Kleinhaus AL, Prichard JW: Differential action of tetrodotoxin on identified leech neurons. *Comp Biochem Physiol* 1983;74C:211–218.
29. Lent CM: Retzius cells neuroeffectors controlling mucus release by the leech. *Science* 1973;179:693–696.
30. Lent CM: Serotonin-containing neurons within the segmental nervous system of the leech, in Osborne N N (ed): *The Biology of Serotonergic Transmission.* London, Wiley, 1982.
31. Lent CM, Dickinson MH: Serotonin integrates the feeding behavior or the medicinal leech. *J Comp Physiol* 1984;154:457–471.
32. Mason A, Sunderland AJ, Leake LD: Effects of leech Retzius cells on body wall muscles. *Comp Biochem Physiol* 1979;63C:359–361.
33. McAdoo DJ: The Reszius cell of the leech *Hirudo medicinalis,* in Osborne N N (ed): *Biochemistry of Characterized Neurons,* Oxford, Pergamon, 1978.
34. McAdoo DJ, Coggeshall RE: Gas chromatographic mass-spectrometric analysis of biogenic amines in identified neurons and tissues of *Hirudo medicinalis. J Neurochem* 1976;26:163–167.
35. Nicholls JG, Baylor DA: Specific modalities and receptive fields of sensory neurones in the leech CNS. *J Neurophysiol* 1968;31:740–756.
36. Ritchie MJ, Greene NM: Local anesthetics, in Goodman LS, Gilman A (eds): *The Pharmacological Basis of Therapeutics.* New York, Macmillan, 1980,pp 300–320.
37. Ritchie JM, Greengaard P: On the active structure of local anesthetics. *J Pharmacol Exp Ther* 1961;133:241–245.
38. Ritchie JM, Rogart RB: The binding of saxitoxin and tetrodotoxin to excitable tissue. *Rev Physiol Biochem Pharmacol* 1977;79:1–50.
39. Rude S: Monomine-containing neurons in the central nervous system of the leech *Hirudo medicinalis. J Comp Neurol* 1969;136:249–372.
40. Sargent PB, Yau K, Nicholls JG: Extrasynaptic receptors on cell bodies of neurons in central nervous system of the leech. *J Neurophysiol* 1977;40:446–452.
41. Sawada M, Coggeshall RE: A central inhibitory action of 5-hydroxytryptamine in the leech. *J Neurobiol* 1976;7:477–482.
42. Seeman P, Chen SS, Chau-Wong M, Staiman A: Calcium reversal of nerve blockade by alcohols, anesthetics, tranquilizers, and barbiturates, *Can J Physiol Pharmacol* 1974;52:526–534.

43. Staiman A, Seeman P: Different sites of membrane action for tetrodotoxin and lipid-soluble anesthetics. *Can J Physiol Pharmacol* 1975;53:513–524.
44. Ulbricht W: Kinetics of drug action and equilibrium results at the node of Ranvier. *Physiol Rev* 1981;4:785–828.
45. Wagner HH, Ulbricht W: Saxitoxin and procaine act independently on separate sites of the sodium channel. *Pleugers Arch Ges Physiol* 1967;364:65–70.
46. Washizu Y: Procaine on smooth muscle. *Comp Biochem Physiol* 1968;27:121–126.
47. Willard AL: Effects of serotonin on the generation of the motor program for swimming by the medicinal leech. *J Neurosci* 1981;1:9:936–944.
48. Yang J, Kleinhaus AL: Procaine actions on leech Retzius cells. *Soc Neurosci Abstr* 1982;8:252.
49. Yang J, Kleinhaus AL: Effects of tetraethylammonium chloride and divalent cations on the afterhyperpolarization following repetitive activity in leech neurons. *Brain Res* 1984;311:380–384.
50. Yang J, Lent CM: Calcium depletion produces Na-dependent, sustained depolarizations of Retzius cell membranes in the leech CNS. *J Comp Physiol A* 1983;150:499–507.
51. Yang J, Johansen J, Kleinhaus AL: Procaine actions on tetrodotoxin senstive and insensitive leech neurons. *Brain Res* 1984;302:297–304.

8

Effects of Isoflurane on some Passive and Active Electrical Properties of Trigeminal Sensory Neurons in Vivo

E. Puil

ABSTRACT. The endotracheal administration of isoflurane to decerebrate guinea pigs resulted in a reduction in the impedance magnitude functions (impedance transfer functions) recorded from trigeminal ganglion neurons. This decrease in the case of neurons, which could be modeled by a simple parallel RC circuit, could be attributed to an increase in input conductance and input capacitance. Most neurons presented active (voltage- and time-dependent) membrane properties reflected as a resonating bandpass characteristic in their impedance magnitude functions, which was decreased by endotracheal administrations of isoflurane. The isoflurane-induced changes in resonance were quantified by curve-fitting the complex impedance with a fitting function based on a linearized Hodgkin–Huxley model or an equivalent electrical circuit model incorporating a series resistance and inductor. Increases in resting conductance and less consistently in input capacitance were observed after administration of 4% isoflurane. The alterations in electrical susceptance (1/reactance) of the neuronal membrane in neurons with the resonating bandpass characteristic in their impedance transfer functions have important implications for phasic signal transfer by synaptic transmission in neurons of the central nervous system during the general anesthetic state. The isoflurane-induced changes in the membrane electrical properties of trigeminal ganglion neurons also would serve to impair the generation of action potentials.

Since the turn of the century, much attention has been focused on efforts to correlate the apparent potencies of substances to induce the general anesthetic state in the whole animal, with certain physicochemical properties, particularly with the ability of these agents to dissolve in lipids.[20,23] One plausible hypothesis, which has emerged from *in vitro* investigations on artificial membranes as well as on biological membranes, is that general anesthetics may act directly on the hydrophobic portions of integral protein(s) which are associated with the lipid components of the neuronal membrane, thereby perturbing its normal functions, e.g., excitability.[9,16,28,31] An assumption implicit in the hypotheses that correlate physicochemical properties with anesthetic action is that the derangement in excitability directly results in a reduction of neuronal activity within neurological arousal systems.

E. Puil • Department of Anaesthesia and Department of Pharmacology and Therapeutics, Faculty of Medicine, The University of British Columbia, Vancouver, British Columbia V6T 1W5, Canada.

Possible mechanisms of general anesthesia of these neurons include alterations in neuronal function which occur (a) presynaptically to retard[18,24,25,33] or to enhance[21] transmitter release, (b) postsynaptically to alter excitability,[2,6,22] including changes in responsiveness either to inhibitory[12,17,26] or to excitatory[3,29] transmitter agonists, or (c) intracellularly to alter metabolism, thereby changing presynaptic release of, and/or postsynaptic responsiveness to, neurotransmitters.[15]

A major difficulty in such electrophysiological investigations is the inability to distinguish the effects of drugs on passive membrane properties, e.g., resting or input resistance and capacitance, from their direct effects on active responses, i.e., potential- and time-dependent opening and closing of ion-selective channels. Indeed, the effects of drugs, in general, on the input capacitance of the membrane have received little attention. The series of experiments described here represents a new approach to this problem; our recently developed computerized methodology based on Fourier transformations[11,24a] permits observations of the separate effects of anesthetics on the active and passive components of the electrical behavior of neuronal membranes.

The trigeminal root ganglion of guinea pigs *in vivo* was selected for study because neurons within this structure are geometrically simple and have not yet been found to possess synaptic impingements on their perikarya. These features, especially their approximately spherical shapes, confer electrical behavior that may be interpreted in a straightforward way. The advantage of the *in vivo* preparation was considered to be important for its relevance to mechanisms of the general anesthetic state in spite of the inherent difficulties in obtaining stable recording conditions and controlling certain influential variables such as blood pressure.

METHODS

The experiments were performed on unanesthetized guinea pigs that were rendered decerebrate while under prior isoflurane anesthesia. The animal initially was anesthetized with 4% isoflurane in a 3-liter chamber. After loss of the righting reflex, which occurred after 40–90 sec, anesthesia was maintained by mask. The two cerebral hemispheres rostral to the colliculi were removed, and the trigeminal ganglion regions in the middle fossa of the cranium were exposed during endotracheal administration of 2% isoflurane. Anesthesia then was discontinued for at least 2 hr before observations of anesthetic effects were made on membrane properties. Arterial blood pressure was monitored from a cannulated carotid artery during measurement of the effects of anesthetics on neuronal membrane properties. Carotid arterial blood pressure may have been slightly elevated or slightly depressed during anesthesia, but, for the most part, was not always significantly affected by the administration of 0.5–2% isoflurane. However, blood pressure usually was reduced by 20–40% in animals administered 4% isoflurane, and the possibility must be considered that the observed changes in membrane properties during the administration of isoflurane may be a consequence of isoflurane-induced changes in blood flow, etc. (cf. ref. 34). Stable conditions for prolonged intracellular recording often were achieved by paralyzing the animal with pancuronium bromide and employing mechanical ventilation. End tidal P_{CO_2} was monitored during this time. High-frequency (900/sec) ventilation was employed in some of the experiments such that arterial P_{CO_2} was maintained at about 40 mm Hg and P_{O_2} values at 100–200 mm Hg.

Complex impedance measurements of trigeminal ganglion neurons were obtained using intracellular recording and current injection, with single microelectrodes filled with 3 M KCl (tip resistances—R_e = 10–60 MΩ) and a computerized methodology involving digital to analogue conversion of a function into sinusoidal current of a rapidly increasing frequency in

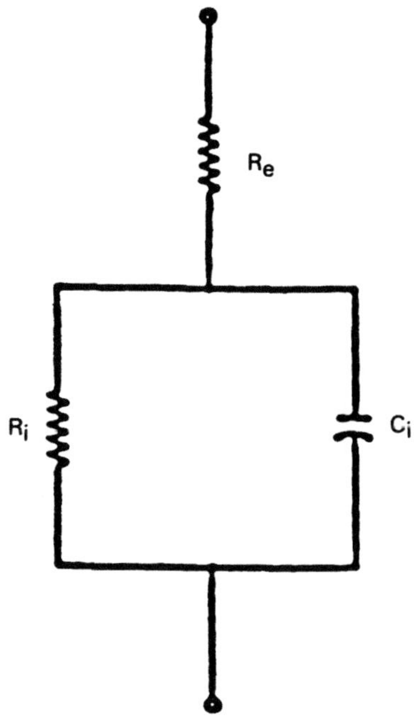

Figure 1. Equivalent electrical circuit of model that behaves according to single time constant process. R_e represents electrode resistance in series with input resistance (R_i) which is in parallel with input capacitance (C_i). The theoretical impedance transfer function of this model, which was used for a fit of experimental data as in Fig. 2, is given by

$$Z = R_e + \frac{R_i}{1 + \omega^2 R_i^2 C_i^2} - j\frac{\omega R_i^2 C_i}{1 + \omega^2 R_i^2 C_i^2}$$

where $\omega = 2\pi f$ and $j = \sqrt{-1}$.

the range of 5–1250 Hz and the subsequent injection (WPI amplifier, Model 701) of such waveform currents *via* the recording electrode into a neuron.[11,24a] The impedance amplitude profile of the electrode–neuron system was computed on-line as a transfer function from the ratio of fast Fourier transforms of the output voltage response from the neuron (sampled at a rate of 5 kHz with an analogue-to-digital converter) and of the injected input current. The duration of each input current waveform that was used to generate the transfer functions was about 200 msec. The amplitude of the input current was adjusted such that perturbation of the membrane potential (E_m) was very small, and mostly within the range of 2–5 mV (peak to peak). The data usually were expressed either as a transfer function representing the absolute value of complex impedance or as a real versus imaginary plot (impedance locus diagram) which reflects changes in phase angle as a function of frequency of input current.[14] Each data point in this diagram represents the maximum voltage response of a neuron to a given frequency of the current input.[14] By utilizing a model based on an equivalent electrical circuit (Figs. 1 and 3) or linearized Hodgkin–Huxley equations[13] which accurately represent the measured transfer function in absolute (e.g., Fig. 2) or complex form (Fig. 4), the response of the electrode–neuron system could be characterized. This was accomplished by using a nonlinear least-squares curve-fitting technique[1] to determine the model parameters that best fit the measured data. The interpretation of the data then could be related to the electrical and biological properties of the system.

RESULTS

Observations of the effects of 28 endotracheal applications of isoflurane were made on 21 neurons in 15 guinea pigs. However, the data presented here were obtained from 12 of these neurons for which recording conditions were very stable. In about half the neurons tested, the administration of isoflurane reduced the rates of rise and fall and amplitude of action potentials evoked by intracellular depolarizing current. The effects of the anesthetic on the firing threshold were variable in tests in seven cells; in four of these neurons the

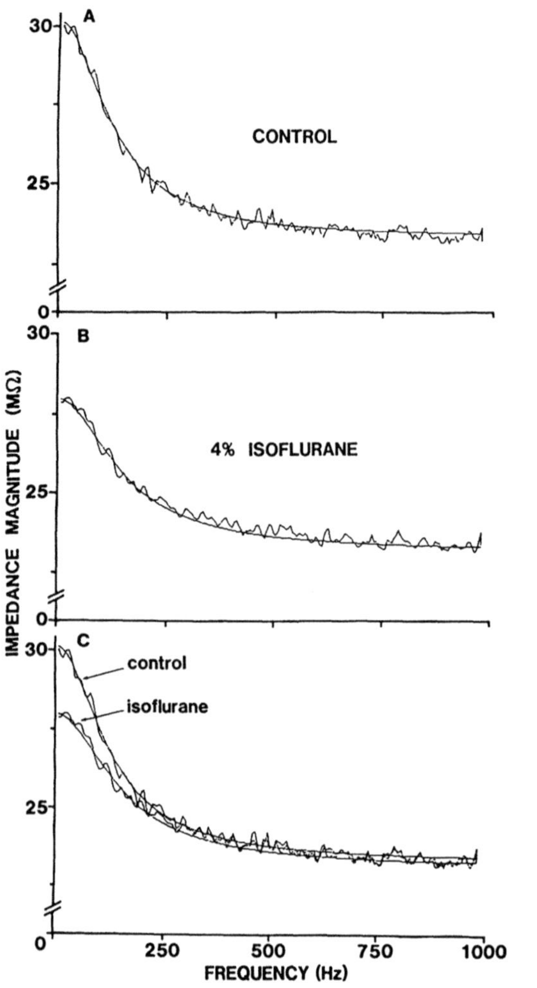

Figure 2. Effects isoflurane anesthesia on impedance transfer functions of trigeminal ganglion neuron. Smooth curves in A, B, and C represent a nonlinear least-squares fit of the theoretical transfer function for an electrode resistance in series with a parallel RC circuit model to each experimentally derived transfer function comprised of 60 averages. Electrical behavior of neuron may be said to conform to a single time constant model. (A) Control transfer function taken during a 30 sec "control period" prior to induction of anesthesia with isoflurane. The nonlinear least-squares fit provided values of $R_e = 23.4$ MΩ, $R_i = 6.8$ MΩ, and $C_i = 189$ pF during this control period. (B) Transfer function obtained about 2 min after the onset of 4% isoflurane anesthesia and for a 30-sec period. The nonlinear least-squares fit provided values of $R_e = 23.2$ MΩ, $R_i = 4.8$ MΩ, and $C_i = 221$ pF during anesthesia. (C) Superimposition of transfer functions of A and B reveals that the anesthetic-induced changes mostly occur in the low-frequency region (<250 Hz) of experimentally derived transfer function.

endotracheal administration of 2–4% isoflurane for 1–2 min (see Methods) resulted in an increase in the threshold for the generation of an action potential evoked by intracellular injection of a depolarizing current pulse. In addition to the changes in spike shape, directionally consistent, large changes in E_m were not observed during isoflurane anesthesia. Both hyperpolarizing and depolarizing responses of membrane potential (<10 mV) were observed during anesthesia, in about an equal number of cells. In five cells (5/12), E_m remained unchanged, while decreases in R_i and increases in C_i were observed during the anesthesia.

The voltage responses to injections of square-pulse hyperpolarizing currents which were consistent with a model comprised of an R_e in series with an RC parallel circuit were observed in about 25% of the neurons at their initial resting potential (near −60 mV; cf. ref. 11; Fig. 1); that is, the exponential nature of the hyperpolarizing voltage response verified this linear behavior.

An example of a transfer function of a neuron that exhibited such simple electrical behavior in the time domain is shown in Fig. 2. A well-approximated fit (smooth line, Fig. 2) of the theoretical transfer function computed for a resistance in series with an RC parallel circuit model (Fig. 1) to the transfer function measured in such neurons provided effective estimates for electrode resistance (R_e), input resistance (R_i), and input capacitance (C_i).

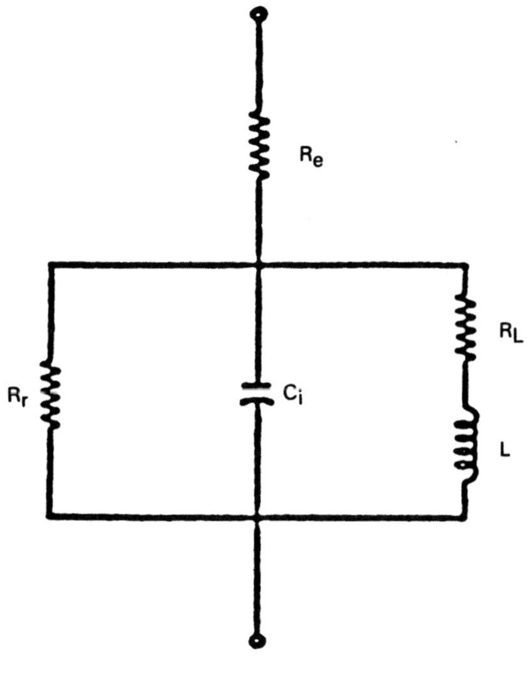

Figure 3. Equivalent electrical circuit of five-parameter model described by general form of fitting equation in text and which has interpretations that are equivalent to linearized Hodgkin–Huxley model. The theoretical impedance transfer function of this model is represented by

$$Z = R_e + \frac{1}{\frac{1}{R_r} + j\omega C_i + \frac{1}{R_L + j\omega L}}.$$

R_e represents electrode resistance in series with resting resistance (R_r). The input capacitance (C_i) is in parallel with R_r and the series resistance (R_L) and an inductor (L).

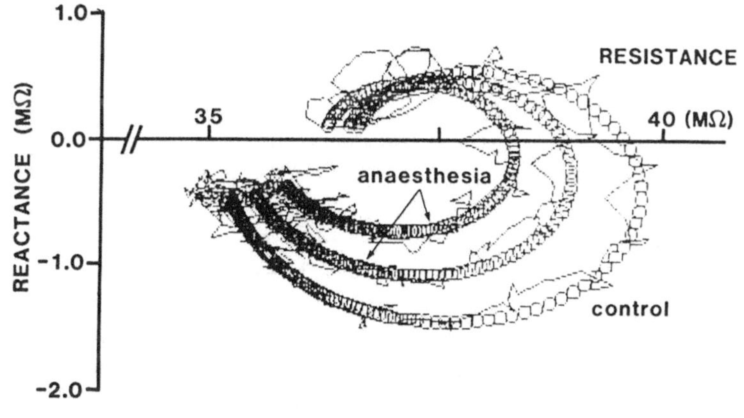

Figure 4. Impedance locus diagram or Cole–Cole plot (40 averages) shows effect of anesthesia on a trigeminal ganglion neuron that exhibited an inductive (positive) reactance. In the time domain, this neuron exhibited anomalous (inward) rectification when injected with step currents of 50 msec duration (not shown). Squares represent a nonlinear least-squares fit of impedance fitting function based on linearized Hodgkin–Huxley equations. Computer plotted points in a continuous manner starting at lowest frequency (5 Hz) in the positive half-plane such that lines were drawn between points. Systematic experimental "noise" then can be seen as "loops" or peaking in the impedance dispersion. Control curve crosses over into capacitive (negative) reactance half-plane at about 170 Hz and <40 MΩ. Control curve with fit that provided parameter values of R_e = 36.0 MΩ, R_r = 7.5 MΩ, and C_i = 364 pF was taken during a 30-sec period prior to isoflurane anesthesia. Middle curve fit provided parameter value of R_e = 36.0 MΩ, R_r = 4.8 MΩ, C_i = 331 pF (during 4% isoflurane for 2 min). Smallest spiral representing fit of data taken at about 10 sec after discontinuing anesthesia administration for a period equivalent in time to control period yielded values of R_e = 35.5 MΩ, R_r = 3.0 MΩ, and C_i = 451 pF.

The endotracheal administration of 4% isoflurane for periods of time that previously caused loss of the righting reflex in the intact animal resulted in a 5–35% diminution of R_i and a less consistent increase of C_i in cells that exhibited behavior ascribable to the single time constant model (Fig. 2B,C). Recovery was observed after 8 min in two of these neurons. Recovery from (1–4%) isoflurane was not observed in all other neurons and apparently would have required more than 10 min after discontinuing its administration.

The majority of trigeminal ganglion neurons seem to exhibit, at their resting membrane potential, a pronounced hump in the computed transfer function at low frequencies (50–250 Hz). This resonant behavior is reflected in the impedance locus diagram (Fig. 4, control) where the locus crosses into the inductive half-plane. Each point in this impedance locus diagram or Cole–Cole plot derives from the maximum voltage response of the neuron to a given frequency of input current. Therefore, each phasor (complex vector) represents the magnitude of the impedance at a particular phase angle (θ) formed with the origin. Because this angle becomes positive at low frequencies (less than about 170 Hz in Fig. 4), the voltage response does not lag, in time, the current in the electrode–neuron system. This existence of a phase lag suggests an inductive behavior of the membrane. Since the electrode properties can be determined and in this case were described by a single time constant process, the electrical behavior of the neuronal membrane may be considered analogous to that of an inductor. This interpretation should not be taken literally, but as a mnemonic in utilizing an electrical model (Fig. 3) which represents accurately (with a series electrode resistance) the impedance locus diagram (Fig. 4). In this way, the system's response may be characterized. Modifications in both capacitive and inductive reactances of the nerve cell with the impedance locus diagram illustrated in Fig. 4 occurred during isoflurane anesthesia. The inductive reactance reflects the resonant behavior in its corresponding transfer function (not shown). The inductive reactance was not evident in many neurons, and in such instances, the impedance locus dispersion resembled a distorted semicircle only in the capacitive half-plane. Such neurons also showed resonance in their transfer functions. Irrespective of whether the resonance in the transfer function might be seen as a distortion in the capacitive half-plane, or as a positive reactance, the impedance locus dispersion may be well approximated by a fit with an electrical RLC-circuit model or equivalently by a function derived from linearized Hodgkin–Huxley equations[13]; this linearization can be justified mathematically and will appear elsewhere.[24a] Both types of fit have the general form

$$Z = R_e + \frac{1}{G_r + j\omega C + \dfrac{g_L}{1 + j\omega\tau_u}}$$

where *Z is the impedance; C* is membrane capacitance, G_r (reciprocal of R_r) is the resting conductance, g_L is the membrane conductance activated by the 2–5 mV (peak-to-peak) perturbation of E_m by the input current, and τ_u is the membrane conductance relaxation time; j and ω have their usual meanings of $\sqrt{-1}$ and $2\pi f$, respectively. An electrical interpretation is that the fitting function represents an equivalent circuit comprised of a resistor (R_L) in series with an inductor (L) which are in parallel with the simple parallel RC circuit model (Fig. 3). The reciprocal of R_L is g_L, whereas L/R_L in the linearized Hodgkin–Huxley interpretation is the relaxation time constant, τ_u. In summary, the first-order differential equation which relates the voltage, as a function of time, across a series g_L circuit to the input current has the same form as the rate process described in the Hodgkin–Huxley formulation[13] for the n parameter associated with potassium conductance.[8] This similarity means that, for a constant voltage condition, the differential equation for the model represented in Fig. 3 also describes, at least kinetically, the linearized Hodgkin–Huxley potassium conductance. Therefore, changes in g_L

and related time constant τ_u would represent changes in voltage activation of an ionic conductance.

Administration of isoflurane increased both the resting conductance and input capacitance of the neuronal membrane in most of the neurons tested. The voltage-dependent conductance activated by the perturbation of membrane potential has been observed to be reduced or, in some cases, increased in these neurons. The input conductance (G_r and g_L) of all neurons showing resonant behavior was increased as in the case of neurons exhibiting single time constant behavior in their transfer functions. In the other neurons that showed anomalous rectification when injected with hyperpolarizing step currents, the resonant characteristic in their transfer functions was decreased during isoflurane anesthesia (not shown).

DISCUSSION

The majority of trigeminal neurons that were examined in the time domain exhibited anomalous rectification or "sagging" voltage responses to injected hyperpolarizing step currents. Our investigations have shown for the first time that these same neurons exhibited corresponding resonance in the frequency domain. In voltage-clamp studies of dorsal root ganglion neurons, inward (anomalous) rectification, activated by membrane hyperpolarization, is believed to be mediated by an inwardly directed "mixed current" involving Na^+ and K^+.[19] In the case of trigeminal neurons, the inductive reactance or the distortion of the semicircular dispersion of the impedance locus evident in the capacitive half-plane of the Cole–Cole plot is probably a manifestation of the activation of similar, voltage- and time-dependent ionic channels. In squid axon, the inductive behavior is attributable to linear time- and voltage-dependent K^+-conductance.[4] This type of nonlinear membrane behavior may be modeled by a linear time-invariant RLC-circuit which includes a conductance in series with an inductance[5,8] or, as in our investigation of trigeminal neurons, by linearized Hodgkin–Huxley equations.[13] The latter interpretation makes the assumption that all the ionic conductances are constant except one for an unknown ion. Therefore, the anesthetic-induced changes in resting conductance and voltage-activated conductance do not provide any information as to what ions may be involved.

The present investigation has provided preliminary evidence that such active as well as passive membrane properties of trigeminal sensory neurons may be affected during general anesthesia with isoflurane. The effects that we have observed with isoflurane on trigeminal ganglion cells are similar to those observed by Scott and Edwards[30] in their studies of cultured dorsal root ganglion neurons exposed chronically to ethanol. An increase in C_i was also apparent in *in vitro* studies of the effects of ethanol on hippocampal neurons (ref. 10; Carlen *et al.*, this volume, Chapter 5) and of other general anesthetics on lipid bilayers.[27,35] Although the question as to how isoflurane increases C_i is difficult to answer because of the complicated interpretation of what measurements of membrane capacitance represent,[32] the excitability of the neuron, e.g., its ability to generate an action potential, would be expected to be compromised by a rise in C_i. As in the case of the dorsal root ganglion studies cited previously, the resting conductance of trigeminal ganglion neurons was increased by application of isoflurane while the changes in E_m were variable. This suggests that isoflurane increases membrane permeability for more than one ion, a feature that also would serve to reduce neuronal excitability.

The use of such modeling and a resolution of neuronal membrane properties into passive and active components by frequency domain analysis permits a well-defined assessment of anesthetic action on neurons exhibiting active responses to a small perturbation of their membrane potential, which in trigeminal ganglion neurons are apparent in the form of resonant behavior in the transfer function. Although isoflurane-induced changes in spike threshold, action potentials, etc. were not investigated systematically in this series of preliminary experiments on the effects of isoflurane, it is likely that the changes in passive and active membrane properties observed here as changes in reactance would have functional consequences for the excitability of neurons, particularly in neurons of the central nervous system, which also exhibit "cable properties." The reduction in amplitude of the resonant peak and broadening of the resonant bandpass induced by isoflurane administration also suggests the possibility for neurons of the central nervous system that anesthetic-induced changes in electrical susceptance (reciprocal of reactance) of the membrane may alter phasic synaptic transmission at pre- and postsynaptic membrane sites. The finding that anesthetics may promote changes in membrane capacitance as well as resting ionic conductance suggests the possibility that these non-receptor-mediated derangements may result from the addition of hydrophobic anesthetic molecules that are charged, or have a net dipole moment, to proteinacious lipids (cf. ref. 7), some of which may be associated with ionic channels.

ACKNOWLEDGMENTS. The author is grateful to the Medical Research Council of Canada for support.

REFERENCES

1. Bevington PR: *Data Reduction and Error Analysis for the Physical Sciences.* New York, McGraw-Hill, 1969; pp 204–246.
2. Brooks CM, Eccles JC: A study of the effects of anaesthesia and asphyxia on the monosynaptic pathway through the spinal cord. *J Neurophysiol* 1947;10:349–360.
3. Catchlove RFH, Krnjević K, Maretić H: Similarity between effects of general anesthetics and dinitrophenol on cortical neurons. *Can J Physiol Pharmacol* 1972;50:1111–1114.
4. Chandler WK, Fitzhugh R, Cole KS: Theoretical stability properties of a space-clamped axon. *Biophys J* 1962;2:105–127.
5. Cole KS: *Membranes, Ions and Impulses.* Berkeley, University of California Press, 1972,pp 12–113.
6. Eccles JC: Synaptic potentials of motoneurones. *J Neurophysiol* 1946;9:87–120.
7. Fernandez JM, Taylor RE, Benzanilla F: Induced capacitance in squid axon. Lipophilic ion displacement currents. *J Gen Physiol* 1983;82:331–346.
8. Fishman HM, Moore LE, Poussart D: Squid axon K conduction: Admittance and noise during short- versus long-duration step clamps, in Adelman W J, Goldman D E (eds): *The Biophysical Approach to Excitable Systems.* New York, Plenum Press, 1981,pp 65–95.
9. Franks NP, Lieb WR: Molecular mechanisms of general anaesthesia. *Nature* 1982;300:487–493.
10. Friedman EB, Carlen PL, Durand D, Blaxter TJ: Computer aided analysis of drug effects on neuronal electrotonic properties using short intracellular current pulses; preliminary results with ethanol. *Can Physiol* 1983;145:105.
11. Gimbarzevsky B, Miura RM, Puil E: Impedance profiles of peripheral and central neurons. *Can J Physiol Pharmacol* 1984;62:460–462.
12. Heyer EJ, Macdonald RL: Barbiturate reduction of calcium-dependent action potential: Correlation with anesthetic action. *Brain Res* 1982;236:157–171.
13. Hodgkin AL, Huxley AF: A quantitative description of membrane current and its application to conduction and excitation in nerve. *J Physiol* 1952;117:500–544.
14. Jack JJB, Noble D, Tsien RW: *Electric Current Flow in Excitable Cells.* Oxford, Clarendon Press, 1983,pp 5–24.
15. Krnjević K: Central actions of anaesthetics, in Halsey MJ, Millar RA, Sutton JA (eds): *Molecular Mechanisms in General Anaesthesia.* Edinburgh, Churchill Livingstone, 1974,pp 65–89.
16. La Bella FS: Is there a general anesthesia receptor? *Can J Physiol Pharmacol* 1981;59:432–442.
17. Larson MD, Major MA: The effect of hexobarbital on the duration of the recurrent IPSP in cat motoneurons. *Brain Res* 1970;21:309–311.
18. Matthews EK, Quilliam JP: Effects of central depressant drugs upon acetylcholine release. *Br J Pharmacol* 1964;22:415–440.

19. Mayer ML, Westbrook GL: A voltage-clamp analysis of inward (anomalous) rectification, in mouse spinal sensory ganglion neurones. *J Physiol* 1983;340:19–45.
20. Meyer H: Zur theorie der alkoholnarkose. *Arch exp Path Pharmak* 1899;42:109–118.
21. Nicoll RA: The effects of anaesthetics on synaptic excitation and inhibition in the olfactory bulb. *J Physiol* 1972;223:803–814.
22. Nicoll RA, Madison DV: General anaesthetics hyperpolarize neurons in the vertebrate central nervous system. *Science* 1982;217:1055–1057.
23. Overton E: *Studien uber die Narkose*. Jena, Fischer, 1902.
24. Potashner SJ, Lake N, Langlois EA, Plouffe L, Lecavalier D: Pentobarbital: Differential effects on amino acid transmitter release, in Fink BR (ed): *Molecular Mechanisms of Anesthesia*. New York, Raven Press, 1980, pp 469–476.
24a. Puil E, Gimbarzevsky B, Miura RM: Quantification of membrane properties of trigeminal root ganglion neurons in guinea pigs. *J. Neurophysiol,* in press.
25. Quastel, DMJ, Hackett JT, Okamoto K: Presynaptic action of central depressant drugs: Inhibition of depolarization–secretion coupling. *Can J Physiol Pharmacol* 1972;50:279–284.
27. Reyes J, Latorre R: Effect of the anesthetics benzyl alcohol and chloroform on bilayers made from monolayers. *Biophys. J* 1979;28:259–280.
28. Richards CD, Keightley CA, Hesketh TR, Metcalfe JC: A critical evaluation of the lipid hypotheses of anesthetic action, in Fink BR (ed): *Molecular Mechanisms of Anesthesia*. New York, Raven Press, 1980,pp 337–351.
29. Sastry BR: Antagonism of substance P induced facilitation of central neuronal activity by 2,4-dinitrophenol and depressant agents. *Can J Physiol Pharmacol* 1978;56:527–530.
30. Scott BS, Edwards BAV: Effect of chronic ethanol exposure on the electric membrane properties of DRG neurons in cell culture. *J Neurobiol* 1981;12:379–390.
31. Seeman P: The membrane actions of anesthetics and tranquillizers. *Pharmacol Rev* 1972;2:583–655.
32. Taylor RE, Fernandez JM, Benzanilla F: Squid axon membrane low-frequency dielectric properties in Adelman W J, Goldman D E (eds): *The Biophysical Approach to Excitable Systems*. New York, Plenum Press, 1981,pp 97–106.
33. Weakly JN: Effects of barbiturates on ''quantal'' synaptic transmission in spinal motoneurones. *J Physiol* 1969;204:63–77.
34. Wittmers LE, Pozos RZ, Stocker K: Structure of the microvasculature in the dorsal root ganglion of the cat. *Soc Neurosci Abstr* 1978;4:572.
35. Yoshida T, Mori T, Ueda I: Giant planar lipid bilayer capacitance and its biphasic response to inhalation anaesthetics. *J Coll Interface Sci* 1983;96:39–47.

II

POSTSYNAPTIC ACTIONS OF ANESTHETICS

9

An Approach to Anesthetic Action from Studies of Acetylcholine Receptor Function

Palmer Taylor, Paul Culver, Robert Dale Brown, Jeffrey Herz, and David A. Johnson

ABSTRACT. The comparatively detailed knowledge of the structure, functional states, and sensitivity to noncompetitive antagonists has made the nicotinic acetylcholine receptor a useful system for understanding how anesthetics affect chemosensitive membrane channels. The receptor exists in multiple states which differ in affinity for agonists and propensity for channel opening. Anesthetics and other noncompetitive inhibitors bind to sites allosteric to the sites of agonist and antagonist association and thus act as heterotropic ligands. Since the affinities of the anesthetics are not equivalent for the individual receptor states, an alteration in their relative populations is achieved through anesthetic association with the receptor. However, a unitary theory of anesthetic mechanism which correlates anesthetic potency with state changes induced in receptors does not emerge from these studies. For example, local and general anesthetics will increase the population of receptors in the desensitized state, yet not all local anesthetics effect this state change and the sensitivity to individual local anesthetics depends on the tissue source of the receptor. Moreover, certain simple amines and peptide antibiotics such as the polymyxins with little or no anesthetic activity are remarkably effective in desensitizing receptors. Thus, a complete analysis of anesthetic action on receptor function will require delineating the sites of anesthetic association on the receptor as it appears that more than a single site is involved for these noncompetitive inhibitors.

INTRODUCTION

Although experimental handles to uncover the molecular basis of anesthetic action are limited by the low potency of the anesthetics and the absence of chemical antagonists, lipid partition theories and pressure reversal of anesthesia have provided a framework for implicating the perturbation of membrane structure and volume in the mechanism of anesthesia. Given this basic tenet, it becomes important to define how anesthetic molecules partitioning into hydrophobic environments can affect specific membrane proteins responsible for cellular excitability. To this end, four distinct approaches to the question of anesthetic mechanisms might be developed. First, a search could be embarked upon for anesthetic-sensitive sites within the central nervous system (CNS). Second, we might consider how anesthetics

Palmer Taylor, Paul Culver, Robert Dale Brown, Jeffrey Herz, and David A. Johnson • Division of Pharmacology, University of California, San Diego, La Jolla, California 92093. Present address of D. A. J.: Division of Biomedical Sciences, University of California, Riverside, California 92521.

block the receptors or other proteins involved in excitatory pathways and responses. Third, anesthetic enhancement of inhibitory receptor responses and inhibitory pathways might be considered. Fourth, structurally well-defined membrane proteins containing ion channels might be selected since from their sensitivity to anesthetics a commonality in response to structural perturbations might emerge.

Our fragmentary knowledge of CNS pathways responsible for the conscious state coupled with the lack of identification of the neurotransmitters and neuromodulators intrinsic to these pathways has proven limiting to the first approach. The predominance of excitatory pathways is believed to rely on the dicarboxylic acids glutamate and aspartate. Unfortunately, the absence of suitable antagonists has restricted the characterization of these systems. The situation improves for the inhibitory transmitter systems, and, in fact, considerable evidence has accumulated to suggest that the antianxiety and sedative properties of the benzodiazepines and barbiturates are related to their ability to augment γ-aminobutyric acid binding to its receptor (see Gage *et al.*, this volume, Chapter 12, and Olsen *et al.*, this volume, Chapter 14). Whether this mechanism extends to general considerations of anesthesia remains an open question (see Macdonald *et al.*, this volume, Chapter 2). Finally, studies of anesthetic action on the acetylcholine receptor have provided the well-defined system necessary to extend studies of anesthetic-induced perturbations in structure beyond the domain of a simple lipid bilayer. In this case structural perturbations can be monitored in the membrane-associated protein and correlated with the block of function. Since studies on the structure and functional behavior of the acetylcholine receptor have developed to an advanced stage and the receptor can be isolated in large quantities, it presents itself as a unique system for examining anesthetic action. If we assume that common structural features exist among receptor-linked ion channels, then this system may reveal essential molecular details of how perturbations in membrane and protein structure affect the function of such channels. Such an approach is taken in this chapter and the two following chapters by Cohen *et al.*, Chapter 10, and Miller *et al.*, Chapter 11.

THE NICOTINIC ACETYLCHOLINE RECEPTOR

A convergence in findings of several laboratories over the last 3–4 years has led to a relatively detailed understanding of the structure of the nicotinic acetylcholine receptor. It is composed of four distinct subunits in the stoichiometry α_2, β, γ, δ, where all five subunits in this pentamer are believed to form the outer perimeter of an internal channel.[6,16,28] The individual subunits show considerable sequence homology[21] and all span the membrane. The α-subunits carry the primary sites for the binding of agonists and competitive antagonists, and occupation of both α-subunits is required for channel opening. The peptide α-toxins such as cobra α-toxin and α-bungarotoxin show no preference for binding to the two α-subunits. In contrast, agonists, the reversible competitive antagonists such as *d*-tubocurarine, and the irreversible diterpenoid antagonist lophotoxin exhibit a preference for one of the two α-subunits.[8,28] Occupation of only one of the subunits by an antagonist is sufficient to block the response to agonist completely (c.f. Fig. 1).[28] Thus, hybrid receptor species with one of their α-subunits occupied with agonist and the other with antagonist are nonfunctional.[28]

At least three distinct receptor states have been defined by both electrophysiological and chemical binding studies, and the schemes below provide a suitable framework for examining the influence of anesthetics and other noncompetitive inhibitors on receptor function.

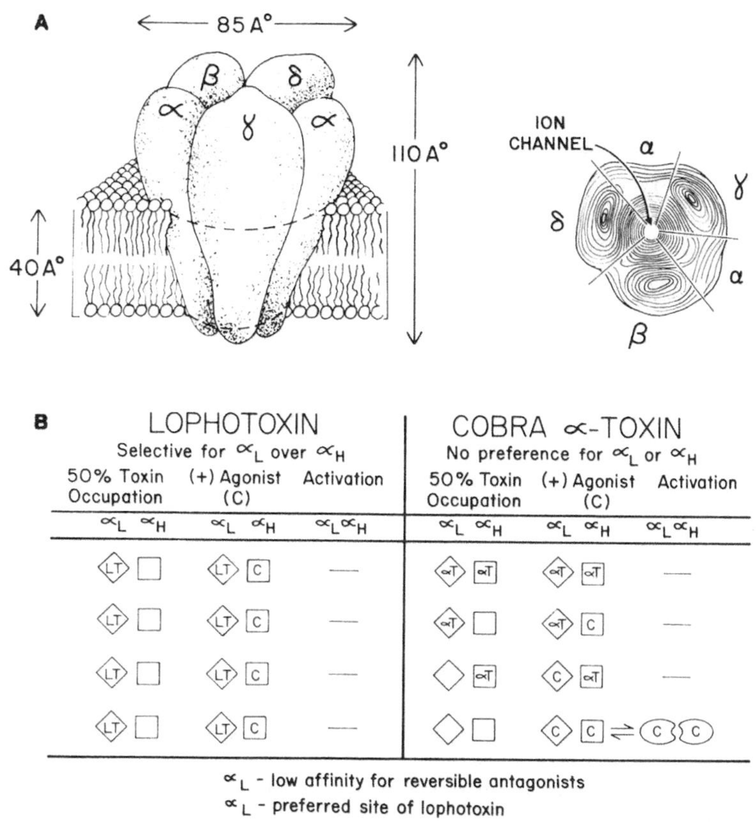

Figure 1. Structure of the nicotinic acetylcholine receptor and a description of the requirements for ligand occupation to activate, irreversibly inhibit, and competitively antagonize receptor function.

(A) The five subunits (two α, β, γ, δ, with apparent molecular weights of 40, 50, 60, and 65 kilodaltons, respectively), which are partly homologous in sequence, are arranged to form the perimeter of an internal cavity which is believed to be the channel (cf. refs. 6,10,16,28). Each of the subunits has an extracellular and intracellular exposure with the bulk of the peptide chain existing on the extracellular side. The α-subunits carry the recognition sites for agonists and antagonists, while noncompetitive or heterotropic ligands presumably bind to the internal channel, interdigitate between α-helical domains that span the membrane, or associate with the boundary lipid on the outer perimeter of the receptor.

(B) Requirements for activation and antagonism of the receptor. Shown are the α-subunits represented by the dimers. Both agonist (represented by C) and the antagonists cobra α-toxin (αT) and lophotoxin (LT) associate with the two α-subunits on the receptor molecule. The transition to the open state only occurs with the association of agonist molecules on both the α-subunits. Occupation of one of the two subunits by competitive or irreversible antagonist is sufficient to block the response. Cobra α-toxin and lophotoxin differ in that cobra α-toxin is not selective for one of the two α-subunits, whereas lophotoxin shows a selectivity. Reversible competitive antagonists also show a selectivity for one of the two subunits, but that appears to be the opposite of lophotoxin. Occupation of one of two of the α-subunits by competitive antagonists is also sufficient to antagonize receptor function.

$$\begin{array}{ccccccc} L + RR & \overset{K_R/2}{\rightleftharpoons} & LRR & \overset{2K_R}{\rightleftharpoons} & LRRL & \underset{\alpha}{\overset{\beta}{\rightleftharpoons}} & LR^*R^*L \\ M \updownarrow & & \updownarrow & & \updownarrow & & \\ L + R'R' & \overset{K_{R'}/2}{\rightleftharpoons} & LR'R' & \overset{2K_{R'}}{\rightleftharpoons} & LR'R'L & & \end{array} \qquad \text{(scheme I)}$$

L represents the agonist, R the activable receptor state, R^* the active or open channel state, and R' the desensitized state or state of a refractory channel. The symbol RR designates the two binding sites on the receptor for agonist which gives rise to potentially cooperative interactions. M, the allosteric contant, equals $R'R'/RR$ and defines the ratio of receptor in the desensitized to activable states in the absence of agonist. K_R and $K_{R'}$ are the dissociation constants for the activatable and desensitized states. α and β are the closing and opening rates for the channel.

If we incorporate LR^*R^*L into $LRRL$, then a state function reflecting the activatable and desensitized receptor states may be defined by

$$\bar{R} = \frac{\Sigma R \text{ states}}{\Sigma(R' + R) \text{ states}} = \frac{\left(1 + \frac{L}{K_R}\right)^2}{\left(1 + \frac{L}{K_R}\right)^2 + M\left(1 + \frac{L}{K_{R'}}\right)^2} \quad \text{(equation 1)}$$

Moreover, the ligand saturation function for (scheme I) can be defined by

$$\bar{Y} = \frac{\frac{L}{K_R}\left(1 + \frac{L}{K_R}\right) + \frac{ML}{K_{R'}}\left(1 + \frac{L}{K_{R'}}\right)}{\left(1 + \frac{L}{K_R}\right)^2 + M\left(1 + \frac{L}{K_{R'}}\right)^2} \quad \text{(equation 2)}$$

Since $M < 1$ and $K_R' < K_R$, the binding of L will increase the population of desensitized receptors. The corresponding values of K_R and $K_{R'}$ for the various agonists do not vary substantially between the receptor in the Torpedo and mammalian receptor from BC3H-1 cells (Table I). By contrast, $M = 0.1–0.2$ in the isolated Torpedo receptor[1,4,23,31] while in intact mammalian cells $M \simeq 0.0001$ (Table I).[24,25] As we shall observe, the different set point of the allosteric constant M becomes a critical factor in quantitating the behavior of anesthetics on the receptor.

THE INFLUENCE OF HETEROTROPIC INHIBITORS ON RECEPTOR STATE

Functionally, heterotropic inhibitors such as anesthetics that bind to sites allosteric to the agonist–antagonist site have two general modes of action (scheme II). First, they show a propensity for binding to the open-channel state of the receptor (LR^*R^*L) which causes channels to flicker, hence reducing opening intervals. This has best been studied by the electrophysiological means of single-channel recording or voltage-current relaxation.[17,20] Second, these inhibitors bind preferentially to the desensitized or R' receptor state.[18,29] Hence, in the absence of agonist, M is increased by these inhibitors,[3,7,25,27,31] and in the presence of agonist, the agonist-elicited conversion to the desensitized state is facilitated. Thus (scheme I) in the presence of anesthetic, A, expands to

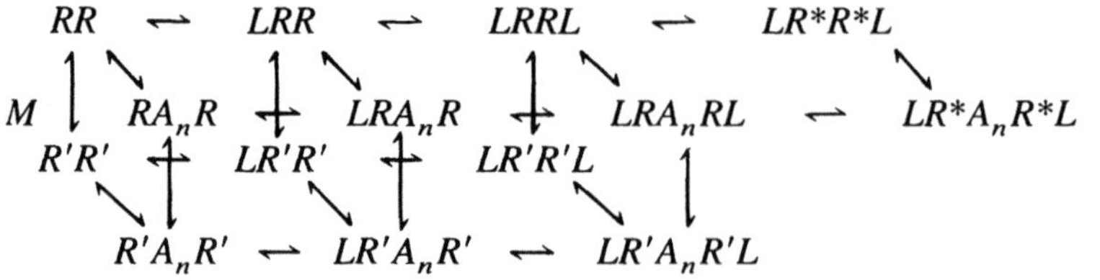

(scheme II)

Table I. Dissociation and Isomerization Constants for Agonist and Heterotropic Inhibitor Binding to the Acetylcholine Receptor from Torpedo and BC3H-1 Cells[a]

Ligand	Constant	*Torpedo* (electric organ)[b,c,d]	BC3H-1 (murine) cells[e]
–	M	0.1–0.2	1×10^{-4}
Carbamylcholine[b]	K_R	2×10^{-5} M	6×10^{-5} M
	$K_{R'}$	4.5×10^{-8} M	2×10^{-7} M
	$K_R/K_{R'}$	400	300
Acetylcholine[b]	K_R	6.1×10^{-7} M	—
	$K_{R'}$	2.3×10^{-9} M	—
	$K_R/K_{R'}$	270	—
Suberyldicholine[b]	K_R	1.3×10^{-7} M	5.1×10^{-7} M
	$K_{R'}$	4.0×10^{-10} M	1.6×10^{-9} M
	$K_R/K_{R'}$	325	350
Dibucaine[c]	$K_{R,A}$	8.0×10^{-5} M	1.0×10^{-3} M
	$K_{R',A}$	7.5×10^{-6} M	5.0×10^{-6} M
	$K_{R,A}/K_{R',A}$	11	200
Histrionicotoxin[d]	$K_{R,A}$	9×10^{-7} M	1.1×10^{-4} M
	$K_{R',A}$	4×10^{-7} M	1.1×10^{-6} M
	$K_{R,A}/K_{R',A}$	2.3	100

[a]Specified constants apply to schemes I and II.
[b]Refs. 30 and 31. Close agreement is found between these values and those reported in Refs. 1, 4, and 23.
[c]Ref. 3.
[d]Ref. 12.
[e]Refs. 24 and 25.

In this case

$$M' = M \left(\frac{1 + \frac{A}{K_{R',A}}}{1 + \frac{A}{K_{R,A}}} \right)^n \qquad \text{(equation 3)}$$

M remains as the allosteric constant in the absence of ligand while $K_{R',A}$ and $K_{R,A}$ are the dissociation constants for the heterotropic ligand, A, binding to the allosteric site of the receptor in the R' and R states, respectively. n represents the number of binding sites per receptor unit. This scheme enables us to see that although the binding of agonists and anesthetics occurs at disparate loci, the sites are intimately coupled through these equilibria. Thus, anesthetics act as heterotropic, allosteric ligands. Agonists have the capacity to increase the affinity of anesthetics, and anesthetics, in turn, increase agonist affinity for the receptor. Scheme II represents the coupled equilibria.

With this general scheme in place, we can now turn our attention to individual anesthetics and other allosteric inhibitors. It is particularly instructive to compare the behavior of anesthetics on mammalian muscle cells in culture and on isolated *Torpedo* membranes. In both cases, the action of the individual inhibitors can be adequately described by scheme II. Since anesthetics exhibit higher affinity for the desensitized relative to activatable state of the receptor (i.e., $K_{R',A} < K_{R,A}$), they effectively increase the value of M. Then M' becomes greater than M [cf. equation (3)].

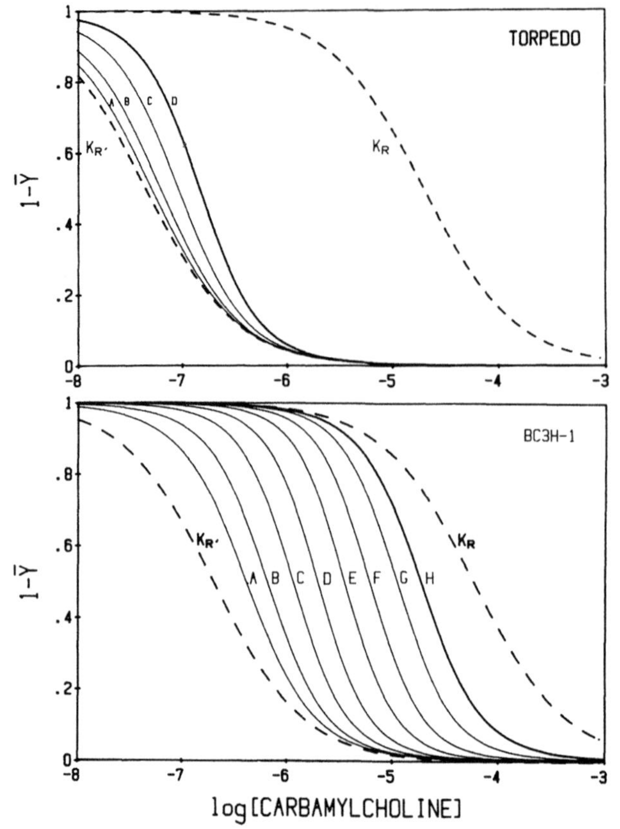

Figure 2. Predicted behavior of allosteric ligands on agonist occupation to the nicotinic acetylcholine receptor. Shown are computer-simulated occupation functions for carbamylcholine. (Top) The *Torpedo* receptor ($K_R = 2 \times 10^{-5}$, $K_{R'} = 4.5 \times 10^{-8}$, $M = 0.1$); (bottom) the receptor in BC3H-1 cells ($K_R = 6 \times 10^{-5}$, $K_{R'} = 2 \times 10^{-4}$, $M = 10^{-4}$). The dotted lines show simple occupation functions described by K_R and $K_{R'}$ and the bold solid lines (D in *Torpedo* and H in BC3H-1 cells) are the equilibrium occupation function of agonist in the absence of local anesthetic. Curves designated by the letters show the influence of M on the agonist occupation curves: (Top) *Torpedo* membranes: A, $M = 3.0$; B, $M = 1.0$; C, $M = 0.3$; D, $M = 0.1$. (Bottom) BC3H-1 cells: A, $M = 0.3$; B, $M = 0.1$; C, $M = 0.03$; D, $M = 0.01$; E, $M = 0.003$; F, $M = 0.001$; G, $M = 0.0003$; H, $M = 0.0001$. The equilibria between species are shown in (scheme II) where two states of the receptor RR and $R'R'$ exist and both agonists (L) and heterotropic ligands (A) show preferred binding to the $R'R'$ state (i.e., $K_{R'} < K_R$ and $K_{R',A} < K_{R,A}$). Data show that K_R and $K_{R'}$ as well as their ratio are very similar in the receptor from *Torpedo* and BC3H-1 cells; however, the allosteric constant M differs substantially (Table I). When M is small (10^{-4}), as in the BC3H-1 cells, the concentration dependence of equilibrium occupation function for agonists is not shifted greatly from a binding function described by K_R. Heterotropic ligands have the capacity by increasing M to effect large changes in agonist occupation. As M becomes greater than $K_{R'}/K_R$, in the limit agonist occupation will be described by $K_{R'}$. In *Torpedo*, M is much larger; so even in the absence of heterotropic ligand, the occupation function is well shifted toward the function described by $K_{R'}$ alone. Hill coefficients are related to the ratio of M to $K_{R'}/K_R$ and can be predicted by this equation (cf. Ref. 15).

In Fig. 2 we show the influence of heterotropic ligands such as anesthetics on agonist occupation of the receptor from *Torpedo* and BC3H-1 cells, as would be predicted by equations (2) and (3). The respective dissociation constants of agonists for the R and R' states are very similar for receptors in *Torpedo* and BC3H-1 cells; however, the ratio of receptor in the two states in the absence of any ligand (designated by M) differs substantially (Table I). In the absence of ligand, approximately 10% of the *Torpedo* receptor exists in the R' state while 0.01% exists in the R' state for the mammalian receptor. Since receptors from both sources predominate in the R state and the rate of transition to R' is slow, the concentration dependence of occupation of agonists immediately following their addition will be described closely by K_R; i.e., $\bar{Y} = L/(L + K_R)$. However, if the system is allowed to come to equilibrium, $\bar{Y}$ will be described by equation (2), and we see a very different relationship between the equilibrium occupation curves for the two receptor sources. When M is very small, the occupation curve approaches an occupation curve described by K_R, as seen in the

BC3H-1 cells. On the other hand, when M increases, the curve shifts to the left toward a limiting value described by $K_{R'}$. In *Torpedo* with $M = 0.1$, the occupation curve is appreciably shifted toward the limiting value.

The heterotropic ligands will cause the agonist occupation curves to shift to lower agonist concentrations by increasing M and in the limit $\bar{Y} = L/(L + K'_R)$. Since M for the *Torpedo* receptor is relatively large, the equilibrium occupation curve is close to that described by the K'_R. This limits the "window" in which effects of heterotropic ligands can be seen to a factor of approximately 5. Since M is much smaller in the case of the BC3H-1 cell, this factor is greater than 100.

That this prediction is borne out by data is shown in Fig. 3. Histrionicotoxin and the local anesthetics dibucaine and tetracaine all show preferential binding to the R' state and substantially enhance agonist affinity. A comparison of these data obtained in BC3H-1 cells[25] with previous studies in *Torpedo* shows that qualitatively similar but smaller shifts in agonist concentration dependencies are elicited by local anesthetics in the *Torpedo* system.[7,27]

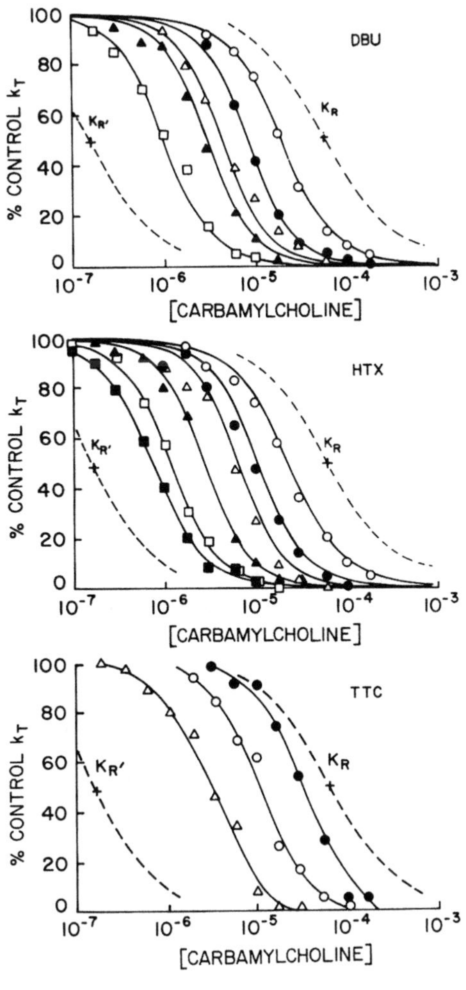

Figure 3. Influence of dibucaine, histrionicotoxin, and tetracaine on the equilibrium occupation function of carbamylcholine. BC3H-1 cells were incubated for 20 min in buffer containing the specified concentrations of carbamylcholine with dibucaine, histrionicotoxin, or tetracaine. (Top panel) Dibucaine concentration: ○, 0; ●, 10 μM; △, 18 μM; ▲, 30 μM; □, 60 μM. (Middle panel) Histrionicotoxin concentrations: ○, 0; ●, 1 μM; △, 3 μM; ▲, 10 μM; □, 30 μM; ■, 60 μM. (Botton panel) Tetracaine concentrations: ○, 0; ●, 10 μM; △, 100 μM. Following the incubation period, medium was aspirated, and the initial rate of [^{125}I]cobra α-toxin binding was measured over a 30-sec interval in an identical buffer containing agonist and the noncompetitive inhibitor. The dotted curves show occupation functions dependent on K_R or $K_{R'}$ only. With increasing concentrations of local anesthetic, M' becomes larger. As M' increases, the saturation isotherm for carbamylcholine will approach the theoretical limit of being described by only $K_{R'}$ [cf. equation (2)]. The solid curves represent the best fits to equation (2) for each competition isotherm for agonist binding, assuming K_R = 60 μM and $K_{R'}$ = 0.2 μM and allowing M to vary. (Some of the data from the top two panels are from ref. 25. Other considerations on the concentration dependence of local anesthetics in affecting the receptor state transition are detailed in that study.)

In addition to this general quantitative difference for the two receptor systems, certain heterotropic ligands, typified by histrionicotoxin and tetracaine, show properties unique to the receptor from one animal species. In comparing the relative dissociation constants $K_{R,A}$ and $K_{R'A}$ for histrionicotoxin in *Torpedo*, only a small difference between $K_{R,A}$ and $K_{R',A}$ is evident (~2),[7,12] while in the BC3H-1 cell $K_{R,A}/K_{R',A}$ ~100 (Table I). In *Torpedo*, tetracaine shows anomalous behavior among the local anesthetics in that its affinity for the *R* state is actually greater than that for the *R'* state. Hence, tetracaine does not effect a conversion of receptor to the *R'* state (ref. 3; Cohen *et al.*, this volume, Chapter 10). Yet in the BC3H-1 cell, the conversion to the *R'* state of the higher agonist affinity is evident (Fig. 3), and tetracaine appears not to differ appreciably in this capacity from other local anesthetics.

Concomitant with the conversion to the state with a higher affinity for agonists induced by the local anesthetic is the enhanced desensitization of the receptor and the block of its functional permeability response. Figure 4 shows the influence of prior agonist exposure on the permeability response in the presence and absence of anesthetic. The enhanced desensitization is reflected in the lower concentration of carbamylcholine that is required to block the permeability response.

It is noteworthy that the acetylcholine receptor from the two sources shows minimal differences in its binding of agonists and competitive antagonists or in the capacity of these ligands to effect state changes in the receptor. Yet considerable differences are observed when the behavior of the receptors from the two sources toward the noncompetitive, heterotropic ligands is compared. Thus, the specificity of the receptor toward agonists and the capacity of an agonist to effect a state change appear to be better conserved through evolution than is receptor specificity and response to allosteric ligands. The *Torpedo* receptor is present in a densely packed array in junctional areas, while there is no evidence for such receptor clustering in the BC3H-1 cell. It is also possible that this difference in sensitivity to heterotropic ligands may arise from the different structural organization of the receptor molecules.

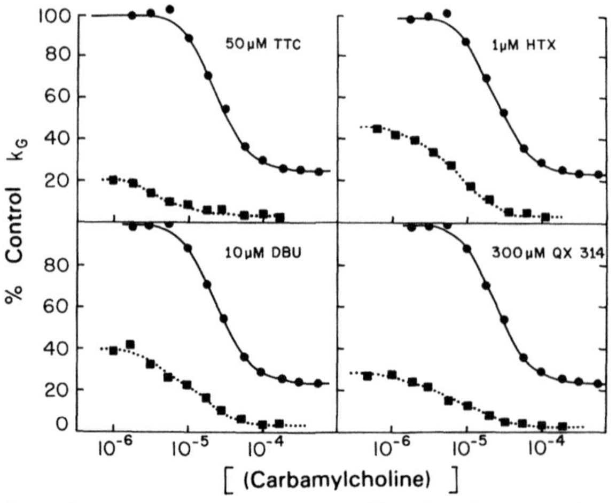

Figure 4. Ability of the heterotropic antagonists to affect carbamylcholine-elicited permeability and desensitization of the acetylcholine receptor in BC3H-1 cells. Cells were exposed to the specified concentrations of carbamylcholine and the indicated noncompetitive inhibitor for 30 min. Following this, the initial rate of Na^+ influx in the presence of 1.0 mM carbamylcholine and the noncompetitive inhibitor was measured over a 15-sec interval. The permeability rate constant, k_G, was estimated from the equations in ref. 24. The permeability at limiting agonist dilution reflects the agonist-elicited permeability in the absence of prior exposure to agonist, while the concentration required to achieve half-maximal desensitization (K_{des}) may be estimated from the concentration dependence. K_{des} has been correlated with the concentration dependence for carbamylcholine occupation of the receptor (see ref. 25). In each panel, the symbols shown by the closed circles represent data obtained in the absence of anesthetic. Data obtained for dibucaine (DBU), histrionicotoxin (HTX), and QX314 (a quaternary local anesthetic) are from ref. 25. ●-●, Carbamylcholine elicited permeability in the absence of heterotropic ligand; ■ - - -■, permeability in the presence of the indicated heterotropic ligand (local anesthetic) concentration.

PEPTIDE ANTIBIOTIC BLOCK OF RECEPTOR FUNCTION

It has long been recognized that certain antibiotics will augment the action of neuromuscular blocking agents, and the BC3H-1 cells in culture provide a well-suited system for analyzing the action of these agents on an intact cell devoid of the influence of presynaptic release of transmitter. Of the antibiotics studied, polymyxin B is clearly the most potent, and a large window exists between the concentration blocking at the agonist–antagonist site and the concentration required to enhance the binding of agonist (Fig. 5).[5] The noncompetitive block by polymyxin is analogous to that found for histrionicotoxin and the great preponderance of local anesthetics and general anesthetics[3,7,12,17,19,25,27,32] in revealing a capacity to convert the acetylcholine receptor to the high-affinity or desensitized state (Fig. 6). As with the other noncompetitive ligands, the permeability response is inhibited and capacity of agonists to promote desensitization enhanced by the polymyxins (Fig. 6).[5] In fact, polymyxin with a K_I of 10^{-7} M appears to have a potency equal to or greater than the local anesthetics in effecting a state change. It will be of interest to determine whether polymyxin has a site of binding mutually exclusive with histrionicotoxin and the local anesthetics. All these heterotropic ligands enhance agonist binding (presumably to the R' state of the receptor) and block agonist-elicited Na^+ permeability noncompetitively, yet the structure of the polymyxins differs substantially from that of the known local or general anesthetics. The former possess a cationic peptide head esterified to a seven- to eight-carbon fatty acid tail. These surface-active antibiotics will perturb membrane structure by positioning their fatty acid group well into the bilayer enabling the polycationic portion to associate with the polar moiety of the phospholipid or with charged domains on the receptor.[26] The polymyxins show selectivity for anionic lipid domains.[2]

Multiple classes of allosteric sites on the receptor involved in the binding of heterotropic noncompetitive inhibitors of the receptor should be considered. The presence of multiple sites provides an explanation for the wide structural variation of the noncompetitive inhibitors. Moreover, it would appear that the site(s) and stoichiometries of anesthetic binding will

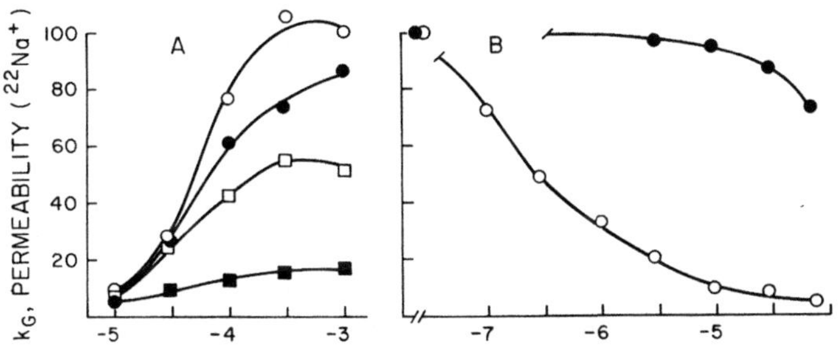

Figure 5. The influence of polymyxin B on the functional properties of the nicotinic acetylcholine receptor. (A) Noncompetitive inhibition by polymyxin B of the carbamylcholine-elicited Na^+ permeability. ○-○, no antibiotic; ●-●, 10^{-7} M; □-□, 3×10^{-7} M; ■-■, 3×10^{-6}M (cf. ref. 5). (B) Concentration dependence of polymyxin B for (○-○) inhibition of carbamylcholine-stimulated $^{22}Na^+$ influx and (●-●) inhibition of the initial rate of [^{125}I]cobra α-toxin binding. Monolayer cultures of BC3H-1 cells were equilibrated with polymyxin B for 20 min. The solution was replaced by a solution containing the same concentration of antibiotic plus a fixed concentration of either [^{125}I]α-toxin (10 nM) or 100 μM carbamylcholine and tracer $^{22}Na^+$. Initial rates of α-toxin binding and carbamylcholine-stimulated $^{22}Na^+$ influx were measured over 20-sec intervals. Results are expressed in terms of percentage of the control in the absence of antibiotic. Reproduced with permission from Brown and Taylor.[5]

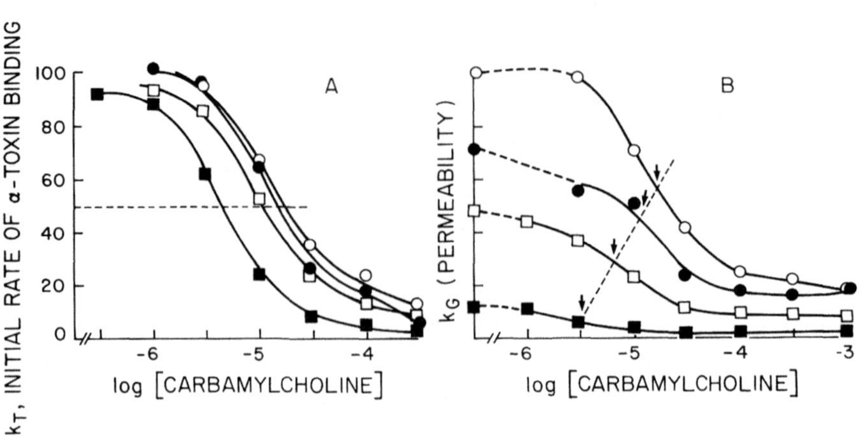

Figure 6. The influence of polymyxin on the equilibrium occupation of carbamylcholine (A) and on the capacity of the prior exposure of carbamylcholine to block permeability and to promote desensitization of the permeability response (B). In both experiments, BC3H-1 cells were incubated with the indicated concentrations of carbamylcholine and (○-○) no polymyxin; (●-●) 1×10^{-7}M polymyxin; (□-□) 3×10^{-7} polymyxin; (■-■) 1×10^{-6}M polymyxin for 20 min. Assay conditions are identical to those in Figs. 3 and 4.

have to be delineated for a rigorous analysis of the acetylcholine receptor as a model for anesthetic action.

Several approaches might be developed for elucidating the sites of binding of the anesthetics and noncompetitive inhibitors. The first would involve site-specific irreversible labeling of the allosteric sites, and several attempts have been initiated[14,22] (see Cohen *et al.*, this volume, Chapter 10). While irreversible inhibitors of different structure have been shown to label distinct subunits, these studies will take on additional significance in the future when the predictions of peptide chain folding and domains inserted in the membrane[9,10] can be related to the sites modified by irreversibly reacting heterotropic ligands. It is likely that the predictions of structure can be further refined when sites of glycosylation and disulfide linkages are assigned. Knowing whether anesthetics label membrane-spanning or more hydrophilic peptide sequences will be of considerable importance. Moreover, it should be possible to ascertain which peptides orient toward the inner perimeter of the channel versus the outer boundary of the receptor. It has been proposed that 20 to 24 membrane-spanning sequences from the five subunits circumscribe the channel in the receptor.[9–11] Thus, a plethora of potential sites for local anesthetic association can be envisioned should local anesthetics interdigitate between such helical segments. Small alterations in the pitch of the helices may be sufficient to alter channel opening or promote the transition to the desensitized state. Anesthetics might also be expected to bind at the outer perimeter of the receptor displacing associated boundary lipid.

The second approach to determining the loci of noncompetitive inhibitor association involves estimation of intersite distances by spectroscopic methods. With the receptor possessing a diameter of 85–90°, fluorescence energy transfer is well suited for this purpose, and studies have also been initiated in this area.[13] As a frame of reference, intersite distances were measured between receptor bound α-toxins selectively labeled at a single amino acid residue. The α-toxins are advantageous in the initial studies since they have the requisite affinity and known stoichiometry of binding. An ambiguity in interpreting the data in relation to overall receptor structure is inherent in not knowing the precise orientation of binding of these larger peptide toxins on the receptor surface. However, the extension of this approach to lower-molecular-weight fluorescent agonists and competitive antagonists and subsequently to local anesthetics should enable one to obtain a clearer picture of the loci of binding of these allosteric inhibitors. An associated advantage of the fluorescence energy transfer method is that the fluorescent ligands could well yield information on the conformational changes that accompany their binding.

The existing fund of knowledge on the structure of the acetylcholine receptor renders this system of value in studying anesthetic association with receptor-linked channels. Considering the diversity of ligands that can affect state changes in the receptor, more than a single site appears to be involved in the allosteric actions of the noncompetitive inhibitors. Delineation of the location of these sites and their coupling to receptor function becomes critical to analyzing general characteristics of anesthetic action on receptor-linked channels.

ACKNOWLEDGMENTS. Supported in part by USPHS Grant GM-24437and fellowships to P. C., J. H., R. D. B., and D. A. J.

REFERENCES

1. Barrantes FD: Agonist-mediated changes of the acetylcholine receptor in its membrane environment. *J Membr Biol* 1978;124:1–26.
2. Bearer EL, Friend DS: Anionic lipid domains: Correlation with functional topography in a mammalian cell membrane. *Proc Natl Acad Sci USA* 1980; 77:6601–6605.
3. Blanchard SG, Elliott J, Raftery MA: Interaction of local anaesthetics with *Torpedo californica* membranes bound acetylcholine receptor. *Biochemistry* 1979; 18:5880–5885.
4. Boyd ND, Cohen JB: Kinetics of binding of [^{3}H]acetylcholine and [^{3}H]carbamylcholine to Torpedo postsynaptic membranes. *Biochemistry* 1980;19:5344–5353.
5. Brown RD, Taylor P: The influence of antibiotics on agonist occupation and functional states of the nicotinic acetylcholine receptor. *Mol Pharmacol* 1982;23:8–16.
6. Changeux J-P: The acetylcholine receptor: An allosteric membrane protein. *Harvey Lect* 1981;75:85–254.
7. Cohen JB, Boyd ND, Shera NS: Interactions of anesthetics with nicotinic postsynaptic membranes isolated from Torpedo electric tissue, in Fink BR (ed): *Progress in Anesthesiology,* Vol. 2. New York, Raven Press, 1980,pp 165–174.
8. Culver P, Fenical W, Taylor P: Lophotoxin irreversibly inactivates the nicotinic acetylcholine receptor by preferential association at one of the two primary agonist sites. *J Biol Chem* 1984;259:3763–3770.
9. Devillers-Thiery A, Girandal J, Benraboulet M, Changeux, J-P: Complete mRNA coding sequence of the acetylcholine binding α-subunit of *Torpedo marmorata* acetylcholine receptor: A model of the transmembrane organization of the polypeptide chain. *Proc Natl Acad Sci USA* 1983;80:2067–2071.
10. Fairclough RM, Finer-Moore J, Love RA, Kristofferson D, Desmeules PJ, Stroud RM: Subunit organization and structure of an acetylcholine receptor. *Cold Spring Harbor Symp Quant Biol* 1983;48:9–20.
11. Guy MR: A structural model of the acetylcholine receptor channel based on partition energy and helix packing calculations. *Biophys J* 1983;45:249–261.
12. Heidmann T, Oswald RE, Changeux J-P: Multiple sites of action for noncompetitive blockers on acetylcholine receptor rich membrane fragments from *Torpedo marmorata. Biochemistry* 1983;22:3112–3127.
13. Johnson DA, Voet JG, Taylor P: Fluorescence energy transfer between cobra α-toxin molecules bound to the acetylcholine receptor. *J Biol Chem* 1984;259:5717–5725.
14. Kaldany R-R, Karlin A: Reaction of quinacrine mustard with the acetylcholine receptor from *Torpedo californica:* Functional consequences and sites of labeling. *J Biol Chem* 1983;258:6232–6238.
15. Karlin A: On the application of "a plausible model" of allosteric proteins to the receptor for acetylcholine. *J Theor Biol* 1967;16:306–320.
16. Karlin A: Molecular properties of nicotinic acetylcholine receptors, in Poste G, Nicolson GL, Cotmann CW (eds): *The Cell Surface and Neuronal Function.* Amsterdam, Elsevier, 1980,pp 191–260.
17. Koblin DR, Lester H: Voltage-dependent and voltage-independent blockade of acetylcholine-receptors by local anesthetics. *Mol Pharmacol* 1979;15:559–580.

18. Magazanik LG, Vyskocil F: Desensitization at the neuromuscular junction, in Thesleff S (ed): *Motor Innervation of Muscle.* New York, Academic Press, 1976,pp 151–176.
19. Miller K, Sauter JF, Braswell LM: A stereoselective pentobarbital binding site in cholinergic membranes from *Torpedo californica. Biochem Biophys Res Commun* 1982;105:669–660.
20. Neher E, Steinbach JH: Voltage-dependent and voltage-independent blockade of acetylcholine-receptors by local anesthetics. *Mol Pharmacol* 1979;15:559–580.
21. Numa S, Noda M, Takahashi H, Tanabe T, Toyosato M, Furutani Y, Kikyotani S: Molecular structure of the nicotinic acetylcholine receptor. *Cold Spring Harbor Symp Quant Biol* 1983;48:57–70.
22. Oswald RE, Changeux JP: Selective labeling of the δ subunit of the acetylcholine receptor by a covalent local anesthetic. *Biochemistry* 1981;20:7166–7172.
23. Quast U, Schimerlik M, Lee T, Witzemann V, Blanchard S, Raftery M A: Ligand-induced conformation changes in *Torpedo californica* membrane-bound acetylcholine receptors. *Biochemistry* 1978;17:2405–2415.
24. Sine SM, Taylor P: The relationship between agonist occupation and the permeability response of the cholinergic receptor revealed by bound cobra α-toxin. *J Biol Chem* 1980;255:10144–10156.
25. Sine S, Taylor P: Local anesthetics and histrionicotoxin are allosteric inhibitors of the acetylcholine receptor. *J Biol Chem* 1982;257:8106–8114.
26. Storm DR, Rosenthal AS, Swanson PE: Polymyxin and related peptide antibiotics. *Annu Rev Biochem* 1977;46:723–763.
27. Taylor P, Weiland G, Sine S, Chignell CF, Brown BD: Cholinergic receptor state transitions and local anesthetic action, in Fink BR (ed): *Progress in Anesthesiology,* Vol. 2. New York, Raven Press, 1980,pp 175–188.
28. Taylor P, Brown RD, Johnson DA: The Linkage between Ligand Occupation and Response of the Nicotinic Acetylcholine Receptor, in Martin BR, Kleinzeller A (eds): *Current Topics in Membranes and Transport,* Vol. 18. New York, Academic Press, 1983,pp 407–444.
29. Terrar DA: Influence of SKF-525A congeners, strophanthidin and tissue-culture media on desensitization in frog skeletal muscle. *Br J Pharmacol* 1974;51:259–268.
30. Weiland G, Taylor P: Ligand specificity of state transitions in the cholinergic receptor: Behavior of agonists and antagonists. *Mol Pharmacol* 1979;15:197–212.
31. Weiland GA, Georgia B, Lappi S, Chignell CF, Taylor P: Kinetics of agonist-mediated transitions in state of the cholinergic receptor. *J Biol Chem* 1977;252:7648–7656.
32. Young AP, Sigman D: Allosteric effects of volatile anesthetics on the membrane-bound acetylcholine receptor protein. *Mol Pharmacol* 1981;20:498.

10

Interactions of Local Anesthetics with Torpedo Nicotinic Acetylcholine Receptors

Jonathan B. Cohen, Leslay A. Correll, Evan B. Dreyer, Ilmar R. Kuisk, Daniel C. Medynski, and Nina P. Strnad

ABSTRACT. Local anesthetics, as well as a number of membrane active compounds, act as potent noncompetitive antagonists of the nicotinic acetylcholine (AcCh) receptor. The nicotinic postsynaptic membranes isolated from the electric organ of the marine ray, *Torpedo*, provide a preparation ideally suited for the direct analysis of the binding of radiolabeled antagonists and their effect on receptor function. Based on studies of the equilibrium binding of [^{3}H]histrionicotoxin (HTX) and [^{3}H]tetracaine, we conclude that each AcCh receptor contains one binding site for noncompetitive antagonists that is distinct from the binding site for agonists and competitive antagonists. This noncompetitive antagonist site is linked allosterically to the AcCh site, and certain drugs such as proadifen bind to the site to cause receptor desensitization while other drugs including tetracaine occupy the site without such effect. Structure–activity relationships of this allosteric antagonist site are analyzed in terms of the potent amine noncompetitive antagonists as well as other compounds (barbiturates, alcohols, detergents) that inhibit the binding of [^{3}H]HTX. Studies of the kinetics of association and dissociation of [^{3}H]HTX and [^{3}H]tetracaine reveal unusual aspects of the binding that suggest that the allosteric antagonist site is not normally accessible on the surface of the AcCh receptor. To localize the allosteric antagonist site within the structure of the receptor, a novel affinity label, [^{3}H]meproadifen mustard ([^{3}H]methyl-chloroethylaminoethyl-2,2-diphenylpentanoate), has been developed. Alkylation of the AcCh receptor α-subunit by meproadifen mustard is inhibited by reversible noncompetitive antagonists and not by agonists or competitive antagonists, a result indicating that the α-subunit contains a portion of the binding site. These studies of ligand binding are discussed with reference to the known effects of noncompetitive antagonists on AcCh receptor function.

INTRODUCTION

Local anesthetics as a class of drugs are identified by their capacity to block impulse conduction along sensory pathways. When applied locally to nerve, they prevent activation of the voltage-dependent Na^+ channels that are involved in the propagation of the action potential in all neurons. Although most local anesthetics in common use are aromatic amines such as procaine and lidocaine, a wide variety of other membrane-active compounds can also act as local anesthetics. These compounds have in common the capacity to partition indis-

Jonathan B. Cohen, Leslay A. Correll, Evan B. Dreyer, Ilmar R. Kuisk, Daniel C. Medynski, and Nina P. Strnad • Department of Anatomy and Neurobiology, Washington University School of Medicine, St. Louis, Missouri 63110.

criminately into all biological membranes and act as general perturbants of membrane structure and function. For a variety of membrane responses, a good correlation is observed between the membrane concentration of the drugs necessary to perturb function and their oil–water (or membrane–buffer) partition coefficients.[39] This correlation provides the basis for the hypothesis that the compounds act primarily as "membrane stabilizers" inhibiting the function of specific membrane proteins as a consequence of a perturbation of lipid structure. Alternatively, local anesthetics could interact directly with the relatively hydrophobic domains that must be a structural feature not only of the Na^+ channel but of all integral membrane proteins. These alternative hypotheses need not be mutually exclusive, but it has been difficult to develop experimental protocols to evaluate the relative contributions of these two "sites" of anesthetic action.

In this chapter we present studies of the interactions of local anesthetics with a specific model system, the nicotinic postsynaptic membranes that can be isolated from the electric organ of the marine ray, *Torpedo*. In terms of structure and function, the nicotinic acetylcholine (AcCh) receptor of *Torpedo* electric tissue is very similar to that of vertebrate neuromuscular junction. In the electric organ, as at the vertebrate neuromuscular junction, receptors are present at very high density ($\sim$10,000 receptors/μ^2) in the region of plasma membrane underlying the nerve terminals. While the endplate region constitutes only a very small fraction of the surface area of skeletal muscle fibers, cholinergic nerve terminals cover as much as 50% of the ventral surface of each *Torpedo* electrocyte. It is possible to isolate from homogenates of electric tissue the highly specialized postsynaptic membrane in which the AcCh receptor constitutes about 50% of the membrane protein. These membranes have been used in extensive studies of AcCh receptor structure and function.[3,11,24,38] Membrane suspensions containing micromolar concentrations of AcCh receptor are readily available, and this preparation is thus ideal for the quantitative analysis of the ligand binding properties of the AcCh receptor in its membrane environment. In addition, the isolated membranes exist as vesicles (diameter $\sim$0.3 μ) sealed to solutes such as $^{22}Na^+$, and filtration techniques can be used to measure agonist-induced ion fluxes into or out of the vesicles.

Electrophysiological studies at the level of isolated muscle fibers provide direct information about the functional properties of individual receptors (for recent reviews, see refs. 2 and 37). The primary permeability response associated with the binding of AcCh is the opening of a cation-selective ion channel. Individual channels open within a fraction of a millisecond of the binding of AcCh, and once open, a channel remains so for several milliseconds. That primary permeability response is not the result of the equilibrium interaction between AcCh and its receptor, since exposure to AcCh for seconds or longer produces a progressive, reversible decrease of the number of open channels (desensitization).

While many quaternary ammonium compounds act either as agonists or competitive antagonists of the action of AcCh, tertiary aromatic amines including the local anesthetics procaine, tetracaine, and lidocaine act as reversible, noncompetitive antagonists. Two apparently distinct mechanisms are involved in their actions. First, they alter the kinetics of channel opening and closing and of ion transport through open channels as if they bind to a site within the structure of the channel.[1,29,43] Second, many of the aromatic amine noncompetitive antagonists enhance the rate and extent of desensitization produced by agonist,[4,27] an effect that could result from binding to a distinct regulatory site. However, in addition to local anesthetics and other aromatic amines, a wide range of membrane-active compounds including general anesthetics, alcohols, fatty acids, steroids, and detergents all act as reversible noncompetitive antagonists having many of the effects of the aromatic amines (reviewed in refs. 37 and 42). Thus, at the nicotinic postsynaptic membrane as well as at the axon, it is possible that local anesthetics inhibit the function of specific membrane transport proteins as a consequence of their capacity to partition into all membranes.

FUNCTIONAL PROPERTIES OF TORPEDO ACCH RECEPTOR

Functional properties of AcCh receptors in the *Torpedo* vesicles can be measured in terms of the agonist-induced flux of ions into or out of the vesicles. To quantify receptor-mediated ion transport, it is necessary to measure ion transport on a time scale of milliseconds. The use of rapid-mixing and quenched-flow techniques[22,31] makes it possible to measure ion transport when vesicles are exposed to known concentrations of agonist (or antagonist) for times as short as 10 msec. One can thus define the rate constant for agonist-induced ion transport (a parameter proportional to the number of open channels), as well as the rate constant for desensitization of the response. The concentrations of AcCh and carbachol that produce half maximal responses (K_{ap}) are 70 μM and 800 μM, respectively,[31] concentrations similar to those necessary to activate receptors in muscle cells. Efficient channel opening requires the binding of two AcCh molecules to each receptor, and occupancy of one site by antagonist is sufficient to block the response.

With the high concentrations of AcCh receptor in the *Torpedo* membranes, equilibrium binding of [^{3}H]AcCh and]^{3}H]tubocurarine can be measured readily by ultracentrifugation assays.[30] [^{3}H]AcCh is bound with high affinity (K_{eq} = 10 nM) to a number of sites equal to the number of α-bungarotoxin binding sites (i.e., two per receptor monomer). Tubocurarine binds to the same sites, although tubocurarine binds to one of the sites with high affinity (K_{eq} = 30 nM) and to the other with low affinity (K_{eq} = 7μM). The nonequivalence of the cholinergic binding sites seen for tubocurarine is also seen for most competitive antagonists.

The discrepancy between the low concentration of AcCh necessary for receptor occupancy at equilibrium (K_{eq} = 10 nM) and the high concentration required for channel activation (K_{ap} = 70 μM) is striking, but it must be remembered that channel activation is a preequilibrium response and that the observed equilibrium binding must reflect interactions with desensitized receptors. Studies of agonist binding kinetics, whether by the use of radiolabeled α-neurotoxins (for example, ref. 44), [^{3}H]AcCh,[6,7] or by fluorescence techniques,[13,20] establish that the receptor exists in multiple interconvertible conformations. Agonists are bound transiently with low affinity, and slow conformational transition results in final high-affinity binding. Although numerous kinetic models have been proposed (reviewed in ref. 11), a two-state cyclic reaction mechanism accounts for many of the features of [^{3}H]AcCh and [^{3}H]carbachol binding.[6,7]

$$\begin{array}{ccc} D + R & \overset{K}{\rightleftharpoons} & DR \\ \uparrow\downarrow & & \uparrow\downarrow \\ D + R' & \underset{K'}{\rightleftharpoons} & DR' \end{array} \tag{1}$$

The receptor contains a single class of AcCh binding sites that exists in the absence of AcCh (D) in two interconvertible conformations: R and R', binding AcCh with low (K = 1 μM) and high (K' = 2 nM) affinity, respectively. For unoccupied *Torpedo* receptors, 85% of receptors exist in the low-affinity state (R/R' = 8), while at equilibrium, greater than 99% of occupied receptors exist in the high-affinity conformation.

This two-state model, originally proposed by Katz and Thesleff[25] to account for desensitization at the neuromuscular junction, is compatible with many features of agonist interactions with *Torpedo* receptors. The high-affinity conformation defined by [^{3}H]AcCh binding is a functionally desensitized receptor,[33] since the rate of desensitization as well as the rate of recovery upon removal of agonist are accounted for quantitatively by the rate constants for the conformational transitions ($R \rightleftharpoons R'$, $DR \rightleftharpoons DR'$). In addition, the model accounts for many of the effects of local anesthetics on the binding of cholinergic ligands by AcCh receptors in *Torpedo* vesicles[8,9,21] and in muscle cells[40] (see also Taylor *et al.*, this volume, Chapter 9). However, the model is an oversimplification. The transient low-affinity binding of AcCh (K = 1 μM) that is measured on a time scale of seconds occurs at agonist concentrations 50-fold lower than necessary to open channels (K_{ap} = 70 μM). This discrepancy between binding affinity and response indicates that the transient low-affinity binding cannot be directly related to the open-channel conformation of the receptor and probably reflects binding by an inactive receptor conformation distinct from the conformation (R') predominant at equilibrium.[22,31,33] Despite this complexity, the two-state model provides a useful first model that accounts for many of the effects of local anesthetics and other noncompetitive antagonists on AcCh receptors in *Torpedo* vesicles.

EQUILIBRIUM BINDING OF REVERSIBLE NONCOMPETITIVE ANTAGONISTS

Direct evidence for the presence in *Torpedo* membranes of a specific binding site for amine noncompetitive antagonists was first obtained in studies of the equilibrium binding of radiolabeled derivatives of potent noncompetitive antagonists: the spiropiperidine alkaloid [^{3}H]perhydrohistrionicotoxin (HTX[14,18]) (prepared by catalytic tritiation of histrionicotoxin), and the aromatic amines [^{14}C]trimethisoquin[41] and [^{14}C]meproadifen[26,33] (prepared as [^{14}C]methyliodide salts of dimethisoquin and proadifen). Each of these ligands bound with moderately high affinity (K_{eq} ~ μM) to a site distinct from the AcCh binding site. The ligands were not displaced by concentrations of agonist or competitive antagonists sufficient to occupy the AcCh site, but binding was inhibited by a variety of amine noncompetitive antagonists including procaine and tetracaine. Initial estimates of the number of sites per AcCh (or α-bungarotoxin) site varied from 0.25 to 2. We discuss below more recent studies of site stoichiometry, but from the outset, several observations indicated that the noncompetitive antagonist site(s) was probably contained within the AcCh receptor itself. (1) High-affinity [^{14}C]meproadifen binding was present in highly purified *Torpedo* postsynaptic membranes and not in other membrane fractions isolated from *Torpedo* electric tissue.[26] (2) The noncompetitive antagonist site was linked allosterically to the AcCh site. Ligands such as proadifen, dimethisoquin, and dibucaine were bound an order of magnitude more tightly with the AcCh site occupied (by agonist or competitive antagonist) than in the absence of cholinergic ligands.[5,26] Tetracaine appeared as an exception since it was bound 100-fold more weakly in the presence of carbachol than in its absence.[5] To emphasize this functional linkage with the AcCh site, we refer to the noncompetitive antagonist site as an *allosteric antagonist site*. (3) The noncompetitive antagonist site was retained in *Torpedo* postsynaptic membranes after removal of nonreceptor proteins by alkaline extraction.[32]

Several general features of the equilibrium binding of noncompetitive antagonists can be illustrated by [^{3}H]histrionicotoxin and [^{3}H]tetracaine as examples. For membrane suspensions in physiological saline at 4°C, [^{3}H]HTX is bound to the same number of sites in the presence or absence of cholinergic ligands (Fig. 1), but the affinity of binding is dependent on the ligand occupying the AcCh site. For [^{3}H]HTX, K_{eq} = 0.4 μM in the presence of carbachol, K_{eq} = 0.7 μM in the presence of tubocurarine, K_{eq} = 5 μM in the absence of cholinergic ligands, and K_{eq} = 7 μM in the presence of α-bungarotoxin. The number of

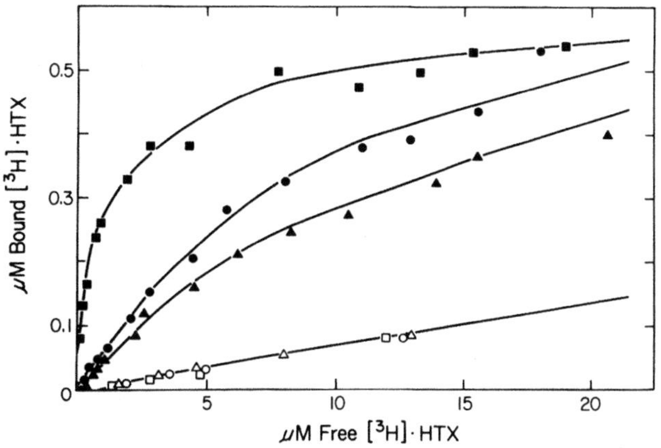

Figure 1. Equilibrium binding of [^{3}H]HTX to *Torpedo* membranes in physiological saline at 4°C. A filtration assay was used to measure binding in the presence of 30 μM carbachol (■,□), in the presence of α-bungarotoxin (▲,△), or in the absence of added cholinergic ligand (●,○). Filled symbols represent total binding and open symbols represent nonspecific binding determined in the presence of 180 μM nonradioactive HTX. Membrane suspensions contained 0.8 μM α-bungarotoxin sites. From Medynski.[28]

[^{3}H]HTX sites equals 0.5 ± 0.05 (n = 11) times the number of [^{3}H]AcCh or [^{125}I]α-bungarotoxin sites. In contrast, [^{3}H]tetracaine is bound with highest affinity in the absence of cholinergic ligands (K_{eq} = 0.4 μM) while K_{eq} = 0.5 μM in the presence of α-bungarotoxin, K_{eq} = 1 μM in the presence of tubocurarine, and K_{eq} ~ 45 μM in the presence of carbachol.[10] For [^{3}H]tetracaine, the number of sites is equal to 0.44 ± 0.1 (n = 16) times the number of [^{3}H]AcCh sites. Thus for [^{3}H]tetracaine as well as [^{3}H]HTX, there is one high-affinity noncompetitive antagonist site per AcCh receptor. Furthermore, despite the fact that the two ligands are bound preferentially by different receptor conformations, they bind in a mutually exclusive manner as judged by displacement of [^{3}H]HTX by tetracaine[5] and the inhibition of [^{3}H]tetracaine binding by HTX (data not shown). These results indicate the presence of a single high-affinity noncompetitive antagonist site per receptor monomer, a conclusion in agreement with studies based on the binding of another noncompetitive antagonist, [^{3}H]phencyclidine.[21]

At equilibrium, [^{3}H]HTX binding affinity is increased when receptor sites are occupied by either carbachol or tubocurarine; hence, it is apparent immediately that reversible agonists and competitive antagonists modify the binding of [^{3}H]HTX by an *allosteric* mechanism. For [^{3}H]tetracaine, equilibrium binding affinity is reduced in the presence of agonist, a result that might result from a competitive rather than allosteric inhibition. The fact that tetracaine binding affinity is only slightly reduced when all AcCh sites are occupied by α-bungarotoxin or tubocurarine establishes, however, that carbachol acts as an allosteric modifier of tetracaine binding.

The effects of cholinergic ligands on the equilibrium binding of [^{3}H]HTX and [^{3}H]tetracaine can be discussed in terms of the simplified two-state model of receptor function [equation (1)] (see also Taylor *et al.*, this volume, Chapter 9). In the presence of carbachol more than 99% of receptors are converted to the desensitized conformation (R'), and binding of HTX or tetracaine measured in the presence of carbachol reflects binding to the desensitized conformation. In *Torpedo* in the absence of cholinergic ligands, 85% of receptors exist in a nondesensitized conformation (R), and the observed equilibrium binding more closely reflects the binding properties of that conformation. Within the two-state model, the observed equilibrium dissociation constant (K_{eq}) of any allosteric antagonist (I) will be determined by three parameters: K_I, the dissociation constant for binding to the R conformation; K_I', the dissociation constant for binding to the R' conformation, and M, the allosteric parameter (equal to R/R'). The effect of agonist or competitive antagonist is to alter the value of M and not to alter the binding properties of either conformation:

$$K_{eq} = \frac{K_I'(1 + M)}{1 + \frac{K_I'}{K_I}M} \tag{2}$$

Values of the allosteric parameter in the absence of any cholinergic effector ($M = 8$) or with receptor sites occupied by carbachol ($M < 0.01$) have been determined from the analysis of the kinetics of binding of [^{3}H]AcCh.[6] Values of K_I' and K_I can thus be determined from the observed equilibrium binding in the absence and presence of carbachol. This simple model readily accounts for an initially puzzling observation such as the effect of tubocurarine on the binding of tetracaine and HTX. Tubocurarine produces a partial desensitization of the AcCh receptor ($M = 0.7$),[8] and thus for [^{3}H]HTX ($K_I > K_I'$) in the presence of tubocurarine, it is predicted that $K_{eq} \sim K_I'$ $(1 + M)$ or 1.7 times the dissociation constant observed in the *presence* of carbachol, while for tetracaine ($K_I < K_I'$), the predicted $K_{eq} \sim K_I\,(1 + M)/M$ or 2.4 times the dissociation constant observed in the *absence* of carbachol. While the two-state model provides a useful starting point for the analysis of noncompetitive antagonist mechanisms, we discuss below results indicating significant limitations of the model.

Structure–activity relations of the allosteric antagonist site can be readily defined by measuring the inhibition of binding of [^{3}H]HTX or other suitable radioligand by nonradioactive ligands. A wide variety of lipophilic secondary and tertiary aromatic amines, including phencyclidine, phenothiazines, and tricyclic antidepressants, can bind to the allosteric site with dissociation constants in the micromolar range (see, for example, refs. 15 and 21), and most ligands are bound with highest affinity by receptors equilibrated with agonist. Tetracaine, however, is by no means a unique exception to this generality, and we present in Table I selected examples of structure–activity relations based on the inhibition of [^{3}H]HTX binding in the absence or presence of carbachol.

A first group of compounds includes a series of phenyl acetic acid esters of diethylaminoethanol where the structure of the aromatic portion is varied. Within this series, the diphenylacetic acid ester (adiphenine) is bound with equal affinity in the presence and absence of carbachol, while the 2,2-diphenylpentanoic acid ester (proadifen) is bound with 10-fold higher affinity in the *presence* of carbachol, and the 2-phenylbutyric acid ester (butethamate) is bound with eight-fold higher affinity in the *absence* of carbachol. Clearly, small changes in structure can determine whether or not a ligand is bound preferentially by the desensitized receptor, and the analysis of series of compounds with related structure will provide a means to determine the structural determinants for high-affinity interactions with different receptor conformations.

A more general concern is to identify the range of structures that bind competitively with [^{3}H]HTX. Inhibition of [^{3}H]HTX binding provides initial evidence of binding to the allosteric antagonist site, although extensive studies are required to distinguish between allosteric and competitive inhibition of binding. In addition to a wide variety of aromatic amines, other noncompetitive antagonists also inhibit [^{3}H]HTX binding. Amobarbital displaces [^{3}H]HTX with an $IC_{50} = 10\ \mu M$ in the absence of cholinergic ligands and is 1000-fold weaker ($IC_{50} = 6$ mM) in the presence of carbachol (Table I and Fig. 2) (see Miller *et al.*, this volume, Chapter 11, for additional studies of barbiturate interactions with the *Torpedo* vesicles). Triton X-100 inhibits [^{3}H]HTX binding with $IC_{50} = 1.6\ \mu M$, and studies of [^{3}H]Triton X-100 binding by *Torpedo* membranes reveal a saturable binding component.[21] Inhibition of [^{3}H]HTX binding by 2-propanol is characterized by $IC_{50} \sim 1M$. This concentration is considerably higher than the concentrations that block axonal conduction ($C_{50} = 0.3$ M) or the hemolysis of erythrocytes ($C_{50} = 0.4$ M).[39] However, when [^{3}H]HTX binding is measured at 4°C in the absence of carbachol, 2-propanol at concentrations between

Table I. Effect of Receptor Densensitization on the Binding of Nicotinic Noncompetitive Antagonists

		IC_{50}[a] (μM) +Carb/−Carb	Ratio IC_{50} (−Carb)/ IC_{50} (+Carb)
A.	R_1—C(C_6H_5)(R_2)—$COOCH_2CH_2N(C_2H_5)_2$		
	Proadifen (R_1 = C_6H_5, R_2 = C_3H_7)	0.6/3	5
	Adiphenine (R_1 = C_6H_5, R_2 = H)	5/4	0.8
	Benactyzine (R_1 = C_6H_5, R_2 = OH)	36/400	11.1
	Butethamate (R_1 = C_2H_5, R_2 = H)	32/4	0.12
B.	Procaine	2000/1500	0.75
	Tetracaine	100/0.4	0.003
C.	Amobarbital	5700/12	0.002
	Triton X-100	1.6/9	6
	2-Propanol	1 M/1 M[b]	1

[a] IC_{50} = Concentration of competing ligand reducing by 50% specific binding at 4°C of [^{3}H]HTX by *Torpedo* membranes in the presence of 0.1 mM carbachol or in its absence (−Carb).
[b] 2-Propanol increases [^{3}H]HTX binding affinity at lower concentrations (see text).

0.05 and 0.5 M produces a dose-dependent increase in [^{3}H]HTX binding affinity (Fig. 2), and it is only at higher concentrations that the presence of the alcohol causes a reduction of binding. The enhancement of [^{3}H]HTX binding affinity is a consequence of the fact that 2-propanol on its own stabilizes the desensitized receptor conformation, i.e., the conformation (R') binding HTX (and agonist) with highest affinity.[8,9] Desensitization of the receptor by 2-propanol results neither from binding to the AcCh site nor from binding to the HTX site. The enhancement of [^{3}H]HTX binding affinity is observed when receptor sites are occupied by α-bungarotoxin, and it occurs at concentrations of 2-propanol that increase [^{3}H]AcCh binding

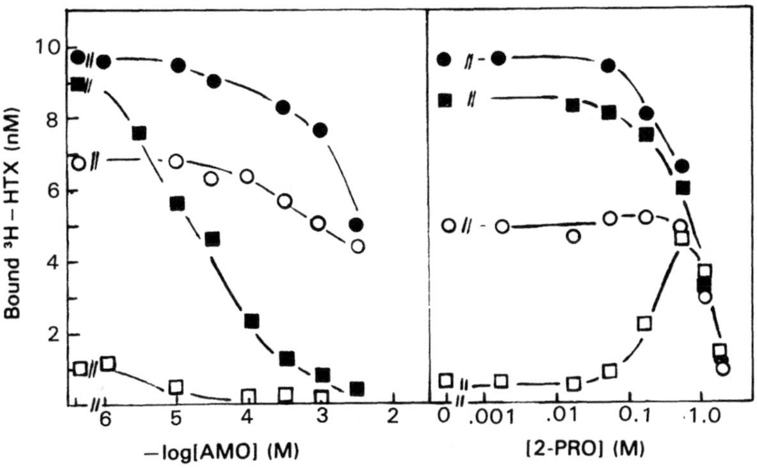

Figure 2. Inhibition of [^{3}H]HTX binding by amobarbital and 2-propanol. Membrane suspensions (0.2 μM α-bungarotoxin sites), were equilibrated with 5 nM [^{3}H]HTX at 4°C (□,○) or at 20°C (■,●) in the presence of 0.1 mM carbachol (○,●) or in its absence (■,□) and the indicated concentrations of amobarbital or 2-propanol. Equilibrium binding was determined by filtration.

affinity. As we discuss elsewhere,[8,9] the potentiation of [^{3}H]HTX binding (and receptor desensitization) caused by aliphatic alcohols probably occurs as a consequence of a nonspecific perturbation of membrane structure, and it is thus important to design experiments to determine whether the inhibition of [^{3}H]HTX binding that occurs at yet higher concentrations results as a consequence of a weak interaction with the HTX site or as a nonspecific modification of the structure of the postsynaptic membranes.

KINETICS OF BINDING OF [^{3}H]HTX AND [^{3}H]TETRACAINE

To understand the mechanism of antagonism by tetracaine or other nicotinic noncompetitive antagonists, it is important to define not only equilibrium binding parameters but also the rate constants of antagonist association and dissociation from the AcCh receptor. More generally, since the AcCh receptor exists in multiple, interconvertible conformations, it is important to relate ligand binding to the functional state of the receptor defined from analysis of agonist action (closed channel, open channel, desensitized). Tetracaine binds preferentially in the absence of agonist, but does tetracaine stabilize AcCh receptor in a resting, closed-channel state, in an open-channel (blocked) state, or perhaps in a novel state induced by tetracaine?

Filtration assays, which have been used to measure the kinetics of binding of [^{3}H]AcCh and competitive antagonists,[6,7] can be used to measure the kinetics of binding of radiolabeled noncompetitive antagonists on a time scale of seconds.[16,17] Before specific data concerning the kinetics of binding of [^{3}H]tetracaine and [^{3}H]HTX are discussed, several comments about the assay are appropriate. In this assay, the time course of ligand binding is determined by filtering at desired intervals aliquots of a reaction mixture to determine the bound ligand at each time (B'_t). Membranes and bound ligand are retained on the filter, while the filtrate contains free ligand. However, there is a substantial interaction of all nicotinic noncompetitive antagonists with the glass filters, and specific interactions with AcCh receptors can be determined only by the incorporation of a postfiltration wash with buffer (~5 ml). As a consequence, the assay can detect binding to sites from which the ligand dissociates slowly relative to the duration of the wash (several seconds). In control experiments, with [^{3}H]HTX and [^{3}H]tetracaine at concentrations to 20 μM, no significant difference was observed when specific binding at equilibrium was determined by filtration and ultracentrifugation. It should be noted that this result is itself somewhat surprising since the ligands bind relatively weakly ($K_{eq} \sim 0.3$ μM), at least in comparison with [^{3}H]AcCh that is bound by the high-affinity (desensitized) receptor conformation with $K' = 2$ nM and for which the half-time of dissociation at 4°C is 15 sec.[7] If tetracaine or HTX binds with a bimolecular association rate constant similar to that of AcCh, the half-time of dissociation would be less than 0.2 sec, and a substantial fraction of the bound ligand would dissociate during the wash!

An example of ligand association data is presented (Fig. 3) for the kinetics of binding of [^{3}H]tetracaine to a *Torpedo* membrane suspension at 4°C. At a concentration of [^{3}H]tetracaine of 0.6 μM, the half-time of the association reaction is ~7 min, and the half-time decreases with increasing tetracaine concentrations. When the data are replotted as a semilogarithmic plot of log $B'_\infty - B'_t$ versus time, for each tetracaine concentration the plots are linear for at least 90% of the reaction. The pseudo first-order rate constants (k_{obs}) determined from the slope of these plots vary linearly with tetracaine concentration, and the tetracaine binding at 4°C in the absence of cholinergic ligands can be fit by a bimolecular association rate $k_+ = 7 \times 10^2\ M^{-1}\ sec^{-1}$ and a dissociation rate constant $k_- = 3 \times 10^{-4}\ sec^{-1}$ (Table II). The dissociation rate constant is equivalent to a half-time of dissociation of 39

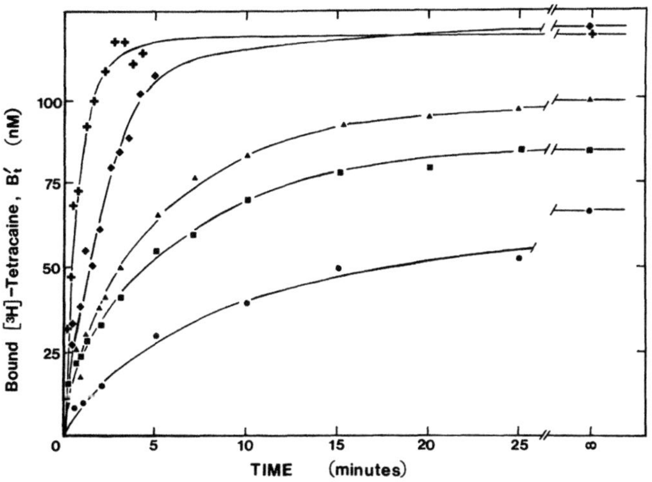

Figure 3. Kinetics of specific binding of [^{3}H]tetracaine by *Torpedo* postsynaptic membranes. Reaction mixtures contained 0.25 μM α-bungarotoxin binding sites in physiological saline (4°C) and 0.6 (●), 1.2 (■), 2 (▲), 4 (♦), or 15 (+) μM [^{3}H]tetracaine. Aliquots were filtered on glass fiber filters at the indicated times after mixing. Specific binding was determined from the difference between the total binding and the binding observed in the presence of 200 μM nonradioactive tetracaine. (N. P. Strnad and J. B. Cohen, unpublished observations).

min, a value that is also observed when the kinetics of [^{3}H]tetracaine dissociation are measured directly.

The slow association and dissociation rates are not unique for tetracaine, since for [^{3}H]HTX in the absence of cholinergic ligands, the observed association and dissociation data at 23°C can also be reasonably well accounted for by a simple bimolecular reaction model with $k_+ = 1 \times 10^3$ M^{-1} sec^{-1} and $k_- = 3 \times 10^{-4}$ sec^{-1}.[28] The apparent bimolecular rate constants for [^{3}H]tetracaine and [^{3}H]HTX are five orders of magnitude smaller than the rate constants for the binding of [^{3}H]AcCh and [^{3}H]tubocurarine to the AcCh site, and the rate constants of dissociation are three orders of magnitude smaller (Table II)! These very small values of k_+ and k_- (as well as those for [^{3}H]phencyclidine[36]) suggest that for these noncompetitive antagonists, the observed binding must be preceded by an initial low-affinity complex and that the observed kinetics reflect a slow, rate-limiting conformational transition of the receptor.

In addition to the analysis of ligand binding kinetics in the absence of cholinergic ligands, it is possible to measure the effects of agonists and competitive antagonists on noncompetitive antagonist kinetic parameters. In view of the strong allosteric linkage revealed in the analysis of equilibrium binding, substantial effects on rate constants are to be expected. As an example (Fig. 4), at 23°C in the absence of cholinergic effector, [^{3}H]HTX dissociates with a half-time of 20 min, while for receptors equilibrated with carbachol the half-time is 10 min. Since there are no open channels either in the absence of agonist or for receptors equilibrated with agonist, it is especially interesting to examine the effects of

Table II. Ligand Interactions with Membrane-Bound Torpedo AcCh Receptor

$L + R \underset{k_-}{\overset{k_+}{\rightleftharpoons}} LR$	k_+ (M^{-1} sec^{-1})	k_- (sec^{-1})	K_d (μM)
[^{3}H]AcCh[a]	6×10^7	0.2	0.003
[^{3}H]dTC	(7×10^6)[b]	0.07	0.01
[^{3}H]HTX (20°C)	1×10^3	3×10^{-4}	0.3
[^{3}H]Tetracaine (4°C)	7×10^2	3×10^{-4}	0.4

[a]Binding at 20°C to high-affinity conformation.[7]
[b]Calculated from k_- and K_d at 4°C.

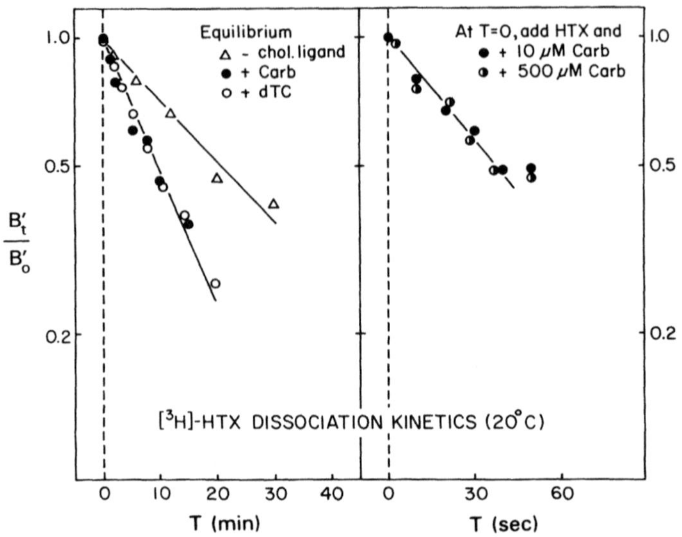

Figure 4. Kinetics of dissociation of [^{3}H]HTX from membrane-bound *Torpedo* AcCh receptor. The rate of dissociation was determined as the rate of exchange of [^{3}H]HTX for nonradioactive HTX (20 μM). A membrane suspension in physiological saline (20°C) was incubated with 0.4 μM [^{3}H]HTX for 2 hr, and then the exchange reaction was initiated by addition of HTX. The exchange reaction was terminated by filtration at the indicated times. B'_t = the amount of [^{3}H]HTX bound specifically at time *t;* B'_0 the specific binding at zero time prior to the addition of nonradioactive HTX. (Left) Dissociation rates were measured in the absence of cholinergic ligand (△), and for membrane suspensions equilibrated with 0.1 mM carbachol (●) or tubocurarine (○). (Right) Membrane suspensions were equilibrated with [^{3}H]HTX in the absence of other cholinergic effector and mixed at zero time with a solution containing nonradioactive HTX and carbachol [10 (●) and 500(◑) μM final concentrations]. Medynski.[28]

agonist under the preequilibrium conditions where the binding of agonist results in channel opening. The effect of transient exposure to agonist is seen in a simple experiment (Fig. 4) in which a membrane suspension is first equilibrated with [^{3}H]HTX and then exposed to a solution containing carbachol and nonradioactive HTX. While [^{3}H]HTX dissociates with a half-time of 20 min in the absence of agonist or with a half time of 10 min from receptors equilibrated with carbachol, upon transient exposure to carbachol, [^{3}H]HTX dissociates with a half-time of 30 sec! Transient exposure to agonist accelerates the kinetics of dissociation by 20-fold, and it is as if the sudden exposure to agonist causes HTX to "pop out" of the AcCh receptor. In addition, transient exposure to agonist accelerates by orders of magnitude the kinetics of association of [^{3}H]HTX[16] and [^{3}H]phencyclidine.[17,36] Although it is tempting to speculate that this effect of agonist is related to channel activation, this appears unlikely since the maximal effects of carbachol occur at concentrations lower than those expected from the known concentration dependence of channel activation.

[^{3}H]MEPROADIFEN MUSTARD, AN ALKYLATING NONCOMPETITIVE ANTAGONIST

To identify within the structure of the AcCh receptor the functionally important ligand binding sites, it is necessary to develop alkylating analogues (affinity labels) of reversible agonists and antagonists. For the AcCh binding site itself, affinity labels have been developed that exploit the presence of a readily reduced disulfide within the binding site. Compounds such as maleimidobenzyltrimethylammonium and bromoacetylcholine bind reversibly to the AcCh site in the native receptor, but after disulfide reduction, these reagents bind to and alkylate a cysteine within the α-subunit of the AcCh receptor.[24,45] There is as yet no information about the amino acid composition of the allosteric antagonist site in the AcCh receptor, and to identify this site we synthesized a reactive analogue of meproadifen that we call meproadifen mustard (chloroethylmethylaminoethyl-2,2-diphenylpentanoate). At physiological pH, this ligand exists with a reactive aziridinium group at the positively charged nitrogen, and this ligand could thus alkylate the site within the receptor that interacts with the positive charge of the amine noncompetitive antagonists.

To identify sites of alkylation, a *Torpedo* membrane suspension is reacted with $[^3H]$meproadifen mustard either in the absence of other ligands or in the presence of any desired concentration of reversible cholinergic agonist, competitive antagonist, or noncompetitive antagonist. After termination of the reaction, the suspension is denatured, and the constituents are resolved by polyacrylamide gel electrophoresis. After staining of the gel to visualize protein, the gel is sliced, and radioactivity in the slices can be determined by liquid scintillation counting. Meproadifen mustard alkylation of the AcCh receptor α-subunit (and of no other component) is inhibited by meproadifen, tetracaine, HTX, and other noncompetitive antagonists (Fig. 5), and the concentrations inhibiting the alkylation reaction correlate well with their reversible binding to the allosteric antagonist site.[12] In addition, it is possible to react *Torpedo* membranes with meproadifen mustard, and then to determine the number of high-affinity $[^3H]$HTX sites or $[^3H]$AcCh sites remaining in the suspension. Alkylation of the α-subunit of the AcCh receptor is associated with a corresponding reduction in the number of high-affinity $[^3H]$HTX sites (and no reduction in the concentration of $[^3H]$AcCh sites), and the irreversible reduction in $[^3H]$HTX binding by meproadifen mustard can be prevented by protection with HTX.[12] These results lead us to conclude that a region of the AcCh receptor α-subunit contains a portion of the allosteric antagonist site. Studies with other alkylating noncompetitive antagonists result in a more complex picture. Studies with

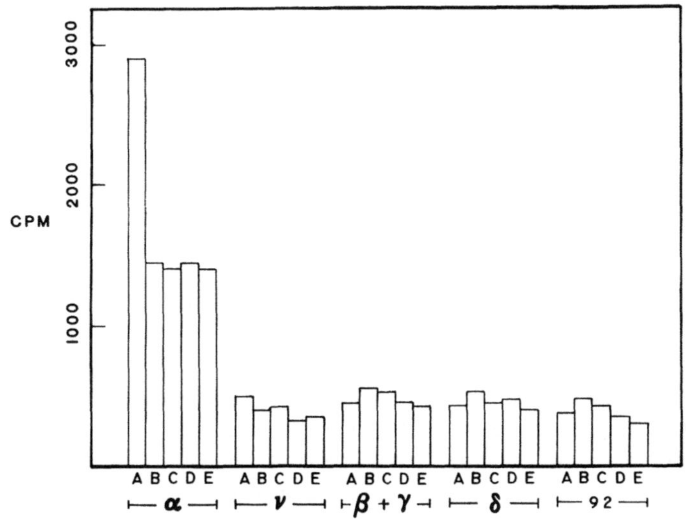

Figure 5. Alkylation of *Torpedo* postsynaptic membranes by $[^3H]$meproadifen mustard: inhibition by reversible noncompetitive antagonists. Membrane suspensions were reacted with $[^3H]$meproadifen mustard in the presence of 1 mM carbachol (A), or 1 mM carbachol and 0.1 mM HTX (B), 0.1 mM meproadifen (C), 0.1 mM proadifen (D), and 0.1 mM trimethisoquin (E). Membrane polypeptides were resolved by SDS-polyacrylamide gel electrophoresis, and the gel was stained for protein with Coomassie blue. Receptor subunits (α, β + γ, δ), as well as 43K protein (ν) and a nonreceptor polypeptide of Mr92000 (92), were cut from each gel track, and incorporated radioactivity was determined by liquid scintillation counting. From Dreyer.[12]

quinacrine mustard[23] and azidophencyclidine[19] provides evidence that AcCh receptor α- and β-subunits are involved in the binding of noncompetitive antagonists, while azidotrimethisoquin alkylates the δ-subunit.[35] Further studies will be necessary to determine whether in fact the allosteric antagonist site represents an interface between receptor subunits and also to identify the alkylation sites within the primary structure (amino acid sequences) of the receptor subunits.

FUNCTIONAL CONSEQUENCES OF BINDING OF NONCOMPETITIVE ANTAGONISTS

The binding studies reviewed here provide strong evidence that a specific binding site for noncompetitive antagonists exists within the AcCh receptor. Ligands bind to this site in the absence of cholinergic agonists or competitive antagonists, and they also bind to the site when receptors are equilibrated with agonist (desensitized). Ligands undoubtedly bind to the site in the preequilibrium condition when the presence of agonist results in open ion channels. The relative affinity of any particular drug for the individual conformational states of the AcCh receptor will determine the effect of the drug on the agonist-induced permeability response. In view of the multiplicity of effects of noncompetitive antagonists on receptor function as well as the complex structure of the AcCh receptor, it is important to identify the pharmacological effects that can be accounted for by the identified binding site.

Equilibrium binding properties of AcCh receptor can be most directly related to receptor desensitization. For the *Torpedo* receptor it can be shown that the same desensitized receptor conformation is stabilized by cholinergic agonist, by some drugs binding to the allosteric antagonist site, and also by general membrane perturbants such as the aliphatic alcohols (refs. 8, 9, and 21 and Miller *et al.*, this volume, chapter 11). Proadifen is representative of the noncompetitive antagonists that are bound preferentially by the desensitized receptor conformation, and proadifen in the absence of any agonist converts as much as 90% of the receptors to the desensitized conformation.[8] That effect of proadifen is a consequence of its binding to the allosteric antagonist site, since it can be blocked by other noncompetitive antagonists (such as adiphenine and HTX) that bind to the same site but do not increase the amount of receptor in the desensitized conformation. Adiphenine or HTX, however, cannot block desensitization produced by agonist or by nonspecific membrane perturbants such as propanol. These results establish that desensitization can occur by three mechanisms: (1) occupancy of the AcCh site; (2) occupancy of the allosteric antagonist site; and (3) general perturbation of membrane structure.

At this time it is less certain that the allosteric antagonist site can also account for the effects detected in electrophysiological studies in skeletal muscle that are described as "channel blockade" (see Gage *et al.* and McLarnon *et al.*, this volume, Chapters 12 and 13). It is significant that alcohols as well as aromatic amines produce similar effects, and there is no compelling reason to conclude that the drugs act by physical occlusion of the channel as opposed to an allosteric modulation of channel structure. Although directly comparable studies have not been performed on *Torpedo* AcCh receptors, the structure of the AcCh receptor is highly conserved,[34,38] and there is no reason to expect more significant differences in details of the binding sites for noncompetitive antagonists than for agonists and competitive antagonists. Drugs such as procaine and lidocaine that act as "channel blockers" all bind competitively with [^{3}H]tetracaine and [^{3}H]HTX, and it is thus likely that there is only a single allosteric antagonist site. Although it will be difficult to determine unambiguously whether the site is contained within the structure of the ion channel, with the availability of appropriate affinity labels, it will be possible to identify the allosteric antagonist site as well as the AcCh site within the primary structure of the AcCh receptor.

ACKNOWLEDGMENTS. This research was supported in part by USPHS Grant NS 19522 and by a grant from the Muscular Dystrophy Association. D.C.M. was a recipient of a predoctoral fellowship from the Pharmaceutical Manufacturers Association; E.B.D. and N.P.S. were supported by predoctoral training grants GM-07306 and GM-07809, respectively.

REFERENCES

1. Adams PR: Voltage jump analysis of procaine action at frog endplate. *J Physiol* 1977;268:291–318.
2. Adams PR: Acetylcholine receptor kinetics. *J Membr Biol* 1981;58:161–174.
3. Anholt R, Lindstrom J, Montal M: The molecular basis of neurotransmission: structure and function of the nicotinic acetylcholine receptor. in Martonosi A (ed.): *Enzymes of Biological Membranes*. Vol. 3, New York, Plenum Press, 1985,pp 335–401.
4. Anwyl R, Narahashi T: Comparison of desensitization and time-dependent block of the acetylcholine receptor response by chlorpromazine, cytochalasin B, Triton X-100 and other agents. *Br J Pharmacol* 1980;69:99–106.
5. Blanchard S, Elliott J, Raftery MA: Interaction of local anesthetics with *Torpedo californica* membrane-bound acetylcholine receptor. *Biochemistry* 1979;18:5880–5885.
6. Boyd ND, Cohen JB: Kinetics of binding of [^{3}H]acetylcholine and [^{3}H]carbamylcholine to *Torpedo* post-synaptic membranes: slow conformational transitions of the cholinergic receptor. *Biochemistry* 1980a;19:5344–5353.
7. Boyd ND, Cohen JB: Kinetics of binding of [^{3}H]acetylcholine to *Torpedo* post-synaptic membranes: Association and dissociation rate constants by rapid-mixing and ultrafiltration. *Biochemistry* 1980b,19:5353–5358.
8. Boyd ND, Cohen JB: Desensitization of membrane-bound *Torpedo* acetylcholine receptor by amine non-competitive antagonists and aliphatic alcohols. *Biochemistry* 1984;23:4023–4033.
9. Cohen JB, Boyd ND, Shera NS: Interactions of anesthetics with nicotinic post-synaptic membranes isolated from *Torpedo* electric tissue, in Fink BR (ed): *Molecular Mechanisms of Anesthesia*. New York, Raven Press, 1980,pp 165–174.
10. Cohen JB, Medynski DC, Strnad NP: Interactions of local anesthetics with nicotinic acetylcholine receptors, in Covino BG, Fozzard HA, Rehder K, Strichartz G (eds): *Effects of Anesthesia*. Baltimore, Williams & Wilkins, 1985, 53–64.
11. Conti-Tronconi BM, Raftery MA: The nicotinic cholinergic receptor. *Annu Rev Biochem* 1982;51:491–530.
12. Dreyer EB: Affinity labelling studies of the acetylcholine receptor (Ph.D. Thesis), Harvard University, Cambridge, MA, 1984.
13. Dunn S, Blanchard S, Raftery MA: Kinetics of carbamylcholine binding to membrane-bound acetylcholine receptor monitored by fluorescent changes of covalently bound probes. *Biochemistry* 1980;19:5645–5651.
14. Eldefrawi AT, Eldefrawi ME, Albuquerque EX, Oliveira AC, Mansour N, Adler M, Daly JW, Brown GB, Burgermeister W, Witkop B: Perhydrohistrionicotoxin: A potential ligand for the ion conductance modulator of the acetylcholine receptor. *Proc Natl Acad Sci USA* 1977;74:2172–2176.
15. Eldefrawi AT, Miller R, Murphy DL, Eldefrawi ME: [^{3}H]Phencyclidine interactions with the nicotinic acetylcholine receptor channel and its inhibition by psychotrophic, antipsychotic, opiate, antidepressant, antibiotic, antiviral, and antiarrhythmic drugs. *Mol Pharmacol* 1982;22:72–81.
16. Eldefrawi M, Aronstam R, Bakry N, Eldefrawi A, Albuquerque E: Activation, inactivation and desensitization of acetylcholine receptor channel complex detected by perhydrohistrionicotoxin binding. *Proc Natl Acad Sci USA* 1980a;77:2309–2313.
17. Eldefrawi ME, Eldefrawi AT, Aronstam RS, Maleque MA, Warnick JE, Albuquerque EX: [^{3}H]-phencyclidine: A probe for the ionic channel of the nicotinic receptor. *Proc Natl Acad Sci USA* 1980b;77:7458–7462.
18. Elliott J, Raftery MA: Binding of perhydrohistrionicotoxin to intact and detergent-solubilized membranes enriched in nicotinic acetylcholine receptor. *Biochemistry* 1979;18:1868–1877.
19. Haring R, Kloog Y, Sokolovsky M: Localization of phencyclidine binding site on α and β subunits of the nicotinic acetylcholine receptor from *Torpedo ocellata* electric organ using azidophencyclidine. *J Neurosci* 1984;4:627–637.
20. Heidmann T, Changeux JP: Fast kinetic studies on the interaction of a fluorescent agonist with the membrane-bound acetylcholine receptor from *Torpedo marmorata*. *Eur J Biochem* 1979;94:255–279.

21. Heidmann T, Oswald R, Changeux JP: Multiple sites of action for non-competitive blockers on acetylcholine receptor rich membranes from *Torpedo marmorata*. *Biochemistry* 1983;22:3112–3127.
22. Hess GP, Pasquale EB, Walker JW, McNamee MG: Comparison of acetylcholine-receptor controlled cation flux in membrane vesicles from *T. californica* and *Electrophorus electricus:* chemical kinetic measurement in the millisecond region. Proc Natl Acad Sci USA 1982;79:963–967.
23. Kaldany R-R, Karlin A: Reaction of quinacrine mustard with the acetylcholine receptor from *Torpedo californica*. *J Biol Chem* 1983;258:6232–6242.
24. Karlin A: Molecular properties of nicotinic acetylcholine receptors, in Cotman C, Poste G, Nicolson G (eds): *The Cell Surface and Neuronal Function* Amsterdam, North Holland, 1980,pp 191–260.
25. Katz B, Thesleff S: A study of the desensitization produced by acetylcholine at the motor end-plate. *J Physiol* 1957;138:63–80.
26. Krodel EK, Beckman RA, Cohen JB: ldentification of a local anesthetic binding site in nicotinic post-synaptic membranes isolated from *Torpedo marmorata* electric tissue. *Mol Pharmacol* 1979;15:294–312.
27. Magazanik LG, Vyskocil F: Desensitization at the neuromuscular junction, in Thesleff S (ed): *Motor Innervation of Muscle*. New York, Academic Press, 1976;pp 151–176.
28. Medynski DC: Histrionicotoxin and aromatic amine interactions with nicotinic acetylcholine receptors from *Torpedo*. Ph.D. Thesis, Harvard University, Cambridge, MA, 1983.
29. Neher E, Steinbach JH: Local anesthetics transiently block currents through single acetylcholine channels. *J Physiol* 1978;277:153–176.
30. Neubig RR, Cohen JB: Equilibrium binding of [^{3}H]tubocurarine and [^{3}H]acetylcholine by *Torpedo* post-synaptic membranes: Stoichiometry and ligand interactions. *Biochemistry* 1979;18:5464–5475.
31. Neubig RR, Cohen JB: Permeability control by cholinergic receptors in *Torpedo* post-synaptic membranes: agonist dose response relations measured at second and millisecond times. *Biochemistry* 1980;19:2770–2779.
32. Neubig RR, Krodel EK, Boyd ND, Cohen JB: Acetylcholine and local anesthetic binding to *Torpedo* nicotinic post-synaptic membranes after removal of non-receptor peptides. *Proc Natl Acad Sci USA* 1979;76:690–694.
33. Neubig RR, Boyd ND, Cohen JB: Conformations of *Torpedo* acetylcholine receptor associated with ion transport and desensitization. *Biochemistry* 1982;21:3460–3467.
34. Noda M, Furutani Y, Takahashi H, Toyosato M, Tanabe T, Shimizu S, Kikyotani S, Kayano T, Hirose T, Inayama S, Numar S: Cloning and sequence analysis of calf cDNA and human genomic DNA encoding α-subunit precursor of muscle acetylcholine receptor. *Nature* 1983;305:818–823.
35. Oswald RE, Changeux JP: Selective labeling of the δ subunit of the acetylcholine receptor by a covalent local anesthetic. *Biochemistry* 1981;20:7166–7174.
36. Oswald RE, Heidmann T, Changeux JP: Multiple affinity states for non-competitive blockers revealed by [^{3}H]phencyclidine binding to acetylcholine receptor rich membrane fragments from *Torpedo*. *Biochemistry* 1983;22:3128–3136.
37. Peper K, Bradley RJ, Dreyer F; The acetylcholine receptor at the neuromuscular junction. *Physiol Rev* 1982;62:1271–1340.
38. Popot J-L, Changeux JP: Nicotinic acetylcholine receptor: structure of an oligomeric integral membrane protein. *Physiol Rev* 1984;64:1162–1239.
39. Seeman P: The membrane actions of anesthetics and tranquilizers. *Pharmacol Rev* 1972;24:583–655.
40. Sine SM, Taylor P: Local anesthetics and histrionicotoxin are allosteric inhibitors of the acetylcholine receptor. *J Biol Chem* 1982;257:8106–8114.
41. Sobel A, Heidmann T, Cartaud J, Changeux JP: Reconstitution of a functional acetylcholine receptor. *Eur J Biochem* 1980;110:13–33.
42. Spivak CE, Albuquerque EX: Dynamic properties of the nicotinic acetylcholine receptor ionic channel complex: Activation and blockade, in Hanin I, Goldberg A (eds): *Progress in Cholinergic Biology: Model Cholinergic Synapses*. New York, Raven Press, 1982,pp 323–357.
43. Steinbach AB: A kinetic model for the action of xylocaine on receptors for acetylcholine. *J Gen Physiol* 1968;52:162–180.
44. Weiland G, Georgia B, Lappi S, Chignell CF, Taylor P: Kinetics of agonist-mediated transitions in state of the cholinergic receptor. *J Biol Chem* 1977;252:7648–7656.
45. Weill CL, McNamee MG, Karlin A: Affinity-labeling of purified acetylcholine receptor from *Torpedo californica*. *Biochem Biophys Res Commun* 1974;61:997–1003.

11

General Anesthetics Act Both Specifically and Nonspecifically on Acetylcholine Receptors

Keith W. Miller, Leon M. Braswell, Leonard L. Firestone, Barbara A. Dodson, and Stuart A. Forman

ABSTRACT. In acetylcholine receptor-rich membranes isolated from the electroplaques of *Torpedo*, general anesthetics are shown to exert a variety of effects which require more than one mechanism to be invoked in explanation.

All general anesthetics examined (volatiles, alcohols, carbamates, and barbiturates) enhance acetylcholine's affinity for its receptor, an effect we have previously shown to be pressure reversible. The pharmacology of this effect thus resembles that of general anesthesia, and spectroscopic studies with spin-labeled lipids show an excellent correlation between it and the ability to perturb the lipid bilayer, particularly near the bilayer center.

The barbiturates at concentrations lower than those at which they enhance desensitization *decrease* acetylcholine binding. We conclusively show with [^{14}C]amobarbital that this is achieved by binding to an allosteric site. The negative heterotropic interaction between the agonist and the barbiturate sites is consistent with a simple two-state model in which barbiturates bind preferentially to the resting state. Ion flux studies suggest that the open-channel state binds barbituate with even higher affinity than the resting state.

Long-chain alcohols and barbiturates inhibit agonist-stimulated ion flux by a noncompetitive mechanism. In contrast, ethanol causes a marked shift of the agonist dose–response curve to the left. This increase in apparent affinity of the agonist is not consistent with models based on either changes in dielectric constant or binding to an allosteric alcohol site. Its mechanism remains to be elucidated.

INTRODUCTION

General anesthetics are a diverse group of molecules that share the property of being lipid soluble. However, many subclasses of anesthetic (for example, the barbiturates and steroids) also have quite well-defined chemical structures including rigid rings. Since general anesthetics work at high concentrations in the body, it is not then surprising that, in addition to their nonspecific general anesthetic action, they also exert relatively selective side effects. Thus some agents cause convulsions at subanesthetic concentrations yet are anesthetics at higher concentrations.[17] Since some of these specific effects may be desirable (for example,

Keith W. Miller and Leonard L. Firestone • Departments of Pharmacology and Anaesthesia, Harvard Medical School and Massachusetts General Hospital, Boston, Massachusetts 02114. *Leon M. Braswell and Barbara A. Dodson* • Department of Anesthesia, Massachusetts General Hospital, Boston, Massachusetts 02114. *Stuart A. Forman* • Department of Biophysics, Harvard University and Harvard Medical School, Boston, Massachusetts 02115.

the anticonvulsive action of phenobarbital), while others may be undesirable (for example, potentiation of catecholamine-induced ventricular arrhythmias),[16] an understanding of their mechanisms is of as much interest as that of general anesthetia *per se*.

Many traditional studies of anesthetic mechanisms have naturally focused on the unitary hypothesis, yet such a view is clearly limited to nonspecific, general anesthetic actions and does not necessarily encompass the more selective actions. A simple working hypothesis would be that general anesthetic actions are related to lipid solubility, and selective actions arise from direct anesthetic–protein interactions.

However, pharmacological specificity alone cannot be relied on to distinguish between lipid- and protein-mediated actions of anesthetics, for although lipid solubility correlates with anesthetic potency,[15] perturbation of lipid structure may be a complex function of both lipid composition and anesthetic structure.[22] Indeed, Haydon *et al.* (this volume, Chapter 22) explain qualitatively different effects of alcohols and hydrocarbons on sodium channels by invoking differential distribution of these agents within the lipid bilayer. On the other hand, while even volatile anesthetics usually exhibit considerable selectivity for proteins (the acetylcholine esterases providing a clear example),[4] in a few cases, notably the luciferases (see Franks and Lieb, this volume, Chapter 26), little or no specificity has been demonstrated, and the binding site appears, at least to a limited range of anesthetics, to resemble a lipid bilayer. Thus, only detailed structural and functional studies on a case-by-case basis can unequivocally determine the mechanism by which agents that are general anesthetics exert each of their many pharmacological actions.

In this article we present studies carried out on the acetylcholine receptor-rich postsynaptic membranes from the electric tissue of *Torpedo*. Although general anesthetics are thought to act at synapses (see Krnjević, this volume, Chapter 1), the main reason for choosing this system was the experimental advantages presented by the high purity and abundance of this receptor, as emphasized in the studies of Taylor *et al.* and Cohen *et al.* (this volume, Chapters 9 and 10). Our studies show that general anesthetics modulate the function of the acetylcholine receptor in several ways, each requiring a different mechanism. Only one of these mechanisms has the same pharmacological profile as general anesthesia.

METHODS

Electric organs from freshly killed *Torpedo* were homogenized and acetylcholine receptor-rich membranes obtained by differential and sucrose density gradient centrifugation. This method was adapted from Cohen *et al.*[7] and has been described in detail elsewhere.[5] Purified membranes were typically 0.5–1.5 μmole acetylcholine sites/gram of protein and were stored at −80°C until used.

Binding of [^{3}H]acetylcholine to membranes (30–80 nM in acetylcholine-binding sites) was determined after incubation (30 min, room temperature) either alone or with anesthetic in *Torpedo* Ringer's solution (in mM: NaCl 250, KCl 5, $CaCl_2$ 3, $MgCl_2$ 2, Na_2HPO_4–NaH_2PO_4 5; pH 7.0). Bound ligand was separated from free ligand by centrifugation (100,000 × *g*, 90 min) or by ultrafiltration (Whatman, GF/F glass filters) for agents with low vapor pressure. The amount of [^{3}H]acetylcholine bound was the difference between the total ligand added to the reaction mixtures and the amount remaining in the supernatant (or filtrate) and was corrected for nonspecific binding using an irreversible inhibitor specific for the nicotinic acetylcholine receptor, α-bungarotoxin.

Torpedo membranes were spin-labeled for electron spin resonance spectroscopy by gently shaking (6 hr, 4°C) with a spin probe—either 5-doxylpalmitate (5-DP) or 12-doxylstearate (12-DS). These probes partition into membranes and report on the motion and orientation of the neighboring native lipid molecules.[13,14] Final concentrations of spin probes were less than 1 mole % of the membrane lipid. Anesthetics in buffer were added to spin-labeled membrane suspensions, which were then sealed in glass capillary tubes. After spectroscopy (Varian E-109 spectrometer, 9.3 GHz; 21.1 ± 0.1°C), the aqueous concentrations of anesthetics were confirmed by gas chromatography. Membrane lipid-order parameter was determined by measurement of spectral splittings (interpeak distances from first derivative absorption spectra) as described by Hubbell and McConnell[14] and correcting the hyperfine splittings for solvent polarity.[10]

Binding of [^{14}C]amobarbital (6 μM) to membranes (2–6 μM in acetylcholine binding sites) was determined after incubation (30 min, room temperature) either alone or with barbiturates in *Torpedo* Ringer's solution.[19] Bound ligand was separated from free ligand by centrifugation (100,000 × *g*, 60 min). Maximal displaceable binding was defined as the difference in the amount of [^{14}C]amobarbital bound to the receptor in the presence and absence of excess unlabeled barbiturates (e.g., 1 mM amobarbital). Specific binding was expressed as % of maximum displaceable binding.

Agonist-activated net efflux of $^{86}Rb^+$ from sealed acetylcholine vesicles at 4°C was measured after a 10-sec exposure to carbamylcholine. Acetylcholine receptor (3–7 μM) vesicles were incubated overnight at 4°C with $^{86}RbCl$ (100–200 μCi/ml) and desired stoichiometric amounts of α-bungarotoxin to block spare receptors. Extravesicular $^{86}Rb^+$ was removed by gel filtration (Sephadex G-50, Pharmacia, 0.5 cm × 20 cm), and then vesicles were incubated for 20 min, during which $^{86}Rb^+$ in "leaky" vesicles equilibrated with the extravesicular buffer.[20] These baseline counts were reduced by rapid passage over a cation exchange column (Dowex 50W, 20–50 mesh). An efflux experiment was initiated when an aliquot of $^{86}Rb^+$-loaded vesicles was added to a tube containing *Torpedo* Ringer with or without carbachol and/or anesthetics. After 10 sec, the contents of the test tube (1 ml) were filtered, and the filtrate was counted. Carbachol-stimulated flux was corrected for slow, time-dependent leakage of $^{86}Rb^+$ from sealed vesicles (see legend to Fig. 7).

RESULTS AND DISCUSSION

All General Anesthetics Desensitize the Acetylcholine Receptor

Previous studies have shown that alcohols and volatile anesthetics desensitize the acetylcholine receptor.[6,8,25,27] The resultant increase in affinity of acetylcholine binding has a steeper concentration dependence than that induced by local anesthetics and appears to occur by an independent mechanism (see Cohen *et al.*, this volume, Chapter 10). All the general anesthetics share this property, and representative data are shown in Fig. 1 and Table I for a structurally diverse group of anesthetics. The anesthetic concentration causing half-maximal effect on acetylcholine binding (EC_{50}) is compared to the median concentration causing general anesthesia (ED_{50}) in amphibians (see Table I). For the five anesthetics shown, the EC_{50} varies 199-fold and the ED_{50} varies 552-fold, yet the multiple of the anesthetic ED_{50} at which a half-effect on desensitization occurs varies only 4.5-fold. Thus there is a fairly good correlation between the ability to cause general anesthesia and desensitization. Although desensitization has been suggested as a mechanism for general anesthesia,[27] it is clear that at anesthetic concentrations, it would cause relatively little inhibition at this peripheral synapse.

The above relationship is supported by studies on some two dozen general anesthetics (unpublished data). Although this receptor is somewhat less sensitive to anesthetics than those associated with general anesthesia, all effects are reversible and an understanding of the mechanisms involved should be useful. We therefore proceeded to test the lipid hypoth-

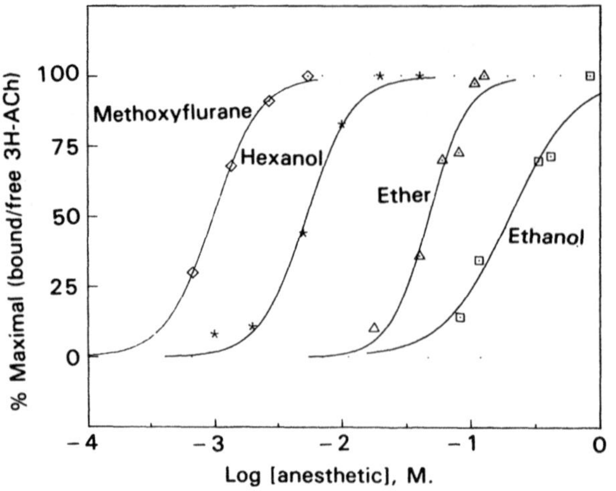

Figure 1. Effect of anesthetics on equilibrium high-affinity [^{3}H]acetylcholine binding. Equilibrium binding was performed with acetylcholine receptor-rich membranes from *Torpedo* at room temperature as described in Methods. Membranes were preincubated with anesthetics before addition of [^{3}H]acetylcholine. Corrections for nonspecific binding using α-bungarotoxin were 4% of the free [^{3}H]acetylcholine. Data are represented as the ratio of bound to free [^{3}H]acetylcholine and plotted on a scale of percent maximal increase in bound/free [^{3}H]acetylcholine to normalize variations in efficacy among anesthetics.

esis by examining the ability of anesthetics to perturb lipid order. (This particular parameter was chosen because of its experimental accessibility.)

The control-order parameters reported by 5-DP and 12-DS in *Torpedo* membranes at 21°C were 0.670 and 0.450, respectively, which reflects the decreasing order of lipid acyl chains toward the hydrophobic core of the bilayer.[14] The presence of a general anesthetic significantly reduced the membrane lipid-order parameter reported by each spin probe. Figure 2 shows typical electron spin resonance spectra obtained with increasing concentration of *n*-hexanol. From such spectra, order parameters were obtained that showed a linear relationship between order parameter and the anesthetic concentration in the buffer which was determined by gas chromatography. The slopes of these lines are given in Table I. These changes were reversible.

The anesthetics perturbed the probe deepest in the membrane's lipid bilayer (12-DS) about threefold (ΔS), or fivefold (percent change), more effectively than that nearer the surface (5-DP). The anesthetics involved in this comparison are alcohols and ethers which

Table I. Comparison of Ability to Desensitize with Anesthetic Potency in Tadpoles and with Lipid Disordering in Acetylcholine Receptor-Rich Membranes

Agent	EC_{50}[a] (mM)	ED_{50}[b] (mM)	EC_{50}/ED_{50}	5DP[c] $-\Delta S$/mM[e]	5DP[c] $-\Delta S^{EC_{50}}$[f]	12DS[d] $-\Delta S$/mM[e]	12DS[d] $-\Delta S^{EC_{50}}$[f]
Methoxyflurane	1.0	0.21	4.8	0.0071	0.0071	0.0170	0.0170
n-Hexanol	5.3	0.7	7.6	0.0020	0.0106	0.0052	0.0210
Urethane	75	14	5.4	0.00004	0.0030	N.D.	N.D.
Diethylether	47	24	2.0	0.00014	0.0071	0.00070	0.0350
Ethanol	199	116	1.7	0.000019	0.0059	0.000051	0.0156
Range	199X	522X	4.5X	374X	3.5X	333X	2.2X
Mean percent change in S					1.0		4.9

[a]EC_{50}'s were computed from a nonlinear least-squares fit of data plotted in Fig. 1 and are half-effect concentrations. Results were obtained from single experiments; data points were the mean of triplicates.
[b]Median doses, ED_{50}'s, for general anesthesia in tadpoles, from the sources cited in the Appendix, this volume.
[c]5-Doxylpalmitate.
[d]12-Doxylstearate. N.D. = not determined.
[e]Single experiments as described in Methods; free aqueous anesthetic concentrations were measured by gas chromatography.
[f]The change in lipid-order parameter associated with the half-effect on binding ($\Delta S^{EC}{}_{50}$) is the product of (ΔS/mM) × (EC_{50}).

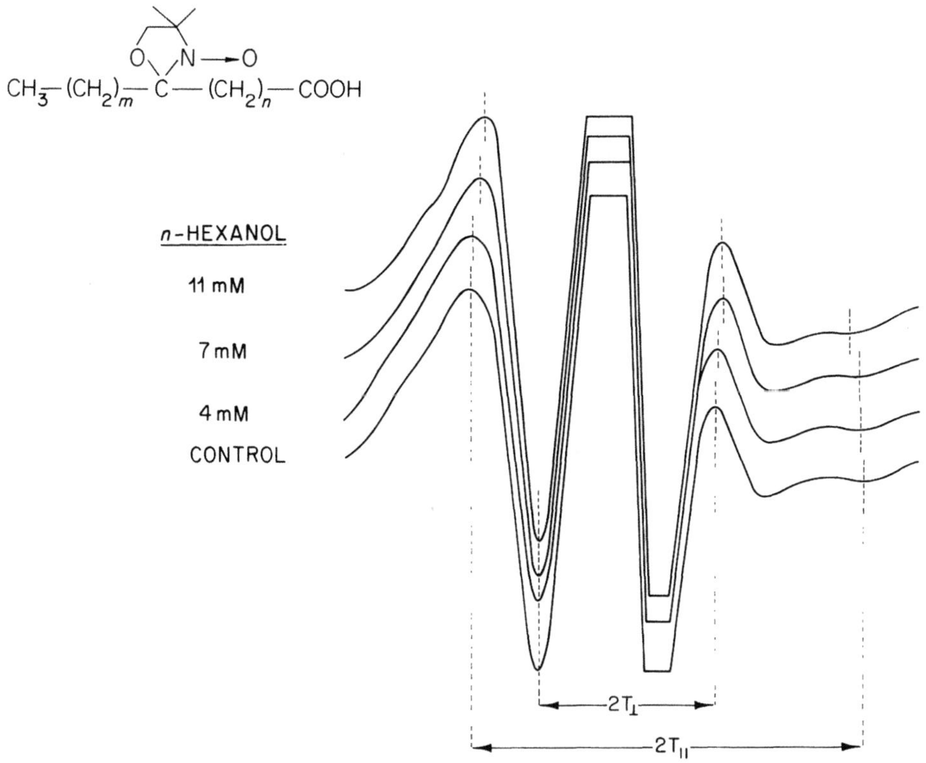

Figure 2. Effect of *n*-hexanol on spin-labeled *Torpedo* nicotinic membranes. Membranes were spin-labeled with 12-DS (insert, $m = 5$, $n = 10$) as described in Methods. Spectra obtained from membrane samples containing *n*-hexanol are aligned above a control for comparison. Note that $2T_{\parallel}$ decreases and $2T_{\perp}$ increases, as a function of *n*-hexanol concentration. This indicates that *n*-hexanol disorders the bulk lipid of *Torpedo* nicotinic membranes.

are polar enough to spend more time on average in the lipid surface than in the hydrophobic core. Such a distribution would tend to create energetically unfavorable voids deep in the membrane which would have to be filled by increasing the disorder of the acyl chains at this depth. There is a small residual variation in the degree of extra perturbation at depth induced by each anesthetic. These differences presumably reflect minor hetereogeneity in anesthetic polarity and shape that gives rise to a variable distribution and perturbing ability within the bilayer.

At a fixed concentration in the buffer, the anesthetics differed more than 300-fold in their ability to perturb the order reported by the spin probes, but when compared at concentrations causing equal degrees of desensitization (EC_{50}), this variation was reduced to 3.5- and 2.2-fold at the fifth and twelfth acyl carbon, respectively.

Others have speculated that the ability of general anesthetics to desensitize the acetylcholine receptor is related to lipid solubility as estimated by octanol/water partition coefficients.[27] Our data show a strong parallel between an anesthetic's ability to disorder the bulk lipid of the acetylcholine receptor and its ability to desensitize it (see Fig. 3 and Table I). Such a relationship could of course arise by chance, as could the relationship between lipid disorder and general anesthesia, but the advance here is that the perturbation of lipid structure and protein function are being measured in the same membrane so that more detailed mechanistic studies can be made in the future.

It is in fact unlikely that changes in bulk lipid order of 1–5% (Table I) would drive an

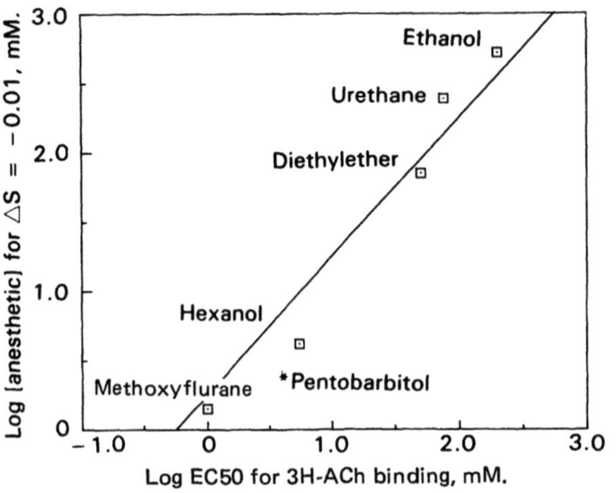

Figure 3. Correlation between anesthetic effects on cholinergic binding and lipid order. The five anesthetics from Table I are represented. Pentobarbital is plotted at the concentration associated with a steep increase in [^{3}H]acetylcholine binding, but is not included in the regression analysis (see text for details). The disordering effect concentration is that which produces a reduction in the lipid-order parameter reported by 5-DP by 0.01 unit. The line represented has a slope of 1. Unconstrained linear least-squares analysis yields a slope of 1.16 and a correlation coefficient of 0.99. A similar plot of binding effect versus disordering reported by 12-DS yields a line with a slope of 1.06 and a regression coefficient of 0.98.

increase in desensitization. Indeed, changes of 5% achieved by removing the labile of two pools of cholesterol from these membranes resulted in no change in acetylcholine binding.[18] However, the nonexchangeable cholesterol pool remaining in the membrane may well be closely associated with the acetylcholine receptor,[9,21] and if this is the case, the local lipid environment of the receptor may not have been perturbed by partial cholesterol depletion. This test suggests that attention should be redirected from the bulk lipid perturbations to the immediate lipid environment of the receptor. To a first approximation, such lipid will interact as strongly with anesthetics as does bulk lipid, but the structural and functional ramifications could be more profound.

Barbiturates Act Both Allosterically and Nonspecifically

At low concentrations, pentobarbital, unlike volatile anesthetics and alcohols, was reported to paradoxically decrease acetylcholine binding.[25] Subsequent studies revealed a [^{14}C]pentobarbital binding site which might be responsible for this effect, but the dissociation constant of the site was so high that accurate work on it was impossible.[19]

A number of barbiturates were surveyed in order to find a better ligand for this site. We found that amobarbital had both the highest apparent affinity and the highest efficacy for reducing acetylcholine binding. It thus offered the best opportunity for studying the relationship between barbiturate binding and acetylcholine receptor function. Figure 4 illustrates the range of effects found on acetylcholine binding. The data for amobarbital and pentobarbital can be described by a binding site model with dissociation constants of 20 and 200 μM, respectively. Secobarbital either does not bind this putative site or does so with an efficacy of zero. If it binds the specific site, one would predict secobarbital would antagonize amobarbital's effect. Indeed, in receptors in which acetylcholine binding had already been decreased by 0.5 mM amobarbital, increasing concentrations of secobarbital successively overcame the effect of amobarbital (Fig. 4). This is consistent with a competitive interaction between amobarbital and secobarbital for a common site.

Using the centrifugation assay, we next examined [^{14}C]amobarbital binding to the acetylcholine receptor. Figure 5 shows a typical experiment. At 10 μM [^{14}C]amobarbital approximately 490 nmole of [^{14}C]amobarbital was bound per gram of acetylcholine receptor protein. Increasing concentrations of unlabeled amobarbital decreased the amount of [^{14}C]amobarbital bound in a dose-dependent saturable fashion to a plateau of approximately 200 nmoles [^{14}C]amobarbital bound per gram of acetylcholine receptor protein, representing

Figure 4. Barbiturate-dependent changes in [^{3}H]acetylcholine binding. The effect of increasing concentrations of secobarbital (★), pentobarbital (■), and amobarbital (×) on the ratio of ([^{3}H]acetylcholine bound to acetylcholine receptor) / (free [^{3}H]acetylcholine) was determined by the filtration assay described in Methods. Both amobarbital and pentobarbital decreased [^{3}H]acetylcholine binding in a dose-dependent, saturable fashion. Secobarbital, itself, had no effect on [^{3}H]acetylcholine binding. The effect of secobarbital in the presence of 0.1 mM amobarbital (◇) was, however, consistent with a competitive interaction between amobarbital and secobarbital.

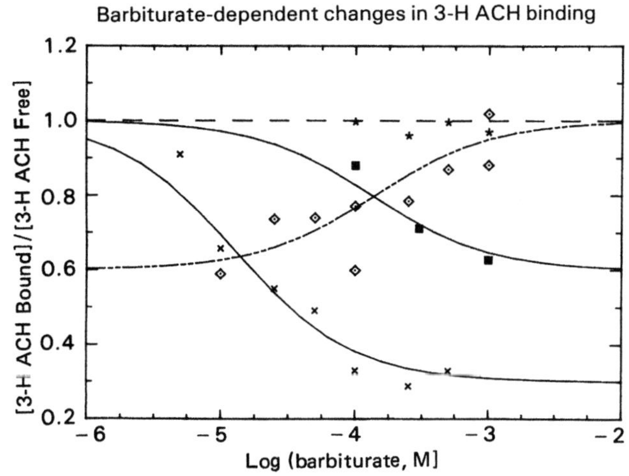

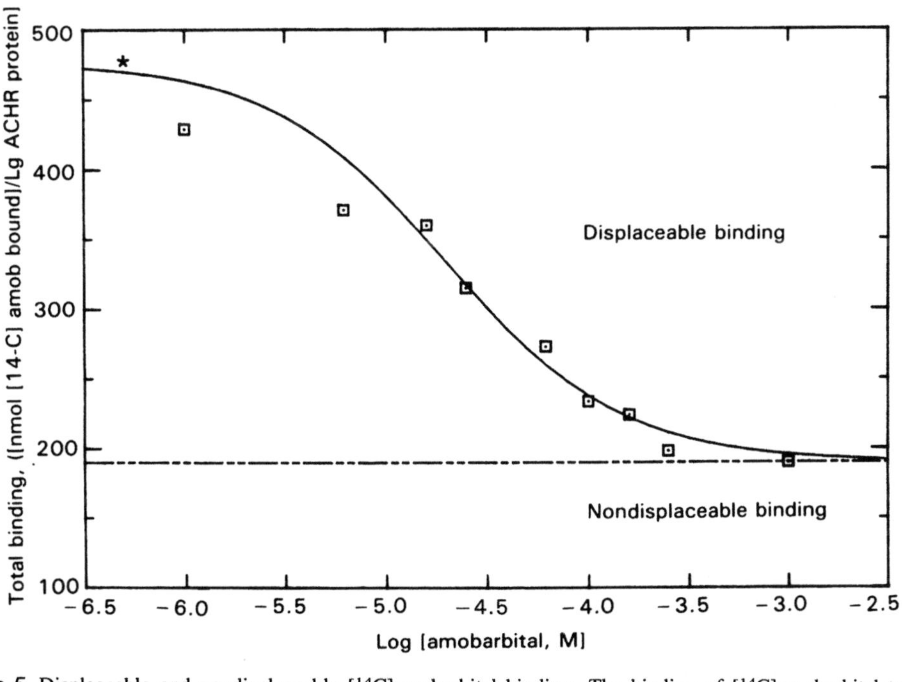

Figure 5. Displaceable and nondisplaceable [^{14}C]amobarbital binding. The binding of [^{14}C]amobarbital to the acetylcholine receptor in the presence and absence of unlabeled amobarbital was determined by the centrifugation assay described in Methods. In the absence of unlabeled amobarbital, there was 477 nmol [^{14}C]amobarbital bound per gram of receptor protein (★). This represents total bound [^{14}C]amobarbital. Increasing concentrations of unlabeled amobarbital decreased this binding in a dose-dependent manner to a maximal displacement with 300 μM amobarbital. Excess (1 mM) amobarbital failed to displace the remaining 190 nmol [^{14}C]amobarbital bound per gram of receptor protein. This represents nondisplaceable binding. Specific, or displaceable, binding is the difference between total and nondisplaceable binding.

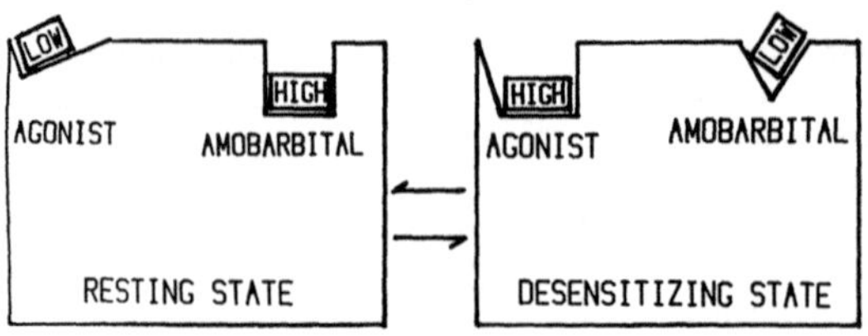

Figure 6. Model for barbiturate–acetylcholine interactions. This simple two-state model represents a possible mechanism for the effect of amobarbital on the acetylcholine receptor. Amobarbital binds with high affinity to the resting state of the receptor, whereas acetylcholine binds with higher affinity to the desensitized state. The concentration of acetylcholine and/or amobarbital would then determine the respective conformation state of the receptor.

nondisplaceable binding. The ratio of displaceable to nondisplaceable binding in our assay ranged from 2 to 2.5 depending on the specific membrane preparation used.

Both amobarbital and secobarbital displaced [^{14}C]amobarbital in a dose-dependent saturable fashion with half-maximal displacements, (IC_{50}s) of 26 and 94 μM, respectively. The IC_{50} for amobarbital is comparable to its apparent K_d for decreasing acetylcholine binding, while that for secobarbital is consistent with its ability to antagonize amobarbital's effect (Fig. 4).

Cholinergic agonists also decreased [^{14}C]amobarbital binding.This cholinergic effect was blocked by preincubation with α-bungarotoxin. However α-bungarotoxin itself had no effect on [^{14}C]amobarbital binding.

The above data support the existence of a barbiturate binding site on the acetylcholine receptor protein, but not the acetylcholine binding site itself, which can allosterically regulate acetylcholine binding to the acetylcholine receptor and is itself allosterically regulated by acetylcholine. Although the acetylcholine receptor has been hypothesized to exist in multiple conformational states (refs. 2 and 6, and see Taylor *et al.* and Cohen *et al.*, this volume, Chapters 9 and 10), the interaction of amobarbital and acetylcholine on the acetylcholine receptor can be explained by a simple two-state model (Fig. 6). Acetylcholine binds with higher affinity to the desensitized than to the resting state, whereas amobarbital binds with higher affinity to the resting state. Increasing concentrations of either of the two ligands will shift the receptor conformation into that ligand's preferred high-affinity state, thus decreasing the affinity of the other ligand for the receptor.

Although this allosteric mechanism satisfactorily accounts for our barbiturate data, it raises an important question concerning the nonspecific desensitization caused by all the other general anesthetics discussed in the preceding section. If barbiturates do not cause a similar effect, then the mechanism cannot be general! In fact, at concentrations higher than those considered so far, barbiturates do cause nonspecific desensitization. Thus, pentobarbital at concentrations between 30 μM and 1 mM gradually decreases acetylcholine binding (Fig. 4), but at concentrations above 1 mM it causes a steep increase in acetylcholine binding (not shown). Such a steep increase is reminiscent of the actions of the alcohols and volatile anesthetics. Unfortunately, the limited solubility of pentobarbital (about 9 mM under our conditions) prevented determination of the maximum effect attainable; so quantitative comparison with the other anesthetics is difficult. However, pentobarbital caused no change in lipid order measured with 5-DP at concentrations where allosteric effects were observed, but caused a linear decrease above 1 mM. These data are plotted in the correlation on Fig. 3 and are consistent with a lipid model. This suggests that this effect of high barbiturate concentrations is the same as that caused by all other general anesthetics.

Thus our data inescapably point to the conclusion that barbiturates exert both a specific, allosteric, and a nonspecific, general anestheticlike, action on the acetylcholine receptor.

The former effect is analogous to that exerted at the GABA receptor (see Olsen *et al.* and Ticku *et al.*, this volume, Chapters 14 and 15).

Anesthetic Effects on Agonist-Induced Cation Flux

The previous sections of this chapter have described equilibrium ligand binding studies with the acetylcholine receptor. In order to elucidate the mechanism(s) of anesthetic action on this protein, its functional states must also be examined. We have therefore initiated studies on the effects of octanol, pentobarbital, and ethanol on agonist-induced cation flux in sealed acetylcholine receptor-rich *Torpedo* vesicles.

Electrophysiological studies at the neuromuscular junction have shown that octanol[12] and pentobarbital[1] shorten acetylcholine receptor channel lifetimes. This action of octanol was observed at buffer concentrations of 0.5–1.0 mM. The acceleration of miniature end-plate current (MEPC) decay by a number of barbiturates (pentobarbital not included) was observed at 50–150 μM, and equilibrium dissociation constants for binding to acetylcholine receptor were calculated to be in the 1 to 10-μM range. On the other hand, a number of experimenters have observed that ethanol (and other short-chain alcohols) cause a *decrease* in MEPC decay rate and an *increase* in single-channel lifetimes from noise analysis.[3,11,23]

Two shortcomings of the electrophysiological approach can be overcome by studying agonist-induced flux in *Torpedo* vesicles. First, in studies of channel dynamic parameters, agonist dose–response data cannot be reliably acquired. Second, drug exposure times are generally on the order of tens of minutes in electrophysiological experiments in order to ensure equilibration. These long perfusions are rate-limiting for data acquisition and may allow irreversible changes to occur, which would not be seen in a short flux assay. However, to study the pharmacology of the activated acetylcholine receptor ionophore in *Torpedo* vesicles requires special techniques because of both desensitization and the presence of spare receptors in the vesicle preparation.

Agonist-induced desensitization is a slow process, characterized by an agonist-dependent time constant of 0.004–0.10 sec^{-1} [2,26] and can thus be circumvented by a relatively rapid ion flux assay.

The presence of spare receptors indicates that within the time frame of a flux assay, a submaximal stimulus (i.e., one that opens less than 100% of the activatable channels) will elicit a maximal response (release of all tracer ions in vesicles with activatable channels). Agonist and antagonist dose–response data under these conditions will vary depending on the amount of spare receptors in a given vesicle preparation. Elimination of spare receptors can be achieved either by using a very rapid flux assay, such that initial flux rates can be monitored, or by reducing the number of active channels.[20] For these experiments, we assayed net $^{86}Rb^+$ efflux from *Torpedo* vesicles during a 10-sec exposure to carbamylcholine and anesthetic agents. Spare receptors were removed by titration with purified α-bungarotoxin until a maximum stimulus (exposure to 5 mM carbamylcholine for 10 sec) produced a submaximal response.

In the absence of spare receptors, a typical carbamylcholine dose–response curve (see Fig. 7) exhibits a C_{50} of 128 ± 6 μM and a Hill coefficient (n_H) of 1.6 ± 0.1. Both octanol and pentobarbital block maximally activated efflux (F_{max}, 5 mM carbamylcholine) in a dose-dependent manner with IC_{50}'s of 16 ± 1 μM and 23 ± 3 μM, respectively. When concentrations of octanol (10 μM) and pentobarbital (16 μM) that inhibit maximum efflux by 30% are present in the agonist dose–response experiment (Fig. 7), little or no change in the C_{50} for carbamylcholine is observed (C_{50} = 126 ± 8 μM with octanol and 155 ± 30 μM with pentobarbital).

In contrast, ethanol up to 3.0 M causes only a 15% reduction in F_{max}, with little or no

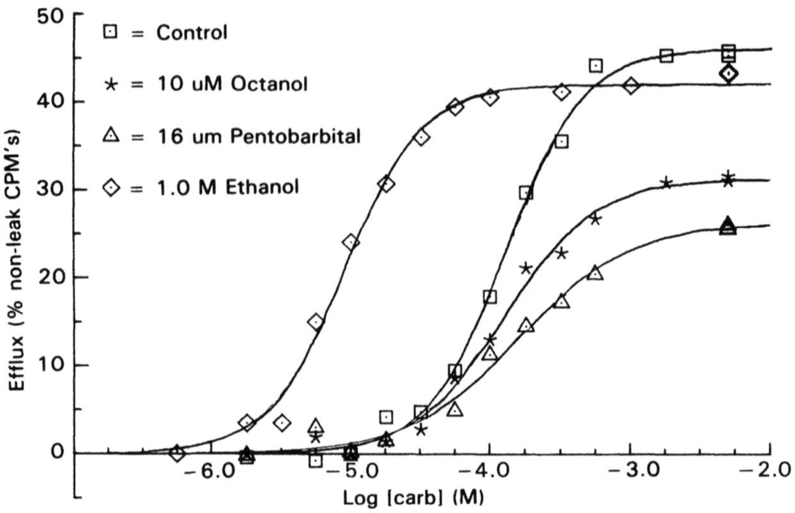

Figure 7. Ten second efflux of $^{86}Rb+$ from *Torpedo* vesicles without spare receptors at 4°C is shown as a function of [Carb]. Ordinate values are calculated as the percent of entrapped tracer ions released, corrected for time-dependent leakage:

$$\text{Efflux [Carb]} = 100\% \times \frac{\text{CPM [Carb]} - \text{CPM (leak)}}{\text{CPM (total)} - \text{CPM (leak)}} \quad (1)$$

The dose–response data were fit by nonlinear least squares to the equation

$$\text{Efflux [Carb]} = F_{\max} \times \frac{[\text{Carb}]^n}{[\text{Carb}]^n + [C_{50}]^n} \quad (2)$$

The experiment was performed in the absence of anesthetics (□), $F_{max} = 46.0 \pm 0.7\%$, $C_{50} = 128 \pm 6$ μM; and in the presence of 10 μM Octanol (★), $F_{max} = 31.2 \pm 0.6\%$, $C_{50} = 126 \pm 8$ μM; 16 μM pentobarbital (△), $F_{max} = 26.8 \pm 1.3\%$, $C_{50} = 155 \pm 30$ μM; and 1 M ethanol (◇), $F_{max} = 42.0 \pm 0.6\%$, $C_{50} = 9.1 \pm 0.5$ μM.

dose dependence above 300 mM. At lower carbamylcholine concentrations, however, ethanol enhances stimulated $^{86}Rb^+$ efflux, by causing a leftward shift of the carbamylcholine dose–response curve (Fig. 7). Each 0.3-M increment in ethanol concentration causes an approximately twofold reduction in C_{50}. This leftward shift effect is nonsaturable up to 3.0 M ethanol.

The mechanisms by which octanol, pentobarbital, and ethanol alter observed acetylcholine receptor channel dynamics and flux parameters are still uncertain. A number of models, based on electrophysiological observations, have been proposed. Gage *et al.*[12] suggested that octanol accelerates channel closing by fluidizing the surrounding membrane. The concentrations at which channel dynamics are affected are 10- to 50-fold higher than those at which flux is blocked in the absence of spare receptors, although they are similar to those that cause altered equilibrium [^{3}H]acetylcholine binding (see previous discussion). Thus, although rapid channel closing may be caused by nonspecific lipid perturbations, flux blockade by octanol apparently acts at a more sensitive site. Bradley *et al.*[3] have suggested a hydrophobic "patch" in the ion channel as the site of alkanol action. This model is consistent with our data, which simply suggest a noncompetitive blockade of carbamylcholine-induced ion flux by octanol and pentobarbital.

A number of hypotheses for the action of ethanol have also been proposed. Gage *et al.*[11] suggest that short-chain alkanols reduce the membrane dielectric constant. This causes a change in the transmembrane electric field encountered by a dipole, resulting in longer channel lifetime and greater ion flux per channel. Our results contradict this model, because ethanol enhances flux from vesicles that have no transmembrane potential. Furthermore, this model predicts that F_{max} should increase in the presence of ethanol, which we do not observe. Pennefather and Quastel (ref. 23, and this volume, Chapter 13) argue that ethanol may slow channel closing by increasing agonist affinity for the open channel (i.e., a slower off rate for acetylcholine), while Bradley *et al.*, [3] using quantitative iontophoresis, report a leftward shift in the acetylcholine dose–response curve with ethanol. Our data are in agreement with a mechanism whereby ethanol enhances the agonist affinity of the open-channel state, but an additional effect is required to explain the observed decrease in F_{max}.

Finally, in one experiment, we have started to test the hypothesis that all alcohols interact with a single discrete hydrophobic "patch" on the receptor. Given this model, we would expect both saturating dose–response relationships and competition between alcohols. We therefore combined 10 μM octanol, which has only a blocking effect, with 1 M ethanol, which has almost exclusively a leftward-shifting effect, and looked at the carbamylcholine dose–response curve (data not shown). A leftward shift in the carbamylcholine dose–response curve was seen, which was identical to that caused by 1 M ethanol alone. This result indicates that 10 μM octanol did not displace ethanol from its "binding" site. However, F_{max} in the combination experiment was increased relative to F_{max} in the presence of 10 μM octanol alone, indicating that ethanol antagonized the octanol blocking effect. We believe that the most likely explanation for these paradoxical results is that 1 M ethanol reduces the partitioning of octanol into a hydrophobic site. In support of this hypothesis, it is known that the membrane/buffer partition coefficient of tetrahydrocannabinol is reduced by a factor of 2 in the presence of 1 M ethanol.[24] Thus, octanol and ethanol appear to exert their actions independently of each other, and a model involving a single site or patch is not consistent with our data. On the other hand, if the free energy cost of transferring an apolar amino acid side chain from a hydrophobic environment to an aqueous phase containing high ethanol concentrations were to be reduced compared to that for transfer to water, this could have marked consequences on the structure of the receptor.

CONCLUSIONS

The acetylcholine receptor-rich membranes from *Torpedo* are enabling anesthetic mechanisms to be examined in a degree of detail not previously achieved in an excitable membrane protein. Such a situation was predictable given the high specific activity of the preparation. What is remarkable, however, is that our studies have revealed at least three actions of general anesthetics. The first of these, the ability to stabilize the high-affinity, desensitized form of the receptor, is brought about by all the classes of general anesthetic examined to date (volatiles, alcohols, barbiturates, and carbamates) and is pressure reversible.[5] The second appears to be specific to low concentrations of barbiturates and the third to the action of small polar molecules, such as ethanol, on the ionophore. Although more detailed work will be needed to establish mechanisms, certain working models can be perceived at this point.

The actions of barbiturates at low concentrations are most readily rationalized if one assumes that the barbiturate binding site is available on all conformations of the receptor, but that each barbiturate's affinity varies with the protein's conformational state. This model is very similar in principle to the modulated receptor model for voltage-sensitive sodium channels (see this volume, Section III). Our flux data suggest that pentobarbital has a dissociation constant for the open state of ~20 μM, while our binding data suggest dissociation constants of 200 μM and >1 mM for the resting and desensitized states, respectively.

Since the selectivity for the open state over the resting state is about 10-fold, this model is consistent with the notion of Adams[1] that at low barbiturate concentrations only open channels combine with the drug.

The mechanism by which all general anesthetics induce desensitization is not established, but equal degrees of anesthetic-induced desensitization are associated with equal degrees of lipid perturbation. This is suggestive of some involvement of lipids, perhaps at the lipid–protein interface, but other models cannot be ruled out. Significantly, this effect of anesthetics can be pressure reversed.[5]

Our studies on the paradoxical effects of ethanol on the ionophore appear to rule out some existing models, but more detailed work will be required before we can approach alternative models.

Clearly, the acetylcholine receptor will continue to provide a procrustean bed on which the concepts of a simpler era may be refined.

ACKNOWLEDGMENTS. This work was supported by a grant to the Harvard Anaesthesia Center from the National Institute of General Medical Sciences (GM-15904). L.L.F. and B.A.D. were supported by GM-07592 and S.A.F. by GM-07753. We thank Jenny Fell for preparing the manuscript.

REFERENCES

1. Adams PR: Drug blockade of open end-plate channels. *J Physiol (London)* 1976;260:531–552.
2. Boyd ND, Cohen JB: Kinetics of binding of [^{3}H]acetylcholine to *Torpedo* postsynaptic membranes: Association and dissociation rates by rapid mixing and ultrafiltration. *Biochemistry* 1980;19:5353–5358.
3. Bradley RJ, Stertz R, Peper K: The effects of alcohols and diols at the nicotinic acetylcholine receptor of the neuromuscular junction. *Brain Res* 1984;295:101–112.
4. Braswell LM, Kitz RJ: The effect in vitro of volatile anesthetics on the activity of cholinesterases. *J Neurochem* 1977;29:665–671.
5. Braswell LM, Miller KW, Sauter J-F: Pressure reversal of the action of octanol on postsynaptic membranes from *Torpedo. Br J Pharmacol* 1984;83:305–311.
6. Changeux J-P, Devillers-Thiery A, Chemouilli P: Acetylcholine receptor: An allosteric protein. *Science* 1984;225:1335–1345.
7. Cohen JB, Weber M, Huchet M, Changeux J-P: Purification from *Torpedo mamorata* electric tissue of membrane fragments particularly rich in cholinergic receptor protein. *FEBS Lett* 1972;26:43–47.
8. Cohen JB, Boyd ND, Shera NS: Interactions of anesthetics with nicotinic postsynaptic membranes isolated from *Torpedo* electric tissue, in Fink RB (ed): *Molecular Mechanisms of Anesthesia, Progress in Anesthesiology,* Vol. 2. New York, Raven Press, 1980.
9. Ellena JF, Blazing MA, McNamee MG: Lipid protein interactions in reconstituted membranes containing acetylcholine receptor. *Biochemistry* 1983;22:5523–5535.
10. Gaffney BJ: Practical considerations for the calculation of order parameters for fatty acid or phospholipid spin labels in membranes, in Berliner LJ (ed): *Spin Labeling, Theory and Applications.* New York, Academic Press, 1976,pp 567–571.
11. Gage PW, McBurney RN, Schneider GT: Effects of some aliphatic alcohols on the conductance change caused by a quantum of acetylcholine at the toad end-plate. *J Physiol (London)* 1975;244:409–429.
12. Gage PW, McBurney RN, Van Helden D: Octanol reduces end-plate channel lifetime. *J Physiol (London)* 1978;274:279–298.
13. Griffith OH, Jost P: Lipid spin labels in biological membranes, in Berliner LJ (ed): *Spin Labeling, Theory and Applications.* New York, Academic Press, 1976,pp 411–454.
14. Hubbell W, McConnell H: Molecular motion in spin-labeled phospholipids and membranes. *J Am Chem Soc* 1971;93:314–326.
15. Janoff AS, Pringle MJ, Miller KW: Correlation of general anesthetic potency with solubility in membranes. *Biochim Biophys Acta* 1981;649:125–128.
16. Johnson RR, Eger EI III, Wilson C: A comparative interaction of epinephrine with enflurane, isoflurane, and halothane in man. *Anesth Analg* 1976;55:709–712.

17. Koblin DD, Eger EI III, Johnson BH, Collins P, Terrell RC, Speers L: Are convulsant gases also anesthetics? *Anesthesiology* 1981;60:464–470.
18. Leibel W, Firestone L, Braswell LM, Miller KW: Two pools of cholesterol in acetylcholine receptor-rich membranes from *Torpedo. Soc Neurosci Abstr* 1983;9:511.
19. Miller KW, Sauter J-F, Braswell LM: A stereoselective pentobarbital binding site in cholinergic membranes from *Torpedo californica. Biochem Biophys Res Commun* 1983;105:659–666.
20. Neubig RR, Cohen JB: Permeability control by cholinergic receptors in *Torpedo* postsynaptic membranes: Agonist dose-response relations measured at second and millisecond times. *Biochemistry* 1980;19:2770–2779.
21. Ochoa ELM, Dalziel AW, McNamee MG: Reconstitution of AChR function in lipid vesicles of defined composition. *Biochim Biophys Acta* 1983;727:151–162.
22. Pang K-YY, Miller KW: Cholesterol modulates the effect of membrane perturbations in phospholipid vesicles and biomembranes. *Biochim Biophys Acta* 1978;511:1–9.
23. Pennefather P, Quastel DMJ: Actions of anesthetics on the function of nicotinic acetylcholine receptors, in Fink BR (ed): *Molecular Mechanisms of Anesthesia, Progress in Anesthesiology,* Vol. 2. New York, Raven Press, 1980,pp 45–48.
24. Roth SH, Williams PJ: The non-specific membrane binding properties of Δ^9-tetrahydrocannabinol and the effects of various solubilizers. *J Pharm Pharmacol* 1979;31:224–230.
25. Sauter J-F, Braswell LM, Miller KW: Action of anesthetics and high pressure on cholinergic membranes in Fink BR (ed): *Molecular Mechanisms of Anesthesia, Progress in Anesthesiology,* Vol. 2. New York, Raven Press, 1980,pp 199–207.
26. Walker JW, Takeyasu K, McNamee MG: Activation and inactivation kinetics of *Torpedo californica* acetylcholine receptor in reconstituted membranes. *Biochemistry* 1982;21:5384–5389.
27. Young AP, Sigman DS: Allosteric effect of volatile anesthetics on the membrane-bound AChR. *Mol Pharmacol* 1981;20:498–505.

12

The Influence of Anesthetics on Postsynaptic Ion Channels

Peter W. Gage, David McKinnon, and Brian Robertson

ABSTRACT. A wide variety of general anesthetics reduce the time constant of decay of endplate currents at concentrations in the clinical range, producing depression of the amplitude of endplate potentials. This effect has been shown to be caused by a reduction in the time for which endplate channels remain open following activation by acetylcholine, either because channels close more rapidly in a more fluid membrane or because open channels are blocked by molecules of anesthetic. The characteristics of channels activated by acetylcholine in denervated rat muscle in the presence of pentobarbitone cannot be explained simply by the latter mechanism, and it is suggested that pentobarbitone may bind to an allosteric site on the receptor–channel complex, making the open state unstable. An opposite effect of several anesthetics has been found at inhibitory synapses on pyramidal cells in the rat hippocampus: pentobarbitone, halothane, and ketamine increase the time constant of decay of inhibitory postsynaptic currents. It is proposed that a fundamental action of general anesthetics is to affect the stability of receptor–channel complexes, either decreasing channel open time at excitatory synapses or increasing channel open time at inhibitory synapses. Such effects would depress excitatory synaptic transmission.

INTRODUCTION

Although the molecular basis of anesthesia is still hotly disputed, there now seems to be a consensus that general anesthetics produce anesthesia by depressing transmission at excitatory synapses. More than 60 years ago, Winterstein showed that synaptic transmission in the spinal cord is more susceptible to the effects of most narcotics than the conduction process in nerve fibers.[16] Sherrington also proposed that anesthetics depress synaptic transmission more than axonal conduction.[44] The classical work of Larrabee and Posternak[28] demonstrated that general anesthetics affected synaptic transmission at low, "clinical" concentrations while action potentials in nerve axons were unaffected. Ether and barbiturates were found to depress monosynaptic reflexes in the spinal cord[46,47,52] and halothane and barbiturates to depress excitatory synapses in the cuneate nucleus[22] and dorsal horn.[27] Over the past 10–15 years, Richards and co-workers have demonstrated that general anesthetics also depress excitatory transmission and the response to glutamate in isolated olfactory cortex and hippo-

Peter W. Gage, David McKinnon, and Brian Robertson • Department of Physiology, John Curtin School of Medical Research, Australian National University, Canberra City A.C.T. 2601, Australia.

campus at concentrations that do not affect conduction velocity of action potentials in very fine axons (for a review, see ref. 41; see also Krnjević, this volume, Chapter 1).

There are many ways in which general anesthetics might depress excitatory transmission. We have been exploring two of these: the depression of excitatory postsynaptic potentials and the potentiation of inhibitory postsynaptic potentials produced by general anesthetics. A useful model for the first has been the neuromuscular junction where it is possible to employ a variety of biophysical techniques to examine the properties of ion channels activated by acetylcholine. In order to examine the effects of general anesthetics on inhibitory transmission mediated by γ-aminobutyric acid (GABA), we have been using a rat hippocampal slice preparation.

Many types of drugs can inhibit synaptic transmission. Tetrodotoxin and local anesthetics block the sodium channels that generate action potentials; divalent ions such as magnesium, manganese, cobalt, and cadmium can block the movement of calcium across cell membranes, and specific antagonists, such as *d*-tubocurarine at nicotinic synapses, can compete with neurotransmitters for their binding sites on the receptors.

It is now realized that drugs that change the duration of synaptic currents, without necessarily changing their amplitude, may have important effects on synaptic transmission. The reason for this is that a synaptic current must generate a potential across a membrane that has capacitive as well as resistive components. Thus, the potential change generated by ionic currents takes time to develop. Synaptic currents that last only a fraction of the membrane time constant will not charge the membrane capacity as much as synaptic currents of equal amplitude that persist for a longer time. This is shown graphically in Fig. 1. The model here is a toad skeletal muscle fiber with a time constant of 8 msec.[14] Rectangular currents (broken lines) of equal amplitude but increasing duration (*a, b, c, d:* 1, 2, 5, 10 msec) produce the potential changes of increasing amplitude (*a, b, c, d*) shown in B. The time course of a synaptic current at the motor endplate consists of a rapid surge of inward current followed by a slower exponential decay. The simulated current with decay time constant of 2 msec (downward trace in A) produces the potential change indicated by the solid waveform in B. Clearly, the change in membrane potential would be less for briefer currents. This is illustrated in Fig. 2. Again, the model is a skeletal muscle fiber (but the same principles apply in

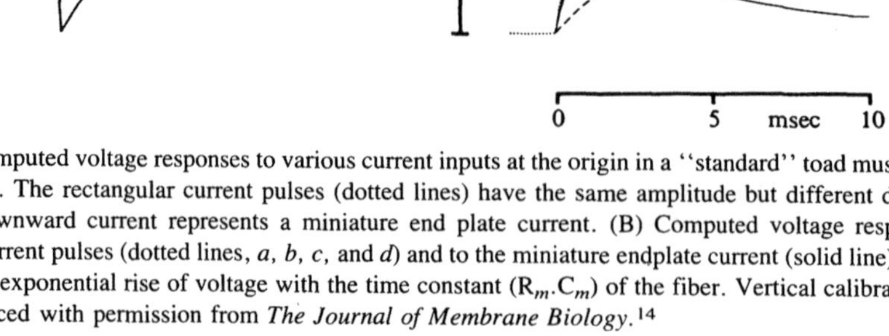

Figure 1. Computed voltage responses to various current inputs at the origin in a "standard" toad muscle fiber. (A) Input currents. The rectangular current pulses (dotted lines) have the same amplitude but different durations (see text). The downward current represents a miniature end plate current. (B) Computed voltage responses to the rectangular current pulses (dotted lines, *a*, *b*, *c*, and *d*) and to the miniature endplate current (solid line). The broken line shows an exponential rise of voltage with the time constant ($R_m.C_m$) of the fiber. Vertical calibration: 5 nA, 1 mV. Reproduced with permission from *The Journal of Membrane Biology*.[14]

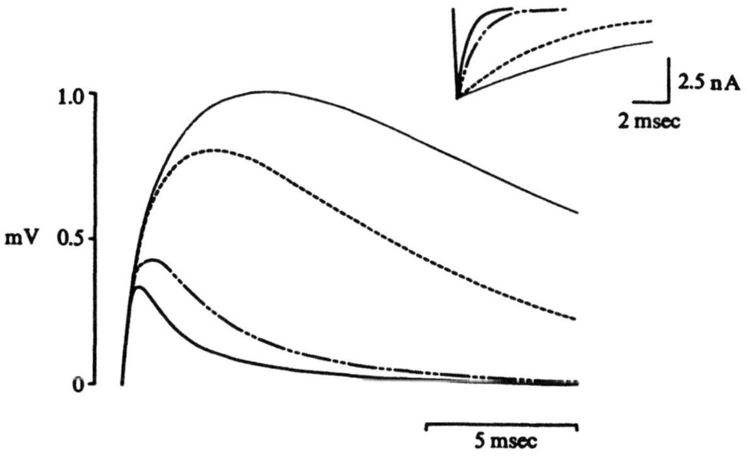

Figure 2. Voltage responses computed for standard input currents (inset) with varying decay time constants (0.5, 1, 5, and 10 msec). Calibrations for inset, 2.5 nA and 2 msec. Long horizontal calibration, 5 msec. Reproduced with permission from *The Journal of Membrane Biology.*[14]

nerve cells). Currents of increasing duration produced depolarizations of increasing amplitude.

If drugs were to modify the time course of currents, similar changes could occur. Specifically, if a drug made synaptic currents decay more rapidly, the amplitude of postsynaptic potentials could be depressed below threshold for an action potential, and transmission would be blocked. Conversely, if the decay of a synaptic current became slower, the potential change it generated could increase in amplitude. Evidence that general anesthetics affect synaptic transmission in both these ways will be presented.

METHODS

Miniature Endplate Currents and Potentials

For recording miniature endplate potentials and currents simultaneously, an extracellular microelectrode was maneuvered at an endplate region in isolated nerve–muscle preparations (toad sartorius, rat or mouse sternomastoid or diaphragm) until extracellular miniature endplate currents were detected. An intracellular electrode was then inserted into the same muscle fiber nearby. (For further details, see ref. 12.)

Single-Channel Currents

Single channels activated by acetylcholine were recorded from denervated (12–16 days) extensor digitorum longus muscles of 300- to 400-g male Wistar rats. Isolated muscles were treated in a 0.3% collagenase solution for 1.0–1.5 hr to digest the collagen matrix surrounding individual muscle fibers. Currents were recorded from isolated fibers using the intact patch technique.[25] The current was low-pass filtered (2 kHz corner frequency, 4th order Bessel) and sampled at 10 kHz. The membrane potential of the muscle fiber was monitored with an intracellular electrode, and the potential across the patch was clamped at −100 mV for all the experiments described here. Patch pipettes contained 200–400 nM acetylcholine chloride. For testing the effects of pentobarbitone, the drug was added to both the pipette and bath solutions. Single-channel records were analyzed using a semiautomatic computer program that identified channel openings with a midlevel threshold crossing routine and displayed the identified event on a graphics terminal. The distributions of open and closed times were calculated directly from the measured transition times. The distribution of burst lengths was calculated using a procedure similar to that of Magleby and Pallotta.[30] The mean open and closed times were corrected to take the limited frequency response of the recording system into account.[34] Average values were obtained from three to five experiments at each concentration of pentobarbitone. Predictions for the sequential blocking model (broken lines in Figs. 7 and 8) were calculated from average values obtained with 50 μM pentobarbitone.

Hippocampal Inhibitory Currents

Spontaneous inhibitory postsynaptic currents were recorded in CA1 neurons in 400-μm-thick rat hippocampal slices using a single microelectrode voltage clamp (ASF1, Axon Instruments Ltd). Slices were maintained on a nylon mesh at an interface between a standard solution containing, in mM: NaCl 124; KCl 5; $NaHCO_3$ 26; Na_2HPO_4 1.25; $MgSO_4$ 2; $CaCl_2$ 2; D-glucose 10, in an atmosphere of 95% O_2 plus 5% CO_2. Intracellular microelectrodes contained KCl or CsCl; occasionally electrodes containing Cs_2SO_4 were used. Clamp switching rates were between 10 and 16 kHz, and adequate settling of the voltage trace prior to voltage sampling was ensured at all times by continuously monitoring the headstage voltage. The anesthetics were freshly added to the standard solution for each experiment. The gaseous anesthetics halothane and nitrous oxide were administered both in solution and in the gas phase.

RESULTS

Depression of Excitatory Transmission

About 20 years ago, it was noticed that ethanol increased the amplitude of miniature endplate potentials.[10] Later, while reexamining the effects of aliphatic alcohols at the neuromuscular junction, Gage, McBurney, and Schneider[18] found that short-chain alcohols (methanol to pentanol) increased the amplitude of miniature endplate potentials by making endplate currents decay more slowly. These observations were confirmed by Quastel and Linder.[38] (See also McLarnon *et al.*, this volume, Chapter 13.)

Close examination of the influence of alcohol chain length and concentration revealed that there was an exponential relationship between alcohol concentrations and the rate of decay of endplate currents, and that the relative effectiveness of these short-chain alcohols was closely correlated with their membrane-buffer partition coefficient.[18] Two independent observations led to the suggestion that alcohols might be influencing the polarizability (dielectric constant) of the environment of the receptor–channel complex. First, the open time of channels was affected by membrane potential, and this may reflect a difference in dipole moment of the receptor–channel complex when channels are open or closed.[31] Second, the rate of a reaction involving a change in dipole moment is exponentially related to the dielectric constant of the medium.[23] It was proposed[18] that ethanol partitions into and changes the dielectric constant of the lipid in close proximity to the dipole. This makes the open state of the channel, associated with an increase in dipole moment, more stable so that open time is increased. In support of this hypothesis, it has been found that ethanol has an opposite effect on channel open time at excitatory synapses where there is probably a decrease in dipole movement when channels open. For example, at glutaminergic neuromuscular junctions in crab and crayfish where channel open time decreases at more positive potentials, postsynaptic currents decay more rapidly and the amplitude of postsynaptic potentials is depressed in the presence of ethanol.[3,24,51]

Surprisingly, longer-chain alcohols (e.g., octanol) had an opposite effect: miniature endplate potentials were depressed in amplitude because the underlying currents decayed more rapidly,[17,20,33] and it was suggested that this effect might be due to an increase in membrane fluidity.[17] As it had been suggested that general anesthetics might produce their effects by increasing membrane fluidity also,[50] Hamill, Torda, and I proceeded to test the effects of a variety of general anesthetics on endplate potentials and currents.[11–13,48,49]

All the anesthetics tested (including barbiturates, ether, halothane, chloroform, enflurane, steroid anesthetics, diazepam, and long-chain aliphatic alcohols) depressed the amplitude of miniature endplate potentials. They shared the common property of making endplate currents decay more rapidly than normal and, in this way, depressed the amplitude of the postsynaptic depolarization.

The effect of ether (10 mM) on a toad sartorius neuromuscular junction[11] is illustrated in Fig. 3. There is clearly a depression of the amplitude of the miniature endplate potential (upper trace) in the presence of ether compared with the control. The underlying traces show the miniature endplate currents recorded simultaneously (see Methods). Although the amplitude of the currents was the same, the more rapid decay in the presence of the ether produced a smaller miniature endplate potential. Similar results obtained at a mouse neuromuscular junction with halothane[12,13] are shown in the lower part of Fig. 3. Again, the faster decay of the currents in the presence of halothane produced depression of the amplitude of the postsynaptic potential. All the other general anesthetics listed previously made miniature endplate currents decay more rapidly so that the amplitude of postsynaptic potentials was depressed.

The effectiveness of the anesthetics in reducing the decay time constant of miniature endplate currents increased with their oil/water partition coefficients. In Fig. 4, the concentrations of four anesthetics (ether, enflurane, chloroform, and halothane) that produced a 30% reduction in decay time constant of endplate currents in mouse diaphragm are plotted against their oil/water partition coefficients.[12] The concentration of anesthetics at some hydrophobic site appears to determine their efficacy. For comparison, the aqueous concentrations of the same anesthetics that produce clinical anesthesia (calculated from minimum alveolar concentrations) are plotted on the same graph (squares, Fig. 4). A similar correlation is observed in accord with the Meyer–Overton rule.

It is now recognized that the decay of endplate currents is determined by the open time of channels activated by acetylcholine; the time constant of decay is determined by the average channel open time.[5] The general anesthetics must depress the amplitude of endplate potentials by reducing the time for which channels activated by acetylcholine remain open. Direct evidence for this conclusion has come from analyzing "current noise" generated by applying acetylcholine to endplates[20] or from recording the behavior of single channels after exposure to general anesthetics (ref. 29; and see below). The decrease in channel open time produced by ether can be seen from the shift of the power density spectrum to the right in the presence of the anesthetic in Fig. 5.[20]

A Possible Molecular Mechanism—The Blocking Model

The molecular basis of the reduction in channel open time produced by general anesthetics at the neuromuscular junction has not been determined. Some suggestions have been that they "fluidize" the lipid in close proximity to the receptor–channel complex so that chan-

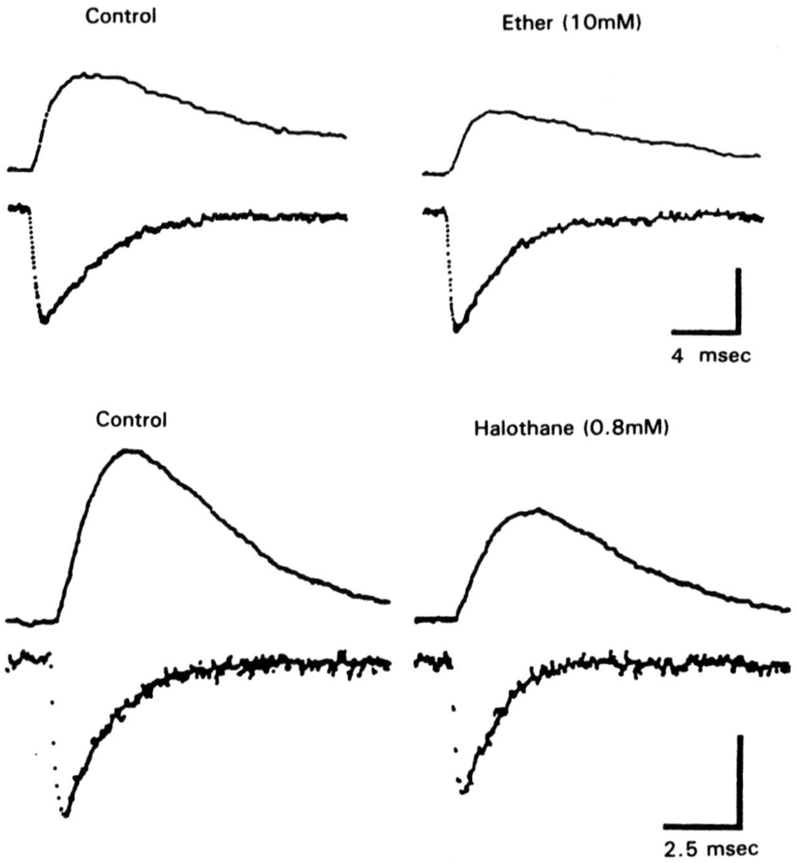

Figure 3. Intracellular records of miniature endplate potentials (upper traces) and the miniature endplate current recorded simultaneously with an extracellular electrode (lower traces). The records before and after exposure to ether (10 mM) were obtained at a toad sartorius endplate at 15°C. The vertical calibration denotes 0.5 mV for the potentials and 0.1 mV for the current. The records before and after exposure to halothane (0.8 mM) were obtained at an endplate in a mouse diaphragm at 20°C. The vertical calibration denotes 1 mV for the potentials and 0.25 mV for the currents. Horizontal calibrations show 4 msec above, 2.5 msec below.

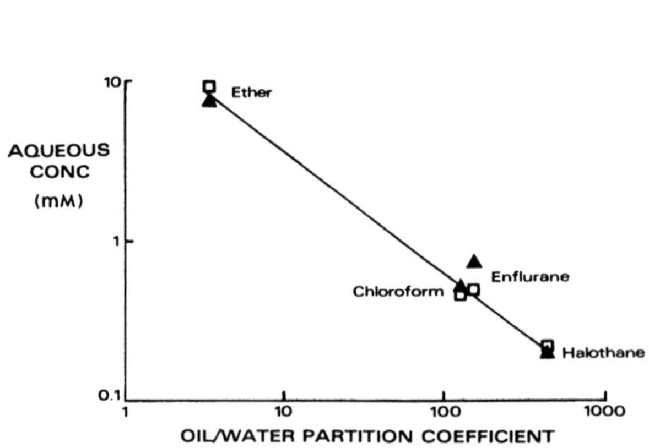

Figure 4. The aqueous concentrations (mM) of four anesthetics producing clinical levels of anesthesia (□) or 30% reduction of the decay time constant of miniature endplate currents in the toad (▲) plotted against oil–water partition coefficients. Note the logarithmic axes. Reproduced with permission from *The British Journal of Pharmacology*.[12]

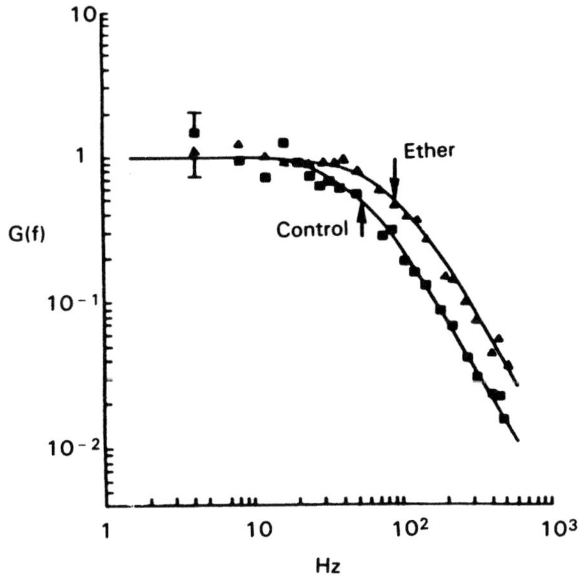

Figure 5. Ether (70 mM) shifts the power spectrum of acetylcholine-induced current noise recorded at an endplate in toad sartorius muscle to higher frequencies. The solid lines show Lorentzians of best fit to the points. The arrows show the half-power frequencies, 52 Hz in control (■) and 88 Hz in the presence of the ether (▲) giving average channel lifetimes of 3.1 msec and 1.8 msec, respectively. Reproduced with permission from *The Journal of Physiology*.[20]

nels relax to a closed state more rapidly,[11] or they may "block" open channels so that their open time is reduced.[1,2]

The advent of the patch clamp technique has allowed closer examination of the latter idea. In its simplest form, the "sequential blocking model" can be represented by

$$C \underset{\alpha}{\overset{\beta}{\rightleftharpoons}} O \underset{b}{\overset{fc}{\rightleftharpoons}} B$$

where C, O, and B are closed, open, and blocked states, respectively, of the channel, α, β, f, and b are first-order rate constants, and c is anesthetic concentration. In essence, this model proposes that, when channels open, they may flicker between an open and blocked state before closing.

The normal behavior of single channels in denervated rat muscle is illustrated in Fig. 6A. The record shows two downward currents that were generated during two openings of a channel. No transient closings can be seen during the open times. (Transient closings are occasionally seen but are very infrequent.) This behavior is in accord with the simplest model,

$$C \underset{\alpha}{\overset{\beta}{\rightleftharpoons}} O$$

where channels may be either open or closed. (More recent results have demonstrated that this model is too simple and that there must be more than one closed state,[45] but the simple model will suffice here.) After exposure to pentobarbitone (50 μM), the open state of a channel is interrupted by transient brief closed states (Fig. 6B). These open periods interrupted by transient closings are commonly called "bursts." The transient closed periods during a burst could conceivably be caused by the movement of blocking molecules into the channel. However, analysis of the characteristics of these events reveals that this is probably not the explanation.

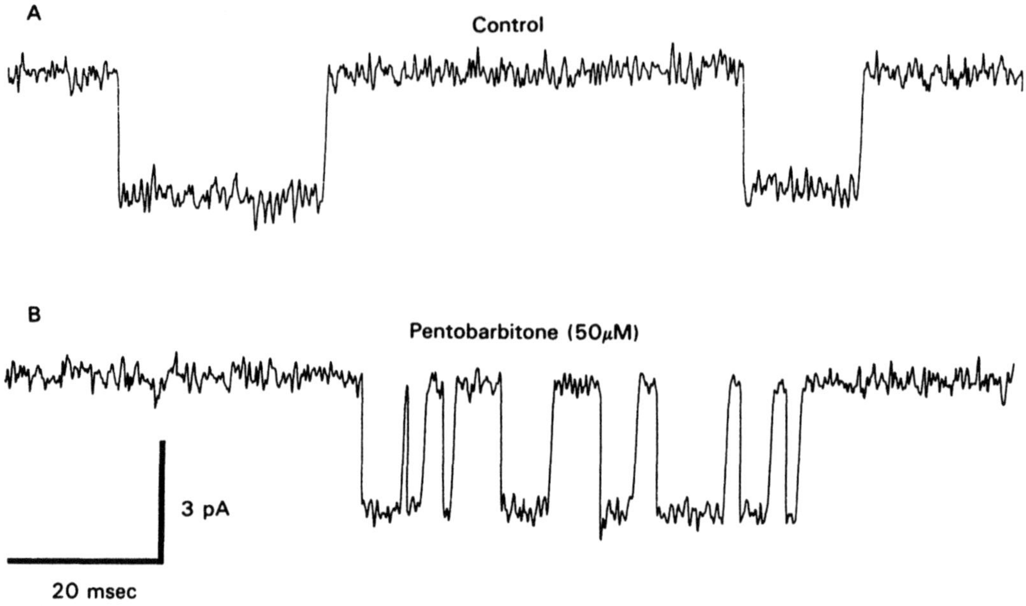

Figure 6. Single-channel currents recorded from a denervated rate extensor digitorum longus muscle before (A) and after (B) exposure to pentobarbitone (50 μM). Records were filtered with a Bessel, fourth-order filter with corner frequency set at 2 kHz. Membrane potential, −100 mV. Temperature, 16°C. The vertical calibration denotes 3 pA, the horizontal 20 msec.

The mean lifetime of any state is given by the inverse of the sum of the rate constants away from that state. Thus, the model above predicts that (1) the mean open time ($1/(fc+\alpha)$) should decrease with increasing drug concentration, and (2) the blocked time ($1/b$) should be independent of blocking drug concentration.

Further predictions of the model are that (3) the average number of "blocked" episodes per burst (fc/α) should increase linearly with drug concentration, and (4) the average burst length ($[1+fc/b]/\alpha$) should also increase linearly with drug concentration (see ref. 7 for derivation of equations).

The mean open time did decrease with pentobarbitone concentration (Fig. 7A). However, the mean transient closed time was not independent of pentobarbitone concentration. The average mean transient closed time is plotted against pentobarbitone concentration in Fig. 7B, and it is clear that the mean transient closed time increased with drug concentration. Other inconsistencies with the model were the dependence of the number of transient closed episodes per burst (Fig. 8A) and average burst length (Fig. 8B) on pentobarbitone concentration. These observations are also inconsistent with several other models of pentobarbitone action, e.g., cyclical models in which blocked channels can revert to closed channels without opening. A successful model must predict the increase in the number and average duration of the transient closed times with increasing drug concentration.

It would appear, therefore, that pentobarbitone does not simply block open endplate channels. Indeed, the idea that drugs such as procaine act solely by blocking open endplate channels according to the above scheme now seems unlikely also.[15,21,34] It is more probable that barbiturates increase the lifetime of an intermediate, nonconducting (closed) state of the receptor–channel complex that normally is not seen because it has a very brief lifetime. This could be the result of an interaction of the drug with a hydrophobic site on the protein or with

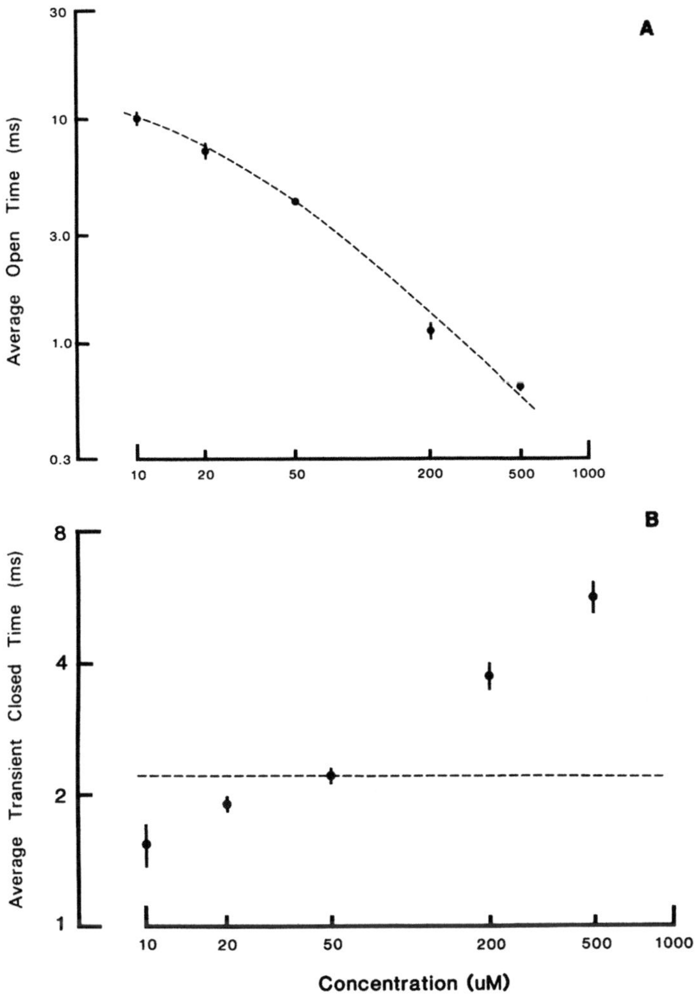

Figure 7. Plots of the averages of the mean open times (A) and mean transient closed times (B) against pentobarbitone concentration. The vertical bars denote ± 1 S.E.M. The broken lines show the predictions of the sequential blocking model (see text).

the lipid in the immediate vicinity of the protein. Alternative models in which the blocked state of a channel passes on to a "desensitized" state can also be made to fit the data.

Potentiation of Inhibitory Transmission

The second way in which anesthetics might depress transmission of action potentials is to potentiate inhibitory transmission. In fact, inhibitory potentials in pyramidal cells in the hippocampus are potentiated by barbiturates.[37] It has also been shown that barbiturates increase the open time of channels activated by GABA in cultured neurons[39,40] (for reviews, see ref. 32 and Owen *et al.*, this volume, Chapter 3). However, in view of the uncertainties introduced by use of cultured tissues in which newly synthesized proteins may be very different from normal, it was thought worthwhile to evaluate the effects of barbiturates on noncultured mammalian neurons.

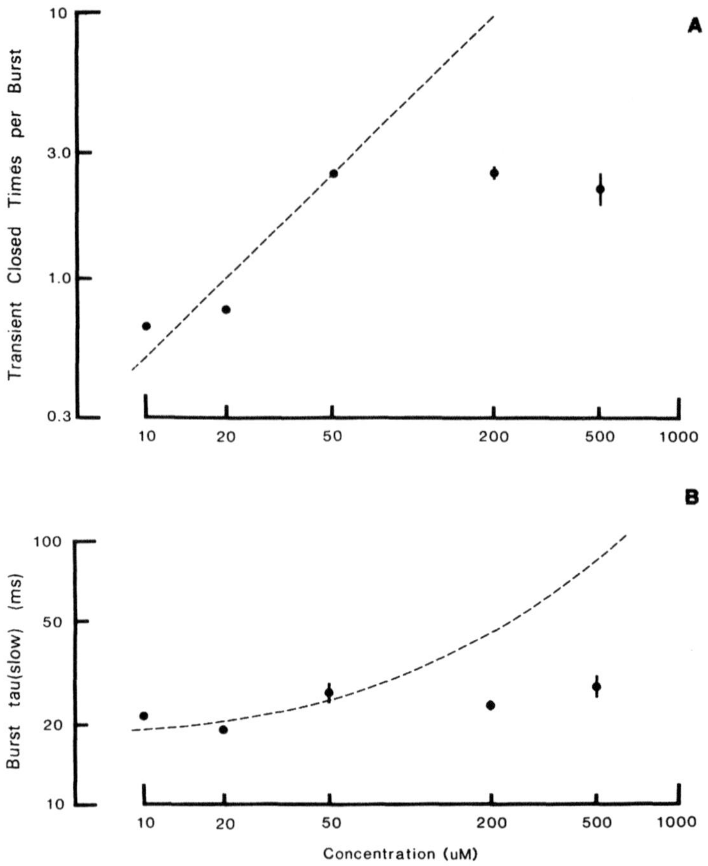

Figure 8. Plots of the averages of the mean number of transient closed times per burst (A) and mean burst length (B) against pentobarbitone concentration. The vertical bars denote ± 1 S.E.M. The broken lines show predictions of the sequential blocking model.

We have examined the effects of several anesthetics on inhibitory currents in CA1 neurons in rat hippocampal slices. A schematic diagram of a hippocampal slice is shown in Fig. 9 with an exploded view of a CA1 neuron and an inhibitory neuron with inputs to the somas of CA1 neurons. Spontaneous inhibitory postsynaptic potentials are known to occur in CA1 neurons,[4] and this offered an ideal opportunity to record the underlying currents free from contamination by excitatory inputs. Using a single electrode voltage clamp, we have recorded inhibitory currents that are blocked by bicuculline and are affected by the transmembrane chloride gradient.[6] These currents are presumably generated by GABA. They have an exponential decay like endplate currents, and the time constant of decay, τ_D, probably reflects the average open time of channels, as in cultured neurons (see ref. 32). However, it was noticeable that the average time constant of decay of currents in CA1 neurons in hippocampal slices (~11 msec) was significantly less than in cultured hippocampal cells (~20 msec).

Low, clinical[53] concentrations (50–100 μM) of pentobarbitone caused a dramatic slowing of the decay of these currents, as illustrated in Fig. 10, but their amplitude and growth time were relatively unaffected. The decay remained exponential, supporting the idea that the increase in τ_D is associated with an increase in channel open time.

Halothane also prolonged the decay phase of the inhibitory currents. The anesthetic was

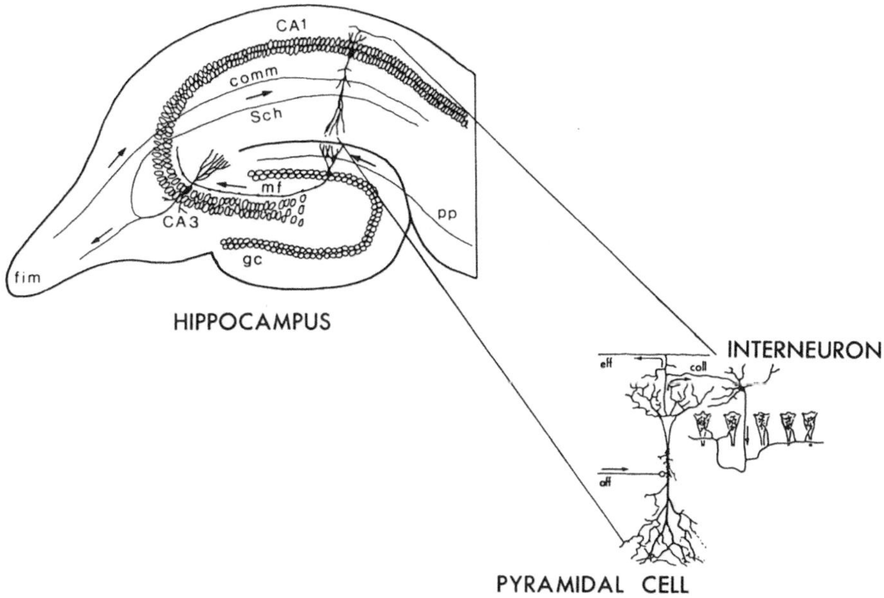

Figure 9. A diagrammatic representation of a cross section of the hippocampus showing the CA1 cell layer and synaptic connection between inhibitory interneurons and pyramidal cells.

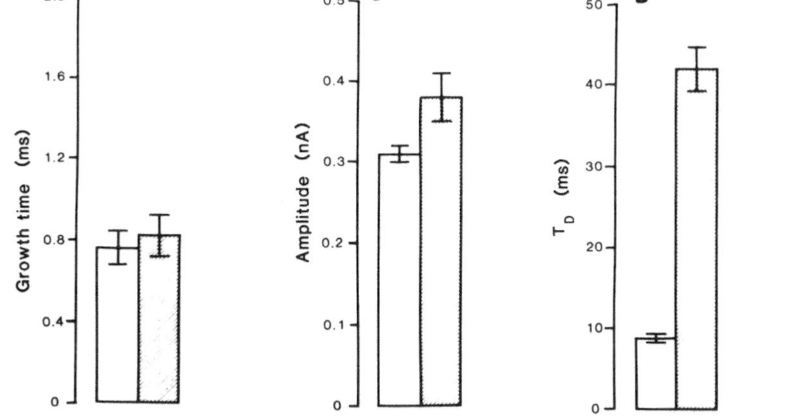

Figure 10. Prolongation of the decay of IPSC's in the presence of pentobarbitone. The IPSC on the left in A was recorded before exposure to pentobarbitone, the other two after exposure to 50 μM pentobarbitone. The average growth time (B), amplitude (C), and τ_D (D) recorded in another cell before (open bars) and after (hatched bars) exposure to the 100 μM pentobarbitone show that the drug causes a large increase in τ_D but no significant change in average growth time or amplitude. The vertical lines at the top of the bars show ± 1 S.E.M. Reproduced with permission from *The Journal of Physiology*.

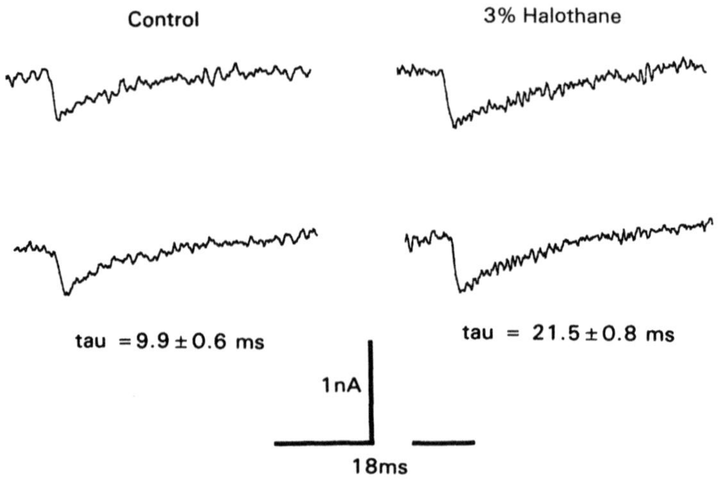

Figure 11. Lengthening of inhibitory postsynaptic currents produced by halothane in a voltage-clamped CA1 neuron. The two currents on the left were recorded before (CONTROL), and the two on the right after, exposure to 3% halothane. The average time constant of decay in this experiment increased from 9.9 ± 0.6 msec to 21.5 ± 0.8 msec (mean ± 1 S.E.M.). Clamp potential, −54 mV. Temperature, 27°C.

applied directly in the atmosphere around the slice, and the effect on the decay of currents was seen within several seconds. Currents were recorded more than 5 min after exposure to halothane when the effect appeared to be stable. Two currents recorded before and two after exposure to 3% halothane are shown in Fig. 11 (note the difference in time scale) and illustrate the slower decay of the currents produced in the presence of the anesthetic. In this neuron, 3% halothane increased the average decay time constant from 9.9 msec to 21.5 msec. The effect was rapidly reversible and repeatable.

Ketamine also slowed the decay of inhibitory currents. The effect of 0.5 mM ketamine is illustrated in Fig. 12. Two typical currents recorded in the presence of this anesthetic clearly decay more slowly than control currents. In this cell, the average decay time constant increased from 11.5 to 20.8 msec, a factor of 1.8. In two other cells exposed to the same

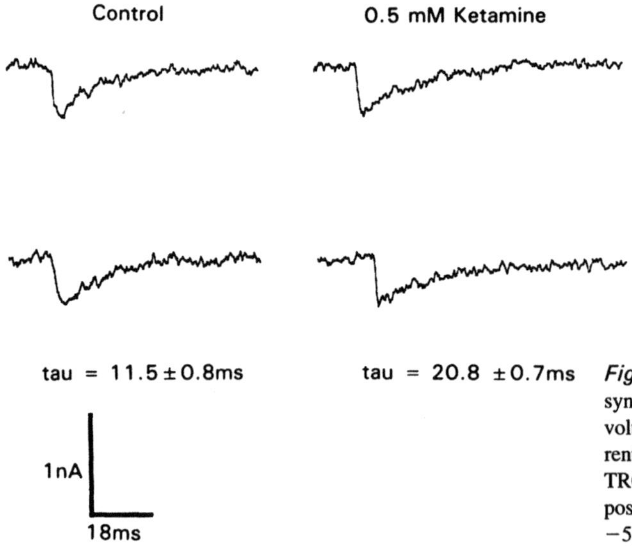

Figure 12. Lengthening of inhibitory postsynaptic currents produced by ketamine in a voltage-clamped CA1 neuron. The two currents on the left were recorded before (CONTROL), and the two on the right after, exposure to 0.5 mM ketamine. Clamp potential, −53 mV. Temperature, 25°C.

concentration of ketamine, the average decay time constant increased by factors of 1.6 and 1.7.

The slower decay of inhibitory currents produced by pentobarbitone, halothane, and ketamine probably explains the potentiation of inhibitory potentials that has been reported previously.[35,37,42,43] In contrast, neither nitrous oxide (75–80%) nor ethanol at concentrations as high as 200 mM had any perceptible effect on the decay of the inhibitory currents. The lack of effect of ethanol was particularly interesting as ethanol might have been expected to change the rate of decay of currents that have a voltage-dependent decay time constant (but see ref. 35). The results also suggest a way in which nitrous oxide and ethanol may produce anesthesia differently from halothane and ketamine: different ion channels are influenced in varying ways by different anesthetics.

DISCUSSION

The results demonstrate that general anesthetics, at clinical concentrations, can modify the open time of postsynaptic channels activated by neurotransmitters. The change in channel open time (or burst length) influences the decay of postsynaptic currents, and this is expressed in changes in the amplitude (and time course) of postsynaptic potentials.

It may be significant that some of the anesthetics tested have opposite effects at excitatory and inhibitory synapses, the end result in both cases being depression of the excitatory signal. There is also evidence that some general anesthetics may affect presynaptic inhibition.[8,9,36] Neither nitrous oxide nor ethanol affected the inhibitory postsynaptic currents. Such differences in modes and sites of action may underlie the well-recognized differences between the drugs. For example, pentobarbitone and other barbiturates produce anesthesia that is quite different from that produced by nitrous oxide. Studies at other synapses may provide clues to the different actions of the general anesthetics.

The effects produced by pentobarbitone at excitatory (and inhibitory) synapses cannot be explained by the simple blocking model. It is also difficult to imagine that pentobarbitone might increase membrane fluidity at an excitatory synapse but decrease fluidity at an inhibitory synapse. This does not exclude the possibility that an increase in fluidity might underlie the effect at an excitatory synapse but that another action is more important at inhibitory synapses. The slower decay of inhibitory currents could have been due to inhibition of uptake of GABA by pentobarbitone, but it does not seem likely: the voltage sensitivity of the current decay was not changed by the barbiturate, and the concentration was at a level that is thought not to affect GABA uptake in the hippocampus.[26]

Whatever the molecular mechanism of action of anesthetics turns out to be, the results obtained so far suggest that these drugs may have an important influence on excitatory and inhibitory synapses by modulating the time for which channels activated by neurotransmitters remain open.

REFERENCES

1. Adams PR: Drug blockade of open end plate channels. *J Physiol* 1976;260:531–552.
2. Adams PR: Voltage jump analysis of procaine action at frog end-plate. *J Physiol* 1977;268:291–318.
3. Adams DJ, Gage PW, Hamill OP: Ethanol reduces the amplitude of excitatory postsynaptic currents at a crustacean neuromuscular junction. *Nature* 1977;266:739–741.
4. Alger BE, Nicoll RA: Spontaneous inhibitory postsynaptic potentials in hippocampus: Mechanism for tonic inhibition. *Brain Res* 1980;200:195–200.
5. Anderson CR, Stevens CF: Voltage clamp analysis of acetylcholine produced end-plate current fluctuations at frog neuromuscular junction. *J Physiol* 1973;235:655–691.

6. Collingridge GL, Gage PW, Robertson B: Inhibitory post-synaptic currents in rat hippocampal CA1 neurones. *J Physiol* 1984;356:551–564.
7. Colquhoun D, Hawkes AG: On the stochastic properties of bursts of single ion channel openings and of clusters of bursts. *Phil Trans R Soc Lond B* 1982;300:1–59.
8. Eccles JC, Malcolm JL: Dorsal root potentials of the spinal cord. *J Neurophysiol* 1946;9:139–160.
9. Eccles JC, Schmidt RF, Willis WD: Pharmacological studies on presynaptic inhibition. *J Physiol* 1963; 168:500–530.
10. Gage PW: The effect of methyl, ethyl and n-propyl alcohol on neuromuscular transmission in the rat. *J Pharmacol Exp Ther* 1965;150:236–243.
11. Gage PW, Hamill O: General anaesthetics: Synaptic depression consistent with increased membrane fluidity. *Neurosci Lett* 1975;1:61–65.
12. Gage PW, Hamill O: Effects of several inhalation anaesthetics on the kinetics of postsynaptic conductance changes in mouse diaphragm. *Br J Pharmacol* 1976a;57:263–272.
13. Gage PW, Hamill O: Postsynaptic effects of general anaesthetics at mammalian neuromuscular junctions. *Proc Aust Physiol Pharmacol Soc* 1976b;7:6P.
14. Gage PW, McBurney RN: An analysis of the relationship between the current and potential generated by a quantum of acetylcholine in muscle fibres without transverse tubules. *J Membrane Biol* 1973;12:247–272.
15. Gage PW, Wachtel RE: Some effects of procaine at the toad end-plate are not consistent with a simple channel-blocking model. *J Physiol* 1984;346:331–339.
16. Winterstein H: *Die Narkose*. Berlin, Springer, 1919.
17. Gage PW, McBurney RN, Van Helden D: Endplate currents are shortened by octanol: Possible role of membrane lipid. *Life Sci* 1974b;14:2277–2283.
18. Gage PW, McBurney RN, Schneider GT: Effects of some aliphatic alcohols on the conductance change caused by a quantum of acetylcholine at the toad end-plate. *J Physiol* 1975;244:409–429.
19. Gage PW, McBurney RN, Van Helden D: Octanol reduces end-plate channel lifetime. *J Physiol* 1978;274:279–298.
20. Gage PW, Hamill OP, Van Helden D: Dual effects of ether on end-plate currents. *J Physiol* 1979;287:353–369.
21. Gage PW, Hamill OP, Wachtel RE: Sites of action of procaine at the motor end-plate. *J Physiol* 1983;335:112–137.
22. Galindo A: Effects of procaine, pentobarbital and halothane on synaptic transmission in the central nervous system. *J Pharmacol Exp Ther* 1969;169:185–195.
23. Glasstone S, Laidler KJ, Eyring H: *The Theory of Rate Processes*. New York, McGraw-Hill, 1941.
24. Hamill OP, Adams DJ, Gage PW: An effect of ethanol at a glutamate synapse. *Clin Exp Physiol* 1978;5:275–276.
25. Hamill OP, Marty A, Neher E. Sakman B, Sigworth FG: Improved patch-clamp techniques for high-resolution current recording from cells and cell-free membrane patches. *Pflugers Arch* 1981;391:85–100.
26. Jessell TM, Richards CD: Barbiturate potentiation of hippocampal i.p.s.p.s is not mediated by blockade of GABA uptake. *J Physiol* 1977;269:42–44P.
27. De Jong RH, Robles R, Heavner JE: Suppression of impulse transmission in the cat's dorsal horn by inhalation anesthetics. *Anesthesiology* 1970;32:440–445.
28. Larrabee MG, Posternak JM: Selective action of anesthetics on synapses and axons in mammalian sympathetic ganglia. *J Neurophysiol* 1952;15:91–114.
29. Lechleiter J, Gruener R: Halothane shortens acetylcholine receptor channel kinetics without affecting conductance. *Proc Natl Acad Sci USA* 1984;81:2929–2933.
30. Magleby KL, Pallotta BS: Burst kinetics of single calcium-activated potassium channels in cultured rat muscle. *J Physiol* 1983;344:605–623.
31. Magleby KL, Stevens CF: A quantitative description of end-plate currents. *J Physiol* 1972;223:173–197.
32. Mathers DA, Barker JL: Chemically induced ion channels in nerve cell membranes. *Int Rev Neurobiol* 1982;23:1–34.
33. McBurney RN, Gage PW, Van Helden D: The effect of aliphatic alcohols on end-plate conductances caused by acetylcholine: II. Heptanol, octanol and decanol. *Proc Aust Physiol Pharmacol Soc* 1974;5:92.
34. Neher E: The charge carried by single-channel currents of rat cultured muscle cells in the presence of local anaesthetics. *J Physiol* 1983;339:663–678.
35. Nicoll RA: The effects of anaesthetics on synaptic excitation and inhibition in the olfactory bulb. *J Physiol* 1972;223:803–14.
36. Nicoll RA: Presynaptic action of barbiturates in the frog spinal cord. *Proc Natl Acad Sci USA* 1975;72:1460–1463.
37. Nicoll RA, Eccles JC, Oshima T, Rubia F: Prolongation of hippocampal inhibitory postsynaptic potentials by barbiturates. *Nature* 1975;258:625–627.

38. Quastel DJM, Linder TM: Pre and postsynaptic actions of central depressants at the mammalian neuromuscular junction, in Fink BR (ed): *Molecular Mechanisms of Anesthesia, Progress in Anesthesiology*. New York, Raven Press, 1975, pp 157–165.
39. Ransom BR, Barker JL: Pentobarbital modulates transmitter effects on mouse spinal neurones grown in tissue culture. *Nature* 1975;254:703–705.
40. Ransom BR, Barker JL: Pentobarbital selectively enhances GABA mediated post-synaptic inhibition in tissue cultured mouse spinal neurones. *Brain Res* 1976;114:530–535.
41. Richards CD: The action of anaesthetics on synaptic transmission. *Gen Pharmacol* 1978;9:287–293.
42. Scholfield CN: A barbiturate induced intensification of the inhibitory potential in slices of guinea-pig olfactory cortex. *J Physiol* 1978;275:559–566.
43. Scholfield CN: Potentiation of inhibition by general anaesthetics in neurones of the olfactory cortex *in vitro*. *Pflugers Arch* 1980;383:249–255.
44. Sherrington CS: *The Integrative Action of the Nervous System*. New Haven, CT, Yale University Press, 1906.
45. Sine SM, Steinbach JH: Activation of a nicotinic acetylcholine receptor. *Biophys J* 1984;45:175–185.
46. Somjen GG: Effects of ether and thiopental on spinal presynaptic terminals. *J Pharmacol Exp Ther* 1963; 140:396–402.
47. Somjen GG, Gill M: The mechanism of the blockade of synaptic transmission in the mammalian spinal cord by diethyl ether and by thiopental. *J Pharmacol Exp Ther* 1963;140:19–30.
48. Torda TA, Gage PW: Effect of barbiturates on synaptic currents. *Anaesth Intensive Care* 1976;4:199–202.
49. Torda TA, Gage PW: Postsynaptic effect of I.V. anaesthetic agents at the neuromuscular junction. *Br J Anaesth* 1977;49:771–776.
50. Trudell JR, Hubbell WL, Cohen EN: The effect of two inhalation anesthetics on the order of spin-labeled phospholipid vesicles. *Biochim Biophys Acta* 1973;291:321–327.
51. Wachtel RE: Aliphatic alcohols increase the decay rate of glutamate-activated currents at the crayfish neuromuscular junction. *Br J Pharmacol* 1984;83:393–397.
52. Weakly JN: Effect of barbiturates on quantal synaptic transmission in spinal motoneurones. *J Physiol* 1969;204:63–77.
53. Willow M, Johnston GAR: Pharmacology of barbiturates: Electrophysiological and neurochemical studies. *Int Rev Neurobiol* 1983;24:15–49.

13

Mechanism of Nicotinic Channel Blockade by Anesthetics

James G. McLarnon, Peter Pennefather, and David M. J. Quastel

ABSTRACT. At the mouse neuromuscular junction the *n*-alkanols hexanol, heptanol, and octanol induce a biphasic decay of miniature end-plate currents (MEPCs), consistent with blockade of endplate channels by these agents, in the same way as seen with barbiturates and local anesthetics. Shorter-chain alkanols (propanol to pentanol) merely reduce MEPC amplitude and prolong decay. We now find that at reduced temperature pentanol also appears to be a channel blocker, and consideration of the modulation of potency with *n*-alkanol chain length leads to the conclusion that channel blockade can account for the action of short- as well as long-chain alkanols. Patch-clamp recording from BC3H1 cells in culture confirms that butanol induces channel transitions consistent with this hypothesis. Thermodynamic analysis, based on the temperature sensitivity of channel blockade by heptanol and octanol (and procaine and scopolamine), shows channel blockade to involve large positive values of entropy and high activation energies for the channel-blocking rate constant, suggesting that channel blockade is associated with conformational change in a segment of the channel protein, induced by a hydrophobic interaction between the agent and an intrachannel site. Some similarities in the action of volatile anesthetics to that of the alkanols suggest that channel blockade may also underly the receptor-channel blocking action of inhalation anesthetics.

INTRODUCTION

A diverse group of "nonspecific" agents including local and general anesthetics, barbiturates, and some *n*-alkanols act at the mammalian neuromuscular junction with kinetics that accord with blockade by the drug of the endplate channel that has been opened as a consequence of acetylcholine (ACh) combination with nicotinic receptor[2–4,6,12,13,18] with little if any interaction of the drug with the inactivated receptor, i.e., with the closed state of the receptor-channel complex.[13] Formally, the scheme is equivalent to "uncompetitive" blockade[14] rather than "noncompetitive" blockade, with ACh remaining bound to the receptor-ionophore in the blocked state. However, it remains unclear whether there exists only one or a diversity of mechanisms for "channel blockade," and whether some drugs might act via channel blockade although they do not induce the changes in the decay phase of miniature

James G. McLarnon and David M. J. Quastel • Department of Pharmacology and Therapeutics, Faculty of Medicine, The University of British Columbia, Vancouver, British Columbia V6T 1W5, Canada. *Peter Pennefather* • Department of Neurobiology and Behavior, State University of New York, Stony Brook, New York 11794.

endplate currents (MEPCs) characteristic of known "channel blockers." In particular, the *n*-alkanols, hexanol, heptanol, and octanol cause a "biphasic" MEPC decay typical of channel blockade, while shorter-chain alkanols (propanol to pentanol) cause reduction of MEPC amplitude with prolongation of a monophasic decay. In the present work we have determined the effect of changing temperature on the action of the *n*-alkanols from pentanol to octanol. With slowing of blocking and unblocking rate constants at low temperature, measurable with hexanol to octanol, it also becomes evident that pentanol, which at room temperature appears to act like propanol and butanol to prolong duration and diminish amplitude[13] of MEPCs, is in fact a channel blocker like hexanol, heptanol, and octanol. Consideration of the modulation with chain length of the potency of the various *n*-alkanols leads to the conclusion that propanol and butanol also may act via channel blockade. Thermodynamic analysis applied to the temperature dependence of blocking and unblocking rate constants for heptanol and octanol, and also those for channel blockade caused by the positively charged local anesthetics procaine and scopolamine, suggests that channel blockade involves an interaction of the "perturbing" substance with a hydrophobic portion of the receptor-channel. With appropriate rate constants, the channel blockade model can also account for the action of volatile anesthetics on the nicotinic receptor-ionophore.

METHODS

All experiments were carried out on mouse diaphragm muscle using a two-microelectrode point voltage clamp, a superperfusion system allowing rapid changes in solutions reaching the clamped region of the endplate, and a Peltier temperature control apparatus. MEPCs were recognized and stored on line using a PDP-12 computer, and subsequent calculations were made from averages of groups of at least 20 MEPCs.[8–11] In most experiments the preparation was exposed for 4 min to paraoxon (4 μM) to inhibit irreversibly acetylcholinesterase and thereby prolong the action of ACh.[9] In terms of the scheme shown below, this is essentially equivalent to a reduction of k_{-1}; this results in an increase of amplitude and prolongation of the slow phase, when there exists a biphasic MEPC decay, and facilitates estimation of the various rate constants.

The time course of decay of biphasic MEPCs was analyzed in terms of the cyclic channel blocking model given below, which predicts a time course of decay expressible as the sum of two exponentials, with rate constants depending on the values of k_{-1}, k_2, k_{-2}, and k_3. The rate constants for channel blocking $k_2[D]$, and for channel unblocking k_{-2}, can be obtained if it is assumed that k_{-1} is independent of drug concentration.[10] A reduction of k_{-1} would be manifest in an increase in the ratio of MEPC area to height. In this scheme any appreciable k_{-4} is reflected in a reduction of MEPC height, while k_3 is manifest in a reduction of the ratio of MEPC area to MEPC peak height, which occurs with most channel blockers.[13] It should be noted that it is implicit in this scheme that ACh does not dissociate from the receptor-channel when the latter is in the blocked state, except via the route designated by k_3, and that the latter route, as well as the k_1/k_{-1} route, may well involve intermediate steps in which ACh remains bound to the receptor.

$$
\begin{array}{ccccl}
2\mathrm{ACh} + R + D & \underset{k_{-1}}{\overset{k_1}{\rightleftharpoons}} & \mathrm{ACh}_2R^* + D & & \text{(channel open)} \\
k_4 \upharpoonleft\!\downharpoonright k_{-4} & & k_{-2} \upharpoonleft\!\downharpoonright k_2 & & \\
2\mathrm{ACh} + RD & \underset{k_3}{\overset{k_{-3}}{\rightleftharpoons}} & \mathrm{ACh}_2RD & & \text{(channel blocked)}
\end{array}
$$

The thermodynamic parameters for the transition state were found as follows.[11,20] The activation energy E_a was determined from the slope of Arrhenius plots (ln rate constant versus T^{-1}), and the enthalpy ($\Delta H^{\ddagger}$) and free energy ($\Delta G^{\ddagger}$) were found using $\Delta H^{\ddagger} = E_a - RT$ and $\Delta G^{\ddagger} = -RT \ln k_r + RT \ln(kT/h)$, where k_r is the rate constant, k is Boltzmann's constant, R is the gas constant, and h is Planck's constant. The entropy of activation ($\Delta S^{\ddagger}$) was then determined from the relation $T\Delta S^{\ddagger} = \Delta H^{\ddagger} - \Delta G^{\ddagger}$.

RESULTS

Channel Blockade by n-Alkanols

At 18°C the *n*-alkanols from octanol to pentanol all cause biphasic (split) MEPC decays (Fig. 1). As listed in Table I, from hexanol to octanol there occurs a progressive reduction of k_{-2} and an increase in k_2 as chain length is increased, with approximately two- to threefold changes in magnitude per methylene group. As previously reported,[13] the shorter-chain *n*-alkanols propanol, butanol, and pentanol (at 28°C) show only a prolonged time course of decay (see also ref. 7) with a diminished amplitude of the MEPC. However, it seems that the two obvious effects of alkanols on MEPC decay, splitting and prolonging, may be manifestations of a single action since pentanol causes a biphasic MEPC decay when the temperature is decreased (Fig. 1). It may readily be calculated that if k_{-2} were sufficiently high, the rapid phase of a split MEPC would escape detection, and the slow phase would appear as a single prolonged exponential decay, with reduced initial amplitude; throughout the MEPC, each activated channel would fluctuate rapidly between open and blocked states. In this case the MEPC amplitude would be halved at a concentration of drug where $k_2[D] = k_{-2}$. As shown in Table I and Fig. 2, the *n*-alkanols show a steady progression of values for concentrations causing half-height MEPCs (propanol to pentanol) or k_{-2}/k_2 (hexanol to octanol) at room temperature (26°–28°C). It is apparent that the data form a smooth sequence, with no break at any point, and are consistent with the shorter-chain *n*-alkanols acting in the same way as the longer-chain agents (i.e., to block endplate channels), with the onward and off-rate constants modulating regularly with chain length. The average decrease in k_{-2}/k_2 is about sixfold for each addition of a CH_2 group, with a rather larger change from propanol to butanol and less change from heptanol to octanol.

A high value for the unblocking rate constant k_{-2}, with an agent such as butanol or pentanol, would be expected to be manifest in a rapid, random flickering between open and blocked states of the ionophore, apparent as highly noisy low-amplitude current pulses. In order to test this prediction, the effects of 20 mM butanol on single channels in cultured mouse BC3H1 cells were studied using the patch clamp technique. The patch clamp record in Fig. 3 (from a patch containing 50–100 channels) clearly shows that the introduction of butanol was associated with a large increase in the apparent noisiness of channels with frequent transitions from the open state to a nonconducting level. It should also be noted in the patch clamp recording that butanol acts to increase the rate of channel opening by ACh. It has previously been reported that modification of MEPCs by ethanol appears to involve an increase in the rate constant for ACh–receptor combination and channel opening,[9] and an action of butanol to increase net conductance change produced by carbachol (ref. 16, and see Discussion) is also consistent with such an action.

Thermodynamics of Channel Blockade

In order to understand better what kind of molecular interactions between drug and channel might be involved in channel blockade, the temperature dependence of the rate

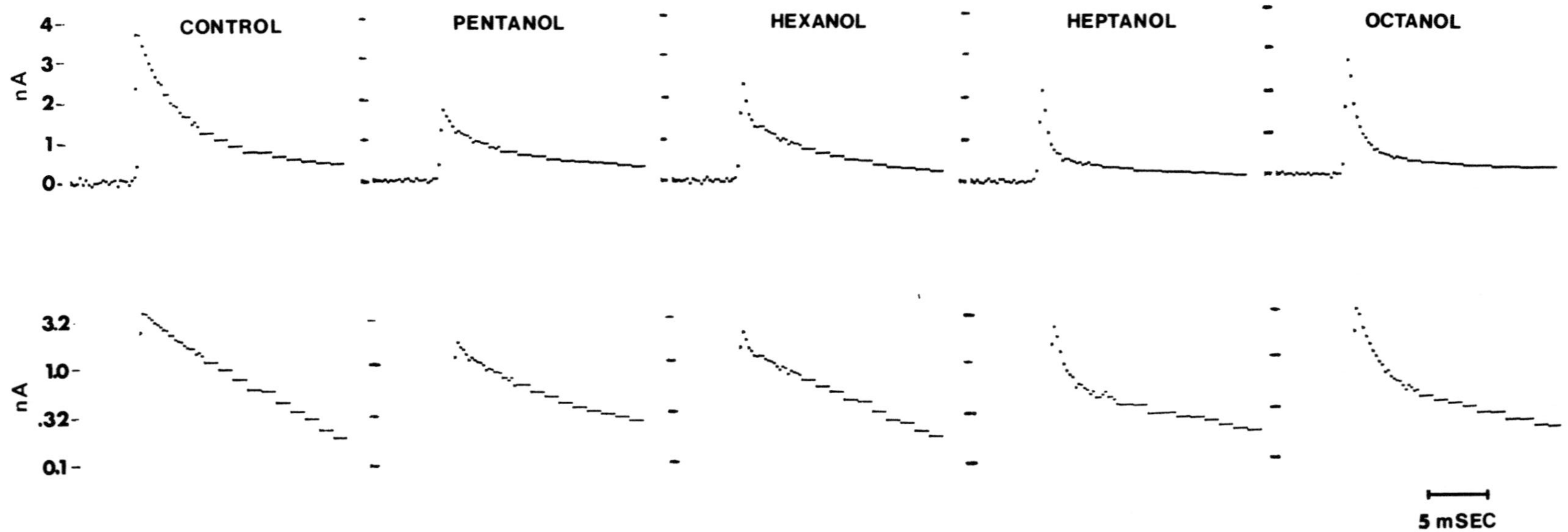

Figure 1. Modification of MEPCs by *n*-alkanols. The MEPCs are plotted linearly (above) and semilog (below) versus time. All records are from the same junction in preparation pretreated with paraoxon (4 μM for 5 min before washoff). Each plot is average of 25–40 individual MEPCs, recorded at a holding potential of −80 mV and at 18°C. The alkanol concentrations were 5 mM pentanol, 0.5 mM hexanol, 0.25 mM heptanol, and 0.1 mM octanol.

Table I. Values of k_{-2}/k_2 or Half-Height Concentration for n-Alkanols

n-Alkanol	k_2 ($10^6\ M^{-1}\ sec^{-1}$)	k_{-2} ($10^3\ sec^{-1}$)	$K = k_{-2}/k_2$ (10^{-3} M)	C = half-height concentration (10^{-3} M)	Log_{10} (K or C)
Octanol	17	0.55	0.03	—	−4.5
Butanol	9.2	1.0	0.11	—	−4.0
Hexanol	3.6	2.3	0.64	—	−3.2
Pentanol	—	—	—	4.8	−2.3
Butanol	—	—	—	35	−1.5
Propanol	—	—	—	373	−0.4

constants k_2 and k_{-2} was determined for heptanol and octanol and for the positively charged local anesthetic agents procaine and scopolamine (at concentrations high relative to those sufficient for muscarinic blockade), which, in contrast to neutral agents, show voltage dependence of both rate constants.[13] From Arrhenius plots (ln rate constant versus 1/T), the activation energies associated with the transition states were estimated, and the other thermodynamic parameters were then determined using the equations given in Methods. The results (Table II, Fig. 4) show that the activation energies associated with both the channel blocking rate constant, k_2, and the channel unblocking rate constant, k_{-2}, are in excess of 10 kcal/mol, higher than expected for simple diffusion. Channel blockade is also associated with large increases in entropy (both for the transition state associated with k_2 and for the equilibrium state) for all the agents; the increase caused by either of the two alcohols is significantly greater than that caused by either of the two positively charged local anesthetics. It appears that the magnitude of the rate constant, k_{-2}, correlates inversely with oil : water partition ratio and is related to the ability of the agent to form hydrophobic bonds, presumably with some site involved in the permeation pathway. An involvement of hydrophobic effects on the actions of various agents and organic ions to alter ionic conductance has previously been noted.[1,4,6,10] The large entropy changes are consistent with hydrophobic interactions leading to increased disorder in the system, involving alterations in hydrogen bonding and perhaps unfolding of a segment of the channel protein.

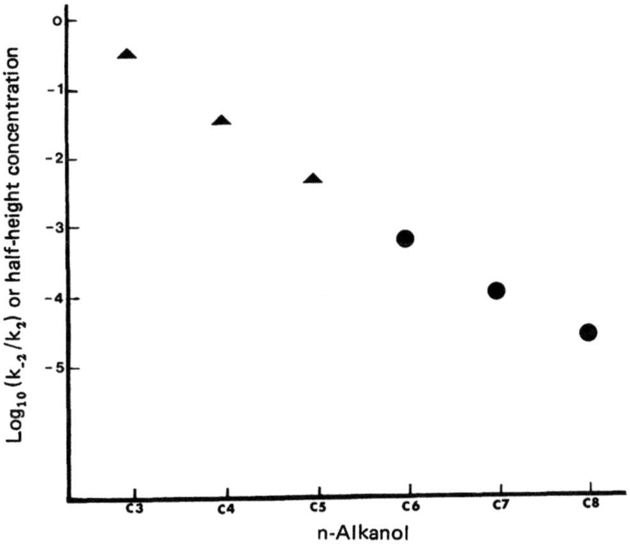

Figure 2. Graph, versus chain length, of log (half-height concentration) for the shorter-chain *n*-alkanols propanol, butanol, and pentanol (▲), and log (k_{-2}/k_2) values for the longer-chain alkanols hexanol, heptanol, and octanol (●).

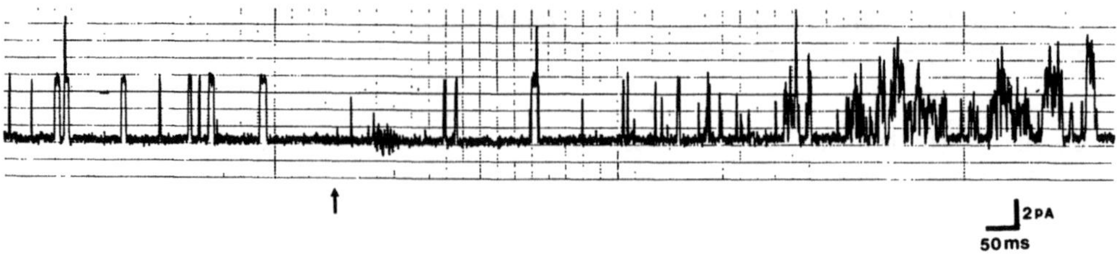

Figure 3. Patch clamp record obtained from cultured mouse BC3H1 cells in the presence of a 1 μM concentration of ACh. A concentration of 20 mM butanol was applied at the time indicated by the arrow.

DISCUSSION

Action of Alkanols on Endplate Channels

Blockade of open endplate channels[2,3] appears to occur with a wide diversity of agents. Although channel blockade is usually associated with biphasic MEPC decays (i.e., a decay phase that fits the sum of two exponentials), a combination of diminished MEPC height and prolonged decay, such as is observed with the intermediate-chain-length *n*-alkanols, is also consistent with a blocking model; the fast phase will escape detection if there is a large value for k_{-2}, the unblocking rate constant. In such a case the observed time course represents the slow phase of the MEPC decay. In support of the applicability of this model to the *n*-alkanols propanol to pentanol, the concentrations of these agents required to reduce MEPC height by half fit on a smooth curve with the concentrations of hexanol, heptanol, and octanol that block open channels by 50%, as determined from the magnitudes of the onward and off-rate constants. Moreover, with pentanol at reduced temperature, biphasic MEPCs are found, and with patch clamp recording, it appears that butanol does indeed cause rapid fluctuations between open and shut (possibly blocked) channel states. It therefore seems likely that all the *n*-alkanols from propanol to octanol act by the same mechanism on the nicotinic receptor–channel complex. However, it should be pointed out that methanol and ethanol act to prolong MEPCs (and channel duration) without any major effect on MEPC amplitude.[13] To fit the action of these agents into the same scheme it might be supposed that they induce a state of the receptor–channel that is similar to the blocked state produced by higher alkanols in terms of ACh being held onto the receptor,[9] but different in that there is no interference with permeation of ions. Alternatively, it may be that short-chain alkanols act mainly by reduction of k_{-1}, a channel-blocking action becoming important only at very high concentrations. Conversely, with long-chain alkanols an action to reduce k_{-1} could become unimportant at concentrations causing channel blockade.

Although the interpretation of thermodynamic parameters is necessarily tentative, our results are consistent with channel blockade by alkanols and local anesthetics involving, as a primary step, interaction between the agent and a hydrophobic region of the protein. In particular, the high values for entropy change associated with k_2, or with k_{-2}/k_2, support this hypothesis, as does the correlation of both k_2 and k_{-2} with *n*-alkanol chain length. The equilibrium values for the enthalpy ($\Delta H°$) of octanol and heptanol action, in Table II, are very similar to the equilibrium enthalpies of 7–8 kcal/mole obtained by Aveyard and Mitchell[5] for the transfer of butanol, pentanol, and hexanol from water to the alkanes octane, dodecane, and hexadecane. If, on the other hand, channel blockade by an alkanol were due to transfer of the molecule from the aqueous phase in the endplate channel to a more polar environment, with formation of ionic bonds, then equilibrium enthalpy values near 2 kcal/mole would be expected.[19] The values of $\Delta H°$ close to 2 kcal/mole for procaine and

Table II. Thermodynamic Parameters Associated with Channel Blocking and Unblocking as Determined from Transition State Analysis[a]

Agent	No. of endplates	k_2 (10^6 M^{-1} sec^{-1}) or k_{-2} ($msec^{-1}$)	E_a (kcal/mole)	ΔH^+ (kcal/mole)	ΔG^+ (kcal/mole)	$T\Delta S^+$ (kcal/mole)	ΔS^+ (entropy units)
Transition state: channel plugging (k_2)							
Procaine	6	29.7 ± 2.9	12.0 ± 0.9	11.4 ± 0.9	7.3 ± 0.2	4.1 ± 0.4	13.8 ± 1.5
Scopolamine	3	18.2 ± 2.0	11.6 ± 1.3	11.0 ± 1.2	7.6 ± 0.2	3.5 ± 0.5	11.7 ± 1.8
Octanol	5	17.4 ± 0.9	18.4 ± 0.7	17.8 ± 0.7	7.6 ± 0.2	10.2 ± 0.7	34.2 ± 2.7
Heptanol	5	9.2 ± 0.5	18.1 ± 0.7	17.5 ± 0.7	8.0 ± 0.2	9.5 ± 0.6	31.9 ± 2.2
Transition state: channel unplugging (k_{-2})							
Procaine	6	0.77 ± 0.05	10.0 ± 0.6	9.4 ± 1.0	13.5 ± 0.5	−4.1 ± 0.6	−13.8 ± 2.2
Scopolamine	3	2.36 ± 0.17	10.3 ± 0.9	9.7 ± 0.8	12.9 ± 0.5	−3.2 ± 0.4	−10.7 ± 1.4
Octanol	5	0.55 ± 0.05	10.4 ± 0.8	9.8 ± 0.8	13.7 ± 0.5	−3.9 ± 0.4	−13.1 ± 1.4
Heptanol	5	1.01 ± 0.07	10.6 ± 0.7	10.0 ± 0.7	13.4 ± 0.5	−3.4 ± 0.4	−11.4 ± 1.4
Associated state				$\Delta H°$	$\Delta G°$	$T\Delta S°$	$\Delta S°$
Procaine	6			2.0	−6.2	8.2	27.6
Scopolamine	3			1.3	−5.3	6.7	22.4
Octanol	5			8.0	−6.1	14.1	47.3
Heptanol	5			7.5	−5.4	12.9	43.3

[a]Associated state values were calculated from the differences between the transition state parameters for channel blocking and unblocking. Rate constants are values for $T = 25°C$ with $V = -80$ mV. Each tabulated value is mean ± S.E.M. of values obtained separately for each endplate studied. Reproduced with permission from McLarnon and Quastel,[11] copyright of Society for Neuroscience.

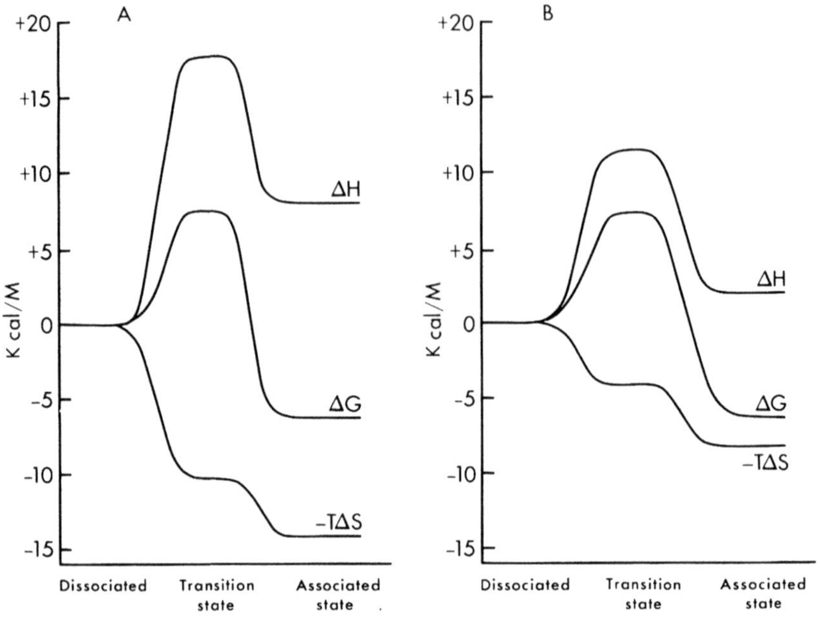

Figure 4. Energy profiles of enthalpy (ΔH), free energy (ΔG), and temperature × entropy ($-T\Delta S$), for octanol (A) and for procaine (B). To account for the temperature dependence of both on- and off-rate constants, it is assumed that the formation of the product state (blocked receptor–channel complex) requires an intermediate transition state with free energy change $\Delta G^{\ddagger}$. The height of the energy barrier associated with the transition state is a measure of the energy that must be acquired by the reactant molecules in order for the reaction to proceed. As shown here, the data indicate that the reaction occurs because formation of products results in an increase in entropy, rather than a decrease in enthalpy. Note that the diagram is highly schematic since the data do not exclude the existence of more than one step in the overall reaction. Reproduced with permission from McLarnon and Quastel,[11] copyright of Society for Neuroscience.

scopolamine (Table II) are consistent with the possibility that, in addition to a hydrophobic interaction, the local anesthetic is stabilized at its binding site by hydrogen bonding. The formation of polar bonds by the charged local anesthetics could also account for the reduced activation energies (E_a) associated with the blocking rate constant k_2, compared with the values determined for the *n*-alkanols (Table II). Another point of comparison with the work of Aveyard and Mitchell[5] is the latter's measurement of an average incremental change, with increase in chain length, of −0.79 kcal/mole per methylene group for the free energy change associated with the incorporation of the intermediate-chain-length *n*-alkanols from water into the alkanes. The values of k_{-2}/k_2 and concentrations that reduce MEPC height by half for the shorter-chain alkanols, in Table I, correspond to a change of $\Delta G°$ of −1.1 kcal/mole per methylene group for each step between butanol and heptanol, of −1.4 kcal/mole between propanol and butanol, and of −0.7 kcal/mole from heptanol to octanol. It is evident from the data in Table II that the decrease in free energy ($\Delta G°$) from heptanol to octanol is associated with a larger $\Delta S°$ for the longer-chain *n*-alkanol. As pointed out to us by N. Franks and W. Lieb (personal communication), the high enthalpies measured for the blocking rate constant, for both heptanol and octanol, are larger than would be expected if the energy barrier consisted solely of the stripping of water from the agents, and with all the agents, including the charged local anesthetics, the high entropy ($\Delta S^{\ddagger}$) for k_2, in association with the high activation energy, also suggests that channel blockade may require a conformational change in a portion of the channel protein. Such a change could account for the relative inability of

ACh to leave its site on the receptor protein (i.e., $k_3 < k_{-1}$) until the channel becomes unblocked.

Action of General Anesthetics on Endplate Channels

In terms of action on MEPC height and time course, a variety of general anesthetics and other central depressants fall into two categories: (1) those that cause an obvious split of the decay phase into two components, the action of which may be attributed to channel blockade (e.g., barbiturates, diphenylhydantoin, chlorpromazine), and (2) those that cause a reduction of MEPC height without any obvious split, or large prolongation, of the decay phase.[13,15] The latter group, which includes volatile anesthetics, is further characterized by a failure to diminish responses to bath-applied carbachol,[13,16] despite a profound depression of MEPC amplitude. This group can also be considered to include short-chain alkanols (propanol to pentanol) if we disregard the prolongation of MEPCs caused by these agents; in the presence of alkanols (ethanol to pentanol were tested), the response to carbachol was enhanced nearly in proportion to the prolongation of MEPCs.[16] At the same time there is with volatile anesthetics a reduction in the "noise" produced by carbachol or ACh, i.e., an apparent reduction of single-channel conductance.[13,17] Since with positively charged local anesthetics that clearly act in accord with a channel-blocking model, with fluctuation of channels between open and blocked states[12] noise analysis shows a similar reduction of apparent single channel conductance,[18] it is clear that the action of this second group of agents can be accommodated to the cyclical model for channel blockade, with the following provisos: (1) there occurs with the drug an effective increase in k_1 (which includes several steps), as with ethanol,[9] (2) values of k_{-2} are usually so high as to preclude resolution of the fast component of the decay phase, and (3) values of k_3 are sufficiently high that there may be little or no net prolongation of the MEPC (except with the alkanols). The route designated by k_3 in the model may tentatively be identified with the development of "desensitized" receptors in a state with a high affinity for ACh.[21]

ACKNOWLEDGMENTS. This work was supported by grants from the Muscular Dystrophy Association of Canada, and the British Columbia Health Care Foundation. We thank Dr. J. P. Dilger, Department of Medicine and Dentistry, University of New Jersey, for his assistance with the patch clamp experiments.

REFERENCES

1. Adams DJ, Nonner T, Dwyer T, Hille B: Block of endplate channels by permeant cations in frog skeletal muscle. *J Gen Physiol* 1981;78:593–615.
2. Adams PR: A model for the procaine end-plate current. *J Physiol (London)* 1975;246:61–63P.
3. Adams PR: Drug blockade of open end-plate channels. *J Physiol (London)* 1976;260:531–552.
4. Adler M, Albuquerque E, Lebeda F: Kinetic analysis of endplate currents altered by atropine and scopolamine. *Mol Pharmacol* 1978;14:514–529.
5. Aveyard R, Mitchell RW: Distribution of *n*-alkanols between water and n-alkanes. *Trans Faraday Soc* 1969;65:2645–2653.
6. Farley J, Yeh J, Watanabe S, Narahashi T: Endplate channel block by quinidine derivatives. *J Gen Physiol* 1981;77:273–293.
7. Gage PW, McBurney RN, Schneider GT: Effects of some aliphatic alcohols on the conductance change caused by a quantum of acetylcholine at the toad end-plate. *J Physiol (London)* 1975;244:409–429.
8. Linder TM, Quastel DMJ: A voltage clamp study of the permeability change induced by quanta of transmitter at the mouse endplate. *J Physiol (London)* 1978;281:535–556.

9. Linder TM, Pennefather P, Quastel DMJ: The time course of endplate currents and its modification by receptor blockade and ethanol. *J Gen Physiol* 1984;83:435–468.
10. McLarnon J, Quastel DMJ: Postsynaptic effects of magnesium and calcium at the mouse neuromuscular junction. *J Neurosci* 1983;3:1626–1633.
11. McLarnon J, Quastel DMJ: Thermodynamic parameters of endplate channel blockade. *J Neurosci* 1984;4:939–944.
12. Neher E, Steinbach JH: Local anaesthetics transiently block currents through single acetylcholine-receptor channels. *J Physiol (London)* 1978;277:153–176.
13. Pennefather P, Quastel DMJ: Actions of anesthetics on the function of nicotinic acetylcholine receptors, in Fink BR (ed): *Molecular Mechanisms of Anesthesia, Progress in Anesthesiology,* Vol. 2. New York, Raven Press, 1980, pp 45–57.
14. Pennefather P, Quastel DMJ: Modification of dose-response curves by effector blockade and uncompetitive antagonism. *Mol Pharmacol* 1982;22:369–380.
15. Quastel DMJ, Linder TM: Pre- and postsynaptic actions of central depressants at the mammalian neuromuscular junction, in Fink BR (ed): *Molecular Mechanisms of Anesthesia, Progress in Anesthesiology,* Vol. 1. New York, Raven Press, 1975, pp 157–168.
16. Quastel DMJ, Pennefather P: Receptor blockade and synaptic function. *J Neural Transmission* 1983;Suppl. 18:61–81.
17. Richter J, Landau EM, Cohen S: Effect of fluorinated ethers on channel conductance in a cholinergic synapse. *Isr J Med Sci* 1977;13:533–534.
18. Ruff RL: A quantitative analysis of local anesthetic alteration of miniature end-plate currents and end-plate current fluctuations. *J Physiol (London)* 1977;264:89–124.
19. Tanford C: *The Hydrophobic Effect,* ed. 2. New York, Wiley, 1980.
20. Weiland G, Molinoff P: Quantitative analysis of drug-receptor interactions: Determination of kinetic equilibrium properties. *Life Sci* 1981;29:313–330.
21. Young AP, Brown FF, Halsey MJ, Sigman DS: Volatile anesthetic facilitation of *in vitro* desensitization of membrane-bound acetylcholine receptor from *torpedo californica. Proc Natl Acad Sci USA* 1978;75:4563–4567.

14

Barbiturate Enhancement of γ-Aminobutyric Acid Receptor Binding and Function as a Mechanism of Anesthesia

Richard W. Olsen, James B. Fischer, and Thomas V. Dunwiddie

ABSTRACT. Anesthetic barbiturates and related depressants modulate radioactive ligand binding at γ-aminobutyric acid, benzodiazepine, and convulsant drug receptor sites in the γ-aminobutyric acid receptor–chloride ion channel complex of neuronal membranes with a specificity that correlates well with enhancement of GABA postsynaptic membrane actions in rat hippocampal slices. Barbiturates and the related depressant drugs, the pyrazolopyridines, etomidate, and LY81067, at 1–300 μM allosterically enhance the binding of γ-aminobutyric acid and benzodiazepine receptor agonists ([^{3}H]GABA or muscimol; [^{3}H]diazepam or flunitrazepam) and inhibit the binding of the GABA antagonist ([^{3}H]bicuculline methochloride) and benzodiazepine inverse agonist ([^{3}H]β-carboline-3-carboxylate methyl ester) to rat brain membranes. These effects are blocked by picrotoxin and related convulsant γ-aminobutyric acid receptor-ionophore antagonists, and all the active depressants and convulsants inhibit radioactive picrotoxin ligand binding ([^{3}H]α-dihydropicrotoxinin or [^{35}S]*t*-butyl bicyclophosphorothionate) at similar concentrations. The interactions are dependent on the presence of physiological levels of chloride or other anions that can permeate γ-aminobutyric acid-regulated chloride channels, suggesting that these *in vitro* effects involve receptor-ion channel sites. An identical chemical specificity [5-ethyl-5-(1,3-dimethyl-butyl) barbituric acid > pentobarbital > hexobarbital > mephobarbital > phenobarbital] and stereospecificity [(−)pentobarbital, mephobarbital, N^1-methyl-5-phenyl-5-propyl barbituric acid, and N^1-ethyl-5-methyl-5-piperidinobarbituric acid more active than the (+) isomers, and (+)hexobarbital and etomidate more active than the (−) isomers] were observed for all the allosteric binding interactions, and also in physiological assays of GABA function (enhancement of γ-aminobutyric acid-activated chloride flux in rat hippocampal slices, monitored with $^{36}Cl^-$, and GABAergic inhibitory synaptic transmission, monitored by electrophysiological techniques).

Abbreviations used in this chapter: GABA, γ-aminobutyric acid; BMC, bicuculline methochloride; βCCM, β-carboline-3-carboxylate methyl ester; TBPS, *t*-butyl-bicyclophosphorothionate; DHP, α-dihydropicrotoxinin; DMBB, 5-ethyl-5-(1,3-dimethylbutyl) barbituric acid; CHEB, 5-ethyl-5-cyclohexylidene-ethyl barbituric acid; EMPB, N^1-ethyl-5-methyl-5-piperidino barbituric acid; MPPB, N^1-methyl-5-phenyl-5-propyl barbituric acid.

Richard W. Olsen and James B. Fischer • Department of Pharmacology, School of Medicine, and Brain Research Institute, University of California, Los Angeles, California 90024. *Thomas V. Dunwiddie* • Department of Pharmacology, University of Colorado Health Sciences Center, Denver, Colorado 80262.

INTRODUCTION

Among the numerous effects that barbiturates have on the vertebrate central nervous system that could contribute to the anesthetic action of these drugs is an enhancement of inhibitory synaptic transmission mediated by the neurotransmitter γ-aminobutyric acid (GABA).[15,31] Barbiturates at hypnotic concentrations show a relatively potent and chemically specific prolongation and/or enhancement of postsynaptic responses to GABA,[6,8,10,14,27,29,30,36–38,40,52,56,58] apparently by increasing the lifetime of GABA-activated chloride channels (see Macdonald *et al.*, Owen *et al.*, and Gage *et al.*, this volume, Chapters 2, 3, and 12).

Studies in our laboratory have provided a biochemical correlate of this barbiturate enhancement of GABA function. Mutual allosteric interactions between three classes of receptor sites suggest that the GABA receptor sites, benzodiazepine receptor sites, and picrotoxin–barbiturate receptor sites are part of a three-receptor, chloride ion channel complex,[31] which can be solubilized intact and purified substantially with retention of all activities.[34,39,43] Barbiturates inhibit the binding of radioactive picrotoxin and cage convulsants to membrane sites related to GABA-regulated chloride channels[19,35,42,49] and allosterically enhance the binding of GABA and benzodiazepine receptor agonist ligands.[22–25,32,34,37,41,46,48,53,55,59] The chemical specificity of these *in vitro* effects of barbiturates and related depressants on the GABA receptor complex correlated well with the drug effects on neurons assayed by electrophysiological techniques. We have studied a few key compounds for ability to enhance GABA-mediated inhibitory synaptic transmission in the hippocampal slice. In addition, we have developed a brain slice assay of GABA receptor function monitored by radioactive chloride ion flux and found the same barbiturates and related depressants to be active in increasing basal and GABA-activated chloride permeability.[58]

METHODS

Receptor Binding Assays

GABA receptor ligand binding was assayed by the centrifugation technique, using thoroughly washed, frozen, and thawed membranes from rat cortex as previously described.[32,33] Briefly, the tissue was homogenized at 0–4°C in 0.32 M sucrose by 10 passes of a motor-driven Teflon pestle in a glass tube, centrifuged for 10 min at 1000 × *g* and the pellet discarded, then centrifuged for 40 min at 200,000 × *g* and the supernatant discarded. The pellet (crude mitochondrial, synaptosomal, and microsomal pellet) was resuspended in 20 volumes of distilled water by sonic disruption (5 sec at 40% of maximal speed with Ultra-Turrax, Tekmar Co., Cincinnati, OH), which produces a hypoosmotic shock, and the membranes were collected by centrifugation at 200,000 × *g* for 30 min. The pellet was resuspended in buffer and frozen at −20°C for 1–120 days. After thawing, the sample was centrifuged (30 min at 200,000 × *g*) and the pellet resuspended and centrifuged again, then resuspended in assay buffer, all at 0–4°C.

Binding of [^{3}H]GABA (4-amino-[2,3-^{3}H]*n*-butyric acid, 4 nM, 50 Ci/mmole, Amersham, Arlington Heights, IL) was measured with a 20-min incubation at 0–4°C in 10 mM potassium phosphate, pH 7.5, plus 0.1 M potassium chloride; binding of [^{3}H]BMC ([*N*-methyl-^{3}H]bicuculline methochloride, 2 nM, 72.7 Ci/mmole, New England Nuclear, Boston, MA) was assayed in a similar manner but with 0.1 M potassium thiocyanate substituted for chloride.[33,59] Nondisplaceable background was determined in the presence of 0.1 mM GABA. The samples were then centrifuged (13,000 × *g* for 20 min), the supernatants discarded, and the pellets rinsed twice superficially with 0.1 M KCl. The pellets were then dissolved in 0.2 ml of Soluene 350 (Packard Instrument Co., Downers Grove, IL), and 1 ml of scintillation fluid (Cytoscint; Westchem Products, San Diego, CA) was added and radioactivity measured in a Beckman model LS 3800 scintillation counter, with an efficiency of 38%. The background was about 25% with [^{3}H]GABA and 50% with [^{3}H]BMC.

Binding of benzodiazepine receptor ligands and cage convulsant were assayed by filtration.[19,25,59] Membranes were prepared as described above with omission of the freezing and thawing step. The assay buffer for all three ligands was 0.2 M sodium chloride, 10 mM sodium phosphate buffer, pH 7.0. [*N*-methyl-^{3}H]Diazepam (1 nM, 83.5 Ci/mmole), [^{3}H]βCCM (β-carboline-3-carboxylate methyl ester, 2 nM, 76 Ci/mmole), and [^{35}S]TBPS (*t*-butyl bicyclophosphorothionate, 10 nM, 105.7 Ci/mmole) were purchased from New England Nuclear. The benzodiazepine ligands were incubated at 0–4°C for 60 min and [^{35}S]TBPS for 90 min at 22°C, with backgrounds determined in the presence of 10 μM diazepam and 10 μM picrotoxin, respectively. Following equilibration, the samples were vacuum filtered on Whatman GF/B glass fiber filters and washed twice with 2.5 ml of 0.2 M NaCl. The filters were counted for radioactivity in 6 ml of Cytoscint (efficiency, 34%), with background amounting to 10–20%.

GABA-activated $^{36}Cl^-$ Flux in Brain Slices

The activation of $^{36}Cl^-$ efflux by GABA and barbiturates was assayed according to our published method.[58] Rat hippocampus was sliced (230 μm) and placed in sieve-bottom baskets and incubated at 37° in oxygenated saline. These were loaded with $^{36}Cl^-$ and the efflux rate measured by sequential transfers to nonradioactive buffer every minute. After the efflux rate stabilized (about 15 min), GABA and/or drugs were introduced and the relative efflux rate was determined (amount of $^{36}Cl^-$ released in a given time period divided by the tissue content of $^{36}Cl^-$ at that time).

Electrophysiological Measurement of Recurrent Inhibition in Hippocampal Slices

Hippocampal brain slices were maintained in a recording chamber under constant superfusion with control medium or medium containing various concentrations of barbiturate drugs.[28] The stimulation protocol involved either stimulation pulses delivered to the Schaffer/commissural input to the CA1 region or paired stimuli consisting of antidromic stimulation of the pyramidal neurons (to activate the recurrent GABAergic inhibitory pathway to the pyramidal neurons) immediately followed by stimulation of the Schaffer/commissural input to these cells. The stimulation intensity was set so as to elicit a 3-mV population spike response to stimulation of the synaptic pathway alone (control). At this point, the interval between the antidromic and synaptic stimuli was varied so as to determine the interpulse interval between the pulses required for the synaptically evoked population spike to recover to 50% of control values. This 50% interpulse interval (IPI_{50}) was then used as an index as to the *duration* of the recurrent inhibition elicited by antidromic stimulation. During and after perfusion with barbiturates, similar tests were run to determine the effects of drugs on this parameter. In the case of barbiturates that directly affected population spike amplitude (1,3-dimethylbutyl barbituric acid and occasionally pentobarbital increased the population spike, whereas high concentrations depressed it), the stimulation voltage was changed slightly to compensate for the direct barbiturate action so that the amplitude of the control population spike remained at 3 mV.

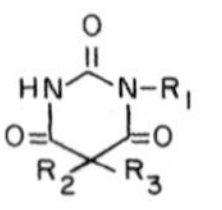

BARBITURATE	RING SUBSTITUTION		
	R_1	R_2	R_3
DMBB	H	$-CH_2CH_3$	$-CH(CH_3)CH_2CH(CH_3)_2$
SECOBARBITAL	H	$-CH_2-CH{=}CH_2$	$-CH(CH_3)CH_2CH_2CH_3$
PENTOBARBITAL	H	$-CH_2CH_3$	$-CH(CH_3)CH_2CH_2CH_3$
MEPHOBARBITAL	CH_3	$-CH_2CH_3$	
HEXOBARBITAL	CH_3	$-CH_3$	
AMOBARBITAL	H	$-CH_2CH_3$	$-CH_2CH_2CH(CH_3)_2$
MPPB	CH_3	$-CH_2CH_2CH_3$	
CHEB	H	$-CH_2CH_3$	$-CH_2-CH{=}$ 
PHENOBARBITAL	H	$-CH_2CH_3$	
METHARBITAL	CH_3	$-CH_2CH_3$	$-CH_2CH_3$
BARBITAL	H	$-CH_2CH_3$	$-CH_2CH_3$
EMPB	C_2H_5	$-CH_3$	

Figure 1. Structures of barbiturates described in the text.

RESULTS

Structure–Activity Relationships for Barbiturate Allosteric Interactions with GABA–Benzodiazepine Receptor Binding

Numerous barbiturates, including stereoisomers, whose chemical structures are given in Fig. 1, and selected nonbarbiturate central nervous system depressants were compared for potency and maximal effect in allosterically modulating GABA and benzodiazepine receptor binding *in vitro*. These hypnotic agents enhanced the binding of [^{3}H]diazepam to benzodiazepine receptors and [^{3}H]GABA to GABA receptors in rat brain membranes in a chloride-dependent manner which was sensitive to picrotoxin and related convulsants.[22,24,25,32] Likewise, these compounds allosterically inhibited in a picrotoxin-sensitive manner the binding of the GABA receptor antagonist [^{3}H]BMC and the benzodiazepine receptor inverse agonist [^{3}H]βCCM (57), as well as inhibiting the binding of the picrotoxinlike convulsant [^{35}S]TBPS.[19] These results are summarized in Table I for barbiturates and Table II for other compounds.

We developed the radioactive picrotoxin analogue [^{3}H]α-dihydropicrotoxinin (DHP) as a probe for the site of action of this convulsant GABA synaptic antagonist.[51] [^{3}H]DHP binding sites were inhibited by appropriate concentrations of biologically active picrotoxin analogues[51] and cage convulsants,[50] as well as by convulsant barbiturates at 0.1–10 μM and depressant barbiturates at 10–1000 μM.[49] The [^{3}H]DHP binding assay is technically poor, because a low ligand affinity produces a low signal : noise ratio of 10–20% displaceable binding in a centrifugation assay; the filter assay method was unsuitable because of the rapid ligand dissociation rate.[51] A new ligand for the picrotoxin receptor sites, *t*-butyl bicyclophosphorothionate ([^{35}S]TBPS), has been developed recently by Squires and associates.[42] We have used this ligand to determine potencies of inhibition by a series of barbiturates and related drugs (Tables I and II). These barbiturate-sensitive picrotoxin/convulsant binding sites have been shown to be linked to the GABA receptor–ionophore complex by virtue of allosteric interactions between the different drug binding sites.

Table I. Relative Potencies of Barbiturates to Allosterically Modulate Cage Convulsants, Benzodiazepine, and GABA Receptors

Compound	IC_{50} (μM) ([^{35}S]TBPS)	EC_{50} (μM)			EC_{20} (μM) ([^{3}H]βCCM)
		[^{3}H]GABA	[^{3}H]BMC	[^{3}H]Diazepam	
(−)DMBB[a]	N.D.[b]	N.D.	N.D.	50	N.D.
(±)DMBB	50	150	50	80	40
(±)Secobarbital	15	300	200	100	110
(−)Pentobarbital	90	200	N.D.	100	N.D.
(±)Pentobarbital	100	300	210	130	170
(+)Pentobarbital	130	400	N.D.	250	N.D.
(+)Hexobarbital	460	300	260	150	250
Amobarbital	N.D.	450	N.D.	350	N.D.
(−)MPPB	300	400	280	100	800
(−)Mephobarbital	140	400	310	50	220
CHEB	210	N.D.	300	500	N.D.
(−)EMPB	600	2000	N.D.	2000	N.D.
(+)EMPB	230	>5000	N.D.	>5000	N.D.
Phenobarbital	4000	>1000	550	(200)[c]	2000
(+)Mephobarbital	200	>1000	950	>1000	440
(−)Hexobarbital	840	>1000	>1000	>1000	>1000
(+)MPPB	2200	>1000	>1000	>1000	2000

[a]For structures, see Fig. 1. For binding assay methods, see text.
[b]N.D. = not determined.
[c]Phenobarbital at 1 mM does not enhance *equilibrium* binding but alters *kinetics* of [^{3}H]diazepam binding (EC_{50} = 200 μM) and reverses pentobarbital enhancement.

The pyrazolopyridine class of anxiolytic agents, such as cartazolate (SQ 65396), etazolate (SQ 20009), and tracazolate (ICI 136753), were observed to enhance, rather than inhibit, benzodiazepine receptor binding[25,45] and GABA receptor binding.[46] Etazolate has been shown to enhance GABA responses on neurons.[6,7] The chloride dependence and picrotoxin sensitivity of the enhancement by etazolate of benzodiazepine receptor binding[45] led to our demonstration that the hypnotic barbiturates likewise showed these interactions[24]; picrotoxin and related convulsants inhibit these interactions at concentrations that inhibit radioactive picrotoxin binding, and the depressants inhibit radioactive picrotoxin binding at concentra-

Table II. Relative Potencies of Nonbarbiturate Depressant Drugs to Allosterically Modulate Benzodiazepine and GABA Receptors[a]

Compound	IC_{50} (μM) ([^{35}S]TBPS)	EC_{50} (μM)			EC_{20} (μM) ([^{3}H]βCCM)
		[^{3}H]GABA	[^{3}H]BMC	[^{3}H]Diazepam	
Etazolate	2	2	3	1	8
Cartazolate	0.2[b]	1	2	0.5	N.D.
Tracazolate	2[b]	3	2	2	N.D.
(+)Etomidate	7	10	10	4	28
(−)Etomidate	155	>100	100	>100	>1000
LY 81067	3	1[c]	N.D.	1[c]	N.D.

[a]N.D. = not determined. Etazolate and cartazolate were the gifts of E. V. Squibb and Sons, Inc. Tracazolate was from Stuart Pharmaceuticals; etomidate from Janssen Pharmaceuticals; and LY 81607 from Eli Lilly Co.
[b]Ref. 42.
[c]Ref. 57.

tions similar to those active in enhancing GABA or benzodiazepine receptor binding. Another depressant, etomidate, was shown to enhance GABA[54] and benzodiazepine receptor binding in a stereospecific manner,[5] and recently another compound with yet a different structure, LY 81067, was found to enhance both GABA and benzodiazepine receptor binding activities and to inhibit the binding of [^{35}S]TBPS.[57]

The enhancement of benzodiazepine and GABA receptor binding by barbiturates and pyrazolopyridines, etomidate, and LY 81067 requires the presence of physiological concentrations of chloride ions or certain other anions.[24,32,45] The anions that will support these interactions turn out to be exactly those anions that can permeate the chloride channels in neurons that mediate inhibitory synaptic transmission,[3,13] chloride channels that are regulated by GABA receptors and sensitive to enhancement by barbiturates.[30]

The enhancement of GABA action by barbiturates has been proposed to correlate well with hypnotic[8,37] and/or anticonvulsant potency.[27,38,56] Likewise, the interactions of barbiturates with GABA receptor complex binding sites *in vitro* correlate reasonably well with hypnotic action[9,24]; e.g., secobarbital > pentobarbital > (+)hexobarbital > amobarbital > (−)hexobarbital > phenobarbital. Certain barbiturates are considered to be primarily excitatory agents, e.g., DMBB (1,3-dimethylbutyl barbituric acid), CHEB (cyclohexylidene-ethyl barbituric acid), and (+)pentobarbital.[2,11,12,18,21,37,48,49,55,56] These compounds inhibit [^{35}S]TBPS with varying potency, but all of them *enhance* GABA and benzodiazepine receptor binding like the depressants and do not inhibit the enhancement by depressants, as seen with convulsants like picrotoxin. However, DMBB, CHEB, and (+)pentobarbital all do *enhance* GABA actions, *in addition to* their neuronal excitatory activity.[2,6,11,16,18,37,56,58] The enhancement of GABA action by barbiturate stereoisomers that have an asymmetrical carbon in the C^5-alkyl group, e.g., DMBB, secobarbital, and pentobarbital, is characterized by activity for both isomers, with a modest two- to threefold greater activity for the (−) isomers.[11,18] The same specificity is seen for enhancement of benzodiazepine and GABA binding by isomers of DMBB and pentobarbital (Table I). On the other hand, the excitatory activity of these compounds is stereospecific with greater activity for the (+) isomers,[11,18] with (±)DMBB being excitatory overall and (±)pentobarbital and (±)secobarbital being depressant.

Looking at another series of stereoisomers, the N^1-methyl compounds, which are asymmetrical at C^5, one finds that stereospecificity also is observed. For example, (+)hexobarbital is more active in hypnosis than the (−) isomer, and in the case of mephobarbital, the (−) isomer appears to be more potent (although this estimate is complicated by the metabolism of both isomers *in vivo* to phenobarbital, a weak hypnotic.[21] Two other examples, N^1-ethyl-5-methyl-5-piperidino barbituric acid (EMPB) and N^1-methyl-5-phenyl-5-propyl barbituric acid (MPPB), show excitatory activity for one isomer and hypnotic activity for the other.[20,21] In this case, both isomers inhibit [^{35}S]TBPS binding but only the hypnotic (−) isomers of EMPB and MPPB allosterically modulate benzodiazepine and GABA receptor binding.

In summary, all the compounds, both barbiturate and related nonbarbiturate depressant drugs, that allosterically enhance benzodiazepine and GABA receptor agonist binding *in vitro* have been found to enhance GABA action at the cellular level[6,7,16,18,28,30,37,38,56] or in the hippocampal slice studied by electrophysiological methods or by radioactive tracer ion flux as described below. No exceptions to this correlation have been found, but some of the barbiturates that show a weak but definite activity on GABA action and as hypnotics, e.g., phenobarbital and (+)mephobarbital, are almost inactive in enhancing benzodiazepine binding affinity at equilibrium.[22] They do, however, alter the kinetics of benzodiazepine binding and, when present in excess, can inhibit the equilibrium enhancement seen by active agents such as pentobarbital.[22] Thus they appear to act at the same barbiturate binding site on the GABA receptor complex. The interactions of the "excitatory" barbiturates with GABA–benzodiazepine receptor binding appear to be related to their *depressant* GABA-enhancing activity, and we propose that their excitatory effects on neurons involve a separate non-GABA mechanism.

Structure-Activity of Barbiturates to Enhance GABA Function

We developed a test tube assay for GABA function with $^{36}Cl^-$ efflux from preloaded hippocampal slices,[58] based on the glutamate-activated $^{22}Na^+$ flux assay described by Luini and colleagues.[26] GABA receptor agonists show an antagonist-sensitive stimulation of $^{36}Cl^-$ efflux rate, and this is enhanced and mimicked by barbiturates (near millimolar levels).[58] Interestingly, etazolate and etomidate [the (+) isomer but not the (−) isomer] and numerous barbiturates (DMBB > secobarbital ≥ pentobarbital > (+)hexobarbital) were able to increase $^{36}Cl^-$ efflux from rat and mouse brain slices (hippocampus, cortex, and striatum were tested). The so-called excitatory compound DMBB, which acts like pentobarbital (only more potently) in enhancing GABA/benzodiazepine binding *in vitro,* also acts like pentobarbital, but more potently, in enhancing $^{36}Cl^-$ efflux. Furthermore, the appropriate stereospecificity was seen with (+)hexobarbital enhancing $^{36}Cl^-$ flux and (−)hexobarbital inactive (up to 1 mM). EMPB isomers gave very weak responses, with the (−) isomer more effective at 1 mM. Phenobarbital up to 3 mM was inactive in this assay. However, in the tissue slice, all drug effects were very weak, apparently owing to the low sensitivity of the method, diffusion barriers, drug uptake, and/or metabolism, etc.[58] This method requires for a detectable signal the responses of a great many cells, some of which are not in rapid equilibrium with the bathing medium. Therefore, the dose–response curves in this assay probably cannot be compared to other biological assays. Nevertheless, the rank order of potencies of active compounds and the lack of activity of other compounds are probably relevant. We have preliminary evidence[60] that mouse cortex slices give a more potent response to GABA agonists and barbiturates, and we are attempting to assay such responses in cultured mouse spinal neurons.

The rat *in vitro* hippocampal slice has also been employed for electrophysiological measurements of barbiturate action on GABA-mediated inhibition. The duration of presumed GABAergic recurrent inhibition was measured in hippocampal pyramidal neurons using a well-characterized protocol.[28,52] Barbiturates caused a dose-dependent prolongation of recurrent inhibition (Fig. 2). This was readily distinguished from effects on excitatory neurotransmission (release or postsynaptic potentials) or on nerve conduction (effects that were seen with higher concentrations of some of the compounds). Although DMBB caused an increase in the amplitude of the EPSP's, it was very effective, like pentobarbital, in shifting the duration of recurrent inhibition. Dose–response curves (Fig. 2) reveal a potency order of DMBB > pentobarbital > mephobarbital > phenobarbital. Phenobarbital was not active at concentrations under 0.5 mM. Etazolate and etomidate [(+) isomer better than (−) isomer] were more effective than pentobarbital, and the (−) isomers of mephobarbital and MPPB were more active than the (+) isomers. In these assays, the effective concentrations of barbiturates (e.g., 10–100 μM DMBB and pentobarbital) agree with those effective in the test tube binding assays. In both the $^{36}Cl^-$ flux measurements and microelectrode studies, an identical drug specificity is observed for barbiturates and related substances as that seen for the allosteric interactions with the GABA/benzodiazepine/cage convulsant receptor binding

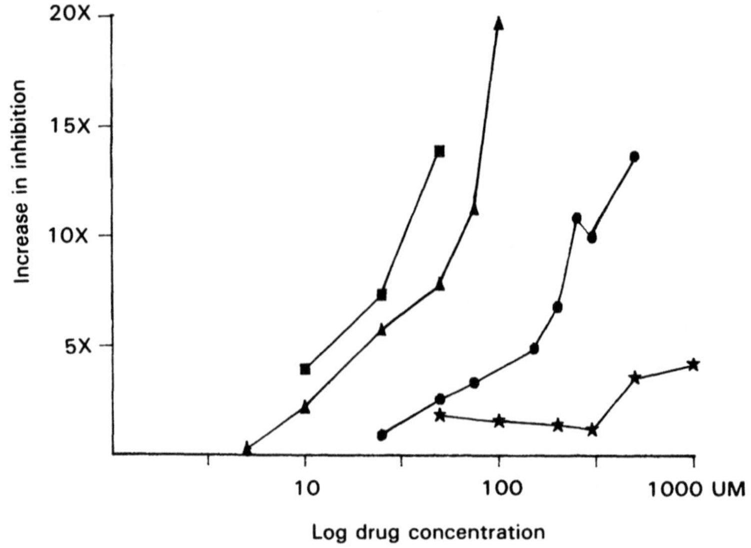

Figure 2. Effect of barbiturates on recurrent inhibition in the rat hippocampal slice. The IPI_{50} was determined for the recurrent inhibitory pathway as described in Methods; slices were then perfused with the indicated concentration of drug, and new IPI_{50} intervals were determined. The average fold increase in the IPI_{50} is shown for different concentrations of drug. Each point represents the mean increase for between two and eight slices recorded from in the presence of DMBB (■), pentobarbital (▲), mephobarbital (●), and phenobarbital (★).

assays. This demonstrates that the barbiturate binding sites involved in the GABA receptor complex assays are related to the action of barbiturates to enhance GABA receptor–chloride ion channel function.

DISCUSSION

The chemical specificity, including stereospecificity, chloride ion dependence, and sensitivity to antagonists such as picrotoxin, cage convulsants, and bicuculline of barbiturate interactions with GABA/benzodiazepine receptor binding sites *in vitro* suggest a relationship to the physiological enhancement of GABA function by barbiturates.[31] These *in vivo* and *in vitro* interactions of depressant and excitatory drugs with the GABA receptor–chloride ion channel have led us to propose a "donut" model of the GABA receptor complex[31] (Fig. 3). The biochemical isolation of this complex with all of its activities intact confirms its existence.[34,39] The question here is whether enhancement of GABA binding/function plays a role in the anesthetic activity of barbiturates or other compounds. It is well established that barbiturates and a few other hypnotic agents can enhance GABA-mediated inhibition, and this correlates with an enhancement of GABA/benzodiazepine receptor binding, as described previously. Some general anesthetics also enhance GABAergic inhibition (Gage *et al.*, this volume, Chapter 12) and receptor binding (S. C. Cheng, personal communication).

Because the barbiturates exert their anesthetic action at relatively high concentrations (over 10 μM for pentobarbital and possibly over 100 μM for some analogues (see Macdonald *et al.*, this volume, Chapter 2), it is difficult to pinpoint their mechanism by using potency data for molecular and cellular assays, because several mechanisms are suggested by relatively similar potencies for modulation by barbiturates. Deciding which of the many possible (and all relatively weak) actions of barbiturates is related to anesthesia requires quantitative comparisons of dose–response curves for a series of compounds including

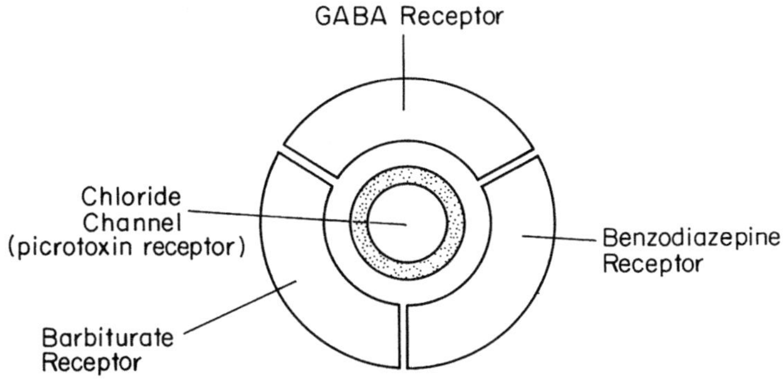

Figure 3. Hypothetical model of the GABA receptor–chloride ion channel complex. Separate binding sites on the same protein complex are present for the neurotransmitter (GABA receptor) for the depressant drugs (benzodiazepine receptor and barbiturate receptor), and for the picrotoxin-type convulsant drugs (chloride channel).

steroisomers, as well as a consideration of the anatomical structures that may be particularly important and/or sensitive in the function that is altered by the drug.

Thus barbiturates are known not only to enhance and mimic GABAergic inhibition via opening or prolonging the open time of chloride channels,[44] but also to inhibit several excitatory ion channels in neurons. The nicotinic acetylcholine receptor–ionophore at the neuromuscular junction is blocked (noncompetively with respect to acetylcholine) by barbiturates (e.g., ref. 1). Central glutamate receptor responses (Na^+ channels) are blocked by barbiturates.[36,47] Several voltage-regulated ion channels are blocked by barbiturates including the Ca^{2+} channels involved in neurotransmitter release[17,37,56] (and possibly in excitation–contraction coupling in muscle and/or in other hormonal responses). Barbiturates inhibit potassium conductances, including the calcium-dependent potassium channels and those that may be regulated by certain neurotransmitter/hormone receptors[1,6]; barbiturates also inhibit voltage-regulated sodium channels of the classical tetrodotoxin-sensitive type[6] (Roth *et al.*, this volume, Chapter 4). Other ion channels may also be susceptible, and this suggests that barbiturates have a relatively specific effect on ion channels; on the other hand, the fact that barbiturates affect *numerous* ion channels suggest either that barbiturate action might be mediated by an interaction with the lipid bilayer (to which all ion channels share a sensitivity owing to analogous function) or that all the ion channel proteins share some structural property that makes them sensitive to these hydrophobic agents.

However, when one applies rigorous criteria of chemical specificity, certain potential mechanisms for barbiturate anesthesia probably can be ruled out. Although barbiturate action correlates well with lipid solubility, the stereospecificity of anesthetic action suggests that a membrane–lipid interaction is not sufficient and that protein targets are involved. The relationship of anesthetic potency to physical properties including pK_A appears primarily relevant to considerations of pharmacokinetics,[12,37,56] rather than to the actual barbiturate "receptor site." Comparisons of chemical specifities for barbiturate inhibition of central excitatory postsynaptic potentials mediated by amino acids like L-glutamic acid do *not* correlate with anesthetic activity; for example, phenobarbital ~ pentobarbital ~ (+)secobarbital = (−)secobarbital[47] and (+)hexobarbital ~ (−)hexobarbital (V. Teichberg, personal communication). However, glutamate receptor heterogeneity leaves open the possibility of a subtype that does show the correct specificity.

The blockade of neuromuscular and electric organ transmission by barbiturates seems unlikely to mediate anesthesia on physiological grounds, and the chemical specificity does not correlate.[1,37,56] Nicotinic acetylcholine receptor–ionophores in the central nervous system presumably show a similar specificity (but compare acetycholine receptors in BC3H1 cells and *Torpedo* electroplaques, Taylor *et al.*, this volume, Chapter 9). Blockade of repetitive neuronal firing by barbiturates, enhancement of spike frequency adaptation, and other interactions of barbiturates with voltage-regulated sodium and potassium channels appear to show improper specificity for anesthesia (e.g., phenobarbital = pentobarbital)[61] (Roth *et al.*, this volume, Chapter 4; P. Carlen, personal communication). This leaves the voltage-regulated calcium channels and the GABA-regulated chloride channels as the most appropriate potential mechanisms on chemical specificity grounds. Of the two ion channels that appear to show the appropriate barbiturate specificity for anesthetic activity, the calcium channel site requires higher concentrations than those involved in enhancement of the GABA-activated chloride channel, and the sedative–hypnotic potencies of the barbiturates correlate best with enhancement of GABA[16,58] (Macdonald *et al.*, and Owen *et al.*, this volume, Chapters 2 and 3).

Barbiturates have been reported to show two qualitatively different effects on the GABA system, both in binding studies, as described previously, and in electrophysiological studies. Barbiturates both enhance and mimic GABA responses on neurons[27,38] and reverse the action of GABA antagonists.[8,40] The enhancement of GABA action by barbiturates is more potent than the direct mimicking action[6,27] (Owen *et al.*, this volume, Chapter 3) or the antagonist reversal,[16] and, as in hypnotic action, pentobarbital is distinctly more potent than phenobarbital[6,16] (Fig. 2). A similar specificity is seen in the binding assays.[22,53] Furthermore, when phenobarbital finally has an effect on GABA/benzodiazepine receptor binding, it is somewhat different from that of pentobarbital, modulating kinetics but not equilibrium binding.[22,53] The greater potency of pentobarbital than phenobarbital and the lower concentrations required suggest that enhancement of GABA binding and functional responses show the best correlation with hypnotic activity. The same argument suggests that the GABA-enhancing mechanism cannot be equally important in the anticonvulsant action of barbiturates (but see Macdonald *et al.*, this volume, Chapter 2).

On the other hand, benzodiazepine receptors are clearly linked to GABA receptors,[31] and these drugs enhance GABA function, but do not mimic GABA.[6,15,40,44] The benzodiazepines have anticonvulsant, anxiolytic, and sedative activities like the barbiturates, but they are not anesthetic. This difference from barbiturates might be explained by a brain regional variation in drug receptor–GABA receptor coupling, or in heterogeneity of GABA receptors.[23,37,44,56] It remains possible that the anesthetic action of the barbiturates requires an additional mechanism besides the sedative–hypnotic enhancement of GABA. Obviously, one cannot rule out a multiple mechanism theory of anesthesia, and it is possible that one of the new ion conductances being discovered almost daily in the neuronal membrane may mediate anesthesia. Nevertheless, given that sedation appears to involve enhancement of GABA action, a higher *degree* of enhancement of GABAergic inhibition in crucial brain regions appears to be the most likely mechanism for barbiturate anesthesia.

ACKNOWLEDGMENTS. This work was supported by NSF Grant BNS 80-19722 and NIH Grants RO1-NS 20704 and AM 07310. We thank Drs. L. M. F. Leeb-Lundberg, V. I. Teichberg, E. H. F. Wong, J. Yang-Ransom, G. Stauber, and R. King for helpful discussions, and Dr. J. Knabe for the gift of barbiturate isomers.

REFERENCES

1. Adams PR, Banks FW: Actions of anesthetics and anticonvulsants on synaptic channels, in Fink BR (ed): *Molecular Mechanisms of Anesthesia*. New York, Raven Press, 1980, pp 95–109.

2. Andrews PR, Jones GP, Lodge D: Convulsant, anticonvulsant, and anaesthetic barbiturates. 5-Ethyl-5-(3′-methyl-but-2′-enyl)-barbituric acid and related compounds. *Eur J Pharmacol* 1979;55:115–120.
3. Araki T, Ito, M, Oscarsson O: Anion permeability of the synaptic and nonsynaptic motoneurone membrane. *J Physiol* 1961;159:410–435.
4. Asano T, Ogasawara N: Chloride-dependent stimulation of GABA and benzodiazepine receptor binding by pentobarbital. *Brain Res* 1981;225:212–216.
5. Ashton D, Geerts R, Waterkeyn C, Leysen JE: Etomidate stereospecifically stimulates forbrain, but not cerebellar, [^{3}H]diazepam binding. *Life Sci* 1981;29:2631–2636.
6. Barker JC, Gratz E, Owen DG, Study RE: Pharmacological effects of clinically important drugs on the excitability of cultured mouse spinal neurons, in Bowery NG (ed): *Actions and Interactions of GABA and Benzodiazepines*. 1984, Raven Press, New York, pp 203–216.
7. Barnes DM, White WF, Dichter NA: Etazolate (SQ20009): Electrophysiology and effects on [^{3}H]flunitrazepam binding in cultured cortical neurons. *J Neurosci* 1983;3:762–772.
8. Bowery NG, Dray A: Reversal of the action of amino acid antagonists by barbiturates and other hypnotic drugs. *Br J Pharmacol* 1978;63:179–215.
9. Butler TC: The delay in onset of action of intravenously injected anaesthetics. *J Pharmacol* 1942;74:118–128.
10. Curtis DR, Lodge D: Pentobarbitone enhancement of the inhibitory action of GABA. *Nature* 1977;270:543–544.
11. Downes H, Perry RS, Ostlund RE, Karler R: A study of the excitatory effects of barbiturates. *J Pharmacol Exp Ther* 1970;175:672–679.
12. Dundee JW: Molecular structure-activity relationships of barbiturates, in Halsey MJ, Millar RA, Sutton JA (eds): *Molecular Mechanisms in General Anesthesia* London, Churchill-Livingston, 1974, pp 16–31.
13. Eccles J, Nicoll RA, Oshima, T, Rubia FJ: The anionic permeability of the inhibitory postsynaptic membrane of hippocampal pyramidal cells. *Proc R Soc Lond B* 1977;198:345–361.
14. Evans RH: Potentiation of the effects of GABA by pentobarbitone. *Brain Res* 1979;171:113–120.
15. Haefely W, Polc P, Keller HH, Pieri L, Mohler H: Facilitation of GABA-ergic transmission by drugs, in Krogsgaard-Larsen P, Scheel-Kruger J, Kofod H (eds): *GABA-Neurotransmitters*. Copenhagen, Munksgaard, 1979, pp 357–373.
16. Harrison NL, Simmonds MA: Two distinct interactions of barbiturates and chlormethiazole with the $GABA_A$ receptor complex in rat cuneate nucleus in vitro. *Br J Pharmacol* 1983;80:387–394.
17. Haycock JW, Levy WB, Cotman CW: Pentobarbital depression of stimulus-secretion coupling in brain: Selective inhibition of depolarization-induced calcium-dependent release. *Biochem Pharmacol* 1977;26:159–161.
18. Huang LY, Barker JL: Pentobarbital: Stereospecific actions of the (+) and (−) isomers revealed on cultured mammalian neurons. *Science* 1980;207:195–197.
19. King RG, Olsen RW: Solubilization of convulsant/barbiturate binding activity on the GABA/benzodiazepine receptor complex. *Biochem Biophys Res Commun* 1984;119:530–536.
20. Knabe J, Buch HP, Reinhardt J: Barbitursaurederivate, 32. Zur zentralnervosen Wirkung racem. und optisch aktiver basisch substituierter Barbitursauren. *Arch Pharm (Weinheim)* 1982;315:832–839.
21. Knabe VJ, Rummel W, Buch HP, Franz N: Optisch aktive Barbiturate. Synthese, Konfiguration und pharmakologische wirkung. *Arzneimittel-Forsch* 1978;28:1048–1056.
22. Leeb-Lundberg F, Olsen RW: Interactions of barbiturates of various pharmacological categories with benzodiazepine receptors. *Mol Pharmacol* 1982;21:320–328.
23. Leeb-Lundberg F, Olsen RW: Heterogeneity of benzodiazepine receptor interactions with GABA and barbiturate receptor sites. *Mol Pharmacol* 1983;23:315–325.
24. Leeb-Lundberg F, Snowman A, Olsen RW: Barbiturate receptors are coupled to benzodiazepine receptors. *Proc Natl Acad Sci USA* 1980;77:7468–7472.
25. Leeb-Lundberg F, Snowman A, Olsen RW: Perturbation of benzodiazepine receptor binding by pyrazolopyridines involves picrotoxinin/barbiturate receptor sites. *J Neurosci* 1981;1:471–477.
26. Luini A, Goldberg O, Teichberg VI: Distinct pharmacological properties of excitatory amino acid receptors in the rat striatum: study by Na^+ efflux assay. *Proc Natl Acad Sci USA* 1981;78:3250–3254.
27. Macdonald RL, Barker JL: Different actions of anticonvulsant and anesthetic barbiturates revealed by use of cultured mammalian spinal cord neurons: a common mode of anticonvulsant action. *Brain Res* 1978;167:323–326.
28. Mueller AL, Hoffer BJ, Dunwiddie TV: Noradrenergic responses in rat hippocampus: Evidence for mediation by alpha and beta receptors in the in vitro slice. *Brain Res* 1981;214:113–126.
29. Nicoll RA, Wojtowicz JM: The effects of pentobarbital and related compounds on frog motoneurons. *Brain Res* 1980;191:225–237.

30. Nicoll RA, Eccles JC, Oshima TC, Rubia F: Prolongation of hippocampal inhibitory post-synaptic potentials by barbiturates. *Nature* 1975;258:625–627.
31. Olsen RW: GABA-Benzodiazepine-Barbiturate receptor interactions. *J Neurochem* 1981;37:1–13.
32. Olsen RW, Snowman AM: Chloride-dependent barbiturate enhancement of GABA receptor binding. *J Neurosci* 1982;2:1812–1823.
33. Olsen RW, Snowman AM: [^{3}H]Bicuculline-methochloride binding to low affinity GABA receptor sites. *J Neurochem* 1983;41:1653–1663.
34. Olsen RW, Wong EHF, Stauber GB, Murakami D, King R, Fischer JB: Biochemical properties of the GABA/barbiturate/benzodiazepine receptor–chloride ion channel complex, in Kito S, Segawa T, Kuriyama K, Yamamura, HI, Olsen RW (eds): *Neurotransmitter Receptors: Mechanisms of Action and Regulation.* New York, Plenum Press, 1984, pp 205–219.
35. Ramanjaneyulu R, Ticku MK: Binding characteristics and interactions of depressant drugs with [^{35}S]*t*-butylbicyclophosphorothionate, a ligand that binds to the picrotoxinin site. *J Neurochem* 1984;42:221–229.
36. Ransom BR, Barker JL: Pentobarbital modulates transmitter effects on mouse spinal neurones grown in tissue culture. *Nature* 1975;254:703–705.
37. Richter JA, Holtman JR: Barbiturates: Their *in vivo* effects and potential biochemical mechanisms. *Prog Neurobiol* 1982;18:275–319.
38. Schulz DW, Macdonald RL: Barbiturate enhancement of GABA-modulated inhibition and activation of chloride ion conductance: Correlation with anticonvulsants and anesthetic actions. *Brain Res* 1981;209:177–188.
39. Sigel E, Barnard EA: A γ-aminobutyric acid/benzodiazepine receptor complex from bovine cerebral cortex. Improved purification with preservation of regulatory sites and their interactions. *J Biol Chem* 1984;259:7219–7223.
40. Simmonds MA: Distinction between the effects of barbiturates, benzodiazepines and phenytoin on responses to γ-aminobutyric acid receptor activation and antagonism by bicuculline and picrotoxinin. *Br J Pharmacol* 1981;73:739–747.
41. Skolnick P, Rice KC, Barker JL, Paul SM: Interaction of barbiturates with benzodiazepine receptors in the central nervous system. *Brain Res* 1982;233:143–156.
42. Squires RF, Casida JE, Richardson M, Saederup E: [^{35}S]*t*-Butylbicyclophosphorothionate binds with high affinity to brain-specific sites coupled to γ-aminobutyric acid-A and ion recognition sites. *Mol Pharmacol* 1983;23:326–336.
43. Stephenson RA, Olsen RW: Solubilization by CHAPS detergent of barbiturate-enhanced benzodiazepine-GABA receptor complex. *J Neurochem* 1982;39:1579–1586.
44. Study RE, Barker JL: Diazepam and (−)pentobarbital: Fluctuation analysis reveals different mechanisms for potentiation of γ-aminobutyric acid responses in cultured central neurons. *Proc Natl Acad Sci USA* 1981; 78:7180–7184.
45. Supavilai P, Karobath M: Action of pyrazolopyridines as modulators of [^{3}H]flunitrazepam binding to the GABA/benzodiazepine receptor complex of the cerebellum. *Eur J Pharmacol* 1981;70:183–193.
46. Supavilai P, Mannonen A, Karobath M: Modulation of GABA binding sites by CNS depressants and CNS convulsants. *Neurochem Int* 1982;4:259–268.
47. Teichberg VI, Tal N, Goldberg O, Luini A: Barbiturates, alcohols, and the CNS excitatory neurotransmission: Specific effects on the kainate and quisqualate receptors. *Brain Res* 1984;291:285–292.
48. Ticku MK: Interaction of depressant, convulsant and anticonvulsant barbiturates with [^{3}H]diazepam binding site at the benzodiazepine-GABA-receptor-ionophore complex. *Biochem Pharmacol* 1981;30:1573–1579.
49. Ticku MK, Olsen RW: Interaction of barbiturates with dihydropicrotoxinin binding sites related to the GABA receptor-ionophore system. *Life Sci* 1978;22:1643–1651.
50. Ticku MK, Olsen RW: Cage convulsants inhibit picrotoxinin binding. *Neuropharmacology* 1979;33:1135–1138.
51. Ticku MK, Ban M, Olsen RW: Binding of [^{3}H]α-dihydropicrotoxinin, a γ-aminobutyric acid synaptic antagonist, to rat brain membranes. *Mol Pharmacol* 1978;14:391–402.
52. Tsuchiya T, Fukushima H: Effects of benzodiazepines and pentobarbitone on the GABAergic recurrent inhibition of neurons. *Eur J Pharmacol* 1978;48:421–424.
53. Whittle SR, Turner AJ: Differential effects of sedative and anticonvulsant barbiturates on specific [^{3}H]GABA binding to membrane preparations from rat brain cortex. *Biochem Pharmacol* 1982;31:2891–2895.
54. Willow M: A comparison of the action of pentobarbital and etomidate on [^{3}H]GABA binding to crude synaptosomal rat brain membranes. *Brain Res* 1981;220:427–431.
55. Willow M, Johnston GAR: Enhancement by anesthetic and convulsant barbiturates of GABA binding to rat brain synaptosomal membranes. *J Neurosci* 1981;1:364–367.
56. Willow M, Johnston GAR: Pharmacology of barbiturates: electrophysiological and neurochemical studies. *Int Rev Neurobiol* 1983;24:15–49.

57. Wong DT, Rathbun RC, Bymaster FP, Lacefield WB: Enhanced binding of radioligands to receptors of γ-aminobutyric acid and benzodiazepine by a new anticonvulsive agent, LY81067. *Life Sci* 1983;33:917–923.
58. Wong EHF, Leeb-Lundberg LMF, Teichberg VI, Olsen RW: γ-Aminobutyric acid activation of $^{36}Cl^-$ flux in rat hippocampal slices and its potentiation by barbiturates. *Brain Res* 1984a;303:267–275.
59. Wong EHF, Snowman AM, Leeb-Lundberg LMF, Olsen RW: Barbiturates inhibit GABA antagonist and benzodiazepine inverse agonist binding. *Eur J Pharmacol* 1984b;102:205–212.
60. Yang-Ransom J, Fischer JB, Olsen RW: γ-Aminobutyric acid receptors in mouse brain: Comparison of binding and chloride channel activation. *Abstr Soc Neurosci* 1984;10:418.
61. Zbicz KL, Wilson WA: Barbiturate enhancement of spike frequency adaptation in Aplysia giant neurons. *J Pharmacol Exp Ther* 1981;217:222–227.

15

Barbiturate-Sensitive Sites in the Benzodiazepine–GABA Receptor–Ionophore Complex

Maharaj K. Ticku and Santosh K. Rastogi

ABSTRACT. The site of action of barbiturates at the GABA receptor complex is discussed. Results of equilibrium and dissociation kinetics studies with [^{35}S]*t*-butylbicyclophosphorothionate (TBPT), which interacts with the picrotoxin (convulsant) site, and the stereoisomers of (±)*N*-methyl-5-phenyl-propyl-barbituric acid (MPPB) [(+) a convulsant, (−) a depressant)] provide possible proof for a fourth "depressant" site in the GABA complex, distinct from the convulsant site. Furthermore, a study of rats *in vivo* to determine whether pentobarbital's anticonvulsant and/or hypnotic effects are mediated via the GABA system demonstrates that facilitation of GABAergic transmission is the mechanism responsible for the anticonvulsant effect of pentobarbital. The hypnotic effect of this drug appears to be independent of the anticonvulsant effect and may involve additional mechanisms.

INTRODUCTION

The GABA synapse serves as a site of action for a variety of centrally acting anxiolytic, convulsant, anticonvulsant, and depressant drugs. The postsynaptic receptor–ionophore complex, with which these drugs interact to enhance or inhibit GABAergic transmission, is composed of multiple, interacting binding sites. Three of these sites have been characterized *in vitro* by radioligand assays, using [^{3}H]GABA agonists for GABA recognition sites; [^{3}H]benzodiazepine (BZ) agonists and antagonists for the BZ sites; and [^{3}H]α-dihydropicrotoxinin (DHP) and [^{35}S]t-butylbicyclophosphorothionate (TBPT; also referred to as TBPS in the literature) for the picrotoxin (convulsant) sites.[5,14,28,29]

Barbiturates have a broad spectrum of activity. Their effect may be sedative, hypnotic, relaxant (muscle), anticonvulsant, or anesthetic. Many racemic barbituates have dual excitatory and depressant effects, and some barbiturates are pure convulsants.[1,3,6,19] Barbiturates can also affect a variety of biological processes, and these have been reviewed in the

Abbreviations used in this chapter: GABA, γ-aminobutyric acid; BZ, benzodiazepine; DHP, α-dihydropicrotoxinin; TBPT, *t*-butylbicyclophosphorothionate; MPPB, *N*-Methyl-5-phenyl-propyl-barbituric acid.

Maharaj K. Ticku and Santosh K. Rastogi • Department of Pharmacology, The University of Texas Health Science Center, San Antonio, Texas 78284.

literature.[12,19] Several lines of electrophysiological evidence indicate that the depressant barbiturates, such as pentobarbital, facilitate GABAergic transmission.[2,4,9,11–13,20] Investigators have shown that at low concentrations (10–150 μM), and in various preparations, pentobarbital enhances GABAergic responses[2,12]; at higher concentrations (> 200 μM), it directly activates GABA receptor-coupled chloride conductance.[11,12] Both these effects are blocked by the GABA antagonists picrotoxin and bicuculline.

In the following discussion concerning the site(s) of action of barbiturates at the GABA receptor complex, we review the effects of barbiturates on BZ, GABA, and TBPT binding in brain membrane and, on the basis of several investigations, propose an argument for the existence of a distinct "depressant" receptor site in the GABA receptor complex. Finally, on the basis of a study in rats, we examine whether or not the anticonvulsant and/or hypnotic effects of pentobarbital are mediated by a GABAergic mechanism.

EFFECTS OF BARBITURATES ON BINDING AT THE POSTSYNAPTIC GABA COMPLEX

BZ Binding

Barbiturates have a variable effect on BZ binding. Several laboratories have shown that depressant barbiturates enhance [^{3}H]BZ agonist binding in brain membranes.[7,8,22,24,26,27,30] For example, pentobarbital enhancement of [^{3}H]BZ binding is dose-dependent, additive with GABA, and blocked by bicuculline, picrotoxin, and the cage convulsants. Moreover, this enhancing effect is stereospecific and anion-dependent. Barbiturates apparently act allosterically, increasing the affinity of BZ agonists for their binding sites, presumably by decreasing the rate of the dissociation.[7,8] However, all barbiturates do not give the same maximal enhancement of [^{3}H]BZ binding, and many exhibit a dual effect, first stimulatory then inhibitory .[27] Furthermore, although certain anesthetics, such as (+)etomidate,[26] enhance BZ binding, other sadative/hypnotics, such as methaqualone, glutethimide, and thalidomide, inhibit it.[5] Thus, enhancement of BZ agonist binding is not a universal feature of all barbiturates.

GABA Binding

Depressant barbiturates enhance [^{3}H]GABA binding in brain membranes in a concentration-dependent manner.[14,15,25,26] This enhancement is blocked by picrotoxin but not by RO15-1788,[21] indicating that the interaction of pentobarbital with the GABA receptors does not require activation of BZ receptors. Nevertheless, the picrotoxin site does appear to be critical. Pentobarbital apparently increases the number of both high- and low-affinity GABA receptors.[15,26] Although radioligand binding studies have demonstrated the presence of high ($K_d \sim$ 10–20 nM) and low ($K_d \sim$ 100–200 nM) affinity sites, there must exist a super-low-affinity site, since the concentration of GABA required to produce a conductance change or to enhance BZ binding is ≥1 μM. *In vitro,* pentobarbital and (+)etomidate cause the number of both high and low affinity GABA receptor sites to increase.[15,26] On the basis of these observations, it has been speculated that barbiturates shift these super-low-affinity sites to other affinity states.[14] The net result of this shift would be an increase in the B_{max} values of the higher affinity sites. The ability of pentobarbital to shift the affinity of GABA to a higher-affinity site is consistent with electrophysiological studies, since one observes a leftward shift of GABA dose–response curves in the presence of pentobarbital[9] (Owen *et al.*, this volume, Chapter 3). It is tempting to speculate that this effect of pentobarbital may be responsible for the potentiation of GABAergic transmission. Thus, since the depressant barbiturates enhance both GABA and BZ binding, the depressant barbiturates must bind to a separate site.

TBPT Binding

A specific binding site for the picrotoxinlike cage convulsants has been identified and characterized in brain membranes. This binding site was characterized initially with a biologically active analogue of picrotoxin, [^{3}H]DHP,[32] and more recently with [^{35}S]TBPT.[16,17,23] Studies with [^{35}S]TBPT have suggested that several classes of convulsants, depressants, and anxiolytics that modulate GABA transmission may interact with the picrotoxin (convulsant) site.[16,17,23,28,29,31] TBPT binds to rat brain membranes at a single site with an apparent K_d value of 20–30 nM, and a B_{max} of ~1.35 pmole/mg protein. The binding of TBPT is dependent on Cl^- concentrations, and maximal binding occurs in buffers containing 200 mM Cl^-. Table I lists the IC_{50} values of barbiturates and other depressants on [^{35}S]TBPT binding. The stereoselective nature of barbiturate inhibition of [^{35}S]TBPT binding is similar to that observed for the enhancement of GABA and BZ binding,[15,28] and is in agreement with the behavioral effects. Table II shows that the IC_{50} values of (+)etomidate and (±)pentobarbital on [^{35}S]TBPT are identical to their respective EC_{50} values on [^{3}H]GABA and [^{3}H]diazepam binding.[26] Thus, it appears that all these effects of pentobarbital and (+)etomidate may be mediated through the same site.

TBPT binding is coupled with both GABA and BZ receptors. GABA agonists displace TBPT binding, allosterically,[10,29] in a bicuculline-sensitive manner and with the following order of potency: muscimol > GABA > THIP.[31] The inhibition of TBPT binding by the GABA agonists appears to be mediated by the low- and/or super-low-affinity sites, since inhibition is observed in the presence of 50 mM ammonium thiocyanate, which eliminates the high-affinity sites.[31] This noncompetitive inhibition of TBPT binding by GABA agonists is consistent with an allosteric inhibition.[31]

Table I. IC_{50} Values of Barbiturates and Other Depressants on [^{35}S]TBPT Binding[a]

Drug	IC_{50} (μM)
Cartazolate	0.5
Etazolate	5
(+)Etomidate sulfate	9
(−)Secobarbital	40
(+)Secobarbital	70
(−)Pentobarbital	70
(±)Pentobarbital	100
Methaqualone	100
(−)Etomidate	100
(−)MPPB	116
(+)Pentobarbital	125
(+)Hexobarbital	145
(±)Hexobarbital	220
(−)Hexobarbital	380
Phenobarbital	480
(+)MPPB	760

[a]From Ramanjaneyulu and Ticku.[16]

Table II. Comparison of the Effect of Etomidate and Pentobarbital on Binding at the Benzodiazepine–GABA Receptor Ionophore Complex

	Binding at the benzodiazepine–GABA receptor complex (μM)		
Ligand	[^{35}S]TBPT (IC_{50})	[^{3}H]GABA (EC_{50})	[^{3}H]Diazepam (EC_{50})
(+)Etomidate	9	8	8
(±)Pentobarbital	100	100	100

A DISTINCT SEDATIVE/HYPNOTIC SITE AT THE BZ–GABA RECEPTOR COMPLEX

It has recently become apparent, on the basis of structural differences and the effects of depressant barbiturates on equilibrium and dissociation kinetics studies of TBPT, that depressant agents, including the barbiturates, interact with the convulsant (TBPT) binding sites, allosterically.[10,29] The hypothetical topography of the picrotoxinin (convulsant) site at the BZ–GABA receptor–ionophore complex has recently been postulated.[29] From this model, it is clear that all convulsants (picrotoxin, TBPT, Tetrazoles, bemegride, RO5-3663) that inhibit GABA responses noncompetitively and inhibit TBPT binding possess an alkyl moiety corresponding to the β-position on γ-butyrolactone.[29] γ-Butyrolactone, which lacks this substituent at the β-position, does not inhibit [^{35}S]TBPT binding up to 1 mM (Ticku, unpublished observation). Barbiturates also lack this corresponding alkyl moiety.

Furthermore, whereas the convulsants inhibit TBPT binding competitively, barbiturates, (+)etomidate, and etazolate inhibit it noncompetitively.[31] Convincing evidence for a distinct allosteric site of action for depressants is provided from our recent dissociation studies with TBPT.[10] Convulsants, such as TBPT, picrotoxinin, and pentamethylenetetrazole, cause TBPT to dissociate from cortical membranes in a monophasic and nondistinguishable pattern, which suggests that they bind to the same site[10] (Fig. 1). Conversely, GABA, (−)*N*-methyl-5-phenylpropyl barbituric acid (MPPB, a depressant), and etazolate accelerate the dissociation of TBPT from its binding sites (Fig. 1).

Studies with MPPB

We have used the stereoisomers of MPPB in our laboratory to determine whether or not they bind to the same or different sites. Behaviorally, R(−)MPPB produced a transient loss of righting reflex and ataxia, whereas S(+)MPPB produced dose-related convulsions. Furthermore, S(+)MPPB was proconvulsant with subconvulsive doses of picrotoxin. The convulsant effects of S(+)MPPB were blocked by R(−)MPPB and pentobarbital (Rastogi and Ticku, unpublished observations). Thus, the two isomers of MPPB exhibited opposite effects. It may be pointed out that although some convulsant barbiturates have been reported to enhance BZ and/or GABA binding,[7,8,33] (+)MPPB enhances binding of neither GABA nor BZ.[28] Moreover, (+)MPPB, like picrotoxin and the cage convulsants, antagonizes the enhancing effect of (−)MPPB and GABA agonists on BZ binding.[27]

The isomers of MPPB also exhibited a differential profile in the dissociation of TBPT[10] (Fig. 1). The depressant R(−)MPPB accelerated the dissociation of TBPT the most and the convulsant S(+)MPPB the least. The $t_{1/2}$ value of the slower phase of [^{35}S]TBPT dissocia-

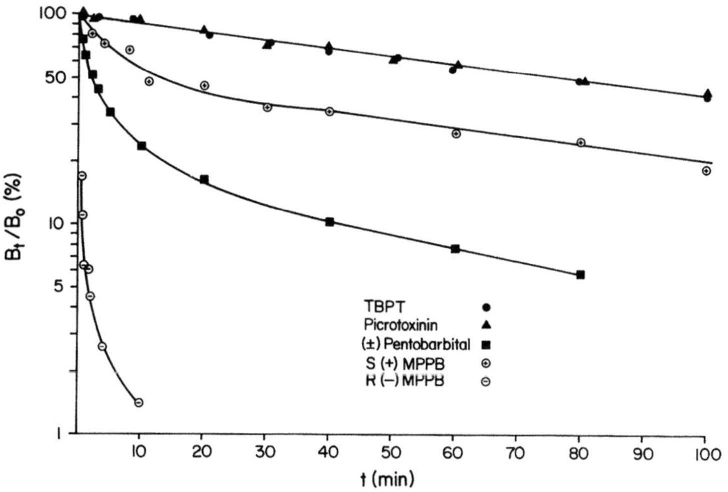

Figure 1. Dissociation plots of (^{35}S)TBPT binding by various ligands in freeze-thawed and washed cortical membranes. Modified from Maksay and Ticku.[10]

tion elicited by S(+)MPPB (76 ± 8 min) was similar to that obtained in the presence of nonradioactive TBPT (72 ± 6 min). It has been postulated that barbiturates may bind to the convulsant sites if they have long, flexible substituents at the 5-position, which can interact with the allosteric hydrophobic regulatory site.[29] On this basis, it could be argued that S(+)MPPB binds predominantly to the convulsant site, whereas R(−)MPPB must bind to a different (depressant) site. Of note, racemic pentobarbital, which possesses a dual depressant–excitatory activity, had an intermediate accelerating potency.[10] These dissociation studies with TBPT have provided a model for differentiating the depressant and convulsant drugs that affect GABAergic transmission.[10]

All these factors, namely, the structural differences between the depressant and convulsant barbiturates, the noncompetitive inhibition of TBPT binding by pentobarbital, and the different dissociation profiles exhibited by depressants versus convulsants, argue strongly that depressant barbiturates bind to a distinct site. This (depressant) site appears to be coupled with the convulsant (picrotoxin), GABA, and BZ sites of the macromolecular complex. The possibility remains, of course, that the depressants and convulsants may bind to the same site in a negatively cooperative manner; however, pentobarbital and other depressants that facilitate GABAergic transmission do not exhibit negative cooperativity in electrophysiological responses. Figure 2 shows a hypothetical version of the four potential sites where GABA and other modulators may bind.

IS AN ANTICONVULSANT AND/OR HYPNOTIC EFFECT OF PENTOBARBITAL MEDIATED VIA THE GABA SYSTEM?

In Vivo Studies

To elucidate the possible involvement of the GABA complex in pentobarbital actions, we have investigated the effect of pentobarbital on convulsions [produced by either an impairment of GABAergic transmission or maximal electroshock seizures (MES)] and the loss of righting reflex in rats.

Pentobarbital protected against electroshock seizures in a dose-related manner (Table III). Significant anticonvulsant action was observed at 20–40 mg/kg. At lower doses, no

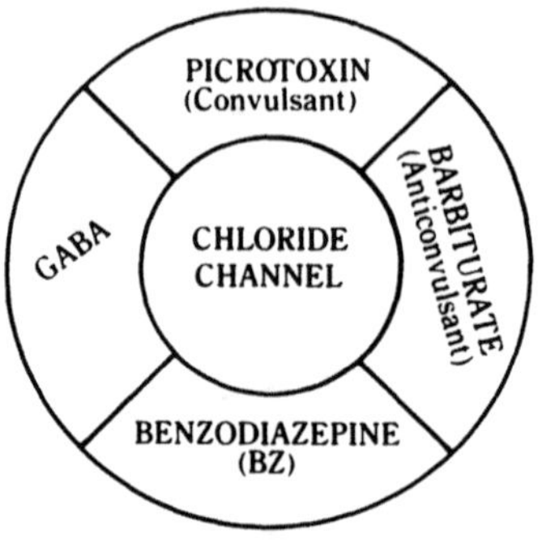

SITES OF DRUG ACTION

GABA RECEPTORS	BZ RECEPTORS	PICROTOXIN SITE	BARBITURATE SITE
GABA	BZ Agonists	Picrotoxin	Barbiturates
Muscimol	BZ Antagonists	TBPS	(+)Etomidate
Bicuculline		PTZ	Pyrazolopyridines
R-5135 (?)	β-Carbolines (?)	R05-3663	Cyclopyrazolones (?)

Figure 2. A fourth "an iconvulsant" receptor site at the BZ–GABA receptor–ionophore complex.

Table III. Effect of Pentobarbital Alone or in Combination with Other GABA-Mimetic Agents on Maximal Electroshock (MES) Seizures and Loss of Righting Reflex in Rats[a]

Treatment	MES Seizures: Protection (%)	MES Seizures: Recovery (min ± S.D.)	Loss of Righting Reflex: Onset (min ± S.D.)	Loss of Righting Reflex: Recovery (min ± S.D.)
Pentobarbital (mg/kg)				
5	0	—	—	—
10	0	—	—	—
20	20	60	7.6 ± 1.5	14.6 ± 2.5
30	60[b]	60	8.3 ± 1.5	45.0 ± 5.0
40	100[c]	90	2.5 ± 0.6	99.0 ± 16.5
Diazepam (40 mg/kg) + pentobarbital (mg/kg)				
5	20	60	4.0 ± 1.0[d]	23.0 ± 3.0
10	100[d]	90	4.0 ± 1.0[d]	68.0 ± 8.0[d]
20	100[d]	120	3.0 ± 1.0[d]	115.0 ± 5.0[d]
Ethanol (0.5 g/kg) + pentobarbital (mg/kg)				
5	80[c]	60	—	—
10	100[d]	60	5.0 ± 1.0[d]	30.0 ± 5.0[d]
20	100[d]	120	2.3 ± 0.6[d]	38.0 ± 2.5[d]

[a]Pentobarbital alone or in combination with diazepam or ethanol was administered simultaneously 30 min before electroshock seizures were induced. Percent protection against the seizures was observed every 30 min for 120 min. Sleep time was recorded as the duration between loss and recovery of the righting reflex. Onset and recovery of righting reflexes represent the mean ± S.D. of five to seven animals, in minutes.

[b]$p < 0.05$.

[c]$p < 0.01$.

[d]$p < 0.001$, as compared to control values of pentobarbital, by Fisher's exact probability test for MES and one-way analysis of variances for loss of righting reflex. The degree of significance was calculated by student Newman–Keuls procedure. A dash indicates the absence of convulsions or detectable loss of righting reflex. Modified from Rastogi and Ticku.[18]

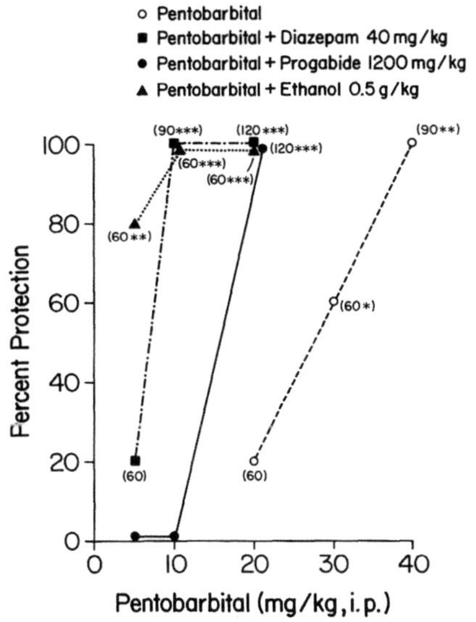

Figure 3. Effect of pentobarbital alone and in combination with GABA modulators on maximal electroshock-induced seizure protection and duration of protection (parenthesis). Subprotective doses of pentobarbital (5, 10, and 20 mg/kg) injected i.p. simultaneously with a single dose of diazepam (40 mg/kg), ethanol (0.5 g/kg), and progabide (1200 mg/kg) 30 min (except progabide; 60 min) before electroshock. Results were obtained from five to seven animals. * $p < 0.05$; ** $p < 0.01$; *** $p < 0.001$, in comparison with control values of pentobarbital. Note that pentobarbital alone produced a dose-related protection against MES seizures. Combination of single doses of various GABA produced significant protection in lower doses of pentobarbital and shifted the dose–response curve to the left.

correlation was found between the hypnotic and anticonvulsant action of this drug, since recovery from loss of righting reflex occurred earlier than recovery from the anticonvulsant effect. At higher doses (40 mg/kg), however, this differential effect was not observed[18] (Table III).

Other prototype agents like diazepam and ethanol, which facilitate GABAergic transmission, also protected against electroshock seizures, but at higher doses.[18] To study their potentiating action, we looked at the effect of different subprotective doses of pentobarbital, in combination with a single subprotective dose of diazepam (40 mg/kg) or ethanol (0.5 mg/kg). These drug combinations caused a significant potentiation of the loss of righting reflex and anticonvulsant effect of pentobarbital (Table III). However, recovery from loss of righting reflex occurred earlier (as it did with pentobarbital alone) than did recovery from the anticonvulsant action (Table III). Thus, these results also suggest that more than one mechanism may be involved in the anticonvulsant and hypnotic effect. Figure 3 shows that, in the presence of diazepam or ethanol, the dose–response curve of pentobarbital is shifted to the left. These behavioral results are consistent with other evidence that diazepam, pentobarbital, and ethanol all facilitate GABAergic transmission.

To determine the potential involvement of the GABA and/or BZ receptor sites in the anticonvulsant effect of pentobarbital, we studied the effect of their respective antagonists. Table IV shows that the anticonvulsant effect of pentobarbital against electroshock seizures was antagonized by bicuculline (2 mg/kg) and, at higher doses, by RO15-1788 (>100 mg/kg), but not by picrotoxin (5 mg/kg). Higher doses of picrotoxin could not be tested, since they would have caused convulsions. Of interest, the antagonists of both GABA and BZ in combination with pentobarbital caused a significant increase in the loss of righting reflex (Table IV). These results provide evidence that different mechanisms may be operating simultaneously to produce the anticonvulsant and hypnotic effects of pentobarbital.

The anticonvulsant profile of pentobarbital alone and in combination with diazepam and ethanol against convulsions induced by various chemoconvulsants has also been investigated in our laboratory. Table V compares the anticonvulsant profile of pentobarbital against picrotoxin, bicuculline, and RO5-4864. Pentobarbital prevented convulsions produced by all three GABA antagonists. In terms of doses, pentobarbital was more effective against convulsions produced by RO5-4864 than those of bicuculline or picrotoxin. The tonic component

Table IV. Effect of GABA and BZ Antagonists on Pentobarbital's Protection against Maximal Electroshock Seizures (MES) and Loss of Righting Reflex[a]

		MES seizures		Loss of righting reflex	
Treatment (mg/kg)		Protection (%)	Recovery (min ± S.D.)	Onset (min ± S.D.)	Recovery (min ± S.D.)
Pentobarbital	(40)	100	90	2.5 ± 0.6	99 ± 17.0
+ Bicuculline	(2)	0	30	2.6 ± 0.9	117 ± 5.1[b]
+ Picrotoxin	(5)	100	90	6.6 ± 1.5[c]	114 ± 4.1[b]
+ RO15-1788	(50)	10	90	2.5 ± 0.8	180 ± 6.0[c]
+ RO15-1788	(100)	40	60	2.6 ± 0.9	163 ± 5.7[c]
+RO15-1788	(200)	10[c]	90	2.3 ± 0.1	150 ± 5.0[c]

[a]Bicuculline, picrotoxin, and RO15-1788 were administered, 60 min before, whereas pentobarbital (40 mg/kg) was administered 30 min before exposure to electroshock.
[b]$p < 0.05$.
[c]$p < 0.001$, in comparison with pentobarbital treatment.

was blocked at lower doses, whereas the clonic component was blocked at higher doses of pentobarbital. Pentobarbital increased the duration of the seizures and reduced the mortality associated with all three convulsants.

The results with pentobarbital alone and in combination with diazepam or ethanol

Table V. Anticonvulsant Profile of Pentobarbital against Chemoconvulsants[a]

	Time to response after administration of convulsant (min ± S.D.)			
Treatment (mg/kg, i.p.)	Onset (clonic)	Severity (tonic)	Total seizure latency	% Mortality
RO5-4864 (40 mg/kg)	3.0 ± 0.7	6.5 ± 1.3	16.5 ± 4.2	80
+ Pentobarbital				
5	3.2 ± 0.4	6.8 ± 0.8	47.0 ± 5.7[b]	0
10	4.4 ± 1.1	—	39.0 ± 2.2[b]	0
20	4.6 ± 1.1	—	45.0 ± 3.6[b]	0
40	—	—	—	—
Bicuculline (4 mg/kg)	1.8 ± 0.8	—	6.4 ± 1.4	100
+ Pentobarbital				
2.5	2.3 ± 0.5	4.0 ± 1.0	17.0 ± 1.5[c]	106
5	2.3 ± 0.5	4.3 ± 0.5	17.3 ± 5.0[b]	100
10	2.3 ± 0.5	4.6 ± 0.5	21.6 ± 2.8[b]	80
20	4.3 ± 0.6[b]	—	—	0
Picrotoxin (10 mg/kg)				
+ Pentobarbital				
5	5.6 ± 1.4	16.6 ± 8.5	14.6 ± 2.9	100
10	6.6 ± 1.4	21.6 ± 3.5[b]	26.1 ± 2.0[c]	100
20	17.2 ± 3.0[b]	29.0 ± 2.4[b]	47.0 ± 4.7[b]	20
40	16.3 ± 1.5[b]	—	75.0 ± 2.0[b]	0

[a]Pentobarbital was given 30 min before administration of the chemoconvulsant. Each value represents the mean ± S.D. of three to five animals, in minutes.
[b]$p < 0.001$.
[c]$p < 0.005$, in comparison with control values of respective convulsants by one-way analysis of variance using student Newman–Keuls procedure. A dash indicates the absence of convulsions.

against MES and GABA antagonist-induced seizures support the hypothesis that facilitation of GABAergic transmission is responsible for the anticonvulsant effect. The hypnotic effect does not appear to play a role in the anticonvulsant action of pentobarbital. Furthermore, the different time courses of recovery from anticonvulsants and loss of righting reflex and the inability of GABA and BZ antagonists to affect pentobarbital protection against loss of righting reflex suggest that one or more different mechanisms are involved in the hypnotic effect produced by pentobarbital.

CONCLUSIONS

The GABA receptor complex, with which various facilitators and inhibitors of GABAergic transmission interact, is composed of four interacting binding sites. Dissociation and equilibrium binding studies with [^{35}S]TBPT suggest that depressants like pentobarbital—which facilitate GABAergic transmission, enhance GABA and BZ binding, and inhibit TBPT binding—bind to a distinct "depressant" site (Fig. 2). Behavioral studies indicate that pentobarbital's anticonvulsant effect against electroshock seizures is mediated by facilitation of GABAergic transmission. This is further supported by the behavioral effects of stereoisomers of MPPB, which produce opposite pharmacological effects. The convulsant effects of S(+)MPPB are blocked by R(−)MPPB and pentobarbital. However, S(+)MPPB fails to antagonize the loss of righting reflex produced by R(−)MPPB or pentobarbital. These results suggest that the anticonvulsant but not the hypnotic effect of barbiturates appears to involve facilitation of GABAergic transmission.

ACKNOWLEDGMENTS. This work was supported, in part, with funds from NIH (NS-15339) and NIAAA (AA-04090). We thank Drs. R. Ramanjaneyulu and G. Maksay for their helpful discussion, Mrs. R. Thyagarajan for technical assistance, and Mrs. M. Wilson for secretarial and editorial assistance.

REFERENCES

1. Downes H, Perry RS, Ostlund RE, Karler R: A study of the excitatory effects of barbiturates. *J Pharmacol Exp Ther* 1970;175:692–699.
2. Evans RH: Potentiation of the effect of GABA by pentobarbitone. *Brain Res* 1979;171:113–120.
3. Grossman W, Jurna I, Theres C: The site of action of the optical isomers of 1-methyl-5-phenyl-5-propylbarbituric acid. *Nanunyn-Schmeideberg's Arch Pharmacol* 1974;282:367–377.
4. Haefely W, Polc P, Keller HH, Pieri L, Möhler H: Facilitation of GABAergic transmission by drugs, in Krogsgaard-Larsen P, Scheel-Kruger J, Kofod H (eds): *GABA—Neurotransmitters*. Copenhagen, Munksgaard, 1979, pp 357–373.
5. Johnston GAR: Regulation of GABA receptors by barbiturates and related sedative-hypnotic and anticonvulsant drugs, in Enna SJ (ed): *GABA Receptors*. Clifton, NJ, Humana Press, 1983, pp 107–108.
6. Knabe J, Rummel W, Büch HP, Franz N: Optical aktive barbiturate: Synthese konfiguration und pharmakologische, wirkung. *Arzneimittel-Forschung* 1978;28:1048–1056.
7. Leeb-Lundberg F, Olsen RW: Interaction of barbiturates of various pharmacological categories with benzodiazepine receptors. *Mol Pharmacol* 1980;21:320–328.
8. Leeb-Lundberg F, Snowman A, Olsen RW: Barbiturate receptor sites are coupled to benzodiazepine receptors. *Proc Natl Acad Sci USA* 1980;77:7462–7472.
9. Macdonald RL, Barker JL: Different actions of anticonvulsant and anesthetic barbiturates revealed by use of cultured mammalian neurons. *Science* 1978;200:775–777.
10. Maksay G, Ticku MK: Dissociation of [^{35}S]t-butylbicyclophosphorothionate binding differentiates convulsant and depressant drugs that modulate GABAergic transmission. *J Neurochem* 1985;44:480–486.
11. Nicoll RA: Pentobarbital: Action on frog motoneurons. *Brain Res* 1975;96:119–123.
12. Nicoll RA: Sedative-hypnotics: Animal pharmacology, in Iversen LL, Iversen SD, Snyder SH (eds): *Handbook of Psychopharmacology: A Generation of Progress*, Vol. 12. New York, Plenum Press, 1980, pp 187–234.

13. Nicoll RA, Wojtowicz JM: The effects of pentobarbital and related compounds on frog motoneurons. *Brain Res* 1980;191:225–237.
14. Olsen RW: Drug interactions at the GABA receptor-ionophore complex. *Annu Rev Pharmacol Toxicol* 1982;22:245–277.
15. Olsen RW, Snowman AM: Chloride-dependent enhancement by barbiturates of gamma-aminobutyric acid receptor binding. *J Neurosci* 1982;2:1812–1823.
16. Ramanjaneyulu R, Ticku MK: Binding characteristics and interactions of depressant drugs with [^{35}S]t-butylbicyclophosphorothionate, a ligand that binds to the picrotoxinin site. *J Neurochem* 1984a;42:221–229.
17. Ramanjaneyulu R, Ticku MK: Interactions of pentamethylenetetrazole and tetrazole analogues with the picrotoxinin site of the benzodiazepine-GABA receptor-ionophore complex. *Eur J Pharmacol* 1984b;98:337–345.
18. Rastogi SK, Ticku MK: Involvement of GABAergic mechanism in anticonvulsant effect of pentobarbital against maximal electroshock-induced seizures in rats. *Pharm Biochem Behav* 1985;22:141–146.
19. Richter JA, Holtzman JR: Barbiturates: Their *in vivo* effects and potential biochemical mechanisms. *Prog Neurobiol* 1982;18:275–319.
20. Simmonds MA: Distinction between the effects of barbiturates, benzodiazepines, and phenytoin in response to γ-aminobutyric acid receptor activation and antagonism by bicuculline and picrotoxin. *Br J Pharmacol* 1981;73:739–747.
21. Skerritt JH, Johnston GAR, Katsikas T, Tabar J, Nicholson GM, Andrews PR: Action of pentobarbitone and derivatives with modified 5-butyl substituents on GABA and diazepam binding to rat brain synaptosomal membranes. *Neurochem Res* 1983;8:1337–1350.
22. Skolnick P, Rick KC, Barker JL, Paul SM: Interaction of barbiturates with benzodiazepine receptors in the central nervous system. *Brain Res* 1982;233:143–156.
23. Squires RF, Casida JE, Richardson M, Saederup E: [^{35}S]t[Butylbicyclophosphorothionate binds with high affinity to brain specific sites coupled to γ-aminobutyric acid-A and ion recognition sites. *Mol Pharmacol* 1983;23:326–336.
24. Supavilai P, Karobath M: Differential modulation of [^{35}S]TBPT binding by the occupancy of benzodiazepine receptors with its ligands. *Eur J Pharmacol* 1983;91:145–146.
25. Supavilai P, Mannonen A, Karobath M: Modulation of GABA binding sites by CNS depressants and CNS convulsants. *Neurochem Int* 1982;4:259–268.
26. Thyagarajan R, Ramanjaneyulu R, Ticku MK: Enhancement of diazepam and GABA binding by (+)etomidate and pentobarbital. *J Neurochem* 1983;41:575–585.
27. Ticku MK: Interaction of depressant, convulsant, and anticonvulsant barbiturates with [^{3}H]diazepam binding sites at the benzodiazepine-GABA receptor-ionophore complex. *Biochem Pharmacol* 1981;30:1573–1579.
28. Ticku MK: Benzodiazepine-GABA receptor-ionophore complex: Current concepts. *Neuropharmacology* 1983;22(No. 12B):1459–1470.
29. Ticku MK, Maksay G: Convulsant/depressant site of action at the allosteric benzodiazepine-GABA receptor-ionophore complex. *Life Sci* 1983;33:2363–2375.
30. Ticku MK, Olsen RW: Interaction of barbiturates with dihydropicrotoxinin binding sites related to the GABA receptor-ionophore system. *Life Sci* 1978;22:1643–1652.
31. Ticku MK, Ramanjaneyulu R: Differential interactions of GABA agonists, depressant, and convulsant drugs with [^{35}S]t-butylbicyclophosphorothionate binding sites in cortex and cerebellum. *Pharm Biochem Behav* 1984; 21:151–158.
32. Ticku MK, Ban M, Olsen RW: Binding of [^{3}H]α-dihydropicrotoxinin, a γ-aminobutyric acid synaptic antagonist, to rat brain membranes. *Mol Pharmacol* 1978;14:391–402.
33. Willow M, Johnston GAR: Enhancement by anesthetic and convulsant barbiturates of GABA binding to rat brain synaptosomal membranes. *J Neurosci* 1981;1:364–367.

III

ANESTHETICS AND SODIUM CHANNELS

16

An Overview of the Mechanisms of Local Anesthetic Action Past, Present, and Future

J. Murdoch Ritchie

ABSTRACT. It is exactly 100 years since Freud in "Ueber Coca" proposed cocaine as a potentially clinically useful local anesthetic agent and Koller used it in ophthalmic surgery. Many theories have been adduced concerning the active structure. These originally focused on the free base, then on the cationic form, and finally on both forms. Interference with the neurochemistry of conduction, membrane expansion, and surface charge effects have all been proposed. Currently the mechanism of local anesthetic action is envisaged as an interaction with a specific membrane receptor in the sodium channel whose properties are modified when this interaction occurs (Hille's "modified receptor hypothesis"). In spite of a century of research, several questions remain outstanding. Is there a true differential sensitivity of different fibers to local anesthetics? Is there more than one site of local anesthetic action? Can other molecular structures yield useful local anesthetic agents?

This chapter will trace the early history of our ideas on the mechanism of action of local anesthetic drugs, will briefly delineate the present state of the subject, and finally will suggest some areas for future research.

PAST

Exactly 100 years ago, in the summer of 1884, Sigmund Freud[10] published his first paper on cocaine, Ueber Coca, which was to open up the whole modern field of local anesthesia. In his monograph Freud mentioned various therapeutic uses for cocaine. It could be employed as a central nervous stimulant for use in wartime, on journeys, during mountain climbing, and on other expeditions as well as a stimulant for diverse kinds of psychic debility. Freud discussed the use of cocaine for digestive disorders, as an agent to treat cachexia, as an agent useful in the treatment of morphine and alcohol addiction, and as a drug that might be useful in the treatment of asthma. Only finally, after a discussion on the use of cocaine as an aphrodisiac, did Freud in a brief paragraph note that cocaine had a marked anesthetizing effect on the skin and mucous membranes, and he suggested that

J. Murdoch Ritchie • Department of Pharmacology, Yale University School of Medicine, New Haven, Connecticut 06510.

"indeed the anesthetizing properties of cocaine should make it suitable for a good many further applications." How sound this suggestion was! This aspect was taken up by his colleague Koller in the Ophthalmology department, and local anesthesia—first with cocaine but subsequently with safer related drugs such as procaine—entered the mainstream of modern medicine.

Exactly how local anesthetics act was to remain obscure for some time. Much of the early thinking in this field focused on the fact that there was an apparent increase of potency in local anesthetic solutions as they were made more alkaline. This fact had long been known to the Indians of Peru who, when chewing the coca leaf, have mixed with it lime, or with ash from burnt saplings, to make the content of their mouths more alkaline so as more effectively to extract the cocaine from the coca leaf and more effectively to absorb it. This effect of alkalinity was studied systematically in 1929 by Trevan and Boock, who showed that for a variety of local anesthetic agents the minimum effective concentration falls as the pH of the solution increases. Since increasing the pH of the solution increases the amount of local anesthetic in the form of the uncharged free base, Trevan and Boock in 1929[33] concluded that "the only active constituent of a solution of local anesthetic is the free base and not the ion or the undissociated salt."

Progress in the area was slow. In the first edition of Goodman and Gilman in 1941,[15] for example, the section on the mechanism of action of local anesthetics begins with the statement: "Little is known of the mechanism of action of the local anesthetics." It was noted that there was an apparent differential sensitivity to local anesthetics, smaller-diameter fibers being blocked before larger-diameter fibers; and (following Trevan and Boock) it was stated that "it is the free base that acts on nerve tissue" to produce conduction block. By 1955, at the time of the second edition of Goodman and Gilman,[16] there was still remarkably little progress. Thus, the section on the mechanism of action of local anesthetics began, as in 1942, by saying: "Little is known of the basic mechanisms of action of the local anesthetics" and, it went on to add, "not too much is gained by discussion of the present inadequate theories." However, a substantial advance in this second edition was that the importance of the un-ionized free base for penetration of the local anesthetic to its site of action was first clearly discussed, as well as the possibility that once the local anesthetic got to the active site, the cationic form might well be ultimately responsible for the local anesthetic action. Indeed, for the first time, the suggestion was made that perhaps "anesthetic drugs act intracellularly in the cationic form." It had taken 70 years to get to this point. But after that, progress was rapid.

By 1959, Taylor[32] had shown in experiments on a voltage-clamped single squid giant axon that local anesthetics act by decreasing the voltage-dependent sodium ion conductance that underlies the generation of the action potential, and the idea was adumbrated that the local anesthetic, in its cationic form, acts by blocking the sodium channel.

As far as the importance of the cation is concerned, only two experiments need be cited. Figure 1[26] shows the effect of exposing a desheathed rabbit vagus nerve to dibucaine at an alkaline pH. When the local anesthetic is first applied, there is only a slight fall in the size of the action potential of the nonmyelinated fibers. Switching to anesthetic-free solution at neutral pH, however, causes a paradoxical marked increase in the degree of conduction block. Return to the anesthetic-containing solution at alkaline pH leads to recovery in the size of the action potential. The explanation is clear, namely, that while being exposed to the anesthetic (at alkaline pH) most of the anesthetic is in the uncharged form whereas in the anesthetic-free solution (at neutral pH) some of the anesthetic that is already trapped in the nerve is converted to the cationic form. It seems, therefore, from this experiment that the

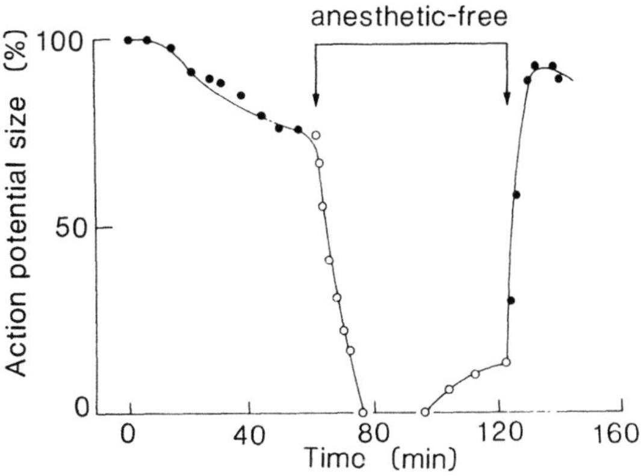

Figure 1. The amplitude of the compound action potential of the nonmyelinated fibers in a rabbit desheathed vagus nerve. From 0–60 min and from 120 min onward the nerve was exposed to 5 μm dibucaine (●) at pH 9.2. During the intervening period it was exposed to anesthetic-free solution at pH 7.2 (○). The amplitude is expressed as a percentage of its original value in anesthesic-free Locke solution. Modified from Ritchie *et al.*[26]

uncharged free base is relatively inactive whereas most of the activity resides in the cationic form.

The second kind of critical experiment that supported this conclusion was done by Narahashi and his colleagues[9,21] using the quaternary form of local anesthetic agents. These quaternary analogues (usually of lidocaine) can never exist as uncharged free amines. Nevertheless, they were shown to be potent local anesthetic agents—provided they were applied internally.

PRESENT

By the beginning of the 1970s it was generally accepted that local anesthetic activity mostly resides in the cationic form of the molecule, although the possibility was not excluded that activity also resides in the nonionized form.[23] There were probably more structure/activity studies done for local anesthetic drugs than for any other drug up to that time, and theories of local anesthesia were abundant. The various theories then current[23,29,30] have for the most part been discarded or greatly modified.

The modern era of the mechanism of local anesthetic action began in 1973 when Strichartz, using quaternary forms of local anesthetics, showed that the degree of block of the sodium channels depended on their previous history. Figure 2, taken from Strichartz's paper,[31] show[5] the sodium currents that flow in response to a depolarizing test pulse in a voltage-clamped node of Ranvier that had previously been exposed to a concentration of

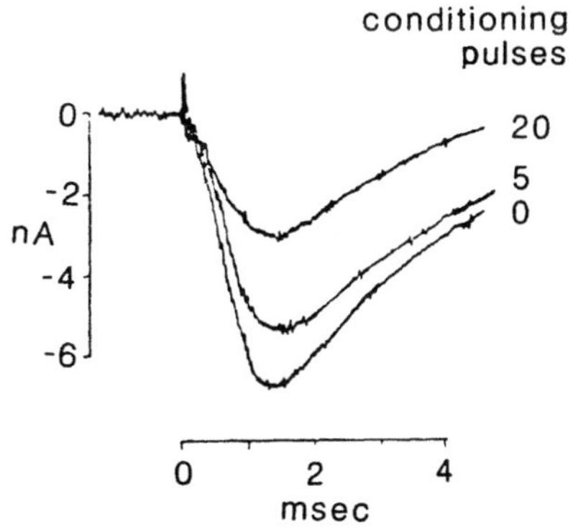

Figure 2. The dependence on previous history of the inhibition of the sodium currents in a node of Ranvier of a single isolated frog nerve fiber. The current in record labeled "0" was elicited by switching the membrane potential from a resting value of −75mV to −45mV. The records labeled "5" and "20" were results of similar procedures preceded, however, by 5 and 20 5-msec depolarizations, respectively, to +75mV, each 1 sec apart. Modified from Strichartz.[31]

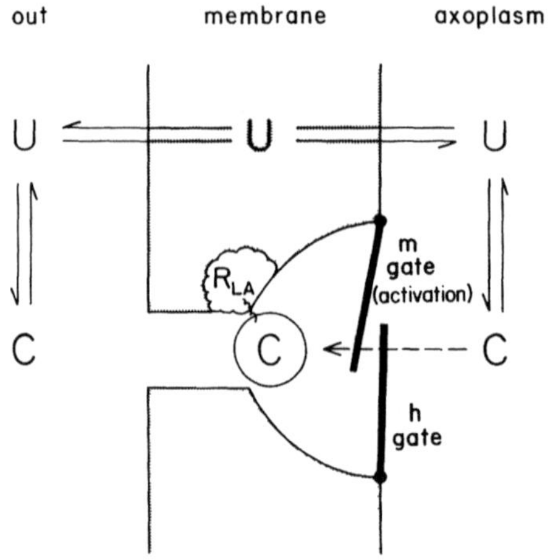

Figure 3. A diagrammatic representation of the interaction of a local anesthetic with the sodium channel. The uncharged form of the anesthetic (U) in the extracellular space, which is in equilibrium with the charged form (C), crosses the lipid axonal membrane into the axoplasm where a new equilibrium between charged and uncharged forms is set up. Provided the *m* and *h* gates are open, the charged molecule gains access to the receptor (R_{LA}) in the membrane to bind to it and block conduction.

local anesthetic that reduced, but did not abolish, the sodium currents. The response labeled "0" shows the sodium current that flowed in response to an occasional infrequent depolarizing pulse. In the record labeled "5" the same depolarizing pulse is applied to the preparation but after five conditioning depolarizing pulses 1 sec apart had been previously applied. Clearly, this previous treatment with depolarizing conditioning pulses reduced the sodium current elicited by the constant test pulse. Indeed, if the current flowing to the depolarizing pulse was tested after 20 prior depolarizing pulses, the sodium current was even less. In other words, the sodium current that flows, the ability of the nerve to conduct impulses, and the degree of block by the local anesthetic agent, all depend on the previous history of the nerve.

Strichartz's interpretation of these and other experiments was extremely important, and it was that the local anesthetic gains access to the receptor with which it interacts on the nerve membrane (and indeed on recovery escapes from this receptor) only when the activation (*m*) and inactivation (*h*) gates controlling the sodium currents are open. This is shown diagramatically in Fig. 3 for a conventional amine local anesthetic. Local anesthetic is present in the extracellular space in both uncharged and charged forms. The uncharged molecule readily penetrates the lipid membrane to the axoplasm, where some is then protonated depending on the prevailing pH. Then, provided the *m* and the *h* gates gates are open, the charged molecule gains access to the receptor in the membrane, binds to it, and conduction is blocked.

Shortly therafter, Courtney[6] in 1975 made the finding, later confirmed in 1977 by Hille,[17] that the mere act of binding of the local anesthetic to the receptor causes a change in the properties of the receptor. Thus the *h* infinity curve, which measures the availability of the sodium-carrying system, is shifted to the left—in the hyperpolarizing direction. The interpretation of this experiment was that the inactivated form of the sodium channel has a higher affinity for the local anesthetic than the resting form, or, alternatively, binding of the local anesthetic to the receptor favors the inactivated form. This and other findings led Hille[17] to propose in 1977 the well-known modified receptor model (Fig. 4) in which the drug-free sodium channel is envisaged as existing in three states in equilibrium with each

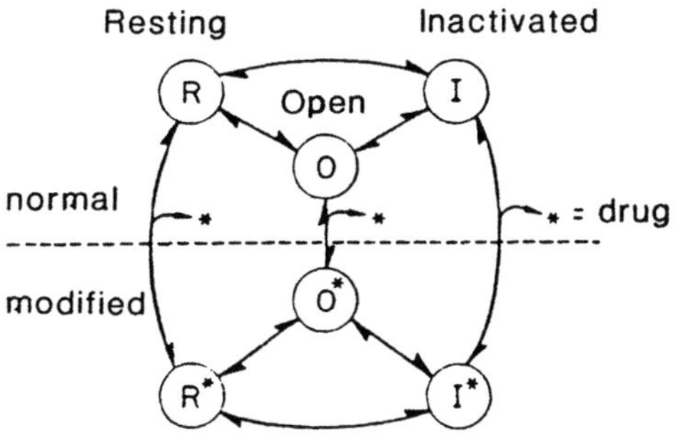

Figure 4. Hille's modified receptor hypothesis. R, O, and I are drug-free resting, open, and inactivated states, respectively, in a dynamic equilibrium determined by voltage-dependent interconversion rate constants. The starred conversions (*) represent the corresponding states when bound to a local anesthetic. For further details, see Hille. Reproduced from *The Journal of General Physiology*,[17] by copyright permission of The Rockefeller University Press.

other: the resting, open, and inactivated states. These states become modified to three other states (resting star, open star, and inactivated star) when the channel interacts with the local anesthetic.

This model is now fairly generally accepted and will be referred to in subsequent chapters. Among other things, the model accounts for the shift in inactivation observed by Courtney[6] and by Hille.[17] Figure 5 shows diagrammatically how a local anesthetic molecule, having gained access to the receptor and having then bound to it, produces a conformational change in the protein to affect the *h* gating subunits so as to lock them in the closed position.

FUTURE

Present ideas on the mechanism of local anesthetic action are embodied in the subsequent chapters of this volume. Clearly, prediction of the future is difficult (and dangerous). However, there are three areas that seem worthwhile suggesting.

The first area concerns whether or not local anesthetics can produce a differential block. The question is not whether or not there is a differential sensitivity of different kinds of fiber type, for it is now becoming reasonably well established on the basis of work by Covino, Gissen, and their colleagues[12–14] that under appropriate circumstances large-diameter myelinated A-type fibers can be consistently blocked at lower concentrations of local anesthetic than the slower-conducting, smaller-diameter, nonmyelinated C-type fibers. The myelinated autonomic B fibers may have an intermediate sensitivity to local anesthetics. Rather, the

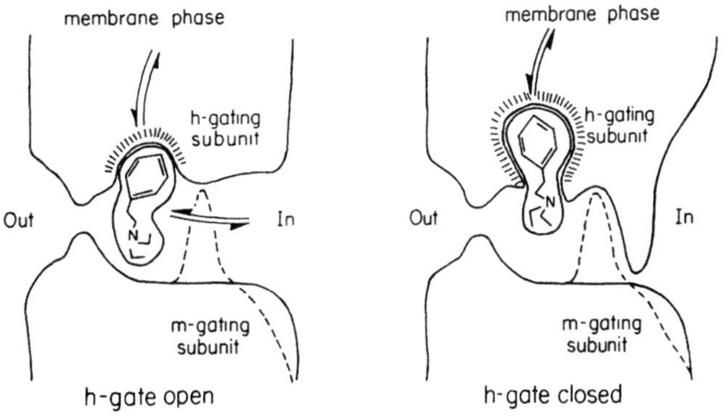

Figure 5. Diagram of a local anesthetic molecule binding in the pore of a sodium channel in a manner that promotes inactivation. In its cationic form the local anesthetic gains access to the binding site when the *m* and *h* gates are open. In its uncharged form (and also perhaps in its charged form in the case of certain lipophilic local anesthetic agents), it can also reach the site from the membrane phase. The binding site has an important hydrophobic component (shading), and closure of the *h* gate enhances the hydrophobic interaction. Taken from Hille. Reproduced from *The Journal of General Physiology*,[17] by copyright permission of Rockefeller University Press.

question is whether within a given fiber type, for example the somatic myelinated A fibers, a differential sensitivity exists.

Gasser and Erlanger[11] in 1929 showed that under their experimental conditions small myelinated A fibers (the small A gamma and A delta fibers) were more readily blocked by local anesthetic agents than the large A alpha fibers. This conclusion was confirmed by Nathan and Sears[22] in 1961, who, however, made the additional point that the fiber size/sensitivity relationship did not extend across fiber types; indeed, some of the smallest myelinated A delta fibers were more sensitive to local anesthetics than the yet smaller nonmyelinated C fibers.

However, Franz and Perry[8] in 1974 showed, quite convincingly, that although under certain circumstances small myelinated axons are indeed blocked more quickly than large myelinated axons, this differential effect cannot be accounted for by differences in minimal effective anesthetic concentrations required to block electrical activity at the nodes of Ranvier in different-sized fibers but rather depends on differences in the initial length of axons that must be exposed to a given concentration to produce block. Indeed, as far as the absolute sensitivity to local anesthetics is concerned, all the axons studied by Franz and Perry seemed to have the same sensitivity.

This conclusion of a lack of absolute differential sensitivity is borne out by computer simulations of the action potential (Fig. 6) showing how the conduction velocity, the spike duration, and the spike height in a 20-μm-diameter frog nerve at 20°C would be expected to change when the relative sodium conductance is decreased, as it would be by a local anesthetic. As the relative sodium conductance falls (or, by reinterpretation, as the degree of local anesthetic action progresses), conduction velocity falls; and finally, at about one-sixth of the normal sodium conductance the impulse fails; i.e., there is a safety factor of 6. In principle, all fibers should be equally sensitive to such a reduction of the sodium conductance by a local anesthetic, if myelinated fibers are scaled down in a regular way,[27] so that, for example, both the thickness of the myelin and the internodal distance are proportional to fiber diameter. However, in the peripheral nervous system, it seems that this relationship of internodal distance to fiber diameter may break down for fiber diameters below about 4 μm.[24] As a result, the safety factor, which is constant at a value of about 6 for all fibers greater than 4 μm in diameter, falls and reaches a value of unity at about a diameter of 1 μm. Smaller fibers cannot conduct impulses, and conduction block necessarily occurs, which is probably why myelinated fibers smaller than 1 μm in diameter are rarely found in the peripheral nervous system. On theoretical grounds, therefore, a differential sensitivity to

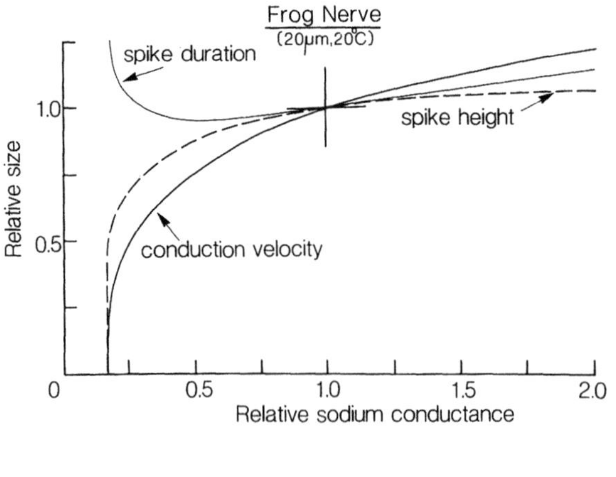

Figure 6. Computer simulation of the dependence on maximum sodium conductance (G_{Na}) of the conduction velocity, spike amplitude, and spike duration in a 20-μM-diameter fiber at 20°C. The basic parameters are those used by Ritchie,[24] which were essentially those of Frankenhauser and Huxley.[7] All values are expressed as fractions of the normal values.

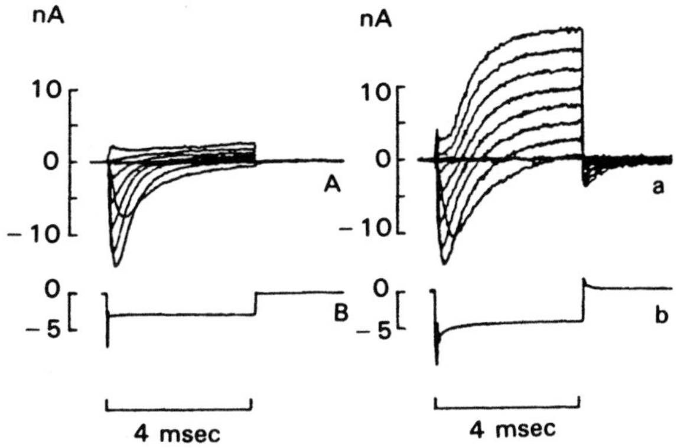

Figure 7. Ionic currents in a rabbit node of Ranvier before (left-hand records) and after (right-hand records) acute treatment to disrupt the myelin. The upper records show currents in response to a series of depolarizing pulses from an initial holding potential of −80 mV in 15 mV increments. The lower pair show the currents obtained in response to a hyperpolarization to −125 mV. Note that the sudden appearance of outward current about 30 min after beginning the myelin-loosening treatment is associated with a large increase in the nodal capacitive transient current. Modified from Ritchie and Chiu.[25]

local anesthetics might be expected for the smaller-diameter fibers. Unfortunately, these are the very fibers that are most difficult to study experimentally.

However, there is another reason why a possible absolute differential sensitivity to local anesthetics should be further explored at this time. It has become clear in the last few years that the mammalian myelinated fiber does not simply behave like a scaled-down version of a squid giant axon that happens to be covered with an electrical insulator over most of its length, except at the nodes of Ranvier where the myelin is interrupted, for voltage-clamp studies show that unlike the squid giant axon or amphibian myelinated fibers, mammalian myelinated fibers exhibit relatively little, voltage-dependent outward potassium current in response to a membrane depolarization; i.e., potassium channels are absent from the mammalian node.[2,5,18]

However, the mammalian fiber is not devoid of the capacity of generating potassium currents. When a rabbit sciatic nerve fiber is subjected to various treatments to produce a paranodal demyelination (in the particular case of Fig. 7 by treating with collagenase), there is no consistent change in the inward sodium current, but there is a dramatic appearance of a large outward potassium current, indicating that potassium channels do exist in the internode even though they are absent from the node itself.[3] There is, in fact, a complementary distribution of sodium and potassium channels in mammalian myelinated fibers[25]: sodium channels are present in high density at the node, where they are needed, with none in the internode; on the other hand, potassium channels are absent from the node but distributed throughout the internode.

A variety of physiological functions are speculated to be subserved by this complementary distribution,[4,25] one of which is relevant to the question of differential blocking by local anesthetics. The internodal potassium channels seem to be important because they support an internodal membrane potential. This in turn is important because if this potential were absent, the internode, which is electrotonically coupled to the node, would place a substantial resistive loading on the node.[4] Indeed, complete abolition of the internodal potential would depolarize the node of Ranvier in a 20-μm-diameter fiber by just under 10 mV.

Such a depolarization would, however, depend on fiber size. Thus, although a 30-μm-diameter fiber might be depolarized by about 5 mV and a 20-μm-diameter fiber by under 10 mV, a 5-μm-diameter fiber would be depolarized by 20 mV. By inactivating the sodium currents, these depolarizations would reduce the safety factor from the normal value of 6 to some smaller value, and this effect would be much greater for smaller fibers than larger fibers. Thus, even with fibers whose diameters are above 4 μm, the possibility of differential sensitivity to local anesthetic agents exists. For example, in the presence of a glycoside, the potassium concentration in the nodal region might not change much from normal because the latter is well irrigated by the extracellular fluid, but potassium might well accumulate in the more protected periaxonal space between the myelin and the internodal axolemma, producing an internodal depolarization and consequently causing (because of the argument just presented) a nodal depolarization, a decrease in safety factor, and hence an increased sensitivity to local anesthetics which would be more prominent in smaller than in larger fibers. Thus, an absolute differential sensitivity to local anesthetics even of moderately sized fibers may well be found under pathophysiological conditions.

A second area warranting further study is the question of whether or not there is a single mechanism of action of local anesthetic agents. Hille's elegant modulated receptor hypothesis is generally interpreted in terms of a single receptor with which all local anesthetics interact, whether charged or uncharged. However, a critical test of such a conclusion seemed to suggest the opposite.[20] The principle involved in testing the hypothesis of one or more sites is quite simple. If there is only one receptor for all local anesthetics, charged or uncharged, then it follows that if two solutions of these agents are equipotent in producing block at equilibrium, any mixture of these two solutions must at equilibrium have the same potency as each of the two components. However, experiments using the action potential of myelinated nerve as the index gave a different result.[20] With equipotent solutions of lidocaine and of the uncharged benzocaine, it was found that a 50/50 mixture of the two is more potent than either of its components, which is incompatible with the idea of a single receptor for the two local anesthetics lidocaine and benzocaine. There must therefore be at least two receptors or sites of action with a differential sensitivity for these two agents.

A similar conclusion, based on quite different grounds, was reached by Huang and Ehrenstein.[19] Looking at how benzocaine and a quarternary analogue of lidocaine (QX572) influence the influx of sodium into neuroblastoma cells through batrachotoxin-activated sodium channels, they showed that the equilibrium dissociation constant for each of these drugs was *reduced* in the presence of the other: classical theory absolutely requires these constants to be increased if the two drugs are competing for the same site.

However, experiments carried out by Schmidtmayer and Ulbrict,[28] studying sodium currents in voltage-clamped frog nerve fibers rather than the action potential, led to the conclusion that only a single site is present. Mixing equipotent solutions of benzocaine and lidocaine seemed to produce the same reduction in sodium current as either above. An idealized form of experiments of Schmidtmayer and Ulbricht is illustrated in Table I. For the sake of argument, it is first assumed that there is only a single site for which lidocaine has an equilibrium dissociation constant of 0.05 mM, whereas benzocaine has an equilibrium dissociation constant of 0.125 mM. Either 0.1 mM lidocaine or 0.25 mM benzocaine will thus separately each produce a block of about 67% (as indeed Schmidtmayer and Ulbricht observed with these concentrations). Furthermore, a mixture of the two (namely, 0.05 mM lidocaine plus 0.125 mM benzocaine) would also produce a 67% fall in the sodium current.

The experimental findings were so close to these predictions that the conclusion was drawn that indeed there is only a single site for lidocaine and benzocaine. However, the voltage-clamp experiments may not have been sufficiently critical to decide the question. Suppose, for example, that there are in fact two independent receptors on the sodium channel, either of which when occupied by a local anesthetic will produce conduction block. If, for the sake of argument, the first receptor is sensitive to both lidocaine and benzocaine with equilibrium dissociation constants of 0.05 and 0.25 mM, respectively, and the second receptor is sensitive only to benzocaine with an equilibrium dissociation constant of 0.5 mM, exposure to either 0.1 mM lidocaine or 0.5 mM benzocaine will cause a 67% decrease in sodium current. With a 50/50 mixture (0.05 mM lidocaine plus 0.25 mM benzocaine), the

Table I. Effect of Lidocaine and Benzocaine on the Basis of Single-Site and Double-Site Models

	Equilibrium dissociation constants for lidocaine (K_L) and benzocaine (K_B) at their site(s) of action		Fractional block produced by 0.1 mM lidocaine (L), by 0.25 mM benzocaine (B), or by 0.05 mM lidocaine plus 0.125 mM benzocaine (mix)		
	Site 1	Site 2	L	B	Mix
Single-site model	$K_L = 0.05$	—	0.67	0.67	0.67
	$K_B = 0.125$	—			
Double-site model	$K_L = 0.05$	—	0.67	0.67	0.68
	$K_B = 0.25$	$K_B = 0.50$			

fall in the sodium current is different from that produced by either component alone, but it is so slightly different, being 68%, that it is doubtful whether any voltage-clamp experiment would detect it. The incorrect conclusion might thus be drawn that this is only one site. The reason that the earlier results with the action potential as index gave a different result probably resulted from the fact that the height of the action potential is a very nonlinear function of sodium conductance, and in the region near block of conductance, the amplitude of the action potential is very sensitive indeed to small conductance changes (Fig. 6).

The question of the number of sites is by no means settled, which is one reason why further study is warranted. However, another important reason is that if there are two or more kinds of local anesthetic receptor site, a new field opens up as far as the use of mixtures of the two kinds of agent is concerned. In general, one might expect mixtures of equipotent solutions of two agents acting at quite different sites to produce a much greater degree of block. Since the concentration of each of the components is necessarily reduced in a mixture (each by factor of two in a 50/50 mixture), the toxic effects of each component will be reduced; and if the spectrum of toxicities of the two drugs is different, one might substantially improve the therapeutic index.

This discussion leads to the third area of future research, namely, the need for finding newer forms of local anesthetic agents. For the reason mentioned previously, it is worthwhile looking for agents that may block conduction in quite a different way from the presently available amino kind of local anesthetic. This would be important, not merely insofar as the toxicity of the agents might be reduced, but there is a clear need at the moment for local anesthetic agents of very long duration, not agents acting for just a few hours, but agents that act nondestructively for a period of several weeks. For example, tetrodotoxin and saxitoxin are of great current interest in this respect. Since these are much more specific and potent than the conventional amine local anesthetics, they might well serve as prototypes for new chemical classes of local anesthetics. Indeed, in animal experiments a combination of saxitoxin and conventional local anesthetic has been shown to produce a nerve block of longer duration than either agent alone.[1] Perhaps with such agents, which are so much more potent and so much more specific than normal local anesthetic agents, it might be possible to produce alkylating analogs that might produce long-lasting anesthesia of the desired kind.

Acknowledgments. This work was supported in part by grants NS 08304 and NS 12327 from the USPHS and by a grant RG 1152 from the National Multiple Sclerosis Society.

REFERENCES

1. Adams HJ, Blair MR, Takman BH: The local anesthetic activity of saxitoxin alone and with vasoconstrictive and local anesthetic agents. *Arch Int Pharmacodyn Ther* 1976;224:274–282.
2. Brismar T: Potential clamp analysis of membrane currents in rat myelinated nerve fibres. *J Physiol (London)* 1980;298:171–184.
3. Chiu SY, Ritchie JM: Evidence for the presence of potassium channels in the internodal region of acutely demyelinated mammalian single nerve fibres. *J Physiol (London)* 1981;313:415–437.
4. Chiu SY, Ritchie JM: On the physiological role of internodal potassium channels and the security of conduction in myelinated nerve fibres. *Proc R Soc Lond B* 1984;220:415–422.
5. Chiu SY, Ritchie JM, Rogart RB, Stagg D: A quantitative description of membrane currents in rabbit myelinated nerve. *J Physiol (London)* 1979;292:149–166.
6. Courtney KR: Mechanism of frequency-dependent inhibition of sodium currents in frog myelinated nerve by lidocaine derivative GEA-968. *J Pharmacol Exp Ther* 1975;195:225–236.
7. Frankenhauser B, Huxley AF: The action potential in the myelinated nerve fibre of *Xenopus laevis* as computed on the basis of voltage clamp data. *J Physiol (London)* 1964;171:302–315.
8. Franz DN, Perry RS: Mechanisms for differential block along single myelinated and non-myelinated axons by procaine. *J Physiol (London)* 1974;236:193–210.
9. Frazier DT, Narahashi T, Yamada M: The site and active form of local anesthetics II. Experiments with quaternary compounds. *J Pharmacol Exp Ther* 1970;171:45–51.
10. Freud S: Ueber Coca. *Centralblat, ges Ther* 1884;2:289–314.
11. Gasser HS, Erlanger J: The role of fiber size in the establishment of a nerve block by pressure or cocaine. *Am J Physiol* 1929;88:581–591.
12. Gissen AJ, Covino B, Gregus J: Differential sensitivity of mammalian nerves to local anesthetic drugs. *Anesthesiology* 1980;53:467–474.
13. Gissen AJ, Covino BC, Gregus J: Differential sensitivity of fast and slow fibers in mammalian nerve. II. Margin of safety for nerve transmission, *Anesth Analg* 1982a;61:561–569.
14. Gissen AJ, Covino BC, Gregus J: Differential sensitivity of fast and slow fibers in mammalian nerve. III. Effect of etidocaine and bupivacaine on fast/slow fibers. *Anesth Analg* 1982b;61:570–575.
15. Goodman L, Gilman A: *The Pharmacological Basis of Therapeutics*, ed 1. New York, Macmillan, 1941.
16. Goodman L, Gilman A: *The Pharmacological Basis of Therapeutics*, ed 2. New York, Macmillan, 1955.
17. Hille B: Local anesthetics: Hydrophilic and hydrophobic pathways for the drug-receptor interaction. *J Gen Physiol* 1977;69:497–515.
18. Horackova M, Nonner W, Stämpfli R: Action potentials and voltage-clamp currents of single rat Ranvier nodes. *Proc Int Union Physiol Sci* 1968;7:198.
19. Huang L-YM, Ehrenstein G: Local anesthetics QX 572 and benzocaine act at separate sites in the batrachotoxin-activated sodium channel. *J Gen Physiol* 1981;77:137–153.
20. Mrose HE, Ritchie JM: Local anesthetics: Do benzocaine and lidocaine act at the same single site? *J Gen Physiol* 1978;71:223–225.
21. Narahashi T, Frazier DT, Yamada M: The site of action and active form of local anesthetics. I. Theory and pH experiments with tertiary compounds. *J Pharmacol Exp Ther* 1970;171:32–44.
22. Nathan PW, Sears TA: Some factors concerned in differential nerve block by local anesthetics. *J Physiol (London)* 1961;157:565–580.
23. Ritchie JM: The mechanism of action of local anesthetic agents, in Lechat P (ed): *Local Anesthetics*. New York, Pergamon Press, 1971, pp 131–166.
24. Ritchie JM: On the relation between fibre diameter and conduction velocity in mammalian nerve fibres. *Proc R Soc Lond B* 1982;217:29–35.
25. Ritchie JM, Chiu SY: Distribution of sodium and potassium channels in mammalian myelinated nerve. *Adv Neurol* 1981;31:329–342.
26. Ritchie JM, Ritchie BR, Greengard P: The active structure of local anesthetics. *J Pharm Exp Ther* 1965; 150:152–159.
27. Rushton WAH: A theory of the effects of fibre size in medullated nerve. *J Physiol (London)* 1951;115:101–122.
28. Schmidtmayer J, Ulbricht W: Interaction of lidocaine and benzocaine in blocking sodium channels. *Eur J Physiol* 1980;387:47–54.
29. Skou JC: Local anesthetics. I. The blocking potency of some local anesthetics and of butyl alcohol determined on peripheral nerve. *Acta Pharmacol Toxicol* 1954;10:281–291.
30. Skou JC: Relation between the ability of various compounds to block nervous conduction and their penetration into a monomolecular layer of nerve tissue lipids. *Biochim Biophys Acta* 1958;30:625–629.

31. Strichartz GR: The inhibition of sodium currents in myelinated nerve by quaternary derivatives of lidocaine. *J Gen Physiol* 1973;62:37–57.
32. Taylor RE: Effect of procaine on electrical properties of squid axon membrane. *Am J Physiol* 1959;196:1071–1078.
33. Trevan JW, Boock E: The relation of hydrogen ion concentration to the action of the local anesthetics. *Br J Exp Pathol* 1929;8:307–315.

17

Lidocaine Block of Sodium Channels in Heart Cells

Bruce P. Bean, Charles J. Cohen, Rosemarie C. Tan, and Richard W. Tsien

ABSTRACT. Lidocaine block of cardiac sodium channels was studied in rabbit Purkinje fibers voltage-clamped with two microelectrodes and in single atrial cells from bullfrog hearts, voltage-clamped with a patch pipette. In Purkinje fibers, dose–response experiments showed that lidocaine blocks the channel by binding one-to-one, with a K_d that depends on holding potential. At a very negative holding potential, where inactivation was completely removed, the K_d was >300 μM, while at a depolarized holding potential, where inactivation was nearly complete, the K_d was ~10 μM. In kinetic experiments, lidocaine induced a component of reactivation much slower ($\tau \sim$ 1–2 sec) than normal; the fraction of slowly recovering channels increased with lidocaine concentration and was about half at 10 μM lidocaine. The experiments are consistent with tight binding of lidocaine ($K_d \sim$ 10 μM) to the inactivated state of the Na channel and much weaker binding ($K_d \sim$ 400 μM) to the resting state. Other kinetic experiments suggested that at therapeutic concentrations of lidocaine, block of open channels is insignificant compared to block of inactivated channels. In single-cell experiments, nonstationary fluctuation analysis experiments showed that lidocaine blocks Na channels in an all-or-none manner and not by changing open-channel conductance. The fluctuation analysis also showed that lidocaine reduced both the number of functional Na channels and the probability of a channel being open at the time of peak Na current; the results suggest that lidocaine blocks and unblocks with a time constant of about 1 sec.

INTRODUCTION

Lidocaine is widely used in cardiology as an antiarrhythmic drug, especially against the ventricular arrhythmias that frequently follow myocardial infarctions. Lidocaine is effective as an antiarrhythmic at much lower concentrations (5–20 μM) than are required for its local anesthetic action on nerve (0.1–1 mM). Despite the difference in effective doses, however, evidence has accumulated suggesting that the action of lidocaine and some other antiarrhythmics on heart muscle is at least partly due to block of Na channels, as for the inhibition of nerve action potentials. The development of new voltage clamp methods for recording currents through cardiac Na channels has opened the way to analyzing how lidocaine and

Bruce P. Bean and Rosemarie C. Tan • Department of Physiology and Biophysics, University of Iowa, Iowa City, Iowa 52242. *Charles J. Cohen* • Miles Institute, New Haven, Connecticut 06509. *Richard W. Tsien* • Department of Physiology, Yale University, New Haven, Connecticut 06510. Present address of B. P. B.: Department of Neurobiology, Harvard Medical School, Boston, Massachusetts 02115.

other drugs block Na channels in heart tissue. Using a rabbit Purkinje fiber preparation with a two-microelectrode voltage clamp,[8,9] we have studied Na channel block by lidocaine over a wide concentration range, with a particular emphasis on block by low concentrations and the question of whether the local anesthetic receptor of cardiac Na channels is fundamentally different than in nerve or skeletal muscle.

In nerve, the experimental phenomenology of local anesthetic block of Na channels is complicated, with block being modulated by steady membrane potential and the frequency of stimulation (see ref. 14). A critical advance was Hille's[14] modulated receptor hypothesis, which, building on earlier work, especially that by Strichartz,[26] Courtney,[10] and Khodorov and collaborators,[19] explained how the complex phenomenology could be understood by a simple idea: that local anesthetics bind to different states of the Na channel with different affinities. A similar model for cardiac muscle was presented by Hondeghem and Katzung,[16] based on observations of voltage-dependent and frequency-dependent block of cardiac action potentials.[4,5,18,27]

Since the work of Hodgkin and Huxley,[15] it has been known that Na channels can exist in at least three states (diagrammed in Fig. 1), with conversions between states promoted by changes in membrane potential. At negative resting potentials, channels are mostly in the resting state (R), where they are closed but available to be opened if the membrane becomes depolarized by an action potential or voltage clamp pulse; upon depolarization, the channels briefly enter the open state (O) and allow Na ions to enter the cell, but after 10 msec or so of a maintained depolarization the channels close again, this time to the inactivated state (I), where the channels are nonconducting and also incapable of being opened by a further depolarization; the membrane must then be repolarized in order to allow the channels to be reprimed by entering the resting state again. The modulated receptor hypothesis supposes that local anesthetics bind more tightly to channels that are in the open or inactivated states than to resting channels, and also that the rates of binding and unbinding can be much slower than normal channel opening and closing kinetics, giving rise to the slow relaxations in Na current size that can be observed in the presence of drugs. For the sake of simplicity, the distinction between the charged and uncharged forms of the lidocaine molecule has been ignored in Fig. 1 (see ref. 23).

The design of the experiments described in this chapter was based on the modulated receptor model in Fig. 1, with a principal aim being to test the hypothesis quantitatively by determining as directly as possible the strength of lidocaine binding to each of the resting, open, and inactivated channels. Consistent with the hypothesis, we find that lidocaine binds very tightly to the inactivated state of cardiac Na channels, with dose–response curves indicating 1 : 1 binding with a dissociation constant of about 10 μM; in contrast, the drug

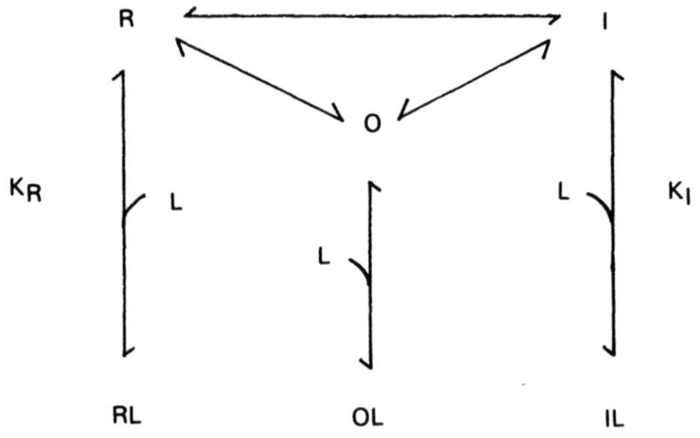

Figure 1. Modulated receptor model for lidocaine (L) binding, after Hille.[14] Na channels are presumed to interconvert between resting (R), open (O), and inactivated (I) states. Lidocaine binds with a different dissociation constant to each state.

binds much more weakly ($K_d \sim 400$ μM) to resting channels. Other experiments suggest that binding to open channels is not significant at therapeutic concentrations of lidocaine. In addition to these results with rabbit Purkinje fibers, preliminary results using fluctuation analysis to analyze Na current from single cardiac cells indicate that lidocaine blocks channels in an all-or-none fashion, with no change in the single-channel current.

METHODS

Rabbit Purkinje Fibers

Rabbit Purkinje fibers were cut into short pieces and voltage-clamped using two microelectrodes.[1,6,7,9] Na currents were recorded at 17°C in external solutions that contained concentrations of NaCl and choline-Cl that totaled 150 mM, 4 mM KCl, 1.8 mM $CaCl_2$, 3.6 mM $MnCl_2$ (to block Ca currents), 0.5 mM $MgCl_2$, 5.0 mM glucose, and 10 mM HEPES, at pH 7.0. To avoid artifacts due to loss of spatial voltage control, Na currents were kept small either by using low Na concentrations (4–10 mM) or by holding the preparation at a membrane potential where most Na channels are inactivated. In experiments where very negative membrane potentials were imposed to ensure that all the channels were in the resting state, 10 mM CsCl was included in the external solution to block currents through the inward rectifier channel.

Single Bullfrog Heart Cells

Single atrial cells were isolated from bullfrog hearts by the method of Hume and Giles,[17] and Na currents were studied at 22°C using the whole-cell variant of the patch clamp technique.[13] The internal solution consisted of 120 mM CsCl, 10 mM Cs EGTA, 5 mM $MgCl_2$, and 10 mM HEPES at pH 7.45; the external solution was 150 mM NaCl, 4 mM KCl, 2 mM $CaCl_2$, 3 mM $CoCl_2$ (to block Ca currents), 10 mM glucose, and 10 mM HEPES, at pH 7.40. To improve the speed of the clamp and to avoid filtering high-frequency fluctuations, pipettes were relatively large (0.5–1.5 MΩ), and series-resistance compensation was used. An initial series of experiments tested for series resistance or cable errors in membrane voltage by looking for shifts in the peak current-voltage curve when currents were reduced by partial inactivation or tetrodotoxin treatment; shifts were always <4 mV (usually <2 mV) for peak unblocked currents of 4 nA or less.

RESULTS

Use-Dependent Block

Figure 2 shows the effect of two concentrations of lidocaine on Na currents in a rabbit Purkinje fiber. The voltage clamp protocol was to first establish a very negative potential (−105 mV) for a long time and then give a train of depolarizing voltage clamp pulses. The top three panels in Fig. 2 show, superimposed, Na current (I_{Na}) during the 1st and 12th pulses in a train of 500-msec pulses given at 1 Hz. With no drug present, the 1st and 12th pulses are exactly the same. With 20 μM lidocaine, the 1st pulse of the train elicits a current almost unchanged from the control; however, as the train continues, currents become smaller, and by the 12th pulse, peak I_{Na} is reduced by about 40%. With 100 μM lidocaine, the current elicited by even the 1st pulse is reduced by about 25% compared to control, and the additional block during the train reduces the current by an additional 60% or so. The "use-dependent" or "frequency-dependent" block during the trains of stimuli is similar to that

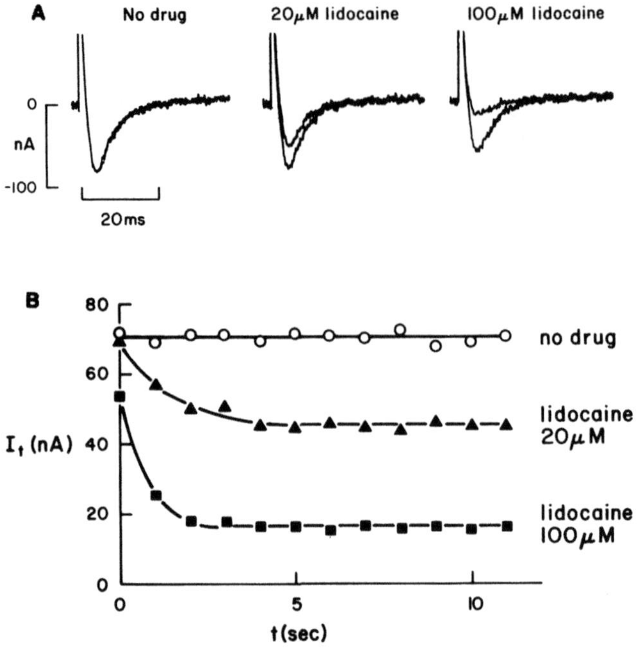

Figure 2. Use-dependent effect of lidocaine. I_{Na} was measured during trains of 500-msec pulses from −105 to −35 mV at 1.0 Hz. (A) The membrane currents were measured on the 1st and 12th pulses in (from left to right) 0, 20, and 100 μM lidocaine. The rest period before each train was long enough for effects of previous trains to be removed. (B) Peak sodium current amplitudes were measured for each of the pulses. The decrease in current magnitude has been fitted by an exponential curve, with τ = 1.3 sec in 20 μM lidocaine and τ = 0.7 sec in 100 μM lidocaine. Preparation C74-3: 10 mM Na, pH 7.0, 17°C. Reproduced from *The Journal of General Physiology*,[1] by copyright permission of the Rockefeller University Press.

previously seen in voltage clamp experiments in nerve and skeletal muscle using higher drug concentrations and is qualitatively similar to results with action potential trains in non-voltage-clamped heart muscle. The experiment with 20 μM lidocaine is particularly interesting because it shows that a clinical concentration of the drug has almost no effect on I_{Na} after a long resting period, but produces a marked reduction under conditions mimicking a train of cardiac action potentials at a physiological frequency.

Binding to Inactivated Channels

The fact that at 20 μM lidocaine has no effect on I_{Na} after a long resting period implies that there is no significant binding to resting Na channels at this concentration (Fig. 2). The accumulation of block during the train of depolarizing pulses must mean that the repriming of channels during the interval between pulses is incomplete in the presence of lidocaine. Figure 3 shows directly how lidocaine changes Na channel repriming after a depolarizing pulse. A long depolarization (conditioning pulse) is followed by a return to the negative holding potential for a variable length of time and then by a second depolarization (test pulse) that tests the completeness of repriming that occurred during the return to the holding potential. With no drug present, repriming occurs with a roughly exponential time course, with a time constant of 30 msec. With 10 μM lidocaine, repriming is no longer exponential but instead occurs in two distinct phases; about half the channels recovery quickly, as in the absence of drug, but the other half of the channels recover much more slowly, with a time constant of about 2 sec. With 200 μM lidocaine, almost all the channels recover with the slow time course.

The modulated receptor hypothesis (Fig. 1) provides an interpretation of the experiment in Fig. 3 as follows: during the long (5 sec) conditioning depolarization, channels open transiently and then become inactivated; the inactivated channels then equilibrate with lidocaine according to the equilibrium constant K_I. Upon repolarization, the channels in state I

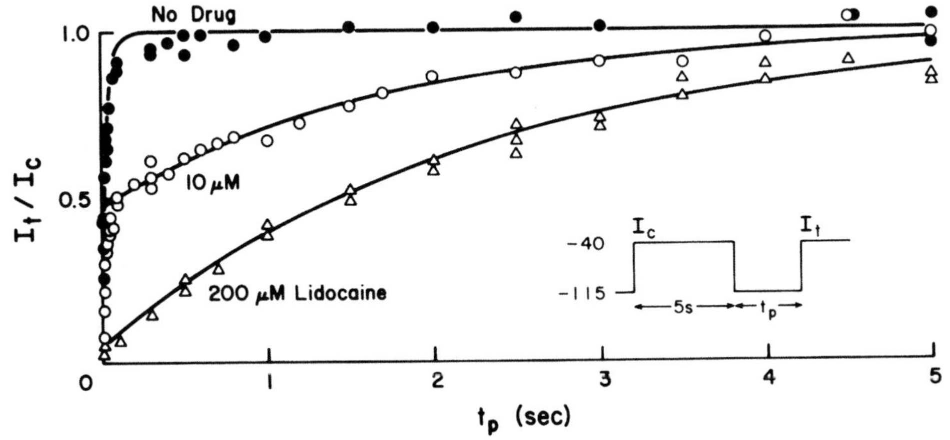

Figure 3. Slowing of sodium channel repriming by lidocaine. Solid curve through control points is drawn according to 1 - exp($-t/\tau$), with τ = 30.1 msec. Curves through the two sets of lidocaine points are drawn according to 1 - A exp($-t/\tau$); for 10 µM, τ = 1.70 sec and A = 0.53; for 200 µM, τ = 2.26 sec and A = 0.95. These parameters were determined by a least-squares fit to points for $t \geq 0.5$ sec. In 200 µM lidocaine, I_c was unchanged when the holding potential was changed to −125 mV, which verifies that removal of inactivation was complete. Preparation C103: 6 mM Na, pH 7.0, 16.5°C.

(inactivated but with no drug bound) reactivate quickly into state R. However, the channels in state IL (inactivated with drug bound) must first unbind the drug molecule (by hypothesis a slow process) before returning to state R, and thus reactivate much more slowly. According to the model, then, the partition of repriming into distinct fast and slow phases represents the partition of inactivated channels between drug-free and drug-bound states. By quantitating the distribution of channels between the slow and fast phases of repriming (which can be done by fitting the slow phase with an exponential and extrapolating to time zero, as in Fig. 3), the strength of lidocaine binding to the inactivated state can be determined; this approach follows that employed by Khodorov *et al.*[19] in their analysis of local anesthetic block of Na channels in frog nerve. Figure 4 shows collected data from 13 experiments of this kind; the data are well fit by a dose–response curve drawn according to 1 : 1 binding of drug to inactivated channels with a dissociation constant (K_I in Fig. 1) of 10 µM. These results help confirm the hypothesis[14] that local anesthetic molecules bind to a specific receptor on sodium channels; if, instead, local anesthetic molecules affected Na channels by an indirect generalized perturbation of membrane lipid structure, one would expect the reactivation of all the Na channels to be slowed more or less equally and not the crisp division into quickly and slowly reactivating channels that is actually seen at low concentrations.

The modulated receptor scheme in Fig. 1 suggests another, more direct method of determining the strength of lidocaine binding to the inactivated state. If the membrane is held at a potential where most of the channels are in the inactivated state, the model predicts that lidocaine will block current elicited by depolarizations with a potency that reflects binding to the inactivated state. The method depends on the fact that resting and inactivated channels are in relatively rapid ($\tau \sim$ 10–100 msec) equilibrium. Lidocaine binding to inactivated channels is in itself electrically silent, since these channels are already nonconducting and incapable of being opened by a depolarization, but because of the equilibrium between resting and inactivated channels, decreasing the number of channels in state I will cause a proportionate decrease in the number in state R and thus lead to a smaller Na current elicited by a depolarization. Such an experiment is shown in Fig. 5 (left curve). The membrane was held at −65 mV, where almost all (~99%) of the Na channels were inactivated. Na current

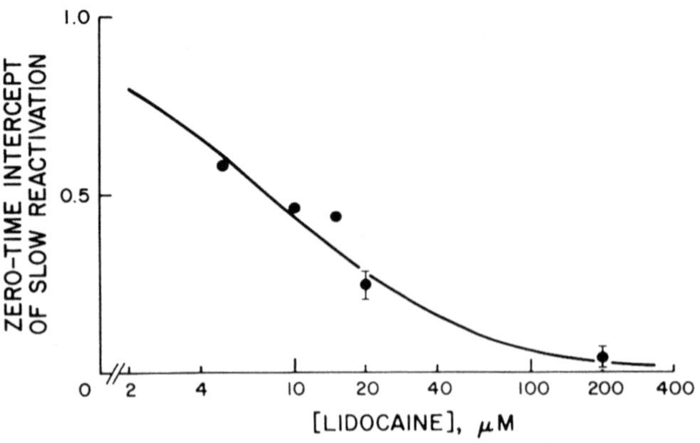

Figure 4. Zero-time intercept of slow repriming versus lidocaine. Points for 5, 10, and 15 μM were from a single fiber (the same as in Fig. 2). Data for 20 μM and 200 μM, plotted as means ± S.E.M., are collected data from four and six experiments, respectively. In each experiment, the conditioning pulse was long enough (2–5 sec) to produce a maximal effect, and the holding potential was negative enough (−105 to −135 mV) to ensure complete removal of inactivation before the conditioning pulse: both conditions are necessary for a simple interpretation by the model in Fig. 3. Each zero-time intercept was obtained from a least-squares fit to the equation $1 - A\exp(-t/\tau)$ to the slow component of repriming ($t > 0.3$ sec). The curve is a least-squares fit to $[1 + [L]/K_d]^{-1}$ with $K_d = 10$ μM. 6.5–10 mM Na, pH. 7.0, 16.5–17.5°C. Reproduced from *The Journal of General Physiology*,[1] by copyright permission of the Rockefeller University Press.

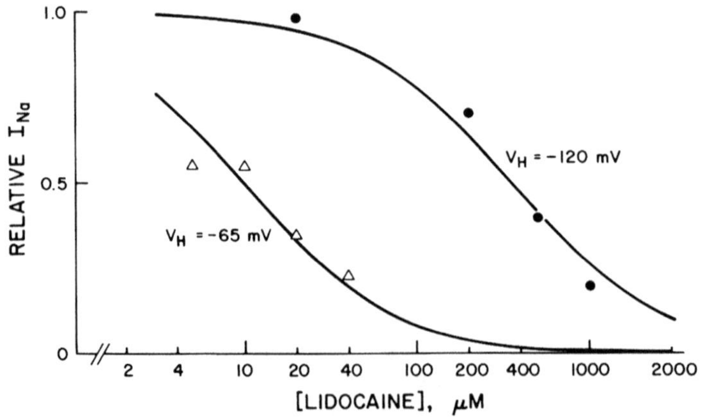

Figure 5. Dose–response curves for block at holding potentials of −120 (●) and −65 mV (△). For block at −120 mV, I_{Na} was measured using test pulses to −40 mV. Sequence of solutions and actual peak currents: control, 23.3 nA; 400 μM lidocaine for 11 min, 10.0 nA; 1 mM lidocaine for 8 min, 5.4 nA; washout for 10 min, 29.1 nA; 200 μM lidocaine for 10 min, 20.5 nA; washout for 14 min, 29.4 nA; 20 μM lidocaine for 12 min, 28.8 nA. Hyperpolarizing to −131 mV did not increase the current size, even in 1 mM lidocaine. Preparation C95-3: 8.5 Na, pH 7.0, 17.5°C. For block at −65 mV, test pulses to −45 mV were used. Sequence of solutions and actual peak currents: control, 94 nA; 10 μM lidocaine for 5.5 min, 54 nA; 20 μM lidocaine for 6 min, 35 nA; washout for 8.5 min, 107 nA; 5 μM lidocaine for 12 min, 61 nA; 40 μM lidocaine for 9 min, 27 nA; washout for 6 min, 119 nA. Preparation C92-1: 155 mM Na, pH 7.0, 17.0°C. For both experiments, each solution was applied long enough for I_{Na} to reach a steady state. The fiber was rested for at least 15 sec before each test pulse. In both experiments, currents were normalized assuming a linear drift of peak I_{Na} in the control solution. Reproduced from *The Journal of General Physiology*,[1] by copyright permission of the Rockefeller University Press.

was elicited by infrequent pulses to −45 mV; although this current is from the only ~1% of channels that is not inactivated, it was measurable because the bathing solution contained 155 mM Na instead of the usual 4–10 mM Na. Various concentrations of lidocaine were applied, and block was allowed to proceed to equilibrium in each solution. Under these conditions, the half-blocking dose was about 10 μM; the solid line corresponds to 1 : 1 binding with an apparent dissociation constant of 9.7 μM. Even 5 μM lidocaine, a dose that is marginally effective in treating arrhythmias, has a substantial blocking effect. The estimate of 10 μM for half-binding to the inactivated state obviously agrees very well with the estimate made from the very different approach in Figs. 3 and 4.

Block of Resting Channels

The strength of lidocaine binding to the resting state of the Na channel can be determined straightforwardly, by applying lidocaine at a very negative holding potential, where all channels are in the resting state, and using infrequent pulses to avoid use-dependent block. Figure 5 (right-hand curve) shows the results of such an experiment, where Na current was elicited by infrequent pulses (1/min) from a holding potential of −120 mV. The data are well fit by a 1 : 1 binding curve with a half-blocking concentration of 353 μM. Other experiments gave values closer to 400 μM. Thus, lidocaine binds about 40 times more weakly to the resting state than the inactivated state, in agreement with the modulated receptor hypothesis.

Determining the strength of binding to the open state of the Na channel is more difficult. Since the open state is transient and is occupied for only a few milliseconds during a depolarization, it is virtually impossible to obtain drug equilibration with open channels under normal circumstances. One strategy would be to look at open-channel block in channels that have had inactivation removed enzymatically or pharmacologically. However, even with normally operating channels, one can do experiments to determine whether open-channel block contributes to the overall block seen under a given set of circumstances. Figure 6 shows an experiment that examines whether open-channel block is important for the use-dependent block seen with lidocaine. A conditioning depolarization was given for various lengths of time, and after a 250-msec return to the holding potential, a test pulse was given. Test pulse Na current gives a measure of the lidocaine-induced block that develops during the conditioning pulse; the 250-msec return to rest is long enough so that lidocaine-free channels would have time to reprime almost completely (see Fig. 3). In Fig. 6A, the time course of development of block by 20 μM lidocaine is shown for two conditioning pulses, −69 mV and +31 mV. At both potentials, almost all the channels inactivate completely during the conditioning pulse, but at −69 mV there was no detectable Na current, while at +31 mV, channel opening is maximal. Despite this difference, the development of block is very similar at the two potentials; in particular, there is not the large, sudden drop in test current for very short conditioning pulses at +31 mV that would be expected if block of open channels were important. Instead, the results are consistent with all of the block arising from lidocaine binding to inactivated channels.

If a much higher lidocaine concentration (200 μM) and a higher pH (8.1 instead of 7.0) are used, there is a rapid phase in the development of block at −40 mV (where channels are open during the depolarization), and such a rapid phase is lacking at −60 mV, where channels do not open significantly. Thus, under these conditions, where both the higher pH and the higher lidocaine concentration will increase the concentration of lidocaine inside the cells, there probably is significant open-channel block during a depolarization.

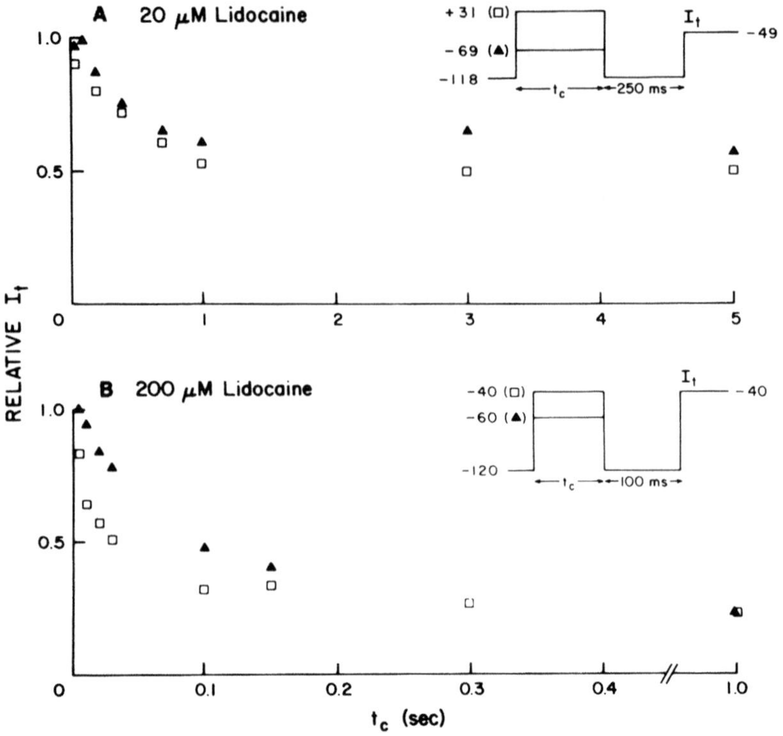

Figure 6. Time course of development of lidocaine block during depolarization. (A) Development of block by 20 μM lidocaine at −69mV (where channel opening was not detectable) and +31 mV (where channel activation is maximal). Preparation C95-2: 10 Na, pH 7.0, 18°C. (B) Development of block by 200 μM lidocaine at −60 mV (no detectable sodium current) and at −40 mV (substantial sodium current) on a faster time scale. Preparation C95-3: 8.5 Na, pH 8.1, 17.5°C. Reproduced from *The Journal of General Physiology,*[1] by copyright permission of The Rockefeller University Press.

All-or-None Channel Blocking

The development of the patch clamp technique[13] will undoubtedly soon lead to much more detailed information about local anesthetic block of Na channels by permitting recording of single-channel behavior in the presence of drug molecules. Many open questions can be addressed by investigating drug block at the single-channel level. For example, the modulated receptor hypothesis implicitly assumes that the single-channel conductance is not affected by local anesthetics; channels are assumed either to be completely blocked or to conduct normally. This prediction has not yet been tested for tertiary or quaternary drugs.

Figure 7 shows an experiment that tests this prediction for lidocaine block of cardiac Na channels, using noise analysis to measure single-channel current. In this experiment, a single cell that was enzymatically isolated from bullfrog atrium (by the method of Hume and Giles[17]) was voltage-clamped using the whole-cell configuration of the patch clamp technique. Under ionic conditions that blocked non-Na currents (internal Cs and EGTA, external Co), whole-cell Na currents were recorded for a series of 117 identical voltage clamp pulses from −100 to −10 mV. Then, using the nonstationary noise analysis technique developed by Sigworth,[24] the records were aligned in time, and the small variations between individual records (due to stochastic opening and closing of the Na channels) were analyzed by calculat-

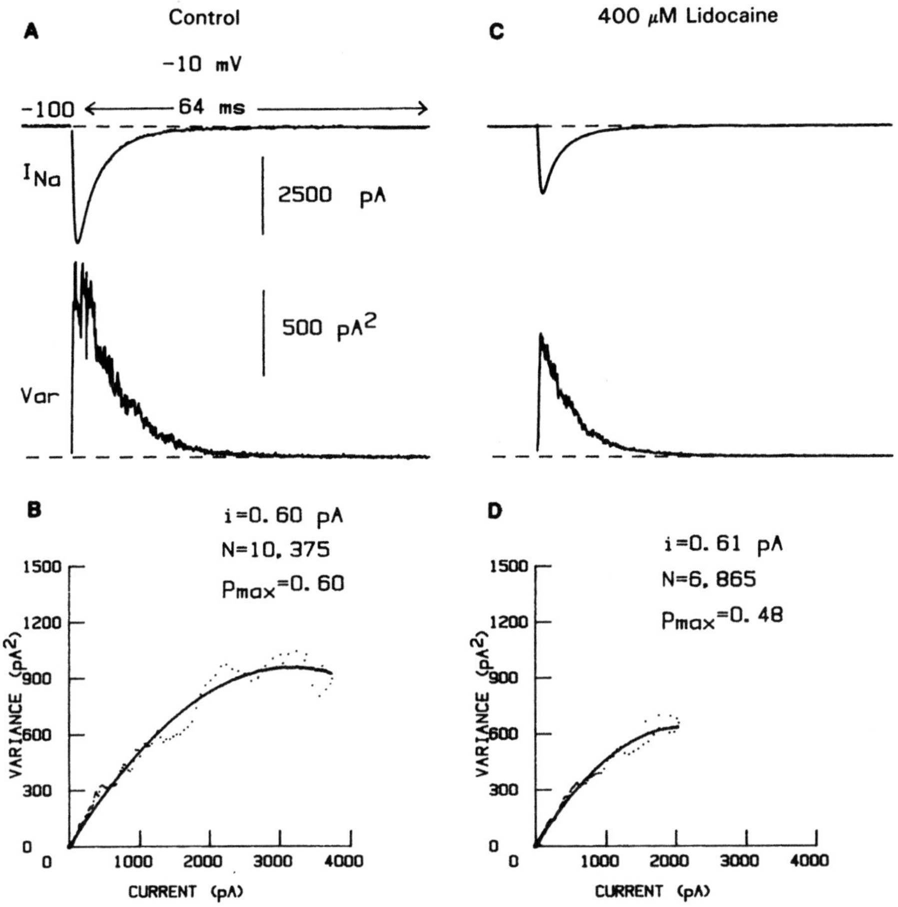

Figure 7. Nonstationary fluctuation analysis of lidocaine block in a single bullfrog atrial cell. (A) Mean and variance for an ensemble of 117 Na currents elicited at a frequency of 1 Hz. (B) Variance versus mean current. Theoretical fit (as described in text) is best (least-squares) fit. (C,D) Sample analysis for an ensemble of 123 sweeps after addition of 400 μM lidocaine to the external solution. 23°C.

ing the variance of the current at each time point over the ensemble of 117 currents. The variance was calculated in a pairwise manner, between each pair of successive currents, and then averaged over the ensemble. Figure 7A shows the time course of the mean current and of the variance for this ensemble of individual Na currents; the variance rises to a peak during the rising phase of the Na current and then falls with a time course roughly paralleling that of inactivation. In Fig. 7B the variance is plotted versus mean current; the relationship can be very well fit by the parabolic relationship

$$\text{Variance} = iI - I^2/N$$

expected for N identical channels with a single-channel current of i pA.[24] The single-channel current derived from the fit is 0.60 pA. In Fig. 7C-D, the noise analysis was repeated in the same cell with 400 μM lidocaine present. With the negative holding potential employed in this experiment, this concentration of lidocaine decreased Na current to about half the control value; however, the noise analysis showed that single-channel current (0.61 pA) was almost unchanged from that in the control. This experiment shows that Na channels are blocked in an all-or-none manner by lidocaine, with unblocked channels retaining a normal single-channel current size.

The ensemble noise analysis also provides information about the kinetics of lidocaine block. With no drug present, the data are fit by assuming that there are about 10,400 channels in the cell and that at the peak of the macroscopic current each channel has a probability of being open (p_{max}) of 0.6. With lidocaine present, the decrease in peak macroscopic current (~50%) is accounted for partly by a decrease in the number of channels (~35%) and partly by a decrease in p_{max} (~20%). This result can be understood if lidocaine blocks and unblocks channels with a time course comparable to the 1-sec interval between the pulses that elicited the currents: if lidocaine bound and unbound very fast compared with the interpulse interval, there would be complete randomization of blocked and unblocked channels between pulses; this would produce a decrease in p_{max}, with no change in N. On the other hand, if lidocaine equilibration with the channels were very slow compared with the interpulse interval, the same channels blocked during one pulse would very likely remain blocked during the next and would not contribute to the current or variance (calculated in a pairwise manner); these channels would be effectively removed from the population of channels giving rise to the current, while those unblocked would behave normally, and the decrease in macroscopic current would be entirely due to a decrease in N, with no change in p_{max} (see ref. 25 for a discussion of fluctuation analysis and time scales of channel block). Since lidocaine reduced both the number of functional channels and p_{max}, the results suggest that lidocaine blocks and unblocks channels with a time constant of about 1 sec, neither much faster nor much slower than the interpulse interval.

DISCUSSION

Tight Binding to the Inactivated State

Our experiments provide direct evidence for the idea that lidocaine binds especially tightly to inactivated channels. Both of the different approaches in Figs. 4 and 5 suggest that the the half-binding concentration for inactivated channels is 10 μM. (A third method, measuring lidocaine-induced shifts in steady-state Na channel availability, yielded the same estimate.[1]) Binding to inactivated channels is about 40 times stronger than to resting channels. Together with the observation that binding to both states appears to be in a 1 : 1 stoichoimetry of drug to receptor, these observations provide direct support for the modulated receptor hypothesis.[14,16]

The agreement between the values of K_I (Fig. 1) estimated from the experiments in Fig. 4, with low Na, and Fig. 5, with full Na, suggests that external Na has no effect on the strength of lidocaine binding to inactivated channels. The work by Cahalan and Almers[3] is frequently cited as evidence that local anesthetic block is antagonized by external Na, but there is no conflict between the results, since their experiments were done with QX-314, the quaternary analogue of lidocaine, for which block of open channels is most important; it is easy to imagine that drug binding in an open channel would be antagonized by Na while binding to closed, inactivated channels would not be, since external cations might have no access to the binding site. Binding to inactivated channels is probably the predominant mechanism at therapeutic concentrations (see below).

The strength of lidocaine binding to inactivated channels helps explain the drug's clinical efficacy against arrhythmias; one kind of arrhythmia following myocardial infarction arises from reentrant pathways produced by damaged, depolarized, slowly conducting tissue (see ref. 22). Probably, this tissue is depolarized enough so that most Na channels are inactivated, and as Fig. 5 shows, the remaining Na current in such tissue would be strongly inhibited by clinical concentrations of lidocaine (5–20 μM) that have almost no effect on the resting channels in normal well-polarized tissue. Thus, the pathological reentrant pathways will be blocked while normal conduction of the action potential remains intact.

Comparison with Nerve and Skeletal Muscle

The experimental protocols we used are somewhat different from those used in work with lidocaine on nerve and skeletal muscle, where local anesthetic block has mostly been studied with large drug concentrations and trains of brief depolarizations (see, however, ref. 19 for experiments with low concentrations of other local anesthetics in nerve). However, it seems likely that lidocaine block is very similar in Na channels in the various tissues. For example, Schwarz *et al.*[23] constructed a model for lidocaine block in frog skeletal muscle with a K_I of 8 μM and K_R of 200 μM, not much different from our values, and the data from frog node of Ranvier[11,14] are quantitatively similar. Thus, although cardiac Na channels are definitely pharmacologically distinct from those in nerve and skeletal muscle in having a very different interaction with tetrodotoxin,[6,21] the local anesthetic receptor may well be the same in Na channels from the different tissues. This conclusion is supported by the similarity of the effects on local anesthetic block produced by changing pH in the different tissues.[1,12,19,23] The much greater potency of lidocaine for combatting arrhythmias in heart muscle than for blocking nerve conduction probably arises because in the first case, lidocaine acts mainly by binding to inactivated channels, while for nerve block, block of resting channels in well-polarized axons is required.

Open-Channel Block

The use-dependent block seen with low, therapeutic concentrations of lidocaine does not seem to involve binding of drug to open channels, but rather cumulative binding to inactivated channels (Fig. 6). This may seem surprising since the early work on use dependence emphasized open-channel block by charged or mostly charged local anesthetics.[10,26] However, the result is in line with the model by Schwarz *et al.*,[23] which predicts that at 20 μM, lidocaine use dependence will be due almost entirely to binding to inactivated channels. The finding that there is little open-channel block by this concentration also fits well with Cahalan's[2] finding that local anesthetic binding to open channels (with inactivation removed) occurs with relatively high half-blocking concentrations of 0.1–1 mM.

One still frequently reads statements that local anesthetics act in the cationic form from the axoplasmic side of the membrane, but the modulated receptor hypothesis[14,23] and the experiments that support it actually show quite clearly that channels can be blocked by both neutral and charged forms. Block of inactivated channels by the neutral form (which may become charged after binding to the channel, as proposed by Schwarz *et al.*[23]) is almost certainly the predominant mechanism at low concentrations of tertiary local anesthetics like lidocaine.

Single-Channel Behavior

The noise analysis results show that lidocaine blocks Na current without altering single-channel current, so that channels are either completely blocked or conducting normally. A similar result was previously obtained for benzocaine block of steady-state Na current in frog nerve treated with sea anemone toxin.[20] The noise analysis also shows that lidocaine block involves moderately rapidly exchanging (about 1 sec) equilibrium drug binding. Under the conditions of the experiment in Fig. 7, block was predominantly due to binding to resting channels; it may well be that if different protocols were used for which binding to inactivated channels is most important, different equilibration rates of drug with channel would be found.

ACKNOWLEDGMENTS. The single-cell work was supported by grants to B.P.B. from the American Heart Association, Iowa Affiliate, and the Pharmaceutical Manufacturers Association Foundation. R.C.T. was supported by a Medical Student Summer Research Fellowship from the University of Iowa.

REFERENCES

1. Bean BP, Cohen CJ, Tsien RW: Lidocaine block of cardiac sodium channels. *J Gen Physiol* 1983;81:613–642.
2. Cahalan MD: Local anesthetic block of sodium channels in normal and pronase-treated squid giant axons. *Biophys J* 1978;23:285–311.
3. Cahalan MD, Almers W: Block of sodium conductance and gating current in squid giant axons poisoned with quaternary strychnine. *Biophys J* 1979;27:57–74.
4. Chen C-M, Gettes LS: Combined effects of rate, membrane potential, and drugs on maximum rate of rise of action potential upstroke of guinea pig papillary muscle. *Circ Res* 1976;38:464–469.
5. Chen C-M, Gettes LS, Katzung B: Effect of lidocaine and quinidine on steady-state characteristics and recovery kinetics of $(dV/dt)_{max}$ in guinea pig ventricular myocardium. *Circ Res* 1975;37:20–29.
6. Cohen CJ, Bean BP, Colatsky TJ, Tsien RW: Tetrodotoxin block of sodium channels in rabbit Purkinje fibers. Interactions between toxin binding and channel gating. *J Gen Physiol* 1981;78:383–411.
7. Colatsky TJ: Voltage clamp measurements of sodium channel properties in rabbit cardiac Purkinje fibers. *J Physiol (London)* 1980;305:215–234.
8. Colatsky TJ, Tsien RW: Electrical properties associated with wide intercellular clefts in rabbit Purkinje fibers. *J Physiol (London)* 1979a;290:227–252.
9. Colatsky TJ, Tsien RW: Sodium channels in rabbit Purkinje fibers. *Nature* 1979b;278:265–268.
10. Courtney K: Mechanism of frequency-dependent inhibition of sodium currents in frog myelinated nerve by the lidocaine derivative GEA 968. *J Pharmacol Exp Ther* 1975;195:225–236.
11. Courtney KR: Comparative actions of mexiletine on sodium channels in nerve, skeletal, and cardiac muscle. *Eur J Pharmacol* 1981;74:9–18.
12. Grant AO, Strauss LJ, Wallace AG, Strauss HC: The influence of pH on the electrophysiological effects of lidocaine in guinea pig ventricular myocardium. *Circ Res* 1980;47:542–550.
13. Hamill OP, Marty A, Neher E, Sakmann B, Sigworth FJ: Improved patch clamp techniques for high-resolution current recording from cells and cell-free membrane patches. *Pflugers Arch* 1981;391:85–100.
14. Hille B: Local anesthetics: hydrophilic and hydrophobic pathways for the drug:receptor reaction. *J Gen Physiol* 1977;69:497–515.
15. Hodgkin AL, Huxley AF: A quantitative description of membrane current and its application to conduction and excitation in nerve. *J Physiol (London)* 1952;117:500–544.
16. Hondeghem LM, Katzung BG: Time- and voltage-dependent interactions of antiarrhythmic drugs with cardiac sodium channels. *Biochim Biophys Acta* 1977;472:373–398.
17. Hume JR, Giles W: Active and passive electrical properties of single bullfrog atrial cells. *J Gen Physiol* 1981;78:19–42.
18. Johnson EA, McKinnon MC: The differential effect of quinidine and pyrilamine on the myocardial action potential at various rates of stimulation. *J Pharmacol Exp Ther* 1957;120:460–468.
19. Khodorov BL, Shishkova L, Peganov E, Revenko S: Inhibition of sodium currents in frog Ranvier node treated with local anesthetics. Role of slow sodium inactivation. *Biochim Biophys Acta* 1976;433:409–435.
20. Neumcke B, Schwarz B, Stampfli R: Block of Na channels in the membrane of myelinated nerve by benzocaine. *Pflugers Arch* 1981;390:230–236.
21. Reuter H, Baer M, Best PM: Voltage-dependent of TTX action in mammalian cardiac muscle, in Morad M (ed): *Biophysical Aspects of Cardiac Muscle.* New York, Academic Press, 1978, pp 129–142.
22. Rosen MR: Cardiac drugs: antiarrhythmic drugs. *Curr Cardiol* 1979;1:259–303.

23. Schwarz W, Palade PT, Hille B: Local anesthetics: Effects of pH on use-dependent block of sodium channels in frog muscle. *Biophys J* 1977;20:343–368.
24. Sigworth FJ: The variance of sodium current fluctuations at the node of Ranvier. *J Physiol (London)* 1980a;307:131–142.
25. Sigworth FJ: The conductance of sodium channels under conditions of reduced current at the node of Ranvier. *J Physiol (London)* 1980b;307:131–142.
26. Strichartz GR: The inhibition of sodium currents by quaternary derivatives of lidocaine. *J Gen Physiol* 1973;62:37–57.
27. Weld FM, Bigger JT: Effect of lidocaine on the early inward transient current in sheep cardiac Purkinje fibers. *Circ Res* 1975;37:630–639.

18

The Kinetic Basis for Phasic Local Anesthetic Blockade of Neuronal Sodium Channels

Gary Strichartz and Ging Kuo Wang

ABSTRACT. The blockade of impulse conduction in nerves by local anesthetics depends on the inhibition of voltage-dependent sodium channels, a reaction that occurs in two modes: a "tonic" block characterizes the reduction of sodium current in a resting nerve, and an additional "phasic" decrease in current occurs during repetitive depolarization. Both modes have been explained previously by Hille's modulated receptor hypothesis (MRH). According to this proposal, local anesthetics bind more tightly to, and thereby stabilize the inactivated form of sodium channels, a nonconducting channel state resulting from membrane depolarization. Two recent results from our laboratory conflict with the MRH, so stated. First, we find that when channel inactivation is prevented, either by a scorpion toxin (αLQ) or by oxidizing reagents [chloramine-T (CT) or *N*-bromoacetamide (NBA)], channels remain sensitive to both tonic and phasic modes of local anesthetic action. Second, depolarizations that are sufficiently long to open (activate) channels, but of too brief a duration to permit most channels to rapidly inactivate, still produce full phasic block by a local anesthetic, a block that develops at the same rate as that from long depolarizations which cause full inactivation. We interpret these results to show that phasic block occurs during the activation process and that it does not require any fast inactivation reactions. In one possible kinetic scheme, equilibrium binding of anesthetics to channels occurs before the conditioning depolarization, which drives the drug receptor to a transient, nonconducting conformation, one that then reverses only slowly to the resting state. Repetitive depolarizations can thus "trap" channels in this high-affinity form, and, upon repolarization, the remaining channels reequilibrate between low-affinity anesthetic-free and anesthetic-bound forms. Because proteolysis by enzymes in the cytoplasmic space removes phasic block for many charged and polar local anaesthetics, yet external αLQ, CT, and NBA do not, the effects of proteolysis on phasic block cannot be due to the removal of fast inactivation *per se*. The enzymes may be digesting regions of a binding site for local anesthetics.

INTRODUCTION AND HISTORICAL REVIEW

Actions of Local Anesthetics on Sodium Channels Defined

The principal impulse-blocking effect of local anesthetics occurs through their action on voltage-dependent sodium channels.[12,27] Many experiments during the past 15 years have investigated this action in detail, using the technique of voltage clamping to measure both the macroscopic ionic currents (I_{Na}) and the assymetrical displacement "gating" currents (I_g) associated with the operation of sodium channels. Each type of current reveals a different

Gary Strichartz and Ging Kuo Wang • Anesthesia Research Laboratory, Brigham and Women's Hospital, and Department of Pharmacology, Harvard Medical School, Boston, Massachusetts 02115.

aspect of channel function; ionic currents show directly the kinetics of the appearance and disappearance of open channels, whereas gating currents reflect the charge movements resulting from motion of the electrically active channel macromolecules, motion that occurs primarily during conformational transitions between resting and open channels.[3]

Two modes of channel inhibition by local anesthetics can be defined functionally. The blockade of ionic currents measured by a single depolarization is called "tonic" block; the additional fall in current amplitude during repetitive depolarization is called "use-dependent" or "phasic" block.[10,26] The degree of this phasic block is larger for more frequent depolarizations, for the block is reversed when the nerve membrane returns to rest. Kinetic analyses of phasic block demonstrate that it must arise from the selective modulation by anesthetic molecules of specific conformational transitions of the channel that result from depolarizations. Tonic block is also potentiated by membrane depolarizations (either constant or from pulses of long duration[10,13] One question about the action of local anesthetics is whether phasic and tonic block occur by the same mechanisms and correspond to anesthetic binding to the same channel sites or channel conformations. In this chapter we wish to survey briefly the published mechanisms for these blocking modes and then to propose a somewhat different concept.

Phasic Block and Inactivation Shifts

Another feature of sodium channel blockade, in addition to phasic block, is the modification of the resting "inactivation" function of sodium channels. Hodgkin and Huxley[14] showed that membrane depolarizations in the voltage range of −90 to −50 mV slowly convert sodium channels to a nonconducting form which cannot be activated or "gated" directly to a conducting state by subsequent, larger depolarizations. Channels in this "inactivated" state (I) can be restored by membrane repolarization to the resting, "recruitable" state (R) without passing through a conducting state. At normal resting potentials of about −80 mV, at equilibrium about 75% of the neuronal Na channels are in the R state and 25% in the I state.

Many local anesthetic compounds (Fig. 1) alter this resting equilibrium to one that favors the I state.[13] The equilibrium shift is anesthetic dose-dependent and at rest appears to be larger for the more lipophilic drugs, like benzocaine, and smaller for the more polar anesthetic molecules.[13] Indeed, when quaternary derivatives of lidocaine, which are permanently cationic, are applied to the nerve, no shift of the inactivation equilibrium is detectable

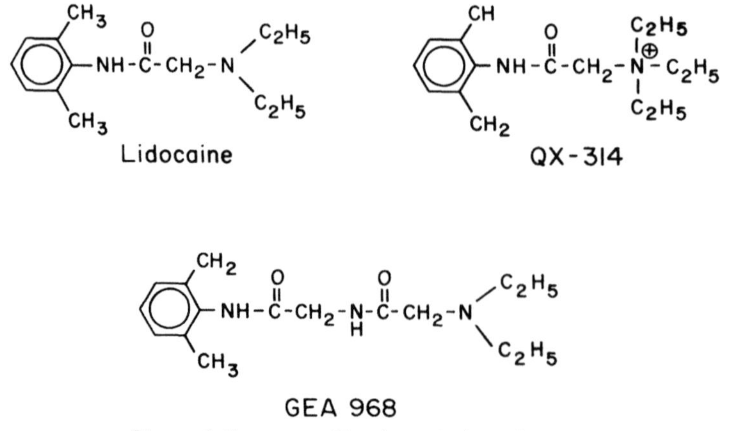

Figure 1. Structures of local anesthetics referred to.

for the channels recruitable from rest.[13,26] Tonic block by tertiary (3°−) amine anesthetics is at least partially antagonized by prolonged membrane hyperpolarization,[10,13] but the tonic block by quaternary (4°−) amine derivatives is affected little by this procedure.[26] Noteworthy, however, is the fact that both the polar 3°-amine compounds and the 4°-amine derivatives display the greatest phasic block. When the extracellular pH is made alkaline, favoring the neutral form of the 3°-amine anesthetics, the phasic effect is small and reverses rapidly; in contrast, there is a large phasic block at neutral pH where many more molecules, on the average, are in the protonated, charged form.[21] And although the 4°-amine derivatives and polar drugs sustain little change in inactivation of resting nerve, they do produce an incremental "accumulation" of inactivation during phasic block.[10] To summarize briefly, the normal availability of sodium channels to be recruited to the ion-conducting state by depolarizing pulses is reduced by local anesthetics. In this manner anesthetics mimic the normal "inactivation" process that occurs during prolonged depolarizations in the absence of drugs. The more lipophilic anesthetics effect such changes without channel openings, in resting membranes, and activation of the channels adds little to the resting inhibition. The more polar and the charged anesthetics, on the other hand, effect less inhibition in resting nerves but produce a sizable incremental inhibition during repetitive membrane depolarization.

The phasic actions of local anesthetics also produce changes in the gating currents which strongly resemble the changes seen during normal inactivation. In the absence of local anesthetic, following a depolarization that "inactivates" the macroscopic ionic current, the gating charge corresponding to the R → O transition is also reduced.[4,15] (The overall R → O transition is almost certainly a multistep process, and the total gating current depends on the amount and rate of charge movement occurring with each step.) A very similar charge "immobilization" is observed during the phasic block by 3°- and 4°-amine anesthetics, and like the phasic block, it requires several seconds to reverse.[5,7,20] Both phasic block and anesthetic-induced immobilization of gating charge are strongly voltage-dependent,[2,8,26,34] and all evidence supports the notion that a phasically blocked channel is also charge-immobilized.[9,33] Channels that have been blocked tonically by benzocaine have their gating charge reduced also, and by an amount proportional to the extent of the I_{Na} reduction.[19]

To integrate the tonic and phasic aspects of block by different types of local anesthetics, Hille[13] postulated a modulated receptor hypothesis (MRH). Figure 2 shows the kinetic

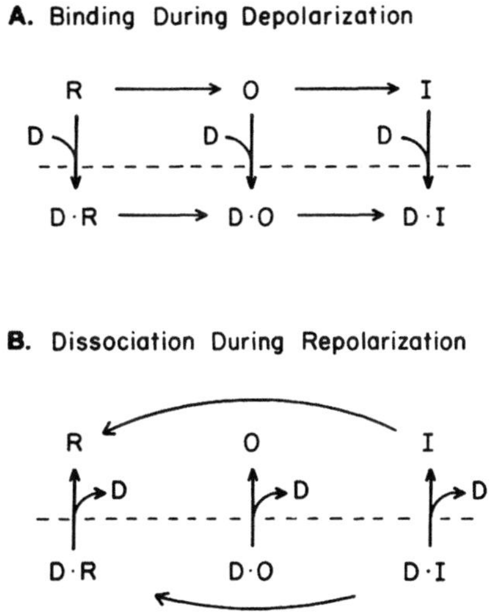

Figure 2. The modulated receptor hypothesis, as proposed by Hille.[13] (A) During membrane depolarization anesthetic molecules (D) bind to voltage-gated channel states, combining more tightly with inactivated states (I) of the channel than with either open (O) or resting (R) states. (B) After repolarization, the return of D·I to D·R is slow, as is the reversal of the binding reactions, accounting for the accumulating fraction of nonactivatable channels that is called "phasic" or "use-dependent" inhibition. Activation of drug-bound channels (D·R → D·O) may also be inhibited.

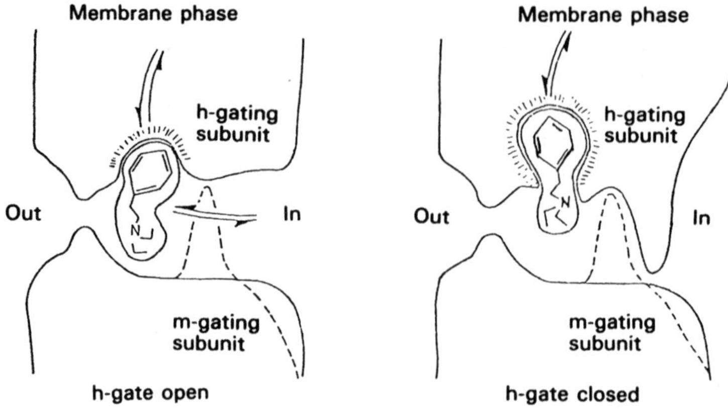

Figure 3. Hydrophobic or hydrophilic pathways permit access of neutral or charged anesthetic molecules, respectively, to a common binding site within the pore of the sodium channel. From Hille,[13] with permission of the author and the Rockefeller University Press.

scheme for the MRH, as originally set forth. The essential features of the scheme are (1) that inactivated channels have a higher affinity than resting channels for anesthetics, accounting for tonic block and the apparent shift in the R → I equilibrium, and (2) that the inactivated state, modified by the binding of polar or charged anesthetics (DI), reverses to the D·R or the I form slowly compared to the other normal gating transitions, thus accounting for the accumulating reduction in the fraction of activatable channels during repetitive depolarizations. To explain the different kinetics of phasic block and the differences in the shift of inactivation at rest, by charged compared to neutral local anesthetics, Hille proposed distinct "hydrophilic" or "hydrophobic" pathways to a single receptor, the first through the channel's aqueous pore, when it was opened by depolarization, and the second through the lipid interior of the membrane (Fig. 3).

The modulated receptor model provided a useful paradigm for the experiments that followed, most of which supported its overall features. The central role of inactivation, in particular, appeared to be confirmed by the finding that removal of the inactivation process, which occurred when proteolytic enzymes (Pronase) were perfused through the axoplasmic space, was accompanied by abolition of the phasic block normally produced by the 4°-amine anesthetics.[6,32] Although not an essential element of the MRH, the physical location of the anesthetic receptor within the aqueous pore of the channel has been supported by results showing that the passage of extracellular permeant cations through the channel lessens anesthetic blocking activity[7,8,20,22] and that extracellular protons increase the duration of phasic block by 3°-amine anesthetics, whereas intracellular protons do not.[18,21] The strong voltage dependence of phasic block by internal 4°-amine anesthetics also has been explained by positing their binding site deep within the channel pore.[26]

Despite the agreement with the MRH by previous experiments, recent findings from our laboratory and others are inconsistent with its details and provoke alterations in both the kinetic and the biophysical aspects of current models of local anesthetic action.

NEW RESULTS

The fast inactivation of sodium channels can be inhibited by procedures other than exposure to internal Pronase. First, specific neurotoxins isolated from scorpion venoms, for example, will prevent complete inactivation,[17,30] apparently by changing the equilibrium

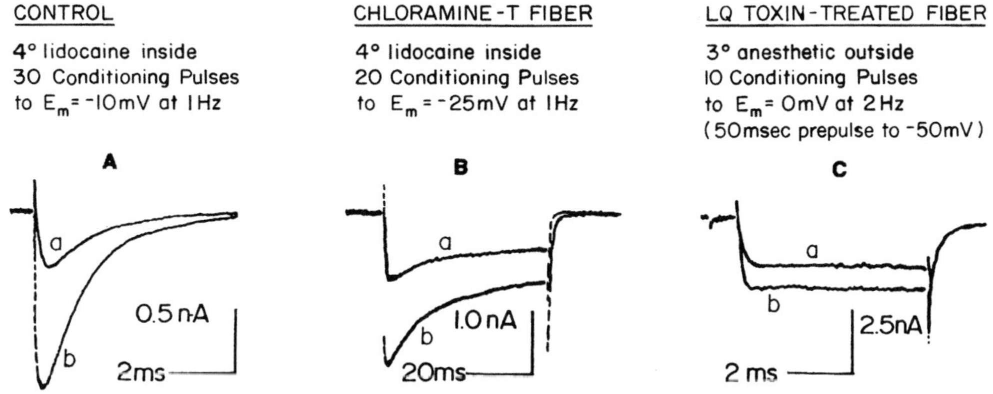

Figure 4. The persistence of phasic inhibition of noninactivating sodium currents in myelinated nerve fibers. Currents before (b) and after (a) a train of conditioning pulses at the test potential noted (E_m). (A) Normally gating currents; QX 314 (1.0 mM) at cut internode. (B) A different, chloramine-T-treated fiber, also with QX 314 (1.0 mM) at cut internode. (C) A different fiber, exposed to 150 nM α-*Leiurus quinquestriatus* scorpion toxin (α-LQIIa), and 0.5 mM GEA 968. In (C) a prepulse of 50-msec duration to −50 mV was applied before each test and conditioning pulse to inactivate all inactivatable channels. A 1-msec repolarization occurred between each prepulse and test or conditioning pulses. Phasic inhibition by GEA 968 normally arises from effects of both prepulse and conditioning pulse.[10,13,21] From Shepley *et al.*[23] with permission of The Physiological Society.

between open and inactivated channels. Second, oxidizing reagents also inhibit Na^+ inactivation; reactions with different channel sites account for the complex mechanisms of these chemical agents.[28,29] Despite their strong interference with channel inactivation, neither method of modification removes the phasic blocking activity of local anesthetics, as shown in Fig. 4.[23] Normal phasic block is illustrated by Fig. 4A, showing sodium current traces before (b) and after (a) depolarization by a train of conditioning pulses. Phasic block persists after treatment by chloramine-T (CT), an oxidizing agent selective for methionine residues (Fig. 4B)[29]; after the conditioning pulses, both the peak (inactivating) currents and the steady-state (noninactivating) currents are reduced. Similar results occur when channel inactivation is inhibited by a toxin (LQIIA) from *Leiurus quinquestriatus* scorpion (Fig. 4C). To produce these traces, a long depolarization (50 msec) followed by a brief repolarization (1 msec) was imposed before the current-eliciting test depolarization; this procedure silences all the inactivatable channels, by definition, and permits measurement of currents through noninactivated channels alone. Regardless of whether a polar 3°-amine anesthetic or a 4°-amine derivative was applied, phasic block is obvious in all three preparations. Clearly, the fast inactivation of channels is not necessary for the expression of phasic anesthetic blockade in amphibian myelinated nerve. Similar responses have also been observed in squid giant axons treated with CT (G. K. Wang, G. Strichartz, D. Eaton, and M. Brodwick, unpublished results).

The average values of tonic and phasic block produced by a 4°-amine derivative of lidocaine (QX-314) on normal and CT-treated nodes are collected in Table I. Tonic blockade of currents, measured either at their peak current or at steady state during one 40-msec depolarization ("I_{Na} maintained"), is reduced by QX-314 in CT-treated nerves as much as the peak current is reduced tonically in untreated nerves. Phasic block is slightly weaker in the CT-treated axons, and the maintained current, measured after 40 msec of depolarization, is slightly less inhibited than that measured at the peak. Thus, noninactivating channels appear to be only slightly less sensitive than normal channels to inhibition by local anesthetics.

Table I. Inhibition of Normal and Noninactivating Sodium Channels[a] by Quaternary Local Anesthetics

	Tonic block		Phasic block[b]	
	Peak I_{Na} inhibition	Maintained I_{Na} inhibition	Peak I_{Na} inhibition	Maintained I_{Na} inhibition
Normal currents[c]	72–81% (n = 2)		60–67% (n = 2)	
Noninactivating currents[d]	73 ± 5% (n = 3)	60 ± 13% (n = 3)	50 ± 1% (n = 4)	45 ± 2% (n = 4)

[a]QX-314 dissolved to 1.0 mM in solution bathing transected internode (cf. ref. 26).
[b]Measured at steady state after 40 pulses to 0–20 mV at 1 Hz frequency.
[c]All currents measured in nodes of Ranvier of sciatic nerve axons from *Rana pipiens*.
[d]Inactivation partially removed (39 ± 5%) by treatment with chloramine-T[(28)]; a small prepulse before each current-testing pulse removed an additional increment of inactivatable currents to yield an amplitude of maintained currents equal to ca · 60% of peak current amplitude, before anesthetic addition.

Is there evidence to support the involvement of states other than the inactivated form in phasic block? If the open state reacts selectively with local anesthetics to produce phasic block, then longer depolarizing pulses, extending throughout the period of open channels, should promote phasic block more effectively than short pulses. In experimental terms, phasic block should develop more rapidly and to a greater steady-state level with increasing duration of conditioning depolarizations, and if only the open state has the higher affinity, then the rate of phasic block development should be proportional to the availability of open channels, the area under the sodium current trace in time.[11] However, for the 3°-amine anesthetic GEA 968 this is not observed (Fig. 5). Short depolarizing conditioning pulses (0.8 msec), too brief to maintain channels open for their normal lifetime, actually induce a slightly greater steady-state phasic block than that produced by pulses long enough for all channels to inactivate. For a range of conditioning depolarization potentials, it was found, in general, that pulses long enough for the macroscopic current to reach its peak value were adequate to induce the largest phasic block.[31] It appears that the most effective anesthetic binding reaction is complete by the time of peak current, even during large depolarizations having peak currents at 100–200 μsec.

When inactivation was removed, the maximum phasic block at short pulses was even more defined than in Fig. 4C, longer pulses becoming only half as effective as short pulses in producing phasic block, even though channels were maintained in the open state for the duration of the pulse. These findings contradict the hypothesis that sodium channels that are open, *per se*, have the highest affinity for GEA 968.[10] Another "state" of the channel must bind the anesthetic molecules more tightly to account for phasic block, or the mechanism for this phenomenon must include processes beyond mere binding of anesthetics to voltage-gated channel conformations.

DISCUSSION

The main concern of this paper is the mechanism for phasic blockade of neuronal channels by local anesthetic agents. In order to evaluate the different affinities of identified states of the sodium channel for local anesthetics, we must consider the kinetic restrictions on binding reactions that occur during voltage clamp experiments. During a conditioning depolarization, a sodium channel passes through several conformations, judging by the time and voltage dependence of ionic and gating currents[3] and by the analysis of current fluctuations[24] and single-channel events.[1,25] The relationship between phasic block development and the duration of channel states will depend strongly on the rates of the various anesthetic

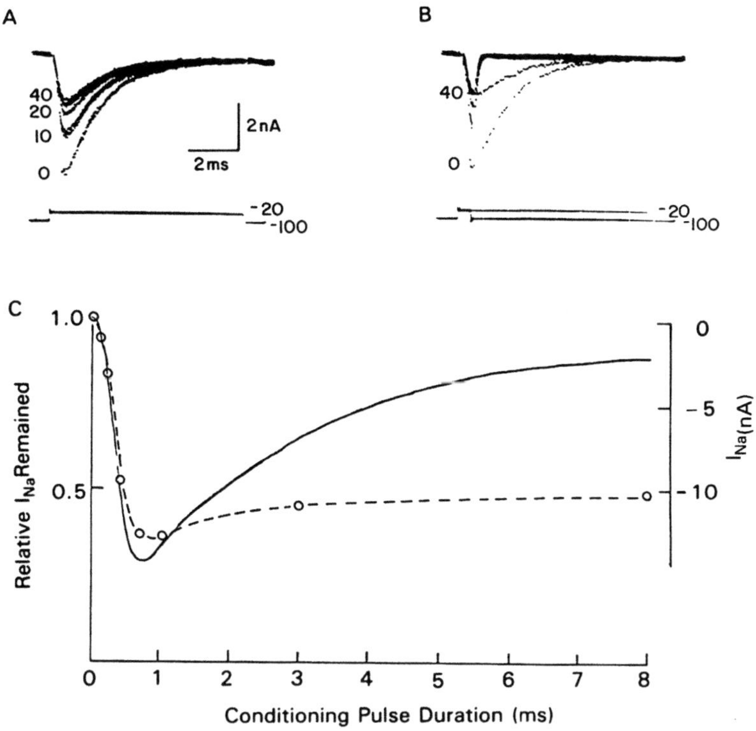

Figure 5. Conditioning pulse duration and phasic inhibition by GEA 968. (A) Sodium currents decrease during repetitive pulses of 8 msec duration in the presence of 1 mM GEA 968. The numbers near the traces identify currents measured before (0) and after 10, 20, and 40 repetitive pulses. (B) Sodium currents in the same fiber also decrease during pulses of 0.5 msec duration. Both the rate and the degree of inhibition are similar to those in A. Full-current traces of 8 msec are shown before (0) and after 40 repetitive pulses only. (C) The effect on another nerve fiber of conditioning pulse duration on the steady state of phasic block is shown; the ratio of peak Na current after to that before repetitive pulses is plotted by the circles and the broken line. The nerve was held at −100 mV and depolarized to −30 mV repetitively at 2 Hz for 40 pulses. The solid line traces the time course of I_{Na} during a single depolarization to the same voltage.

binding reactions.[11] In theory, an increase in the fraction of blocked channels during a train of repetitive depolarizations could arise either from increased drug binding to channel conformations with higher drug affinity from which dissociation occurred as rapidly as from a resting channel (i.e., greater k_{on}), or from drug molecules bound to channels that were converted by voltage to conformations from which dissociation was far slower than from resting channels (i.e., smaller k_{off}). The analysis is further complicated by the possibility that anesthetic binding reactions will be different at rest than during a depolarization and that both of the kinetic mechanisms just mentioned can contribute to phasic block during a pulse train.

The transient appearance of most channel states makes any kinetic analysis of the intact reaction scheme somewhat equivocal. Also, the "states" of a sodium channel that are hypothesized in order to model transitions based on ionic and gating currents are not necessarily identical to real channel conformations that have different affinities for various drugs. Furthermore, although many different drugs can cause a phasic decrement in I_{Na}, all of them do not necessarily accomplish this by the same kinetic or molecular mechanisms.[2,9] In this discussion we shall restrict our comments to traditional local anesthetics and their 4°-amine derivatives.

The results we report here support several conclusions, although they by no means provide a complete explanation for all anesthetic actions. The persistence of phasic block in axons that are free of the fast channel inactivation reaction shows that the inactivated state of the channel is not required for phasic block. And since pulses that are longer than the time-to-peak I_{Na} become less effective in promoting phasic block by GEA 968, in both normally gated and inactivation-modified channels, the more GEA-reactive form(s) of the channel must appear transiently during the conditioning pulse and be converted to a lower-affinity state during a maintained depolarization. Therefore, the open states of the channel are not the highest-affinity conformations for phasic block by GEA 968. A different logical argument also applies to the same question. If anesthetics were bound most tightly to the open-channel conformation, then the binding reaction would have to reach equilibrium very rapidly ($\tau < 1$ msec), for very slow binding reactions ($\tau > 20$ msec) would be promoted by the prolongation of the open state (which they are not, Fig. 5C), and binding reactions with rates that are comparable to channel gating times would change the observed time course of I_{Na}. The latter does occur in normal channels with drugs like yohimbine[16] and *N*-methylstrychnine[6,22] and is very evident in Pronase-treated squid axons with a variety of cationic drugs,[6,34] but in the presence of many traditional local anesthetics or their 4°-amine derivatives, no changes in the I_{Na} time course are detectable. If open channels are bound very rapidly by anesthetics, then equilibrium binding will be reached within the duration of a single conditioning pulse, and the accumulating inhibition that characterizes phasic block must then arise from the incomplete reversal of this binding during the period between conditioning pulses. In order to account for the total contributions to phasic block of the different channel states, experiments must be conducted measuring the development of phasic block by conditioning pulses of different length presented at different repetition frequencies. This procedure will permit the direct evaluation of the *on* and *off* rate constants for anesthetic binding to the reactive channel conformations.

To account for the various results reported so far, we propose a general scheme for phasic block without the explicit involvement of either open or inactivated channel conformations. As diagrammed in Fig. 6, the "anesthetic" molecule, designated D, binds to a receptor in the resting conformation, R_r, an "intermediate" conformation, R_x, or an "activated" conformation, R_a, the last two of which are populated through membrane depolarization. Binding at rest occurs rapidly, but during a conditioning depolarization, binding to R_x or R_a conformations may be slow. It is essential that either the $D{\cdot}R_a$ or the $D{\cdot}R_x$ complex undergoes slow drug dissociation and slow conversion to $D{\cdot}R_r$ in order to produce phasic channel blockade. Although this scheme bears similarities to the original modulated receptor hypothesis, it does not explicitly involve the inactivated or the open state of the channel. The only requirements are that the binding site be modulated by activation, and that the reaction rates and affinities for binding to the three conformations be different for different drugs. The scheme is sufficiently general to account for both the phasic block from repetitive depolarization and the time-dependent decrease in currents during a single depolarization observed with

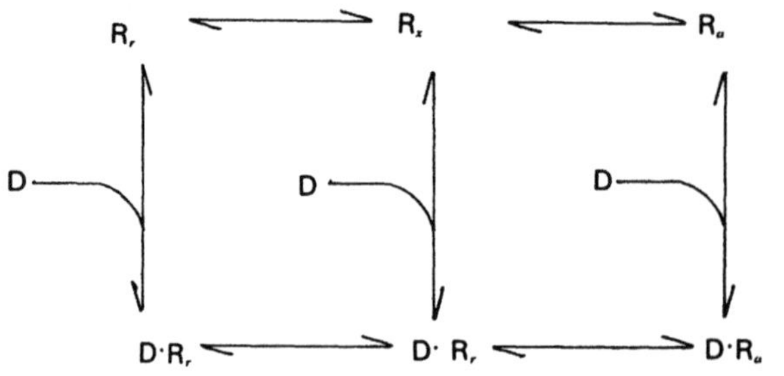

Figure 6. A general kinetic scheme for tonic and phasic inhibition of sodium channels by "local anesthetics." See Discussion for description.

drugs like *N*-methyl strychnine. It can also easily account for the inhibitions (immobilization) of gating current produced by local anesthetics.[5,7,33]

Since fast channel inactivation *per se* is not essential for phasic block, we believe that consideration of this more general model is appropriate. Phasic block probably corresponds to drug binding reactions that alter (and are altered by) any of several activated conformations, and the general concept of the modulated receptor is still valid. And although we have used the terms tonic and phasic *block* throughout this chapter, we do not mean to suggest that local anesthetics are inhibiting currrents by occluding a channel's pore. Any mechanism that prevents complete activation to the open state satisfies the general kinetic scheme.

The actions of Pronase treatment on phasic inhibition also must be reconsidered. Intracellular Pronase may act directly on part of the local anesthetic receptor or on a separate region of the channel to prevent the conformational changes that lead to an increased drug affinity, although only for certain anesthetics, since the actions of some drugs are unaffected by Pronase.[6] A universal mechanism for the actions of all "local anesthetic" drugs may invoke certain types of general kinetic modifications, but must also specify particular conformations of the channel or separate binding reactions with the macromolecule in order to account for the actions of all types of anesthetic molecules.

ACKNOWLEDGMENTS. This work was supported by USPHS grants GM/NS-30160 and GM-15904 and a research grant from the National Multiple Sclerosis Society (RG-1513A1). G.S. thanks Astra Lakemedel, Sodertalje, Sweden, for support during the writing of this chapter and Dr. Bertil Takman, Astra Pharmaceuticals, Worcester, MA, for generously providing experimental drugs.

REFERENCES

1. Aldrich RW, Corey DP, Stevens CF: A reinterpretation of mammalian sodium channel gating based on single channel recording. *Nature* 1983;306:436–441.
2. Almers W, Cahalan MD: Block of sodium channels by internally applied drugs: Two receptors for tertiary and quaternary amine compounds? in Salank J (ed): *Physiology of Excitable Membranes,* Vol 4. New York, Pergamon Press, 1981, pp 67–74.
3. Armstrong CM: Sodium channels and gating currents. *Physiol Rev* 1981;61:644–683.
4. Armstrong CM, Bezanilla F: Currents associated with the ionic gating structures in nerve membrane. *Ann NY Acad Sci* 1975;264:265–277.
5. Armstrong CM, Croop RS: Simulation of Na channel inactivation by thiazin dyes. *J Gen Physiol* 1982;80:641–662.
6. Cahalan M: Local anesthetic block of sodium channels in normal and pronase-treated squid giant axons. *Biophys J* 1978;23:285–311.
7. Cahalan MD, Almers W: Interactions between quaternary lidocaine, the sodium channel gates and tetrodotoxin. *Biophys J* 1979a;27:39–56.
8. Cahalan MD, Almers W: Block of sodium conductance and gating current in squid giant axons poisoned with quaternary strychnine. *Biophys J* 1979b;27:57–74.
9. Cahalan M, Shapiro BI, Almers W: Relationship between inactivation of sodium channels and block by quaternary derivatives of local anesthetics and other compounds, in Fink BR (ed): *Molecular Mechanisms of Anesthesia, Progress in Anesthesiology,* Vol. 2. New York, Raven Press, 1980.
10. Courtney KR: Mechanism of frequency-dependent inhibition of sodium currents in frog myelinated nerve by the lidocaine derivative GEA-968. *J Pharmacol Exp Ther* 1975;195:225–236.
11. Courtney KR, Kendig JJ, Cohen EN: The rates of interaction of local anesthetics with sodium channels in nerve. *J Pharmacol Exp Ther* 1978;207:594–604.
12. Hille B: The common mode of action of three agents that decrease the transient change in sodium permeability in nerves. *Nature* 1966;210:1220–1222.
13. Hille B: Local anesthetics: Hydrophilic and hydrophobic pathways for the drug-receptor reaction. *J Gen Physiol* 1977;69:497–575.

14. Hodgkin AL, Huxley AF: The dual effect of membrane potential on sodium conductance in the giant axon of *Loligo*. *J Physiol* 1952;116:487–506.
15. Keynes RD, Rojas E: Kinetic and steady-state properties of the charged system controlling sodium conductance in the squid giant axon. *J Physiol (London)* 1974;239:393–434.
16. Lipicky RJ, Gilbert DL, Ehrenstein G: Effects of yohimbine on squid axons. *Biophys J* 1978;24:405–422.
17. Meves H, Rubly N, Watt DD: Effect of toxins isolated from the venom of the scorpion *Centruroides sculpturatus* on the Na currents of the node of Ranvier. *Pflugers Arch* 1982;393:56–62.
18. Narahashi T, Frazier D, Yamada M: The site of action and active form of local anesthetics, I. Theory and pH experiments with tertiary compounds. *J Pharmacol Exp Ther* 1970;171:32–44.
19. Neumcke B, Schwarz W, Stampfli R: Block of Na channels in the membrane of myelinated nerve by benzocaine. *Pflugers Arch* 1981;390:230–236.
20. Revenko SV, Khodorov BI, Shapovalova LM: The effect of yohimbine on sodium and gating currents in frog Ranvier node membrane. *Neuroscience* 1982;7:1377–1387.
21. Schwarz W, Palade PT, Hille B: Local anesthetics: Effect of pH on use-dependent block of sodium channels in frog muscle. *Biophys J* 1977;20:343–368.
22. Shapiro BI: Effects of strychnine on the sodium conductance of the frog node of Ranvier. *J Gen Physiol* 1977;69:915–920.
23. Shepley MP, Strichartz GR, Wang GK: Local anesthetics block non-inactivating sodium channels in a use-dependent manner in amphibian myelinated axons. *J Physiol (London)* 1983;341:62P.
24. Sigworth FJ: The variance of sodium current fluctuations at the node of Ranvier. *J Physiol (London)* 1980; 307:97–129.
25. Sigworth FJ, Neher E: Single Na^+ channel currents observed in cultured rat muscle cells. *Nature* 1980; 287:447–449.
26. Strichartz GR: The inhibition of sodium currents in myelinated nerve by quaternary derivatives of lidocaine. *J Gen Physiol* 1973;62:37–57.
27. Taylor RE: Effect of procaine on electrical properties of squid axon membrane. *Am J Physiol* 1959;196:1071–1078.
28. Wang GK: Irreversible modification of sodium channel inactivation in toad myelinated nerve fibres by the oxidant Chloramine-T. *J Physiol (London)* 1984a;346:127–141.
29. Wang GK: Modification of sodium channel inactivation in single myelinated nerve fibers by methionine-reactive chemicals. *Biophys J* 1984b;46:121–124.
30. Wang GK, Strichartz GR: Purification and physiological characterization of isotoxins from venoms of the scorpions *Centruroides sculpturatus* and *Leiurus quinquestriatus*. *Mol Pharmacol* 1983;23:519–533.
31. Wang GK, Strichartz GR: Local anesthetics produce phasic block of sodium channels during activation. *Biophys J* 1984;45:286a.
32. Yeh JZ: Sodium inactivation mechanism modulates QX-314 block of sodium channels in squid axons. *Biophys J* 1978;24:569–574.
33. Yeh JZ, Armstrong CM: Immobilization of gating charge by a substance that simulates inactivation. *Nature* 1978;273:387–389.
34. Yeh JZ, Narahashi T: Kinetic analysis of pancuronium interaction with sodium channels in squid axon membranes. *J Gen Physiol* 1977;69:293–323.

19

Single-Channel Analysis of the Action of Na Channel Blockers 9-Aminoacridine and QX-314 in Neuroblastoma Cells

Jay Z. Yeh, William A. McCarthy, Jr., Fred N. Quandt, and Daisuke Yamamoto

ABSTRACT. Although the Na channel is well known to be the site of action for local anesthetics, the exact mechanism of action of these compounds is still an area being investigated. Single-channel recording methods were applied to this problem in order to determine whether local anesthetic agents reduce current through each Na channel, alter the temporal properties of open channels, or reduce the frequency of channel opening to a conducting state. The actions of 9-aminoacridine, a tertiary amine, and QX-314, a quaternary derivative of lidocaine, were determined. Currents through single Na channels were measured from excised patches of membrane from cultured neurons. 9-Aminoacridine reduced the number of Na channels conducting during a series of depolarizations, but did not alter the current amplitude through individual open channels. Open Na channels "flickered" between conducting and drug-blocked states once opened in the presence of 9-aminoacridine. Flickering was not dependent on an intact Na channel inactivation process. A comparison of the time that the channel burst in the presence of the drug with the normal open time, with or without inactivation, indicated that the channel could become inactivated while blocked. In contrast, QX-314 reduced the frequency of Na channel opening to a conducting state without alterations in either the current amplitude or the temporal properties of the open channel. While the reduction in frequency of openings by these compounds would give rise to "resting" block of Na currents through the neural membrane, block of the open channel by 9-aminoacridine would also contribute to a "use-dependent" reduction in these currents.

INTRODUCTION

By measuring the macroscopic ionic currents in voltage-clamped squid axons, Taylor[20] showed that procaine suppresses Na and K currents. Since the K current is not required for action potential propagation, he concluded that the block of Na current by procaine is responsible for block of nerve conduction. Over the following two decades, it has been fully established that the Na channel is the site of action for all local anesthetics.[8]

Jay Z. Yeh and William A. McCarthy, Jr. • Department of Pharmacology, Northwestern University Medical School, Chicago, Illinois 60611. *Fred N. Quandt* • Department of Physiology, University of Calgary/Medical School, Calgary, Alberta T2N 4N1, Canada. *Daisuke Yamamoto* • Mitsubishie-Kasei Institute of Life Sciences, Tokyo, Japan.

9-Aminoacridine QX-314

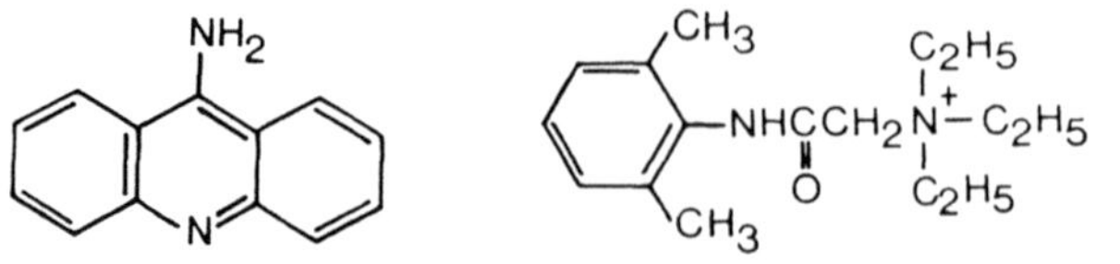

Figure 1. Chemical structure of QX-314 and 9-aminoacridine (9-AA). QX-314 is a quaternary derivative of lidocaine. The ring nitrogen of 9-AA carries the positive charge at pH = 7.3 because of its pKa (9.9).

Since current measurements using traditional voltage clamp techniques only reflect the average behavior of many individual channels opening and closing in response to a step change in the membrane potential, the Na current (I_{Na}) at any time equals the product of the number of open channels open (N) and the single-channel current (i), as follows:

$$I_{Na} = N \cdot i \qquad \text{(equation 1)}$$

According to this equation, a decrease in I_{Na} caused by the presence of a local anesthetic agent could be due either to a mechanism that involves a reduction in the number of channels that are opening as a result of modification of channel gating[6,11] or to a mechanism that involves a reduction in the single-channel current as a result of channel occlusion,[19] or both.[5,8]

Although these various mechanisms cannot be easily distinguished by measuring the average membrane current from the whole cell, they can be directly tested using single-channel recording techniques. The current flowing through a single channel can be directly measured in order to clarify alterations in channel permeability to ions. In addition, the latency to first opening and the channel open and shut times can be measured in order to determine modifications in the gating of the channel.

This study was undertaken in an attempt to answer three questions: (1) Do local anesthetic agents reduce single Na channel current? (2) Do they change the temporal properties of the open Na channels? (3) Do they change the frequency (or number) of channel opening events? The patch clamp technique was used to record single-channel currents from "inside-out" excised patches from cultured N1E-115 neuroblastoma cells. QX-314, a quaternary derivative lidocaine, and 9-aminoacridine (9-AA), a triple-ring tertiary amine (see Fig. 1) were chosen for this study because they share many features of local anesthetic action on nerve membranes: namely, both produced resting and frequency-dependent block. We found (1) that both QX-314 and 9-AA decreased the number of channel openings observed in response to a series of step depolarizations, (2) neither compound altered the single channel conductance, and (3) 9-AA caused the channel to flicker, i.e., to alternate between conducting and nonconducting states during the period it was gated open; however, QX-314 did not induce such flickering behavior.

The blocking effects found for 9-AA are consistent with two mechanisms of action: 9-AA is able to transiently block a channel that has been gated open; in addition, 9-AA will modify the closed state of the channel and reduce the frequency of channel opening. QX-314 shares this latter property.

METHODS

Preparation

All experiments were performed on N1E-115 neuroblastoma cells. These cells were maintained in tissue culture at 37°C in humidified air containing 10% CO_2 and were grown in Dulbecco's modified Eagle medium, supplemented with 10% newborn calf serum. Three to ten days before use, 2% dimethylsulfoxide was added to the cells to enhance neuronal characteristics.[10]

Recording Solutions

The bath solution for the cells was 140 mM NaCl, 5 mM CsCl, 1.8 mM $CaCl_2$, and 5 mM HEPES (pH 7.3). Internal solution (for inside-out patches) for the cells was 145 mM CsF, 1.4 mM NaF, and 5 mM HEPES (pH 7.4). Sucrose (198 mM) was added to the internal solution for isoosmolarity with the high-Na pipette solution. When indicated, 300 μM *N*-bromoacetamide (NBA) was added to the internal solution to eliminate Na inactivation.[16,17] After 10–20 min, the NBA was washed away with normal internal solution to prevent deterioration of the patch of membrane. The pipette solution was 250 mM NaCl, 5 mM CsCl, 1.8 mM $CaCl_2$, and 10 mM HEPES (pH 7.3). All solutions were filtered during use with a 0.2-μm-pore-size membrane filter and were cooled for recording to temperatures indicated in the text.

Patch Clamp

Single-channel currents were recorded from excised membrane patches using the gigohm seal patch clamp technique.[7] All membrane patches were inside-out, as shown in Fig. 2. Patch electrodes were vertically pulled from glass capillary tubes, fire-polished using a glass-coated platinum wire, and coated with red insulating varnish to reduce capacitative transients. The electrode resistances of the saline-filled pipettes were 2–10 MΩ. Gigohm seals were formed as follows: The tip of the pipette was lightly brushed across the cell surface to form a low-resistance mechanical seal, and then gentle suction was applied to the pipette interior until the gigohm seal was formed. For good seals, this typically occurred in less than 10 sec. After seal formation, the bath solution was switched to an internal solution. An inside-out patch was then formed by rapidly pulling the pipette away from the cell.

The patch clamp was designed and built in our laboratory and had three stages of capacitative compensation. The rise time of the clamp response to a capacitatively injected

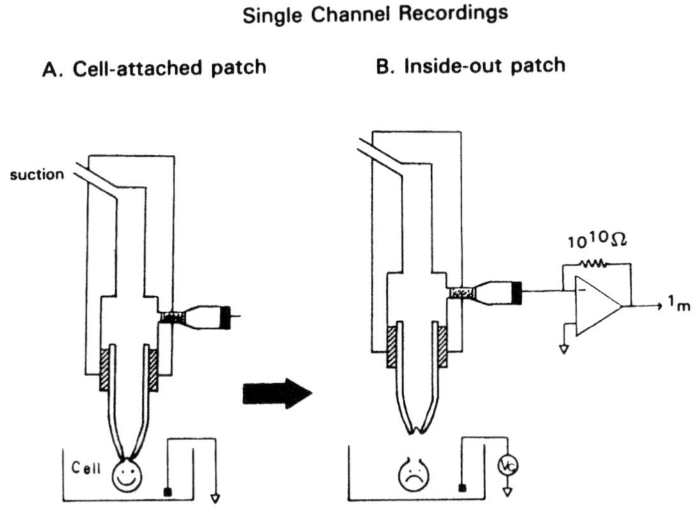

Figure 2. Single-channel recordings from the whole cell-attached patch (A) and from the excised inside-out patch (B). In both configurations, the pipettes were filled with an extracellular solution. The cell was bathed in an extracellular solution in A and in an intracellular solution in B.

triangle wave was set to 120 μsec. The head stage of the patch clamp contained a Burr–Brown 3523L operational amplifier with a 10-GΩ feedback resistor.

The holding potential for all experiments ranged from −90 to −100 mV. Pulses from the holding potential depolarized the patch to membrane potentials of −60–0 mV, as indicated.

The stimulating voltage pulses and the current trace storage were done on-line using a Digital PDP 11/23 or 11/34 computer, with a 14-bit analog-to-digital converter which sampled at 100 μsec. All current traces were low-pass-filtered (using a six-pole Bessel filter) and amplified before being passed to the computer. Traces were stored temporarily on a hard disk and then transferred to a floppy disk for later analysis.

Analysis

In the experiments with 9-AA the transitions between the conducting (open) and non-conducting (shut or blocked) states were determined by using the half-amplitude threshold analysis of Colquhoun and Sigworth.[2] In the experiments with QX-314, current traces were analyzed off-line using the PDP-11 computer. Traces without channels (as determined by the operator) were averaged and used for digital subtraction of capacitative and leakage currents. A semiautomatic search program was used to analyze all channel events. This program used a point histogram to allow it to determine what the single-channel events were and where the baseline was. It then generated a vector of channel status, consisting (for a 2000-sample point trace) of 2000 integers, representing the number of open channels at each point of the trace. This vector was used in conjunction with the actual current trace by the program to determine the mean baseline current and the amplitudes and durations of all single-channel events. All calculated events were displayed for operator approval before they were written either to a line printer or to a file saved on the hard disk. The widths of the peaks of the point histogram could also be manually overriden, to allow the threshold for channel activity to be raised or lowered, depending on the signal-to-noise ratio of the trace. Time constants were fitted to the histograms of durations of channel events using a least-squares fit to one or two exponential distributions. This time constant represents the mean duration of the event.

9-AA hydrochloride was purchased from K and K Laboratories Inc. (Plainview, NY). 9-AA has a pKa of 9.99 and is almost entirely ionized at a pH of 7.3. The ring nitrogen carries the positive charge. QX-314, a quaternary derivative of lidocaine, was a generous gift from Dr. B. Takman of Astra Research Laboratories, Inc., Farmingham, MA. The chemical structures of 9-AA and QX-314 are shown in Fig. 1.

RESULTS

9-AA

When the excised membrane patch was depolarized to more than −55 mV, an inward pulselike event was observed. This pulselike event has been interpreted as ion flow through an individual channel. Figure 3 shows several channel current traces in response to a 40-msec pulse to −26 mV. The control traces (Fig. 3A) show the opening and closing of single channels, the opening being indicated by a downward deflection and the closing by a subsequent upward deflection. Most traces showed an initial upward spike and a symmetrical downward spike at the end of the trace, which represented the capacitative charging of the pipette at the beginning and end of the depolarizing pulse and were not channel events. Some channels did not close by themselves during the depolarizing pulse, but were cut off by the end of the pulse (first and sixth traces from the top). The mean open time in the control traces

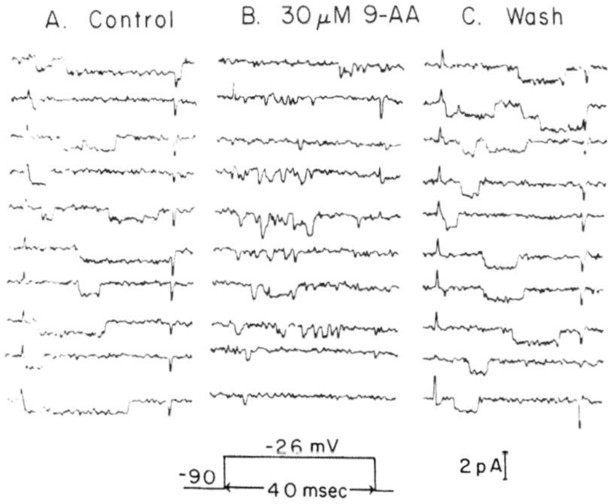

Figure 3. Characteristics of the alterations in temporal properties of the open state of Na channels by 9-AA. Examples of Na channel openings recorded from an inside-out patch, with Na inactivation intact, in response to a 40-msec step depolarization to −26 mV from a holding potential of −90 mV. The depolarization is shown in the bottom of the panel and was given at 3.5-sec intervals. 9-AA caused the channel to burst; i.e., the channel made alternating rapid transitions between conducting and nonconducting states. The pipette contained 125 mM Na, in contrast to 250 mM Na, which was used in the remaining experiments. Temperature = 5.5°C.

was quite long (10.3 msec), owing to the low temperature at which these experiments were performed (5.5°C). Introduction of 30 μM 9-AA to the cytoplasmic side of the membrane had two main effects on the Na channel recordings: First, it reduced the mean frequency of channel opening to an average of 63 ± 9% (n = 5); second, the single-channel current no longer appeared as a simple square wave event, but showed a bursting or flickering behavior (Fig. 3B). Following washout of the 9-AA, the mean frequency of opening recovered to the control value, and the flickering behavior disappeared (Fig. 3C).

In order to simplify the kinetic analysis of the efect of 9-AA, Na channel inactivation was removed by perfusing the cytoplasmic side of the patch with *N*-bromo acetamide (NBA). The flickering observed in the presence of 9-AA with Na inactivation present could also been seen if Na inactivation was removed by NBA. Figure 4 shows two single-channel traces in response to a 400-msec depolarizing pulse to −40 mV, after the membrane patch had been treated with 300 μM NBA. The control trace (Fig. 4A) has a single-channel event, lasting 75 msec. The extreme length of this opening was due to removal of Na inactivation by NBA. When 30 μM 9-AA was added to the bathing solution (Fig. 4B), flickering appeared. Thus, flickering can be seen with or without Na inactivation present.

The interpretation of the temporal effect of 9-AA on the channel is best described using a statistical analysis of the data. Figure 5 is a pictorial representation of what an ideal single-channel recording might look like, in the presence of a drug that causes flickering. Three periods of rapid channel opening and closing (bursts) were present, in which the channel

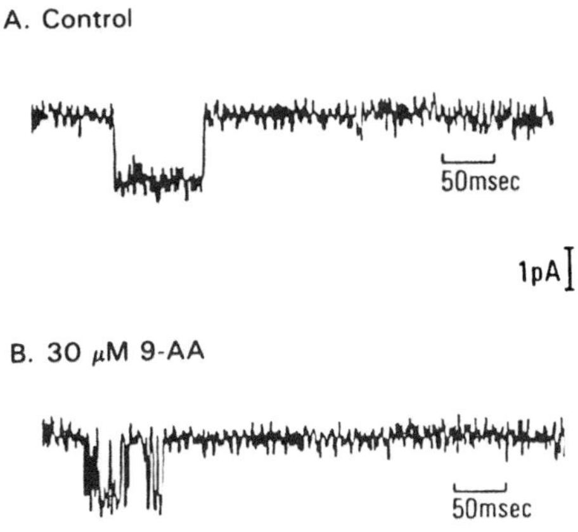

Figure 4. Examples of single sodium channel currents recorded before (A) and during (B) perfusion of 30 μM 9-AA. Both A and B were obtained from a single inside-out membrane patch pretreated with 300 μM NBA. The Na channel currents were obtained by depolarizing the membrane to −40 mV from a holding potential of −100 mV. 9-AA transiently blocked the Na channel causing the open channel to burst. Temperature = 6°C.

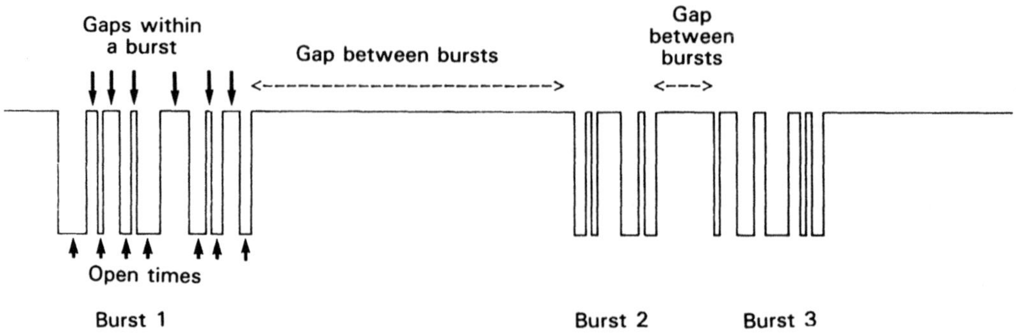

Figure 5. Pictorial representation of idealized bursting behavior in the presence of a drug that causes flickering. Three bursts are shown here: Burst 1 and burst 2 are clearly distinct because they are separated by a long gap with no channel activity. However, burst 2 and burst 3 are not so obviously distinct. A criterion is required to determine whether a gap belongs within a burst or is a gap between bursts. The criterion used is that a gap which is 3.5 times greater than the mean gap within an unambiguous burst (for example, burst 1) will be treated as a gap between bursts. Using this criterion, the probability that the gap within a burst will be mistaken for a gap between bursts is calculated to be 3%.

flickered rapidly between the open and closed states, separated by relatively long intervals of no channel activity. Each burst was composed of several open times and several gaps when the channel was nonconducting. The open times and the gap times within a burst can be analyzed in terms of certain kinetic models to yield information concerning the rate constants governing the blocking and unblocking actions of the drug. For this analysis, it is important to be able to distinguish between the gaps within a burst and the gaps between bursts; i.e., if two bursts are too close together, it is possible that they represent a single burst, separated by an unusually long gap within a burst. In Fig. 5, bursts 1 and 2 were clearly distinct; however, bursts 2 and 3 were not so obviously distinct. Further statistical analysis would have to be done in this case to determine how unlikely it would be to have such a long gap within a burst.

To this end, we made a cumulative gap time histogram (Fig. 6), consisting of all periods

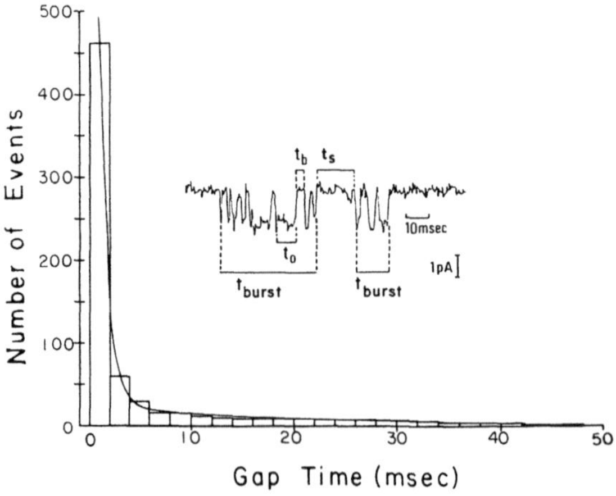

Figure 6. The probability distribution of gap times for Na channels recorded with 9-AA added to the internal perfusate indicates two underlying events. The period of time between inward currents through Na channels exposed to 30 μM 9-AA was measured. The number of events (n) that have gap durations longer than the indicated time bins is plotted. The data were best fitted by the sum of two exponential distributions, as shown by the curve which plots $n = 150 \exp(-t/0.83) + 30 \exp(-t/19.4)$. The inset depicts the interpretation of the origin of the two exponentials. The first term in the equation, having a smaller time constant, represents the distribution of gap times for events within the burst. This event appears to be related to the time the channel is in the blocked state (t_b). The second term, with a longer time constant, gives the distribution of the gap between bursts, which is the time that the channel is in the closed, unblocked state (the shut state, t_s).

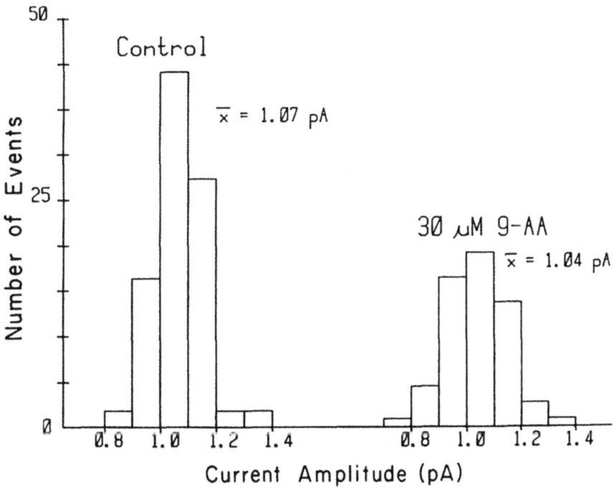

Figure 7. 9-AA does not reduce the amplitude of single Na channel currents. The current amplitude was measured only from single-channel opening events. In the presence of 9-AA, the current amplitude was measured from the open states within the burst.

of time between open-channel events (gaps between bursts and gaps within bursts). The first bin contains all gaps that were at least 0 msec long; the second bin contains all gaps that were at least 2.0 msec long, etc. The smooth curve in Fig. 6 was a sum of two exponentials. These two exponentials can be interpreted to arise from two kinetically distinct events. The small time constant (0.83 msec) represents the short-lived gap within a burst, and the much larger time constant (19.4 msec) represents the much longer-lived gap between bursts.

Although 9-AA had an extreme effect on the temporal properties of single channels, it had no measurable effect on the amplitude (and hence the conductance) of single channels. Figure 7 shows current amplitude histograms for the control case and in the presence of 30 μM 9-AA. The number of single-channel events was much lower in the presence of 9-AA, but the mean current amplitude was not significantly different: 1.07 pA in control conditions versus 1.04 pA in the presence of 9-AA.

From Fig. 7, it is clear that the current amplitude of the channels is not affected by the presence of 9-AA. From Figs. 3 and 4 flickering behavior is seen only in the presence of 9-AA. A simple model which is consistent with these two findings is an open channel block model:

$$\text{Shut} \underset{\beta}{\overset{\alpha}{\rightleftharpoons}} \text{Open} \underset{l}{\overset{k[B]}{\rightleftharpoons}} \text{Block} \qquad \text{(scheme I)}$$

Here α is the apparent opening rate constant, β is the rate constant governing the transition from open to the last shut state prior to the open state of the sodium channel, $[B]$ is the concentration of the blocking molecule, and k and l are the blocking and unblocking rate constants. In this model, the gap between bursts represent the time when the channel is in the shut state, and the gap within a burst represents the time when the channel is being blocked by the drug. This model was first proposed by Cahalan[1] and Yeh[24] to explain the blocking action of 9-AA on microscopic sodium current in squid axons. This model actually predicts three different types of channel blocking behavior, depending on the relative magnitudes of the rate constants governing the blocking and unblocking reactions to the closing rate constant, β. If $k[B]$ and l are relatively larger than β, the channel will undergo several blocking and unblocking reactions before closing. Channel events will exhibit classic flickering behavior (rapid opening and closing events, open events having the same amplitude as the control). This is the case for 9-AA action on single Na channels[18,21] and for QX-222 block of ACh-receptor channels.[14] If k[B] and l are much larger than β, the blocking and unblocking events of the drug will occur much too quickly to be resolved given the bandwidth limitations of the patch clamp. This is called "ultrafast" flickering. The single-channel events will appear to have a reduced conductance, owing to the time integration (averaging) of the blocked and unblocked states of the bursts. This is probably the case for block of Na channels by tetramethylammonium ions[9] and Ca^{2+} ions.[22] Finally, if $k[B]$ and l are comparable to or are slower than β, flickering again will not be seen. The drug will block the channel and stay there. Single-channel events will have the same amplitude as the control, but will have shorter open times.

The two rate constants governing the blocking and unblocking reactions ($k[B]$ and l) can be accurately estimated from the mean open time and from the mean blocked time. This assumes that we can clearly identify all bursts (not confuse a single burst that has an abnormally long gap within it for two different bursts). Even though the blocked and the shut state are nonconducting, since this model assumes that they cannot communicate they can be distinguished from each other if they have very different mean durations. This was indeed the case for 9-AA, as illustrated in Fig. 6. The mean shut time was 20 times larger than the mean blocked time.

According to this model, the duration of the channel in the open state is exponentially distributed with a mean duration equal to the reciprocal of the sum of the rate constants leaving that state (see ref. 2). In other words, the reciprocal of the mean open time (t_o) is

$$1/t_o = \beta + k[B] \qquad \text{(Equation 2)}$$

Similarly, the duration of the channel in the blocked state is exponentially distributed with the mean duration equal to the reciprocal of the unblocking rate constant. In other words, the reciprocal of the mean blocked time (t_b) is

$$1/t_b = l \qquad \text{(Equation 3)}$$

Figure 8 shows the cumulative histograms for the open times and for the blocked times. These histograms were constructed from data obtained with a NBA-pretreated single patch during perfusion of 60 μM 9-AA. The cumulative histograms exhibit a single exponential distribution. A fit of each distribution to an exponential by a least-squares method gave the mean open time and the mean blocked time as 1.4 msec and 1.0 msec, respectively.

According to equation 2, the reciprocal of the open time is linearly concentration-dependent. Equation 3 predicts that the reciprocal of the mean blocked time is not concentration-dependent. The data in Fig. 9A showed that the blocked time had virtually no significant concentration dependence. The slight variation in $1/t_b$ over concentration ranges from 30 to 100 μM was within the experimental error. The data in Fig. 9B demonstrated that the values for $1/t_o$ increased linearly with increasing drug concentration. The blocking rate constant, k, and the channel closing rate constant, β, were obtained from this figure using equation 2 since the slope of the line relating $1/t_o$ to the drug concentration, [B], corresponds to k, and its intercept at the ordinate corresponds to β. The values of k are 5.8×10^6 $M^{-1}sec^{-1}$ at -25 mV and $5.9 \times 10^6 M^{-1}sec^{-1}$ at -40 mV, and values for β are 123 sec^{-1} and 135 sec^{-1} at -25 and -40 mV, respectively.

Another prediction that the sequential open channel block model makes is that the mean burst time in the presence of the blocker should be longer than the mean open lifetime in the control, since the channel can go from the open state to either the blocked state or the closed state in the presence of the blocker. Since the channel closing rate constant, β, was considerably less than either the blocking rate, k[B], or the unblocking rate constant, l, the reciprocal mean burst duration could be approximated as follows:

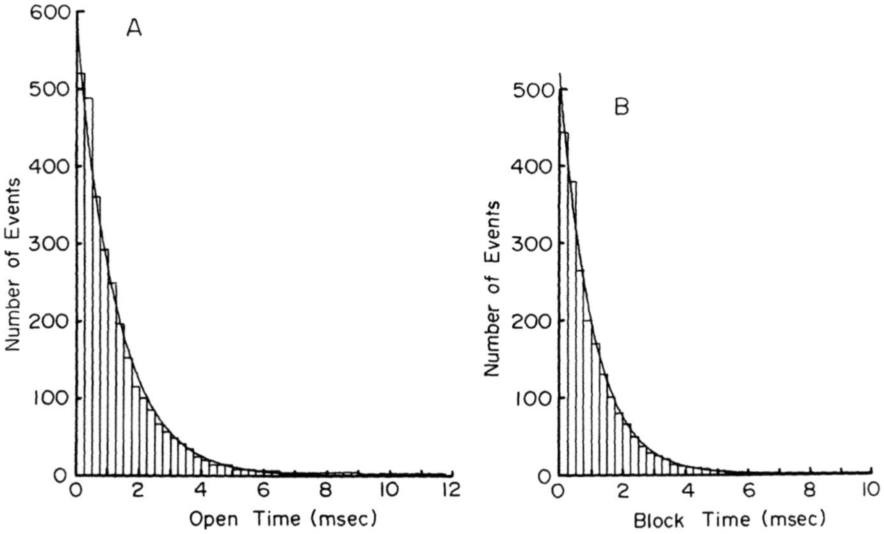

Figure 8. Cumulative histograms of time intervals for the open time (A) and the block time (B) in the presence of 60 μM 9-AA. The single-channel currents were activated by depolarizing the membrane to −40 mV from a holding potential of −100 mV. The membrane patch was pretreated with 300 μM NBA. The number (n) of events having a duration longer than that indicated on the abscissa is plotted against the duration (t). The curves were drawn by the equation $n = 600 \exp(-t/1.4)$ for the open time (A) and $n = 578 \exp(-t/1.0)$ for the blocked time (B). Temperature = 16°C. Reproduced from *The Journal of General Physiology*,[21] by copyright permission of the Rockefeller University Press.

$$t_{\text{burst}} = \beta^{-1}(1 + k[B]/l) \qquad \text{(Equation 4)}$$

Figure 10 shows that the mean lifetime of open channels was 6.7 msec. In the presence of 30 μM 9-AA, the burst time was 11.7 msec. On the average, the burst time was 80% longer than the normal open channel lifetime.

In four experiments on patches with intact Na inactivation and exposed to 30 μM 9-AA, the burst time was found to be greater than the normal open channel lifetime by only 36%

Figure 9. Concentration dependence of the reciprocal of the mean blocked time, t_b^{-1} (A) and the reciprocal of the mean open time, t_o^{-1} (B) at −25 mV and at −40 mV. Lines are the linear regression with the equation $t_o^{-1} = k[B] + \beta$, where *[B]* is the concentration of 9-AA (μM), β is the rate constant (sec^{-1}), at 0 μM concentration of 9-AA, and k is the slope. The values are $\beta = 135\ sec^{-1}$, $k = 5.9 \times 10^6\ M^{-1}sec^{-1}$ (○), $\beta = 123\ sec^{-1}$, $k = 5.8 \times 10^6 M^{-1}$ (△). The reciprocal of the mean blocked time has very little concentration dependence. Temperature = 6°C. Reproduced from *The Journal of General Physiology*,[21] by copyright permission of the Rockefeller University Press.

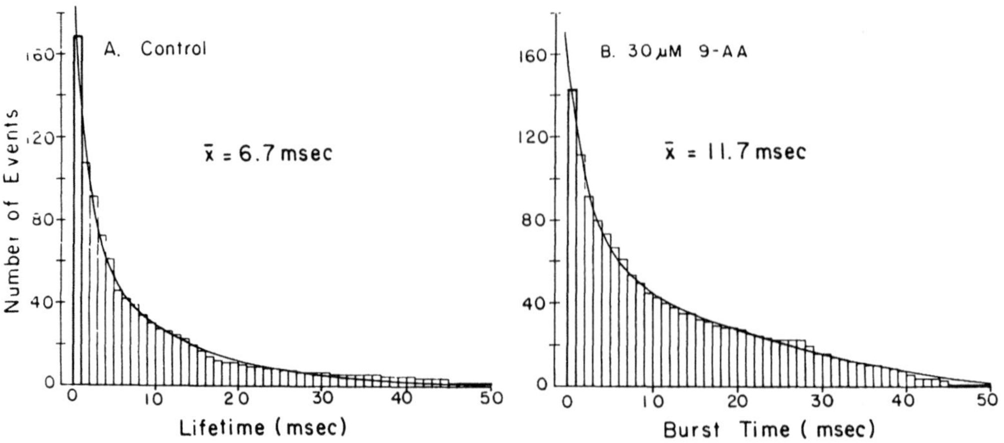

Figure 10. The burst time of the Na channel following exposure to 9-AA is greater than the normal open time of the channel. The patch of membrane was pretreated with NBA to remove Na inactivation. The duration of the open time of single Na channels under control conditions was measured, as well as the burst time of channels that were observed during exposure to 30 μM 9-AA. The number of events (n) having durations longer than the time on the abscissa is plotted. The curve fitting the open time histogram was drawn by the equation $n = 245 \exp(-t/1.8) + 75 \exp(-t/12.1)$. The curve describing the burst time was drawn according to the equation $n = 145 \exp(-t/2.6) + 87 \exp(-t/33)$. The average open time duration (control conditions) for all events was 6.7 msec, while the average burst duration upon exposure to 9-AA was 11.7 msec. Membrane potential, −40 mV. Temperature = 14°C.

(±25%). However, in two experiments burst time was greater by 200–300%. In an additional case, burst time was smaller than the normal open time by 88%. These experiments were performed between −40 and −20 mV, and at temperatures between 4 and 5°C. Mean open time in the presence of 9-AA was 1.33 msec (±0.36 msec) and mean block time was 1.90 msec (±0.51 msec) for these experiments.

QX-314

In marked contrast to 9-AA, QX-314 applied to the membrane did not cause flickering of the current through the Na channel. Figure 11 shows several consecutive traces (interpulse interval of 5 sec) for the control and with 100 μM QX-314 present. In Fig. 11A, the control traces each show several channel events, with many of them overlapping (two or three channels open at the same time). Following addition of QX-314 to the cytoplasmic face of the membrane, the frequency of channel opening events was markedly reduced (Fig. 11B). This was evident from the reduced number of channel events per trace. Only one trace showed overlapping openings (third trace from the top) and 5 of the 12 traces had no channel events at all. Furthermore, no flickering could be seen in any of the QX-314 traces. This is especially evident in the third and twelfth traces from the top.

Current amplitude histograms were constructed to determine the effect of QX-314 on the single-channel conductance. As can be seen in Fig. 12, most of the single-channel events were normally distributed about a mean value, with a small subpopulation being centered at a higher value. This subpopulation represents the events in which two channels were open at the same time and would be expected to be centered about a value twice the mean of the unitary channel events. The control histogram (Fig. 12A) reveals a greater number of multiple channel events than that indicated by the QX-314 histogram (Fig. 12B), as well as a greater number of total events. The means of the amplitudes of the two populations were

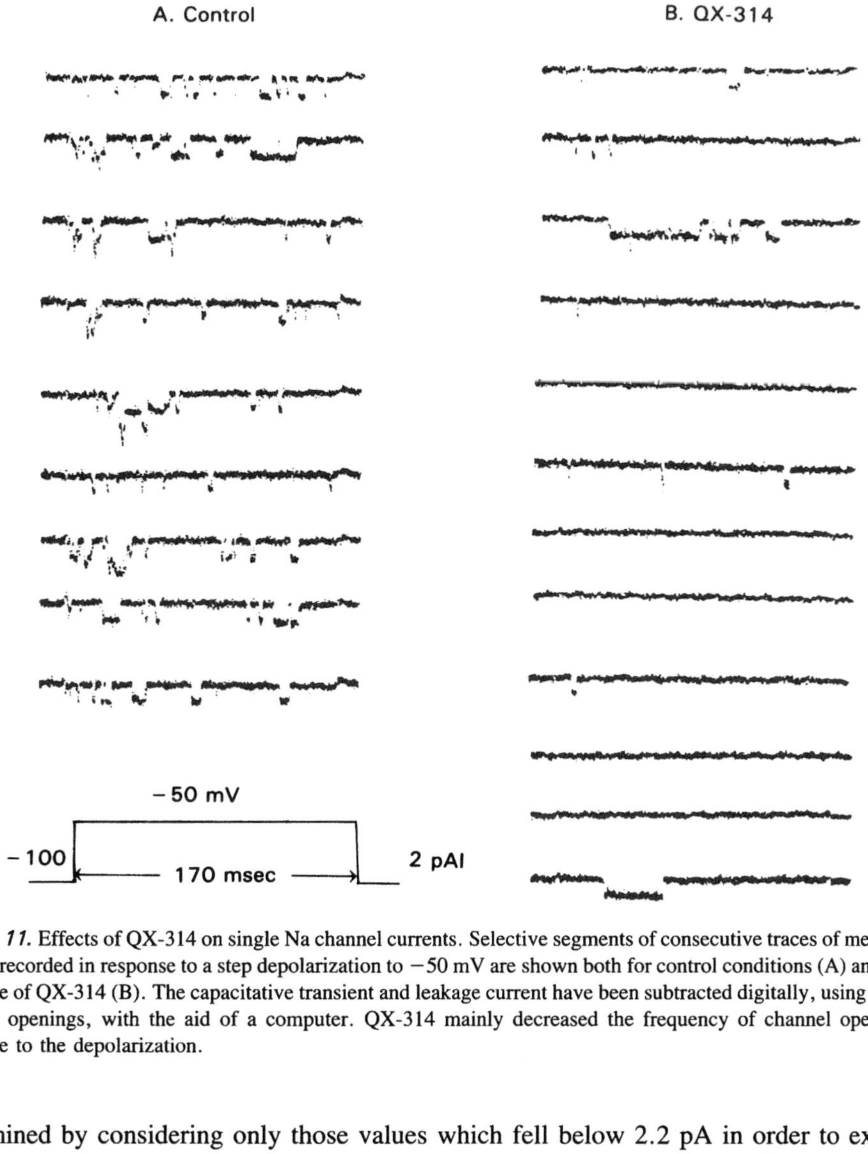

Figure 11. Effects of QX-314 on single Na channel currents. Selective segments of consecutive traces of membrane current recorded in response to a step depolarization to −50 mV are shown both for control conditions (A) and in the presence of QX-314 (B). The capacitative transient and leakage current have been subtracted digitally, using records without openings, with the aid of a computer. QX-314 mainly decreased the frequency of channel opening in response to the depolarization.

determined by considering only those values which fell below 2.2 pA in order to exclude currents due to multiple channel openings. The means were 1.56 ± 0.22 pA for the control histogram and 1.46 ± 0.25 pA for events recorded in the presence of QX-314. Thus, although there seems to be a possible reduction of the single-channel conductance by QX-314, it is not statistically significant.

The cumulative open time histograms for the control case and following exposure to QX-314 are shown in Fig. 13. The control open time cumulative histogram (Fig. 13A) was well described by a single exponential, which was fitted using a least-squares method, and has a time constant of 2.9 msec. In contrast, the QX-314 cumulative open time histogram (Fig. 13B) was best fitted by a sum of two exponentials. The time constants are 2.2 msec and 12.8 msec. The smaller time constant is very similar to the control and presumably represents channels that have not been affected by the drug. The larger time constant, indicating channels that are open for much longer than in the control, could possibly represent drug-modified channels. This question is currently under investigation in our laboratory.

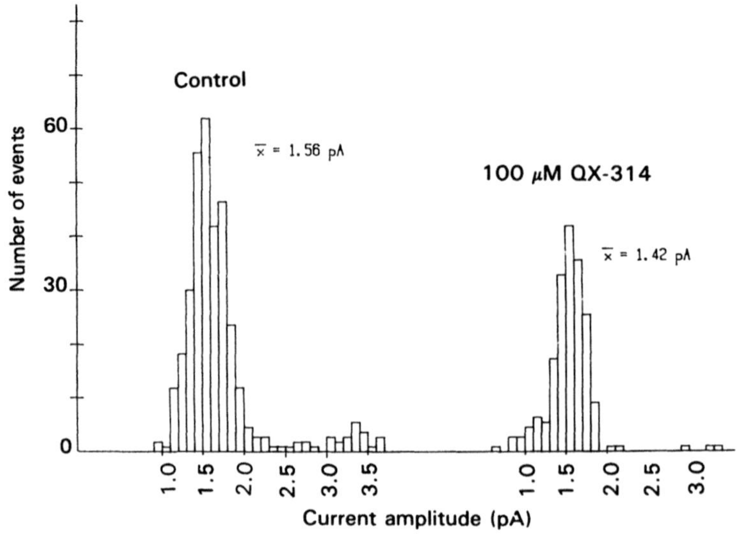

Figure 12. QX-314 does not reduce the amplitude of single Na channel currents. The current amplitude was measured from the zero current level, using traces similar to those shown in Fig. 11. Each histogram includes events with two simultaneous channel openings, which have amplitudes around 3 pA. To avoid including these multiple openings, the means were calculated from values below 2.2 pA. The average single-channel current was not significantly altered by QX-314.

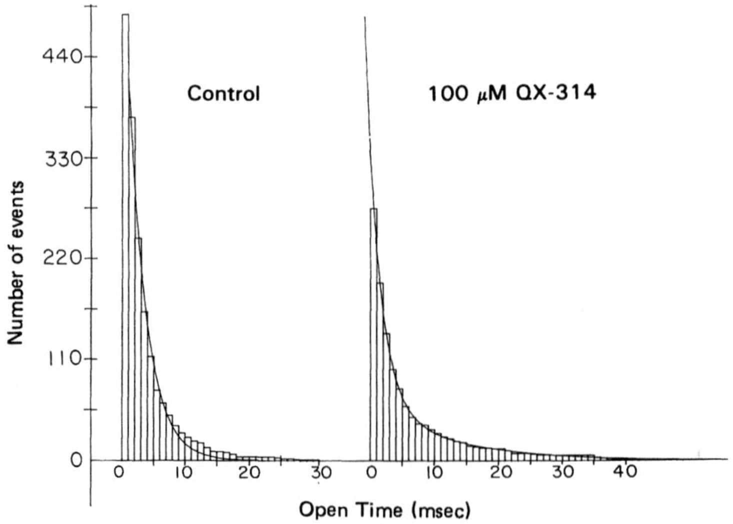

Figure 13. Comparison of the cumulative open time histogram for Na channels under control conditions with that during application of 100 μM QX-314. The single-channel currents were activated by depolarizing the membrane to −50 mV from a holding potential of −100 mV. The number of events (n) having a duration longer than that indicated on the abscissa is plotted against the open time (t). The curves were drawn by the equation $n = 590 \exp(-t/2.9)$ in the control case and $n = 330 \exp(-t/2.2) + 30 \exp(-t/12.8)$ for the data obtained during exposure to QX-314.

DISCUSSION

This chapter describes block of single Na channels by 9-AA and QX-314 in neuroblastoma cells. Single-channel studies have revealed a striking difference between 9-AA and QX-314 in their effects on the temporal properties of single-channel currents, even though they share many features of blocking action on Na channels from previous studies of macroscopic current.

Several interesting features of the 9-AA block of Na channels were revealed in this study. First, 9-AA caused the channel to flicker; i.e., the channel exhibited rapid transitions between fully conducting (open) and nonconducting (blocked) states. This rapid transition produced a bursting behavior. Second, the mean open time within a burst became shorter as the concentration was increased, whereas the mean blocked time was concentration-independent. Third, after removal of inactivation, the mean burst time in the presence of the drug was longer than the mean lifetime of an open channel in the absence of the drug. Fourth, the conductance of the open-unblocked state was not reduced by 9-AA. All the above features of 9-AA block of single Na channels are compatible with a sequential model for open channel block. In this model, 9-AA molecules can bind to, and dissociate from, Na channels with Na inactivation intact only while these channels are open. The bound channel has zero conductance and cannot close until it has freed the drug molecule from the site in the channel at which block occurs. Thus 9-AA represents the first agent described that causes the Na channel to flicker and presents itself as a new member to the "flicker" club.[3,4,12–15] Moreover, this result supports the notion that the channel occlusion model is a mechanism by which local anesthetic agents could block Na currents.

With respect to the burst time, equation (3) predicts that in the presence of 30 μM 9-AA, the burst time should be greater than the control open channel lifetime by 67% at −30 mV.[21] Indeed, in the NBA-treated patches exposed to 30 μM 9-AA, the burst time was greater than the control open lifetime by 80%. However, we found that in many patches not treated with NBA the burst time was greater than the control open time by only 36%. One possible modification in the above model (scheme 1) for untreated patches that would account for the shorter burst time would be to include inactivated states accessible from either open or blocked states. This alternative is particularly likely, since 9-AA causes a "use-dependent" block of Na^+ currents recorded from squid axon membrane, which is dependent on the presence of an inactivation gate.[1,24] The presence of such a state would cause the burst time to be less than that predicted by the above model, since the effect is to reduce the number of interruptions that return to the open state. In addition, a closed-blocked state would contribute to the reduction in frequency of opening of Na channels with intact inactivation after exposure to 9-AA.

In contrast, Na channels did not show flickering in the presence of QX-314. The failure to see flickering might be due to the possibility that the blocking and unblocking events of QX-314 occur much too quickly to be resolved by our recording system. Based on results in which macroscopic currents were measured, Yeh[25] proposed a kinetic model that could account for both resting and frequency-dependent block caused by QX-314. In this model, the rate constants governing the blocking and unblocking reactions with an open channel are fast, being on the order of 10 and 16 $msec^{-1}$, respectively. These rate constants are in a range that could produce "ultrafast" flickering. In this case QX-314 would be predicted to reduce single-channel conductance. This reduction would depend on the concentration of QX-314. However, single-channel recordings show that QX-314 did not decrease single-channel conductance at concentrations between 30 and 100 μM. This result indicates that QX-314 is not an "ultrafast" blocker, as has been proposed in the kinetic model.[25] The result also illustrates the ability of single-channel studies to provide information capable of ruling out certain kinetic schemes that could equally account for the macroscopic current data.

In summary, when the channel is open and conducting, neither 9-AA nor QX-314 has an effect on the amplitude of the single-channel current. Thus, a decrease in single-channel current is not the basis for 9-AA or QX-314 block of macroscopic currents. Both 9-AA and QX-314 decrease the frequency of channel opening. This may be the cause of the resting block of Na channels observed in measurements of macroscopic current with voltage clamp techniques.[1,19,23,24] In the case of 9-AA, open channel block would further decrease the macroscopic currents because the channel opening is interrupted into short openings. When currents due to individual channels are summed, these short openings would provide a smaller total current flow than under control conditions because the time channels are conducting in the presence of the drug is not synchronous as in the control case. This effect would contribute to the phasic (use-dependent) block caused by 9-AA.

ACKNOWLEDGMENTS. We thank Rae Barolet for secretarial assistance. This study was supported by NIH 31458, the Medical Research Council of Canada, and the Alberta Heritage Foundation for Medical Research.

REFERENCES

1. Cahalan MD: Local anesthetic block of sodium channels in normal and pronase-treated squid giant axons. *Biophys J* 1978;23:285–311.
2. Colquhoun D, Sigworth FJ: Fitting and statistical analysis of single-channel records, in Sakmann B, Neher E (eds): *Single-Channel Recording*. New York and London, Plenum Press, 1983, pp 191–263.
3. Coronado R, Miller C: Decamethonium and hexamethonium block K^+ channels of sarcoplasmic reticulum. *Nature (London)* 1980;288:495–497.
4. Coronado R, Miller C: Conduction and block by organic cations in a K^+-selective channel from sarcoplasmic reticulum incorporated into planar phospholipid bilayers. *J Gen Physiol* 1982;79:529–547.
5. Courtney DR: Mechanism of frequency dependent inhibition of sodium currents in frog myelinated nerve by the lidocaine derivative GEA 968. *J Pharmacol Exp Ther* 1975;195:225–236.
6. Gilbert DL, Lipicky RJ: A model of drug-channel interaction in squid axon membrane, in, Adelman WJ Jr, Goldman DE (eds): *The Biophysical Approach to Excitable Systems*. New York, Plenum Press, 1981, pp 213–251.
7. Hamill OP, Marty A, Neher E, Sakman B, Sigworth FJ: Improved patch-clamp techniques for high-resolution current recording from cell and cell-free membrane patches. *Pflugers Arch* 1981;391:85–100.
8. Hille B: Local Anesthetics: Hydrophilic and hydrophobic pathways for the drug-receptor reaction. *J Gen Physiol* 1977;69:497–515.
9. Horn R, Patlak J, Stevens CF: The effect of tetramethylammonium on single sodium channel currents. *Biophys J* 1981;36:321–327.
10. Kimhi Y, Palfrey C, Spector I, Barak Y, Littauer VZ: Maturation of neuroblastoma cells in the presence of dimethylsulfoxide. 1980. *Proc Natl Acad Sci USA* 1976;73:462–466.
11. Lipicky RJ, Gilbert DL, Ehrenstein G: Effects of yohimbine on squid axons. *Biophys J* 1978;24:405–422.
12. Miller C: Bis-quaternary ammonium blockers as structural probes of the sarcoplasmic reticulum K^+ channel. *J Gen Physiol* 1982;79:869–891.
13. Neher E: The charge carried by single-channel currents of rat cultured muscle cells in the presence of local anesthetics. *J Physiol (London)* 1983;339:663–678.
14. Neher E, Steinbach JH: Local anaesthetics transiently block currents through single acetylcholine receptor channels. *J Physiol (London)* 1978;277:153–176.
15. Ogden DC, Siegelbaum SA, Colquhoun D: Block of acetylcholine-activated ion channels by an uncharged local anaesthetic. *Nature (London)* 1981;289:596–598.
16. Oxford GS, Wu CH, Narahashi T: Removal of sodium channel inactivation in squid giant axons by *N*-bromoacetamide. *J Gen Physiol* 1978;71:227–247.
17. Patlak J, Horn R: Effect of *N*-bromoacetamide on single sodium channel currents in excised membrane patches. *J Gen Physiol* 1982;79:333–351.

18. Quandt FN, Yeh JZ, Narahashi T: Contrast between open and closed block of single Na channel currents. *Biophys J* 1982;37:319a.
19. Strichartz GR: The inhibition of sodium currents in myelinated nerve by quaternary derivatives of lidocaine. *J Gen Physiol* 1973;62:37–57.
20. Taylor RE: Effect of procaine on electrical properties of squid axon membrane. *Am J Physiol* 1959;196:1071–1078.
21. Yamamoto D, Yeh JZ: Kinetics of 9-aminoacridine block of single Na channels. *J Gen Physiol* 1984;84:362–377.
22. Yamamoto D, Yeh JZ, Narahashi T: Voltage-dependent calcium block of normal and tetramethrin-modified single sodium channels. *Biophys J* 1984;45:337–344.
23. Yeh JZ: Sodium inactivation mechanism modulates QX-314 block of sodium channels in squid axons. *Biophys J* 1978;24:569–574.
24. Yeh JZ: Dynamics of 9-aminoacridine block of sodium channels in squid axons. *J Gen Physiol* 1979;73:1–21.
25. Yeh JZ: A pharmacological approach to the structure of the Na channel in squid axon, in Haber B *et al.* (eds): *Progress in Clinical and Biological Research,* Vol. 79, *Proteins in the Nervous System: Structure and Function.* New York, Alan R. Liss, Inc., 1982, pp 17–49.

20

Inhibition of Voltage-Sensitive Sodium Channels by Local Anesthetics and Anticonvulsants

A Biochemical and Electrophysiological Analysis

Max Willow, Tohru Gonoi, Sidney W. Postma, and William A. Catterall

ABSTRACT. The inhibitory action of a number of clinically effective local anesthetics and anticonvulsants on voltage-sensitive sodium channels in cultured neuroblastoma cells and rat brain synaptosomes is examined. These drugs inhibit the binding of [^{3}H]batrachotoxinin A 20-α-benzoate (BTX-B) to a receptor site responsible for the activation of sodium channel ion flux by alkaloid neurotoxins. Scatchard analysis of BTX-B binding to synaptosomes shows that diphenylhydantoin (DPH), carbamazepine (CBZ), and local anesthetics reduce receptor affinity without altering maximal binding capacity. The reduction in receptor affinity in the presence of these agents appears to be due to an increased rate of ligand dissociation from the receptor–ligand complex, suggesting an indirect allosteric mechanism for inhibition of BTX-B binding. CBZ, like DPH and lidocaine, acts as a competitive inhibitor of sodium channel activation by the full agonist batrachotoxin, but produces mixed inhibition of veratridine-activated channels. This finding supports the conclusion that these drugs act as allosteric inhibitors of neurotoxin-activated sodium channels. Electrophysiological experiments demonstrate that bupivacaine, DPH, and CBZ reduce peak sodium conductance without altering the current–voltage relationship. In the presence of DPH and CBZ, steady-state inactivation is markedly shifted in the hyperpolarizing direction while bupivacaine produces only a very small negative shift. All three drugs show marked voltage-dependent effects and appear to stabilize inactivated states of the channel. Bupivacaine and DPH exert frequency-dependent effects when cells are stimulated at a rate of 2 Hz. At this frequency, the action of CBZ is nearly unaffected. These results are consistent with different rates of drug dissociation which indicate that CBZ dissociates rapidly upon membrane hyperpolarization while DPH and bupivacaine are relatively slow in their time course for dissociation. These results are consistent with the voltage- and time-dependent actions of local anesthetics and related compounds predicted by the modulated receptor hypothesis.

INTRODUCTION

As we have seen in the preceding chapters, there is now considerable evidence to suggest that local anesthetics and some classes of anticonvulsants exert their pharmacological actions by interacting with voltage-sensitive sodium channels in electrically excitable tissues.[19,26,27,29]

Max Willow, Tohru Gonoi, and William A. Catterall • Department of Pharmacology, University of Washington, Seattle, Washington 98195. *Sidney W. Postma* • Departments of Pharmacology and Anesthesiology, University of Washington, Seattle, Washington 98195. Present address of M. W.: School of Medicine, McGill University, Montreal, Quebec H3G 1Y6, Canada.

In nerve, the action of local anesthetics and pharmacologically related compounds appears to be dependent on variables such as membrane potential and frequency and duration of sodium channel activation.[8–11,17,18,20,21,28] These observations have led to the development of the modulated receptor hypothesis,[17,18] which postulates that (1) local anesthetics bind to a receptor site in close association with the voltage-dependent sodium channel, and (2) the affinity of the drug receptor is modulated by membrane potential or channel state in which resting, open, and inactivated channels can have differing rates of association and dissociation with local anesthetics. The objective of the work described in this chapter was to address the postulates of the modulated receptor hypothesis in order to gain further insight into the nature of the receptor site(s) for local anesthetics and certain anticonvulsants that interact with voltage-sensitive sodium channels.

Our initial experiments have examined the interaction of local anesthetics and anticonvulsants with radiolabeled neurotoxins that bind to specific receptor sites associated with sodium channels. Neurotoxins act at several distinct receptor sites associated with sodium channels in pinched-off nerve terminals (synaptosomes) and electrically excitable neuroblastoma cells.[4,25] Receptor site 1 binds the inhibitors tetrodotoxin and saxitoxin.[4,25] Receptor site 2 binds grayanotoxin and the alkaloids veratridine, batrachotoxin, and aconitine. These toxins cause persistent activation of sodium channels.[25] Receptor site 3 binds the polypeptides α-scorpion toxin and sea anemone toxin. These toxins slow or block sodium channel inactivation and enhance persistent activation of sodium channels by alkaloid toxins acting at receptor site 2 through an allosteric mechanism.[25] The interaction of neurotoxins with each of these receptor sites can be studied by ligand binding or ion flux experiments on electrically excitable neuroblastoma cells[1–3] or synaptosomes isolated from rat brain.[6,7,24]

In order to study the time- and voltage-dependent actions of local anesthetics and anticonvulsants on transmembrane sodium channels, we have employed the whole cell–patch clamp recording technique[16] to analyze sodium currents in cultured neuroblastoma cells. This approach affords a more precise control of membrane potential and time resolution of channel events than is possible in biochemical experiments.

METHODS

Biochemical Experiments

Preparation of Synaptosomes. Synaptosomes were prepared from whole rat brain according to a modification[30] of the method of Gray and Whittaker.[15] The synaptosomes were divided into 0.5-ml aliquots and slowly frozen over 30 min on dry ice prior to storage in liquid nitrogen. Previous studies have shown that there is no appreciable loss of toxin binding activity or $^{22}Na^+$ flux activity using this procedure.[30]

Measurement of Radiolabeled Neurotoxin Binding in Rat Brain Synaptosomes. Specific binding of ^{125}I-labeled scorpion toxin, [^{3}H]saxitoxin, and [^{3}H]batrachotoxinin A 20-α-benzoate ([^{3}H]BTX-B) was measured as described previously.[6,7,24] Prior to use, an aliquot of frozen synaptosomes was thawed at 36° for 5 min and then stored on ice. The synaptosomes were diluted fourfold in standard incubation medium to a final protein concentration of 1.10 mg/ml in the assay. ^{125}I-labeled scorpion toxin binding was measured by incubating synaptosomes for 10 min at 36° in the standard incubation medium consisting of 130 mM choline chloride, 5.4 mM KCl, 0.8 mM $MgSO_4$, 5.5 mM glucose, and 50 mM HEPES/TRIS, at pH 7.4, plus 0.1 nM ^{125}I-labeled scorpion toxin, 1 μM tetrodotoxin, and 1 mg/ml bovine serum albumin (BSA) in the absence or presence of batrachotoxin (1.25 μM) and varying concentrations of added drug. In some experiments, synaptosomes were depolarized by increasing the KCl concentration and decreasing choline chloride accordingly. Nonspecific binding of ^{125}I-labeled scorpion toxin was determined in the presence of 100 nM unlabeled scorpion toxin. Binding reactions were initiated by the addition of 25 μl of the synaptosomal suspension to 175 μl of the reaction mixture. Samples were rapidly mixed, and following incubation for 10 min, the reaction was stopped by the addition of 3 ml of ice-cold wash medium consisting of 163 mM choline chloride, 5 mM HEPES/TRIS (pH 7.4), 1.8 mM $CaCl_2$, 0.8 mM $MgSO_4$, and 1 mg/ml BSA. The synaptosomes were immediately collected on glass-fiber filters (Whatman GF/C) under vacuum and washed three times with 3 ml of wash medium. ^{125}I bound to filters was determined in a gamma scintillation spectrometer.

[^{3}H]Saxitoxin binding was measured in a similar manner to that described above. The reaction mixture contained standard incubation medium plus [^{3}H]saxitoxin (1.0 nM), BSA (1 mg/ml), and varying concentrations of added drugs. Binding was measured for 10 min at 36°. Nonspecific binding of [^{3}H]saxitoxin was determined in the presence of tetrodotoxin (1 μM). Filters containing trapped synaptosomes were suspended in liquid scintillation fluid, and ^{3}H bound was determined in a liquid scintillation spectrometer.

[^{3}H]BTX-B binding was studied by incubating synaptosomes for 30 min at 36° in the standard incubation medium plus [^{3}H]BTX-B (10 nM), tetrodotoxin (1 μM), scorpion toxin (0.3 or 2 μM), BSA (1 mg/ml), and varying concentrations of added drugs. Nonspecific binding of [^{3}H]BTX-B was determined in the presence of 0.3 mM veratridine. [^{3}H]BTX-B bound was determined as described previously.

Measurement of Batrachotoxin-Activated $^{22}Na^+$ Flux in Rat Brain Synaptosomes. Prior to use, synaptosomes were thawed at 36° for 5 min and then stored on ice. Synaptosomes (0.62 mg of protein) were incubated with batrachotoxin (1–2 μM) for 20 min at 36° in 50 μl of standard binding medium. After 20 min, 150 μl of the $^{22}Na^+$ uptake solution was added and mixed rapidly by vortexing. The uptake solution contained the same toxin concentration used in the preincubation plus 2 mM NaCl, 128 mM choline chloride, 5.4 mM KCl, 0.8 mM $MgSO_4$, 5.5 mM glucose, 50 mM HEPES/TRIS (pH 7.4), and BSA (1 mg/ml). The mixture was rapidly filtered under vacuum through an Amicon cellulose filter disk (0.45 μm), and the filter disk and trapped synaptosomes were washed twice with 3 ml of the wash solution. Radioactivity was determined by liquid scintillation spectrometry. Specific $^{22}Na^+$ influx attributable to activation of the voltage-sensitive sodium channel was determined as the difference in $^{22}Na^+$ uptake in the absence and presence of 1 μM tetrodotoxin added to both preincubation and uptake solutions. The data presented are corrected for nonspecific uptake and nonspecific binding of $^{22}Na^+$ to filter discs.

Measurement of Batrachotoxin- and Veratridine-Activated $^{22}Na^+$ Flux in Cultured N18 Neuroblastoma Cells. In experiments in which batrachotoxin (0.02–1.25 μM) was used to activate sodium channels, neuroblastoma cell cultures were incubated with the alkaloid and varying concentrations of drugs as indicated in the figures for 30 min at 36° in 0.25 ml of medium containing 135 mM KCl, 50 mM HEPES/TRIS (pH 7.4), 5.5 mM glucose, 0.8 mM $MgSO_4$, and BSA (1 mg/ml). After 30 min, the medium was removed by aspiration, and the cells were incubated for 30 sec in assay medium containing the same concentrations of batrachotoxin and drugs, and 5.4 mM KCl, 120 mM choline chloride, 10 mM NaCl, 5 mM ouabain, 50 mM HEPES/TRIS (pH 7.4), 5.5 mM glucose, 0.8 mM $MgSO_4$, and ^{22}NaCl (1.0 μCi/ml). The assay was terminated by washing the cells three times with ice-cold wash medium identical with that used in the synaptosomal flux procedures described previously. In some experiments, scorpion toxin (0.3 μM) was included in the incubation and assay media.

In experiments in which sodium channels were activated with veratridine (0.01–0.2 mM), cell cultures were incubated for 30 min at 36° with varying concentrations of drugs in standard incubation medium. The $^{22}Na^+$ flux solution contained varying concentrations of veratridine and drugs. In all other respects, the procedures were identical with those used in the batrachotoxin experiments. Previous experiments in this laboratory have shown that the procedures described here provide an accurate estimate of sodium ion permeability at constant membrane potential.[2,3]

Other Methods. Synaptosomal and cellular protein concentration was measured according to the method of Peterson.[22] Except where mentioned, the data presented are the results of a single experiment that are representative of three to four similar experiments. Each data point represents the mean of three to six replicate determinations. In all experiments, standard errors of the mean were less than 10% of the mean value observed. Linear segments of Scatchard plots and double reciprocal plots of neurotoxin activation of sodium channels were fitted by computer-assisted least-squares regression analysis. IC_{50} determinations were calculated by log-probit analysis according to the method of Finney.[14] For anticonvulsants/local anesthetics that act as competitive inhibitors of batrachotoxin action, IC_{50} values were converted to K_I values according to the relationship

$$K_I = \frac{IC_{50}}{1 + [L]/K_L}$$

where K_L is the equilibrium dissociation constant for batrachotoxin, $[L]$ is the concentration of batrachotoxin used, and IC_{50} is the concentration of drug that produces 50% inhibition of batrachotoxin-activated $^{22}Na^+$ flux.

Electrophysiological Experiments

Cell Culture. N18 neuroblastoma cells were cultured in 100-mm-diameter Petri dishes for 5–8 days until cells became confluent. Details of the culture techniques have been described previously.[1] Three to eight hours prior to recordings, cells were harvested by trituration in Ca^{2+}–Mg^{2+} free-phosphate-buffered saline and reseeded to enable us to record from isolated spherical cells as required for voltage clamp measurements.

Voltage Clamp Recording. All recordings were performed at room temperature (20±1°C) in mammalian Ringer's solution (150 mM NaCl, 5 mM KCl, 1.5 mM $CaCl_2$, 1.0 mM $MgCl_2$, 5 mM glucose, and 5 mM Na^+–HEPES, pH 7.4). The voltage clamp was based on the one-pipette, gigaseal whole-cell recording technique.[16] Glass hematocrit microtubes were pulled in two steps, and tips of the pipettes were fire-polished. The tip resistances were from 0.2 to 1 MΩ in mammalian Ringer's solution. Gentle suction applied after the pipette was in contact with the cell surface sufficed to form a seal and rupture the patch of membrane in the orifice. After 10–15 min the total input resistance of pipette and cell rose to 10^8–10^{10} Ω. The internal pipette medium contained 160 mM CsF and 10 mM NaF. This solution blocked potassium currents without altering sodium currents and allowed stable recordings up to 3 hr. The voltage clamp circuit used no electrodes other than a single bath electrode and the patch pipette. It therefore included a circuit for compensation of up to 1 MΩ series resistance in the current-passing pathway. Leak currents were subtracted electronically. The membrane capacitances of the cells studied ranged from 30 to 185 pF, and the time constants of decay of capacitive currents were typically 50–150 μsec before the series resistance compensation was applied. The cell was maintained at a holding potential of −75 mV. For measurement of peak sodium currents, I_{Na}, the membrane was hyperpolarized in a prepulse to −105 mV for 90 msec to remove inactivation of sodium channels and then depolarized to test potentials from −60 mV to +105 mV for 10 msec. For measurement of the voltage dependence of sodium channel inactivation, the prepulse potential was varied from −135 mV to −30 mV. The test pulse was always +7.5 mV. In experiments examining the voltage dependence of drug action on peak sodium currents, the holding potential was varied from −135 mV to −30 mV. Frequency- or use-dependent effects of local anesthetics and anticonvulsants were assessed as described in the figure legends.

RESULTS

Anticonvulsant Inhibition of [³H]BTX-B Binding to Synaptosomes

In a previous study, it was found that diphenylhydantoin and lidocaine were competitive inhibitors of batrachotoxin action at neurotoxin receptor site 2, but had no effect on saxitoxin or scorpion toxin binding to neurotoxin receptor sites 1 and 3 on sodium channels in neuroblastoma cells.[5] Recently, methods have been developed to measure specific binding of [^{3}H]BTX-B to neurotoxin receptor site 2 on sodium channels in synaptosomal membranes.[7] In this binding assay scorpion toxin is used to decrease the K_d for BTX-B from approximately 1 μM to 70 nM through the allosteric interaction between neurotoxin receptor sites 2 and 3.[4] Sodium-free medium and tetrodotoxin are used to block ion transport through sodium channels activated by BTX-B during the binding measurement. Under these conditions, specific binding of [^{3}H]BTX-B to sodium channels comprises 70%–80% of total [^{3}H]BTX-B binding and can be accurately estimated.[7] Since this binding requires scorpion toxin,[7] a highly specific sodium channel ligand,[2,24,30] it is unlikely that binding to calcium channels or other synaptosomal components make a significant contribution to the data.

The effect of a number of anticonvulsants on the binding of [^{3}H]BTX-B (10 nM) to synaptosomes in the presence of 0.3 μM scorpion toxin is illustrated in Fig. 1A and B. Diphenylhydantoin (DPH) and carbamazepine (CBZ), both of which are effective in the management of grand mal and partial seizure disorders, inhibited binding in a concentration-dependent manner with IC_{50} values of 40.8 ± 2.5 μM (n = 4) and 158 ± 14 μM (n = 4), respectively (Fig. 1A). Trimethadione, ethosuximide, and sodium valproate did not affect [^{3}H]BTX-B binding at concentrations up to 500 μM (Fig. 1A). Diazepam (IC_{50} = 152 ± 14 μM, n = 5) and phenobarbital (IC_{50} = 2.60 ± 0.25 mM, n = 5) also inhibited [^{3}H]BTX-B binding (Fig. 1B), but the concentrations required far exceed therapeutic levels.[31]

Local Anesthetic Inhibition of [³H]BTX-B Binding to Synaptosomes

Similarly, local anesthetics inhibited specific binding of [^{3}H]BTX-B without affecting nonspecific binding. Inhibition of specific toxin binding is dose-dependent, with mean values of K_d's ranging from 1.7 μM to 1.78 mM, as listed in Table I. In order to determine whether the inhibition of sodium channel ion conductance by local anesthetics results from the same mechanism of action as block of [^{3}H]BTX-B binding, we have compared (Table I) the K_d values derived from our neurotoxin binding measurements with those derived from electrophysiological experiments under conditions of tonic block.[9] These data indicate that there is a significant correlation of the two parameters over a 1000-fold range of $K_{0.5}$ values.

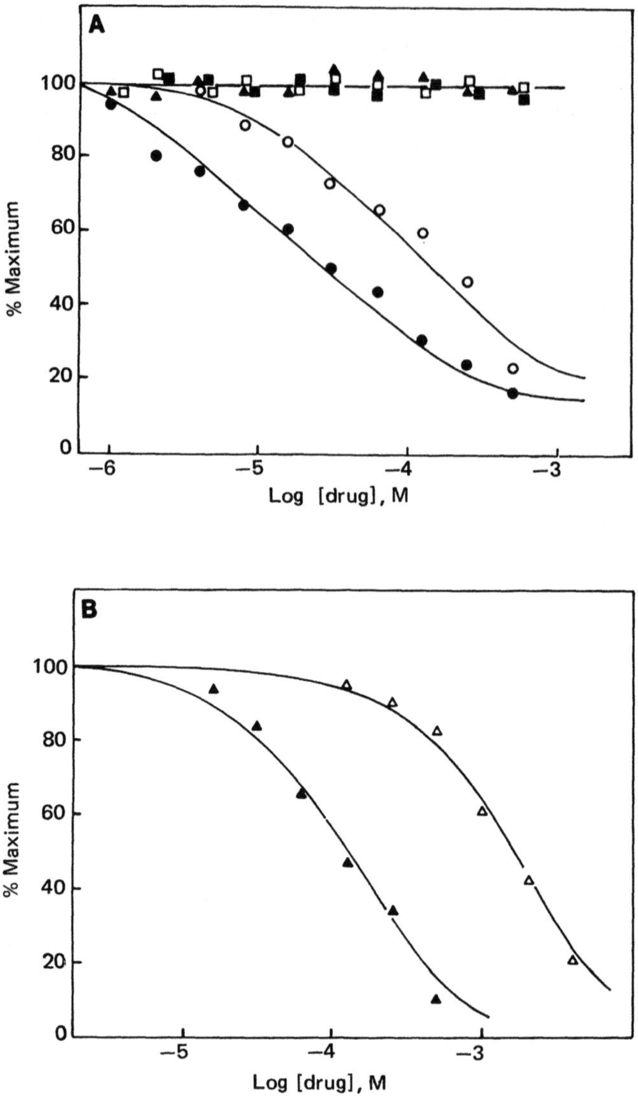

Figure 1. Effect of anticonvulsants on the binding of [^{3}H]BTX-B to synaptosomes. (A) Effect of DPH, CBZ, trimethadione, ethosuximide, and sodium valproate. Synaptosomes were incubated in the presence of 10 nM [^{3}H]BTX-B (in the presence of 0.3 μM scorpion toxin) and varying concentrations of DPH (●), CBZ(○), trimethadione (▲), ethosuximide (□), or sodium valproate (■) for 30 min. Binding was then measured as described in Methods. (B) Effect of diazepam and phenobarbital. Synaptosomes were incubated in the presence of 10 nM [^{3}H]BTX-B (in the presence of 0.3 μM scorpion toxin) and varying concentrations of diazepam (▲) or phenobarbital (△) for 30 min. Binding was then measured as described in Methods. Reprinted with permission from Willow and Catterall.[31]

This conclusion is confirmed by regression analysis of a plot of log K_d for drug binding against log $K_{0.5}$ for channel blockade, which yields a slope of 0.84 and a correlation coefficient, *r*, of 0.86.

Estimation of the affinity for local anesthetic binding and action is complicated by the frequency dependence of these parameters.[17] Local anesthetic inhibition of sodium channels is more rapid and complete if nerve preparations are repetitively depolarized to cause channel activation.[8,28] These effects are thought to result from rapid binding of the drugs to the ion-conducting state(s) of sodium channels together with high-affinity binding to the inactivated state(s).[17] Since sodium channels cannot be repetitively activated during our binding assays, we have compared our binding data with values for tonic inhibition, the extent of channel block at very low frequencies of stimulation. Since a good correlation is observed (Table I), it is likely that tonic block of sodium channels by local anesthetics also prevents [^{3}H]BTX-B binding. Creveling *et al.*[13] observed a close correlation between block of BTX-induced

Table I. Effect of Local Anesthetics on BTX-B Binding

Local anesthetic	K_d (BTX-B) (μM)	$K_{0.5}$ (electrophysiological data)[a] (μM)
Tetracaine	1.7 ± 0.6	0.31
Etidocaine	3.6 ± 1.2	7.1
Bupivacaine	4.5 ± 1.8	31
(±)-Propranolol	15.5 ± 4.5	28
Prilocaine	50.0 ± 3.1	556
Mepivacaine	192 ± 19	140
Procaine	161 ± 73	90
Tocainide	241 ± 41	636
Lidocaine	311 ± 53	140
W 36017	1100 ± 100	426
Benzocaine	1785 ± 86	2300

[a]$K_{0.5}$ was calculated by applying equation (4) of ref. 9 and for benzocaine from ref. 11.

depolarization of brain microsacs and block of [^{3}H]BTX-B binding by local anesthetics, providing additional support for the conclusion that these two effects result from the same mechanism of action.

Effect of DPH, CBZ, and Local Anesthetics on Kinetic and Equilibrium Properties of BTX-B Binding

Scatchard analysis of BTX-B binding over an extended concentration range (Fig. 2A and B) shows that this ligand binds to a single class of high-affinity binding sites, in agreement with earlier studies.[7] DPH (50 μM) and CBZ (200 μM) increase the K_d for [^{3}H]BTX-B from 125 nM to 256 nM or 357 nM, respectively. The maximal binding capacity for BTX-B (2.2 pmoles/mg) is not significantly affected by DPH or CBZ. These results indicate a competitive mechanism for inhibition of BTX-B binding by DPH and CBZ. Bupivacaine (6.2 μM) increased the K_d for [^{3}H]BTX-B from 116 to 312 nM without effect on the maximal binding capacity (data not shown).

Neither 50 μM DPH nor 200 μM CBZ influences the rate of association of [^{3}H]BTX-B binding to the receptor complex. In contrast, both DPH (200 μM) and CBZ (300 μM) accelerate the rate of dissociation of [^{3}H]BTX-B from the steady-state ligand–receptor complex following the addition of 200 μM veratridine (Fig. 3A and B). The dissociation rate constant (k_{-1}) is increased from 0.0086 min^{-1} to 0.014 min^{-1} in the presence of DPH and from 0.0090 min^{+1} to 0.0196 min^{-1} in the presence of CBZ. These kinetic data are not consistent with a simple competitive inhibition mechanism in which the drugs bind at the BTX-B receptor site, since this mechanism predicts a reduction in the rate of binding with no effect on the rate of dissociation of the toxin–receptor complex. The data indicate that DPH and CBZ are indirect allosteric competitive inhibitors of BTX-B binding.

Figure 4 illustrates the dissociation of the [^{3}H]BTX-B/receptor complex in the presence of the competitive ligand aconitine or aconitine plus local anesthetics at concentrations about 10 times greater than their K_d values. A range of dissociation rates is observed with local anesthetics. In each case, local anesthetics significantly increase the dissociation rate over that with a saturating concentration of aconitine alone. Dissociation rate constants and half-lives for these dissociation reactions are summarized in Table II. The increase in dissociation rate constant over that with aconitine ranges from 2.3-fold for tetracaine to 8.5-fold for

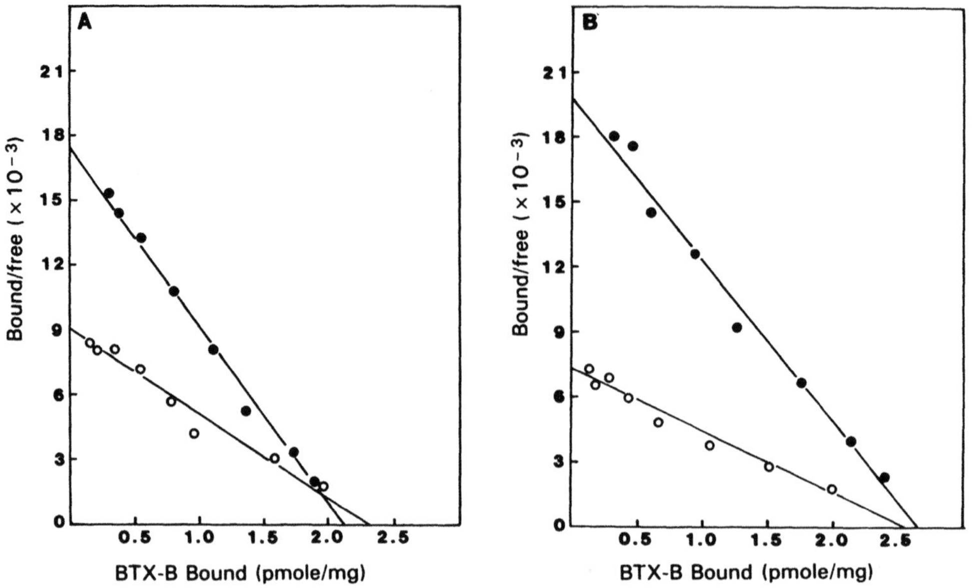

Figure 2. Scatchard analysis of [^{3}H]BTX-B binding to synaptosomes. (A) Binding of 10 nM [^{3}H]BTX-B was measured in standard incubation medium (scorpion toxin concentration = 2 μM) in the absence (●) or presence (○) of 50 μM DPH, with increasing concentrations of unlabeled BTX as described in Methods. (B) Binding of 10 nM [^{3}H]BTX-B was measured in standard incubation medium (scorpion toxin concentration = 2 μM) in the absence (●) or presence (○) of 200 μM CBZ, with increasing concentrations of unlabeled BTX as described in Methods. Reprinted with permission from Willow and Catterall.[31]

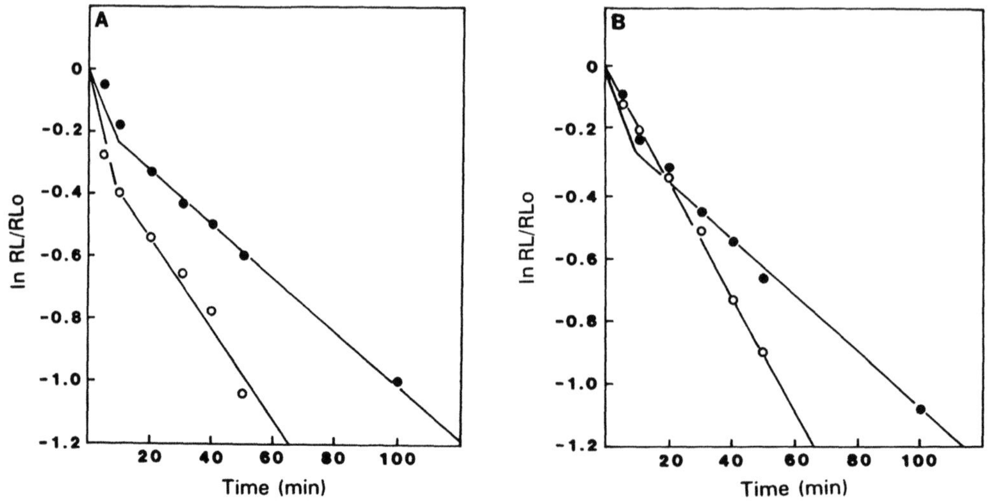

Figure 3. Time course of dissociation of the [^{3}H]BTX-B/receptor complex. (A) Synaptosomes were incubated in standard incubation medium for 30 min with 10 nM [^{3}H]BTX-B (scorpion toxin concentration = 0.3 μM), and at zero time, 200 μM veratridine [in the absence (●) or presence (○) of 50 μM DPH] was added. At the indicated times, samples were removed, and bound [^{3}H]BTX-B was determined as described in Methods. (B) Experimental conditions were similar to those in A, except that, at zero time, 200 μM veratridine [in the absence (●) or presence (○) of 300 μM CBZ] was added. Reprinted with permission from Willow and Catterall.[31]

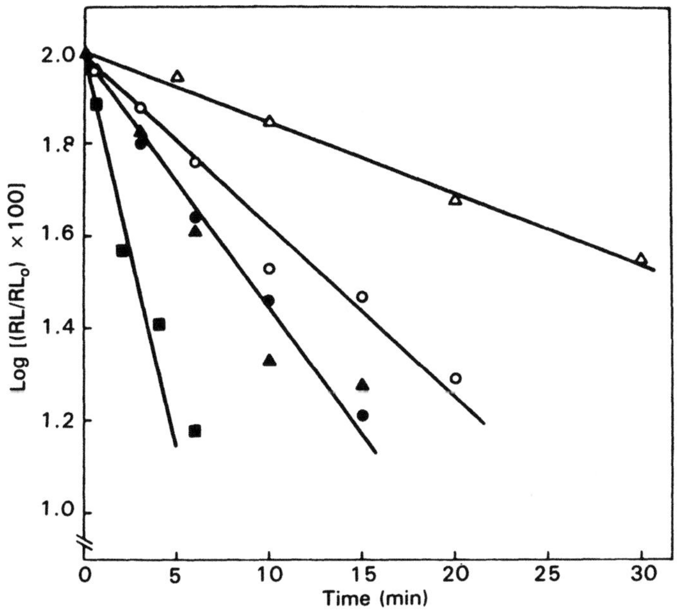

Figure 4. Effect of local anesthetics on the time course of dissociation of the [^{3}H]BTX-B/receptor complex. Synaptosomes were incubated in standard incubation medium for 30 min with 10 nM [^{3}H]BTX-B, and at zero time, 200 μM aconitine alone (△) or 200 μM aconitine plus 13 μM tetracaine (○), 470 μM prilocaine (●), 1.6 mM tocainide (▲), or 2.5 mM lidocaine (■) was added. Reprinted with permission from Postma and Catterall.[23]

lidocaine. These results indicate that local anesthetics also act as indirect allosteric competitive inhibitors of BTX binding.

Effect of DPH, CBZ, and Local Anesthetics on the Binding of Saxitoxin and Scorpion Toxin to Synaptosomes

DPH, CBZ, and local anesthetics have no effect on saxitoxin binding to neurotoxin receptor site 1 at concentrations that are effective in blocking sodium channels.[5,23,31] Binding of scorpion toxin to neurotoxin receptor site 3 on the sodium channel can be measured directly using ^{125}I-labeled scorpion toxin.[24] In synaptosomes, scorpion toxin binding is

Table II. Kinetic Data for the Dissociation of the [^{3}H]BTX-B/Receptor Complex in the Presence of Local Anesthetics and/or Aconitine

Ligand	Concentration (μM)	Dissociate rate constant (k_{-1}) (min^{-1})	Half-life ($t_{1/2}$) (min)
Aconitine	200	0.0156	19.3
Tetracaine	13	0.0354	8.5
Prilocaine	470	0.0516	5.8
Tocainide	1600	0.0519	5.8
Lidocaine	2500	0.133	2.26

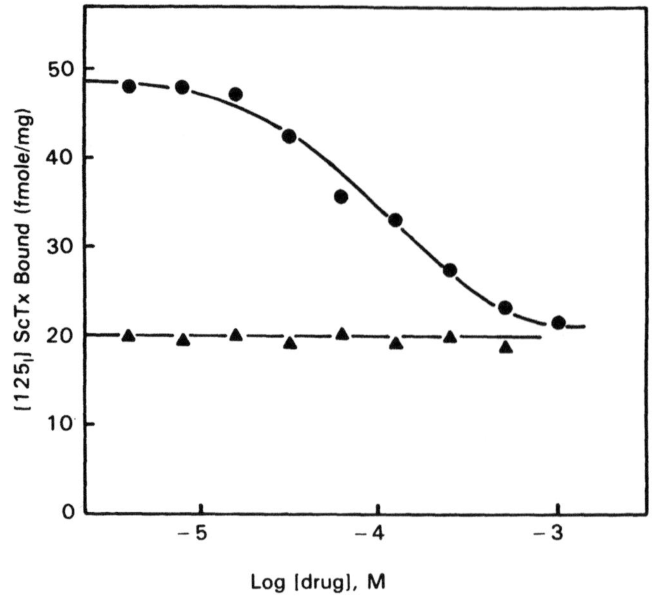

Figure 5. Effect of DPH on ^{125}I-labeled scorpion toxin binding. Synaptosomes were incubated for 10 min in medium containing 0.1 nM ^{125}I-labeled toxin [in the absence (▲) or presence (●) of 1.25 μM unlabeled batrachotoxin] in varying concentrations of DPH. The concentration of KCl in the medium was 25 mM. Binding was then measured as described in Methods. Reprinted with permission from Willow and Catterall.[31]

enhanced by batrachotoxin.[7] The enhancement is greater in depolarized synaptosomes. Thus, in medium containing 25 mM K^+ to partially depolarize synaptosomes, the effect of DPH and CBZ on scorpion toxin binding can be measured directly and compared with the effect of these drugs on the batrachotoxin enhancement of scorpion toxin binding. The results of Fig. 5 show no effect of DPH on specific scorpion toxin binding at concentrations as high as 0.5 mM. In the presence of 1.25 μM batrachotoxin, scorpion toxin binding is enhanced two- to threefold in medium containing 25 mM KCl. DPH produced a concentration-dependent inhibition of that fraction of ^{125}I-labeled scorpion toxin binding which is enhanced by the alkaloid toxin with an IC_{50} = 100 μM (Fig. 5). These data indicate a selective effect of DPH on batrachotoxin binding. Similar results were obtained with CBZ and tetracaine.[23,31]

Effect of Anticonvulsants and Local Anesthetics on Neurotoxin-Activated $^{22}Na^+$ Flux in N18 Neuroblastoma Cells and Synaptosomes

The inhibitory effects of DPH, CBZ, and local anesthetics on [^{3}H]BTX-B binding to synaptosomes predict that these agents should exert similar effects on batrachotoxin-activated $^{22}Na^+$ flux at comparable concentrations. Table III summarizes results showing that DPH and CBZ inhibit batrachotoxin-activated $^{22}Na^+$ flux in both neuroblastoma cells and synaptosomes at concentrations comparable to central nervous system concentrations that protect laboratory animals against maximal electroshock seizures. Diazepam and phenobarbital are, by comparison, relatively poor inhibitors of synaptosomal and neuroblastoma cell $^{22}Na^+$ flux at therapeutically relevant concentrations, in agreement with neurotoxin binding experiments (Fig. 1B). The local anesthetic lidocaine inhibits batrachotoxin-activated $^{22}Na^+$ flux (K_I = 300 μM) at comparable concentrations required to inhibit [^{3}H]BTX-B binding to rat brain synaptosomes (K_I = 311 μM).

In N18 neuroblastoma cells and synaptosomes, batrachotoxin and other alkaloid toxins act as allosteric activators of voltage-sensitive sodium channels.[3] They bind with high affinity to the active state of sodium channels and shift the channel population toward that state according to the laws of mass action. Double reciprocal plot analysis of the block of

Table III. Effects of Anticonvulsants and Local Anesthetics on Neurotoxin-Activated $^{22}Na^+$ Flux in Neuroblastoma Cells and Synaptosomes

Drug	K_I BTX-activated $^{22}Na^+$ influx (cells) (μM)	K_I BTX-activated $^{22}Na^+$ influx (synaptosomes)	Brain levels[a] (pmole/kg)
DPH	35	38	10–20
CBZ	41	22	10–50
Phenobarbital	1200	1300	20–80
Diazepam	51	63	0.5–1.0
Lidocaine	300	—	

[a]Levels achieved during effective prevention of seizures in rats. See ref. 31 for citations.

batrachotoxin activation of sodium channels by lidocaine,[5] DPH,[5] or CBZ (Fig. 6A) reveals a competitive pattern of inhibition. This might result from direct steric inhibition of batrachotoxin binding at neurotoxin receptor site 2 or from an indirect allosteric mechanism in which the anticonvulsants and local anesthetics bind with high affinity to nonconducting states of sodium channels and thereby stabilize those states which have lower affinity for the toxin. The allosteric competitive mechanism of inhibition predicts that activation by full agonists

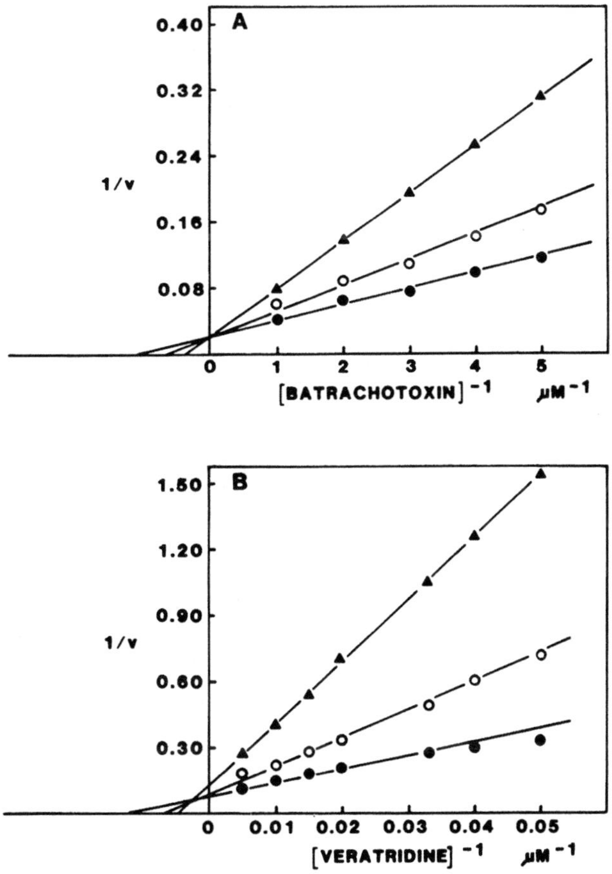

Figure 6. Mechanism of inhibition of neurotoxin activation of sodium channels in N18 cells by carbamazepine. (A) Cell cultures were incubated with the indicated concentrations of batrachotoxin in the absence (●) and presence of 100 μM (○) and 200 μM (▲) carbamazepine. $^{22}Na^+$ flux was measured as described in Methods. (B) Cell cultures were incubated with the indicated concentrations of veratridine in the absence (●) and presence of 100 μM (○) and 200 μM (▲) carbamazepine. $^{22}Na^+$ flux was measured as described in Methods. Reprinted with permission from Willow *et al.*[32]

like batrachotoxin should be blocked competitively, but that activation by partial agonists like veratridine should be blocked in a noncompetitive or mixed manner.[3] Therefore, double reciprocal analysis of drug inhibition of veratridine activation of sodium channels provides an independent test of the inhibition mechanism. For DPH,[5] lidocaine,[5] and CBZ (Fig. 6B), the inhibition pattern is mixed competitive/noncompetitive, confirming the conclusion from BTX-B binding studies that the local anesthetics and anticonvulsants act in an allosteric manner to prevent the binding of neurotoxins at receptor site 2 and activation of sodium channels.

Electrophysiological Characterization of the Inhibitory Effects of DPH, CBZ, and Bupivacaine on Voltage-Sensitive Sodium Channels in N18 Neuroblastoma Cells

Whole-cell voltage clamp procedures[16] offer the opportunity to examine the actions of local anesthetics and anticonvulsants on sodium channels in neuroblastoma cells with finer time resolution and voltage control and to compare the results to biochemical data on the same experimental system. The nature of sodium channel block produced by DPH, CBZ, and bupivacaine in the N18 neuroblastoma cell is summarized in Table IV. The current–voltage relationship for I_{Na} and the reversal potential (E_{Na}) for Na^+ are both unaffected by these agents. Therefore, the reduction in peak I_{Na} by these drugs must reflect reduction in peak Na conductance (G_{Na}).

Steady-state sodium channel inactivation was measured by varying the prepulse potential prior to the test pulse (+7.5 mV, 10 msec). In this study only single prepulses (−135 to −15 mV, 90 msec) were used. Initial observations suggested that in this cell type there is predominantly a single fast time course of inactivation. A slower component of inactivation has been described in other neuroblastoma cells.[20] Steady-state inactivation was described by the midpoint E_h and steepness k_h of the inactivation curve obtained from the following equation:

$$h_\infty = \frac{1}{\left[1 - \exp\frac{(E - E_h)}{k_h}\right]}$$

Both DPH and CBZ produced concentration-dependent shifts in the steady-state inactivation curve in the hyperpolarizing direction. At concentrations that reduce peak inward

Table IV. Effects of Diphenylhydantoin (DPH), Carbamazepine (CBZ), and Bupivacaine (BPV) on the Properties of Voltage-Sensitive Sodium Channels in N18 Neuroblastoma Cells

Drug	E_{Na} (mV)	IC_{50} peak I_{Na} (μM)	Voltage shift in h_∞ curve at IC_{50}[a] (mV)	Voltage dependency	Frequency dependency	Dissociation time from inactive channels[b]
DPH	+70–75	22	15–20	Yes	Yes	5–15 sec
CBZ	+70–75	33	15–20	Yes	Small	≤0.5 sec
BPV	+70–75	5	≤5	Yes	Yes	≥40 sec

[a]Shift in steady-state inactivation is hyperpolarizing.
[b]Estimates of drug dissociation rates from inactive channels are made by depolarizing cells to −45 mV for 3 sec followed by hyperpolarization to −105 mV and application of 0-mV test pulses (5 msec) every 500 msec thereafter.

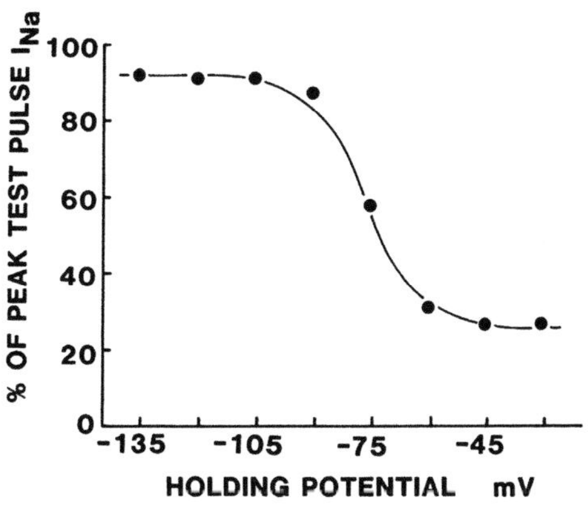

Figure 7. Voltage-dependent inhibition of sodium current in N18 cells by bupivacaine. Voltage-dependent inhibition of peak inward sodium currents evoked in response to 10-msec test pulses (0 mV) was studied by varying the holding potential (900 msec) in the absence and presence of 5 μM bupivacaine. Each test pulse was preceded by a hyperpolarizing prepulse (90 msec, −105 mV).

currents by 50%, these agents changed E_h from −65 ± 5 mV to −80 or −85 mV (Table IV) and also significantly reduced the slope (k_h) of the *h* curve. In contrast, bupivacaine produced only a small (≤5 mV) hyperpolarizing shift in the h_∞ curve at concentrations that produce half-maximal inhibition of peak currents.

All three drugs studied demonstrated marked voltage-dependent effects on sodium channel block (Table IV). The voltage dependency is most prominent between −90 mV and −45 mV. For bupivacaine (5 μM) block of peak inward current varies from 9% at −105 mV to 70% at −45 mV (Fig. 7). Voltage-dependent effects for DPH and CBZ follow a similar pattern. Since inactivation of sodium channels occurs in this voltage range, these results indicate that these three drugs all have high affinity for the inactivated state of sodium channels.

As illustrated in previous chapters of this volume, inhibition of sodium channels by many local anesthetics is enhanced by frequent stimulation of the preparation to activate sodium channels. We have found marked differences in the degree of frequency dependence for DPH, CBZ, and bupivacaine. Bupivacaine (5 μM) produces a marked frequency dependence when prepulses are removed from test-pulse protocols driven at 2 Hz (Fig. 8). Even at the end of 25 pulses, a new level of equilibrium has not been fully established, and upon reintroduction of the prepulse, recovery is slow (Fig. 8). Similar experiments (data not shown) indicate that DPH also has frequency-dependent actions, albeit less pronounced than those of bupivacaine. DPH requires fewer pulses to establish a new equilibrium after removal of the prepulse, and fewer pulses for full recovery upon reintroduction of the prepulse. Under these conditions CBZ produces little frequency-dependent inhibition. These differences among the three drugs may reflect differences in the rate of drug dissociation from inactivated channels. Experiments in which the membrane potential is stepped from −45 mV (3 sec) to −105 mV followed by 5-msec test pulses every 500 msec thereafter reveal that recovery of block in the presence of 5 μM bupivacaine can take 40 sec or more (Table IV). For DPH recovery is somewhat faster (5 sec), and recovery in the presence of CBZ is very rapid (≤0.5 sec).

DISCUSSION

The present study provides evidence that local anesthetics and anticonvulsants bind to a specific receptor site(s) associated with the voltage-sensitive sodium channel to block fast inward tetrodotoxin-sensitive sodium currents. Specificity of drug action is reflected by the lack of effect on saxitoxin (site 1) and scorpion toxin (site 3) binding. On the other hand,

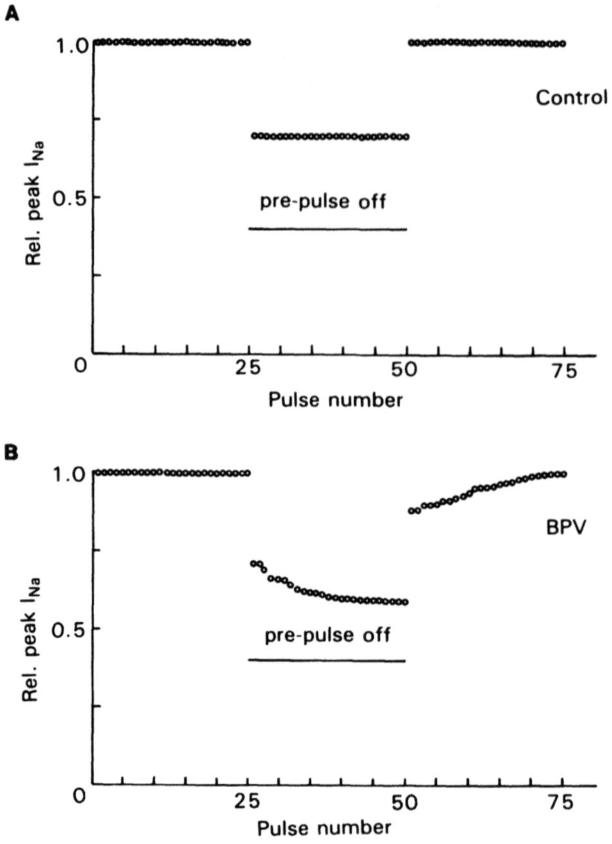

Figure 8. Frequency-dependent inhibition of sodium currents in N18 cells by bupivacaine. Cells are maintained at a membrane potential of −75 mV prior to administration of a 0-mV test pulse (5 msec, 2 Hz) in the absence and presence of a hyperpolarizing prepulse (90 msec, −105 mV). Peak sodium currents are reduced if the prepulse is removed. (A) Control currents in the absence of added drug. (B) Currents in the presence of 5 μM bupivacaine (BPV).

both local anesthetics and the anticonvulsants DPH and CBZ inhibit the binding of BTX-B to neurotoxin receptor site 2 at concentrations that (1) inhibit neurotoxin-activated $^{22}Na^{+}$ flux and (2) inhibit sodium currents in neuroblastoma cells. Of further pharmacological interest is the observation that the concentrations of DPH and CBZ required to inhibit sodium channels correlates closely with central nervous system drug levels required to protect laboratory animals against maximal electroshock seizures (cf. ref. 31). With respect to neurotoxin binding, it appears that local anesthetics and anticonvulsants act similarly in reducing the affinity of receptor site 2 for BTX-B. Since a reduction in receptor affinity may reflect changes in association and/or dissociation rate kinetics, it was of interest to examine the effects of these drugs on ligand–receptor on/off rates. The dissociation of BTX-B following addition of excess unlabeled ligand to the receptor–ligand suspension is accelerated by DPH, CBZ, and local anesthetics. In contrast, association rate kinetics remain unaffected in the presence of these agents. This mode of inhibition is consistent with that of an allosteric competitive inhibitor. Further evidence in support of this interpretation is provided from the effects of local anesthetics[5] and anticonvulsants[32] on batrachotoxin- or veratridine-stimulated $^{22}Na^{+}$ flux in N18 cells. These agents are competitive inhibitors of batrachotoxin action but mixed inhibitors of veratridine activation of sodium channels. This pattern of inhibition is consistent with that of an allosteric inhibition mechanism.

Inhibition of sodium currents by DPH and CBZ in N18 cells is associated with large hyperpolarizing shifts in fast steady-state inactivation (h_{∞}). In addition, DPH, CBZ, and bupivacaine all caused a marked voltage-dependent inhibition of sodium channels consistent with preferential binding of these drugs to the inactivated state of the channel.[9,10,17,20,21,28] Thus, the biochemical evidence for allosteric inhibition of neurotoxin activation of sodium channels and the electrophysiological evidence for enhanced inactivation and voltage-dependent binding both point to selective binding to inactivated states of the channel and stabilization of the channel population in those states as major components of the action of local anesthetics and anticonvulsants. Both these sets of data are accommodated within the modulated receptor hypothesis,[17] which is an extension of the ideas developed to explain the properties of allosterically regulated enzymes to analysis of drug action on ionic channels. Frequency-dependent inhibition of sodium channels is observed when drugs that bind preferentially to the active or inactivated states have dissociation rates such that drug bound during one stimulus cannot completely dissociate by the next. Cumulative inhibition is observed until a frequency-dependent steady state is reached. Frequency-dependent inhibition is observed for bupivacaine and DPH when the rate of stimulation is 2 Hz. In these experiments the recovery period between brief depolarizing test pulses (5 msec) is 495 msec. Dissociation of these two drugs at −75 mV during this period would be relatively slow based on the data presented in Table IV. On the other hand, carbamazepine, which is more hydrophobic, can escape rather rapidly during the recovery period. On the basis of its voltage-dependent actions, one may expect to observe frequency-dependent effects for CBZ at higher frequencies. It will be of interest to establish the relative importance of molecular size as opposed to solubility in determining the rates of recovery from drug block and degree of frequency dependence.[9] It seems likely that the voltage-dependent and frequency-dependent actions of anticonvulsants and antiarrhythmic/local anesthetics are of major importance in their pharmacological applications, particularly in repetitively firing, depolarized tissues such as may occur during an arrhythmia or in epileptic foci.

ACKNOWLEDGMENTS. We express our gratitude to Professor B. Hille, Department of Physiology and Biophysics, University of Washington, for his assistance and helpful discussions in whole-cell patch clamp experiments.

REFERENCES

1. Catterall WA: Activation of the action potential sodium ionophore by veratridine and batrachotoxin. *J Biol Chem* 1975a;250:4053–4059.
2. Catterall WA: Membrane potential dependent binding of scorpion toxin to the action potential sodium ionophore: Studies with a toxin derivative prepared by lactoperoxidase-catalyzed iodination. *J Biol Chem* 1975b;252:8660–8668.
3. Catterall WA: Activation of the action potential Na^+ ionophore by neurotoxins: an allosteric model. *J Biol Chem* 1975c;252:8669–8676.
4. Catterall WA: Neurotoxins that act on voltage-sensitive sodium channels in excitable membranes. *Annu Rev Pharmacol Toxicol* 1980;20:15–43.
5. Catterall WA: Inhibition of voltage-sensitive sodium channels in neuroblastoma cells by antiarrhythmic drugs. *Mol Pharmacol* 1981;20:356–362.
6. Catterall WA, Morrow CS, Hartshorne RP: Neurotoxin binding to receptor sites associated with sodium channels in intact, lysed, and detergent-solubilized brain membranes. *J Biol Chem* 1979;254:11379–11387.
7. Catterall WA, Morrow CS, Daly JW, Brown GB: Binding of batrachotoxinin A 20-α-benzoate to a receptor site associated with sodium channels in synaptic nerve ending particles. *J Biol Chem* 1981;256:8922–8927.
8. Courtney KR: Mechanism of frequency-dependent inhibition of sodium currents in frog myelinated nerve by the lidocaine derivative GEA 968. *J Pharmacol Exp Ther* 1975;195:225–236.
9. Courtney KR: Structure-activity relations for frequency-dependent sodium channel block in nerve by local anesthetics. *J Pharmacol Exp Ther* 1980;213:114–119.
10. Courtney KR, Etter EF: Modulated anticonvulsant block of sodium channels in nerve and muscle. *Eur J Pharmacol* 1983;88:1–9.

11. Courtney KR, Kendig JJ, Cohen EN: Frequency-dependent conduction block: The role of nerve impulse pattern in local anesthetic potency. *Anesthesiology* 1978a;48:111–117.
12. Courtney KR, Kendig JJ, Cohen EN: The rates of interaction of local anesthetics with sodium channels in nerve. *J Pharmacol Exp Ther* 1978b;207:594–604.
13. Creveling CR, McNeal ET, Daly JW, Brown GB: Batrachotoxin-induced depolarization and [^{3}H[batrachotoxinin A 20-α-benzoate binding in a vesicular preparation from guinea pig cerebral cortex: Inhibition by local anesthetics. *Mol Pharmacol* 1983;23:350–358.
14. Finney DJ: *Probit Analysis.* Cambridge, Cambridge University Press, 1971.
15. Gray EG, Whittaker VP: The isolation of nerve endings from brain: an electron-microscopic study of cell fragments derived by homogenization and centrifugation. *J Anat* 1962;96:79–88.
16. Hamill OP, Marty A, Neher E, Sakmann B, Sigworth FJ: Improved patch-clamp techniques for high-resolution current recording from cells and cell-free membrane patches. *Pflügers Arch* 1981;391:85–100.
17. Hille B: Local anesthetics: Hydrophilic and hydrophobic pathways for the drug-receptor reaction. *J Gen Physiol* 1977;58: 599–619.
18. Hondeghem LM, Katzung BG: Time and voltage dependent interactions of antiarrhythmic drugs with cardiac sodium channels. *Biochim Biophys Acta* 1977;472:373–398.
19. Lipicky RJ, Gilbert DL, Stillman IM: Diphenylhydantoin inhibition of sodium conductance in squid giant axon. *Proc Natl Acad Sci USA* 1972;69:1758–1760.
20. Matsuki N, Quandt FN, Ten Eick RE, Yeh JZ: Characterization of the block of sodium channels by phenytoin in mouse neuroblastoma cells. *J Pharmacol Exp Ther* 1984;228:523–530.
21. McLean MJ, Macdonald RL: Multiple actions of phenytoin on mouse spinal cord neurons in cell culture. *J Pharmacol Exp Ther* 1983;227:779–789.
22. Peterson GL: A simplication of the protein assay methods of Lowry *et al.* which is more generally applicable. *Anal Biochem* 1977;83:346–356.
23. Postma SW, Catterall WA: Inhibition of binding of [^{3}H]batrachotoxinin A 20-α-benzoate to sodium channels by local anesthetics. *Mol Pharmacol* 1984;25:219–227.
24. Ray R, Morrow CS, Catterall WA: Binding of scorpion toxin to receptor sites associated with voltage sensitive sodium channels in synaptic nerve ending particles. *J Biol Chem* 1978;253:7307–7313.
25. Ritchie JM, Rogart RB: The binding of saxitoxin and tetrodotoxin to excitable tissue. *Rev Physiol Biochem Pharmacol* 1977;79:1–51.
26. Schauf CL, Davis FA, Marder J: Effects of carbamazepine on the ionic conductances of Myxicola giant axons. *J Pharmacol Exp Ther* 1974;189:538–543.
27. Schwarz JR, Vogel W: Diphenylhydantoin: Excitability reducing action in single myelinated nerve fibres. *Eur J Pharmacol* 1977;44:241–249.
28. Strichartz G: The inhibition of sodium currents in myelinated nerve by quarternary derivatives of lidocaine. *J Gen Physiol* 1973;62:37–57.
29. Strichartz G: Molecular mechanisms of nerve block by local anesthetics. *Anesthesiology* 1976;45:421–441.
30. Tamkun MM, Catterall WA: Ion flux studies of voltage-sensitive sodium channels in synaptic nerve ending particles. *Mol Pharmacol* 1981;19:78–86.
31. Willow M, Catterall WA: Inhibition of binding of [^{3}H]batrachotoxinin A 20-α-benzoate to sodium channels by the anticonvulsant drugs diphenylhydantoin and carbamazepine. *Mol Pharmacol* 1982;22:627–635.
32. Willow M., Kuenzel EA, Catterall WA: Inhibition of voltage-sensitive sodium channels in neuroblastoma cells and synaptosomes by the anticonvulsant drugs diphenylhydantoin and carbamazepine. *Mol Pharmacol* 1984;25:228–235.

IV

BIOPHYSICAL MECHANISMS OF ANESTHETICS

21

Inside the "Black Box"

Keith W. Miller and Sheldon H. Roth

Abstract. The classical physicochemical approach to characterizing the site of action of anesthesia does not always yield unequivocal results. For example, while the inhibitory action of anesthetics on proteins rarely correlates with anesthetic potency, there are proteins (e.g., luciferases) that are inhibited by anesthetics in a manner which correlates with both anesthetic potency and lipid bilayer solubility. The existence of such contradictory, or degenerate, correlations reinforces the need for a more mechanistic approach. Mechanisms must be sought at physiological, cellular, and molecular levels. Anesthetics can interact with proteins in at least two ways. First, they may bind specifically to a discrete site and cause allosteric modulation of the protein's conformation. Examples include the local anesthetics on the voltage-sensitive sodium channel and the barbiturates at GABA and acetylcholine receptors. Second, they may adsorb to sites or clefts without causing a conformational change. Some sites, such as those to which xenon binds in myoglobin, are "hard" and accept only anesthetics of a particular size. Other sites, such as those in adenylate kinase and luciferase, are more flexible and interact with many anesthetics nonspecifically. The latter type of site binds endogeneous ligands which are foced to compete with the anesthetic for their regulatory site. Lipid bilayers provide an even more plausible site with which all anesthetics interact, but mechanisms describing how bilayer perturbations might change membrane protein function have been slow in coming. This may merely reflect the current level of ignorance of lipid–protein interactions, or it may have more serious implications for theories of anesthesia.

INTRODUCTION

The pharmacological approach to defining the site of action of anesthetics suggests that it is hydrophobic in nature.[4,8] However, this approach is probably incapable of elucidating the site in more detail than has been achieved so far. Environments in lipid bilayers and on protein surfaces or clefts provide plausible models, yet to achieve more precision requires a mechanistic approach at two levels, cellular and molecular. This chapter outlines some of the molecular mechanisms that may underlie anesthetic action and discusses some of the problems inherent in deducing the relationship between pharmacological data in the intact organism and mechanisms obtained in isolated model systems.

Claude Bernard's unitary hypothesis ("It [anesthetic] anesthetizes all the cells, benumbing all the tissues. . . ."[1]), Meyer and Overton's lipid solubility rule ("The narcotizing substance enters into a loose physicochemical combination with the vitally impor-

Keith W. Miller • Departments of Pharmacology and Anaesthesia, Harvard Medical School and Massachusetts General Hospital, Boston, Massachusetts 02114. *Sheldon H. Roth* • Department of Pharmacology and Therapeutics, Faculty of Medicine, The University of Calgary, Calgary, Alberta T2N 4N1, Canada.

tant lipoids of the cell. . . .''[11]), and the critical volume hypothesis ('' . . . anesthesia occurs when the volume of a hydrophobic region expands beyond a critical volume. . . .''[14]) have provided predictive tools of great power. Nonetheless, their approach leaves unopened the ''black box'' whose behavior they model so well. The two great questions that now confront the field are (1) What is the physiological or cellular basis of general anesthesia? And, (2) what is the molecular mechanism by which general anesthesia is produced? The answers to all these questions are likely to be complex and not singular.[16]

The aims of the molecular approach are to discover the key structural features underlying the sensitivity to anesthetics and thus to describe the mechanisms by which excitable membrane function is modified. Once obtained, a knowledge of these underlying mechanisms may enable some important questions about actions on the whole organism to be answered. For example, why do these rather nonspecific drugs act selectively on some membrane proteins while leaving most cellular functions intact at clinical concentrations?[16]

Some of the model systems where molecular mechanisms could now be resolved behave with a pharmacological specificity which mimics that of general anesthesia, while others behave with much more selectivity. Examples of the latter types are the inhibition of acetylcholine esterase or of ethylene's action as a plant growth promoter.[7,10] In both cases, general anesthetics act with relative potencies that bear no relationship to their ability to cause anesthesia. Paradoxically, however, it may prove simpler to relate such pharmacologically selective mechanisms to their physiological functions, which may involve side effects of anesthetics, than to do so for mechanisms having an unselective or general anesthetic pharmacology. This problem arises because often several quite distinct mechanisms give rise to effects that are pharmacologically equally unselective. An example is provided by studies on two model systems which represent opposite schools of thought on how anesthetics act.

Neither of these systems pretends to do anything but model that unknown physiological site of action where general anesthesia is produced. The first model is a simple lipid bilayer composed of phosphatidylcholine and the second a pure protein, luciferase. That the ability of the general anesthetics (volatiles and alcohols) to inhibit the pure firefly luciferase (dissociated from any membrane) (see Franks and Lieb, this volume, Chapter 26) correlates well with their lipid/buffer partition coefficient is illustrated in Fig. 1. This correlation is clearly devoid of any mechanistic meaning. The only conclusion that can be drawn is the rather discouraging one that it is going to take a more critical approach than that provided by pharmacological data to yield unequivocal answers. One could retort that this meaningless

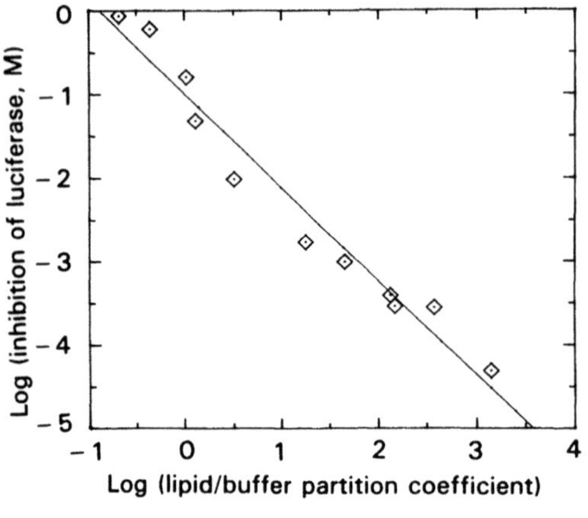

Figure 1. A comparison of the relative potency of 11 anesthetics for inhibiting firefly luciferase (see this volume, Chapter 26) with their lipid/buffer partition coefficients in phosphatidylcholine. Data were taken from the Appendix to this volume where sources may be found. From left to right, the compounds plotted are methanol, ethanol, acetone, propanol, butanol, benzyl alcohol, hexanol, halothane, methoxyflurane, octanol, and nonanol. The regression line is $y = -0.995\ (^{+}0.12) - 1.12\ (\pm 0.072)x$ and $r^2 = 0.964$.

correlation would be broken by using a wider range of compounds, particularly those that are known to not always "follow the rules" (for example, long-chain alcohols and alkanes that lack anesthetic potency) or those with better-defined structural features (for example, barbiturates and steroids that have sterioisomers). One could also ask whether both these models account for the pressure reversal of general anesthesia. It is probable that not all these critical strategies may work; for example, both pressure reversal[8] (and Smith *et al.*, this volume, Chapter 28) and the lack of potency of long-chain alkanols[8,9] (and Franks and Lieb, personal communication) are accounted for by both models in Fig. 1. This problem has been discussed in some detail recently.[13]

Whether or not more pharmacological tinkering can resolve the type of uncertainty expressed by Fig. 1, the mechanistic approach will itself eventually lead to new predictions which will provide further tests that may distinguish between degenerate models. For example, competition between anesthetics and a cofactor for the latter's binding site on luciferase accounts for the observed inhibition of this enzyme (see Franks and Lieb, this volume, Chapter 26). If general anesthesia is produced by an analogous mechanism, one can predict that an effective anesthetic concentration will be dependent on the cofactor concentration.

ANESTHETIC–PROTEIN INTERACTIONS

In addition to action on lipids, general and local anesthetics interact with proteins in a number of ways, some of which have functional consequences. It is in this area where research has come closest to revealing mechanisms.

Allosteric Behavior

Evidence summarized in this volume, Sections II and III, reveals that local anesthetics and barbiturates act as allosteric effectors on voltage-sensitive sodium channels, on GABA receptors, and on acetylcholine receptors. In each case it is possible to demonstrate, either directly with radiolabeled anesthetics or indirectly using other radiolabeled ligands, that the anesthetic binds to a site and that this binding has consequences that are sensed at distinct binding sites for other ligands elsewhere on the protein.

These actions resemble those of allosteric effectors well known in enzymology, and this means that many of the general principles have already been established by studies in that field.[3]

The concept of allosterism in anesthetic action has been elaborated most successfully in the case of the local anesthetics acting on the sodium channel (the modulated receptor hypothesis). The key point is that anesthetics bind to the different conformational states of the protein with different affinities, thus altering the relative number of channels populating the resting, active, and inactivated states.

Although these concepts seem clear, there are hints of a greater level of complexity with more than one allosteric site (Ritchie, this volume, Chapter 16) or with secondary nonspecific mechanisms occuring as well (Miller *et al.*, this volume, Chapter 11).

Adsorption Behavior

Many other instances of anesthetics associating with proteins are known. In some cases anesthetics seek out available hydrophobic pockets within or on the surface of proteins. For example, that the xenon–myoglobin complex consists of at least four different environments for xenon binding has been established by both detailed X-ray diffraction[20] and nuclear magnetic resonance spectroscopy.[19] This work reveals the type of protein pocket available to be rather "hard" and capable of receiving anesthetics of only a limited size range, not at all like the "oil drop" model for the hydrophobic interior of proteins that was once postulated. Furthermore, such binding seems to cause remarkably little functional change.[18]

Resolution of how such binding could have functional consequences without the anesthetic causing any structural changes to the protein comes from studies on adenylate kinase[17] and on the luciferases[12] (and Franks and Lieb, this volume, Chapter 26). In each case the anesthetics adsorb to the surface of a hydrophobic cleft, displacing an endogenous compound.

ANESTHETIC–LIPID INTERACTIONS

The advent of lipid bilayers and of isolated biomembranes in the 1960s led to intensive efforts to develop the Meyer–Overton rule into a mechanistic theory of action. These studies have given a more defined meaning to the old lipid solubility rule—a good correlation can now be demonstrated between solubility in *lipid bilayers* and general anesthetic potency[9]—but *mechanisms* supported by structural studies have been slow in coming. Indeed, two major problems remain: first, the structural changes are small at clinical concentrations, and second, the function of only selected membrane proteins is affected, even though anesthetics dissolve in all bilayers regardless of lipid composition.

The lipid theories can be formulated in a very general way as

$$E_{50} = C_{50} \cdot P_1 \cdot T_p \qquad (1)$$

where E_{50} represents the half-point in some functional change induced by an anesthetic in a membrane protein; C_{50} is the corresponding concentration of the anesthetic in the bilayer; P_1 is any perturbation of the lipid bilayer properties per mole of anesthetic in the bilayer; and T_p expresses the transmission of the lipid perturbation, P_1, to the membrane protein.

The term E_{50} will be a constant independent of the anesthetic. The corresponding concentration in the membrane, C_{50}, usually varies little (Meyer–Overton rule), but in principle it could vary so long as P_1 changes in a compensatory direction.

Lipid theories vary in the meaning they assign to P_1 (for example, lipid fluidity, volume increase, phase transition, for reviews see refs. 4 and 8), but since the bilayer is only two molecules thick, it is artificial to consider one of its properties in isolation from the others. Indeed, one of the qualities that distinguishes the recent work of Haydon *et al.*, reviewed in this volume, Chapter 22, is that changes in several properties of the bilayer are invoked to explain the several effects of different general anesthetics on sodium channels.

Although most properties of lipid bilayers change in a colligative manner on addition of anesthetics (i.e., the change is related to the anesthetic's *concentration* in the bilayer), others depend in addition on some property of the anesthetic molecule itself. One such property is the polarity of the anesthetic, which may cause it to distribute nonrandomly through the amphiphilic asymmetrical environment provided by the lipid bilayer of biological membranes. Several examples occur in the work reported in Section IV of this volume. This recent work on lipid bilayers suggests that the tendency to regard bilayers as microscopic bulk solvents is wrong. Instead they are capable of complex, nonlinear reactions to the intrusion of anesthetics. How far such observations will contribute to the robustness of the lipid approach remains to be seen, but it introduces a new degree of interest.

The question of how a membrane protein reacts to a given lipid perturbation [i.e., the relationship of P_1 to T_p in equation (1)] has barely been explored, yet it is most probably here that the reason why some membrane proteins are more sensitive than others to anesthetics must be sought by the lipid theories. One would expect, for example, that only proteins that are voltage sensitive are likely to respond to a change in potential across the bilayer.

LIPID–PROTEIN INTERACTIONS AND ANESTHETICS

In addition to interacting with lipids and proteins, anesthetics are also expected to interact with the lipid–protein interface. This interface consists of at least two regions: first, the lipid head group–protein interface, and second, the lipid acyl chain–protein interface. An anesthetic that wanders into either region will experience a relatively rigid protein environment on one side and a flexible lipid environment on the other. Since the hydrophobic protein surface is rough,[5] the interaction may be enhanced by favorably shaped pockets on the protein. Given the ambiguity expressed in Fig. 1, such a site may be quite acceptable to the Meyer–Overton rule, but would such an interaction have functional consequences?

This is an area that has received little attention. Haydon *et al.* (this volume, Chapter 22) attempt to address the issue by considering both the general and local lipid environment. Others have made attempts to implicate the lipids immediately neighboring the protein.[6] Recent preliminary studies have probed these lipids spectroscopically, but solutions are far from available.[2]

CONCLUSIONS

The tendency toward more mechanistic studies of anesthetic action has had some success where discrete sites have been found on specific protein models, but the basis of less specific actions on membrane proteins remains unclear. Current rapid progress in elucidating the molecular structure of excitable membrane proteins,[15] together with the availability of powerful spectroscopic tools (see, for example, this volume, Section IV), suggests that significant advances are not far off.

REFERENCES

1. Bernard C: *Lecons sur les Anesthesique et L'Asphyxie*. Paris, Balliere, 1875.
2. Bigelow DJ, Thomas DD: Molecular dynamics and enzymatic activity in sarcoplasmic reticulum: A comparison of the effects of ether and temperature. *Biophysics J* 1984;45:320a.
3. Burgen ASV: Conformational changes and drug action. *Fed Proc* 1981;40:2723–2728.
4. Franks NP, Lieb WR: Molecular mechanisms of general anaesthesia. *Nature* 1982;300:487–493.
5. Guy HR: A structural model of the acetylcholine receptor channel based on partition energy and helix packing calculations. *Biophys J* 1984;45:249–261.
6. Hesketh TR, Smith GA, Houslay MD. McGill KA, Birdsall NJM, Metcalfe JC, Warren GB: Annular lipids determine the ATPase activity of a calcium transport protein complexed with dipalmitoyllecithin. *Biochemistry* 1976;15:4145–4151.
7. Irvine RF, Roth SH, Musgrave A: Ethylene as a plant hormone and anaesthetic. *Experientia* 1974;30:1259–1260.
8. Janoff AS, Miller KW: A critical assessment of the lipid theories of general anaesthetic action, in Chapman D (ed): *Biological Membranes,* Vol 4. London, Academic Press, 1982, pp 417–476.
9. Janoff AS, Pringle MJ, Miller KW: Correlation of general anaesthetic potency with solubility in membranes. *Biochim Biophys Acta* 1981;649:125–128.
10. Kitz RJ, Braswell LM: The effect *in vitro* of volatile anaesthetics on the activity of cholinesterases. *J Neurochem* 1977;29:665–671.
11. Meyer HH: Theories of narcosis, in *Harvey Lectures.* Philadelphia, Lippincott, 1906, pp 11–17.
12. Middleton AJ, Smith EB: General anaesthetics and bacterial luminescence II. The effect of diethyl ether on the *in vitro* light emission of *Vibrio fischeri. Proc Roy Soc Lond B* 1976;193:173–190.
13. Miller KW: The nature of the site of general anaesthesia. *Int Rev Neurobiol* 1985;27:1–61.

14. Miller KW, Paton WDM, Smith RA, Smith EB: The pressure reversal of anaesthesia and the critical volume hypothesis. *Mol Pharmacol* 1973;9:131–143.
15. Numa S, Noda M, Takahashi H, Tanabet T, Toyosato M, Furutani Y, Kikyotani S: Molecular structure of the nicotinic acetylcholine receptor. *Cold Spring Harbor Symp Quant Biol* 1983;58:47–68.
16. Roth SH: Membrane and cellular actions of anesthetic agents. *Fed Proc* 1980;39:1595–1599.
17. Sachsenheimer W, Pai EF, Schultz GE, Schirmer RH: Halothane binds in the adenine-specific niche of crystalline adenylate kinase. *FEBS Lett* 1977;79:310–312.
18. Settle W: Function of the myoglobin molecule as influenced by anaesthetic molecules, in Featherstone RM (ed): *Guide to Molecular Pharmacology and Toxicology,* Part II. New York, Marcel Dekker, Inc., 1973, pp 477–493.
19. Tilton RF Jr, Kuntz ID Jr: Nuclear magnetic resonance studies of xenon-129 with myoglobin and hemoglobin. *Biochemistry* 1982;27:6850–6857.
20. Tilton RF Jr, Kuntz ID Jr, Petsko GA: Cavities in protein: Structure of a metmyoglobin—xenon complex solved to 1.9 A. *Biochemistry* 1984;23:2849–2857.

22

The Action of Nonionic Anesthetic Substances on Voltage-Gated Ion Conductances in Squid Giant Axons

Denis A. Haydon, James R. Elliott, Bruce M. Hendry, and Bernd W. Urban

ABSTRACT. The suppression of sodium currents in squid axons by a wide range of nonionic lipophilic substances has been examined. Each substance acts on the ion channels in several different ways to produce both kinetic and steady-state perturbations of the voltage-gated currents. While, in every instance, gating processes are accelerated by the anesthetics, other effects depend on the polarity or surface activity of the substance. Thus, most hydrocarbons produce hyperpolarizing shifts in both steady-state activation and inactivation. In contrast, polar molecules shift the steady-state activation in the depolarizing sense but (with an exception) have little influence on inactivation. Both the steady-state activation and inactivation effects are major factors in current suppression, and the evidence available suggests that changes in membrane thickness and surface (dipole) potential are responsible. Thickness changes appear to influence both activation and inactivation, while the surface potential seems to affect chiefly the activation.

The effects described could arise through interactions involving primarily the membrane lipid. An exception, however, is found for molecules containing double-bonded oxygen atoms, such as carboxylic esters and ketones where a direct interaction with protein is probably involved. A study of the properties of these molecules suggests a mechanism of action for the local anesthetic benzocaine, which acts in a manner very similar to that of ethyl benzoate and aliphatic carboxylic esters.

INTRODUCTION

It is well known that a wide variety of simple lipophilic or surface-active substances, many of which are general anesthetics, suppress the sodium current in nerve axons.[3,22,24] The concentrations required for 50% suppression of the maximum sodium current range from about 3 to 10 times the minimal alveolar concentrations quoted for various animals. The effects that give rise to the sodium current suppression are, however, still detectable at clinical concentrations, where significant perturbations of the sodium channel function and of the resting

Denis A. Haydon, James R. Elliott, and Bruce M. Hendry • Physiological Laboratory, University of Cambridge, Cambridge CB2 3EG, England, and the Laboratory of the Marine Biological Association, Plymouth PL1 2PB, England. *Bernd W. Urban* • Departments of Anesthesiology and Physiology, Cornell University Medical College, New York, New York 10021.

potential are found. Although the role of sodium channels in general anesthesia is quite unclear, their excitation in postsynaptic membranes is essential for impulse propagation. For this reason, and because there is as yet no unambiguous evidence that any ion channels, whether transmitter-activated or voltage-gated, are more sensitive to general anesthetics than the sodium channel, the contribution of the latter cannot at present be discounted.

From a somewhat different standpoint, there are a number of local anesthetics, such as benzocaine, which act on the sodium channel and which, structurally, are built up of molecular groups which, in themselves, behave as anesthetics. From studies of simple organic molecules it is therefore possible to gain a better appreciation of the significance and probable mechanisms of action of the more complex molecules used clinically.

The most convenient concentrations at which to examine in detail the various changes brought about by the anesthetics are usually those that produce about 50% suppression of the maximum current. Under these conditions, the various effects are comfortably large and the membrane currents still of magnitude sufficient for accurate analysis. At lower concentrations the relative importance of the different perturbations does not necessarily remain the same, but it is reasonable to assume their molecular origins to be identical.

The nerve preparation discussed throughout this chapter is the giant axon of the squid *Loligo forbesi*. There is, as yet, relatively little information as to the detailed responses of other nerve preparations to the various anesthetics considered. The indications are, however, that while the fundamental molecular mechanisms may be similar, different membrane structural features, such as lipid or protein asymmetry, may give rise to differences in the relative importance of these mechanisms in current suppression.

The discussion that follows begins by summarizing briefly earlier results for hydrocarbons and alcohols to show clearly the essential differences between these two types of molecule. The properties peculiar to carboxylic esters will then be described, and finally, evidence will be presented as to the mechanisms of action on the sodium channel of all three classes of anesthetic.

METHODS

The axons were dissected from freshly killed squid (*Loligo forbesi*) and were cleaned of surrounding fibers and connective tissue. The axon diameters ranged from 600 to 1000 μm.

The voltage-clamp and data-acquisition procedures, the impedance measurement techniques and the data analysis were as previously described.[10,11,15] The chamber in which the axons were mounted and the means of introducing and controlling the temperature and flow rate of the bathing solutions were also as described previously.[15]

Particular attention was given throughout to the compensation for the resistance in series with the axon membrane. This was necessary because many of the anesthetics investigated produced shifts in the voltage dependence of the Hodgkin–Huxley parameters of the membrane current. Most of the experiments were carried out in artificial sea water containing only one-quarter of the normal sodium concentration. Under these conditions the currents were sufficiently small that, with compensation applied, errors in the membrane potential should have been less than 1 mV.

After the anesthetics were introduced, the changes in the voltage-clamp currents occurred usually in less than 20 min. After this, there were sometimes very slow drifts but these were sufficiently small that they could be ignored on the time scale of the experiments. The results presented therefore reflect effectively an equilibrium between the bathing solution and the axon. Since the anesthetics were lipid soluble, it may be reasonably supposed that they

Table I. Concentrations That Suppress Reversibly by about 50% the Maximum Inward Current in Voltage-Clamped Intact Axons of the Squid (Loligo forbesi) *at 6°C in Artificial Sea Water*[a]

Substance	Concentration[b] (mM)		Substance	Concentration (mM)
n-Butane	0.68	(1 atm)	*n*-Octyl(oxyethylene)$_3$ alcohol	0.44
n-Pentane	0.28	(0.90)	*n*-Decyl(oxyethylene)$_3$ alcohol	0.035
n-Hexane	0.06	(0.95)	*n*-Decyl(oxyethylene)$_4$ alcohol	0.05
Cyclopropane	7.0	(1.0 atm)	Diethyl ether	75.0
Cyclopentane	0.70	(0.37)	Methoxyflurane	1.0
Cyclohexane	0.19	(0.35)	Isoflurane	3.0
Benzene	2.72	(0.15)	Enflurane	3.6
Methyl benzene	0.93	(0.20)	Methyl pentanoate	4.2
Ethyl benzene	0.30	(0.24)	Methyl octanoate	0.18
Propyl benzene		(0.30)	Methyl butyl ketone	10.0
Dichloromethane	8.0		Methyl heptyl ketone	0.30
Trichloromethane	3.0		Glycerol 1-monooctanoate	1.3
Tetrachloromethane	1.0		Benzyl alcohol	8.0
Halothane	2.0		Methyl benzoate	1.30
n-Pentanol	14.8		Ethyl benzoate	0.40
n-Hexanol	3.52		Benzocaine	0.70
n-Heptanol	0.93		Diheptanoyl phosphatidylcholine	0.52
n-Octanol	0.29		Dioctanoyl phosphatidylcholine	0.15
n-Nonanol	0.068			
n-Decanol	0.022			

[a]For CsF-perfused axons the corresponding concentrations are usually higher by as much as two times.
[b]Values in parentheses are fractional saturation, except for *n*-butane and cyclopropane.

permeated quickly across the axon membranes and were present throughout the preparation. Various observations and calculations support this supposition.

RESULTS

Hydrocarbons and Alcohols

The suppression of excitability in the squid axon by neutral molecules has its origins in several different effects which are not necessarily related to each other. Moreover, while most substances produce all these effects in some degree, there are clear differences in relative importance for the various types of substance. Detailed studies have been carried out for many of the substances listed in Table I, and these results form the basis for the following discussion.

With one or two significant exceptions (which will be considered later), the substances listed exhibit graded behavior, which is bounded by the hydrocarbons at one extreme and the alcohols at the other. These behavioral differences may be conveniently described and quantitatively evaluated by means of the parameters defined by Hodgkin and Huxley.[19] These parameters are derived from studies of voltage-clamped axons and give the degrees of activation and inactivation in the steady state (m_∞ and h_∞), the maximum sodium conductance ($\bar{g}_{Na}$) that can be achieved were inactivation to be absent, and the time constants (τ_m and τ_h) for the activation and inactivation processes.

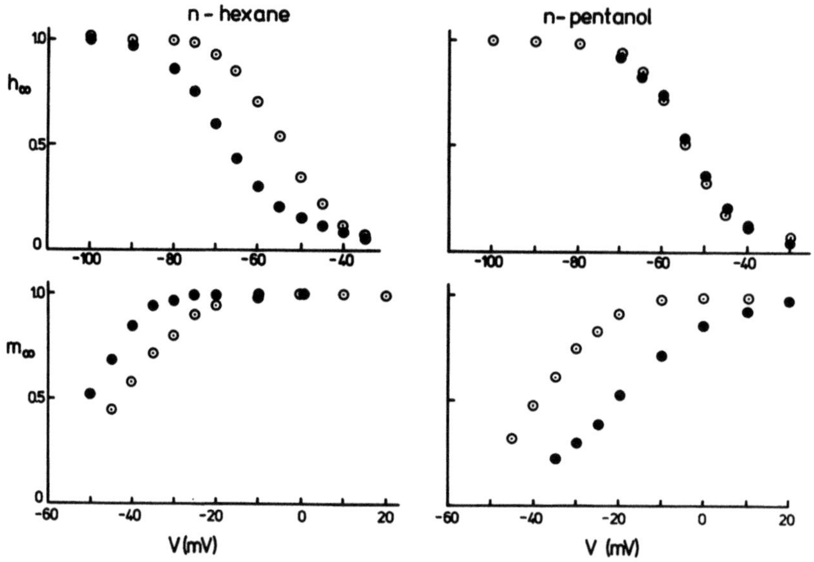

Figure 1. The effects of 61 μM (0.95 saturated) *n*-hexane and 14.8 mM *n*-pentanol on the steady-state inactivation (h_∞) and steady-state activation (m_∞) curves for the sodium current of the squid giant axon. The relationship between m_∞ and the sodium conductance is given in the legend of Fig. 2. ⊙, control; ●, test. The "test" records were taken when the currents had become steady with time about 20 min after exposure to the anesthetic. V is the membrane potential. T = 6 ± 1°C. The perfusion fluid, the bathing solution, and other experimental details were as described in Haydon and Urban.[11–13]

In Fig. 1 can be seen the effects that aliphatic hydrocarbons and alcohols typically produce on the steady-state activation and inactivation curves. Hydrocarbons invariably induce substantial hyperpolarizing shifts in the h_∞ curve and decrease the slope at the midpoint. Alcohols, by contrast, have very little effect.[1,24] For the steady-state activation (m_∞) curve, the alcohols invariably promote depolarizing shifts while the hydrocarbons give varied results. The higher-molecular-weight aliphatic hydrocarbons act in much the same way on m_∞ as on h_∞; i.e., they produce hyperpolarizing shifts, although these are smaller for the former than for the latter parameters. Cyclopropane and the aromatic hydrocarbons (see Fig. 3) shift m_∞ weakly in the depolarizing direction. Figure 2 shows typical effects on the maximum sodium conductance. Both hydrocarbons and alcohols depress $\bar{g}_{Na}$, although the hydrocarbons tend to be much more potent in this respect. The one feature that hydrocarbons and alcohols (and almost all the other substances in Table I) have in common is that, after allowance for voltage shifts, they all reduce the time constants for both activation and inactivation. It will be obvious that for the different substances, the various effects described combine in different ways to reduce the excitability of the axon and that, in general, no single effect is wholly responsible for the anesthetic action.

Carboxylic Esters and Ketones

It was shown by Haydon and Urban[12] that the carboxylic ester, methyl octanoate, influenced the sodium current in much the same way as for other amphipathic molecules except that its effect on the steady-state inactivation curve was more like that for a hydrocarbon than for an alcohol. It has subsequently been found that this effect is apparently general for esters and that it is exhibited also by ketones.[9] Results for a simple aromatic ester, ethyl

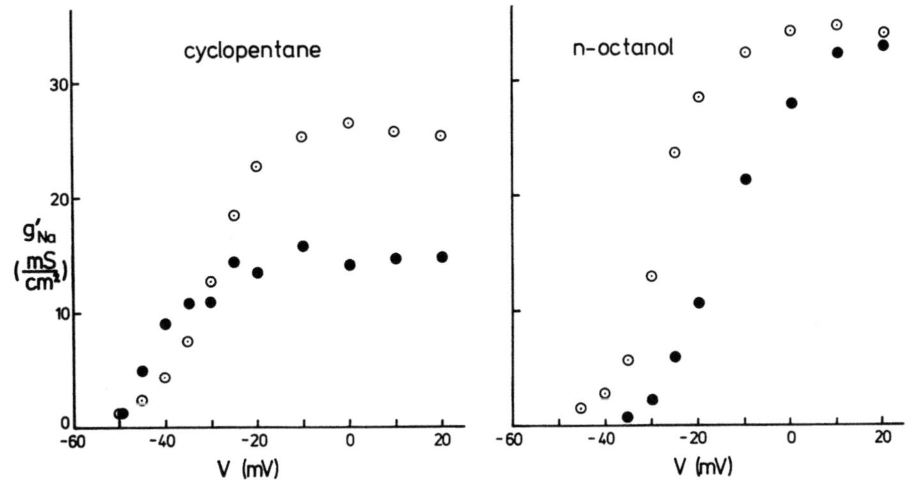

Figure 2. The effects of 1.33 mM (0.7 saturated) cyclopentane and 0.29 mM *n*-octanol on the sodium conductance parameter $\bar{g}_{Na}$ of the squid giant axon. ($\bar{g}_{Na}$ is the conductance per unit area that would be reached at a given potential were inactivation to be absent.) Under the conditions of the experiment, $g_{Na}' = \bar{g}_{Na}' m_\infty^3$, where $\bar{g}_{Na}$ is the maximum sodium conductance. ⊙, control; ●, test. V is the membrane potential. Experimental details as in Fig. 1.

benzoate, are shown in Fig. 4. The shift in m_∞ (for a concentration sufficient to suppress the maximum sodium current by 50%) is similar to that for an alcohol, but the h_∞ curve is appreciably displaced in the hyperpolarizing direction (by 8.5 mV) and its slope is reduced. In fact, the results for ethyl benzoate closely resemble those for benzene (Fig. 3). Benzocaine is merely ethyl benzoate with an amino group in the para position (Fig. 6), and it is not surprising that the h_∞ and m_∞ curves for this anesthetic (Fig. 5) are almost indistinguishable from those for ethyl benzoate. It is interesting that ethyl benzoate is more potent than benzocaine in suppressing the sodium current (Table I). In the squid axon, benzocaine reduces $\bar{g}_{Na}$ as well as shifting the h_∞ and m_∞ curves. The changes in these three parameters together account for nearly all the sodium current suppression as the effects on the time constants τ_m and τ_h are small at the half-suppression concentration.

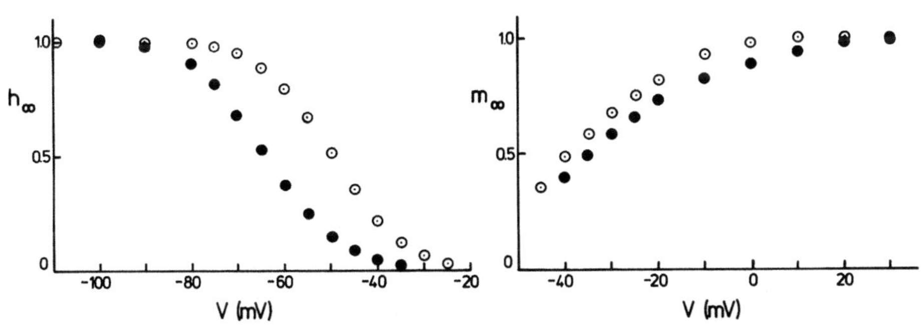

Figure 3. The effects of 3.6 mM (0.2 saturated) benzene on the steady-state inactivation (h_∞) and steady-state activation (m_∞) parameters of the sodium current of the squid giant axon. ⊙, control; ●, test. V is the membrane potential. Experimental details as in Fig. 1.

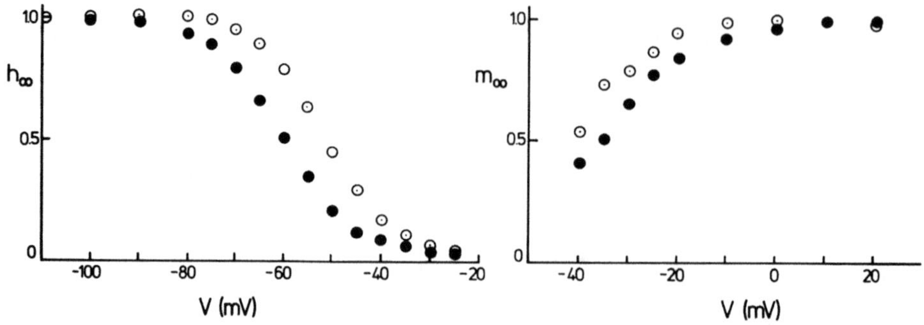

Figure 4. The effects of 0.6 mM ethyl benzoate on the steady-state inactivation (h_∞) and steady-state activation (m_∞) parameters of the sodium current of the squid giant axon. ⊙, control; ●, test. V is the membrane potential. Experimental details as in Fig. 1.

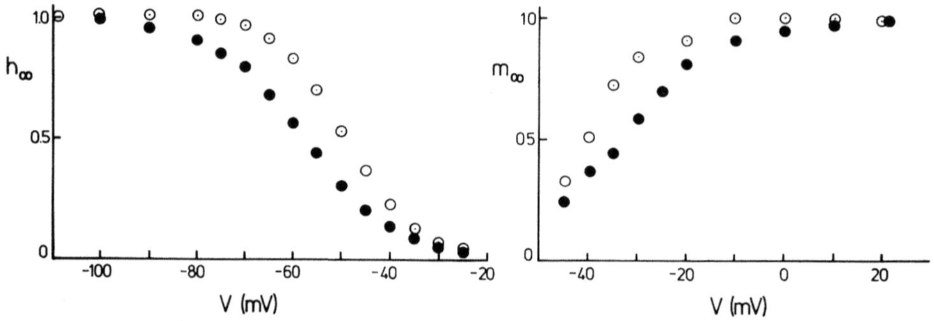

Figure 5. The effects of 1 mM benzocaine on the steady-state inactivation (h_∞) and steady-state activation (m_∞) parameters of the sodium current of the squid giant axon. ⊙, control; ●, test. V is the membrane potential. Experimental details as in Fig. 1.

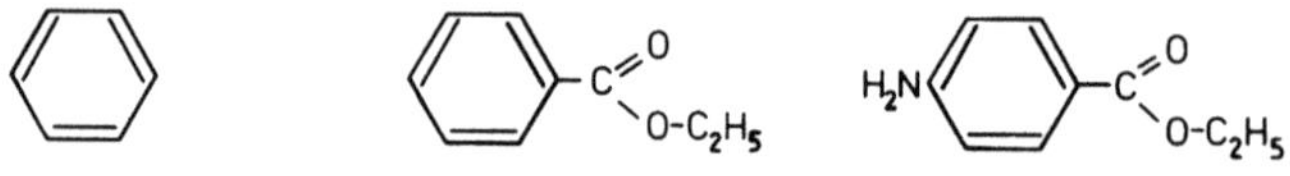

benzene ethyl benzoate benzocaine

Figure 6. The structural features of some of the "test" substances.

DISCUSSION

The Mechanisms of Action of the Hydrocarbons and Alcohols

The hydrocarbons and alcohols influence in varying degrees all the parameters of the sodium currents, i.e., h_∞, m_∞, $\bar{g}_{Na}$, τ_m, and τ_h. Evidence as to the molecular origins of these effects is not uniformly abundant. Thus, the time constants, after allowing for voltage shifts, nearly always decrease in the anesthetic solution, and it is tempting to suppose that increases in the fluidity of the lipid parts of the membrane are responsible. There is, however, no direct evidence that this is the mechanism, and one or two results, such as that for cyclohexane,[11] do not obviously fit in with this idea. In the present systems, the more important effects are those on the steady-state activation and inactivation curves and on the maximum sodium conductance. Fortunately, there is more evidence as to the origins of these effects.

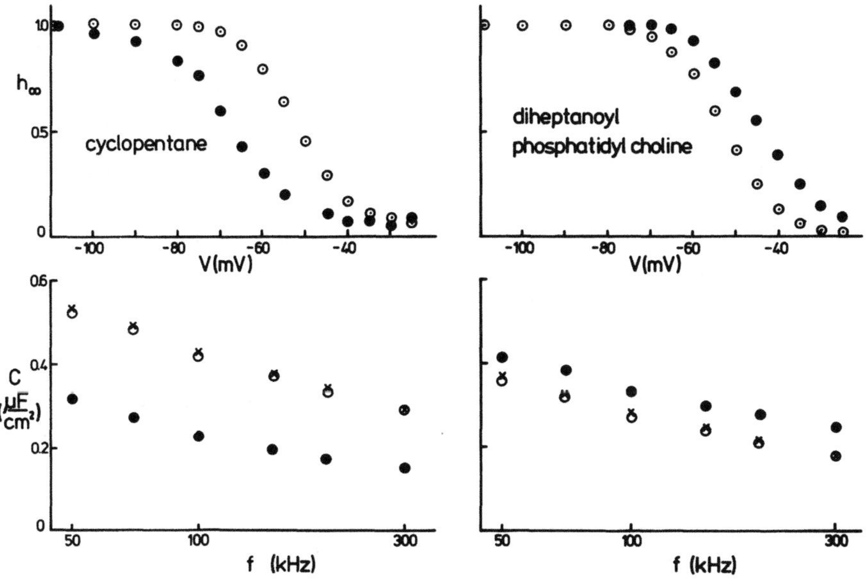

Figure 7. A comparison of the effects on the steady-state inactivation (h_∞) and axonal membrane capacitance per unit area (C) of a hydrocarbon (left-hand diagrams) and a short-chain phospholipid. ⊙, control; ●, test; ×, recovery. The cyclopentane was 1.33 mM (0.7 saturated), and the diheptanoyl phosphatidylcholine was 0.52 mM. V is the membrane potential, and f is the frequency of the signal used in the capacitance measurements. Details of the latter were as given by Haydon, Requena, and Urban[15] and Haydon and Urban.[14]

The possible origin of the large hyperpolarizing shifts and slope decreases of the steady-state inactivation curves caused by hydrocarbons has been discussed in earlier papers.[10,11,16] It was pointed out that nonpolar molecules would tend to adsorb into the inner hydrophobic regions of membranes rather than into their surfaces, and that the membrane would thus become thicker. At constant membrane potential, the internal field in the membrane would then decrease, and if the inactivation voltage sensor were a dipole in this field, a shift in the steady-state inactivation curve in the hyperpolarizing direction would occur. It is also a consequence of this model that the slope of the h_∞ curve should decrease. Various pieces of evidence support this general mechanism.[11,16] First, the model is internally consistent in that, for the hydrocarbons, the thickness changes estimated from the h_∞ curve shifts are in good agreement with those calculated from the slope changes. Second, membrane capacitance measurements at radio frequencies are consistent with the idea that the adsorption of nonpolar molecules produces a thickening.[11,14] Moreover, the thickening calculated from the capacitance per unit area accounts reasonably well for the observed shifts in the inactivation curve. Third, short-chain phosphatidylcholines, which would be expected to *thin* a lipid membrane, have been shown both to shift the h_∞ curve in the *depolarizing* direction and to *increase* the membrane capacitance[12] (Fig. 7). The capacitance increase illustrated in Fig. 7 may not be entirely the consequence of a membrane thinning since the adsorption of the phosphatidylcholine should produce an increase in membrane area which may or may not be "seen" by the measuring electrodes. At least, some 38% of the observed capacitance change would reflect a thickness decrease. An important implication of the phosphatidylcholine experiment is that the thickness changes relevant to the shifts in h_∞ occur in lipid regions of the membrane. Thus, for the hydrocarbon adsorption, it could be argued that protein as well as lipid regions of the membrane would be thickened and that the protein was the more important. For the short-chain phospholipid adsorption, however, it is clear that the lipid regions would become thinner but less obvious that the protein could so do. Finally, the thickening hypothesis is consistent with the observations that alcohols have very little effect on lipid bilayer thickness[5–7,21] and also produce only small shifts in the h_∞ curve. More recently a detailed study has been attempted to show the dependence on hydrocarbon concentration of h_∞ shift, axonal membrane capacitance per unit area, and bilayer capacitance per unit area (D. Needham, J. R. Elliott, B. M. Hendry, and D. A. Haydon, unpublished observation). It has been found that both the h_∞ shifts and the axonal capacitance are more sensitive to hydrocarbons at low concentrations than is the lipid bilayer capacitance. The conclusion would appear to be that, structurally, the nerve membrane is significantly different from that of a lipid bilayer, though whether the important differences lie in the lipid composition or in the presence of the protein, or in some combination of these factors, is as yet unknown.

Shifts in the steady-state activation (m_∞) curve can occur in either the hyperpolarizing or depolarizing directions. For substances, such as alcohols, that do not thicken bilayers and that have little or no effect on h_∞, there are substantial depolarizing shifts in m_∞. The origin of these shifts has been suggested to be an asymmetrical change in the surface dipole potentials of the membrane.[12,16] This proposal is illustrated in Fig. 8. A depolarizing shift in m_∞ suggests that the internal field of the membrane has been increased by the anesthetic. This could occur through an asymmetrical adsorption. Furthermore, if, as the evidence suggests, the anesthetics tend to *lower* the surface potential (see, for example, refs. 2, 21, and 23), it would be necessary for the adsorption to be greater on the inner surface of the membrane. X-ray diffraction studies of nerve myelin[4] have demonstrated the asymmetry of these membranes, and it is well known that the sodium channel proteins are asymmetrical in their response to enzymes such as pronase. The general supposition that anesthetic adsorption is asymmetrical is not therefore unreasonable. An experiment that demonstrates that the axonal membrane is asymmetrical in its response to nonspecific substances has recently been carried out using ions that do not readily permeate the membrane. *N*-alkyltrimethylammonium ions in which the *n*-alkyl residue is eight or less carbon atoms in length perturb the sodium current in much the same way as do *n*-alkanols.[8] These ions are effective, however, only when applied to the inside of the axon (Fig. 9). This result might have arisen from an interaction with the inner part of the sodium channel protein, but since the observed m_∞ shift is characteristic of many other types of molecule, it is more likely that it originates in the membrane lipid adjacent to the protein.

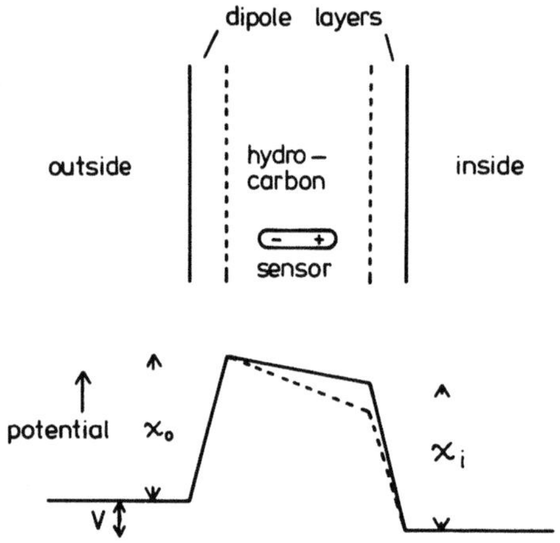

Figure 8. An illustration of how the transmembrane potential is influenced by the presence of oriented dipoles at the membrane surfaces. The potential jump χ_o at the exterior surface is assumed not to be affected by the adsorption of the anesthetic. The potential χ_i at the inner surface is lowered, however, and the internal field is increased (dashed line). A larger depolarization is thus required to activate a given fraction of the sodium conductance; i.e., the m_∞ curve shifts to more positive potentials.

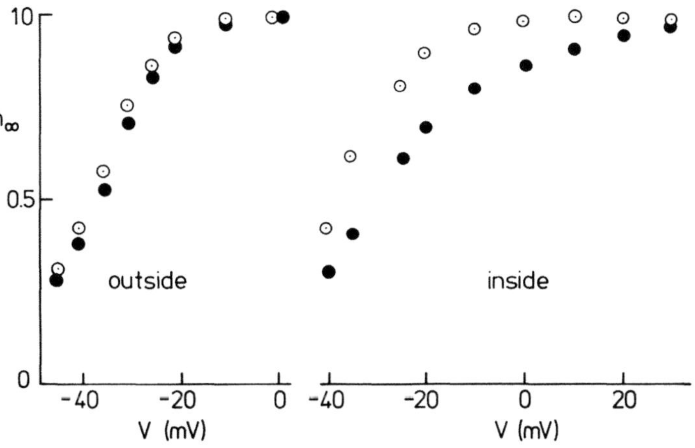

Figure 9. The asymmetrical effect of 0.8 mM *n*-octyltrimethylammonium ions on the steady-state activation parameter (m_∞) of the squid giant axon. The maximum sodium current was approximately half-suppressed for the interior addition. ⊙, control; ●, test. The shift produced by the ions internally is similar in character to that produced by *n*-octanol and other alcohols; i.e., the sodium current suppression becomes smaller for larger depolarizations, in contrast to the effects of cations like QX-222. No significant use dependence was observed.[8]

It has been necessary to invoke two distinctly different mechanisms to account for the h_∞ and m_∞ shifts. As pointed out elsewhere,[11–13,16] while only the thickening effect seems to affect h_∞, both the thickening and the surface potentials appear to affect m_∞. One can only speculate as to why this is so,[16] and there is no need for repetition here of these suggestions. The interesting point is that inactivation and activation seem to originate from voltage sensors that are to some extent independent. This is particularly clear for benzene, the carboxylic esters, and benzocaine, where the shifts in the h_∞ and m_∞ curves are in opposite directions on the voltage axis.

The mechanism of the suppression of the maximum sodium conductance by neutral molecules has not been investigated in the same detail as have the origins of the shifts in h_∞ and m_∞. A theory has, however, been put forward for systems in which significant membrane thickening by hydrocarbons is thought to occur. Thus, a combination of a thickness increase, together with a concomitant tension increase, could destabilize the sodium channel in much the same way as for gramicidin channels in lipid bilayers.[10,11,17]

The Mechanisms of Action of Carboxylic Esters and Benzocaine

The substantial hyperpolarizing shifts in the steady-state inactivation curve produced by some carboxylic esters are out of line with the results for many other polar molecules, e.g., the alcohols and oxyethylene alcohols, diethyl ether, glycerol monooctanoate, and both *n*-alkyl anionic and cationic molecules.[8,12,13] Since the h_∞ shifts produced resemble those of the hydrocarbons both in their direction and in their associated slope changes, it is obviously necessary to consider whether a similar explanation, i.e., in terms of membrane thickening, might be applicable. Polar molecules would not be expected to partition preferentially into the central regions of a lipid bilayer and thus should not produce thickening. However, if there were a specific interaction (involving, perhaps, hydrogen bond formation) between the anesthetic molecule and some part of the sodium channel protein buried in the lipid, then the energetics of the partitioning would be changed. It may be significant that the polar molecules so far known to exhibit the h_∞ shift have ketonic oxygen atoms that could form strong

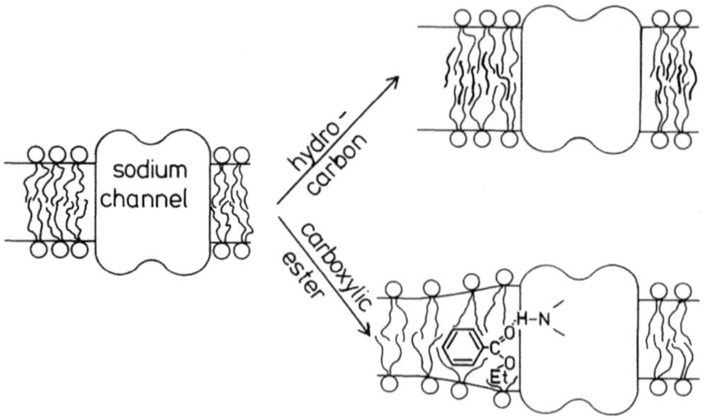

Figure 10. A schematic illustration of the different effects of hydrocarbons and carboxylic esters on the axonal membrane. The hydrocarbon is believed to produce a relatively uniform thickening of the lipid regions of the membrane. The ester, on the other hand, partitions into the central region only through hydrogen-bond formation with groups on the sodium channel protein. The thickening is thus localized to the immediate environment of the channel.

hydrogen bonds with, for example, >NH groups of a peptide chain. The monoglycerides and phospholipids, although also embodying ketonic oxygen atoms, have additional polar groups (hydroxyls in the glyceride and ionic groups in the phospholipid), would be much more strongly anchored in the membrane surface than the simple ester, and therefore would not be expected to partition into the interior, even with the incentive of the hydrogen bond.

The result of this proposed adsorption mechanism would be a local thickening of the membrane adjacent to the sodium channel (Fig. 10). Such a thickening would almost certainly be too restricted to be detected by the capacitance method, which measures only average changes for the whole area of the membrane.

This local thickening mechanism could account for the h_∞ shift produced by benzocaine in squid axons. Interactions with the membrane lipid, so as to produce surface potential changes as described previously, could be the origin of the effects of benzocaine on m_∞. The evidence that benzocaine acts from *inside* the ion pore of the sodium channel is indirect, and unlike lignocaine, it does not exhibit use dependence.[18,20]

REFERENCES

1. Armstrong CM, Binstock L: The effects of several alcohols on the properties of the squid giant axon. *J Gen Physiol* 1964;48:265–277.
2. Bangham AD, Mason W: The effect of some general anaesthetics on the surface potential of lipid monolayers. *Br J Pharmacol* 1979;6:259–265.
3. Bean BP, Shrager P, Goldstein, DA: Modification of sodium and potassium channel gating kinetics by ether and halothane. *J Gen Physiol* 1981;77:233–253.
4. Caspar DLD, Kirschner DA: Myelin membrane structure at 10 Å resolution. *Nature, New Biology (London)* 1971;231:46–52.
5. Ebihara L, Hall JE, MacDonald RC, McIntosh TJ, Simon SA: Effect of benzyl alcohol on lipid bilayers. A comparison of bilayer systems. *Biophys J* 1979;28:185–196.
6. Elliott JR, Haydon DA: The interaction of *n*-octanol with black lipid bilayer membranes. Biochim Biophys Acta 1979;557:259–263.
7. Elliott, JR, Haydon DA: The influence of n-alkanols on the capacity per unit area of planar lipid bilayers. *Biochim Biophys Acta* 1984;773:165–168.
8. Elliott JR, Haydon DA, Hendry BM: The asymmetrical effects of some ionized *n*-octyl derivatives on the sodium current of the giant axon of *Loligo forbesi*. *J Physiol* 1984a;350:429–445.
9. Elliott JR, Haydon DA, Hendry BM: Anaesthetic action of esters and ketones: Evidence for an interaction with the sodium channel protein in squid axons. *J Physiol* 1984b;354:407–418.
10. Haydon DA, Kimura JE: Some effects of *n*-pentane on the sodium and potassium currents of the squid giant axon. *J Physiol* 1981;312:57–70.
11. Haydon DA, Urban BW: The action of hydrocarbons and carbon tetrachloride on the sodium current of the squid giant axon. *J Physiol* 1983a;338:435–450.

12. Haydon DA, Urban BW: The action of alcohols and other non-ionic surface active substances on the sodium current of the squid giant axon. *J Physiol* 1983b;341:411–427.
13. Haydon DA, Urban BW: The effects of some inhalation anaesthetics on the sodium current of the squid giant axon. *J Physiol* 1983c;341:429–439.
14. Haydon DA, Urban BW: The admittance of the squid giant axon at radio frequencies and its relation to membrane structure. *J Physiol* 1985;360:275–291.
15. Haydon DA, Requena J, Urban BW: Some effects of aliphatic hydrocarbons on the electrical capacity and ionic currents of the squid giant axon membrane. *J Physiol* 1980;309:229–245.
16. Haydon DA, Elliott, JR, Hendry BM: Effects of anesthetics on the squid giant axon, in Kleinzeller A, Baker PF (eds): *Current Topics in Membranes and Transport* Vol 22, London, Academic Press, 1984, pp 445–482.
17. Hendry BM, Urban BW, Haydon DA: The blockage of the electrical conductance in a pore-containing membrane by the *n*-alkanes. Biochim Biophys Acta 1978;513:106–116.
18. Hille B: Local anaesthetics: Hydrophilic and hydrophobic pathways for the drug–receptor interaction. J Gen Physiol 1977;69:497–515.
19. Hodgkin AL, Huxley AF: A quantitative description of membrane current and its application to conduction and excitation in nerve. *J Physiol* 1952;117:500–544.
20. Mrose HE, Ritchie JM: Local anesthetics: Do benzocaine and lidocaine act at the same single site? *J Gen Physiol* 1978;71:223–225.
21. Reyes J, Latorre R: Effect of the anesthetics benzyl alcohol and chloroform on bilayers made from monolayers. *Biophys J* 1979;28:259–280.
22. Seeman P: The membrane action of anesthetics and tranquilizers. *Pharmacol Rev* 1972;24:583–655.
23. Simon SA, Bennett PB: Membrane thermodynamics and anesthesia mechanisms, in Fink BR (ed): *Molecular Mechanisms of Anesthesia, Progress in Anesthesiology,* Vol 2. New York, Raven Press, 1980, pp 305–319.
24. Swenson RP, Oxford GS: Modification of sodium channel gating by long-chain alcohols: Ionic and gating current measurements, in Fink BR (ed): *Molecular Mechanisms of Anesthesia, Progress in Anesthesiology,* Vol 2. New York, Raven Press, 1980, pp 7–16.

23

Solubility of Volatile Hydrocarbons in Lipid Bilayers

A New Perspective

Stephen H. White, Glen I. King, and Russell E. Jacobs

ABSTRACT. We review our work of the past several years involving studies of the interaction of hexane with lipid bilayers. Specifically, we summarize neutron diffraction measurements on oriented dioleoylphosphatidylcholine (DOPC) bilayers at 66% relative humidity and nuclear magnetic resonance (NMR) measurements on dimyristoylphosphatidylcholine (DMPC) and DOPC liposomes in excess water. Our results are interesting and surprising. (1) The partial molar volume of hexane in the hydrocarbon core of oriented DOPC bilayers is approximately zero. (2) The phase behavior of DMPC is greatly affected by hexane whereas for DOPC the hexane acts as a simple impurity; the motions of the acyl chains are not affected greatly in either case. (3) The motion of the hexane molecules in both DOPC and DMPC liposomes is fast on the NMR time scale, anisotropic, and such that both ends of the hexane experience the same average environment. We do not interpret these results in terms of mechanisms of general anesthesia. However, we suggest that the question of whether or not general anesthesia is mediated by the lipid bilayer should remain open.

INTRODUCTION

Our laboratory is interested in the behavior of the lipid bilayer as a solvent for hydrophobic molecules. This problem is important for understanding the organization and biosynthesis of cell membranes which are now generally considered to be a quasi-two-dimensional solution consisting of a bilayer "solvent" in which hydrophobic protein "solutes" are dissolved.[19] We are attacking the problem through studies of the interactions of alkanes with bilayers. Alkanes are quintessential hydrophobic molecules available in a wide range of sizes and isomeric and isotopic configurations and are thus ideal for probing the interior of lipid bilayers. This paper summarizes some of our recent work in which we have used neutron

Abreviations used in this chapter: DOPC, dioleoylphosphatidylcholine; DMPC, dimyristoylphosphatidylcholine; ND, neutron diffraction; NMR, nuclear magnetic resonance; SLD, scattering length density; HCC, hydrocarbon core of the bilayer.

Stephen H. White, Glen I. King, and Russell E. Jacobs • Department of Physiology and Biophysics, University of California at Irvine, Irvine, California 92717.

diffraction (ND) to determine the structure of bilayers containing hexane and nuclear magnetic resonance (NMR) to obtain information about the dynamics of the interaction of hexane with bilayers.

The work we describe is clearly relevant to the issue of the mechanism of general anesthesia induced by volatile hydrocarbons. However, we have no position on the question of whether or not anesthesia in such cases is mediated directly by the bilayer. Franks and Lieb[3] have reviewed the situation and concluded that the primary site of action is not the bilayer. Although we cannot disagree with this conclusion, our data suggest that the picture may not yet be complete. Franks and Lieb[3,4] base their conclusion largely on X-ray and neutron diffraction measurements which show no obvious structural changes in bilayers in the presence of high concentrations of volatile anesthetics such as nitrous oxide and cyclopropane. We have made a similar observation[8,9,23]; *protonated* hexane has no apparent effect on the structure of dioleoyl lecithin bilayers at clinical anesthesia levels and beyond. However, its presence and location can be detected using deuterated hexane and neutron diffraction difference structure methods.[23] Furthermore, more complete studies[8,9] show that remarkable things are happening within the bilayer which are not apparent unless one uses methods unique to neutron diffraction. Although Franks and Lieb[3,4] did use neutron diffraction, neither the anesthetic nor the lipids were deuterated, so that the complete power of neutron diffraction methods was not used.

Our results from both ND and NMR measurements are rather surprising. For oriented dioleoylphosphatidylcholine (DOPC) bilayers at 66% relative humidity (RH),[9] hexane enters the hydrocarbon core in such a way that its partial molar volume appears to be approximately *zero!* For bilayers formed from dimyristoylphosphatidylcholine (DMPC) or DOPC in excess water,[6,7] we find that hexane has practically no effect on the motions of the acyl chains in either case. Despite this, the phase behavior of DMPC is altered drastically while for DOPC it acts as a simple impurity. The motion of the hexane molecules in either DOPC or DMPC liposomes is fast on the NMR time scale, anisotropic, and such that both ends of the molecule experience the same average environment.

Our studies thus reveal several peculiar properties of lipid bilayers which are observable with one technique but not another. With protonated hexane, neutron diffraction reveals little about structural changes induced by the hexane. With deuterated compounds and appropriate analyses, however, we find large amounts of hexane entering the bilayer without a significant change in hydrocarbon volume. For DMPC, we find a major effect of hexane on the phase behavior of the lipid using calorimetric methods, but deuterium NMR shows no major effect on acyl chain motions. We conclude that our understanding of bilayers is far from complete and that the question of the bilayer's role in general anesthesia might best be left open for the present.

STRUCTURE OF BILAYERS CONTAINING HEXANE

Methods

The general methods of structural analysis of bilayers by low-angle diffraction methods are described in references 2, 16, and 17. The basic principles for determining the structure of a bilayer using either X rays or neutrons are illustrated in Fig. 1. A lamellar structure such as the bilayer produces a diffraction pattern of evenly spaced spots. The spacing of the spots is determined by the characteristic repeat distance (d-spacing, D) of the structure given by the Bragg condition $2D \sin \theta = h\lambda$. The intensities of the spots are determined by the scattering lengths and spatial distributions of the molecules comprising the bilayer system. The struc-

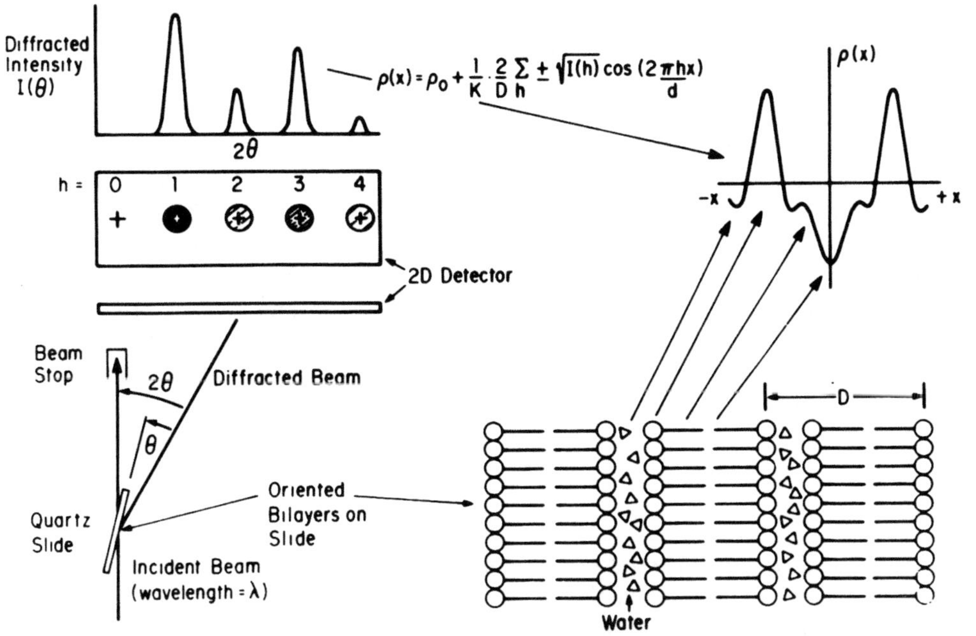

Figure 1. Summary of the method of determining the structure of symmetrical bilayers using X-ray or neutron diffraction. The lipid, DOPC for the experiments described here, is deposited on the quartz slide from a chloroform/methanol solution. ρ(x) is the scattering length density (SLD) of the structure as a function of the distance across the bilayer.

ture of the bilayer can be derived knowing these intensities as follows. Each type of bilayer under a particular set of conditions has a unique diffraction pattern. The structure of the bilayer, in terms of the scattering length density versus thickness [ρ(x)], can be determined from the Fourier transform of the so-called structure factors F(h). F(h) is the mathematical product of two terms—a phase factor and the square root of the intensity of the diffraction spot of order h. Because the bilayer is symmetrical, the phase factor is either +1 or −1. Thus, the structure factors for our system are nothing more than the square roots of the intensities of the spots (corrected in various ways) chosen to have a + or − sign depending on phase. It is important to appreciate the uniqueness of the F(h) for a particular structure. If there is no change in structure, or if the diffraction experiment cannot detect any, then the F(h) will not change either.

The ability of the sample to diffract depends on differences in the scattering lengths of the various atoms of the sample. We are interested in detecting hexane molecules distributed among the acyl chains of the lipid. However, because the hexane is essentially equivalent to acyl chains, there is insufficient contrast to distinguish them from each other unless the hexane or acyl chains can be labeled with atoms that have different scattering lengths. This can be done by deuterating either the hexane or the acyl chains and using neutrons for diffraction. X-ray diffraction is caused by scattering from electrons whereas neutron diffraction is caused by the scattering from nuclei. Neutrons scatter differently from deuterons than from protons, and one can thus distinguish, say, perdeuterated hexane from protonated acyl chains. The distribution of hexane in a bilayer can be obtained by determining the structure of the bilayer in the presence of protonated hexane at a particular vapor pressure and then in the presence of deuterated hexane at the same vapor pressure. Subtracting the two structures from one another nulls out the common features (polar groups and water) and leaves the difference in scattering length density between protonated and deuterated hexane. The resulting difference structure reveals the distribution of the deuterated hexane in the system.

The experiments were performed at the High Flux Beam Reactor at Brookhaven National Laboratory. DOPC was deposited on a quartz slide and placed in a closed chamber containing two reservoirs. One reservoir was filled with a saturated salt solution to maintain the relative humidity fixed at 66% (for the studies reported here). The other reservoir contained a mixture of hexane (protonated or deuterated) dissolved in hexadecane to maintain the vapor pressure of hexane at a value $P = X_H \cdot P_0$ where X_H is the mole fraction of hexane in the hexadecane and P_0 is the vapor pressure of pure hexane. Hexadecane is essentially involatile and forms a nearly ideal mixture with the hexane. The chamber containing the lipid deposited on a quartz slide was placed in the neutron beam. The diffracted beams were measured using a two-dimensional detector and various counting and control circuits. The details of the experiments can be found in White *et al.*[23] and King *et al.*[8,9]

Results

General Structural Features. Figure 2 shows the corrected structure factors (four orders) for DOPC bilayers containing protonated *n*-hexane as a function of X_H up to 0.8. Beyond this value, a new phase of unknown structure appears reversibly. Our studies were confined to values less than 0.8. These data have three striking features. First, there is *no* change in the d-spacing (49.7 ± 0.5A) over the full range of X_H. Second, there is an abrupt structural change between $X_H = 0.4$ and 0.5. Third, above and below this structural change the structure factors are independent of X_H. All these things taken together mean that there is no obvious "swelling" of the structure and that two structures suffice to describe the system. The two structures derived from these structure factors, one for no hexane and one at maximum hexane ($X_H = 0.8$), are shown in Fig. 3. Ignoring the structural change at X_H = 0.4–0.5, these results are not unlike those of Franks and Lieb.[3,4] Were it not for the peculiar structural change, one might conclude either that hexane has no effect on the structure or that it was not even entering the structure.

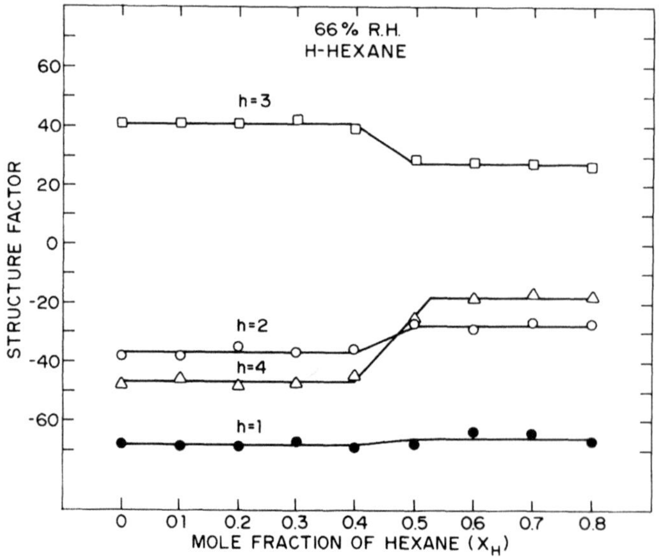

Figure 2. The structure factors of DOPC bilayers containing varying amounts of protonated hexane. The hexane is introduced to the bilayer via the vapor phase from a hexane/hexadecane mixture whose mole fraction of hexane is X_H. The amount of hexane in the bilayer at a particular vapor pressure has been determined using ^{3}H-labeled hexane (see Table I). The d-spacing (*D*) is constant over the entire range of X_H. Note that the structure factors are constant below and above X_H = 0.4–0.5. Data taken from King *et al.*[9]

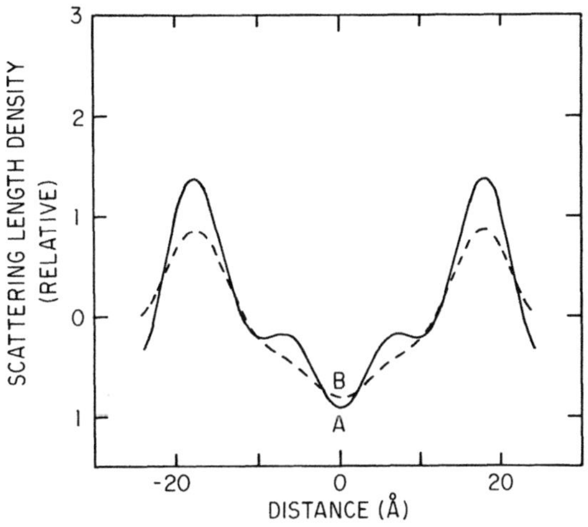

Figure 3. The structures of DOPC bilayers containing protonated hexane. Because the structure factors (Fig. 2) are *constant* below (A) and above (B) X_H = 0.4–0.5, two structures suffice to describe the bilayer at all hexane contents. The d-spacing of the system is independent of the amount of hexane; note that the spacing of the polar groups is unchanged (compare Fig. 1). Also note that the hydrocarbon core (HCC) becomes more disordered and the scattering length density of the head group region decreases for X_H above 0.5. Data from King *et al.*[9]

Two sets of experiments show that the hexane is actually present. Figure 4 shows the results of measurements similar to those of Fig. 2 except that deuterated hexane (which now contrasts strongly with the acyl chains) replaces the protonated hexane. The structural change at X_H = 0.4–0.5 is still apparent, but now one sees, in addition, changes in the F(h) with increases in X_H as expected on theoretical grounds if the hexane enters the bilayer. The amount of hexane in the bilayer can also be measured using ^{3}H-labeled hexane. These results are summarized in the second column of Table I, where it can be seen that about 1.2 hexanes/DOPC enter the bilayer at X_H = 0.8. One might have expected that as hexane entered the bilayer, it would go into the hydrocarbon core causing it to swell resulting in, at

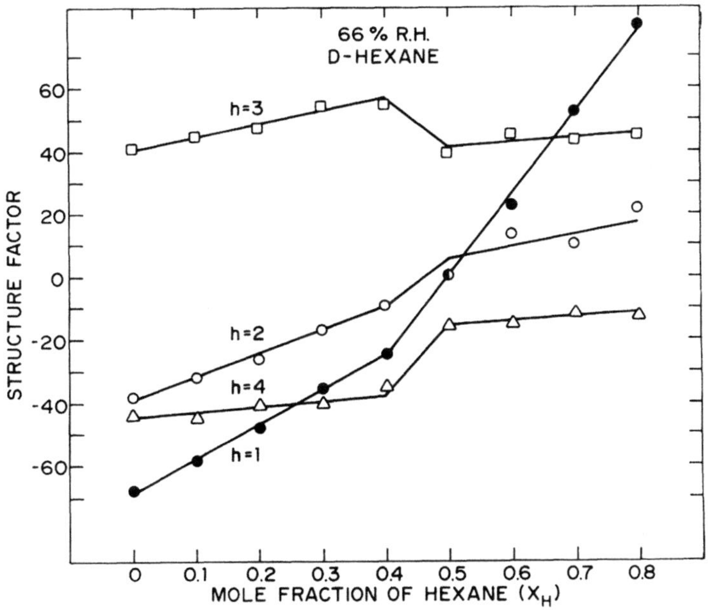

Figure 4. The structure factors of DOPC bilayers containing deuterated hexane. Note that they now change with increases in X_H. The abrupt change in structure between X_H = 0.4 and 0.5 is still apparent (see Fig. 2). The other changes are those expected if deuterated hexane is entering the structure. Data from King *et al.*[9]

Table I. Summary of the Structural Parameters of Oriented DOPC Multilayers Containing Hexane (66% RH, 22.5°C)[a]

X_H	N_H^X	D (Å)	D_h (Å)	S_X (Å²)	V_{hc} (Å³)	V_{hc}^* (Å³)	d_{hc} (g/cc)
0.0	0.00	49.7	28.2	67.4	475	475	0.83
0.1	0.04	49.7	28.2	67.4	475	480	0.83
0.2	0.12	49.7	28.2	67.4	475	489	0.85
0.3	0.22	49.7	28.2	67.4	475	499	0.87
0.4	0.32	49.7	28.2	67.4	475	508	0.88
0.5	0.40	49.7	27.0	72.1	486	522	0.88
0.6	0.61	49.7	26.0	76.2	495	542	0.90
0.7	(0.95)	49.7	26.0	76.2	495	574	0.95
0.8	1.17	49.7	26.0	76.2	495	608	0.99

[a]Hexane is introduced via the vapor phase from a hexane/hexadecane mixture with a hexane mole fraction of X_H. The amount of hexane in the bilayer per phospholipid (N_H^X) was determined using ^{3}H-labeled hexane (S.D. about ± 10%).[9] D is the d-spacing (±0.5 Å), D_h the thickness of the hydrocarbon core (HCC) (±0.6 Å), S_X the area/phospholipid (±1.4 Å²), and V_{hc} the volume of the HCC *per acyl chain* (±20 Å³). V_{hc}^* is the volume of the HCC per acyl chain expected *if* the hexane mixed with the acyl chains in an ideal volumetric way. See the text for the details of the calculations.

least, an increase in d-spacing. This did not happen. The questions that need to be answered are (1) where does the hexane go in the bilayer, (2) why are structural changes not apparent with protonated hexane, and (3) how does the hexane pack into the bilayer?

Distribution of Hexane. The first question is the one most easily answered. Figure 5 shows two bilayer structures obtained at $X_H = 0.3$ for protonated (curve H) and deuterated hexane (curve D). The curves appear to be the same except for the increased scattering length density (SLD) in the center due to the D-hexane. The difference of these two curves gives the residual SLD due to the D-hexane and shows the distribution of the hexane across the width of the bilayer. A series of such curves at various X_H up to 0.3 are shown in Fig. 6. The amount of D-hexane is linearly related to X_H, and it is well localized in the central 10 Å of the bilayer. The "ripples" indicated by the arrows are Fourier "noise" due to the use of only

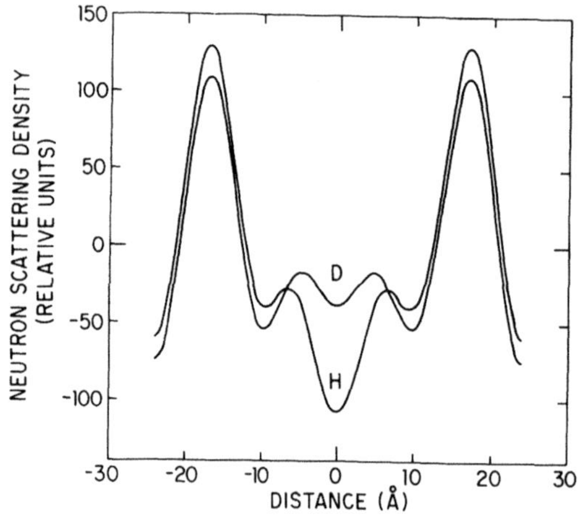

Figure 5. Comparison of the structures of DOPC bilayers containing protonated hexane (curve H) or deuterated hexane (curve D). $X_H = 0.3$ in both cases. The difference of these two curves yields the difference structure which shows the distribution of the deuterated hexane as a function of thickness (see Fig. 6). Taken from White *et al.*[23]

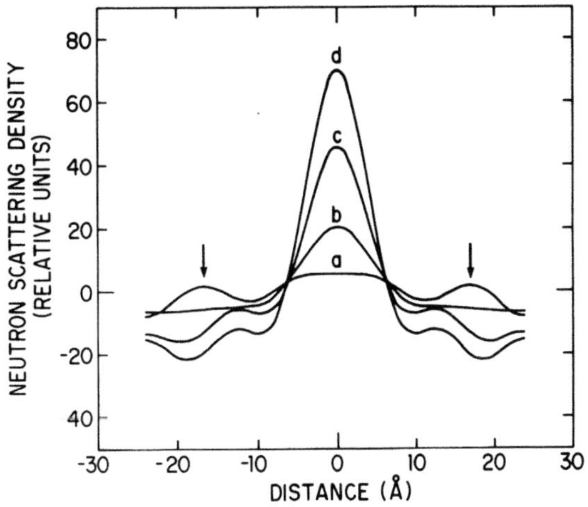

Figure 6. The distribution of deuterated hexane across the bilayer at various values of X_H (a, 0.05; b, 0.1; c, 0.2; d, 0.3). The hexane is well confined to the central 10 Å of the bilayer (see also Fig. 8). The arrows indicate so-called Fourier noise resulting from limited resolution (h = 4) of the measurements. This noise may be ignored. Taken from White *et al.*[23]

four orders of diffraction data and can be ignored at present. This noise is bothersome, however, for detailed analyses and can be gotten around using strip-function modeling procedures described later.

Work on black lipid films formed from glycerol monooleate and various *n*-alkanes suggested that the double bonds of the monooleate defined the space available to the alkane molecules.[21] To examine the location of the hexane relative to the positions of the double bonds, we determined the difference structure of bilayers formed with protonated DOPC and DOPC deuterated at the double-bond carbons. The difference structure shows the residual SLD due to the double bonds across the thickness of the bilayer. Figure 7 shows a superposition of such data with the X_H = 0.3 curve of Fig. 6. The space between the double bonds neatly defines the location of the hexane.

Our data clearly show that hexane enters the bilayer in significant amounts—up to 1.2 hexanes/phospholipid at X_H = 0.8. Yet, there are no significant changes in structure appar-

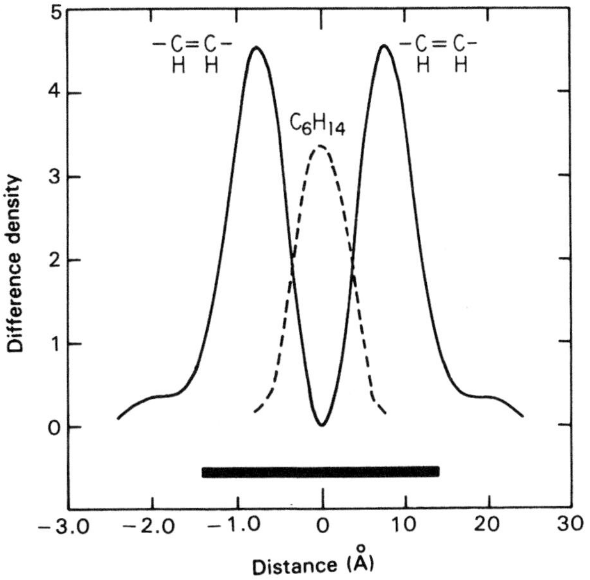

Figure 7. The distribution of deuterated hexane relative to the double bonds of the acyl chains of DOPC. The hexane curve is taken from the data of Fig. 6 (curved). The distribution of the double bonds comes from difference-structure measurements on protonated DOPC and DOPC deuterated at the carbons of the double bonds (unpublished results). The horizontal bar represents the thickness of the hydrocarbon core (HCC) (see Table I).

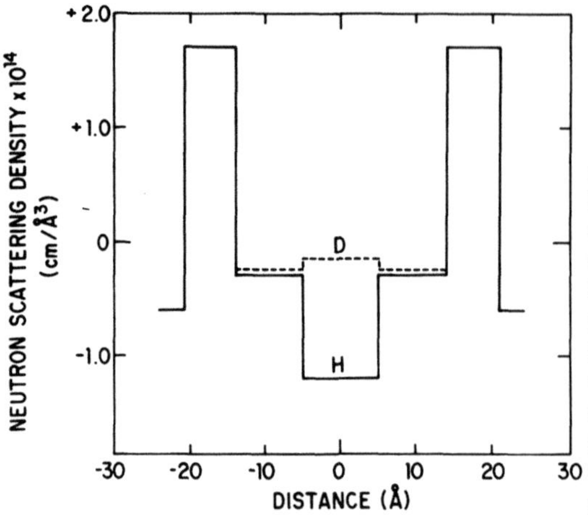

Figure 8. The strip-function structures of DOPC bilayers containing protonated hexane (curve H) and deuterated hexane (curve D) at $X_H = 0.3$. The bilayer is divided into several zones of different average scattering length densities. The widths and densities are varied until the Fourier transforms of the structures (strip-function structure factors) agree to within experimental error with the measured structure factors. This procedure accurately predicts the thickness of the hydrocarbon core (see Fig. 9). Note that the hexane is well contained within the central 10 Å of the bilayer. Taken from White *et al.*[23]

ent for protonated hexane except for that between $X_H = 0.4$ and 0.5. When we began the experiments several years ago, we expected the system to behave as black films which show increases in hydrocarbon core thickness as the activity of the alkane increases.[20,22] On this basis we expected the hydrocarbon thickness to increase by about 7.4 Å for the entrance of one hexane/phospholipid into the bilayer. Such an increase would have resulted in structure factors drastically different from those shown in Fig. 2.[9] Thus, we are led to questions (2) and (3) above concerning the lack of apparent structural changes and the nature of the packing of the hexanes into the hydrocarbon core.

Hydrocarbon Core Structural Analysis. We need some additional structural parameters to answer these two questions. Most important is the thickness of the hydrocarbon core, D_h. The approach we use here to determine D_h is to construct strip-function models that are derived from the structure factors and represent the hydrated bilayer as several contiguous strips with some average scattering length density representing the water, polar groups, and hydrocarbon interior. Such representations for the structures shown in Fig. 5 are presented in Fig. 8. Note that this representation demonstrates clearly the localization of the hexane to the central 10 Å of the bilayer. These strip-functions are derived by varying the widths and average SLDs of the strips until the Fourier transforms of the strip-structures agree to within

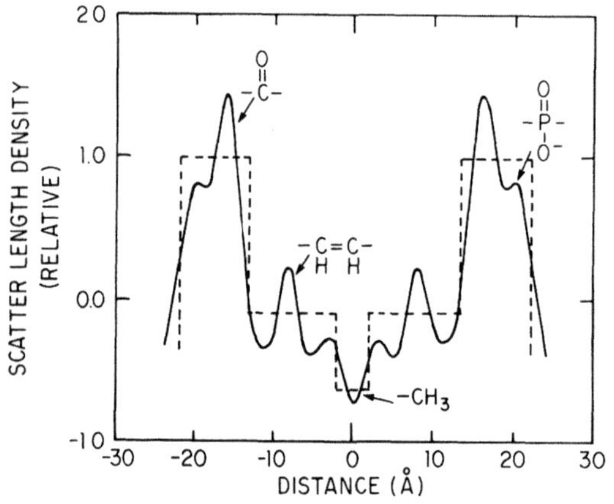

Figure 9. Comparison of the calculated strip-function structure (dashed curve) with the actual structure of a hypothetical DOPC bilayer (solid curve). The strip-function model defines clearly the thickness of the hydrocarbon core, which is generally assumed to contain all of the acyl chain except for the carbonyl group. Based on unpublished data.

experimental error with the observed structure factors. The main question that arises in such analyses is whether or not the strips representing D_h are accurate; that is, do the strips include appropriate parts of the acyl chains?

It is generally accepted that D_h should include all of the acyl chain except for the carbonyl group. The validity of the strip-function approach is shown in Fig. 9. We constructed a more or less arbitrary structure of a DOPC bilayer by assuming the various groups to be located at certain points. We allowed them to have Gaussian distributions and thereby arrived at a "structure" shown as the solid curve. Note that the double bonds and carbonyl groups are clearly visible because of the absence of hydrogens attached to these groups compared to the methylenes and methyls. We then determined the Fourier transform of the structure to arrive at structure factors. These structure factors were then used to construct strip-structures from the first four structure factors according to the protocols used for the real structure factors. The result is the dashed curve shown in Fig. 9. The D_h accurately and appropriately represents the thickness of the hydrocarbon region. We have also measured D_h directly for DOPC bilayers at 66% RH using DOPC labeled with deuterium at the C-2 positions on the acyl chains.[9] The value obtained in this way is in complete agreement with the value calculated from the strip-function model.

Molecular Packing in the Hydrocarbon Core. The D_h calculated from the strip functions are shown in Table I. Note that D_h is constant at 28.2 Å below the structural change at X_H = 0.4–0.5 and constant at 26.0 Å above it. There must also be an increase in the area/phospholipid above the transition because the scattering of the head groups decreases (Fig. 3). Knowledge of the mass density of the hydrocarbon region permits the area/phospholipid (S_X) and consequently the volume and mass density of the hydrocarbon core (HCC) as a function of X_H to be calculated. Preliminary calculations and the well-known fact that the high-angle X-ray scattering of bilayers is the same as bulk alkanes[10] suggest that in the absence of alkane the mass density of the hydrocarbon core should be the same as in bulk. We thus assumed a molecular volume/acyl chain of 475 Å^3.[15] This assumption yields an S_X of 67.4 Å^2 for no hexane. The value of S_X for X_H = 0.8 was then calculated from the zero-hexane value from knowledge of the decrease in the head group SLD. The value arrived at was 76.2 Å^2. From this value, the volume/acyl chain of the HCC was calculated to be only 495 Å^3. This volume *includes* the volume of the 1.2 hexanes/phospholipid but is little different from the hexane-free value. We thus conclude that the partial molar volume of the hexane in the HCC is approximately zero!

Discussion

The consequences of the zero partial molar volume effect are shown in Table I where we show as V_{hc}* the volume/acyl chain expected for the HCC if ideal volumetric mixing occurred. We also show the mass densities of the HCC we expect given the zero partial molar volume. The densities approach (but *do not* exceed) those expected of *solid* alkyl substances. Because the structures seen in Fig. 3 are characteristic of disordered liquidlike structures, we conclude that the HCC must be glasslike at the high hexane concentrations. The entrance of the hexane into the HCC of the bilayer at the highest hexane concentrations is analogous to a liquid-to-solid transition in terms of molecular packing-density. They must be, then, a "latent heat of packing" associated with the process. This latent heat would have to be negative and of a magnitude somewhere between the latent heats of fusion of hexane and heptadecene.

These conclusions violate drastically one's intuition about how the system should behave. However, our intuition is based on the behavior of isotropic bulk systems. Thus, one thinks of the mixing of, say, liquid hexane with liquid hexadecane. There the total mixed volume is very nearly the simple sum of the two starting volumes. Our experimental result *seems* to be equivalent to pouring hexane into a volume of hexadecane and having it "disappear" into preexisting free volume of the hexadecane. The entrance of hexane into the free volume of the hexadecane is favored enthalpically through van der Waals forces but does not occur because there would be an accompanying decrease in entropy which is unfavorable. As a result, no hexane disappears, the average molecular packing-density remains constant, and the mixed volume is about equal to the sum of the starting volume.

Additional insight to what is happening in the system can be gained by considering the *total* volume change of the system for $X_H = 0.8$ compared to $X_H = 0$ rather than the volume change of the HCC. The total volume per phospholipid including lipid, water, and hexane is easily calculated from $S_X \cdot D/2$. For no hexane the value is 1675 Å^3, and for maximum hexane ($X_H = 0.8$) it is 1894 Å^3. The difference is 219 Å^3, which corresponds closely to the molecular volume of liquid hexane (217 Å^3). Therefore, for the *whole* system the volume change is about that expected for ideal volumetric mixed. How can this be explained? The simplest answer is that the organization of the water and polar groups must be changing as the hexane enters the HCC of the structure. One can imagine that as the hexane is increasing the packing-density of the HCC, the organization and packing of the polar groups and water are changing in a thermodynamically favorable way which must, to a large extent, involve an appropriate increase in entropy. Please note the following, however. In this calculation of total volume, we have considered only two values of X_H, zero and 0.8. Examination of Table I shows that the total volume ($S_X \cdot D/2$) is constant between $X_H = 0$ and 0.4 and $X_H = 0.6$ and 0.8. This means that over the full range of hexane concentrations, the partial molar volume of hexane is still behaving peculiarly even when the total volume is considered. Thus, whatever changes occur in the packing of the polar groups and water must be discontinuous ones.

Our data lead us to the following position. If we treat the HCC as an independent isotropic bulk system, we find a situation that is untenable with regard to our past experience with the mixing of bulk alkyl liquids. If, on the other hand, we consider the HCC as only one part of a highly nonisotropic system, the apparent intuitive catastrophe can be avoided to some extent. If we considered the lipid/water system as a "black box" and measured only its total volume at the extremes of hexane concentration, we would not find anything unusual. Our difficulty arises only when we have the ability, as we do, to look inside the box and measure independently a particular part of it. That part in the present experiments is the HCC. We thus conclude that the HCC cannot be considered independently of the interactions between the polar groups and water. It is inappropriate, then, to apply bulk phase analysis to the HCC as has so often been done in "molecular" theories of anesthesia.

DYNAMICS OF HEXANE/BILAYER MIXTURES

Methods

The general principles of calorimetry and NMR spectroscopy are more commonly known than those of neutron diffraction and will not be discussed in much detail here. The reader is referred to several excellent reviews on these subjects[1,5,11,13,18] and to Smith and Butler, this volume, Chapter 25. All of the NMR measurements were made on samples prepared with either the hexane or the lipid acyl chains specifically deuterated or perdeuterated. The principal measurement in all cases was of the quadrupole splitting ($\Delta\nu_q$) which is related to the order parameter S_{CD} of a particular C-D bond by the relation

$$\Delta\nu_q = (3/4)\ (e^2qQ/h)S_{CD}$$

where e^2qQ/h is the static quadrupole coupling constant: 167 kHz for $-CD_2-$ or 175 kHz for $=CD-$. $\Delta\nu_q$ or S_{CD} is a measure of the anisotropy of the group on the NMR time scale. If the motion is isotropic, a single NMR line is observed ($\Delta\nu_q = 0$). If the motion is anisotropic, then a doublet powder pattern is observed with a peak-to-peak separation of $\Delta\nu_q$ (see, for example, Fig. 12). Increases or decreases in the anisotropy will be accompanied by increases or decreases of the peak separation.

The samples for the measurements were prepared by depositing lipid from organic solvent over the inner surface of a glass flask, evaporating the solvent, and then equilibrating with water vapor and a particular hexane vapor pressure. After equilibration, the lipids were dispersed in excess water. Complete details can be found in Jacobs and White.[6,7] All the data presented here can be found in these papers as well.

Results

Phase Behavior: Calorimetry. The addition of hexane to DOPC and DMPC bilayers causes the main lipid phase transition endotherm to shift to lower temperatures and broaden as shown in Fig. 10. The enthalpies of the transitions are essentially unchanged in both cases. There is a second heat absorption event in the DMPC/hexane mixture at lower temperatures, but above 13 mole% hexane, their separation is an equivocal process. The total heat absorbed per mole of DMPC over the entire temperature range does not change appreciably on going from 5 to 26 mole% hexane. It should be noted that the so-called pretransition observed in pure DMPC disappears completely when even a small amount of hexane is added. The second heat absorption event should not be confused with the pretransition event. A plot of the onset and completion temperatures for the DMPC/hexane mixture is shown in Fig. 11. This is essentially the phase diagram of a monotectic mixture. The nearly horizontal

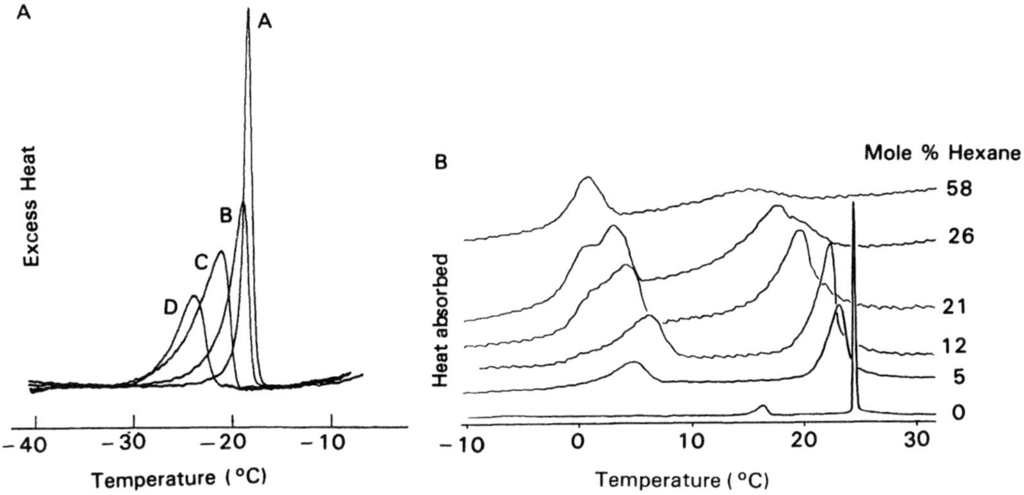

Figure 10. Differential scanning calorimeter (DSC) thermograms of DOPC (A) and DMPC (B) in excess water containing varying amounts of hexane. For A, the mole % of hexane is 0 (A), 3 (B), 27 (C), or 45 (D). For DOPC, the behavior is that expected if the hexane behaves as a simple impurity. The behavior of DMPC is much more complicated (see Fig. 11). A taken from Jacobs and White[7] and B from Jacobs and White.[6] Copyright (1984) American Chemical Society.

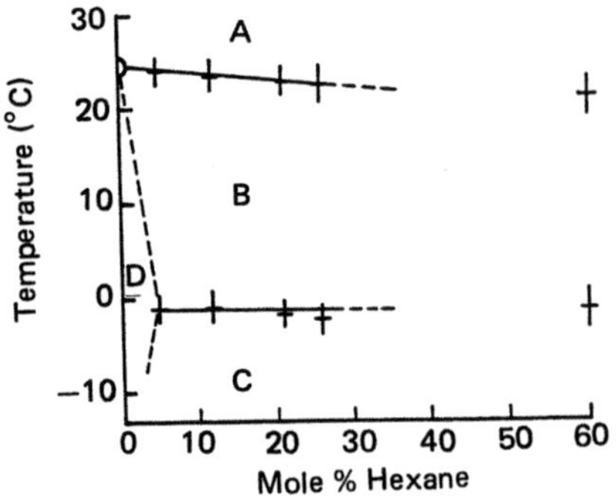

Figure 11. The phase behavior of DMPC/hexane mixtures in excess water. The phase diagram is that expected of a monotectic system. Taken from Jacobs and White.[6]

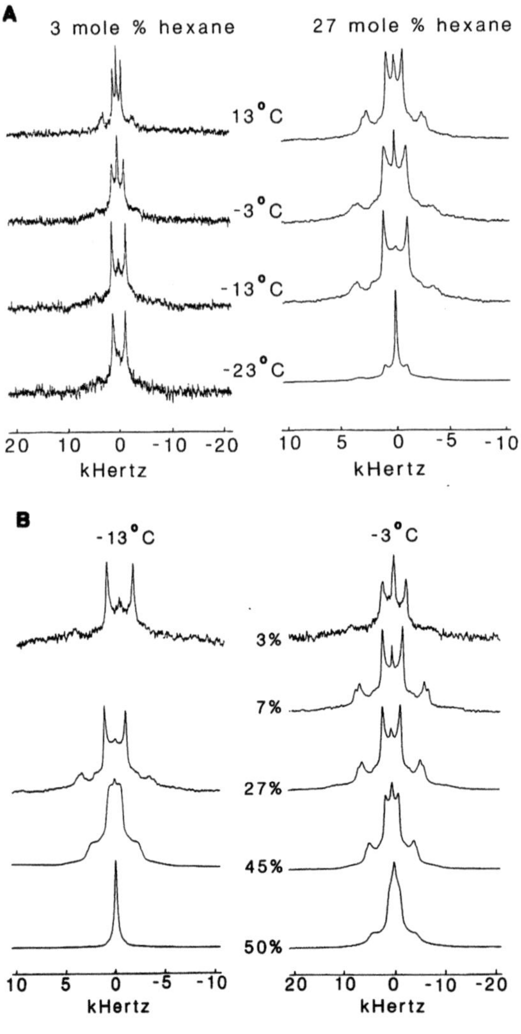

Figure 12. The ^{2}H–NMR spectra of perdeuterated hexane in DOPC liposomes at several temperatures and hexane concentrations. (A) Spectra taken at different temperatures (+13°C to −23°C) for liposomes containing 3 mole % hexane (left) and 27 mole % hexane (right). (B) Spectra made at −13°C (left) and −3°C (right) for liposomes containing (from top to bottom) 3, 7, 27, 45, and 50 mole % hexane. Taken from Jacobs and White.[7] Copyright (1984) American Chemical Society.

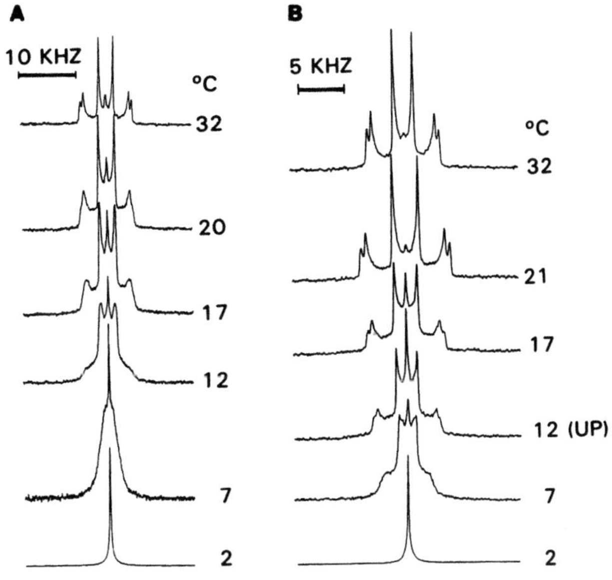

Figure 13. The ^{2}H–NMR spectra of perdeuterated hexane in DMPC liposomes at several temperatures and hexane concentrations. (A) Spectra made at temperatures ranging from 32°C to 2°C (top to bottom) for liposomes containing 3 mole % hexane. (B) Same conditions as in A except hexane present at 26 mole %. The spectrum for 12°C was made after the temperature had been first lowered to 2°C to show reversibility. Taken from Jacobs and White.[6] Copyright (1984) American Chemical Society.

straight line formed by the onset temperatures indicates that the mixture undergoes isothermal melting at about 0°C. This implies that at low hexane concentrations, the DMPC bilayer undergoes a phase separation at about 0°C. Hexane in DOPC bilayers, on the other hand, behaves as an impurity causing the bilayer phase transition endotherm to broaden and shift to lower temperatures. This behavior extends to equimolar amounts of incorporated hexane.

Motions of Hexane. Figures 12 and 13 show ^{2}H–NMR spectra of perdeuterated hexane dissolved in, respectively, DOPC and DMPC liposomal bilayers at a variety of concentrations and temperatures. The hexane ^{2}H–NMR powder patterns give direct information about the average order imposed on the hexane by the bilayer and, by inference, information about the types of anisotropic motion being executed. At relatively low concentrations and high temperatures relative to the pure lipid bilayer phase transition temperature (T_m), the spectra are composed of *three* overlapping powder patterns in an intensity ratio of 3:2:2. The powder pattern with the smallest quadrupole splitting is most intense. In these figures each -CD_2- and -CD_3 group contributes a powder pattern doublet to the spectra. At high hexane concentrations and low temperatures, a central isotropic peak emerges as the dominant spectral feature. Plots of $\Delta\nu_q$ as a function of temperature (Fig. 14) exhibit maxima at approximately the T_m of the pure bilayer. Figure 14 also shows that an increase in hexane content causes a decrease in $\Delta\nu_q$ for the hexane, but the effect is smaller for DOPC than for DMPC.

Motions of Acyl Chains. ^{2}H–NMR spectra of 1,2(,10-$^2H_2$) DOPC bilayers with 0 and 50 mole% hexane are shown in Fig. 15. Four overlapping powder patterns of equal intensity are observed (one from each of the olefinic deuterons), although the outer two patterns have almost the same $\Delta\nu_q$. The temperature dependence of the ^{2}H–NMR spectrum of DMPC with both acyl chains perdeuterated (per-^{2}H–DMPC) plus 26 mole% hexane is shown in Fig. 16. The general appearance of the spectra and the temperature dependence of the $\Delta\nu_q$'s in the deuterated lipid bilayer are *not* greatly affected by the addition of up to 50 mole % hexane to the DOPC bilayer and 26 mole % hexane to the DMPC bilayer. In the

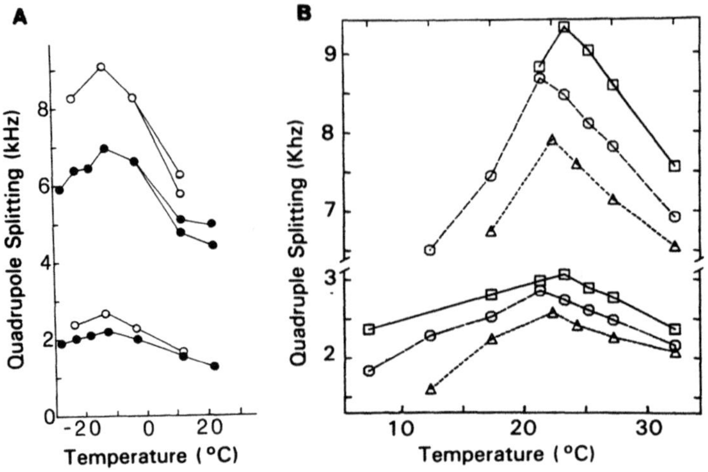

Figure 14. Plots of the innermost (lower traces) and outermost (upper traces) quadrupole splittings of perdeuterated hexane in DOPC (A) and DMPC (B) liposomes as a function of temperature for several hexane concentrations. (A) ○, 5 mole %; ●, 20 mole %. (B) □, 13 mole %; ○, 21 mole %; △, 26 mole %. A taken from Jacobs and White[7] and B from Jacobs and White.[6] Copyright (1984) American Chemical Society.

latter case, larger amounts of hexane result in the collapse of the powder patterns into an isotropic line. For the DOPC case, 50 mole% hexane was the maximum amount that could be introduced using our experimental protocol.

Discussion

Except for the lowest hexane concentrations, the temperature and concentration dependence of the ^{2}H–NMR spectra of perdeuterated hexane incorporated into DOPC and DMPC bilayers follows the same general course. At low concentrations of hexane in DOPC bilayers (less than 7 mole %), relatively sharp ^{2}H–NMR powder patterns are observed for the labeled hexane at all temperatures examined, even well into the gel phase of the bilayer. In the gel phase of the hexane/DMPC bilayer system at low hexane content, almost all of the intensity in the spectrum of the deuterated hexane is in a central isotropic peak. This behavior is

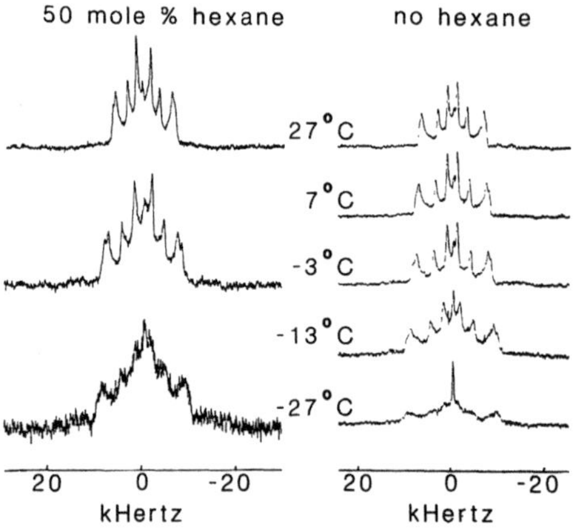

Figure 15. ^{2}H–NMR spectra of DOPC deuterated at the acyl chain double bonds as a function of temperature at two hexane concentrations. The spectra were recorded at temperatures ranging from +27 to −27°C in the presence of hexane (50 mole %, left) and in the absence of hexane (right). Taken from Jacobs and White.[7] Copyright (1984) American Chemical Society.

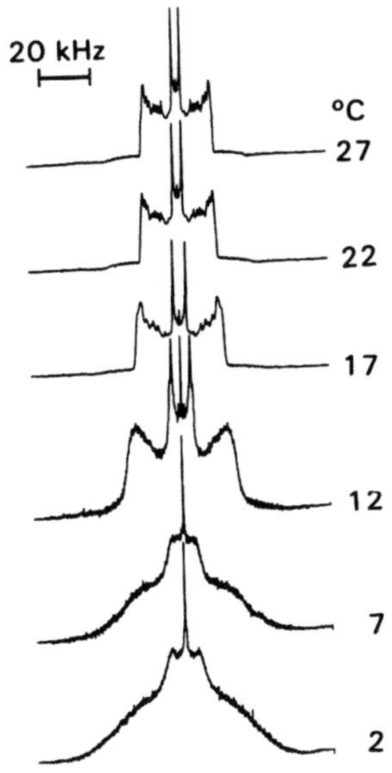

Figure 16. ^{2}H–NMR spectra of DMPC with perdeuterated acyl chains containing 26 mole % of hexane as a function of temperature. Taken from Jacobs and White.[6] Copyright (1984) American Chemical Society.

observed at higher hexane concentrations in the DOPC bilayer where an isotropic peak appears superimposed on the powder patterns as the sample temperature is lowered through its phase transition temperature (see Fig. 12). This effect is reversible. Raising the temperature after the powder pattern disappears causes it to reappear (see, for example, Fig. 14). Figure 12 reveals that at constant temperature the amount of isotropic hexane (as compared to the anisotropic component) increases with the total concentration of hexane. In both bilayers at very high hexane contents there is an appreciable isotropic component even in the liquid crystalline phase of the bilayer. In DMPC bilayers the appearance of a isotropic peak in the ^{2}H–NMR spectra appears to coincide with the phase separation within the bilayer as observed by DSC. This is not the case for the hexane/DOPC bilayer (no phase separation is observed in the DSC thermograms of this mixture), although in both systems the isotropic behavior is prevalent at low temperatures and high concentrations of the hexane.

The ^{2}H–NMR results presented here reveal that the motional characteristics of hexane dissolved in liquid crystalline and gel phase bilayers are quite different. The high temperature disordered phase imposes constraints on the hexane motional freedom resulting in anisotropic motion giving rise to the observed powder pattern lineshape in the ^{2}H–NMR spectra. In the more ordered gel phase, the principal feature of the spectra is a single resonance in the center of the spectra implying isotropic motion of the hexane component. There are two mechanisms by which this could take place.[7] The first mechanism involves rotational diffusion of the hexane molecules within the gel phase. In effect, they would no longer sense the presence of the surrounding anisotropic medium. This is a plausible mechanism considering that the order parameter of the hexane in the liquid crystalline bilayer is small ($\Delta\nu_q$ never larger than 10 kHz) and that the shape of the molecule is probably not far from spherical (the length-to-width ratio of all-*trans* hexane is only 2.4). The second mechanism involves rapid lateral diffusion of hexane through different bilayer orientations with respect to the external magnetic field. Calculations show that the correlation time for lateral diffusion is likely to be small enough in our system for this effect to be observed. We note that these two mechanisms are not mutually exclusive and that increasing the concentration of hexane in the bilayer would tend to increase both the translational and rotational diffusion constants toward those found in the pure alkane.

At temperatures above the T_m it is somewhat surprising that only three powder patterns are observed from per-^{2}H–hexane. If the hexane molecules align themselves parallel to the acyl chains, and if the order parameter gradient found in bilayers[12,14] is imposed on hexane, then one would expect the ^{2}H–NMR spectrum of per-^{2}H–hexane in this environment to contain six overlapping powder patterns. If one uses 1-^{2}H–hexane, only the single inner peak is observed, which identifies this peak as coming from the -CH_3 group. This observation, coupled with the observation of only three powder patterns, implies that both ends of the molecule have the same order parameter. This is probably a result of the general disorder of the bilayer and the hexane.

It is interesting that the phase behavior of hexane/DOPC and hexane/DMPC are so different. Clearly, the presence of the double bond has a large effect on the system behavior. It is also interesting that temperature and hexane concentration have a huge effect on the hexane motions but only small effects on acyl chain motions. If one were to use only acyl chain motions as a criterion for hexane (read anesthetic) effects, one might easily miss phenomena important for understanding the effect of hexane on the system properties.

CONCLUSIONS

The work reviewed here shows that the interaction of small hydrophobic molecules such as hexane and, quite likely, volatile anesthetics have very interesting and complex interactions with lipid bilayers. The most interesting of these is that the hexane can virtually disappear into the hydrocarbon core of the bilayer without any change in the hydrdocarbon volume! We do not yet know how widespread this effect is. Is it true of only certain combinations of lipid and alkane? Is it a peculiarity of only very low hydrations? If it depends on the detailed interaction of polar groups with water, can other molecules at the surface of the bilayer have a similar effect? If the phenomenon is widespread, how might lipid/protein interactions be affected? It seems, then, that more questions have been raised than answered. Until much more work is done, it will be impossible to rule out the participation of the lipid bilayer in general anesthesia. We cannot say at present that any of the phenomena we observe play a role in anesthesia. We can say, though, that the bilayer system is highly anisotropic and that it is inappropriate to treat the hydrocarbon core of the bilayer as an isolated bulk system. Any theory that makes these assumptions should be approached cautiously.

ACKNOWLEDGMENTS. This research was supported by the National Science Foundation and the National Institute of General Medical Sciences. During the early phases of the work, one of us (S.H.W.) was supported by a Research Career Development Award from the National Institute of Neurological Disease, Stroke, and Communicative Disorders. The neutron diffraction experiments were performed at Brookhaven National Laboratory on the High Flux Beam Reactor under the auspices of the Department of Energy with the additional support of the National Science Foundation. We thank Dr. Benno Schoenborn and his staff for their advice and encouragement. The NMR experiments were performed at the Southern California Regional NMR Facility at the California Institute of Technology. We appreciate the valuable assistance of Drs. Muller, Croasmun, and Chan.

REFERENCES

1. Davis JH: The description of membrane lipid conformation, order and dynamics by ^{2}H–NMR. *Biochim Biophys Acta* 1983;737:117–171.

2. Franks NP, Levine YK: Low-angle X-ray diffraction, in Grell E (ed): *Membrane Spectroscopy*. Berlin, Springer-Verlag, 1981, pp 437–487.
3. Franks NP, Lieb WR: Where do general anaesthetics act? *Nature* 1978;274:339–342.
4. Franks NP, Lieb WR: The structure of lipid bilayers and the effects of general anaesthetics. *J Mol Biol* 1979;133:469–500.
5. Jacobs RE, Oldfield E: NMR of membranes. *Prog Nuc Magn Reson Spectrosc* 1981;14:113–136.
6. Jacobs RE, White SH: Behavior of hexane dissolved in dimyristoylphosphatidylcholine bilayers: An NMR and calorimetric study. *J Am Chem Soc* 1984a;106:915–920.
7. Jacobs RE, White SH: Behavior of hexane in dioleoylphosphatidylcholine bilayers: An NMR and calorimetric study. *J Am Chem Soc* 1984b;106:6909–6913.
8. King GI, Chao N-M, White SH: Neutron diffraction studies on incorporation of hexane into oriented lipid bilayers, in Schoenborn BP (ed): *Neutrons in Biology*. New York, Plenum Press, 1984, pp 159–172.
9. King GI, Jacobs RE, White SH: Hexane dissolved in dioleoyl lecithin bilayers has a partial molar volume of approximately zero. *Biochemistry* 1985;24:4637–4645.
10. Levine, YK, Wilkins MHF: Structure of oriented lipid bilayers. *Nature New Biol* 1971;230:69–72.
11. Mabrey S, Sturtevant JM: High-sensitivity differential scanning calorimetry in the study of biomembrane and related model systems, in Korn ED (ed): *Methods in Membrane Biology,* Vol 9. New York, Plenum Press, 1978, pp 237–274.
12. McFarland BG, McConnell HM: Bent fatty acid chains in lecithin bilayers. *Proc Natl Acad Sci USA* 1971;68:1274–1278.
13. Melchior DL, Steim JM: Thermotropic transitions in biomembranes. *Annu Rev Biophys Bioeng* 1976;5:205–238.
14. Oldfield E, Meadows M, Rice D, Jacobs R: Spectroscopic studies of specifically deuterium labeled membrane systems. Nuclear magnetic resonance investigation of the effects of cholesterol in model systems. *Biochemistry* 1978;17:2727–2740.
15. Requena J, Haydon DA: Van derWaals forces in oil-water systems from the study of thin lipid films II. The dependence of the Van der Waals free energy of thinning on film composition and structure. *Proc R Soc Lond* 1975;A347:161–177.
16. Schoenborn BP (ed): *Neutron Scattering for the Analysis of Biological Structures*. Springfield, VA, National Technical Information Service. 1976.
17. Schoenborn, BP (ed): *Neutrons in Biology*. New York, Plenum Press, 1984.
18. Seelig J: Deuterium magnetic resonance: Theory and application to lipid membranes. *Q Rev Biophys* 1977;10:353–418.
19. Singer SJ, Nicolson GL: The fluid mosaic model of the structure of cell membranes. *Science* 1972;175:720–731.
20. Waldbillig RC, Szabo G: Solvent-depleted bilayer membranes from concentrated lipid solutions. *Nature* 1978;272:839–840.
21. White SH: Studies of the physical chemistry of planar bilayer membranes using high-precision measurements of specific capacitance. *Ann NY Acad Sci* 1977;303:243–265.
22. White SH: Mechanism of the compression of black lipid membranes by an electric field. *Biophys J* 1979;25:9a.
23. White SH, King GI, Cain JE: Location of hexane in lipid bilayers determined by neutron diffraction. *Nature* 1981;290:161–163.

24

The Influence of Anesthetics on the Structure and Thermal Properties of Saturated Lecithins

Sidney A. Simon, Thomas J. McIntosh, and Michael L. Hines

Abstract. The gel to liquid crystalline phase transition temperature of lipid bilayers is often measured as a function of anesthetic concentration to obtain information on the membrane–water partition coefficient of the anesthetic. For several small amphiphilic molecules, such as ethanol, methanol, benzyl alcohol, and chlorpromazine, there is a sharp discontinuity in the plot of lipid transition temperature versus anesthetic concentration. In this chapter we show that this discontinuity is caused by the anesthetic inducing a structural change in the gel state bilayer, converting the normal bilayer gel phase to one where the lipid acyl chains from opposing monolayers fully interpenetrate or interdigitate. Our analysis shows that the partition coefficients for these anesthetics are larger for the interdigitated gel phase than for the usual bilayer gel phase. This is due to the fact that the interdigitated phase has twice the number of acyl chains at the bilayer/water interface and thus has more "sites" where amphiphilic anesthetic can bind to the membrane. For the case of ethanol, the partition coefficient in the interdigitated gel phase is larger than in the liquid crystalline phase, and the transition temperature actually increases with increasing anesthetic concentration.

INTRODUCTION

In the late 19th century, it was postulated that anesthetic potency is related to the solubility of the anesthetic in an oil phase.[20] This lipid solubility hypothesis has withstood the test of time, and as Franks and Lieb[4] have recently pointed out, "Although there have been many

Symbols used in this chapter: T, temperature; T_m, lipid gel to liquid crystalline transition temperature; ΔE_{HC}, energy cost exposing lipid terminal methyl groups to water; ΔE_{VdW}, Van der Waals energy per CH_2 group; n, number of CH_2 groups per molecule; ΔE, energy difference between normal gel state bilayer and interdigitated phase; N, number of nearest neighbors; γ, interfacial tension at bilayer–water interface; ΔA, surface area of two methyl groups; N_A, Avogadro's number; α, extent of conversion of lipid from gel to liquid crystalline phase; ΔH_{vH}, van't Hoff enthalpy; ΔH_m, excess calorimetric enthalpy; X_2,c, mole fraction of component 2 in the liquid crystalline phase; $X_{2,g}$, mole fraction of component 2 in the gel phase; K, $X_{2,g}/X_{2,1c}$; M_2, molecular weight of component 2; W_2, concentration of component 2 in mg/ml; K_d, lipid-water partition coefficient in the liquid crystalline phase in mole fraction units; C_{ex}, excess molar heat-capacity; $L\alpha$, liquid crystalline bilayer phase; $L_{\beta'}$, gel bilayer phase.

Sidney A. Simon • Department of Physiology and Anesthesiology, Duke University Medical Center, Durham, North Carolina 27710. *Thomas J. McIntosh* • Department of Anatomy, Duke University Medical Center, Durham, North Carolina 27710. *Michael L. Hines* • Department of Physiology, Duke University Medical Center, Durham, North Carolina 27710.

attempts to correlate potency with other physical properties of the anaesthetic molecule, none of them has been as complete or successful as the correlation with partition coefficients.'' From an analysis of the effect of anesthetics on the lipid gel to liquid crystalline transition temperature, T_m, the membrane water partition coefficient in both the gel and liquid crystalline phases can be obtained.[13] Some anesthetics, such as halothane and chloroform, simply lower T_m with increasing concentration indicating, from freezing point depression analysis, that the anesthetics are more soluble in the liquid crystalline phase than the gel phase.[6] However, it has recently been reported[9,26] that several amphipathic anesthetics, such as methanol, ethanol, and butanol, have a more complicated ''biphasic'' effect on T_m. That is, in plots of T_m versus anesthetic concentration, there are two regions of the curve with different slopes. The reason for the sudden breaks in these biphasic T_m versus anesthetic concentration curves is not apparent from thermodynamic considerations alone, although it might be surmised that an increase in the membrane–water partition coefficient in the gel phase could account for an increase in T_m. In this chapter, we will show by X-ray diffraction analysis that many common amphipathic anesthetics, such as ethanol and chlorpromazine, induce the acyl chains in gel phase saturated lecithins to fully interpenetrate or interdigitate at the same concentrations where the breaks occur in the biphasic T_m versus anesthetic concentration curves. Moreover, we will demonstrate that the structure of this interdigitated lipid phase, where the terminal methyl groups of one monolayer abut the interfacial region of the opposing monolayer, can explain the biphasic nature of the T_m versus anesthetic concentration curves, as well as other properties of these phase transitions.

INDUCTION OF THE INTERDIGITATED PHASE CAUSES BIPHASIC TRANSITION BEHAVIOR

Figure 1A shows a graph of the shift in the main endothermic transition temperature, T_m, of L-α-dipalmitoyl phosphatidylcholine (DPPC) as a function of mole percent of chlorpromazine. These data are taken from the thermograms of Frenzel *et al.*[5] As the mole percent

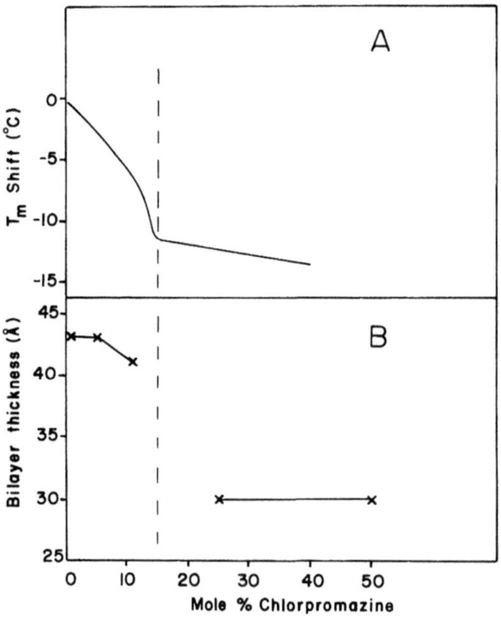

Figure 1. (A) Shift in the transition temperature, T_m, for DPPC dispersions as a function of mole percent chlorpromazine. The calorimetric data were taken from Frenzel *et al.*[5] and are reproduced with permission. (B) Bilayer thickness for gel state DPPC ($T = 20 \pm 1$°C) as a function of mole percent chlorpromazine. The bilayer thickness was obtained from X-ray diffraction data and represents the phosphate-to-phosphate distance as determined from electron density profiles generated from diffraction patterns using the phase angles given in McIntosh *et al.*[19]

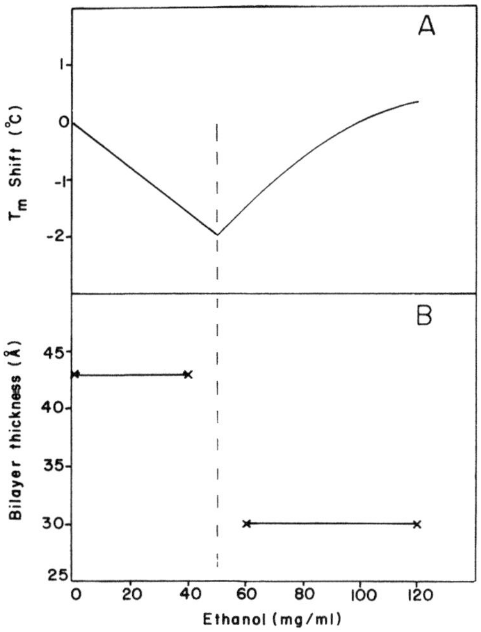

Figure 2. (A) Shift in transition temperature, T_m, of DPPC dispersions as a function of aqueous ethanol concentration. These data were taken from Rowe.[26] (B) Bilayer thickness for gel state DPPC as a function of ethanol concentration. The bilayer thickness is reported as the phosphate-to-phosphate distance.

increases, the transition temperature rapidly decreases until about 15 mole % chlorpromazine, where there is a sudden break in slope in the curve. Figure 1B shows a plot of the bilayer thickness, taken as the lipid head group separation in electron density profiles (see below), for gel phase DPPC in the presence and absence of chlorpromazine. There is a sudden decrease in bilayer thickness at the chlorpromazine concentrations where the change in slope in the T_m curve occurs (Fig. 1A).

Figure 2A shows the shift in T_m for DPPC as a function of ethanol concentration as determined by the absorbance data of Rowe.[26] The transition temperature, T_m, decreases as the ethanol concentration increases until a concentration of about 50 mg/ml (1.1 M) is reached. With the further addition of ethanol, T_m increases linearly until saturation occurs at about 100 mg/ml (2.2 M). Figure 1B shows the dependence of the thickness of gel state DPPC bilayers with ethanol concentration. For ethanol concentrations below 40 mg/ml (0.9 M), the bilayer thickness is about 43 Å, and at 60 and 120 mg/ml (1.3 and 2.6 M, respectively), the bilayer thickness is 30 Å. Thus, for both chlorpromazine and ethanol, a structural change in the gel phase correlates with the break in the plot of T_m versus anesthetic concentration. However, above the transition temperature, in the $L\alpha$ phase, neither ethanol[29] nor chlorpromazine[19] significantly changes the thickness of the bilayer.

To determine the structural changes of the gel phase in the presence of chlorpromazine or ethanol, it is necessary to examine the wide-angle reflections as well as the low-angle reflections. Figure 3 shows a densitometer tracing of X-ray diffraction patterns of DPPC in ethanol solutions of 40 mg/ml ethanol and 60 mg/ml ethanol at 20°C. At 40 mg/ml, there are four lamellar reflections of a repeat period of 64 Å and a double wide-angle reflection consisting of a sharp reflection at 4.21 Å and a broad band at 4.1 Å. The lamellar repeat period and the double wide-angle reflection are identical to those seen in gel phase DPPC in the absence of ethanol.[17,31] Thus, ethanol, at concentrations less than 50 mg/ml, does not affect the structure of DPPC either above or below T_m. The double reflections at 4.1 and 4.21 Å indicate that the acyl chains are tilted about 30° with respect to the bilayer plane and each

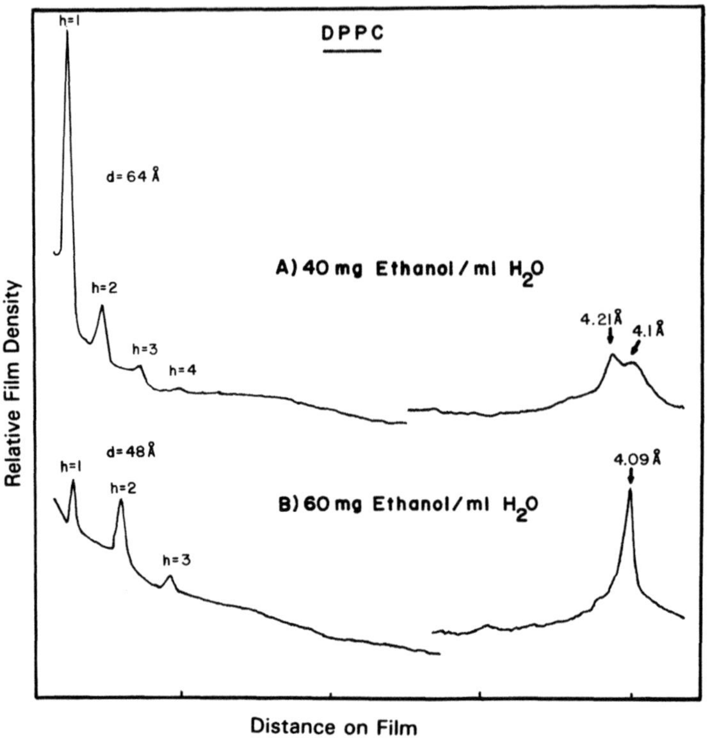

Figure 3. Densitometer traces of X-ray diffraction patterns of DPPC in solutions containing ethanol, (A) 40 mg/ml (0.9 M) and (B) 60 mg/ml (1.3 M). The patterns were recorded at 20°C. In trace A there are four orders of a lamellar repeat period of 64 Å and a double wide-angle reflection at 4.21 Å and 4.10 Å. In trace B there are three orders of a lamellar repeat period of 48 Å and a single wide-angle reflection at 4.09 Å. This figure is taken from Simon and McIntosh[29] and is reproduced with permission.

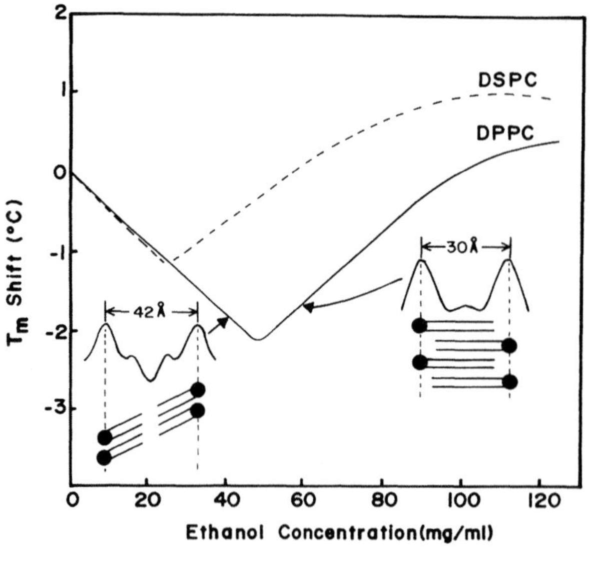

Figure 4. Change in transition temperature for suspensions of distearoylphosphatidylcholine (DSPC) (dashed line) and DPPC (solid line) as a function of ethanol concentration. The break in the DPPC curve occurs at an ethanol concentration of 50 mg/ml (1.1 M). Also shown are electron density profiles for DPPC in the $L_{\beta'}$ phase at 40 mg/ml (0.9 M) ethanol and in the interdigitated phase at 60 mg/ml (1.3 M) ethanol. The electron density profiles were generated from the diffraction patterns given in Fig. 3. Below each profile is a schematic view of the structure, with the black dots representing the lipid headgroups and the straight lines representing the lipid hydrocarbon chains. This figure was taken from Simon and McIntosh[29] and is reproduced with permission.

hydrocarbon chain has four neighbors at 4.86 Å and two neighbors at 4.73 Å.[31] The electron density profiles for gel state DPPC at 40 mg/ml ethanol (Fig. 4) show two high-density peaks, which correspond to the lipid head groups, surrounding a lower-density hydrocarbon chain region. The sharp trough in the center of the profile corresponds to the localization of terminal methyl groups in the geometrical center of the bilayer. The thickness of the bilayer, as measured by the distance between head group peaks, is about 43 Å. This distance is consistent with fully extended chains tilted about 30° relative to the bilayer surface.[31] At an ethanol concentration of 60 mg/ml, which is in the region of the T_m versus ethanol curve that has a positive slope (Fig. 2A), there are three lamellar orders of a 48 Å repeat period and a single sharp wide-angle reflection at 4.09 Å (Fig. 3B). This indicates that each acyl chain is oriented perpendicular to the membrane plane and has six nearest neighbors at 4.72 Å.[31] The electron density profile for gel state DPPC at 60 mg/ml ethanol (Fig. 4) shows the two high-density head group peaks separated by only 30 Å. Moreover, there is no terminal methyl trough in the center of the bilayer. Since the hydrocarbon chains are perpendicular to the plane of the membrane, it follows that the acyl chains must be interdigitated, as shown schematically in Fig. 4 (see also Ranck *et al.*[24,25] and McIntosh *et al.*[19]). Analysis of similar X-ray data shows that the 30-Å-wide bilayer containing chlorpromazine (Fig. 1B) also is an interdigitated phase.[19]

If the chains in DPPC were fully extended and perpendicular to the membrane phase, the bilayer thickness would be about 49 Å.[17,31] In the presence of larger concentrations of chlorpromazine or ethanol, the bilayer thickness is 30 Å. This difference of 19 Å is equivalent to about 15 CH_2 groups in the all trans configuration, since the C–C bond length is 1.25 Å. This means that in the interdigitated phase the acyl chains from one monolayer have penetrated up to the first or second CH_2 group on the opposing monolayer.

MOLECULAR REQUIREMENTS FOR INTERDIGITATION OF DPPC

It is of interest to determine what characteristics molecules must possess for them to be able to induce the interdigitated phase in DPPC. In addition to the two molecules discussed previously, we have shown that the interdigitated phase is induced by methanol, ethylene glycol, glycerol,[16] benzyl alcohol,[1] phenyl butanol, 1-anilino-8-napthalene sulfonate (ANS), and tetracaine.[19] The first point to note is that a molecule's ability to induce the interdigitated phase does not depend on the presence or absence of a formal charge, since ANS is negatively charged, chlorpromazine is positively charged, and benzyl alcohol is uncharged. Second, all these molecules are amphipathic. The addition of molecules that are totally hydrophobic such as hexane, tetradecane, or benzene does not induce the interdigitated phase.[3,15,18,33] Thus, it appears that one requirement for the induction of the interdigitated phase is that the added molecules must reside at the membrane–water interface. However, not all amphipathic molecules will induce chain interdigitation, as the interdigitated phase will not form in the presence of large concentrations of long-chain saturated fatty acids[23] or cholesterol. Consequently, to induce the interdigitated phase, a molecule must not extend too deeply into the acyl chain region. Evidently, the reason that small amphipathic molecules can induce chain interdigitation when they intercalate between the lipid molecules is that they tend to separate the lipid head groups, and since the gel state acyl chains are rigid, this would tend to form voids (or defects) between acyl chains. The formation of voids in hydrocarbons is energetically unfavorable,[8] and consequently the chains must minimize their free energy. In this gel phase the chains could bend cooperatively or else interdigitate. For the several bulky amphiphilic molecules listed previously, the lowest energy phase is the interdigitated phase (as will be discussed in the next section). In the liquid crystalline phase, the fluid acyl chains can accommodate these small molecules without forming voids, and hence the interdigitated phase is not formed.

STABILITY OF THE INTERDIGITATED PHASE

The structural evidence shows that the interdigitated phase is stable in the presence of various anesthetics. In this section we present calculations to show why this phase is energetically favorable. Since the $L\alpha$ phase appears unchanged in the presence of these anesthetics, we must compare the energetics of the usual gel phase, commonly called the $L\beta'$ phase,[31] and the interdigitated phase. In a transition from the $L\beta'$ to the interdigitated phase there is a gain in the Van der Waals energy, since the acyl chains are packed closer together in the interdigitated phase, as evidenced by the wide-angle diffraction patterns seen in Fig. 3. However, there is an energy cost, ΔE_{HC}, going to the interdigitated phase from the $L\beta'$ phase because of the exposure of the terminal methyl groups to water. The energy due to the formation of rotational isomers (or kinks) should be similar in the two phases. Other terms, such as the change in the interaction between the dipoles in the polar head group, should be small compared to the Van der Waals energy and ΔE_{HC}.[22] Thus, the major difference in energy, ΔE, between the $L\beta'$ phase and the interdigitated phases can be written

$$\Delta E = n\Delta E_{\mathrm{VdW}} + \Delta E_{HC} \tag{1}$$

where ΔE_{VdW} is the Van der Waals energy per CH_2 group and n is the number of CH_2 groups per molecule.

From equation (1) it is evident that the stability of the interdigitated phase will depend on the chain length of the lipid molecule since the van der Waals energy is proportional to the number of CH_2 groups per hydrocarbon chain and ΔE_{HC} is independent of the acyl chain length. That is, the longer the lipid chain, the more energetically favorable the interdigitated state becomes. Therefore, the interdigitated phase should form with lower ethanol concentrations for longer-chained lipids. Indeed, the data of Rowe[26] clearly demonstrate this trend (also see Fig. 4).

To calculate ΔE_{VdW}, we follow the analysis of Nagle and Wilkinson[22] and write

$$\Delta E_{\mathrm{VdW}} = \left[\frac{2.03}{4}\right]\left[\frac{N}{6}\right]\left[\left[\frac{4.34}{r}\right]^{25} - 5\left[\frac{4.34}{r}\right]^{5}\right] \tag{2}$$

where N is the number of nearest neighbors at distance r. When r is in Å, E_{VdW} is in kcal/mole per CH_2. Substituting in the values obtained from the wide-angle spacing given previously for both the $L\beta'$ and interdigitated phases, we calculate $\Delta E_{\mathrm{VdW}} = -0.13$ kcal/mole per CH_2. For the particular case of DPPC, n (ΔE_{VdW}) = −4.25 kcal/mole. The second term on the right-hand side in equation (1) can be written $\Delta E_{HC} = \gamma \Delta A\, N_A$, where γ is the interfacial tension at the membrane–water interface, which has been estimated to be between 30 and 50 dynes/cm,[2,8] ΔA is the surface area of two methyl groups, which is about 64 Å^2,[29] and N_A is Avogadro's number. Thus, for all lipids that form the interdigitated phase, ΔE_{HC} is between 2.8 and 4.6 kcal/mole depending on the chosen value of γ.

This simple analysis shows how the induction of the interdigitated phase depends on lipid chain length and, for DPPC at least, that the Van der Waals energy gained is comparable to the energy lost exposing hydrocarbons to water upon interdigitation.

ANALYSIS OF PHASE TRANSITIONS THAT INVOLVE THE INTERDIGITATED PHASE

Induction of the interdigitated phase leads to a break in graphs of T_m versus anesthetic concentration as seen in Figs. 1A and 2A. In this section we will analyze the transition behavior of saturated lecithins in the presence of ethanol using the data of Rowe.[26] We will begin by analyzing the behavior of diheptadecylphosphatidylcholine (diPC17:0), since Dr. Rowe has kindly provided us with her published raw absorbance data of diC17:0 in the presence and absence of ethanol. The analysis presented for diC17:0 is also applicable for the other saturated lecithins.

We will follow the analysis of Sturtevant,[30] who used ideal solution theory to generate excess specific heat curves in the presence of various water-soluble molecules. Thus,

$$\frac{T_m - T}{RT_mT} = \left[\frac{1}{\Delta H_{\mathrm{vH}}} \ln\left[\frac{1-\alpha}{\alpha}\right] - \frac{\ln X_{1,1c}}{\Delta H_m}(1-K)\right] \tag{3}$$

where α is the extent of the conversion of the lipid in going from the gel to the liquid crystalline phase, T_m is the phase transition temperature of the pure lipid at a point where the transition is half completed (i.e., $\alpha = 1/2$), ΔH_{vH} is the van't Hoff enthalpy, ΔH_m is the excess calorimetric enthalpy, $X_{1,1c}$ is the mole fraction of lipid (component 1) in the lipid crystalline phase, and K is the ratio of mole fraction of anesthetic (component 2) in the gel phase, $X_{2,g}$, to mole fraction of anesthetic in the liquid crystalline phase, $X_{2,c}$ (note that $X_{1,1c} + X_{2,1c} = 1$). ΔH_{vH} can be obtained from the raw absorbance data using the equation

$$\Delta H_{\mathrm{vH}} = \frac{\partial \ln\left[\frac{1-\alpha}{\alpha}\right]}{R\partial(1/T)} \tag{4}$$

The value of $X_{2,1c}$ is not a constant but depends on several variables including the concentration of lipid and anesthetic. Previously, Kamaya *et al.*[11] showed that the absolute value of the depression in the transition temperature depends on the lipid concentration. The expression $X_{2,1c}$ is

$$X_{2,1c} = \frac{W_2}{\frac{M_2}{M_1}\left[\frac{55.6\,M_1}{K_d - X_{2,1c}} + \frac{(1-\alpha)KW_1}{1-KX_{2,1c}} + \frac{\alpha W_1}{1-X_{2,1c}}\right]} \tag{5}$$

where M_2 and M_1 are the molecular weights of the anesthetic and lipid, respectively, W_2 and W_1 are the concentration of the anesthetic and lipid (in mg/ml), respectively, and K_d is the lipid–water partition coefficient in the *liquid crystalline phase* in mole fraction units; i.e., $K_d = X_{21c}/X_w$. The range of α is from 0 to 1, with the value 0 meaning that the lipid is entirely in the gel phase and 1 meaning it is entirely in the liquid crystalline phase. Values of α can be calculated from spectroscopic measurements following the analysis outlined by Inoue *et al.*[7]

We shall now proceed to generate the T_m versus ethanol concentration curve for diC17:0 as experimentally determined by Rowe.[26] The experimental data seen in Fig. 5 are shown by x's. The solid line was obtained from equation (3) by using separate curves for the low- and high-ethanol-concentration ranges. We used the parameters listed in the figure caption, all of which were measured except for K. Thus K was manually adjusted to produce a reasonable

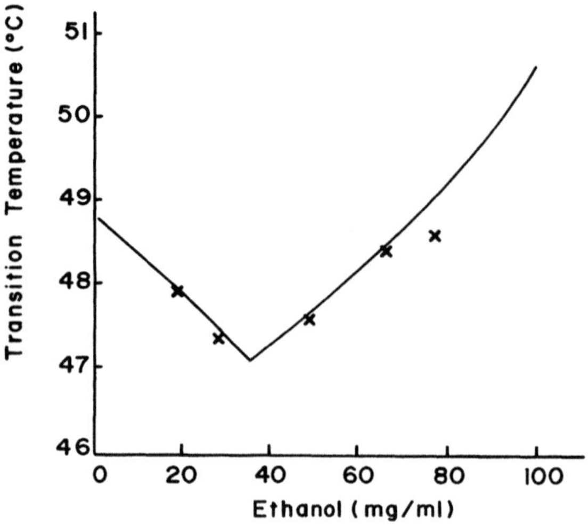

Figure 5. Transition temperature for diheptadecylphosphatidyl choline (diC17:0) as a function of ethanol concentration. The x's represent the data points obtained from the spectrophotometric data of Rowe,[26] whereas the solid lines are theoretical fits to the data generated from equations (3) and (5). The parameters used in generating these curves are: descending limb; T_m = 48°C, ΔH_m = 8800 cal/mole; W_1 = 0.5 mg/ml (from Rowe[26]), $\Delta H_{vH} = 2 \times 10^6$ cal/mole as calculated from the absorbance data using equation (4), M_1 = 762, M_2 = 46, K_d = 16.5 (Katz and Diamond[12]), and K = 0.72. Ascending limb: T_m = 45.9°C, ΔH_m = 10000 cal/mole, W_1 = 0.5 mg/ml, M_1 = 762, M_2 = 46, K_d = 16.5, $\Delta H_{vH} = 2.35 \times 10^6$ cal/mole, and K = 1.24. The value of T_m was obtained by extrapolating the ascending limb to zero ethanol. The value of ΔH_{vH} was obtained from the raw absorbance data using equation (4).

fit to the data. For the decending limb of the curve $K = 0.72$ whereas for the ascending limb $K = 1.24$

With these values of K, the entire T_m versus ethanol concentration curve (Fig. 2A) can now be satisfactorily explained. In the descending region of the curve, the partition coefficient in the liquid crystalline phase is larger than in the gel phase ($K = 0.72$), and a freezing point depression results. However, at a critical concentration of ethanol, the interdigated phase is induced which increases the number of acyl chains per head group at the interface by a factor of 2. Thus, there are more "sites" at the interface for ethanol to "bind," and the partition coefficient of ethanol in the gel state increases. In fact, it increases so much that K becomes greater than it is in the liquid crystalline phase causing the slope of the T_m versus ethanol curve to change sign. At high ethanol concentrations, the T_m versus ethanol curve levels off, probably owing to saturation of the gel state binding "sites." Note that equations (3) and (5) do not include a term to account for the saturation effect, so a discrepancy between theory and experiment is expected at the high concentrations as seen in Fig. 5. Thus, we explain the DPPC–ethanol transition curve in the low ethanol region as a "freezing point depression effect" and in the higher ethanol region, where T_m increases with increasing ethanol concentration, as an increase in K as a result of chain interdigitation. Should this explanation be correct, then one would expect that, for a lipid that normally exists in the interdigitated phase in excess water, the addition of ethanol should monitonically increase T_m. To test this hypothesis, we added ethanol at various concentrations to β,γ-dihexadecylphosphatidyl choline (DHPC), an ether-linked analogue of DPPC that exists in the interdigitated phase in excess water at 20°C.[28] Figure 6 shows that the addition of ethanol to DHPC does indeed increase T_m. The value of T_m for DHPC in water we obtained is 43.2°C, which is in agreement with previously published values that range between 42 and 44.7°C.[14,32] The theoretical line was calculated by using the parameters listed in the figure caption and by varying K. A good fit was obtained with a value of $K = 1.15$. This is in excellent agreement with our value of $K = 1.24$ for the ascending region of the T_m versus ethanol curve for DPPC. Again at high ethanol concentrations, the theoretical curve diverges from the experimental data as a result of saturation of the binding "sites."

The behavior of the DPPC–chlorpromazine phase transition curves seen in Fig. 1 can

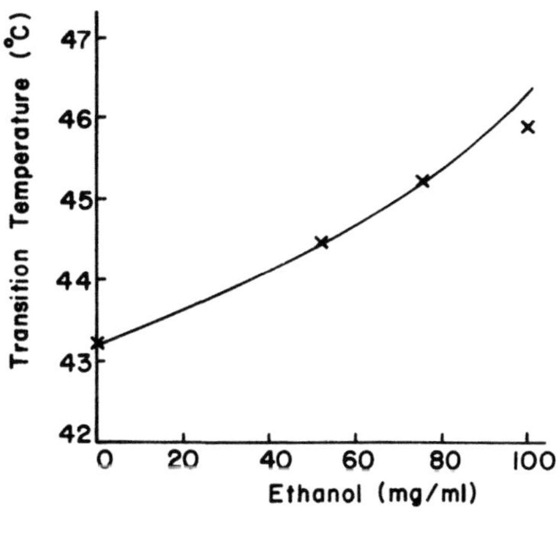

Figure 6. Transition temperature of β,γ-dihexadecylphosphatidylcholine (DHPC) as a function of ethanol concentration. These data (x's) were obtained spectrophotometrically by measuring the absorbance at 400 nm of a 0.5 mg/ml dispersion as a function of temperature. The heating rate was about 1°C/min. The value of T_m for DHPC was 43.2°C. The theoretical curve was obtained using equations (3) and (5) and the following values: T_m = 43.2°C, ΔH_m = 9500 cal/mole (Vaughan and Keough,[32] Lee and Fitzgerald[14]), $\Delta H_{vH} = 2.8 \times 10^6$ cal/mole, K_d = 16.5 (Katz and Diamond[12]), W_1 = 0.5 mg/ml, M_1 = 732, and M_2 = 46.

also be modeled with separate curves for the low- and high-concentration ranges. This fitting procedure was more difficult in the case of chlorpromazine, since neither K_d nor K has been directly measured in DPPC and cannot be uniquely determined from the given data. We therefore selected a reasonable value for K and fit the initial slope by varying K_d. The fit to the shift in transition temperature data is shown in Fig. 7. A reasonable fit for the descending region of the curve was obtained for $K_d = 271$ and $K = 0.8$. This value of K_d is reasonable if one extrapolates the partition coefficient of chlorpromazine in red blood cells reported by Seeman[27] to that of pure lipid bilayers above their transition temperature.[10] In the higher chlorpromazine region of the curve, K_d was kept at 271 and K was adjusted to 0.99 to fit the curve. That is, the change in slope was caused by an increase in binding of chlorpromazine in the interdigitated phase.

ANALYSIS OF EXCESS HEAT CAPACITY

For both chlorpromazine[5] and ethanol[26] the transition width, which is related to the transition cooperativity,[30] is narrower in the high-concentration region. In this section we will calculate the molar excess heat capacity and show that this sharpening of the transition is a direct consequence of the induction of the interdigitated phase.

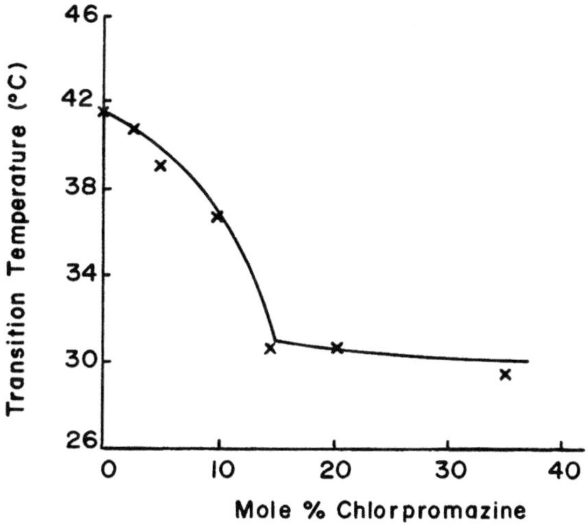

Figure 7. Graph of change in T_m for DPPC in the presence of chlorpromazine. The crosses (x) are the experimental points taken from Frenzel *et al.*[5] The solid lines are obtained from equations (3) and (5). In the region from 0 to 15 mole % chlorpromazine the following parameters were used: T_m = 41.6°C, ΔH_m = 8800 cal/mole, W_1 = 1.0 mg/ml, M_1 = 734, M_2 = 319, $\Delta H_{vH} = 2 \times 10^6$ cal/mole, K_d = 271, and K = 0.8. For the 15–40 mole percent region the following parameters were used: T_m = 31.75°C, ΔH_m = 6000 cal/mole, W_1 = 1.0 mg/ml, M_1 = 734, M_2 = 319, K_d = 271, K = 0.99, and $\Delta H_{vH} = 2.35 \times 10^6$ cal/mole. The value of T_m = 41.75°C was obtained by extrapolating the T_m curve in this region to 0% chlorpromazine. The values of ΔH_m were taken from Frenzel *et al.*[5]

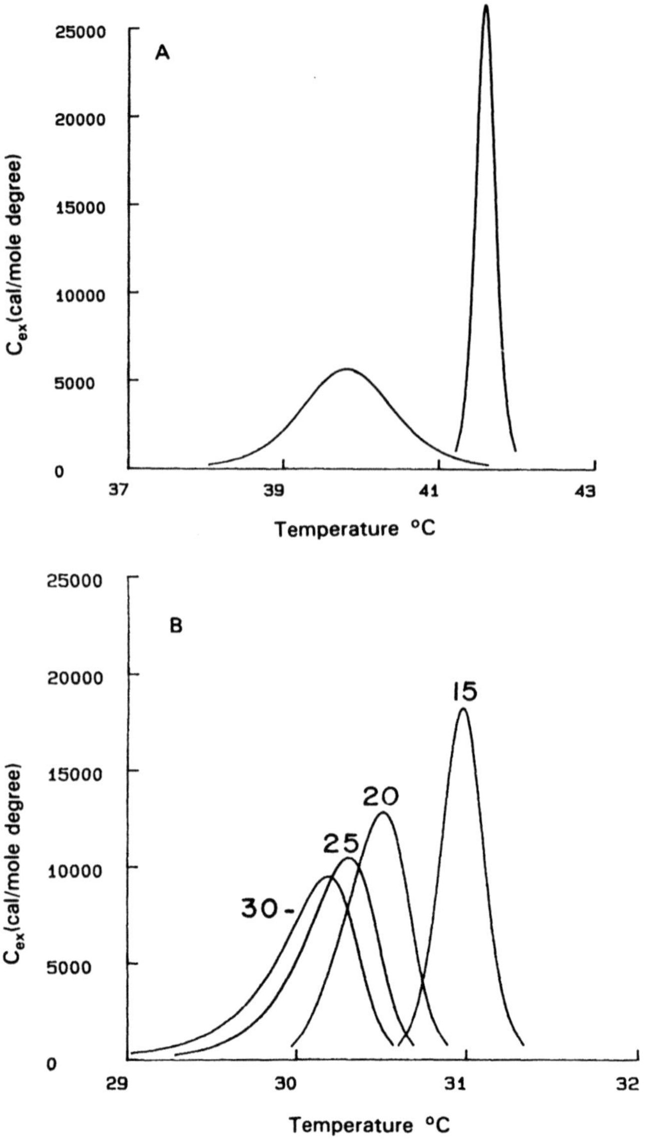

Figure 8. Excess molar specific heat, C_{ex} versus temperature for (A) DPPC in excess water and in the presence of 5 mole % chlorpromazine. C_{ex} for DPPC was calculated from equation (6) using the same values given for the 0–15 mole % region given in Fig. 7. The fit for 5 mole % was obtained by changing ΔH_{vH} from 2×10^6 to 5×10^5 cal/mole. This change in ΔH_{vH} does not change T_m but broadens the transition. (B) DPPC in the presence of 15, 20, 25, and 30 mole % chlorpromazine. The values used to obtain these figures are identical to those given in Fig. 7 in the 15–40 mole % range.

The excess molar heat capacity, C_{ex}, can be obtained[30] using the equation

$$C_{ex} = \Delta H_m \left(\frac{d\alpha}{dt} \right) \tag{6}$$

Graphs of C_{ex} versus temperature for DPPC in chlorpromazine are seen in Fig. 8A, which shows the transition behavior of DPPC in excess water and in the presence of 5 mole % chlorpromazine. As was experimentally found by Frenzel *et al.*[5] the addition of small mole fractions of chlorpromazine broadens and lowers the transition temperature. This was analytically accomplished by reducing ΔH_{vH} from 2×10^6 cal/mole to 5×10^5 cal/mole and keeping the other parameters the same as given in Fig. 7. The reduction in ΔH_{vH} means that the transition cooperativity decreases while ΔH_m remains constant upon the addition of small concentrations of chlorpromazine. A similar behavior was seen by Mountcastle *et al.*[21] when halothane was added to DPPC.

Figure 8B shows the calculated calorimetric behavior of DPPC–chlorpromazine dispersions at 15, 20, 25, and 30 mole % chlorpromazine. The values used to calculate these curves are identical to those given in Fig. 4. In agreement with the experimental data of Frenzel *et al.*,[5] the transition width initially narrows when DPPC melts from the interdigitated phase to the $L\alpha$ phase as seen in Fig. 8B and then broadens at the higher concentrations. Thus, we are able to successfully model the shape of the heat capacity curves using experimental parameters for the normal and interdigitated gel phases.

ACKNOWLEDGMENTS. This work was supported by grants from the National Institute of Health (GM 27278 and NS 11613), and the North Carolina United Way PC. We thank Dr. Elizabeth Rowe for kindly providing her absorbance data on diC17:0. We would like to thank Steffani Webb and Mrs. Gay Blackwell for their excellent job in helping to prepare this manuscript.

REFERENCES

1. Ebihara L, Hall JE, McDonald RC, McIntosh TJ, Simon SA: Effect of benzyl alcohol on lipid bilayers: A comparison of bilayer systems. *Biophys J* 1979;28:185–196.
2. Evans EA, Waugh R: Mechanochemistry of closed vesicular bilayers. *J Colloid Interface Sci* 1977;60:286–298.
3. Franks NP, Lieb WR: The structure of lipid bilayers and the effect of general anaesthetics: An X-ray and neutron diffraction study. *J Mol Biol* 1979;133:469–500.
4. Franks NP, Lieb WR: Molecular mechanisms of general anaesthesia. *Nature* 1982;300:487–493.
5. Frenzel J, Arnold K, Nuhn P: Calorimetric, ^{13}C NMR and ^{31}PNMR studies of the interaction of some phenothiazine derivitives with palmitoyl phosphatidylcholine model membranes. *Biochim Biophys Acta* 1978;507:185–197.
6. Hill MW: The effect of anaesthetic-like molecules on the phase transition in smectic mesophases of dipalmitoyllecithin I. The normal alcohols up to C = 9 and three inhalation anaesthetics. *Biochim Biophys Acta* 1974;356:117–124.
7. Inoue S, Nishimura M, Yasunage T: Studies on the phase transition in the single lamellar liposomes. 3. Kinetic behavior of the phase transition. *J Phys Chem* 1981;85:1401–1405.
8. Israelachvili JN, Marcelja S, Horn RG: Physical principles of membrane organization. *Q Rev Biophys* 1980;13:121–200.
9. Jain MK, Wu NM: Effect of small molecules on the dipalmitoyl lecithin liposomal bilayers: III Phase transition in lipid bilayer. *J Memb Biol* 1977;34:157–201.
10. Janoff AS, Miller KW: A critical assessment of the lipid theories of general anesthesia, in Chapman D (ed): *Biological Membranes,* Vol 4. New York, Academic Press, 1982, pp 417–476.
11. Kamaya H, Kaneshina S, Ueda I: Partition equilibrium of inhalation anesthetics and alcohols between water and membranes of phospholipids with varying acyl chain-lengths. *Biochim Biophys Acta* 1981;646:135–142.
12. Katz Y, Diamond JM: Thermodynamic constants for nonelectrolyte partition between dimyristoyl lecithin and water. *J Memb Biol* 1974;17:101–120.
13. Lee AG: Lipid phase transitions and phase diagrams. II Mixtures involving lipids. *Biochim Biophys Acta* 1977;472:285–344.
14. Lee TC, Fitzgerald V: Phase transitions of alkylether analogs of phosphatidylcholine. *Biochim Biophys Acta* 1980;598:189–192.
15. McDaniel RV, Simon SA, McIntosh TJ, Borovyagin V: Interaction of benzene with bilayers: Thermal and structural studies. *Biochemistry* 1982;21:4116–4126.
16. McDaniel RV, McIntosh, TJ, Simon SA: Nonelectrolyte substitution for water phosphatidylcholine bilayers. *Biochim Biophys Acta* 1983;731:97–108.
17. McIntosh TJ: Differences in hydrocarbon chain tilt between hydrated phosphatidylethanolamine and phosphatidylcholine. *Biophys J* 1980;29:237–246.
18. McIntosh TJ, Simon SA, MacDonald RC: The organization of *n*-alkanes in lipid bilayers. *Biochim Biophys Acta* 1980;597:445–463.

19. McIntosh TJ, McDaniel RV, Simon SA: Induction of an interdigitated gel phase in fully hydrated phosphatidylcholine bilayers. *Biochim Biophys Acta* 1983;731:109–114.
20. Meyer HH: Theories der alkohol narkose. *Arch Exp Path Pharmacol* 1899;42:109–141.
21. Mountcastle DB, Biltonen RL, Halsey MJ: Effect of anesthetics and pressure on the thermotropic behavior of multilamellar dipalmitoylphosphatidylcholine liposomes. *Proc Natl Acad Sci USA* 1978;75:4906–4910.
22. Nagle JF, Wilkinson DA: Lecithin bilayers: Density measurements and molecular interactions. *Biophys J* 1978;23:159–175.
23. Podo, F, Blasie JK: Proton magnetic relaxation studies of mixed phosphatidylcholine/fatty acid and mixed phosphatidylcholine bimolecular bilayers. *Biochim Biophys Acta* 1977;419:1–18.
24. Ranck JL, Tocanne JF: Polymyxin β induces interdigitation in dipalmitoylphosphatidylglycerol lamellar phase with stiff hydrocarbon chains. *FEBS Lett* 1982;143:171–174.
25. Ranck, JL, Keira T, Luzzati V: A novel packing of the hydrocarbon chains in lipids: The low temperature phases of dipalmitoylphosphatidylglycerol. *Biochem Biophys Acta* 1977;488:432–441.
26. Rowe ES: Lipid chain length and temperature dependence of ethanol-phosphatidylcholine interactions. *Biochemistry* 1983;22:3299–3305.
27. Seeman P: The membrane action of anesthetics and tranquilizers. *J Pharmacol Rev.* 1972;24:583–653.
28. Siminovitch DJ, Ruocco MJ, Griffin RG: A differential scanning (DSC), X-ray diffraction and ^{31}P-NMR study of the gel phases of dihexadecyl (DHPC) and dipalmitoylphosphatidylcholine. *Biophys J* 1984;45:197a.
29. Simon SA, McIntosh TJ: Interdigitated hydrocarbon chain packing causes the biphasic transition behavior in lipid/alcohol suspensions. *Biochim Biophys Acta* 1984;773:169–172.
30. Sturtevant JM: The effects of water-soluble solutes on the phase transitions of phospholipids. *Proc Natl Acad Sci USA* 1984;81:1398–1400.
31. Tardieu A, Luzzati V, Raman RC: Structure and polymorphism of the hydrocarbon chains of lipids: A study of lecithin-water phases. *J Mol Biol* 1973;75:711–733.
32. Vaughan DJ, Keough KM: Changes in phase transitions of phosphatidylethanolamine and phosphatidylcholine-water dispersions induced by small modifications in the headgroup and backbone regions. *FEBS Lett* 1974;47:158–161.
33. White, SH, King GI, Cain JE: Location of hexane in lipid bilayers determined by neutron diffraction. *Nature* 1981;290:161–163.

25

Molecular Details of Anesthetic–Lipid Interaction as Seen by Nuclear Magnetic Resonance

Ian C. P. Smith and Keith W. Butler

ABSTRACT. We have used electron spin resonance (ESR) and nuclear magnetic resonance (NMR) to study the effect of local anesthetics on membranes. Both model membranes (lipids in water) and natural membranes have been used. ESR studies with labeled fatty acids in lipids demonstrated that the potency of local anesthetics in reducing molecular order correlated with their pharmacological potency.

The molecular order of the fatty acid region of phospholipids can also be studied by selectively incorporating deuterium in the lipids and observing the deuterium NMR spectra. The headgroup region can be observed by phosphorus NMR. At pH 5.5, tetracaine reduced the order in the fatty acid region of the phospholipid but increased order at the phosphate group.

In addition, the local anesthetics themselves were specifically labeled with deuterium at several different positions. Spectra of the anesthetic in the lipid and in the buffer could be observed simultaneously, allowing partition coefficients to be measured as well as the position of the anesthetic in the bilayer computed. In the case of phosphatidylcholine, at least two binding sites were observed. With phosphatidylethanolamine and bovine spinal cord, only one membrane-bound site for anesthetic was apparent.

There are almost as many theories about the mechanism of anesthesia as there are theorists. This situation exists despite an impressive literature on all aspects from chemistry to pharmacology. Most experimental methods have concentrated on measuring the effects of anesthesia, rather than the molecular events leading to the effects.

Physical chemistry has reached a stage of technological expertise at which direct visualization of molecular components in relevant systems is possible. One of the methods now in use is nuclear magnetic resonance (NMR). It shows the best hope for attainment of precise answers, free from effects due to perturbation by probes or approximate methods of analysis. Our preliminary results on anesthetic–membrane systems were obtained using electron spin resonance (ESR) of nitroxide-labeled components. At present we are concentrating our efforts on the use of NMR.

The nuclei commonly observed in magnetic resonance studies of interactions between

Ian C. P. Smith and Keith W. Butler • Division of Biological Sciences, National Research Council of Canada, Ottawa K1A OR6, Canada

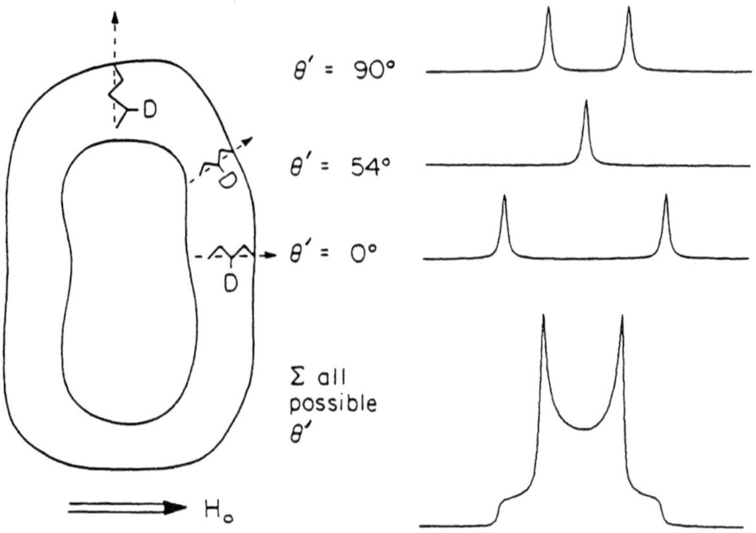

Figure 1. The origin of the ^{2}H NMR powder spectrum. The angle θ′ is between the director (broken line) and H_o.[7]

membranes and local anesthetics are ^{1}H, ^{31}P, ^{13}C, and ^{2}H. The first two, ^{1}H and ^{31}P, are the most abundant isotopes of their elements, approaching 100%, whereas only about 1% of carbon is ^{13}C and only 0.015% of hydrogen exists as ^{2}H. This natural abundance is a major influence on experimental design. No labeling is required with ^{1}H or ^{31}P experiments. On the other hand, all compounds in the sample that contain the element of interest are seen in the spectra. For ^{31}P of membranes, this is not a major problem, but proton spectra of biological systems can be quite complex.

Because of the low natural abundance of deuterium, generally the only resonances observed by deuterium magnetic resonance are of water (because of the high molar concentration of water in most biological systems) and of the deuterium added by the experimenter. In a typical experiment, protons at a particular position on a molecule are replaced by deuterons. In the present context, this molecule can be a local anesthetic or a membrane component which is either biologically incorporated or added to a membrane suspension.

Deuterium differs from the other nuclei mentioned in that it has a spin of 1 whereas the others have a spin of 1/2.[6,14] Deuterium has an unsymmetrical nuclear charge distribution; the resulting nuclear quadrupole interaction causes a splitting of the Zeeman resonance. The spectrum of a deuterated fatty acid in an unoriented liquid crystalline membrane has two distinct peaks whose separation is proportional to the degree of ordering of the particular labeled region (Fig. 1).

If the sample is cooled, a broader component indicative of the more ordered gel phase appears in the spectrum.[15] When the rate of molecular reorientation is rapid, relative to the reciprocal of the ^{2}H quadrupolar coupling constant, the ^{2}H NMR spectrum collapses to a single resonance. This situation is frequently seen for molecules in aqueous solution and for labels in small vesicles.[16]

These nuclei also differ in their relative ease of detection. For a given magnetic field and number of nuclei, the sensitivity of nuclear magnetic resonance is 6.6% for ^{31}P, 1.6% for ^{13}C, and 0.96% for ^{2}H relative to that of protons.

EFFECTS OF LOCAL ANESTHETICS ON MODEL MEMBRANES

In early experiments, electron spin probes, stable free radical derivatives of sterols or of fatty acids, were incorporated into films of lipid to explore the effects of local anesthetics on model membranes.[4,11] These probes, like their deuterium-containing counterparts, give information about the degree of molecular order within membranes. ESR has the advantage of much greater sensitivity over deuterium magnetic resonance, permitting much smaller sample size and faster spectral aquisition (10^{-9} moles of a spin probe in 2–4 10^{-7} moles of lipid may be observed in a few minutes). Deuterium magnetic resonance has the advantages that substituting ^2H for ^1H is much less likely to perturb the system being studied than is the insertion of a comparatively massive nitroxide group,[18] and that deuterium labels are less likely to suffer redox reactions in living systems than are free radical ESR probes.

Planar lipid films, formed from material extracted from the white matter of cattle brains and freed of protein, exhibit a high degree of molecular order.[4] Local anesthetics, particularly at high pH, decreased this order. The relative potency of a series of local anesthetics in disordering films correlated well with their potency as anesthetics. Lowering the cholesterol content of this lipid lowered the order in the films. With these less ordered films, both an ordering and a disordering effect of local anesthetic was observed, depending on the pH and the concentration of local anesthetic.

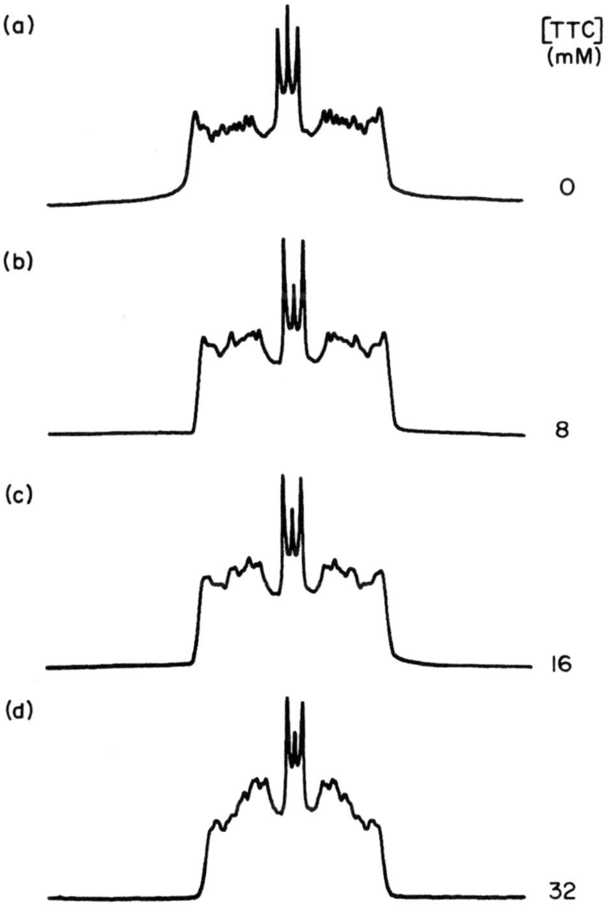

Figure 2. ^2H NMR spectra (46.06 mHz) of 65 mM DPPC-d_{62} multilamellar dispersions in buffer at pH 5.5 with (a) no tetracaine, (b) 8 mM tetracaine, (c) 16 mM tetracaine, and (d) 32 mM tetracaine. The spectra were recorded at 47°C using the quadrupolar echo technique. Reprinted with permission from Boulanger *et al.*[3] Copyright (1981) American Chemical Society.

A combination of ^2H and ^{31}P NMR was used to study the effect of tetracaine on purified lipids.[3] Dispersions of DPPC-d_{62} (lecithin with two perdeuteropalmitate chains) produce characteristic deuterium spectra. The well-defined shoulders result from a number of positions on the fatty acid chain having the same large degree of order (the plateau region) (Fig. 2).

The effects of tetracaine in its charged (pH 5.5) form and its uncharged (pH 9.5) form have been compared. A reduction in intensity of the shoulders is observed at both pH values, indicating a reduction of the region of high order. The order of the fatty acid chains is reduced overall at both pH values. In agreement with the partition constants ($K_p = 660$ at pH 9.5 and 22 at pH 5.5), a lower aqueous concentration of tetracaine is needed to disorder at the higher pH.

Measurements were also made of the effect of tetracaine on the degree of order of the headgroup region. Methylation of the nitrogen of phosphatidylethanolamine (PE) with deuteromethyl groups results in lecithin with nine deutrons (three on each of three carbons) in the headgroup. Exposure to tetracaine increased the degree of order of these deuterons at pH 5.5 and decreased it at pH 9.5.

The effect of tetracaine on the ^{31}P spectra of dispersions of lipids was also observed. The shape of the phosphorus powder pattern has been used to detect the presence of hexagonal, bilayer, or isotropic phases, as well as to measure the degree of order in the headgroup region of the bilayer.[13] The addition of 0.5 mole of tetracaine/mole of lecithin increases the width of the pattern by 8% at pH 5.5 but has no effect at pH 9.[3] These results were interpreted as indicating two different locations within lecithin bilayers for tetracaine. At low pH, where the local anesthetic is positively charged, it can interact with the headgroups, increasing their order. At high pH, the tetracaine is buried more deeply within the bilayer.

Similar experiments were performed with specifically deuterated PE.[10] The labeled PE was mixed 1:1 with either unlabeled PE or phosphatidylserine (PS).

Deuterons in the ethanolamine headgroup produce two quadrupolar splittings which could be assigned to those adjacent to the oxygen (α) and to those adjacent to the nitrogen(β). The addition of tetracaine decreased the quadrupolar splitting of the α-deuterons and increased that of the β-deuterons at both pH 9.5 and pH 5.5. In addition, at pH 5.5 low concentrations of tetracaine resulted in spectra with the superposition of three quadrupolar patterns (Fig. 3).

"De-Paking"[1] is a mathematical procedure by which the spectrum due to the labelled molecules whose directors of motional averaging are aligned perpendicular to the magnetic field of the spectrometer ($\theta = 90°$, Fig. 1) is derived from the observed powder pattern. This procedure simplifies the measurement of order parameters and makes spectral detail more visible, at the expense of a decrease in the signal-to-noise ratio.

The "de-Paked" spectra (Fig. 3, right) plainly show that the resonance due to the α deuteron is divided into two splittings upon exposure to tetracaine. One splitting is essentially unchanged by tetracaine, whereas the other has a reduced quadrupolar splitting. As the concentration of tetracaine increases, the inner deuteron doublet decreases in splitting and increases in intensity while the outer doublet decreases in intensity, eventually disappearing.

The addition of tetracaine at pH 5.5 to PE samples labelled at the 4 or 14 position of the fatty acid chain also divided each doublet into two, one of which had a reduced quadrupolar splitting.[10] At pH 9.5, tetracaine reduced the quadrupolar splitting of deuterons at the position of the headgroup and at the 4 and 14 positions on the fatty acids, but did not split them into two doublets.

The presence of two spectra from a single labeled position indicates that the compound

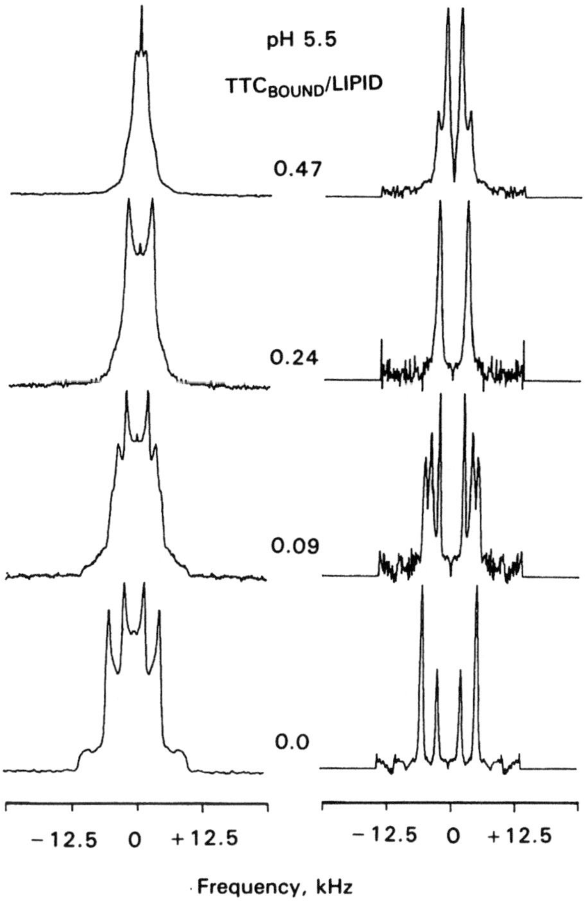

Figure 3. ^{2}H NMR spectra of a 1:1 mixture of 3-(1′-2′-$^2H_4$) DMPE and egg PE at pH 5.5 with varying amounts of tetracaine. (Left) Experimental spectra. (Right) De-Paked spectra. Reprinted with permission from Kelusky and Smith.[10]

experiences two different environments, with interchange between them slow on the NMR time scale. The authors[10] postulated that tetracaine at pH 5.5 undergoes slow lateral diffusion in PE membranes. This slow lateral diffusion leads, at low tetracaine concentrations, to a superposition of spectra from areas of PE influenced by tetracaine and from areas unaffected by tetracaine.

At pH 9.5 the diffusion of tetracaine in PE is sufficiently rapid that two environments are not observed. It is unknown whether, at pH 5.5, it is the tetracaine–PE or the PE–PE interactions that are important for retarding the rate of diffusion of the tetracaine through the PE. The presence of two signals from deuterons at a given position on PE were not observed with PE–PS mixtures.

Both the quadrupolar splitting of the deuterons and the powder pattern width of the ^{31}P NMR spectrum of PE were increased upon exposure to tetracaine. It was concluded that the effect of tetracaine on PE was to alter the conformation of the headgroup and to reduce the order of the fatty acid chains.

STUDIES WITH DEUTERIUM-LABELED LOCAL ANESTHETICS

Specific protons on local anesthetics can be replaced by deuterons. The interaction of these labeled anesthetics with unlabeled lipids can be followed by NMR. The use of deu-

TETRACAINE

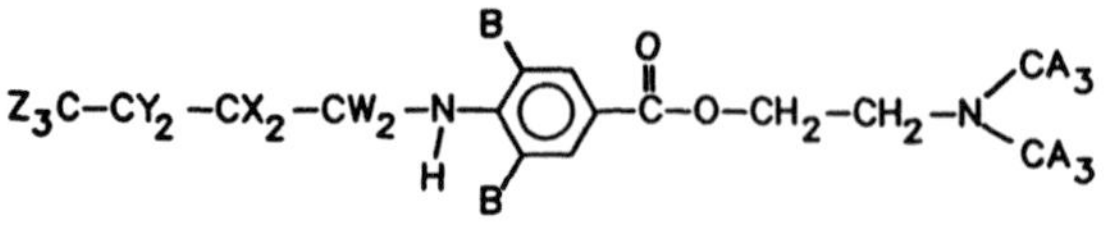

TTC–d_6	A = D	;	B,W,X,Y,Z = H
TTC–d_2	B = D	;	A,W,X,Y,Z = H
TTC–d_3	Z = D	;	A,B,W,X,Y = H
TTC–d_9	W,X,Y,Z = D	;	A,B = H

Figure 4. Structures of the labeled tetracaine species.

terium-depleted water reduces the size of the signal from ^2H at natural abundance in water.

In studies with lecithin[2,9,19] it was found that tetracaine-d_6 produced spectra consisting of the superposition of a quadrupolar pattern and a central resonance. Changing the ratio of lipid to water altered the ratio of the areas of the central resonance and the quadrupolar pattern without changing the width of the quadrupolar pattern. The width of the central resonance was dependent on the lipid:water ratio. These results are consistent with a three-site model: tetracaine exists in a highly ordered site in the bilayer, in slow exchange with all other sites, and in a weakly ordered site near the surface of the membrane, in rapid exchange with tetracaine in aqueous solution (Fig. 5).

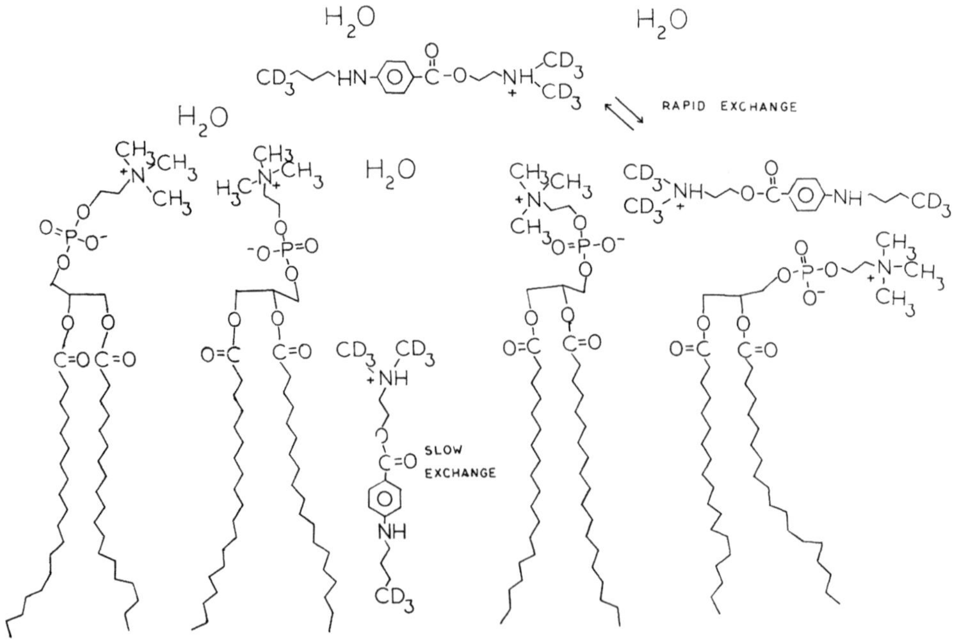

Figure 5. Model for the partitioning of tetracaine between lipids (egg lecithin) and water. The evidence for the location and properties of the strongly bound (slowly exchanging) site is very clear.[3] The nature of the rapidly exchanging site is less so. It is not highly ordered, and is probably at the water–lipid interface.

Freezing and thawing hydrated mixtures of lecithin and deuterium-labelled anesthetics can change the spectra, either eliminating[9,19] or increasing the quadrupolar component, depending on the pH and position of the label.

Deuterium-labelled tetracaine in PE, in contrast to tetracaine in lecithin, showed only two sites, one with a quadrupolar pattern and one free in aqueous solution.[8] These sites are apparently in slow exchange on the NMR time scale. The partition of tetracaine between buffer and lipid was measured by centrifugation and correlated well with the ratios of the areas of the quadrupolar pattern and of the central resonance (aqueous site). Changing the pH had little effect on the width of the quadrupolar pattern. Thus, it appeared that both the charged and uncharged forms of tetracaine bound to similar locations in PE, in contrast to the deeper location of the uncharged form in PC.[2] This was postulated to be due to the complementary wedgelike shapes of tetracaine and PE.[8] Procaine has a lower lipid:water partition coefficient, to the extent that the quadrupolar component of the spectrum is small and hard to see.

STUDIES WITH BIOLOGICAL MEMBRANES

The model membrane studies demonstrated that the position of local anesthetics in model membranes depends on the anesthetic, the lipid, and pH. The partition coefficients measured by NMR were consistent with centrifugation measurements.

Conrad and Singer[5] found the same partition coefficient for chlorpromazine between water and large unilamellar vesicles of soybean lecithin, whether measured by centrifugation or by a filtration technique (hygroscopic desorption). However, they found that the latter technique gave a 1000-fold lower partition coefficient for chlorpromazine between red cells and buffer than did centrifugation. These findings provided an additional impetus for studying local anesthetics in biological membranes, using both labeled membranes and labeled local anesthetics (in separate experiments).

Acholeplasma laidlawii will take up and incorporate intact into its membranes deuterium-labelled fatty acids.[17] Exposure to either procaine or tetracaine at either pH 5.5 or pH 9.5 decreased the quadrupolar splitting observed from deuterons at either the 5 or the 18 (terminal) position of oleic acid[7] (Fig. 6).

Perdeuteropalmitic acid intercalated into the membranes of human red cell ghosts also showed a reduction in quadrupolar splitting when incubated with tetracaine at pH 9.5.

Line broadening, but no quadrupolar pattern, was observed when tetracaine labeled at various positions was added to either red cell ghosts or to unlabeled *A. laidlawii* membranes. From the response of the spectra to changes in temperature, it was deduced that the relaxation time (T_{2e}) of the ordered anesthetic was too short for a quadrupolar pattern to be observed in these systems.

In another series of experiments (M. Pass, K. W. Butler, E. C. Kelusky, and I. C. P. Smith, unpublished observations) spinal cord from freshly killed cattle was bathed in dilute solutions of tetracaine-d_6. Within minutes, a quadrupolar pattern appeared. The total signal strength depended on the volume and tetracaine concentration of the solution in which the spinal cord was soaked before it was placed in the NMR tube. The ratio of the areas of the quadrupolar pattern and the central resonance depended on the ratio of dry weight to water in the NMR tube. Because spinal cord contains water with deuterium present in natural abundance, and this water does not completely wash out with soaking in buffer made up with deuterium-depleted water, a central resonance was observed in the absence of label. This natural abundance spectrum was subtracted from each of the experimental spectra to yield those plotted in Fig. 7.

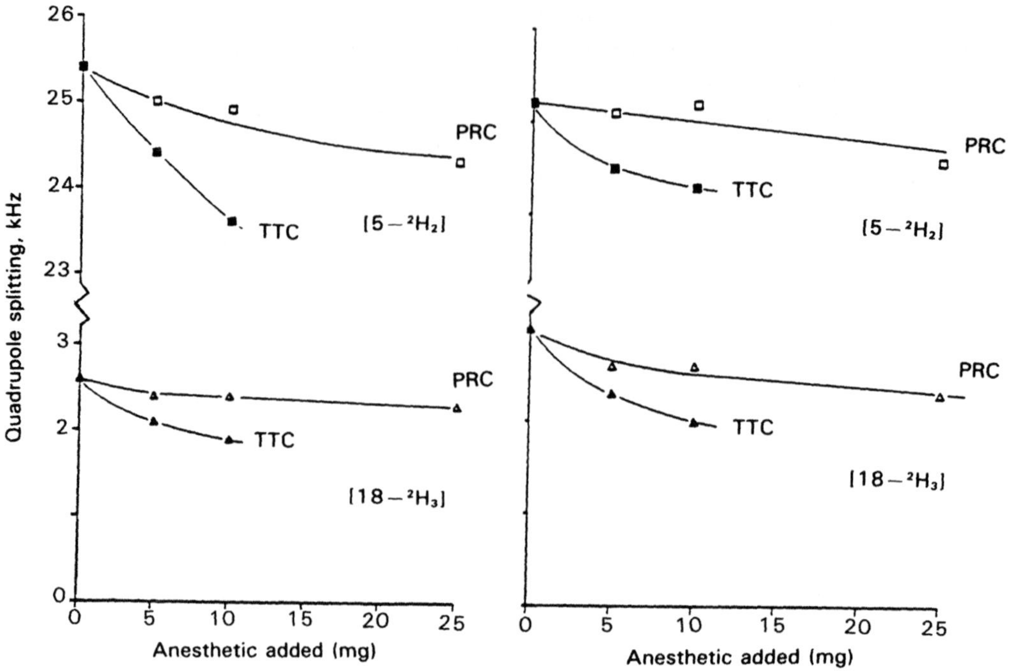

Figure 6. The influence of tetracaine or procaine at pH 5.5 (left) and at pH 9.5 (right) on *A. laidlawii* membranes grown in medium containing oleic acid labeled at the indicated positions.[7]

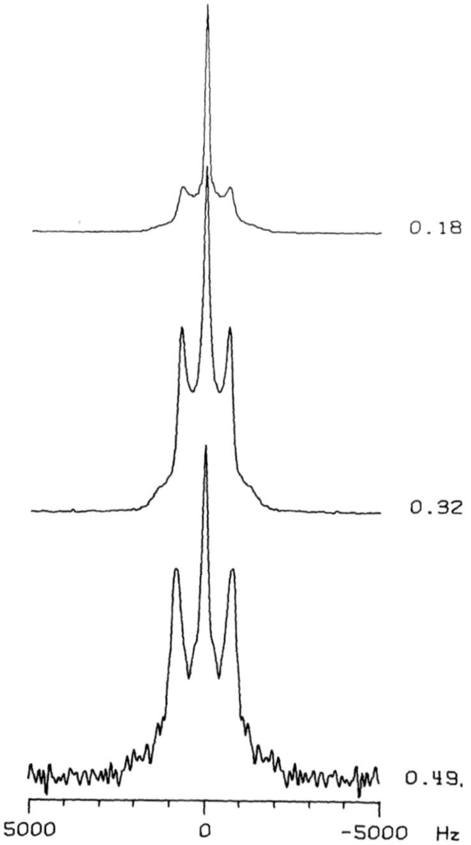

Figure 7. ^{2}H NMR spectra (46.06 mHz) of tetracaine-d_6 in the presence of pH 7.0 buffer and bovine spinal cord at the indicated ratios of dry weight to water. Top spectrum (tetracaine) = 1%; rest (tetracaine) = 0.5%. The natural abundance deuterium spectrum was subtracted from each set of experimental data before plotting.

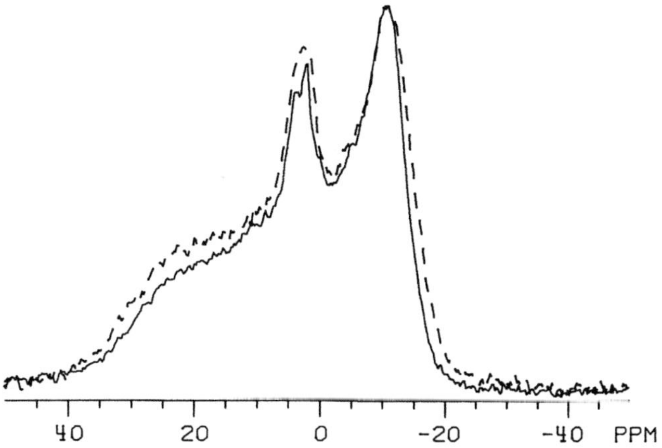

Figure 8. ^{31}P NMR (121 mHz) spectra of bovine spinal cord incubated with pH 7.0 buffer (solid spectrum) or with the same buffer containing 0.5% tetracaine (dashed line). Spectra were acquired at 5°C using a 90° pulse of 7 μsec, a recycle time of 1 sec, gated proton decoupling, and the Hahn echo.[12]

These spectra were then simulated by adding together a theoretical quadrupolar pattern and a single resonance. The change in the ratio of the quadrupolar pattern to single resonance necessary to duplicate an experimental spectrum corresponded to the change in the ratio of dry weight (or of lipid determined by total fatty acid analysis) to the weight of water. The data are consistent with a two-site model, in which tetracaine in the membrane produces a quadrupolar pattern in slow exchange (on the NMR time scale) with the single central resonance due to tetracaine in aqueous solution.

Spectra of labeled tetracaine in aqueous dispersions of lipid extracted from bovine spinal cord were similar to those obtained with intact spinal cord. A number of freeze–thaw cycles extending over several days had no noticeable effect on the spectra.

Phosphorus NMR of spinal cord showed a change in the headgroup region of the phospholipids upon exposure to tetracaine. The same change was noted at 37°C and with 1% tetracaine (Fig. 8).

In summary, the results with biological membranes are membrane-dependent. Deuterium NMR of labeled fatty acids incorporated biologically (*A. laidlawii*) or added exogenously (red cell ghosts) demonstrated a disordering effect of local anesthetics on the fatty acid chains. Deuterium-labeled tetracaine in red cell ghosts and in *A. laidlawii* membranes has too short a relaxation time (T_{2e}) to allow observation of a quadrupolar pattern. Experiments with spinal cord clearly show a strong partitioning of tetracaine into the membrane lipids. We have no data consistent with the proposition of Conrad and Singer that anesthetics partition very differently into model and natural membranes.

CONCLUSION

Anesthetics can interact very strongly with lipids in model and natural membranes, causing strong disordering effects whose magnitudes vary with position in the membrane, nature of the anesthetic, temperature, and pH. These effects can now be quantitated by means of NMR. More data on other anesthetics, lipids, membranes, and proteins (!) are required before mechanistic model building begins, but we are almost ready to start. Extension into whole nerves and even simple animal models is within reach. By the Fourth International Conference on Molecular and Cellular Mechanisms of Anaesthesia we hope to be able to propose a model based on NMR data.

REFERENCES

1. Bloom M, Davis JH, MacKay AL: Direct determination of the oriented sample NMR spectrum from the powder spectrum for systems with local axial symmetry. *Chem Phys Lett* 1981;80:198–202.

2. Boulanger Y, Schreier S, Leitch, LC, Smith ICP: Multiple binding sites for local anesthetics in membranes: Characterization of the sites and their equilibria by deuterium NMR of specifically deuterated procaine and tetracaine. *Can J Biochem* 1980;58:986–995.
3. Boulanger Y, Schreier S, Smith ICP: Molecular details of anesthetic–lipid interaction as seen by deuterium and phosphorus-31 nuclear magnetic resonance. *Biochemistry* 1981;220:6824–6830.
4. Butler KW, Schneider H, Smith ICP: The effects of local anesthetics on lipid multilayers: A spin probe study. *Arch Biochem Biophys* 1973;154:548–554.
5. Conrad MJ, Singer SJ: The solubility of amphipathic molecules in biological membranes and lipid bilayers and its implications for membrane structure. *Biochemistry* 1981;20:808–818.
6. Davis JH: The description of membrane lipid conformation, order, and dynamics by ^{2}H-NMR. *Biochim Biophys Acta* 1983;737:117–171.
7. Kelusky EC: Nuclear magnetic resonance studies of the interaction of local anesthetics with membranes and motions in solid *n*-alkanes, Ph.D Thesis, University of Ottawa, 1983.
8. Kelusky EC, Smith ICP: Characterization of the binding of the local anesthetics procaine and tetracaine to model membranes of phosphatidylethanolamine: A deuterium nuclear magnetic resonance study. *Biochemistry* 1983;22:6011–6017.
9. Kelusky EC, Smith ICP: Anesthetic–membrane interaction: A ^{2}H nuclear magnetic resonance study of the binding of specifically deuterated tetracaine and procaine to phosphatidylcholine. *Can J Biochem Cell Biol* 1984a;62:178–184.
10. Kelusky EC, Smith ICP: The influence of local anesthetics on molecular organization in phosphatidylethanolamine membranes. *Mol Pharmacol* 1984b;26:314–321.
11. Neal MJ, Butler KW, Polnaszek CF, Smith ICP: The influence of anesthetics and cholesterol on the degree of molecular organization and mobility of ox brain white matter lipids in multibilayer membranes: A spin probe study using spectral simulation by the stochastic method. *Mol Pharmacol* 1976;12:144–155.
12. Rance M, Byrd RA: Obtaining high-fidelity spin-1/2 powder spectra in anisotropic media: Phase-cycled Hahn echo spectroscopy. *J Mag Res* 1983;52:221–240.
13. Smith ICP, Ekiel IH: Phosphorus-31 NMR of phospholipids in membranes, in Gorenstein, D (ed): *Phosphorus-31 NMR:Principles and Applications*. New York, Academic Press, 1984, pp 447–475.
14. Smith ICP, Mantsch HH: *Deuterium NMR spectroscopy*. ACS Symposium Series #191, Washington, American Chemical Society, 1982, pp 97–117.
15. Smith ICP, Butler KW, Tulloch AP, Davis JH, Bloom M: The properties of gel state lipid in membranes of *Acholeplasma laidlawii* as observed by ^{2}H NMR. *FEBS Lett* 1979;100:57–61.
16. Stockton GW, Polnaszek CF, Tulloch AP, Hasan F, Smith ICP: Molecular motion and order in single-bilayer vesicles and multilamellar dispersions of egg lecithin and lecithin–cholesterol mixtures. A deuterium nuclear magnetic resonance study of specifically labeled lipids. *Biochemistry* 1976;15:954–966.
17. Stockton GW, Johnson KG, Butler KW, Tulloch AP, Boulanger Y, Smith ICP, Davis JH, Bloom M: Deuterium NMR study of lipid organization in *Acholeplasma laidlawii* membranes. *Nature* 1977;269:267–268.
18. Taylor MG, Smith ICP: Conformation of nitroxide-labeled fatty acid probes of membrane structure as studied by ^{2}H-NMR. *Biochim Biophys Acta* 1983;733:256–263.
19. Westman J, Boulanger Y, Ehrenberg A, Smith ICP: Charge- and pH-dependent drug binding to model membranes. A ^{2}H NMR and light absorption study. *Biochim Biophys Acta* 1982;685:315–328.

26

Do Direct Protein/Anesthetic Interactions Underlie the Mechanism of General Anesthesia?

N. P. Franks and W. R. Lieb

ABSTRACT. In order to test the hypothesis that general anesthetics can act directly on protein, we have studied the inhibition of a pure, soluble, lipid-free protein by a wide range of general anesthetic agents. We purified the luciferase enzyme from lanterns of the North American firefly *Photinus pyralis* using affinity chromatography and obtained a pure crystalline enzyme, completely free of its substrate luciferin. Luciferase activity was extremely sensitive to the presence of anesthetics, and the degree of inhibition was found to depend on the luciferin concentration. When the luciferin concentration was equal to the K_m of the uninhibited enzyme, the concentrations of general anesthetics needed to inhibit enzyme activity by 50% were essentially identical to the concentrations needed to induce general anesthesia in animals. This identity held over a 100,000-fold range of potencies and for a diverse range of simple anesthetic agents: inhalation agents (such as halothane, methoxyflurane, and chloroform), aliphatic and aromatic alcohols, ketones, ethers, and alkanes. The inhibition was competitive in nature, with anesthetic molecules competing with substrate (luciferin) molecules for binding to the protein. We found that the anesthetic-binding site could accommodate only one large, but more than one small, anesthetic molecule. The obvious mechanism suggested by our results is that general anesthetics, despite their structural diversity, may act by competing at specific receptor or enzyme sites for the binding of endogenous ligands.

Compared to the enormous research effort that has been expended on studying lipid/ anesthetic interactions, surprisingly little work has been carried out on the interactions of general anesthetics with proteins. This imbalance is not only unfortunate but remarkable, in view of the central role[8] that protein/anesthetic interactions may play both in the induction of general anesthesia and in many of its clinically important side effects. Much of the reluctance to look for direct effects on proteins has arisen from the widely held feeling that the well-constrained structures and specific binding properties of proteins are unlikely to explain how such a wide range of structurally and chemically dissimilar agents can induce general anesthesia.

In order to test the hypothesis that general anesthetics can act directly on protein, it is essential to study a pure protein, completely free of lipid, so that any effects observed can be interpreted unambiguously in terms of direct protein/anesthetic interactions. We chose[9] to study the luciferase enzyme from the North American firefly *Photinus pyralis*. As we show

N. P. Franks and W. R. Lieb • Biophysics Section, Blackett Laboratory, Imperial College of Science and Technology, London SW7 2BZ, England.

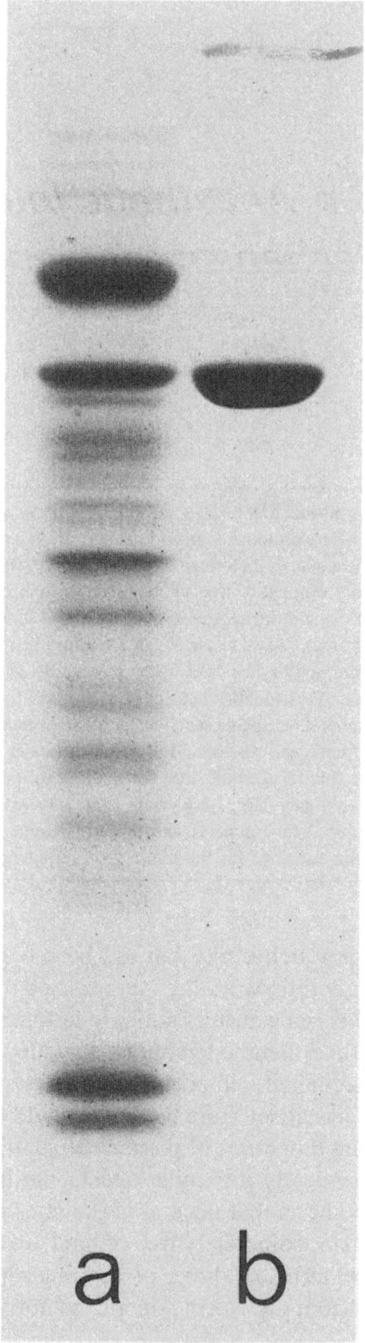

Figure 1. SDS polyacrylamide gels of (a) crude firefly lantern extract and (b) pure firefly luciferase. The gel of the pure enzyme was heavily overloaded to illustrate the high degree of purity.

$$LH_2 + ATP + E \overset{Mg^{2+}}{\rightleftharpoons} E \cdot LH_2\text{-}AMP + PP_i$$

$$E \cdot LH_2\text{-}AMP + O_2 \rightarrow \text{PRODUCTS} + \text{LIGHT}$$

LH_2: FIREFLY LUCIFERIN, M.W. = 280

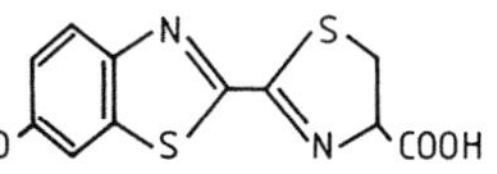

E : FIREFLY LUCIFERASE, M.W. ≈ 100,000 αβ DIMER

Figure 2. Details of the firefly luciferase reaction. See ref. 5 for further details.

below, our results with this protein clearly demonstrate that direct effects of anesthetics on protein form a plausible mechanism for the physiological actions of general anesthetics.

LUCIFERASE INHIBITION MIRRORS ANESTHETIC POTENCIES IN ANIMALS

We purified the luciferase enzyme using affinity chromatography[2] and obtained an extremely pure crystalline protein (see Fig. 1), completely free of the substrate luciferin. Firefly luciferase combines with its substrate luciferin, in the presence of ATP, Mg^{2+}, and O_2, to give a photon of light[5] (see Fig. 2). We constructed a rapid-mixing apparatus with a time resolution in the millisecond range and used this to assay luciferase activity in the presence and absence of a wide variety of general anesthetic agents. When the reactants are rapidly mixed, the light intensity increases rapidly to a maximum and then more slowly decreases. The maximum activity (peak height) was found to be extremely sensitive to the presence of low levels of anesthetics (Fig. 3).

To see to what extent inhibition of the luciferase enzyme mirrored anesthetic potencies

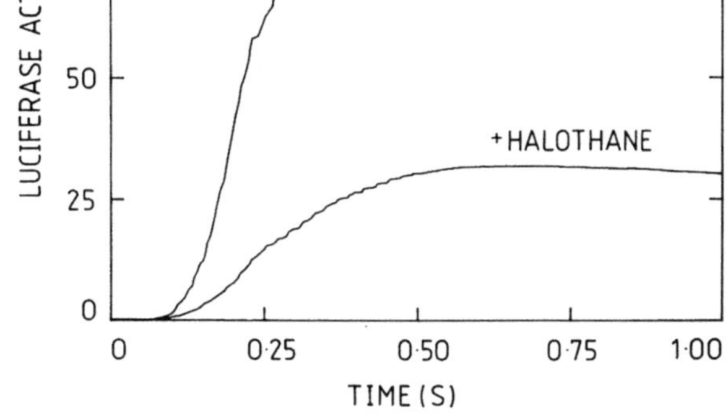

Figure 3. Inhibition of luciferase activity by surgical levels of halothane. The curves show luciferase activity (light intensity) as a function of time, after the rapid mixing of luciferase and luciferin with ATP. For the lower curve, the final concentration of halothane was 0.45 mM, equivalent to a partial pressure of 0.0073 atm. Reproduced from Franks and Lieb.[9] Reprinted by permission from *Nature*. Copyright © 1984 Macmillan Journals Limited.

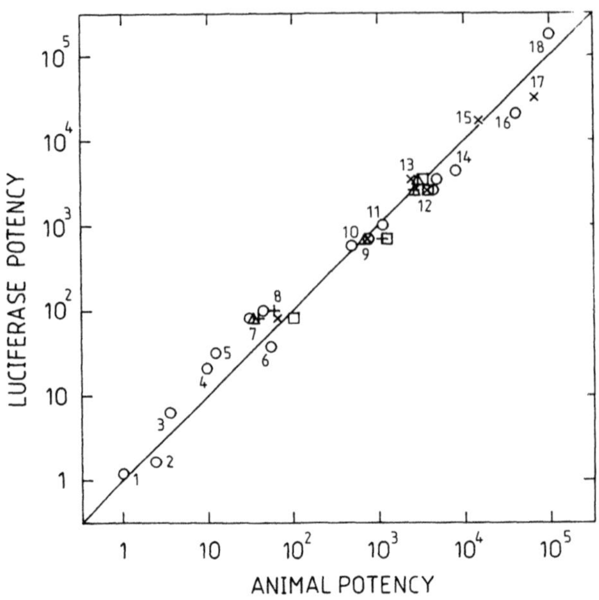

Figure 4. Identity of general anesthetic concentrations needed to anesthetize whole animals and to inhibit luciferase activity by 50%, for a diverse range of simple anesthetics over a 100,000-fold range of potencies. The line is the line of identity. The data are plotted as potencies, defined as reciprocals of aqueous ED_{50} molar concentrations. For the luciferase measurements, the luciferin concentration was equal to the K_m of the uninhibited enzyme. □, man; X, mouse; +, newt; ○, tadpole; △, goldfish. Anesthetics are referred to as follows: 1, methanol; 2, ethanol; 3, acetone; 4, *n*-propanol; 5, butanone; 6, paraldehyde; 7, diethyl ether; 8, *n*-butanol; 9, benzyl alcohol; 10, chloroform; 11, *n*-hexanol; 12, halothane; 13, methoxyflurane; 14, *n*-octanol; 15, *n*-pentane; 16, *n*-nonanol; 17, *n*-hexane; and 18, *n*-decanol. Reproduced from Franks and Lieb.[9] Reprinted by permission from *Nature*, copyright © 1984 Macmillan Journals Limited.

in animals, we determined the concentration of each anesthetic that produced 50% inhibition, under conditions where the luciferin concentration was equal to the K_m of the uninhibited enzyme and where ATP, Mg^{2+}, and O_2 were not rate-limiting. We found that, remarkably, the concentration of each anesthetic necessary to inhibit the enzyme by 50% was essentially identical to that which produces general anesthesia in animals, over a 100,000-fold range (see Fig. 4). This finding is all the more striking in view of the diversity of anesthetics used: inhalational agents such as halothane, methoxyflurane, chloroform, and ether; aliphatic and aromatic alcohols; ketones and alkanes.

MECHANISM OF ANESTHETIC INHIBITION

How can such a chemically and structurally diverse range of compounds cause inhibition? The first clue was that the degree of inhibition was dependent not only on the concentration of the anesthetic but also on the concentration of the substrate luciferin. At a fixed anesthetic concentration, we found that the inhibition was greater at low luciferin levels than at high levels. This strongly suggested that luciferin and anesthetic molecules were competing for binding to the enzyme. This was confirmed by measuring enzyme activities over a range of both luciferin and anesthetic concentrations. The data for halothane are shown in Fig. 5, where we have plotted the reciprocal of the enzyme activity against the reciprocal of the luciferin concentration, for halothane concentrations ranging from 0 to about twice the human minimum alveolar concentration.

Each line in Fig. 5 corresponds to a single halothane concentration and has a slope of K_m^{app}/V_{max}^{app} and a y intercept of $1/V_{max}^{app}$, where K_m^{app} and V_{max}^{app} are the apparent Michaelis constant and maximum activity, respectively. (There has been considerable debate as to which is the most statistically sound way to analyze enzyme kinetic data. We found that essentially identical results were obtained using our data and either distribution-free[3] or weighted least-squares procedures.[3]) V_{max}^{app} was independent of anesthetic concentration and

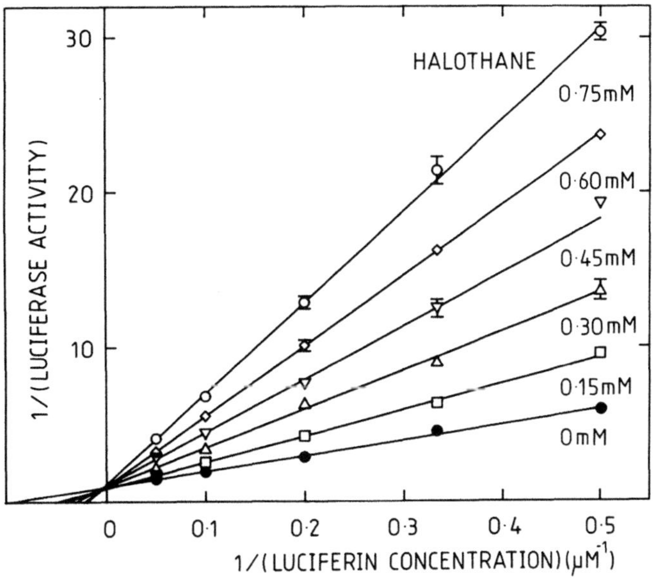

Figure 5. Halothane inhibits luciferase activity by competing with its substrate luciferin. The reciprocal of the maximum rate of light production (the peak height) is plotted, in relative units, against the reciprocal of the luciferin concentration, at various concentrations of halothane: ●, control; □, 0.15 mM; △, 0.30 mM; ▽, 0.45 mM; ◇, 0.60 mM; and ○, 0.75 mM. Reproduced from Franks and Lieb.[9] Reprinted by permission from *Nature*, copyright © 1984 Macmillan Journals Limited.

equal to the V_{max} of the uninhibited enzyme, whereas K_m^{app} progressively increased with increasing anesthetic concentration. This quantitatively confirmed the competitive nature of the inhibition.

ANESTHETIC MOLECULES BIND TO A SITE OF LIMITED SIZE

Further information on the anesthetic/protein interaction is contained in the values of the slopes of Fig. 5. Each slope provides an accurate value for K_m^{app}/V_{max}, and these values (each divided by the slope in the absence of anesthetic) are plotted against anesthetic concentration in Fig. 6. The function $f(A)$, so defined, is the factor by which the uninhibited K_m

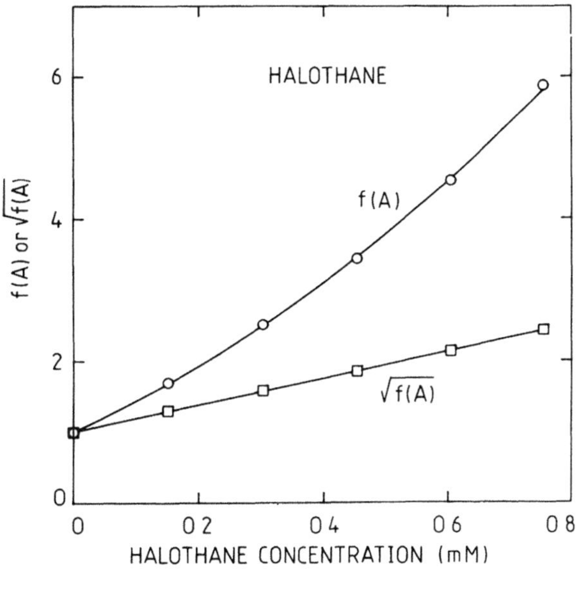

Figure 6. The factor $f(A)$ by which the uninhibited K_m increases in the presence of increasing concentrations of halothane. Reproduced from Franks and Lieb.[9] Reprinted by permission from *Nature*, copyright © 1984 Macmillan Journals Limited.

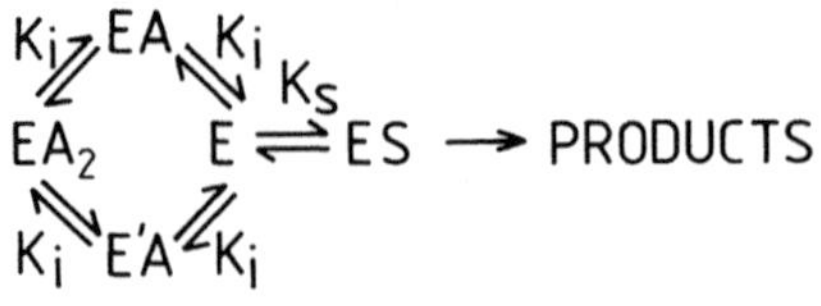

$$\frac{1}{v} = \frac{1}{V_{max}}\left[1 + \frac{K_S}{S}\left(1 + \frac{A}{K_i}\right)^2\right]$$

Figure 7. The simplest inhibition scheme that can explain our data for small anesthetic molecules. The symbols have their usual meanings.

increases in the presence of an anesthetic at a concentration A. [Alternatively, $f(A)$ can be thought of as the ratio of luciferin concentrations that would be required to produce the same activity in both the presence and absence of anesthetic.] Interestingly, the data points do not lie on a straight line ($p < 0.001$) but fit a quadratic equation very accurately. This immediately suggests that two molecules of halothane are involved in the inhibition.

The simplest way to explain our data is to assume that two halothane molecules can bind to the protein but that the presence of only one is sufficient to cause inhibition. If the two anesthetic molecules bind independently and in any order, with dissociation constants K_{i1} and K_{i2}, then it is easy to to show that $f(A) = 1 + (K_{i1} + K_{i2})(K_{i1}K_{i2})^{-1}A + (K_{i1}K_{i2})^{-1}A^2$. If, moreover, the molecules bind with equal affinities, then $K_{i1} = K_{i2} = K_i$ (see Fig. 7), and $f(A)$ takes the very simple form $(1 + A/K_i)^2$, so that a plot of the square root of $f(A)$ versus A should give a straight line with a slope of $1/K_i$. Figure 6 shows that the halothane data do indeed give a straight line when plotted in this way, and from the slope we calculate K_i = 0.54 mM.

The next question was: were the two halothane molecules acting at two well-separated sites, or were both sites contained within a single region of limited size, such as a small

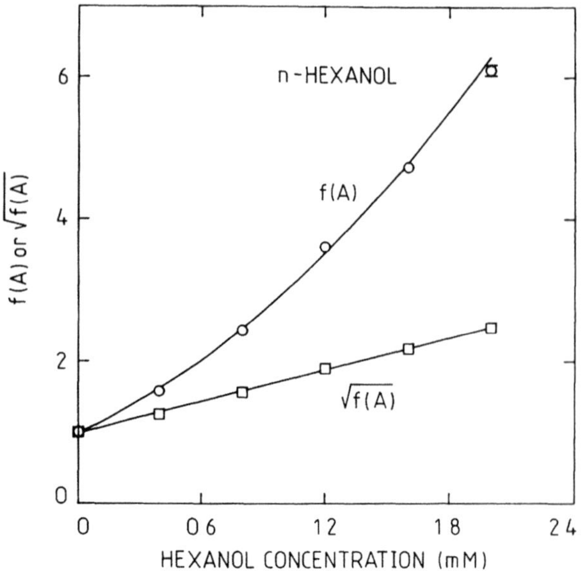

Figure 8. The factor $f(A)$ by which the uninhibited K_m increases in the presence of increasing concentrations of *n*-hexanol. Reproduced from Franks and Lieb.[9] Reprinted by permission from *Nature*, copyright © 1984 Macmillan Journals Limited.

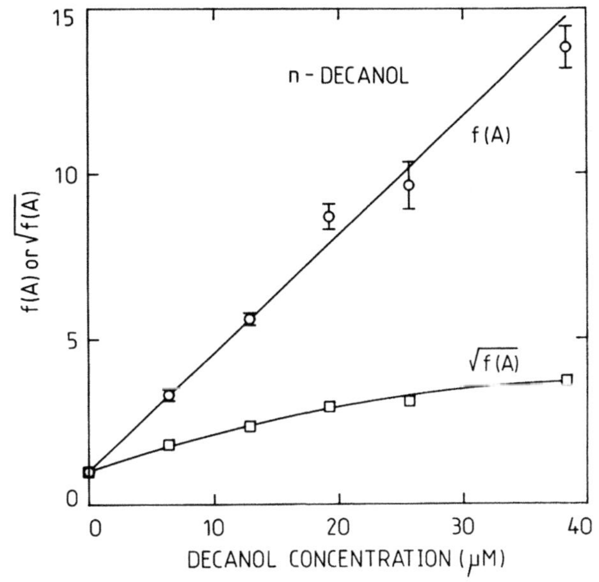

Figure 9. The factor $f(A)$ by which the uninhibited K_m increases in the presence of increasing concentrations of *n*-decanol. Reproduced from Franks and Lieb.[9] Reprinted by permission from *Nature,* copyright © 1984 Macmillan Journals Limited.

hydrophobic pocket? If the latter were true, then one would expect that, as one looked at the effects of progressively larger and larger anesthetic molecules, a point would be reached at which only a single molecule could be accommodated. We therefore studied the effects of the homologous series of *n*-alcohols, from methanol to *n*-decanol. We found that for the lower members of the series, up to and including hexanol, the inhibition resembled that of halothane, in that more than one anesthetic molecule was involved. This is shown for the hexanol data in Fig. 8, where it can be seen that a plot of $\sqrt{f(A)}$ versus A gives a straight line. However, for octanol and decanol, the pattern of inhibition suddenly changed. For these large anesthetics, it was clear that only a single molecule was involved in the inhibition. This can be seen from the data for *n*-decanol, which are shown in Fig. 9. Here the plot of $f(A)$ versus A is a straight line, even up to very high anesthetic concentrations where $f(A) = 14$, whereas the square root plot is no longer linear. This is consistent with a scheme in which only a single anesthetic molecule binds, for which $f(A) = 1 + A/K_i$ (see Fig. 10). It thus appears that the anesthetic molecules bind to the protein at a site of limited size.

$$\mathrm{EA} \underset{}{\overset{K_i}{\rightleftharpoons}} \mathrm{E} \overset{K_S}{\rightleftharpoons} \mathrm{ES} \longrightarrow \mathrm{PRODUCTS}$$

$$\frac{1}{v} = \frac{1}{V_{max}}\left[1 + \frac{K_S}{S}\left(1 + \frac{A}{K_i}\right)\right]$$

Figure 10. The simplest inhibition scheme that can explain our data for large anesthetic molecules. The symbols have their usual meanings.

THE PROBLEM OF ADDITIVITY

For mixtures of simple anesthetic agents acting on animals, it is usually found that additivity is observed.[18] This means, for example, that if half the ED_{50} dose of one anesthetic is added to half the ED_{50} dose of a second anesthetic, then 50% of the animals will be anesthetized. Would additivity be expected for anesthetic inhibition of luciferase? Although we have yet to carry out such experiments, predictions can be made using reaction schemes similar to those shown in Figs. 7 and 10. When the two anesthetic molecules are of similar size (either large or small), then strict additivity is predicted. On the other hand, when a small anesthetic is mixed with a large anesthetic, deviations from perfect additivity would be expected; however, the predicted effects are small and would probably be difficult to establish experimentally.

THE SIZE AND NATURE OF THE ANESTHETIC BINDING SITE

Some idea as to the size of the anesthetic-binding site on luciferase can be obtained from our data. Since the site can accommodate two molecules of hexanol but only one molecule of octanol, the minimum volume of the site must be roughly that of two hexanol molecules. This is consistent with our finding that the site binds two molecules of halothane, whose volume is slightly less than that of hexanol. (Interestingly, two hexanol molecules have a molar volume of 250 ml, which is close to the apparent volume of a strong binding site on β-lactoglobulin.[20,25]) The luciferin binding site appears to accommodate only a single luciferin molecule (since the double-reciprocal plots of Fig. 5 give straight lines), whose volume is close to that of decanol. The similarity between the molecular sizes of two halothane molecules, two hexanol molecules, one decanol molecule, and one luciferin molecule can be seen from the space-filling models shown in Fig. 11. Taking these size considerations together with the competitive nature of the inhibition, our results strongly suggest that the anesthetic molecules are binding within the luciferin site—a conclusion supported by the finding that luciferin binds in a hydrophobic pocket.[4,6]

Such a binding site, which presumably extends to the water/protein surface, might naturally be expected to have both polar and apolar parts, an apparent characteristic of general anesthetic target sites.[7] Indeed, the amphiphilic nature of the anesthetic-binding site on luciferase can be seen by comparing potencies with partition coefficients in simple solvent systems. For example, chloroform and hexanol have very similar potencies for inhibiting luciferase activity (see Fig. 4) and virtually identical partition coefficients between the amphiphilic solvent octanol and water.[7] However, the partitioning[7] of these two anesthetics between a purely apolar solvent (*n*-hexadecane) and water differs by a factor of more than 40.

DISCUSSION

Very little previous work has been published on anesthetic interactions with firefly luciferase. Ueda and Kamaya[23] used a crude luciferase/luciferin extract and concluded that anesthetic inhibition involved a major structural change in the enzyme, a view we have previously[8] criticized. Although more work has been carried out using bacterial luciferases, almost all these studies have been carried out using whole cells,[10,14,24] where mechanistic interpretation is very difficult. Reports on cell-free systems are sparse, but it has been shown that some inhalational agents[1,15] can compete with the binding of an aldehyde. It must be remembered, however, that the mechanisms of the bacterial and firefly enzymes are very

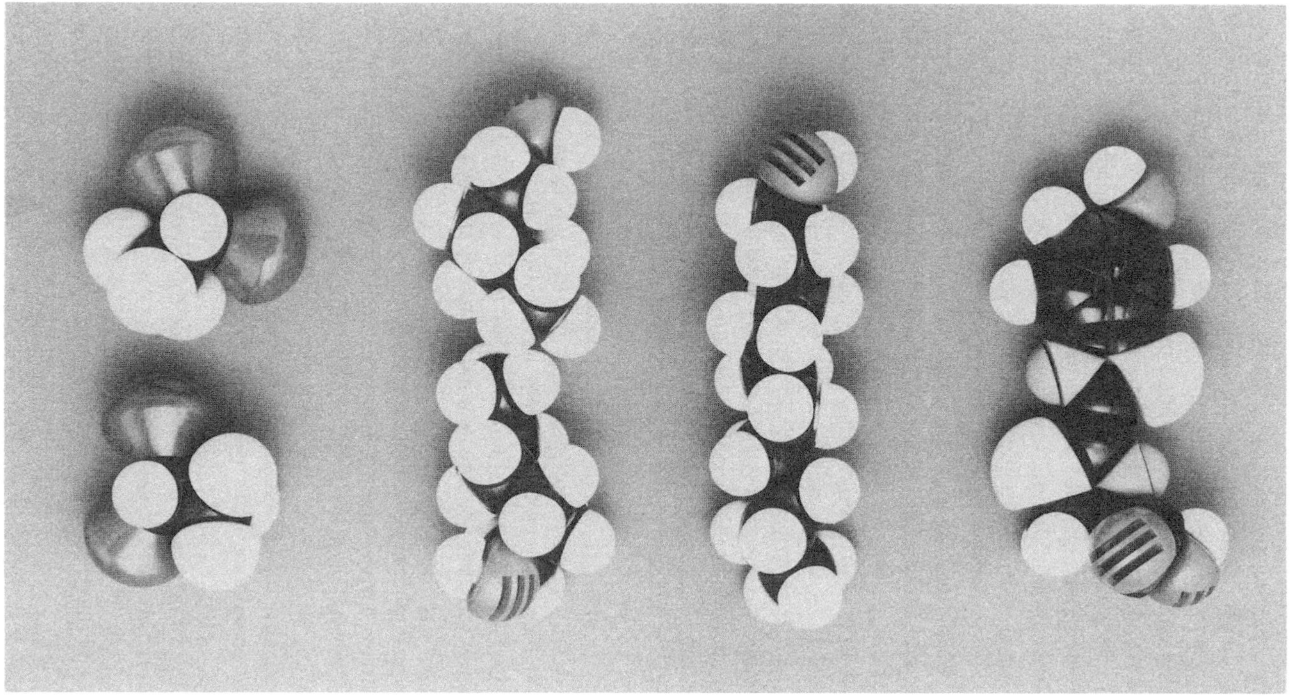

Figure 11. Space-filling models showing, from left to right, the relative sizes of two molecules of halothane, two molecules of *n*-hexanol, one molecule of *n*-decanol, and one molecule of the substrate luciferin. Reproduced from Franks and Lieb.[9] Reprinted by permission from *Nature*, copyright © 1984 Macmillan Journals Limited.

different, and it remains to be seen whether the pattern of inhibition we have observed here is found with pure bacterial enzymes over a more extended range of anesthetics and potencies.

Most proteins are simply insensitive to clinical levels of general anesthetics,[8] while some are inhibited only by molecules that resemble the natural substrates. For example, α-chymotrypsin is inhibited competitively by a range of *aromatic* compounds (including the anesthetic benzyl alcohol) which resemble its natural substrates.[12] However, nonaromatic anesthetics such as ether[16] and *n*-alcohols (from pentanol to decanol[22]) have virtually no effect at concentrations that cause general anesthesia. Thus work with substrate analogues can be misleading in the context of general anesthesia. In addition, certain proteins that are known to bind a wide range of small organic molecules, such as serum albumin and hemoglobin, show reasonable correlations between binding and octanol/water partition coefficients, but the slopes[11] of the log–log plots are much less than the value of unity demanded by a good model for general anesthesia.[7] Moreover, the oxygen-binding properties of hemoglobin are unaffected by clinically relevant levels of halothane and cyclopropane.[17] Thus α-chymotrypsin, serum albumin, and hemoglobin are not good models of the general anesthetic target site. Wishnia and his colleagues,[20,25] however, have presented some interesting results on the binding of hydrophobic ligands to β-lactoglobulin. This protein resembles luciferase in that it can bind two molecules of a small but only one molecule of a large ligand. It will be interesting to see whether β-lactoglobulin can bind the wide range of anesthetics that we have found to inhibit luciferase. Thus, in contrast to what is known about these other proteins, firefly luciferase is not only sensitive to a diverse range of anesthetic agents, but its correlation with general anesthetic potency is essentially an identity (see Fig. 4).

A striking feature of our luciferase data is the demonstration that the luciferin site can not only bind a wide variety of different molecules but that it will bind some anesthetic molecules very much tighter than the luciferin molecule itself, even though they bear no discernible structural similarities to the natural substrate. This would not have been easy to predict *a priori* and clearly shows that a molecule need not resemble an endogenous substrate in order to be a very effective competitive inhibitor.

Finally, what are the implications of these results for possible molecular mechanisms of general anesthesia? Our data show plainly that anesthetics might well act by binding to proteins directly, without involving lipids at all. The functions of these proteins could be affected in a number of different ways,[13,19,21] but our results emphasize the possible importance of competitive interactions. Hydrophobic pockets have been found in a number of soluble proteins, and it seems reasonable to suppose that such pockets will be found in both enzymes and membrane receptors in the central nervous system. If this is so, then general anesthetics might act by competing at these pockets for the binding of endogenous ligands such as neurotransmitters and neuromodulators.

ACKNOWLEDGMENTS. We thank Professor Alan Fersht for help throughout this work, Dr. Roy Burns for running the gels, John Akins for expert technical assistance, Mike Harris and Susan Watts for constructing some of the apparatus, Nick Jackson for the photographic work, and Ian Clark for the artwork. This research was supported by grants from the MRC and the BOC Group, Inc.

REFERENCES

1. Adey G, Wardley-Smith B, White D: Mechanism of inhibition of bacterial luciferase by anaesthetics. *Life Sci* 1976;17:1849–1854.
2. Branchini BR, Marschner TM, Montemurro AM: A convenient affinity chromatography-based purification of firefly luciferase. *Anal Biochem* 1980;104:386–396.

3. Cornish-Bowden A: *Fundamentals of Enzyme Kinetics*. London, Butterworths, 1979.
4. DeLuca M: Hydrophobic nature of the active site of firefly luciferase. *Biochemistry* 1969;8:160–166.
5. DeLuca M, McElroy WD: Kinetics of the firefly luciferase catalyzed reactions. *Biochemistry* 1974;13:921–925.
6. Denburg JL, Lee RT, McElroy WD: Substrate-binding properties of firefly luciferase. I. Luciferin-binding site. *Arch Biochem Biophys* 1969;134:381–394.
7. Franks NP, Lieb WR: Where do general anaesthetics act? *Nature* 1978;274:339–342.
8. Franks NP, Lieb WR: Molecular mechanisms of general anaesthesia. *Nature* 1982;300:487–493.
9. Franks NP, Lieb WR: Do general anaesthetics act by competitive binding to specific receptors? *Nature* 1984;310:599–601.
10. Halsey MJ, Smith EB: Effects of anaesthetics on luminous bacteria. *Nature* 1970;227:1363–1365.
11. Hansch C, Dunn WJ: Linear relationships between lipophilic character and biological activity of drugs. *J Pharm Sci* 1972;61:1–19.
12. Hymes AJ, Cuppett CC, Canady WJ: Thermodynamics of α-chymotrypsin-inhibitor complex formation. *J Biol Chem* 1969;244:637–643.
13. LaBella FS: The general anesthesia receptor. *Perspect Biol Med* 1982;25:322–331.
14. Middleton AJ, Smith EB: General anaesthetics and bacterial luminescence. I. The effect of diethyl ether on the *in vivo* light emission of *Vibrio fischeri*. *Proc Roy Soc B* 1976a;193:159–171.
15. Middleton AJ, Smith EB: General anaesthetics and bacterial luminescence. II. The effect of diethyl ether on the *in vitro* light emission of *Vibrio fischeri*. *Proc Roy Soc B* 1976b;193:173–190.
16. Miles JL, Morey E, Crain F, Gross S, San Julian J, Canady WJ: Inhibition of α-chymotrypsin by diethyl ether and certain alcohols: A new type of competitive inhibition. *J Biol Chem* 1962;237:1319–1322.
17. Millar RA, Beard DJ, Hulands GH: Oxyhaemoglobin dissociation curves in vitro with and without the anaesthetics halothane and cyclopropane. *Br J Anaesth* 1971;43:1003–1011.
18. Richards CD: The mechanisms of general anaesthesia, in Norman J, Whitwam JG (eds): *Topical Reviews in Anaesthesia*, Vol 1. Bristol, John Wright & Sons Ltd., 1980, pp 1–84.
19. Richards CD, Martin K, Gregory S, Keightley CA, Hesketh TR, Smith GA, Warren GB, Metcalfe JC: Degenerate perturbations of protein structure as the mechanism of anaesthetic action. *Nature* 1978;276:775–779.
20. Robillard KA, Wishnia A: Aromatic hydrophobes and β-lactoglobulin A. Thermodynamics of binding. *Biochemistry* 1972;11:3835–3840.
21. Seeman P: The membrane actions of anesthetics and tranquilizers. *Pharmacol Rev* 1972;24:583–655.
22. Smith RN, Hansch C: Hydrophobic interaction of small molecules with α-chymotrypsin. *Biochemistry* 1973;12:4924–4937.
23. Ueda I, Kamaya H: Kinetic and thermodynamic aspects of the mechanism of general anesthesia in a model system of firefly luminescence *in vitro*. *Anesthesiology* 1973;38:425–436.
24. White DC, Dundas CR: Effect of anaesthetics on emission of light by luminous bacteria. *Nature* 1970;226:456–458.
25. Wishnia A, Pinder TW: Hydrophobic interactions in proteins. The alkane binding site of β-lactoglobulins A and B. *Biochemistry* 1966;5:1534–1542.

V

PRESSURE–ANESTHETIC INTERACTIONS

27

Homogeneous and Branching Axons

Differing Responses to Anesthetics and Pressure

Joan J. Kendig and Yoram Grossman

ABSTRACT. A major unresolved question in anesthetic theory is whether pressure reversal of anesthesia represents directly opposing effects of anesthetics and pressure at the same molecular site, or indirect antagonism. Several possibilities for each are outlined, together with the evidence currently available. The target of much current research is the identification of specific neuronal proteins, whose function is altered by anesthetics in a way consistent with a role in anesthesia. One such relatively well-understood protein, the sodium channel that underlies the action potential, displays at least a limited pressure–anesthetic antagonism in some nerves.

Studies on the sodium channel in myelinated nerve reveal a basis for pressure reversal of an anesthetic effect, but fail to discriminate between direct and indirect antagonism. In other types of nonmyelinated axon, however, pressure either does not reverse anesthetic effects or does so only transiently. The conclusion is drawn that responses of defined ion channels to either anesthetics or pressure may be determined by their cellular or membrane environment.

INTRODUCTION

On exposure to hyperbaric pressures of the order of tens of atmospheres, anesthetized animals resume locomotion. Although the phenomenon of pressure reversal of the anesthetic state has been recognized for more than 30 years,[14] its basis in cellular and membrane physiology is not yet established. Much of the research into this problem has been motivated by the assumption that the underlying antagonism is direct at the most fundamental level, that pressure actually does "reverse" a unitary molecular change or set of changes underlying anesthesia for all types of agents. If this assumption is correct, pressure studies have great value as a tool for exploring the cellular and molecular mechanisms of anesthesia. Pressure raises the anesthetic requirement for all the molecularly diverse types of anesthetic agents, including inert gases.[3,10,22] Evidence has been presented, however, that the pressure/anesthetic requirement curve may be nonlinear and display different slopes for different types of agents.[11] If that is the case, it is a finding consistent with an indirect antagonism

Joan J. Kendig • Department of Anesthesia, Stanford University School of Medicine, Stanford, California 94305. *Yoram Grossman* • Department of Physiology, Faculty of Health Sciences, Ben-Gurion University of the Negev, Beer-Sheva 84 105, Israel.

between pressure and essential actions of anesthetic agents, and with differences among agents in the ways in which they bring about anesthesia. Whether the molecular antagonism is direct or indirect is the major unresolved question in pressure–anesthetic relationships.

Since the Second Conference on Molecular and Cellular Mechanisms of Anesthesia was held,[6] little progress appears to have been made in identifying the cellular substratum of consciousness. It is thus not yet possible to ascribe the loss of consciousness produced by anesthesia to a specific cellular effect or set of effects. There has been a considerable advance, however, in the extent to which current work reveals cellular and membrane actions common to several classes of anesthetic agents. One may point to evidence presented in this volume that stabilization of the desensitized state of the acetylcholine receptor is brought about by several classes of anesthetic agents (see Taylor *et al.*, Cohen *et al.*, and Miller *et al.*, this volume, Chapters 9–11); that a hyperpolarizing shift in sodium-channel inactivation voltage dependence, a similar stabilization of an inactive conformation of this channel, appears to be induced by all types of anesthetics (see Haydon *et al.*, this volume, Chapter 22); and that some but not all general anesthetics enhance GABA-mediated inhibition (see Macdonald *et al.*, Gage *et al.*, and Owen *et al.*, this volume, Chapters 2, 3, and 12). The relevance of any of these effects to a mechanism of anesthesia remains to be determined. They do illuminate, however, a number of possibilities for either a direct or an indirect antagonism between hyperbaric pressure and anesthetics. For instance, as has often been proposed, anesthetics may act principally by blocking a postsynaptic membrane receptor–ion channel complex at excitatory synapses. Although at each of the limited number of excitatory synapses that have been studied, hyperbaric pressure depresses excitatory synaptic transmission,[15] it is conceivable that some central synapses may show an opposite response. Indirect evidence has recently been presented in support of such a possibility.[1] If it should be found that anesthetics block while pressure enhances the postsynaptic response to some transmitter, this would constitute a direct antagonism between pressure and anesthetics at the cellular level. Alternatively, pressure may restore normal excitability to an anesthetic-depressed cell by an indirect action. For instance, anesthetics have been reported to selectively depress excitatory as opposed to inhibitory synaptic transmission[23]; it is possible that pressure may exert the opposite effect, selectively depressing inhibition. The result would be to restore an approximation of "normal" excitability to the cell by indirect means. This remains to be demonstrated directly; however, evidence is presented by E. B. Smith *et al.* (this volume, Chapter 28) that at least the convulsant if not the antianesthetic effects of hyperbaric pressure may be mediated through specific depression of inhibition.

At the molecular level of single channels, even less can currently be said. It is possible also at this level that anesthetics and pressure could antagonize each other directly. For example, anesthetics may act to depress chemically gated or voltage-gated ion channels by decreasing mean channel open time. Pressure could antagonize this effect directly, by increasing channel open time, or indirectly, by increasing single-channel conductance. The sole study available to date suggests that pressure affects kinetics, rather than conductance, in the alamethicin channel model.[4] It is probable that more such single-channel studies will be carried out in the near future.

The goal of recent work in our laboratory has been to identify the sites in excitable cell membranes at which pressure and anesthetic agents interact antagonistically, and to try to relate our findings to pressure "reversal" of anesthesia and/or to anesthetic amelioration of the high pressure neurological syndrome. Early findings showed that impulse-conducting axons support pressure–anesthetic antagonism in the form of pressure-induced restoration of amplitude in the partially anesthetic-blocked action potential.[19] Therefore, we have pursued studies on axons and on the voltage-dependent sodium channel, which underlies the action potential, to identify the mechanism for the antagonism and, if possible, to test whether it is direct or indirect. This chapter describes the effects of anesthetics and pressure on two axons, one the node of Ranvier of vertebrate peripheral myelinated nerve, and the other a crustacean axon, which we found to behave quite differently.

METHODS

Axons from frog (*Xenopus laevis*) sciatic nerve were prepared for vaseline gap voltage clamping by our standard procedure.[21] In most cases, sodium and potassium currents were pharmacologically separated by blocking the axon with tetraethylammonium chloride or tetrodotoxin, respectively. The chamber containing the node of Ranvier was placed on a Peltier thermoelectric device within a small pressure chamber. Temperature was maintained at approximately 10° ± 0.25°C while measurements were made; transient excursions during compression and decompression steps were limited to 2°C. The chamber was pressurized by admitting helium from a commercial cylinder at 3500 $lb/in.^2$ Compression was carried out in steps of 3.5 MPa (35 atm).

The bifurcating common excitor motor neuron to the dorsal abdominal extensor muscles of the lobster *Homarus americanus* or *Panulirus interruptus* was isolated according to published methods.[9,24] The axon (Ax) and its two branches (medial and lateral) were prepared for extracellular stimulation and recording in the same pressure chamber used for the node of Ranvier, using modified suction electrodes. A limited number of voltage clamp experiments on the main Ax segment also were carried out in this chamber, using the same techniques as for the node of Ranvier with some modification to accommodate the larger diameter and different specific membrane resistance of this axon. Compression and temperature control were as described previously. A larger chamber, accommodating a remote-controlled micromanipulator, was used to permit simultaneous extracellular recording from the branches and intracellular microelectrode recording from Ax. The preparation was usually maintained at room temperature (20–22°C), with some experiments at lower temperatures.

RESULTS AND DISCUSSION

It has become clear from recent work on voltage-dependent sodium channels (refs. 2, 13, 16, and 21; and Haydon *et al.*, this volume, Chapter 22) that one of the common effects of anesthetic agents is to shift the curve relating membrane voltage to channel inactivation in the hyperpolarizing direction. That is, anesthetics tend to favor the inactive state, reducing the number of available channels. This effect accounts for part of the conduction block associated with these agents. One of our first goals was to test whether hyperbaric nitrogen, an anesthetic at pressures of several atmospheres, behaves like other general anesthetics in shifting inactivation voltage dependence. Exposure of the voltage-clamped node of Ranvier to 7 or to 14 atm of nitrogen, in the anesthetic range, produced a reversible pressure-dependent depression of sodium currents. When the Hodgkin–Huxley h value, representing channel inactivation, is plotted against voltage, nitrogen is seen to shift the curve in the hyperpolarizing direction (Fig. 1). This clearly shows that hyperbaric nitrogen, a classic inert gas anesthetic, does behave like all the clinical anesthetics, such as barbiturates and ether, in favoring the inactive state of the sodium channel. The result supports the assumption that inert gas narcosis and general anesthesia are fundamentally similar.[17] From this demonstration we were led to examine the effects of pressure on the same system, to test the possibility that pressure relief of anesthetic conduction block might be due to effects on inactivation.

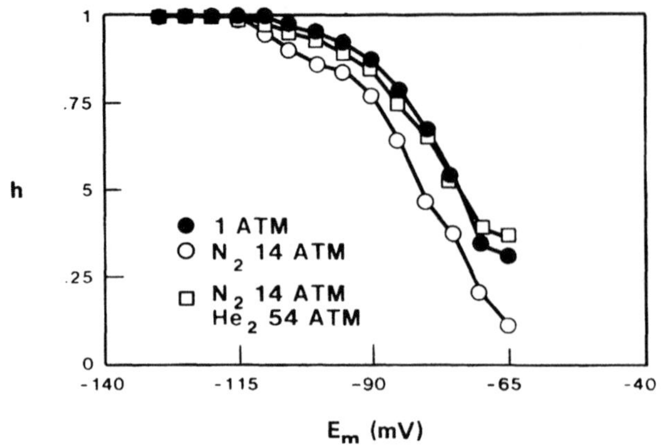

Figure 1. Opposing effects of hyperbaric nitrogen and helium pressure on sodium-channel inactivation in the node of Ranvier. The Hodgkin–Huxley *h* constant is shown on the vertical axis as peak I_{Na} normalized to 1, representing the proportion of available (noninactivated) channels; membrane potential is shown on the horizontal axis as the prepulse potential imposed for 50 msec preceding a test pulse. Nitrogen at 14 atm pressure shifts the curve of inactivation voltage dependence in the hyperpolarizing direction, reducing the proportion of available channels at any given voltage. Addition of helium to bring the total pressure to 68 atm restores the curve to its control (1 atm) position. Reproduced by permission from Kendig.[17]

When the node of Ranvier was exposed to helium pressure at 70 atm, we observed a shift of inactivation voltage dependence in the opposite, depolarizing direction. Therefore, helium pressure exerts an effect on sodium-channel inactivation in a direction opposite to the effect exerted by anesthetic agents. The next step was to interact the two together to see if this would yield pressure antagonism to anesthetic effects. This was the case. In a preparation exposed to 14 atm of nitrogen, we observed the hyperpolarizing shift due to the anesthetic effect (Fig. 1). Addition of helium with the nitrogen still present, to increase the total pressure to 68 atm, restored the curve of inactivation voltage dependence to its 1-atm position (Fig. 1). The opposing effects of anesthetics and hyperbaric pressure on inactivation provide a very plausible basis for the kind of pressure reversal of anesthetic conduction block that we have observed in other axons. In our previous studies, pressure reversal of anesthetic conduction block was found to be partial.[18,19] If anesthetics reduced the compound action potential by an amount much greater than 10% of its control amplitude, pressure did not fully restore the amplitude to control values. These shifts in sodium-channel inactivation will yield exactly that sort of partial antagonsim. Since there is also a basal level of block that is not dependent on inactivation, one would not expect this mechanism to provide complete pressure reversal, but only a partial restoration up to the level where *h* is again equal to 1. Beyond that level, pressure would be ineffective. Therefore, this mechanism provides a rather satisfying explanation in cellular terms for pressure antagonism to anesthetic conduction block.

One would like to pursue this question in molecular terms and ask whether this, in fact, represents a direct antagonism between pressure and anesthetics in their effects on conformation states of the sodium channel. Steady-state inactivation levels do not provide much information about mechanism. One way to get at mechanism is to look at the voltage dependence of the time constant of inactivation, τ_h. In the classic Hodgkin–Huxley analysis, τ_h is made up of two rate constants, one of which predominates at hyperpolarized potentials, the other at depolarized potentials. On examining the voltage dependence of τ_h at hyperbaric pressure, we observed a pressure-induced increase that was very significantly voltage dependent. Pressure appeared to selectively increase τ_h at depolarized potentials, with very little effect at hyperpolarized potentials. This result can be interpreted as a selective slowing effect on the development of inactivation. This result is at least qualitatively consistent with the observed shift in steady-state inactivation at hyperbaric pressure.

Do anesthetics also exert a voltage-dependent effect on τ_h? Evidence had been presented in another preparation that general anesthetics, in fact, act on the time constant of inactivation in a way that could be consistent with a direct antagonism between anesthetics and pressure; that is, anesthetics were observed to increase the rate of development of inactivation,[2] changing it in the direction opposite to the pressure effect we observed. Such a result would be consistent with, although not proof of, a direct opposition on the same molecular conformation change of the channel. Unfortunately, in spite of several attempts on the node of Ranvier, we were never able to satisfy ourselves that any general anesthetic, including diethyl ether, barbiturates, or hyperbaric nitrogen, had a selective effect on one end or the other of the τ_h/voltage curve.[17,21] Therefore, for general anesthetics this preparation provides no evidence on the question of direct versus indirect antagonism at any molecular level. It is probable that evidence at that level is going to have to await single-channel studies. For onc class of conduction-blocking agents, however, the evidence is already quite clear, and it is negative. Some time ago we demonstrated that local anesthetic conduction block is also reversible by pressure.[18] In the case of benzocaine, which exerts one of the largest shifts in steady-state inactivation, the antagonism must be indirect; the shift in steady-state inactivation induced by benzocaine is associated with a change in the rate constant that corresponds to recovery from, rather than development of, inactivation.[21] That is, benzocaine shifts steady-state inactivation by a mechanism that must be different from that which accounts for the opposite effect produced by pressure. For the local anesthetics, then, we have evidence on this system that the apparent antagonism between anesthetics and pressure is produced indirectly, even though both are acting on the same ionic channel.

The demonstration of the apparent basis for pressure reversal of anesthetic conduction block is satisfying in itself. No one, however, has ever been satisfied with arguments that studies on the sodium channel in axons are relevant to anesthesia in the central nervous system, because of the evidence that axons, such as vertebrate peripheral nerve, are very insensitive to anesthetic agents. One way around this argument is to try to find an axon that, in fact, is sensitive to anesthetic agents. Our choice was the motor neuron of the common excitor to the abdominal extensor muscles of the lobster. The axon branches, and at the point of bifurcation, conduction is very labile; at physiological frequencies, quite modest rates of about 30 Hz, conduction in the branches fails.[9,24] This axon therefore represents a case in which the safety factor for conduction is quite low compared with the homogeneous peripheral axons we had used in our previous studies. This axon provided a critical test of the possibility that even such a large axon, 50–100 μ in diameter, might have special properties rendering it sensitive to general anesthetics at or close to the clinical range. This proved to be the case at approximately twice the clinical anesthetic concentration, halothane at 1.7–3.4 mM and ether at 100–200 mM produced frank conduction block in the large-diameter main axon. More interestingly, at very much lower concentrations close to the anesthetic range, there was a marked shift downward in the frequency at which conduction was blocked in the branches. The time required to block at stimulus rates of 50–75 Hz was reduced 10-fold at anesthetic concentrations that did not significantly depress the amplitude of the singly evoked action potential. Low anesthetic concentrations permitted single impulses to pass but markedly altered the pattern of impulse traffic.[7] This axon therefore fulfilled the criterion of sensitivity to anesthetics. The remaining task was to demonstrate that one can antagonize the effect of anesthetics with pressure; one could then propose that if axons with similar sensitivity could be found in the vertebrate central nervous system, the sodium channel would acquire credibility as a possible site for anesthetic action.

On compression of the axon to hyperbaric pressure, the amplitude of the intracellularly recorded action potential in the bifurcating axon did, in fact, increase (Fig. 2). Unfortunate-

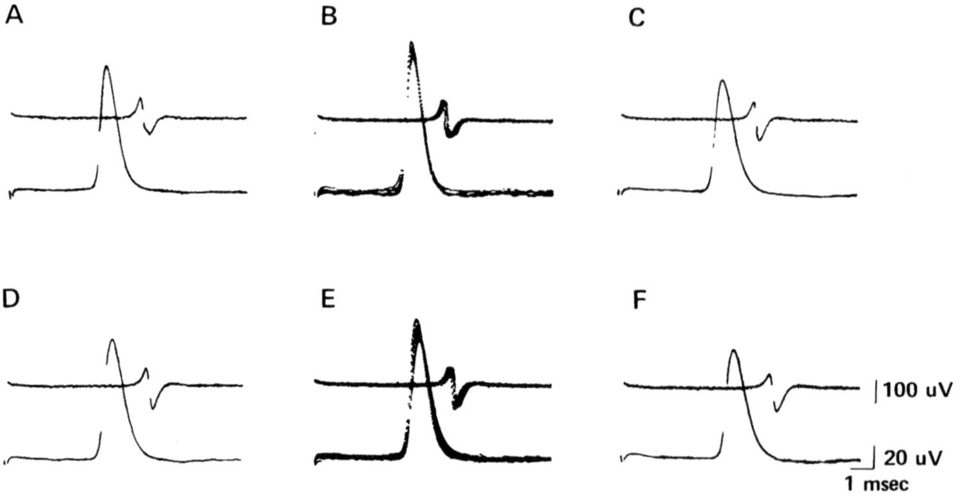

Figure 2. Bimodal effects of compression on action potential amplitude in a crustacean axon. (A) Intracellularly recorded action potential in the main axon (Ax) (lower trace), extracellular response in the medial branch (upper trace) at 35 atm. (B) Responses of the repetitively stimulated axon during compression to 70 atm. Action potential amplitude and conduction velocity increase. (C) After 20 min at the same pressure, the response amplitude decreases to below the initial level and broadens; conduction velocity is slowed. (D–F) The same sequence on compression from 70 atm (D) to 110 atm (E) and 20 min later at 110 atm (F). Initial improvements in function on compression in this axon, followed by delayed depression, were reflected also in the relative refractory period and amplitude of the inward sodium current.

ly, the effect was transient. After 15–20 min, the amplitude of the action potential declined to a level below control. This was the first time we had seen a time-dependent effect of pressure on an axon. It was also the first time we had seen any great effect of pressure in either direction on the conducted action potential in the absence of anesthetics. Extracellularly, we found that in fact pressure transiently improved the ability of this axon to conduct through the bifurcation, but again, the transient improvement was followed by a depression in ability to conduct to below-control levels. On examining the resting potential, we found that in the typical axon exposed to pressures much above 35 atm, there was a decrease in the resting potential of about 10 mV. In an axon treated with ouabain exposed to the same pressures, there was no change in the resting potential. This is evidence that, in this particular axon, pressure seems to be acting to decrease active transport, since pressure-induced depolarization is sensitive to ouabain. In voltage-clamped axons, there was a transient increase in the inward sodium currents upon exposure to 50 atm pressure for 5 min, followed by an eventual decline to below-control levels after 20 min at the same pressure. On this axon, three different recording methods, i.e., extracellular, intracellular microelectrode, and voltage clamp, produced evidence that pressure would have produced only a transient antagonism to anesthetics followed by a delayed additive depressant effect of pressure and anesthetics.[8]

Several other impulse-conducting membranes have been examined at pressure. To sum up the various axonal pressure effects, we and others have seen: Vertebrate peripheral nerve supports opposing shifts in inactivation level induced by anesthetics and pressure. A few crustacean axons exhibit repetitive firing.[20] The branching crustacean axon we have just described does not fire repetitively at pressure, but does exhibit a bimodal effect of pressure with a transient increase, followed by a delayed decrease in its ability to conduct. In voltage-clamped squid giant axon, there is no effect of pressure on steady-state inactivation,[5] nor is there in molluscan cell bodies.[12] In neither of the latter two cases is there evidence for pressure–anesthetic antagonism at the level of channel proteins. The only similarity among different cells so far observed is that pressure slows the time constants of both sodium and potassium currents, but to an extent that differs considerably among axons. It is difficult to account for this range of axon responses to pressure. The sodium channel that governs the action potential is evolutionarily a very conservative channel: in phyla as disparate as molluscs and vertebrates, it seems to display similar properties. Therefore, the variability in the response to pressure of this very stable, very well-defined channel seems to depend on the axonal membrane environment, including sheath components, in which it finds itself. Dismayingly enough, it appears that one cannot predict from studies on one axon how another axon is going to behave in response to pressure with respect to the ability of the axon actually to generate and conduct an action potential. If indeed it is impossible to generalize from one impulse-conducting membrane to another in this fashion, then it follows that no conclusions can be drawn from studies on any peripheral axon about the integrated behavior of central axons in response to anesthetics and pressure. If this is the case, then only studies on the central nervous system, the undoubted site of anesthesia, will produce results of clear-cut relevance to the role played by any particular cellular process in anesthesia and its reversal by pressure.

REFERENCES

1. Angel A, Halsey MJ, Little H, Meldrum BS, Ross JAS, Rostain J-C, Wardley-Smith B: Specific effects of drugs at pressure: Animal investigations. *Phil Trans Roy Soc B* 1984;304:85–94.
2. Bean BP, Shrager P, Goldstein DA: Modification of sodium and potassium channel gating kinetics by ether and by halothane. *J Gen Physiol* 1981;77:233–253.
3. Brauer RW, Hogan PM, Hugon M, MacDonald AG, Miller KW: Patterns of interaction of light metabolically inert gases with those of hydrostatic pressure as such—A review. *Undersea Biomed Res* 1983;9:353–396.
4. Bruner LJ, Hall JE: Pressure effects on alamethicin conductance in bilayer membranes. *Biophys J* 1983;44:39–47.
5. Conti F, Fioravanti R, Segal, JR, Stuhmer W: Pressure dependence of the sodium currents of squid giant axon. *J Membr Biol* 1982;69:23–34.
6. Fink BR, (ed): *Molecular Mechanisms of Anesthesia, Progress in Anesthesiology,* vol 2. New York, Raven Press, 1980.
7. Grossman Y, Kendig JJ: General anesthetic block of a bifurcating axon. *Brain Res* 1982;245:148–153.
8. Grossman Y, Kendig JJ: Pressure and temperature: Time-dependent modulation of membrane properties in a bifurcating axon. *J Neurophysiol* 1984;52:692–708.
9. Grossman Y, Parnas I, Spira ME: Ionic mechanism involved in differential conduction of action potentials at high frequency in a branching axon. *J Physiol* 1979;295:307–322.
10. Halsey MJ, Wardley-Smith, B: Pressure reversal of narcosis produced by anesthetics, narcotics and tranquilizers. *Nature* 1975;257:811–813.
11. Halsey MJ, Wardley-Smith, B, Green, CJ: The pressure reversal of general anesthesia—A multi-site expansion hypothesis. *Br J Anaesth* 1978;50:1091–1097.
12. Harper AA, MacDonald AG, Wann KT: The action of high hydrostatic pressure on the membrane currents of Helix neurons. *J Physiol* 1981;311:325–339.
13. Haydon DA, Urban BW: The action of hydrocarbons and carbon tetrachloride on the sodium currents of the squid giant axon. *J Physiol* 1983;338:435–450.
14. Johnson FH, Flagler EA: Hydrostatic pressure reversal of narcosis in tadpoles. *Science* 1950;112:91–92.
15. Kendig JJ: Anesthetics and pressure in nerve cells, in Fink BR (ed): *Molecular Mechanisms of Anesthesia, Progress in Anesthesiology,* vol 2. New York, Raven Press, 1980, pp 59–68.
16. Kendig JJ: Barbiturates: Active form and site of action at node of Ranvier sodium channels. *J Pharmacol Exp Ther* 1981;218:175–181.
17. Kendig JJ: Nitrogen narcosis and pressure reversal of anesthetic effects in node of Ranvier. *Am J Physiol* 1984;46(*Cell Physiol* 15):C91–C95.

18. Kendig JJ, Cohen EN: Pressure antagonism to nerve conduction block by anesthetic agents. *Anesthesiology* 1977;47:6–10.
19. Kendig JJ, Trudell JR, Cohen EN: Effects of pressure and anesthetics on conduction and synaptic transmission. *J Pharmacol Exp Ther* 1975;195:216–224.
20. Kendig JJ, Schneider TM, Cohen EN: Pressure, temperature and repetitive impulse generation in crustacean axons. *J Appl Physiol* 1978;45:742–746.
21. Kendig JJ, Courtney KR, Cohen EN: Anesthetics: Molecular correlates of voltage and frequency dependent sodium channel block in nerve. *J Pharmacol Exp Ther* 1979;210:446–452.
22. Miller KW, Wilson MW: The pressure reversal of a variety of anesthetic agents in mice. *Anesthesiology* 1978;48:104–110.
23. Richter J, Landau EM, Cohen S: The action of volatile anesthetics and convulsants on synaptic transmission: A unified concept. *Mol Pharmacol* 1977;13:548–559.
24. Spira ME, Yarom Y, Parnas I: Modulation of spike frequency by regions of special axonal geometry and by synaptic inputs. *J Neurophysiol* 1976;36:882–899.

28

New Observations on the Mechanism of Pressure–Anesthetic Interactions

E. Brian Smith, Frank Bowser-Riley, and Stephen Daniels

ABSTRACT. The application of high hydrostatic pressure to animals exposed to general anesthetics can produce a striking change in the level of response. One aspect of this process, the pressure reversal of anesthesia, has long been regarded as a potential clue to the mechanism of action of anesthetics. In this chapter we review some recent experiments that shed new light on pressure–anesthetic interactions.

Evidence is put forward which indicates that the pharmacology of pressure in vertebrates is more similar to that of strychnine than any other centrally acting stimulant. As strychnine is believed to act by blocking the action of the inhibitory transmitter glycine, it is suggested that the effects of pressure may arise, to a large extent, from an interaction with some aspect of this inhibitory process.

It has been observed that crustacea, which are insensitive to strychnine, do not exhibit the pressure reversal of anesthesia characteristic of vertebrates. The extent to which glycinergic processes may be involved in those aspects of general anesthesia that are pressure-reversed is discussed. It is concluded that the effects of pressure and anesthetics may not be entirely independent. There is no direct evidence that glycinergic inhibition plays a part in general anesthesia, but, on the other hand, if it was involved, it would provide a simple explanation of pressure–anesthetic interactions. Evidence from model systems, such as luminous bacteria, show that a single protein, the bacterial luciferase, can provide the basis for a realistic theory of the effects of pressure and anesthetics on the luminosity of these organisms. A protein associated with glycinergic inhibition could perhaps play an equivalent role within the mammalian central nervous system.

INTRODUCTION

The degree to which chemical and physical equilibria and rate processes are affected by pressure is determined by the volume changes accompanying the equilibrium or activation processes involved,[33] ΔV, by

$$\frac{d \ln K}{dP} = \frac{\Delta V}{RT}$$

Here K is a parameter that characterizes the position of equilibrium or the rate of the reaction involved. For this reason, changes in pressure may, like temperature, induce a general

E. Brian Smith, Frank Bowser-Riley, and Stephen Daniels • Oxford Hyperbaric Group, Physical Chemistry Laboratory, Oxford OX1 3QZ, England.

perturbation of the processes that give rise to general anesthesia. Following the discovery that the level of anesthesia in a wide variety of species could be modified by the application of hydrostatic pressure, it was thought that further investigation of the effects of pressure might lead to a clearer understanding of the mechanism of general anesthesia. Unfortunately, this has not proved to be the case, and it seemed as if pressure would contribute as little to our understanding of anesthesia as had studies of the effects of temperature some decades earlier. We have sought, in the light of recent developments, to reevaluate the role of pressure as a key to elucidating aspects of the mode of action of general anesthetics. The following section summarizes some basic ideas on which the subsequent discussion will draw. This will be followed by an outline of recent research that has cast new light on pressure–anesthetic interactions. Finally, some possible consequences that arise from these findings are examined.

BACKGROUND

Physical Site of Action of Anesthetics

In 1961 Pauling[32] and Miller [30] independently proposed that the site of action of general anesthetics lies in the aqueous phases of the central nervous system. Both theories, though differing in detail, sought to relate potency to the stability of gas hydrates that many, but not all, anesthetics may form in aqueous solution. This model was not well supported by physicochemical evidence,[27] since (1) fluorine compounds did not fit the correlation (2) many anesthetics including ether and halothane do not form hydrates, and (3) the potency of mixtures was found to be additive.[10,28] These observations have considerably narrowed the search for a molecular mechanism of anesthetic action.

Some older theories of anesthetic action, with roots in the 19th century, sought to relate potency to solubility in fatty substances. A model, described by Meyer,[25] stated that "narcosis commences when any chemically indifferent substance has attained a certain molar concentration in the lipids of the cell. This concentration depends on the nature of the animal or cell but is independent of the narcotic." Therefore, it was argued that anesthetic potency should be directly proportional to solubility in a solvent selected to match the properties of the site action. This has been tested with many solvents, olive oil and octanol being most favored. The site of action has been characterized more systematically by use of the solubility parameter (δ) theory. This parameter relates solubility to the strength of intermolecular forces and is defined for fluids by $\delta = (-E/V)^{1/2}$, where E is the energy of vaporization of the solvent and V its molar volume. It was found that a solvent with the solubility parameter, δ, equal to 9.1 (cal $cm^{-3})^{1/2}$ was consistent with the anesthetic data. This lipid solubility model enables anesthetic potency to be predicted to within 20% over a range of potency of a factor of 10,000. It has been called one of the most remarkable correlations in science.

The Molecular Nature of the Site of Action

It is now generally accepted that the site of action of anesthetics is hydrophobic, but the molecular nature of the site is still a matter of debate. The traditional view is that anesthetics act in the lipid region of the cellular membrane, to which is often added the belief that the induced perturbation interferes with ion transport through the membrane. However, studies of the effects of anesthetics on membrane processes lend little support to this view.

Membranes are surprisingly resistant to the effects of anesthetics, and the changes that result are such as could be produced by variations of temperature within the physiological range.[11] In 1954, Johnson, Eyring, and Polissar[19] wrote, "... it is not necessary to assume that anaesthetic action involved cell lipids." Indeed, on the basis of the results of experiments on the action of anesthetics on ion transport processes, it was concluded that "the results give no indication that the general perturbation of membrane processes is an essential feature of the mode of action of general anaesthetics at clinical concentrations."[15] There seems no reason to modify this conclusion at the present time.

The hypothesis that anesthetics act at a hydrophobic region of a protein was argued by Johnson *et al.*[19] to be equally plausible. Considerable attention has been paid to evaluating this possibility. For most proteins investigated, no functional change appears to arise from the binding of anesthetics.[11] However, the inhibition of the light output of luminous bacteria by anesthetics, near clinical concentrations, appears to arise from an interaction with luciferase, the enzyme responsible for the light emission.[16,26,35] These findings may suggest a hydrophobic region within a macromolecule as the critical site. This does not conflict with the Meyer–Overton model since the interaction appears to be closely related to solubility in hydrophobic solvents. This has been confirmed by the recent experiments on firefly luciferase by Franks and Lieb.[12] They showed that, over a 100,000-fold range in potency, the anesthetic concentrations required to reduce the activity of purified firefly luciferase by 50% were essentially identical to those which induce anesthesia in mammals. Anesthetics were shown to act competitively with the luciferase substrate. Franks and Lieb[12] concluded that the volatile general anesthetics, despite their diversity, all acted by competing with the endogenous ligands by binding to specific receptors, and it was suggested that the anesthetic binding site can accommodate two small anesthetic molecules but only one for agents larger than hexanol.

Pressure Reversal of General Anesthesia

In 1951, Johnson and Flagler[18] made a remarkable observation. They showed that pressure could restore the luminosity of luminous bacteria exposed to an anesthetic agent and conducted a similar experiment on tadpoles treated with ethanol. Tadpoles exposed to 2.5% ethanol ceased swimming owing to narcosis. Swimming was also inhibited by pressures of 200–300 atm. However, at lower pressures, swimming was restored in ethanol-treated tadpoles, thus indicating an antagonism between the effects of pressure and the anesthetic action of ethanol.

This antagonism has since been observed with other animals using helium as the pressurizing agent.[22] For the Italian great crested newt 34 atm of nitrogen acts as an anesthetic and reduces the ability of the animals to right themselves. As additional pressure is applied, the performance of the animals is restored to normal levels. Similar observations have been made using nitrous oxide as the anesthetic. This antagonism has been observed in mammals, where mice have remained active at 280 atm at concentrations of anesthetic that would normally be lethal.

One hypothesis that accounts for this antagonism implies that the concentration of molecules at the active site is the critical factor determining anesthetic action. Thus it is natural to suppose that pressure reverses anesthesia by "squeezing out" the dissolved molecules, and calculations show that between 300 and 1000 atm would be required to reverse anesthesia. However, these values are high when compared with the experimental value of approximately 100 atm, and the squeezing-out effect cannot be regarded as an adequate theory to account for the pressure reversal of anesthesia.

An alternative model that gives reasonably good prediction of the antagonism is the so-called critical volume hypothesis.[22,29] This assumes that when the anesthetic gas dissolves at an undefined hydrophobic site, it produces a fractional expansion given by

$$\Delta V/V = \gamma_a P_a V_a$$

where γ_a,P_a, and V_a are the solubility coefficient, the partial pressure, and the partial molar volume, respectively, of the anesthetic gas. Hydrostatic pressure will lead to a reduction in volume by

$$\Delta V/V = -\beta P_T$$

where β is the isothermal compressibility and P_T the total pressure. It was supposed that the function of the site was unimpaired if the effects of the anesthetic and the pressure were such as to maintain the site at its original volume. Then

$$\gamma_a P_a V_a = \beta P_T$$

This equation enables us to calculate the conditions under which pressure reversal might be expected to occur. Since the physical chemical evidence points to a hydrophobic site, γ_a, V_a, and β may be estimated with reasonable precision. The converse aspect of this antagonism, the amelioration of the effect of pressure, has also been interpreted using this model, and it has been used to calculate the optimum mixture for divers to breathe.

The High-Pressure Neurological Syndrome

The onset of the high-pressure neurological syndrome (HPNS), in man, is characterized by tremor, dysmetria, fasciculation, myoclonus, and "microsleep." In mice the effects of hydrostatic pressure are uncoordinated tremors with an onset pressure of about 30 atm. At pressures of 90–100 atm convulsions are observed; higher pressures induce respiratory distress, and at pressures of 120–140 atm death is observed. The onset pressures quoted must be regarded as approximate since they are dependent on the rate of compression.[8]

Although many more stages of HPNS in rodents have been reported, different laboratories have not always been able to agree on consistent definitions. In this paper, therefore, we shall simply confine our attention to the seizures, which are the clearest feature of HPNS in rodents.

Early pharmacological investigations of HPNS[8] showed that systemic administration of reserpine, a monoamine-depleting alkaloid, reduced the onset pressures of convulsions in mice by as much as 60%. More recent investigations have employed drugs that selectively inhibit the synthesis of monoamines[21] and the monoamine-depleting neurotoxins.[6] From these studies it was suggested that noradrenaline modified the sensitivity to pressure.[4]

Neuroanatomical studies have shown that pressure has a subcortical site of action[4,5,20] and that the cortex provides a substantial measure of inhibition to counter the convulsive activity that arises from subcortical regions. The interaction between pressure and the monoamine transmitters, when taken together with the results of the neuroanatomical studies, has suggested that pressure has no direct cortical action, and that the descending inhibition originating from the cortex is primarily dependent on noradrenaline.[4] The observation that reserpine abolishes the compression rate dependence of HPNS (and is therefore most effective in lowering convulsion onset pressures, during slow compressions) suggests that the cortical control mechanism may account for the compression rate dependence of HPNS and the remission of signs that frequently occurs on prolonged exposure to pressure. This model is in keeping with the variation of the compression rate dependence among species.[8]

The subcortical action of pressure suggested an analogy with the convulsants pictrotoxin and strychnine. Pictrotoxin is thought to antagonize the actions of the inhibitory neurotransmitter GABA at subcortical sites. It was observed that agents that potentiate the effects of GABA postpone the onset of high-pressure seizures.[2,3,14] However, the discovery that muscimol,[3] a GABA agonist, and baclofen,[4] a GABA analogue, do not confer protection against HPNS indicates that GABA does not play a dominant role in determining HPNS. More recently it has been shown that a number of more specific GABA-enhancing agents do not prevent the behavioral changes associated with high pressure.[17] Experiments with the rat cervical ganglion have shown that pressure does not affect the GABA-mediated transmission in this preparation.[23]

Evidence for the role of glycine in HPNS comes from the close analogy that has now been established between the actions of pressure and those of strychnine.[4,7] These data will be reviewed in the following section.

RECENT ADVANCES

Mechanism of Action of Pressure

Although some evidence has been presented to show that drugs which influence the actions of GABA and other neurotransmitters[1–3] can effect the onset of HPNS, there is at present no evidence that they mediate the primary effect of pressure. However, in the case of glycine, there is more reason to suppose that it plays a major role in determining the effects of pressure on the CNS. In this context the centrally acting muscle relaxants related to mephenesin have proved particularly interesting. We find that

1. The aromatic propandiols mephensin and methocarbamol gave excellent protection against seizures caused by pressure. They are also effective against seizures induced by strychnine.[4]
2. The aliphatic analogues meprobamate and carisoprodol failed to protect against the convulsions associated with HPNS or those due to strychnine.[4]
3. The relative potency of the ortho, meta, and para isomers of mephensin was the same for protection against both pressure- and strychnine-induced convulsions.
4. There appears to be an inverse correlation for the effectiveness of these drugs against seizures induced by metrazol (Table I).

Strychnine is believed to act postsynaptically on glycinergic inhibitory neurons, and this pharmacological evidence suggests that pressure acts in a similar manner.

Further evidence for this point of view can be obtained from the dose–response curves for strychnine measured at high pressures of helium.[7] It was found that the effects of strychnine and pressure in inducing seizures were strictly additive, suggesting that both agents may share a common mechanism in the production of convulsions. In contrast, for picrotoxin highly nonadditive behavior was observed[7] (Fig. 1). Figure 1 also illustrates the effects of simultaneous administration of strychnine and pressure in the presence of mephenesin. The effects of both strychnine and pressure are ameliorated to the same degree by mephenesin, so that the effects of the two agents remain additive, although higher doses are required in the presence of mephenesin to restore the original level of response.[7]

The actions of strychine and pressure are both characterized by extremely steep dose–response curves for which $ED_{33}/ED_{67} = 1.10 \pm 0.05$. It is interesting to note that dose–response curves of this unusual degree of steepness are observed with general anesthetics.

Table I. Action of Anticonvulsant Drugs[a]

Drug	Protection against seizures induced by			
	Pressure	Strychnine	Picrotoxin	Metrazol
Phenobarbitone	↑ ↑	↑ ↑	↑ ↑	—
Valproate	↑ ↑	↑ ↑	↑ ↑	↑ ↑
Diphenylhydantoin	0	0	0	0
Ketamine	↑ ↑	↑ ↑	—	—
Althesin	↑ ↑	↑ ↑	—	—
Mephenesin	↑ ↑	↑ ↑	0	0
Methocarbamol	↑ ↑	↑ ↑	—	0
Meprobamate	0	0	—	↑ ↑
Carisoprodol	0	0	—	↑ ↑
Baclofen	0	0	—	↑ ↑
Diazepam	↑	↑	↑ ↑	↑ ↑
Flurazepam	↑	—	↑ ↑	—

[a] ↑ ↑, Good protection; ↑, moderate protection; 0, little effect; —, no information available.

Species Variation

The fact that glycine is not thought to act as a neurotransmitter in invertebrates[13] provides the possibility of investigating its involvement in the effects of pressure more directly. In particular, we have sought to elucidate its role in the pressure reversal of anesthesia. A comparative study was carried out on the effects of pressure and anesthetics on the electrically stimulated swimming activity of freshwater shrimps (*Gammarus pulex*) and tadpoles (larvae of *Rana temporaria*) under similar experimental conditions.[34] The measure of swimming activity was the fraction of the total population not resting on the bottom when this region was exposed to an electrical stimulus. The anesthetics, chloroform, diethyl ether, ethanol, halothane, and sodium pentabarbitone were employed, and their potencies were found to be very similar for both species. However, the effect of pressure was strikingly different for shrimps and tadpoles. For the tadpoles, pressure clearly reversed the effects of

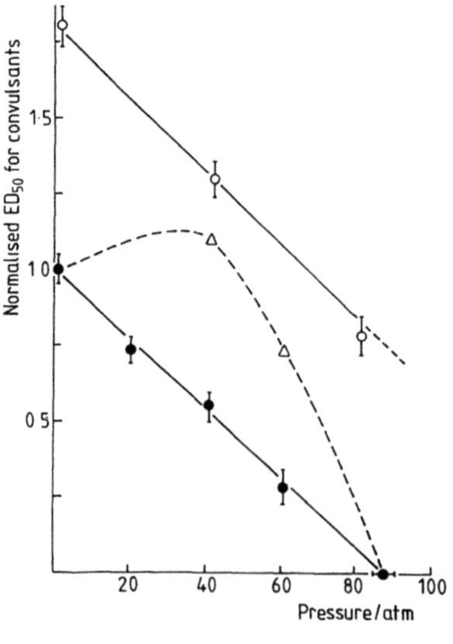

Figure 1. Effect of pressure on the ED_{50} for strychnine (●) and picrotoxin (△) in mice. The ED_{50}'s were determined by probit analysis of the dose–response curves observed for strychnine and picrotoxin administered subcutaneously at fixed pressures of 20, 40, and 60 atm. As shown, the calculated ED_{50}'s are normalized relative to the ED_{50} determined in the same manner at atmospheric pressure. The errors bars show the S.E. of the ED_{50}. The upper line (○) shows the normalized ED_{50}'s for strychnine (s.c.) at pressure following administration of mephenesin (130 mg.kg^{-1};s.c.).[7]

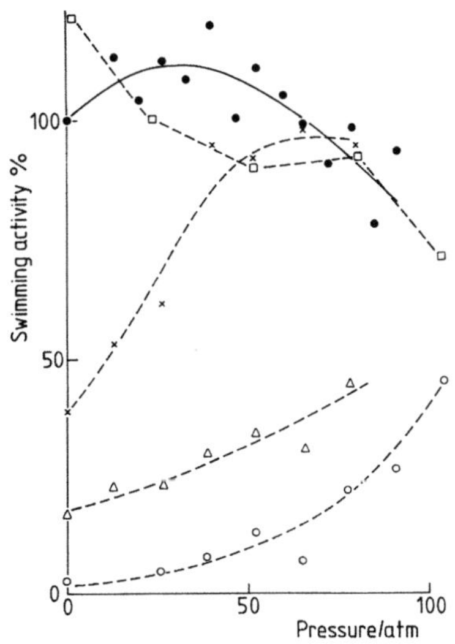

Figure 2. Swimming activity of tadpoles (larvae of *Rana temporaria*) in the presence of ethanol. ●, No ethanol; □, 0.01 M ethanol; x, 0.20 M ethanol; △, 0.265 M ethanol; ○, 0.275 M ethanol.[34]

the anesthetics enhancing the swimming activity (Fig. 2). In contrast, for the shrimps, no pressure reversal was observed, and pressure enhanced the inhibitory effects of the anesthetics (Fig. 3). In light of our previous investigations on the action of pressure, it is possible that the lack of pressure reversal may arise from the absence of glycine as a neutoransmitter in the crustacean central nervous system (CNS).

The specificity of pressure reversal does not appear to be in keeping with the most commonly employed models of pressure reversal which suppose that pressure causes a general perturbation of the cellular membrane. It suggests rather that specific proteins might be involved. Pressure and anesthetics may, of course, act at quite different sites, the final response representing a merely physiological antagonism. However, the results raise the possibility that proteins associated with glycinergic inhibition could play a role in determining those aspects of general anesthesia that are reversed by pressure (including loss of consciousness in mammals) and may indicate that the mechanisms of general anesthesia are more specific than is frequently supposed.

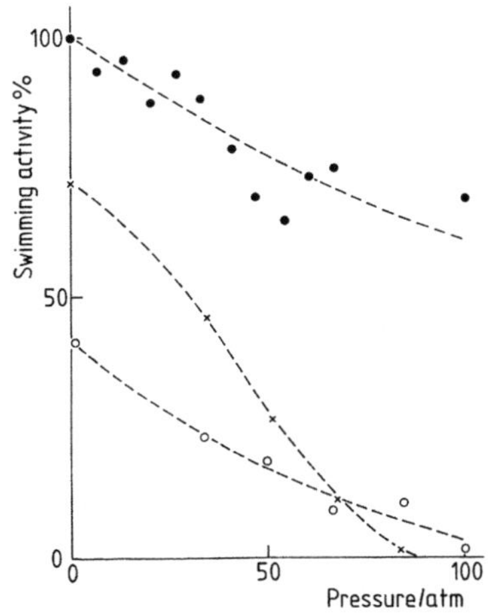

Figure 3. Swimming activity of shrimps (*Gammarus marinus*) in the presence of ethanol. ●, No ethanol; x, 0.15 M ethanol; ○, 0.25 M ethanol.[34]

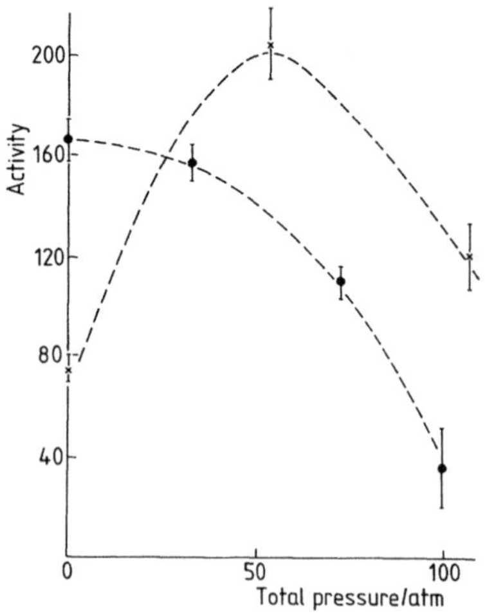

Figure 4. Activity of mice measured in a rocking cage as a function of pressure. No added nitrous oxide (x), with 12 psi lb/in.^{2} (0.82 atm) nitrous oxide (●).[33]

Hyperactivity

In vertebrates hyperactivity can be induced by low doses of anesthetics. In rodents the signs closely resemble those produced by pressures in the range 20–80 atm including tremors, increased grooming activity, bouncing motions, and, in some circumstances, increased translational movements. The phenomena appear to be a stable feature of anesthetic action at low dose levels and not merely a transient phase. In general, the increased activity becomes significant at approximately one-third of the normal anesthetic dose and reaches a maximum at about one-half this dose. This constant ratio may indicate that hyperactivity arises from a mechanism related to that which leads to the inhibitory aspects of anesthetic action.

The effect of increased pressure on the level of hyperactivity induced by anesthetics is of particular interest. Unfortunately, the quantitative measurement of hyperactivity at pressure has not been the object of systematic investigation. However, a preliminary investigation has been carried out using two methods of assessing hyperactivity, one of which employed an activity cage that could respond to fine movement and the other which determined only the extent of gross translational motion. In no case was pressure found to enhance the hyperactivity induced by 12 lb/in.^{2} of nitrous oxide, and in most experiments a substantial reduction in the level of activity was observed (Fig. 4) (P. M. Wolstenholme, S. Daniels, W. D. M. Paton, and E. B. Smith, in press). Pressure was also found to reduce the level of electrically stimulated swimming activity of tadpoles enhanced by low doses of ethanol (Fig. 2).

The results from experiments involving divers are less clear and may be determined more by the effects of compression rate rather than by the equilibrium effect of pressure.[31] However, there is considerable evidence that the addition of low doses of an anesthetic gas (nitrogen) to the oxy–helium mixtures breathed by divers leads to a reduction in the tremors and other aspects of HPNS.

Bacterial Luminescence

Indirect evidence for the view that both pressure and anesthetics could act at the same site has been obtained from the study of model systems. Membrane models have been the most widely investigated, but on the whole, it has proved difficult to relate the actions of anesthetics on such preparations to general anesthesia.[11] On the other hand, investigations employing proteins have proved more interesting. Though most enzyme processes are resistant to the effects of clinical concentrations of anesthetics, the enzymes responsible for luminescence in a number of organisms have been shown to reflect many of the features associated with the actions of general anesthetics on vertebrates. Franks and Lieb[12] have, as described earlier in this volume (Chapter 26), shown that the relative and absolute potencies of anesthetics, when inhibiting firefly luciferase, closely match those required to produce general anesthesia in mammals. Similar results are obtained for bacterial luciferases (from *Photobacterium phosphoreum* and *Aerobacter fischeri*). For these systems the detailed mechanism of action of anesthetics has been elucidated and the effects of high pressures investigated.

The reaction scheme that results in bacterial luminescence is relatively complicated. First, reduced flavin mononucleotide binds to the enzyme. The intermediate that results is oxidized to produce a long-lived species. This, in the presence of cofactor, a long-chain aldehyde, decays producing luminescence with a high quantum yield. When the luminescent process is studied, *in vivo* in the presence of anesthetics, it is observed that the doses of anesthetics required to reduce the luminescence by 50% (ED_{50}) vary through the growth cycle of the subcultures. Relatively high doses are required both early and late in the growth cycle (Fig. 5).[26] However, during the logarithmic growth stage, the ED_{50} is much lower. During the early and late phases of growth, hyperluminosity is observed with low doses of anesthetics. This effect does not occur during the logarithmic phase (Fig. 4). The biphasic response to anesthetics can be shown to arise from two components of anesthetic action which can be resolved both *in vitro* and *in vivo* [26] as: (1) Primary inhibition: this is a reduction in luminosity that occurs on a time scale less than 0.1 sec. This appears to arise from competitive inhibition by the anesthetic for the aldehyde cofactor required for luminescence. (2) Secondary stimulation: this is an enhancement of luminescence, which occurs on a time scale of seconds, that appears to require enzyme turnover and furthermore does not occur in the logarithmic phase of growth. Secondary stimulation may arise from the anesthetics facilitating the binding of the reduced flavin to the enzyme.

The net effect of anesthetic action depends on the relative magnitudes of these two components which are controlled by the level of substrates and cofactor available to the enzyme. Thus conditions that favor a large degree of secondary stimulation lead to hyperluminosity and a high ED_{50}.

The effects of pressure are also dependent on the phase of growth. Pressure reversal of the effects of anesthetics is observed only in the early and late stages of the growth cycle. It closely parallels the occurence of flavin-dependent hyperluminosity (Fig. 5) (M. J. Hallowell and E. B. Smith, in press) and is related to the sensitivity of the organism to anesthetics.

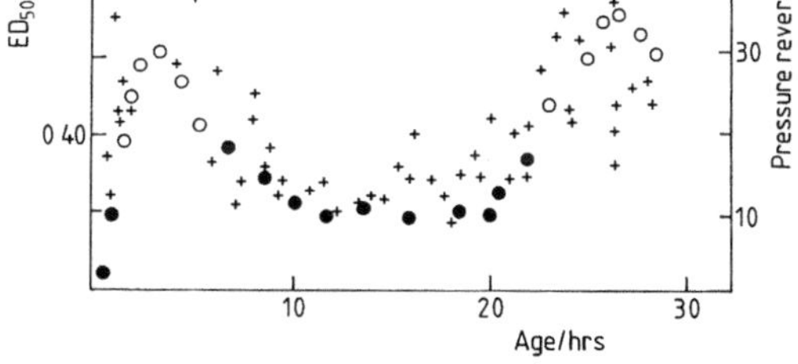

Figure 5. Anesthetic potency of urethane as a function of age of subculture for *Photobacterium phosphoreum* (M. J. Hallowell and E. B. Smith, in press). ED_{50} (○) × 10/M. Occurrence of hyperluminosity at low doses of anesthetic indicated by ●. Crosses (+) represent the percentage recovery of luminescence of the bacteria, in the presence of 0.05 M urethane on the application of 270 atm pressure at 18.5°C.

These observations suggest that both anesthetics and pressure can act on the bacterial system to produce excitation by mechanisms that are probably related through the secondary stimulation process. Although there is no evidence that the luciferases are related to any proteins found within the vertebrate CNS, the findings raise the possibility that an interaction of pressure and anesthetics with a single protein would be sufficient to explain many of the complex features of the effects of pressure and anesthetics on the vertebrate CNS.

DISCUSSION

The Mode of Action of Pressure

The additivity of strychnine and pressure in inducing seizures leads us to suggest that they mediate their actions via the same neurological pathway. The fact that they both have dose–response curves of similar (and unusual) steepness is compatible with the suggestion that they share a common mechanism. The observation that the administration of mephenesin affects, in equal proportions, the dose of either strychnine or pressure required to produce a given response adds further support to this view. If the convulsive action of strychnine arises from an antagonism of the postsynaptic inhibitory processes mediated by the neurotransmitter glycine, it can therefore be suggested that the convulsive action of pressure arises via a similar mechanism.

Interaction of Pressure and Anesthetics

The question of how far the mechanism of general anesthesia is related to that of pressure is less easily resolved. Clearly the two agents can interact, as is most obvious in the pressure reversal of anesthesia. Furthermore, their actions have some features in common. Anesthetics share with strychnine and pressure unusually steep dose–response curves for which $ED_{33}/ED_{67} \sim 1.10$. However, the starting point for any discussion must be that pressure and anesthetics act independently and that pressure reversal represents a physiological rather than a pharmacological antagonism. On the basis of this assumption, we can analyze the species differences between shrimps and tadpoles, and we will refer to this as *Model 1*.

Model 1. *Anesthetics act by the same mechanism in both vertebrates and invertebrates. Pressure acts by an independent mechanism.*

This model is consistent with the fact that the sensitivity to anesthetics of shrimps, tadpoles, and indeed a wide selection of mammals is very similar. However, it also suggests that the dose–response curves for anesthesia in different organisms would be similar. Although this has not been the subject of systematic investigation, we have observed that for the electrically stimulated swimming activity of shrimps $ED_{33}/ED_{67} \sim 1.70$, corresponding to less steep dose–response curves than those reported for vertebrates.

The model suggests that pressure reversal of anesthesia arises from an independent action of pressure in tadpoles by a mechanism that does not occur in shrimps. The mechanism of anesthesia is presumed to be the same in both organisms and, as shown by the shrimps, not sensitive to pressure. Thermodynamic reasoning suggests that the volume change associated with the anesthetic process is close to zero. This is contrary to the commonly employed predictive models which suggest that anesthetics act following absorption at a hydrophobic site, for which the associated volume change would be expected to be 50–100 cm^{-3} $mole^{-1}$.

Both species might be expected to show a similar degree of anesthetic-induced hyperactivity if the same mechanism was involved in both excitatory and inhibitory aspects of anesthetic action. The fact that shrimps do not show such hyperactivity is inconsistent with this view, although a number of alternative explanations are possible. A more substantial difficulty arises from the expectation that the pressure and anesthetic-induced components of hyperactivity (being independent) should add to or at least not diminish each other. In fact, in both tadpoles and mice, there is some evidence that pressure *reduces* the level of anesthetic-induced hyperactivity.

These difficulties suggest that other models may be worthy of attention. Perhaps the most obvious is the following.

Model 2. Anesthesia in shrimps results from a different mechanism than that operating in vertebrates: The actions of pressure and anesthetics in vertebrates are not independent.

On the basis of this view, the similar potencies observed in shrimps and vertebrates would be coincidental. The different slopes of the dose–response curves would be directly explained.

This model allows us to consider the possibility that in vertebrates, pressure and anesthetics act by a linked mechanism. If, for these species, anesthetics bind with a positive volume change, then we have a ready explanation for the observation that both the excitatory and inhibitory effects of anesthetics are reduced by the application of pressure. The principal excitatory effect of pressure could involve a different mechanism from that for anesthesia. On the other hand, some aspects of the action of general anesthetics could involve glycinergic inhibition. The possibility that both pressure and anesthetics could act on the same protein is made more plausible by the results on bacterial luminescence discussed previously.

Direct evidence that anesthetics act on glycinergic processes at higher levels within the CNS is not available. Different investigations of the action of anesthetics at the spinal level lead to conflicting conclusions. Results have been presented which show that ketamine blocks the excitatory responses of spinal neurons, in an anaesthetized cat, to *n*-methyl aspartate but not the inhibitory responses to glycine.[24] However, experiments on a lamprey reticulospinal preparation[9] suggested that ketamine could exert a dose-dependent antagonism of the inhibitory response to glycine thus accounting for the hyperactive anesthetic action of ketamine.

CONCLUSIONS

Research over the past 2 years has indicated that sensitivity to strychnine and pressure is strongly correlated. Since strychnine is believed to block glycinergic processes, this may indicate that postsynaptic glycinergic inhibition plays an important role in determining the effects of pressure. The action of anesthetics and pressure do not appear to be entirely independent, and at the present time, we cannot rule out the possibility that glycine-mediated inhibitory processes may play some part in determining those aspects of general anesthesia, such as loss of consciousness in mammals, that are reversed by pressure.

Irrespective of any conclusions in this matter, the recent results are hard to reconcile with the view that anesthetics act by causing a general perturbation of membrane processes. The fact that *either* anesthesia in crustacea and vertebrates occurs by a different mechanism *or* the mechanism of anesthesia in both species involves a negligible volume change is more consistent with a protein site of action rather than an extensive phase such as a membrane.

At the present time many loose ends remain. But recent findings on the interactions between pressure and anesthetics suggest many further experiments, the results of which may aid the elucidation of the mechanisms that underly general anesthesia.

REFERENCES

1. Angel A, Halsey MJ, Little H, Meldrum BS, Ross JAS, Rostain J-C, Wardley-Smith B: Specific effects of drugs at pressure: Animal investigations. *Phil Trans R Soc Lond* B 1984;304:85–95.
2. Bichard AR, Little HJ: The benzodiazepine antagonist, RO 15-1788, prevents the effects of flurazepam on the high pressure neurological syndrome. *Neuropharmacology* 1982;21:877–880.
3. Bichard AR, Little HJ, Paton WDM: The involvement of GABA in the high pressure neurological syndrome. *Br J Pharmacol* 1984;74:221P.
4. Bowser-Riley F: Mechanistic studies on the high pressure neurological syndrome. *Phil Trans R Soc Lond B* 1984;304:31–41.
5. Bowser-Riley F, Paton WDM, Smith EB: Experiments to investigate the neural origin of the high pressure neurological syndrome. *Br J Pharmacol* 1981;74:820–821.
6. Bowser-Riley F, Dobbie JA, Paton WDM, Smith EB: A possible role for mono-aminergic inhibition in the high pressure neurological syndrome. *Undersea Biomed Res Suppl* 1982;9:32.
7. Bowser-Riley F, Tyers G, Paton WDM, Smith EB: The additive effect of pressure with strychnine but not picrotoxin in evoking convulsions in mice. *J Physiol* 1984;357:19.
8. Brauer RW, Beaver RW, Sheenan ME: Role of monoamine neurotransmitters in the compression rate dependence of HPNS convulsions, in Shilling CW, Bennett MW (eds): *Proceedings of the VIth Symposium on Underwater Physiology*. Bethesda, FASEB, 1978, pp 49–60.
9. Cullen KD, Martin RJ: Effects of injectable anaesthetics on responses to L-glutamate and on spontaneous synaptic activity in *Lamprey* recticulo-spinal neurones. *Br J Pharmacol* 1984;82:659–666.
10. Eger EI, Lundgren C, Miller SL, Stevens WC: Anaesthetic potencies of SF_6, CF_4, chloroform and ethrane in dogs. *Anesthesiology* 1969;30:129–135.
11. Franks NP, Lieb WR: Molecular mechanisms of general anaesthesia. *Nature (London)* 1982;300:487–493.
12. Franks NP, Lieb WR: Do general anaesthetics act by competitive binding to specific receptors. *Nature (London)* 1984;310:599–601.
13. Gerschenfield HM: Transmitters at invertebrate synapses. *Physiol Rev* 1973;53:1–120.
14. Gran L, Coggin R, Bennett PB: Diazepam under hyperbaric conditions in rats. *Acta Anaesth Scand* 1980;24:407–411.
15. Hale J, Keegan R, Smith EB, Snape TJ: The effects of general anaesthetics on active cation transport in human erythrocytes. *Biochim Biophys Acta* 1972;288:107–113.
16. Halsey MJ, Smith EB: Effects of anaesthetics on luminous bacteria. *Nature London)* 1970;277:2363–2365.
17. Halsey MJ, Rostain J-C, Wardley-Smith B: Effect of GABA enhancing drugs on the behavioral response to high ambient pressure in the rat. *J Physiol* 1984;350:25.
18. Johnson FH, Flagler EA: Hydrostatic pressure reversal of narcosis in tadpoles. *Science* 1951;112:92.
19. Johnson FH, Eyring H, Polissar MJ: *The Kinetic Basis of Molecular Biology*. New York, Wiley, 1954.
20. Kaufmann PG, Bennett PB, Farmer JC Jr: Differential effects of pressure on the mammalian central nervous system in Bachrach AJ, Matzen MM (eds): *Underwater Physiology* VII. Bethesda, Undersea Medical Society, 1981, p 371.
21. Koblin DD, Little HJ, Green, AR, Daniels S, Smith EB, Paton WDM: Brain monoamines and the high pressure neurological syndrome. *Neuropharmacology* 1980;19:1031–1038.
22. Lever MJ, Miller KW, Paton WDM, Smith EB: Pressure reversal of anaesthesia. *Nature (London)* 1971;231:368–371.
23. Little HJ: Effects of ketamine and of high pressure on responses to γ-aminobutyric acid to the rat superior cervical ganglion. *Br J Pharmacol* 1982;77:209–216.
24. Lodge D, Anis NA: Effect of ketamine and three other anaesthetics on spinal reflexes and inhibitions in the cat. *Br J Anaesth* 1984;56:1143–1151.
25. Meyer KH: Contributions to the theory of narcosis. *Trans Faraday Soc* 1937;33:1062–1068.
26. Middleton AJ, Smith EB: General anaesthetics and bacterial luminescence. *Proc R Soc Lond B* 1976;193:173–190.
27. Miller KW, Paton WDM, Smith EB: Anaesthetic pressures of certain fluorine compounds. *Br J Anaesth* 1967;39:910–918.
28. Miller KW, Paton, WDM, Smith RA, Smith EB: Physico-chemical approaches to the mode of action of general anaesthetics. *Anesthesiology* 1972;36:339–351.
29. Miller KW, Paton WDM, Smith RA, Smith EB: The pressure reversal of anaesthesia and the critical volume hypothesis. *Mol Pharmacol* 1973;9:131–143.
30. Miller SL: A theory of gaseous anaesthesia. *Proc Natl Acad Sci USA* 1961;47:1515–1524.
31. Naquet R, Lemaire C, Rostain J-C: The high pressure nervous syndrome: psychometric and clinico-electrophysiological correlations. *Phil Trans R Soc London B* 1984;304:95–101.

32. Pauling L: A molecular theory of general anaesthesia. *Science* 1961;134:15–21.
33. Smith EB: The biological effects of high pressure: Underlying principles. *Phil Trans R Soc Lond B* 1984;304:5–16.
34. Smith EB, Bowser-Riley F, Daniels S, Dunbar IT, Harrison CB, Paton WDM: Species variation and the mechanism pressure-anaesthetic interactions. *Nature (London)* 1984;311:56–57.
35. White DC, Dundas CR: Effects of anaesthetics on emission of light by luminous bacteria. *Nature (London)* 1970;226:456–458.

29

Modification of Pressure and Anesthesia Tolerance by Multiday Exposures to High-Pressure Heliox or to Subanesthetic Levels of General Anesthetics

Ralph W. Brauer, William M. Hinson, and Julie Dutcher

Abstract. The effects of prolonged exposures of mice to either high-pressure atmospheres or the action of general anesthetics, on pressure tolerance and on anesthesia susceptibility, have been explored in six of eight possible combinations of factors (cf. Table I). In every case, the results conform qualitatively to the predictions based on the hypothesis of pressure/anesthesia antagonism: Pressure conditioning entails increased pressure tolerance and anesthesia susceptibility, whereas anesthesia conditioning yields the opposite results. More detailed examination of quantitative aspects of these effects, and especially of changes with varying duration of conditioning, indicate that while anesthesia conditioning effects conform acceptably to the predictions of this hypothesis, pressure conditioning effects require the assumption that changes in anesthesia susceptibility and in pressure tolerance proceed independently of one another. These results reinforce the conclusion previously drawn from acute experiments that although the hypothesis of a direct antagonism of pressure and anesthesia at the molecular level may account acceptably for the effects of pressure on anesthesia, it does not do so for the case of modification of pressure effects by anesthesia.

INTRODUCTION

In a recent review,[4] we examined interactions of hydrostatic pressure with a series of anesthetically effective gases and reached two conclusions that appear pertinent to the present subject: First, the total body of information available allowed distinguishing two broad classes of biological effects: One of these conformed in a general way to the predictions of such constructs as the critical volume hypothesis, predicated upon the assumption of a direct antagonism between hydrostatic pressure and anesthetic agents at the molecular level; a second (rather heterogeneous) group comprises effects that deviate widely qualitatively and/or quantitatively from the predictions of such theories. Second, we found that even among the responses which fell into the first group and which on the whole seemed to

Ralph W. Brauer, William M. Hinson, and Julie Dutcher • Institute of Marine Biomedical Research, University of North Carolina at Wilmington, Wilmington, North Carolina 28403.

Table I. Design Matrix for Study of IG/HP Antagonism in Conditioning Experiments[a]

Test parameter (and criterion)	Conditioning parameter A	AP	P
A (ED)	+	?	—
AP (ED)	——	——	(0)
AP (P_c)	0	——	——
P (P_c)	—	0 or ±	+

[a]A, anesthesia; P, pressure; AP, anesthesia at pressure or compression in presence of anesthetic; ED, partial pressure of anesthetic at test endpoint; P_c, total pressure at seizure onset; +, increase in test parameter relative to controls at endpoint; —, decrease in test parameter relative to controls at endpoint; 0, no change in test parameter relative to controls at endpoint; ?, not tested.

represent pressure and anesthesia effects on excitable tissues, the quantitative relations between total pressure and partial pressure of the effective gases for a given level of biological effectiveness varied enough from the predictions of theory to suggest that new experimental approaches to further clarify the basic question as to the nature of their antagonism would be highly desirable.

CHRONIC EXPOSURE TO N_2O—EFFECTS ON SUSCEPTIBILITY TO THE HIGH-PRESSURE NEUROLOGICAL SYNDROME

All experiments involving whole animals reported so far in this general area have involved acute studies. If it were possible to demonstrate that long-term exposures to pressure, or to general anesthetics, were associated with acclimation phenomena (habituation phenomena), then the resulting situations might contribute new data that might shed additional light on the basic question of the nature of the high pressure/inert gas antagonism. A study from which such investigations might take off was published in 1980 by Koblin and coworkers,[6] who reported the results of multiday exposures of mice to subanesthetic levels of nitrous oxide. Such a procedure produced hyperexcitable animals with markedly increased tolerance for several anesthetics. These authors considered the possibility that the observed changes might be a secondary result of biochemical effects of N_2O exposure, such as vitamin B_{12} depletion, but concluded that this possibility was remote, and that their results could confidently be attributed to habituation to the narcotic effects of the gas.

As a first step in the present inquiry, we decided, therefore, to replicate the conditioning procedures of Koblin and co-workers[6] and to determine whether the changes in anesthesia tolerance under these conditions are associated with significant changes in pressure tolerance.

Experimental Procedures

Our conditioning procedures replicated in their essentials those used in ref. 4. Adult CD-1 mice (22–28 g) housed in groups of four per standard 8 × 10 in. plastic mouse cage were exposed to 0.5 ± 0.1 atm of N_2O and 0.4 ± 0.1 atm oxygen, for periods ranging from 3 days to 2 weeks. Food and water were supplied *ad libitum*. At the end of the desired

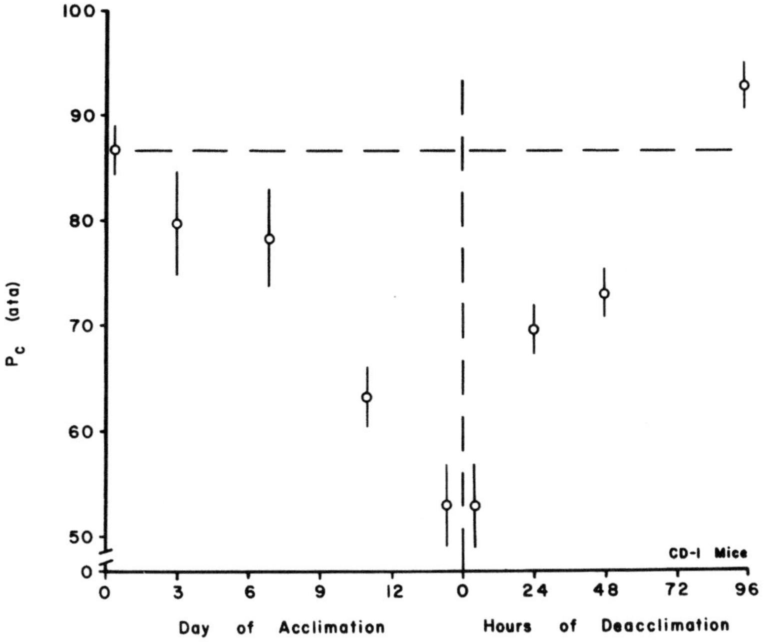

Figure 1. Change in HPNS Type I convulsion threshold pressure after various periods of exposure of CD-1 mice to 0.5 atm N_2O, and during recovery in air following 2 weeks of such exposure.

conditioning period, the animals were removed from the nitrous oxide and kept in room air for periods ranging from 1 to 96 hr. General appearance and body weights of these animals on removal from nitrous oxide closely replicated those reported in ref. 4. Following the selected recovery period in air, four mice at a time were transferred to 15-liter cylindrical pressure chambers equipped with internal blower/CO_2 scrubbers and maintained at a temperature of 34.5 ± 0.5°C. Oxygen partial pressure was brought to 0.5 atm, and chamber pressure was raised at a rate of 100 atm/hr by the injection of helium into the throat of the ventilator. The biological endpoint for each animal was the visually observed onset of Type I high-pressure neurological syndrome (HPNS) seizures.[3] At the end of the procedure the animals were sacrificed by rapid decompression; previous tests had shown this to result in flat electroencephalograms within 10 sec after beginning of decompression. Data are reported in the text ± standard deviation, and significance of differences is assessed throughout in terms of Student's t test with a cutoff value $p < 0.05$. The illustrations show standard error throughout.

Results

Mean convulsion threshold pressure (P_c) 1 hr after removal from the nitrous oxide (Fig. 1, left panel) decreased progressively over the course of 2 weeks' nitrous oxide exposure, falling to a value 60% that for control mice. Mean convulsion threshold pressures in 2-week-acclimated mice were restored to a level indistinguishable from that in the control mice by adding 12% nitrogen to the compression mixture. The partial pressure of nitrogen at convulsion onset was 10.7 atmospheres of air (ata), equinarcotic at the same total pressure with 0.47 atm nitrous oxide.[4] In a sense, therefore, these animals behave as though they were

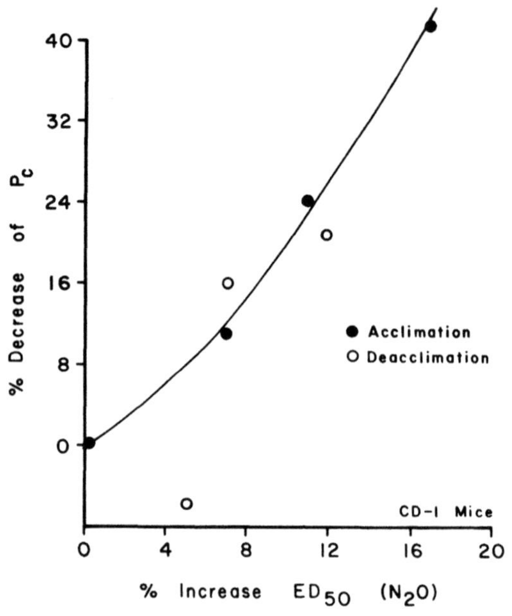

Figure 2. Correlation of convulsion threshold pressures from the present study with values for ED_{50} for N_2O narcosis from Koblin *et al.*[6] for corresponding stages of N_2O conditioning or deconditioning.

suffering a deficiency of anesthetic equivalent to the N_2O partial pressure to which they had been conditioned.

In mice maintained in air for various periods of time after termination of 2 weeks' nitrous oxide acclimation, this sensitization to high pressure regressed progressively. After approximately 75 hr in air, mean convulsion threshold pressures were equal to those of control animals, and by 96 hr mean convulsion threshold pressure was several atmospheres higher than in control animals (Fig. 1, right panel).

Since the age and strain of animals used in the present study, as well as the relevant conditions of N_2O exposure in the present series, closely replicated those reported by Koblin and co-workers,[6] direct comparison of the results appears justifiable. To achieve this, the mean P_c values for any given day of acclimation or deacclimation to N_2O as determined in the present study were plotted against changes in ED_{50} for N_2O narcosis reported by Koblin *et al.* for the same stage of their experiments (Fig. 2). If one excepts the 96h deacclimation point (since the P_c overshoot noted here has no parallel in the anesthesia susceptibility studies of Koblin *et al.*[6]), the two sets of data are linearly correlated; the regression equation is $(P_c^o - P_c)/P_c^o = 2.54\ (ED_{50} - ED_{50}^o)/ED_{50}^o - 0.039$, with a correlation coefficient of 0.97 (0 superscripts designate control values).

ETHANOL HABITUATION EXPERIMENTS

To further test the possibility that the effects of protracted nitrous oxide exposure might reflect specific biochemical changes reportedly associated with nitrous oxide exposures (cf., for example, ref. 7), we have undertaken two additional sets of experiments involving protracted exposures either to 12 ata of nitrogen (approximately equinarcotic with 0.5 atm nitrous oxide at 1 ata), or to ethanol, respectively.

While the results of the nitrogen experiments are not yet in, the results of ethanol habituation are clear cut: Animals prepared according to ref. 8 by 2 weeks' maintenance on a liquid diet containing 7% ethanol were removed from the ethanol regime, held for 3–6 hr to permit elimination of the bulk of the residual ethanol, and tested to determine HPNS Type I convulsion threshold pressures. After 3 hr without alcohol mean convulsion threshold pressure was 73.4 ± 10.2 ata; after 6 hr without alcohol, it was 79.4 ± 5.3 ata, compared with 93.5 ± 10.2 ata in control animals compressed at the same rate (60 atm/hr). Both after 3 and after 6 hr off ethanol, convulsion threshold pressures of ethanol-habituated mice were highly significantly lower than in the control animals ($p < 0.001$). These data thus replicate qualitatively the effects of N_2O conditioning on P_c and lend further support to the view that the observed effects are indeed the results of changes in central nervous system (CNS) function attributable to prolonged exposure to the CNS depressant effect of the agents rather than to any N_2O specific biochemical complications.

Altogether, then, these data reveal a high degree of symmetry in the way pressure and anesthesia susceptibilities are affected by protracted exposure to subanesthetic levels of nitrous oxide. The results thus far, therefore, are compatible with the hypothesis that directly opposed changes at the same cellular and molecular levels might underlie both effects of such acclimation.

EFFECTS OF PROLONGED EXPOSURE TO HIGH-PRESSURE HELIOX

This encouraged us to explore further the possibility that the effects of pressure and of anesthesia exposures might prove to be fully symmetrical by carrying out a series of studies involving protracted exposures of mice to high-pressure heliox, to determine changes in convulsion threshold pressures as well as anesthesia tolerances.

Experimental Procedures

Mice were exposed to heliox atmospheres containing 0.45 ± 0.05 atm 0_2 at pressures from 30 to 100 ata and a temperature of 34.5 ± 0.5°C (the temperature preferendum of CD-1 mice in 50–80 ata heliox) for periods ranging from 3 to 12 days. A vertical 80L cylindrical chamber, equipped with internal scrubber/(brushless) blower system, lighting on a 12hL/12hD cycle, and observation window was used in which the animals were placed, eight at a time, in a cylindrical wire cage 40 cm in diameter, 30 cm high, with food and water *ad libitum* over a mineral oil pan 1-in. deep to isolate urine and feces from the chamber atmosphere. Marked restriction of food intake and weight losses were observed in these experiments. We have shown elsewhere[5] that these changes in food intake and body weight are an intrinsic effect of high-pressure exposure of mice but are not correlated with the changes in CNS excitability which are our concern in this chapter. At the end of the chosen period of pressure conditioning, animals were decompressed at a rate of 8 atm/hr, oxygen partial pressure being maintained throughout between 0.4 and 0.6 atm. On return to the surface animals were weighed, placed in holding cages, and transferred to the pressure test assembly after surface intervals varied from 1 to 48 hr. Tests for pressure tolerance were carried out in the same fashion as for the nitrous oxide conditioning experiments except that a compression rate of 60 atm/hr (rather than 100 atm/hr) was used. Anesthesia tolerance was tested in the same chamber system following the techniques of Brauer and Way,[2] using nitrogen added at a rate of 30 atm/hr as the narcotic agent and the point at which each animal first lost its upright posture as the biological endpoint.

Results

Pressure conditioning resulted in marked increases in HPNS convulsion threshold pressures to a maximum of about 135 ata (Fig. 3). Change in P_c is most rapid during the first 2

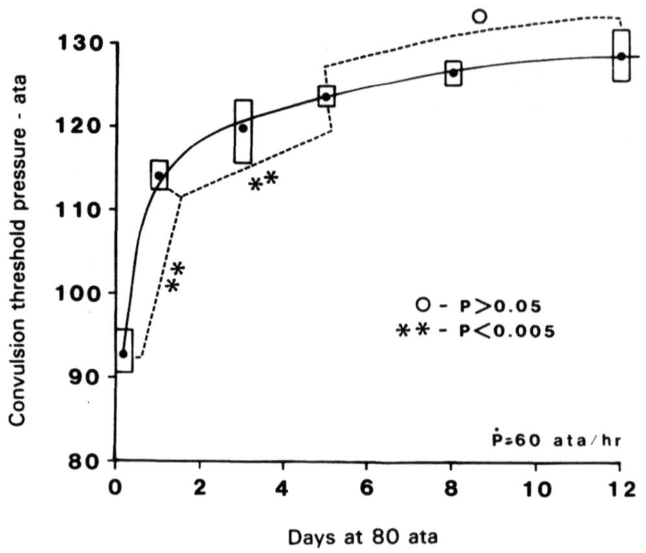

Figure 3. Change in HPNS Type I convulsion threshold pressure after various periods of exposure of mice to heliox at 80 ata.

days of pressure conditioning and progresses slowly thereafter for an additional 8–10 days. Thus, in contrast to the nearly constant rate of decrease in convulsion threshold pressure during nitrous oxide conditioning, the course of pressure conditioning follows a hyperbolic pattern, convex to the time axis.

The degree of increase in convulsion threshold pressure under these conditions is determined not only by time at pressure but also by the pressure at which the animals are being conditioned (Fig. 4): The relation between these parameters is nearly linear up to about 60 ata. There is a slight further rise in P_c at a holding pressure of 80 ata. However, further increase of conditioning pressure to 100 ata (requiring a two-step procedure to avoid premature occurrence of HPNS convulsions) not only does not result in further increase but yields a statistically significant decrease in P_c relative to its value at 80 ata, suggesting an absolute limit to the heliox pressure to which these mice can accommodate.

As in the case of nitrous oxide conditioning, the effect of pressure conditioning is dissipated when the animals are held in air at 1 ata following decompression. This process is rather more rapid than in the case of nitrous oxide, being all but complete within 24 hr.

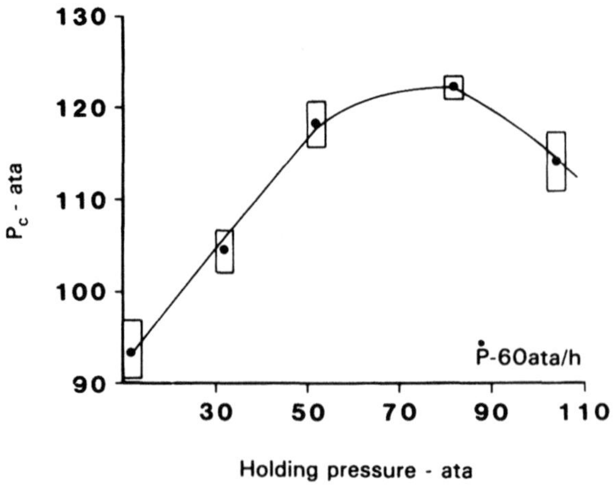

Figure 4. HPNS convulsion threshold pressures in CD-1 mice after 5 days of exposure to heliox at the pressures shown on the abscissa.

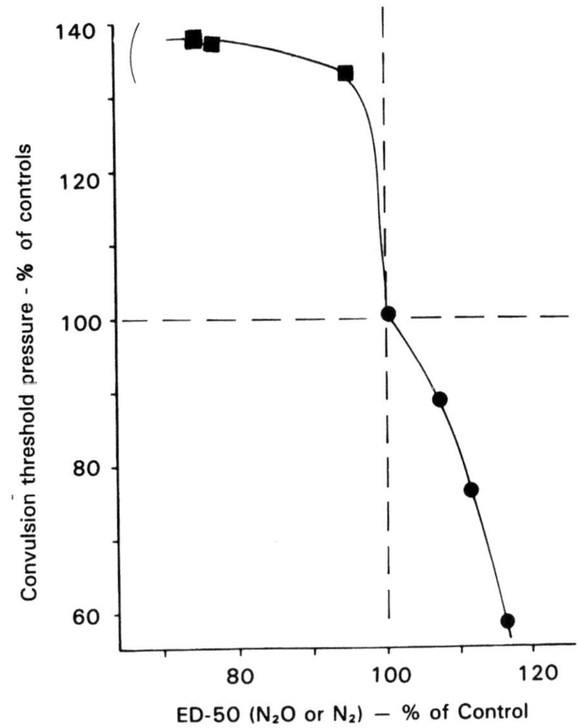

Figure 5. Correlation of changes in P_c and in relative ED_{50} for narcosis by N_2 or N_2O, during conditioning to 80 ata heliox (left of center line) or to 0.5 atm N_2O (right of center line—data from Koblin *et al.*[6] and present study as in Fig. 2).

We have only begun to map out changes in anesthesia tolerance corresponding to these changes in convulsion threshold pressure. However, it is already clear that the time course of evolution of anesthesia resistance here is almost the opposite of that for convulsion resistance: After 5 days at high pressure, the partial pressure of nitrogen required to induce loss of righting reflexes in these animals is reduced below control levels only to a slight and statistically insignificant extent (96 ± 7% of controls). Further exposure to high pressure, on the other hand, entails a relatively rapid decrease in anesthesia tolerance, so that after 10 days at pressure the animals have anesthesia tolerances only 76 ± 4% as great as 1 ata air controls.

As in the case of the nitrous oxide conditioning experiments, one can compare the effects of pressure acclimation on anesthesia resistance with those on pressure tolerance by plotting the changes in P_c against changes in anesthesia tolerance observed at corresponding stages of the acclimation process. Figure 5 shows that the results for the case of acclimation to 80 ata of heliox depart widely from any linear correlation between the two parameters. It seems difficult to interpret these results in any way other than by suggesting that the two effects of prolonged pressure exposure evolve independently of one another. This situation thus differs radically from that obtaining in the case of nitrous oxide conditioning.

EFFECT OF N_2O OR N_2 ADDITION TO HIGH-PRESSURE ACCLIMATION ATMOSPHERES

Either N_2O or N_2, when added to the conditioning atmosphere, substantially reduces the effectiveness of high-pressure conditioning in protecting the mice against HPNS seizures. Thus, after 5 days' conditioning at 80 ata in heliox containing 0.4 atm, N_2O mean convulsion threshold pressures on recompression were 99 ± 6 ata, far below the P_c = 123 ± 4 ata for heliox-conditioned mice at the same stage ($p < 0.001$), and perhaps still perceptibly above that for control mice (93.5 ± 5 ata). Increasing the N_2O concentration to yield $P_{N_2O} = 0.6$ atm in the conditioning mixture had no greater effect on P_c than 0.4 atm. Substituting 9.6 atm N_2 for the N_2O yielded $P_c = 108 \pm 6$ ata after 5 days, significantly below the results for heliox conditioning ($p < 0.01$), but significantly greater than the P_c obtained with 0.4 atm N_2O ($p < 0.01$). Thus, the relative potency of N_2O in this context, if N_2 is taken as 1.0, is significantly greater than 24 (the ratio of partial pressures of the two gases at the same total conditioning pressure), and probably exceeds 50. These values are to be compared with a relative intrinsic narcotic potency for N_2O of 22.6 on the same scale (cf. ref. 1) and a relative intrinsic anti-HPNS convulsant potency of 30–50.

DISCUSSION

Before considering the bearing of the above results on the question of molecular mechanisms of pressure/anesthesia antagonism, brief comment is in order as to factors that might complicate their interpretation. The two families of data under consideration relate to the effects of prolonged exposure to subanesthetic concentrations of nitrous oxide and of prolonged exposure to subconvulsant pressures of heliox atmospheres (cf., for example, ref. 7). With regard to the nitrous oxide exposure, one must consider whether the results are compromised by biochemical effects due to reactivity of nitrous oxide in the organism. Koblin *et al.*[6] reported that mice living in atmospheres containing 0.7 atm partial pressure of N_2O for periods up to 3 weeks, and in atmospheres containing 0.4 atm N_2O for 3 months, showed no apparent behavioral or physical abnormalities. Like the animals of Koblin *et al.*, ours gained weight at rates statistically indistinguishable from control animals of the same age. The inference that the CNS effects observed in these adult animals can confidently be attributed to prolonged exposure to the CNS depressant effects of N_2O is materially strengthened by the observation that qualitatively similar effects can be produced by ethanol habituation. A critical experiment in which nitrous oxide is replaced by a truly inert gas anesthetic is currently underway and should once and for all allow us to resolve this question.

Prolonged high pressures in mice are accompanied by marked weight loss. We have explored this phenomenon in considerable detail[5] and have shown that (1) it is not relieved by substituting a liquid diet for the pelleted diet; (2) its severity can be markedly reduced by substituting a "cafeteria diet" for the standard pelleted diet; (3) the weight loss can be eliminated altogether by including 0.4 atm nitrous oxide in the heliox atmosphere; and (4) changes in weight loss are not correlated with changes in convulsion threshold pressure accompanying the various types of pressure conditioning experiments. We conclude that the weight loss observed under these conditions is probably the result of disturbance of appetite at the central level but that the effects on convulsion threshold pressure and on anesthesia susceptibility here observed proceed largely independently of the ensuing metabolic changes.

Taken together, then, these several considerations seem to us to justify treating the body of data presented here as reflecting primarily changes in CNS function consequent upon prolonged exposure to the anesthetic or the high-pressure environment and turning to the evaluation of these data from the point of view of their bearing on the question of pressure/anesthesia antagonism.

A first overview of the data may be obtained by reducing them to the form of a logic matrix made up of three columns, representing the possible combinations of test parameters, and four rows, representing the corresponding combinations of test parameters and endpoints (Table I). Eliminating meaningless combinations, this table presents eight cells for a test for antagonism between anesthetic and high-pressure effects. For six of these, data have been presented above; the results for the seventh represent a highly likely inference from refs. 1 and 4. In every one of these cases, the qualitative results conform with predictions based on the hypothesis that effects of pressure conditioning are the reverse of anesthesia conditioning, and that, when exhibited in combination, pressure and anesthesia effects (chronic, from conditioning, as well as acute, in the course of the test procedures) tend to annul each other.

The hypothesis of a direct pressure/anesthesia antagonism also accounts well for the pattern of change during N_2O acclimation and deacclimation—P_c and ED_{50} for N_2O at various stages of the conditioning process are directly proportional to one another throughout all but the very late recovery stage. On the other hand, marked discrepancies emerge when one examines the quantitative aspects of the effects of high-pressure conditioning: Most striking among these is the dissociation of the time course of development of increased anesthesia susceptibility and of increased pressure tolerance (Fig. 5). These data are hard to explain unless one is willing to grant that changes in the two characteristics follow mutually independent time courses and hence that whatever changes underlie these effects at the cellular and molecular levels, they must be in large measure distinct from one another. Other discrepancies that may point in the same direction include the lack of linearity of the dose/response relations for reversal of pressure conditioning by N_2O in a partial pressure zone where pressure reversal of anesthesia displays a very satisfactory degree of linearity,[4] and the aberrant potency ratio for N_2 to N_2O in the same context.

All things considered, therefore, we conclude that the effects of conditioning to anesthetics upon anesthesia and upon pressure tolerance could be compatible with the hypothesis of direct antagonism at the molecular level between high pressure and general anesthetics. However, the corresponding effects of high-pressure conditioning bespeak mechanisms which would appear to distinguish sharply between modification of anesthesia tolerance and modification of pressure tolerance. At least with respect to this combination of factors therefore, we conclude that the data currently available are difficult to reconcile with the view that primary sites of attack of pressure and of anesthesia in the central nervous system of mammals are identical and merely affected in opposite directions by the two factors. We suggest that the body of data is far more readily compatible with the interpretation that in intact mammals the effects of prolonged exposure to high hydrostatic pressure reflect an antagonism of overall effects of actions in large part exerted at multiple and distinct sites. These data supplement results previously available bearing on modification of acute effects of high pressure by anesthesia and seem to us to lend further weight to the conclusion that at the whole-organism level a direct antagonism of high pressure effects by anesthetics is irreconcilable with the experimental data. Our present results, on the other hand, are compatible with the conclusion that the modification of anesthesia by high pressure approaches far more closely to the pattern of behavior predicted for a simple one-to-one antagonism between the two agents than the modification of high-pressure effects by anesthetics.

ACKNOWLEDGMENTS. This research has been supported by the Office of Naval Research through Contract N00014-75-C-0468.

REFERENCES

1. Brauer RW, Hinson WM: Effects of variations in the time pattern of nitrogen addition upon development of the high pressure neurologic syndrome in mice. *Undersea Biomed Res* 1984;10(4):13–37.
2. Brauer RW, Way RO: Relative narcotic potencies of hydrogen, helium, nitrogen and of their mixtures. *J Appl Physiol* 1970;29(1):23–31.
3. Brauer RW, Mansfield WM, Beaver RW, Gillen HW: Stages in the development of the HPNS in the mouse. *J Appl Physiol* 1979;46(4):756–765.

4. Brauer RW, Hogan PM, Hugon M, MacDonald AG, Miller KW: Patterns of interaction of the effects of the light metabolically inert gases with those of hydrostatic pressure as such—A review. *Undersea Biomed Res* 1982;9(4):353–396.
5. Hinson WM, Brauer RW: Weight loss associated with prolonged exposure of mice to heliox atmospheres at pressures from 30–80 ata (abst). *Undersea Biomed Res* 1984;11:36.
6. Koblin DD, Eger EI II, Smith RA, Winter PM: Chronic exposure of mice to subanaesthetic concentrations of nitrous oxide, Fink R (ed): *Molecular Mechanisms of Anesthesia.* New York, Raven Press, New Press, 1980, pp 156–162.
7. McKenna B, Weir DG, Scott JM: The induction of functional vitamin B12 deficiency in rats by exposure to nitrous oxide. *Biochim Biophys Acta* 1980;628:314–321.
8. Ritzman RF, Tabakoff B: Body temperature in mice: A quantitative measure of alcohol tolerance and physical dependence. *J Pharmacol Exp Ther* 1976;199:158–170.

30

Pressure Reversal of Halothane's Antiviral Action

Effect of Halothane and Pressure on Measles Virus Protein Synthesis

Elliott Bedows, Peter J. Cohen, and Paul R. Knight

ABSTRACT. In order to study general anesthetic mechanisms, we have developed a simple biological system: the replication of an animal virus whose assembly is highly sensitive to inhalational anesthetic exposure and whose anesthetic-induced inhibition is partially reversed by hyperbaric pressure. This system enables us to monitor a select group of protein molecules as they are synthesized and metabolized within the cell during exposure to anesthetic or pressure. Of all the viruses studied thus far, the replication of measles virus is the most sensitive to halothane exposure. Measles virus replication in cultured monkey kidney cells is completely inhibited by 1.8% halothane at 1 atm of air (ata). While the antiviral effects of halothane are not completely reversed, more than 100 times as much virus is synthesized in the presence of 2.0% halothane at 100 ata (using helium) than at 1 ata. Pressures of 100 ata alone do not significantly lower the amount of measles virus synthesized. However, pressures of 100 ata did lower the amount of herpesvirus synthesized under identical experimental conditions. Halothane's antiviral effect was not reversed when herpesvirus-infected cells were exposed to 2.4% halothane at 100 ata. Rather, the effects of halothane and pressure were additive suggesting that pressure and halothane may be affecting herpesvirus synthesis independently.

Examination of the proteins synthesized by measles virus during exposure to halothane and pressure have revealed that the halothane-induced inhibition of the synthesis of two membrane-associated proteins (H and F) was reversed under 100 ata of helium. The discovery of two viral membrane proteins whose synthesis parallels the sensitivity and recovery of the virus during halothane exposure and application of hyperbaric pressure provides a powerful tool to study the effects that general anesthetics have on the proteins of an intact and functional biological membrane.

INTRODUCTION

Because there is a strong correlation between lipophilicity and anesthetic potency, it is reasoned that general anesthetics interact with either the lipid bilayer of the cell membrane[25] the hydrophobic regions of essential proteins within the membrane,[9,11,28] or both.[30] X-ray and neutron diffraction studies suggest that the lipid bilayer cannot by itself serve as the

Elliott Bedows, Peter J. Cohen, and Paul R. Knight • Departments of Anesthesiology and Epidemiology, University of Michigan, Ann Arbor, Michigan 48109. Present address of P.J.C.: Department of Anesthesia, University of Pennsylvania, Philadelphia, Pennsylvania 19104.

anesthetic site and that proteins are involved.[10] It is likely that the primary action of anesthetics occurs in the lipid bilayer of the biological membrane with secondary effects altering the function of target protein molecules. By nature of their chemical structure or location within the membrane, these target proteins might be affected by comformational changes produced by anesthetics on the lipid phase of the membrane. Consistent with this model for general anesthetic action is the fact that while potency correlates with the partition coefficient of the anesthetic between olive oil and gas, certain molecules (e.g., C_3F_8) appear to be too apolar to serve as an anesthetic.[24] Thus, it appears that an amphiphilic site (i.e., one that has both polar and apolar properties, such as the biological membrane) would serve as an ideal anesthetic binding site and that a primary interaction between the cell membrane(s) and anesthetic molecules might serve to initiate a cascade of events which then have profound effects on cellular structure and function and, subsequently, on the organism as a whole.

One of the strongest arguments implicating the membrane as the primary site of general anesthetic action has been the observation of numerous investigators that many effects of anesthesia are reversed after exposure to hyperbaric or hydrostatic pressure. Hyperbaric pressures up to 100 atmospheres absolute (ata) increase the partial pressures of nitrous oxide and isoflurane required to produce anesthesia[12] and decrease the barbiturate sleeping time in mice.[23] Anesthetics also inhibit the development of high-pressure nervous syndrome in humans, a condition of tremors and seizures associated with hyperbaric pressures.[6,14,22] Recently it was discovered that hyperbaric pressures of 100 ata will partially reverse the inhibition of measles virus replication in cell cultures exposed to 2.0% halothane.[2] The ability of hydrostatic pressure to antagonize anesthetic action is also well documented. Restoration of the light output of anesthetized luminous bacteria by 2000–4000 ata[16] (see E. B. Smith *et al.*, this volume, chapter 28), recovery of swimming by tadpoles treated with 2.5% ethanol (200–300 ata)[15] or 0.5–1.5% halothane (35–171 ata),[13] and reversal of halothane's inhibition of axonal transmission (35–69 ata)[17] are but a few examples. Not all effects attributed to inhalational anesthetics are reversible by pressure, however. For example, pressure fails to reverse the halothane-induced depression of oxygen uptake by isolated rat liver mitochondria.[8] Similarly, the halothane-induced inhibition of synaptic transmission in the rat superior cervical ganglion,[18] nerve cell function in *Aplysia californica*,[26] and the depression of the ciliary beat of *Tetrahymena pyriformis*[27] are not reversed by pressure. This has led us to question which part(s) of the cell are involved in the pressure-reversible and pressure-nonreversible aspects of general anesthesia.

In order to study anesthetic mechanisms, we have developed a simple biological system: the replication of an animal virus whose assembly is highly sensitive to inhalational anesthetic exposure[19] and whose anesthetic-induced inhibition is partially reversed by hyperbaric pressure.[2] This system enables us to monitor the synthesis, biochemical modification, and function[1] of a select group of protein molecules as they are metabolized within the cell. Viruses are obligate intracellular parasites and, as such, require the machinery of the host cell to facilitate their replication. Different families of viruses have varying degrees of sensitivity toward halothane exposure,[4] which is not surprising since different combinations of cellular compartments and organelles are utilized by various viruses during the course of their replication. It is likely that the inhibition of virus replication observed during halothane exposure results from interactions between the anesthetic and sites within the cell vital for virus replication. Of all the viruses studied thus far, the replication of the syncytia-forming paramyxoviruses is the most sensitive to halothane exposure, and measles virus is the most sensitive member of this group.[4] Measles virus replication in cultured monkey kidney cells is completely inhibited by 1.8% halothane at 1 ata.[19] However, in the presence of 2.0% halothane, more than 100 times as much virus is synthesized at 100 ata (using helium) than at 1 ata although the halothane-induced inhibition is not totally reversed.[2]

Measles virus synthesizes six major proteins,[29] which are assembled into infectious virus particles at sites distributed throughout the soluble cellular cytoplasm and membranes,[21] and the synthesis of particular halothane sensitive proteins can be identified. In this chapter we describe experiments in which we used both an indirect immunofluorescent staining technique as well as polyacrylamide gel electrophoresis to identify individual measles virus proteins which were synthesized in the presence of 2.0% halothane at either 1 ata or 100 ata of helium. We report here that of the five major proteins studied, the onset of synthesis of the two membrane-bound measles virus proteins, F and H, were retarded extensively while the synthesis of a third cytoplasmic protein, P, was also inhibited moderately during exposure to 2.0% halothane at 1 ata. However, in cultures exposed to 2.0% halothane and 100 ata, the synthesis of the membrane proteins was nearly normal although the inhibition of the cytoplasmic protein was approximately the same at 100 ata as at 1 ata.

EXPERIMENTAL METHODS

Monkey kidney (Vero) cells were grown on 11-×-22-mm glass coverslips in 60-mm glass Petri dishes in Eagle's modified minimal essential medium supplemented with 10% fetal bovine serum and antibiotics.[19] When cell cultures reached 70–80% confluency, they were infected with the Edmonston vaccine strain of measles virus as follows. Culture medium was removed and replaced with 1.0 ml of buffer containing 0.1 or 1 infectious virus [plaque-forming unit (PFU)] per cell. The virus adsorbed to the cells for 1 hr at 22°C, at which time the virus-containing solution was removed and washed once with TRIS-buffered saline, pH 7.4, after which 5 ml of fresh culture medium was added.

Immediately after the culture medium was replaced, the virus-infected cultures were placed into Parr distruption bombs (Fig. 1). Two percent halothane in 95% air/5% CO_2 was allowed to flow through two of the bombs for 20 min. Two other bombs were prepared identically, except that halothane was omitted. One halothane-containing and one control bomb were then pressurized to 100 ata (10,110 kPa) with helium while the other two bombs (one halothane-containing and one control) remained at normobaric pressure (101.1 kPa).

For experiments involving immunofluorescent staining, the following procedure was followed. At a designated time following virus infection, the pressurized bombs were decompressed over a 20-min period, and the cell-containing coverslips were washed twice in phosphate-buffered saline (PBS) and immersed in acetone at −20°C for 30 sec followed by a second acetone exposure for 10 min at 22°C. After the coverslips were thoroughly dried, they were stored at −70°C for further use. To study viral protein synthesis, the cells were rehydrated by immersion in PBS, and the coverslips were placed face down on a 25-μl drop of mouse ascites fluid which contained a monoclonal antibody against one of the five major measles virus antigens studied. (Monoclonal antibodies directed against measles virus N, M, G, H, or F proteins were a gift of Dr. E. Norrby, Karolinska Institute, Sweden.) In some experiments the coverslips were exposed to polyclonal human serum obtained from a patient with atypical measles and a high measles virus neutralization titer. Following a 40-min incubation at 22°C, the slides were washed twice in PBS and treated with 25 μl of a secondary antibody by the procedure described previously. The secondary antibody used in these experiments was fluorescein-conjugated, sheep antimouse antibody. After staining and washing as described previously, the slides were mounted in 25% glycerol containing 8.4% polyvinyl alcohol and 2.5% 1,4-diazabicyclo(2,2,2)octane (to prevent fading). The mounting medium was adjusted to pH 8.5 prior to use. Slides were examined under a Leitz fluorescent microscope, and the number of cells staining positively for each antigen was recorded.

Measles virus proteins were also analyzed using polyacrylamide gels and the procedure

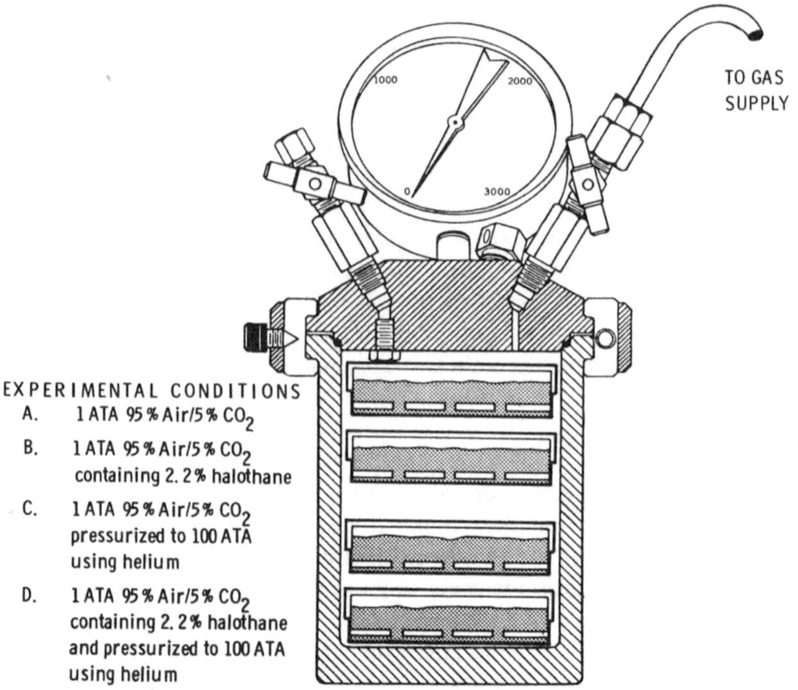

Figure 1. Parr disruption bomb containing Petri dishes used for growth of virus-infected cells in pressure experiments.

of Laemmli.[20] Using a 3% acrylamide stacking gel and an 8.5% separating gel in the presence of 0.1% sodium dodecyl sulfate (SDS) we have identified measles virus-specific proteins from immunoprecipitated preparations. Vero cells were grown to confluency and infected with measles virus at a multiplicity of infection of 0.1 PFU/cell. After virus adsorption, methionineless minimal essential medium supplemented with 300 μCi ^{35}S-methionine plus 5.0 μg/ml nonradioactive methionine was added, after which cultures were exposed to halothane and pressure as described above. For the immunoprecipitation procedure, the culture media was removed, and cells were washed in PBS and lysed in 0.15 M NaCl, containing 0.01 M TRIS (pH 7.8), 1% Triton X-100, 1% deoxycholate, 0.1% SDS, 1 mM phenylmethylsulfonylfluoride, and 100,000 U/ml Aprotinin (Sigma Chemical Co.). The ruptured cell membranes and cell lysate mixture were pelleted at 9000 *g* for 2 min. The supernate was incubated on ice with 1/10th its volume of high titer measles virus antiserum, for 30 min. An equal volume of protein A–Sepharose complex was added to the antigen–antibody mixture and incubated on ice for 15 min. This results in an antigen–antibody complex attached to protein A and connected to the sepharose beads. This complex was pelleted, washed, and centrifuged. The resulting pellet was then suspended in sample buffer (0.6 M TRIS, 2% SDS, 10% glycerol, and 5% 2-mercaptoethanol) and boiled for 3 min (to release the antigens) followed by centrifugation. Samples were applied to gels which ran 4–4 ½ hr at 70 mA. After electrophoresis gels were fixed with 30% methanol containing 10% trichloroacetic acid and 10% glacial acetic acid for 1 hr and immersed in Enhance^R for 1 hr and H_2O for 14 hr. Gels were then dried and exposed to Dupont Cronex 4 film for 5 days and developed.

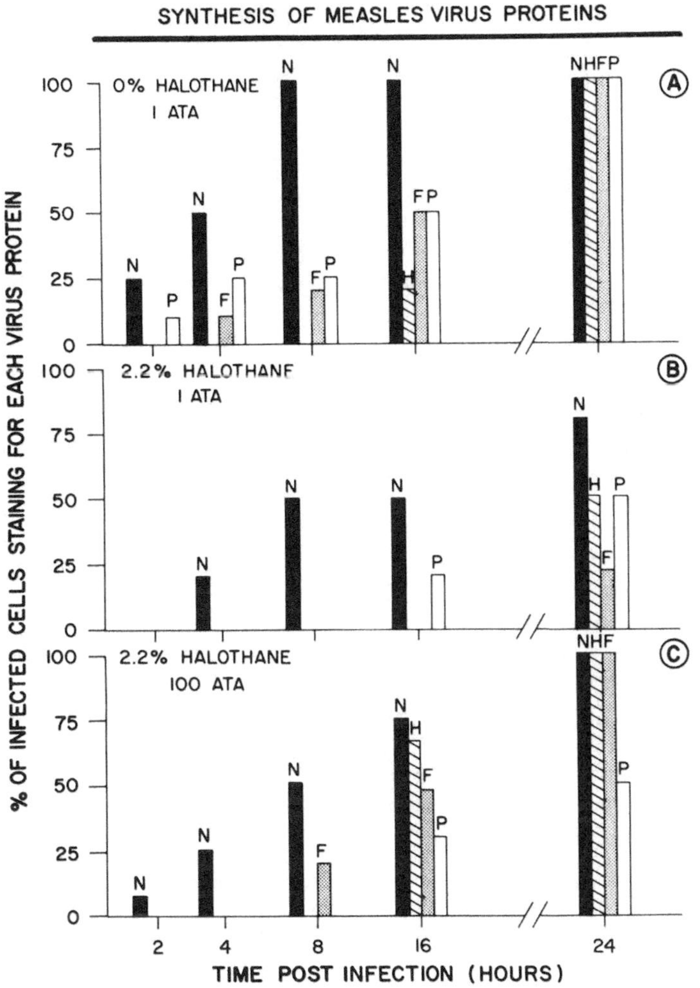

Figure 2. Measles virus proteins synthesis in cells exposed to halothane and pressure as determined by indirect immunofluorescent staining using monoclonal antibodies against specific measles virus antigens. H, P, N, and F are virus proteins as described in Fig. 3 and Table I.

RESULTS

Using our indirect immunofluorescent assay we determined the approximate percentage of cells staining positively for each measles virus protein. Figure 2 summarizes the results of several experiments. Each bar represents observations made on 15–18 random fields of cells with 50–55 cells screened for each determination. Figure 2A shows the time course of measles virus protein synthesis in nonexposed infected Vero cells. Figure 2B shows that exposure to 2.2% halothane dramatically affects the synthesis of certain proteins. The synthesis of three of the measles virus proteins (H, F, and P) was affected by exposure to halothane. H and F are membrane-associated proteins while P is a cytoplasmic protein. Not only did fewer cells stain positively for these antigens, but the time of onset for their synthesis was delayed. Figure 2C demonstrates that the number of cells staining for the two

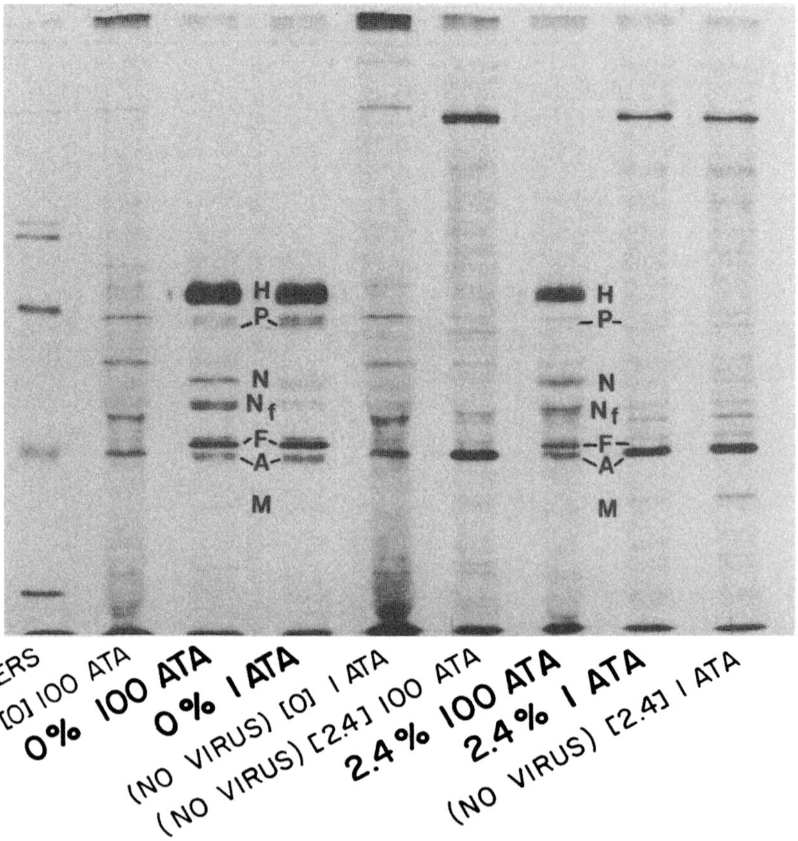

Figure 3. Measles virus protein synthesis in cells exposed to halothane and pressure as determined by polyacrylamide gel electrophoresis. Proteins shown are the virus hemagglutin (H), polymerase (P), nucleocapsid structural (N), fusion (F), and matrix (M) proteins as well as fragmentation products of the N protein (N_f) and cellular actin (A).

Table I. Synthesis of Measles Virus Proteins during Exposure to 2.2% Halothane[a]

Measles virus protein (mol. wt.)	Cellular location of synthesis	Synthesized at 2.2% halothane	
		1 ata	100 ata
L (180,000)	Cytoplasmic	?	?
H (80,000)	Membrane	—	+
P (70,000)	Cytoplasmic	—	—
N (60,000)	Cytoplasmic	+	+
F_1 (41,000)	Membrane	—	+
M (37,000)	Cytoplasmic	+	+

[a]Summarized is the synthesis of the major virus proteins at 1 atm of 95% air/5% CO_2 containing 2.2% halothane (1 ata) or the same gas mixture pressurized to 100^2 ata using helium. Pressure of 100 ata in anesthetic nonexposed cells did not affect the synthesis of any of the measles virus proteins.

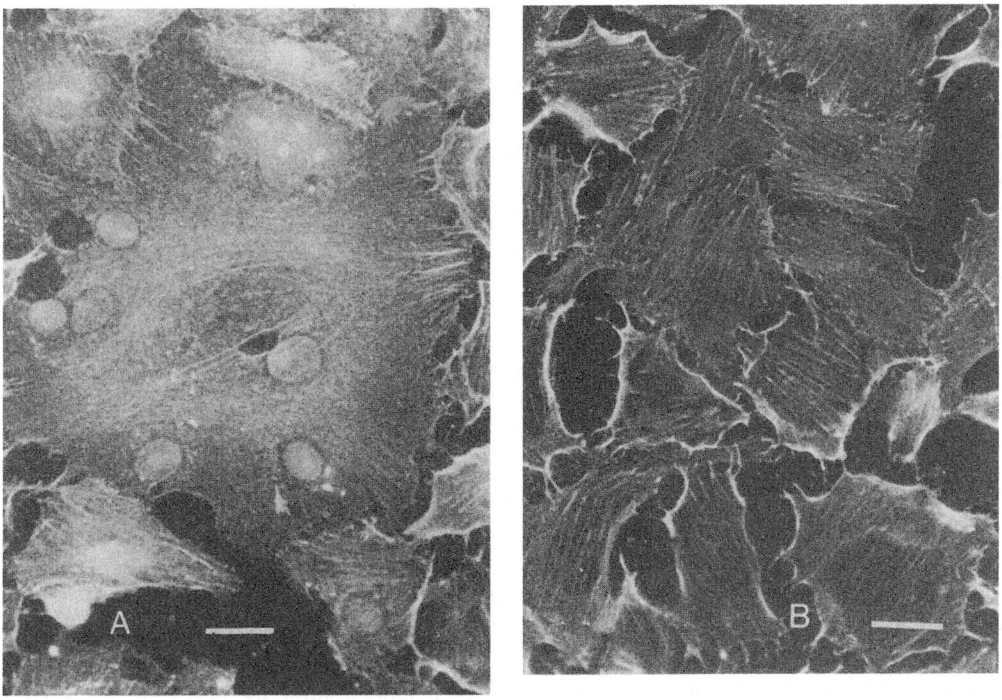

Figure 4. Measles virus cytopathology. Measles virus produces giant, multinucleated cells due to the F-protein-mediated fusion of the plasma membranes of adjacent cells (A). Halothane exposure of 2.2% inhibits this fusion (B) presumably because the F protein is not synthesized (see Figs. 2 and 3). The cells shown in this figure have been stained for F-actin using NBD-phallacidin as previously described.[3] Bar, 20 uM.

membrane proteins in the presence of 2.2% halothane was substantially higher at 100 ata than at 1 ata. However, no significant difference in the number of cells staining for the cytoplasmic P protein was observed at 100 ata as compared to 1 ata.

Figure 3 also demonstrates that the inhibition of synthesis of the membrane-associated proteins was reversed by hyperbaric pressure. In the experiment shown, infected cells were labeled with ^{35}S-methionine immediately prior to halothane exposure. When a control culture of halothane-nonexposed cells grown at 1 ata demonstrated virus specific cytopathology (see Fig. 4), all cells were harvested and lysed, and measles virus proteins were isolated by immunoprecipitation with hyperimmune human serum. The proteins were then separated by polyacrylamide gel electrophoresis. The reversibility of the halothane-induced inhibition of the two membrane-bound proteins (F and H) at 100 ata is readily observable. Also, as was observed by immunofluorescent staining, the halothane-induced inhibition of the cytoplasmic P protein was not reversed by pressure. Table I summarizes the results of our protein synthesis studies.

DISCUSSION

Anesthesia can be reversed by the application of hydrostatic or hyperbaric pressure.[5,9,11,25] However, not all the effects of anesthetic exposure on cells are pressure reversible.[8,18,26,27] (See Kendig and Grossman and Brauer *et al.*, this volume, Chapters 27 and 29.) This implies that certain sites within the cell vital to anesthesia might selectively be identified by virtue of having their anesthetic-induced alterations reversed by elevated ambient pressure.

Our results on the effects of halothane exposure on virus replication demonstrate that the inhibition of the synthesis of halothane-sensitive proteins associated with the plasma membrane is reversible by 100 ata of pressure, while the inhibition of another halothane-sensitive cytoplasmic protein is refractile to increased pressure. We feel that this differential reversibility by pressure may reflect the interaction of halothane with sites within the cell vital for viral protein synthesis, and that the domains occupied by the sensitive proteins might be similar to those responsible for anesthesia.

Two major sites within the cell occur for the translation of messenger RNA's into proteins: the soluble polysomes of the cytoplasm, and the membrane-bound polysomes of the endoplasmic reticulum (ER). The membrane-bound measles virus proteins are synthesized on the rough ER and are membrane associated throughout the course of their synthesis, glycosylation, and insertion into virions. All other measles virus proteins are presumably synthesized in the cytoplasm.[7] The M (matrix) protein is transported to the plasma membrane immediately after its synthesis where it functions in virus assembly while the nucleocapsid structural (N) and the two polymerase (P and L) proteins function in initiating genomic replication and nucleocapsid assembly. The measles virus particle is assembled when the genomic RNA is coated with the N protein as soon as it is synthesized. The P and L proteins are also added to the assembling virus core at this stage. The completed nucleocapsid then migrates to the cell plasma membrane. It attaches to the inner surface of the plasma membrane via the (matrix) M protein and buds from the plasma membrane taking along the viral H and F proteins, and host cell proteins and lipids en route. This results in the deposition of the viral envelope and completes the synthesis of the infectious virus particle.[7,21]

The production of infectious measles virus is blocked in a concentration-dependent manner during halothane exposure.[19] At 1.8% halothane and 1 ata, measles virus core assembly, infectious progeny production, and cytopathology (the F protein is responsible for the fusion type of cytopathology characteristic of the measles virus family) are all totally inhibited in infected Vero or BSC-1 cells. However, more than 100 times as much virus is produced in Vero cells in the presence of 2.0% halothane when cultures are pressurized to 100 ata with helium. Similarly measles-virus-induced cytopathology is observed in the presence of 2.0% halothane and 100 ata of helium.[2]

Hyperbaric pressures of 100 ata alone do not affect measles virus replication in Vero cells,[2] which is somewhat surprising. Most biological and biochemical systems ranging from humans to membrane model systems are affected by high pressures of inert gases[5] including the replication of herpes simplex virus (HSV, a DNA-containing nuclear virus) in Vero cells (Bedows *et al.*, unpublished data). Recent observations from our laboratory reveal that not only is the amount of HSV significantly ($p < 0.001$) lower at 100 ata, but that the antiviral effects of halothane on HSV replication are not reversible. In fact, the inhibitions induced by pressure and anesthetic exposure alone are additive for HSV ($p < 0.05$). This suggests that for the synthesis and assembly of HSV, pressure and anesthesia may be acting independently.

It may not be merely coincidental that the virus(es) most reliant upon the host plasma membrane during its replication is also most sensitive to halothane exposure, or that the halothane sensitivity of this virus is pressure reversible. The further study of the plasma-membrane-associated proteins of measles virus seems warranted as a model system which should provide a great deal of insight into the mechanism(s) by which general anesthetics interact with the proteins of the biological membrane to produce anesthesia.

CONCLUSIONS

1. Several halothane-sensitive measles virus functions including assembly of infectious virus, fusion of cellular plasma membranes, and the synthesis of membrane-bound proteins are restored toward normal values when ambient pressure is increased.

2. In contrast, the inhibitory effects of halothane on herpesvirus replication are not reversed by hyperbaric pressures.

3. The difference between these two virus systems probably reflects different sites in the cell utilized during their replication and provides an excellent tool for the study of membrane protein synthesis and function during anesthetic exposure.

ACKNOWLEDGMENTS. We thank Bruce A. Williams and Bruce A. Davidson for their excellent technical assistance. This work was supported in part by grant GM-28911-04 from the United States Department of Health, Education and Welfare, Public Health Service. P. R. K. is an awardee of Research Career Development Award GMA 100-00477 from the United States Department of Health Education and Welfare, Public Health Service.

REFERENCES

1. Bedows E, Knight PR: Viruses as probes for the study of anesthetic action. *Anesthesiology* 1981;55:A351.
2. Bedows E, Knight PR: Pressure reversal of halothane's antiviral effect. *Anesthesiology* 1983;59:29–32.
3. Bedows E, Rao KMK, Welsh MJ: Fate of microfilaments in Vero cells infected with measles virus and herpes simplex type 1 virus. *Mol Cell Biol* 1983;3:712–719.
4. Bedows E, Davidson BA, Knight PR: Effect of halothane on the replication of animal viruses. *Antimicrob Agents Chemother* 1984;25:719–724.
5. Bennet PB, Simon S, Katz Y: High pressures of inert gases and anesthesia mechanisms, in Fink BR (ed): *Molecular Mechanisms of Anesthesia, Progress in Anesthesiology,* vol 1. New York, Raven Press, 1975, pp 367–403.
6. Brauer RW, Goldman SM, Beaver RW, Sheehan ME: N_2, H_2 and N_2O antagonism of high pressure neurological syndrome in mice. *Undersea Biomed Res* 1974;1:59–72.
7. Choppin PW, Compans RW: Reproduction of paramyxoviruses, in Frankel-Conrat H, Wagner RR (eds): *Comprehensive Virology,* vol 4. New York, Plenum Press, 1975, pp 95–178.
8. Cohen PJ: Effect of hydrostatic pressure on halothane-induced depression of mitochondrial respiration. *Life Sci* 1983;32:1647–1650.
9. Eyring H, Woodbury JW, D'Arrigo JS: A molecular mechanism of general anesthesia. *Anesthesiology* 1973;38:415–424.
10. Franks NP, Lieb WR: Where do general anesthetics act. *Nature* 1978;274:339–342.
11. Franks NP, Lieb WR: Molecular mechanism of general anesthesia. *Nature* 1982;300:487–493.
12. Halsey MJ, Eger EI, Kent DW, Warne PJ: High pressure studies of anesthesia, in Fink BR (ed): *Molecular Mechanisms of Anesthesia, Progress in Anesthesiology,* vol 1. New York, Raven Press, 1975, pp 353–361.
13. Halsey MJ, Wardley-Smith B, Green CJ: Pressure reversal of general anesthesia—A multisite expansion hypothesis. *Br J Anesth* 1978;50:1091–1097.
14. Hunter WL, Bennet PB: The causes, mechanisms and prevention of the high pressure nervous syndrome. *Undersea Biomed Res* 1974;1:1–28.
15. Johnson FH, Flagger EA: Hydrostatic pressure reversal of narcosis in tadpoles. *Science* 1951;112:91–92.
16. Johnson FH, Eyring H, Williams RW: The nature of enzyme inhibitions in bacterial luminescence; sulfanilamide, urethane, temperature and pressure. *J Cell Comp Physiol* 1942;20:247–268.
17. Kendig JJ, Cohen EN: Neural sites of pressure–anesthesia interactions, in Fink BR (ed): *Molecular Mechanisms of Anesthesia, Progress in Anesthesiology,* Vol 1. New York, Raven Press, 1975, pp 421–430.
18. Kendig JJ, Trudell JR, Cohen EN: Effects of pressure and anesthetics on conduction and synaptic transmission. *J Pharmacol Exp Ther* 1975;195:216–224.
19. Knight PR, Nahrwold ML, Bedows E: Anesthetic action and virus replication: Inhibition of measles virus replication in cells exposed to halothane. *Antimicrob Agents Chemother* 1980;17:890–896.
20. Laemmli UK: Cleavage of structural proteins during the assembly of the head of bacteriophage T_4. *Nature (London)* 1970;227:680–685.

21. Matsumoto T: Assembly of paramyxoviruses. *Microbiol Immunol* 1982;26:285–320.
22. Miller KW: Inert gas narcosis, the high pressure neurological syndrome and the critical volume hypothesis. *Science* 1974;185:867–879.
23. Miller KW, Wilson WW: The pressure reversal of a variety of anesthetic agents in mice. *Anesthesiology* 1978;48:104–110.
24. Miller KW, Paton WDM, Smith EB, Smith RA: Physiochemical approaches to the mode of action of general anesthetics. *Anesthesiology* 1972;36:339–351.
25. Miller KW, Paton WDM, Smith RA, Smith EB: The pressure reversal of general anesthesia and the critical volume hypothesis. *Mol Pharmacol* 1973;9:131–143.
26. Parmentier JL, Bennet PB: Hydrostatic pressure does not antagonize halothane effects in single neurons of *Aplysia californica. Anesthesiology* 1980;53:9–14.
27. Pope WDB, Jones AJ, Halsey MJ, Nunn JF: Pressure enhancement of the depressant effect of halothane on cilial beat. *Can Anaesth Soc J* 1978;25:319–322.
28. Richards CD, Martin K, Gregory S, Keightley CA, Hesketh, TR, Smith GA, Warren GB, Metcalf JC: Degenerate perturbations of protein structure as the mechanism of anesthetic action. *Nature* 1978;276:775–779.
29. Rima BK: The proteins of morbilliviruses. *J Gen Virol* 1983;64:1205–1219.
30. Seeman P: The membrane actions of anesthetics and tranquilizers. *Pharmacol Rev* 1972;24:583–655.

VI

TOXICITY OF LOCAL ANESTHETICS

31

Potentially Toxic Actions of Local Anesthetics in Cardiac Tissue

Kenneth R. Courtney

ABSTRACT. Local anesthetics that gain access to the general circulation can produce a number of potentially toxic effects with some of the most important occurring in the heart. Several such actions of a series of amide-linked local anesthetics are described including (1) depression of atrial contractility, (2) reduction of sinoatrial rate, and (3) depression of cardiac excitability (sodium channels). The potency an amide-linked drug has for producing these effects is, in each instance, directly related to the drug's lipid distribution coefficient. In addition, local anesthetics having ether-linked structures are more potent than amide-linked drugs having equivalent solubilities in octanol.

These results have important implications regarding the potential toxicity of these drugs when they gain inadvertent access to the heart during regional anesthesia procedures. For instance, the long-acting local anesthetics etidocaine and bupivacaine are 10–20 times more potent than lidocaine in reducing atrial contractility whereas they are only about 4 times as potent as lidocaine as local anesthetics. Thus they may be relatively cardiotoxic compared to lidocaine. These new structural concepts may also have important implications regarding the intentional application of this class of drugs as antiarrhythmic agents. Both ether-linked and amide-linked local anesthetics which are used as antiarrhythmics (examples mexiletine and tocainide) have potencies for reducing excitability (an antiarrhythmic effect) that are highly correlated with their potencies for reducing cardiac contractility (an unwanted effect).

INTRODUCTION

Local anesthetics can be harmful or toxic when used inappropriately. While these drugs are targeted to block nerve conduction in specific tracts, they can interfere with the proper function of any excitable organ system to which they gain access. The central nervous system and cardiovascular system are two of the most important systems that these drugs can influence (see Feldman *et al.*, this volume, Chapter 33). Seizures and significant electrical and pumping malfunctions of the heart can occur.[1,2] It is important to emphasize that these toxic actions occur at drug concentrations that are substantially lower than levels attained at nerve blocking sites. Often these unwanted responses follow inadvertent systemic administration of the drug during regional anesthesia procedures. Are some local anesthetics more dangerous than others in these regards? Studies reported in this chapter will address this important question by presenting structure/activity relations governing several potentially toxic actions of local anesthetics on excised cardiac preparations.

Kenneth R. Courtney • Palo Alto Medical Foundation, Palo Alto, California 94301.

In order to be successful in clinical practice, a local anesthetic must possess certain characteristics. For instance, it must be able to be easily administered to a patient and have a high intrinsic potency at the neural effector site. It must be able to get to the site of action by penetrating one or more diffusion barriers, and it must have a suitable duration of action. In addition to incorporation of receptor specific groups, such as a tertiary amine, other molecular characteristics of the drug including lipid and aqueous solubility, size, and acid-dissociation characteristics (pKa) are also manipulated in order to achieve the desired actions. Many of these manipulations in drug structure have predictable effects on the actions these drugs have on cardiac tissue. For instance, local anesthetics are given a greater intrinsic potency and greater duration of action by increasing their lipid solubility, such as has been done for etidocaine and bupivacaine. These increases in lipid solubility (and size), while they do enhance beneficial anesthetic actions, may enhance cardiotoxic actions of these drugs, such as their negative inotropic effect.

METHODS

The depressant actions that local anesthetics have on cardiac contractility, automaticity (sinoatrial pacemaker), and excitability (sodium channels) were all measured using excised cardiac tissues as described below. Young adult guinea pigs or rabbits were sacrificed by cervical dislocation, and their hearts were rapidly excised. Preparations were superfused with Tyrode's solution containing (in mM): 137 NaC1, 4.5 KC1, 1.8 $CaC1_2$, 1.1 $MgC1_2$, 12 $NaHC0_3$, and 5.5 glucose oxygenated with a 5% CO_2–95% O_2 gas mixture. Contractility studies were performed at 31–32°C, and studies on automaticity and sodium channel (upstroke velocity) block were performed at 35–36°C. Alprenolol and the amide-linked local anesthetics (lidocaine, tocainide, etidocaine, bupivacaine, prilocaine, and mepivacaine) were supplied by Astra Pharmaceutical Products (Worcester, MA). Other drugs utilized included mexiletine (Beohringer Ingelheim, Ridgefield, CT), procaine, imipramine, propranolol (Sigma), and procainamide (Pronestyl, Squibb).

The distribution coefficient (Q) is used to characterize a drug's capability for distributing into a model hydrophobic (octanol) environment. This parameter includes consideration of both the intrinsic lipid solubility of the drug (log P) and the fraction of neutral drug in the aqueous phase according to

$$\log Q = \log P - \log (1 + 10^{\mathrm{pKa}-\mathrm{pH}})$$

These values were obtained from Hansch and Leo[10] or calculated using their fragment technique[10] from a closely related compound whose log P value was measured.

RESULTS

Local anesthetics that gain access to the general circulation can produce several potentially toxic effects in cardiac tissue. Sinus bradycardia, severe reductions in cardiac contractility, and arrhythmias have all been observed[2] (see Feldman *et al.*, this volume, Chapter 33). Structure/activity relationships that relate to each of these local anesthetic actions are described below. It is useful in this regard to distinguish between drugs which are amide-linked, such as lidocaine, tocainide, and bupivacaine, and those which are not, such as procaine, propranolol, and mexiletine (Fig. 1).

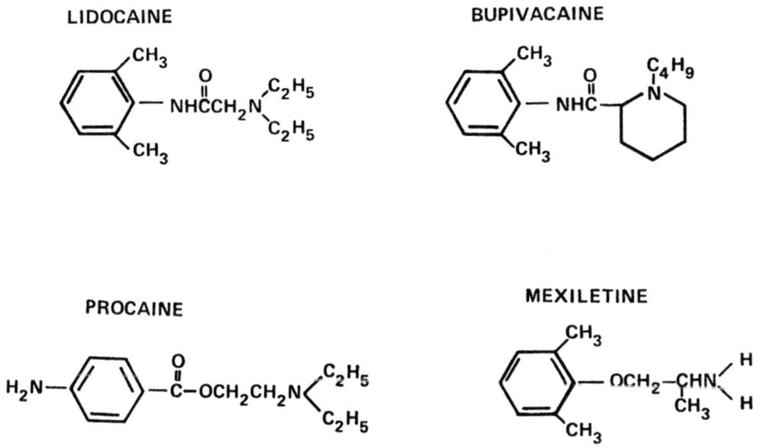

Figure 1. Examples of local anesthetics that are amide-linked include lidocaine and bupivacaine. Those which are not amide-linked include the ester-linked procaine and the ether-linked antiarrhythmic mexiletine.

Effects on Automaticity

Many local anesthetics depress automaticity of cardiac tissues. They do so with widely varying potencies, and, of course, greater depression of automaticity is observed at higher concentrations of drug. A number of amide-linked anesthetics have been compared for their ability to depress sinoatrial automaticity. Sinoatrial rate was monitored by using either

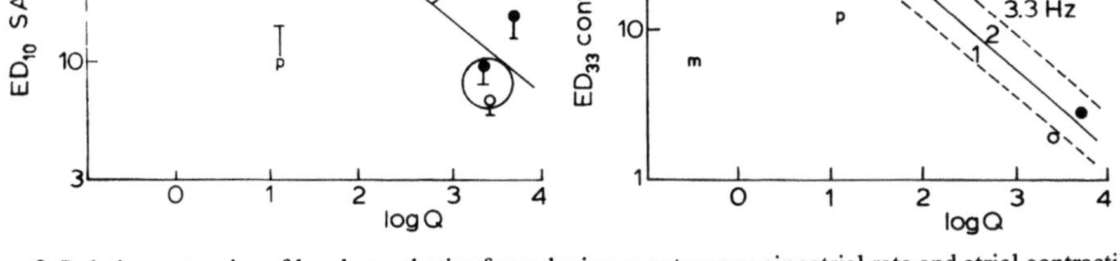

Figure 2. Relative potencies of local anesthetics for reducing spontaneous sinoatrial rate and atrial contractility. (A) Concentration of amide-linked drug (smaller circles) that reduces sinoatrial rate by 10% depends on its lipid distribution coefficient, log Q. Amide-linked drugs include (left to right) tocainide, prilocaine, mepivacaine, lidocaine, bupivacaine, and etidocaine. Larger circles enclose similar results obtained using rabbit and guinea pig preparations for both lidocaine (left) and bupivacaine (right). Standard errors indicated (n = 2–5). Mexiletine (m) and propranolol (p) are ether-linked drugs. (B) Concentration of drug that reduces atrial contractions at 2 Hz by 33% (ED_{33}) also depends on log Q for amide-linked structures (circles and solid best-fit line, from ref. 6). Amide-linked drugs are (open circles, left to right) procainamide, tocainide, lidocaine, and bupivacaine with new result for etidocaine (filled circle at far right). Dotted lines indicate best-fit lines for measurements, omitted in graph, at slower and faster drive rates which produce, respectively, stronger and weaker negative inotropic effects. Propranolol (p) results from Ref. 11.

electrical activity (electrocardiogram) or contractions of adjacent right atrial tissue. It should be understood that the effects that these drugs have on sinoatrial automaticity may not be the same as those that might be observed in other cardiac pacemaker regions. For instance, Purkinje fibers are capable of automaticity, generally at a rate slower than that of the primary pacemaker located in the sinoatrial region, and these Purkinje pacemakers are much more sensitive to K^+ ion levels and sodium channel block by tetrodotoxin than is the sinoatrial pacemaker.

Results using both guinea pig and rabbit sinoatrial tissue are pooled in Fig. 2. This has been done because almost identical results were obtained with both these preparations for lidocaine and bupivacaine. Sinoatrial rate was reduced, on the average, by 10% over all drugs and concentrations utilized. The concentration required to depress rate by 10% was extrapolated in each case in order to compare different drugs with each other regarding their ability to depress sinoatrial automaticity. This was done by assuming first-order kinetics; extrapolation errors were kept to a minimum by using 10% as the target rate reduction. The resulting plot (Fig. 2A) shows that amide-linked local anesthetics have negative chronotropic potencies that increase with increases in their lipid distribution coefficients.

There are two drugs included in Fig. 2A that are not amide-linked; these are mexiletine (Fig. 1) and propranolol, which are ether-linked structures. They are both considerably more potent than amide-linked drugs of equivalent solubility in octanol. As will be shown later, this is a common finding for a variety of local anesthetic actions in both cardiac and nerve preparations.

Effects on Contractility

A previously published report described the negative inotropic actions of several local anesthetics.[6] These results, which are reproduced in Fig. 2B, concern measurements that were made of the concentration of drug required to reduce peak atrial contractions by 33% (excised guinea pig preparations, 2 Hz stimulation). Here again the amide-linked drugs have a potency that increases with increases in their lipid distribution coefficients. In addition, ether-linked drugs appear to be more potent than amides of equivalent solubilities in octanol. Etidocaine (filled circle at right) appears to be very similar to bupivacaine (open circle at right) with regard to its negative inotropic potency.

Figure 2B includes estimates of the effect of drive rate on the negative inotropic potency of the same amide-linked drugs. They all produced greater negative inotropic actions at slower rates of stimulation.

Procaine actually increased atrial contractility slightly in the few experiments where it was tried. Others have reported that procaine and benzocaine can have positive as well as negative inotropic actions in cardiac tissues. The positive inotropic action may be attributable to an adrenergic action of these ester-linked local anesthetics.[14]

Effects on Myocardial Sodium Channels

Maximum rates of rise (upstroke velocities) of myocardial action potentials have been used to assess the changes in sodium channel availability that occur during exposure to local anesthetics. In drug-free conditions, sodium channel availability (hence excitability) recovers within 50 msec or so of the end of a cardiac action potential. However, this changes dramatically with drug treatment. Drug block of sodium channels is modulated by the state of channel (resting, open, or inactivated) and is, as a consequence, substantially greater after an

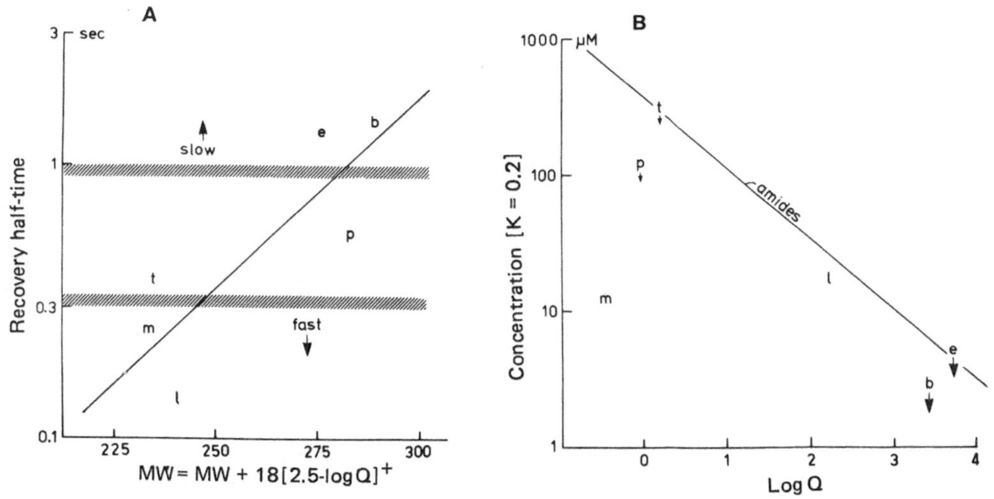

Figure 3. Structure activity relations governing the rate of sodium channel unblocking between action potentials (APs) and governing the amount of channel blocking during APs. (A) Sodium channel availability, as measured by upstroke velocity of APs, is reduced by evoking trains of APs after a period of rest. Half-times (ordinate) which govern rate of recovery from this use-induced block of channels are determined primarily by the size and, to some extent, by the solubility of the drug.[3] Abscissa represents a linear combination of drug size (MW) and solubility (log *Q*) where the molecular weight adjustment (MW') is made for those drugs having log *Q* values less than 2.5. (B) Amount of blocking of channels per cardiac APs can be estimated from blocking behavior during trains of APs if unblocking behavior between APs is known.[5] This blocking parameter kappa is concentration dependent, and the concentration of drug that produces a standard amount of such blocking (kappa = 0.2) depends on log *Q* of amide-linked drug structures (regression line from ref. 5). Ethers (m and p) are again relatively more potent than amides of equivalent octanol solubility.[5] See text for significance of arrows. Local anesthetics, coded by first letter, include lidocaine, procaine, etidocaine, and bupivacaine, as well as the closely related antiarrhythmic drugs tocainide and mexiletine. Guinea pig papillary muscle preparations.

action potential than before[12] (see this volume, Section III). Recovery from these use-induced increases in channel block occurs with widely varying half-times, ranging from a tenth of a second to tens of seconds depending on the drug being studied.[3] Previous results from my laboratory have shown that there is a highly predictable recovery half-time associated with all drugs having capabilities for blocking sodium channels. What is most interesting is that drug size (primarily) and lipid solubility (secondarily) can be used to predict where within this "memory" spectrum a given drug will be found.[3]

Figure 3A indicates the predictions of this size/solubility hypothesis for a number of commonly used local anesthetics. Two local anesthetics not included in Fig. 3A are prilocaine and mepivacaine, which should have fast and intermediate recovery kinetics, respectively. It was clear from the original results[3] that drugs of very low lipid solubility induced slower recovery half-times than expected based on consideration of their size alone; this could be attributable to their retention of hydration shells which may be increasing their effective sizes. More recent results with bupivacaine and etidocaine suggest that only low lipid solubility modulates this exquisitely size-dependent recovery process. Etidocaine and bupivacaine display recovery kinetics (1–2 sec half-times) that are as slow as predicted based on consideration of their size alone.[4] Their exceptionally large lipid solubility does not appear to speed up their recovery kinetics although it will affect their intrinsic potency as described in the next section.

Quantifying Drug Blocking during Action Potentials

Block of sodium channels develops during the open and/or inactive conformations of the channel protein which occur during each cardiac action potential. There is recovery from this use-induced channel block during the diastolic time interval between action potentials when the channel is in its low-affinity state. If the drug is kinetically fast, as is the case for lidocaine, then there will be time for complete recovery between beats. However, channel block can accumulate, even at normal heart rates, if the drug's kinetics are slower. A drug's "potency" for blocking myocardial excitability is then determined both by its intrinsic potency for blocking channels during individual action potentials and by its unblocking kinetics between action potentials.

Structure/activity relations (SAR) regarding drug unblocking between action potentials, when the channels are in a low-affinity state, were reviewed in the previous section. The size/solubility concept described in Fig. 3A involves unblocking rates that are not dose-dependent.[3] Substantially different, dose-dependent SAR govern the blocking of channels that occurs during the high-affinity state(s) that occur with each excitation. In order to complete these SAR, a mathematical analysis was developed that calculated an effective rate of channel blocking (kappa) during each action potential in a train while channel block is accumulating.[5] Intrinsic drug potencies, so estimated, were related to drug lipid distribution coefficients much as observed in Fig. 2. Amide-linked drugs showed a systematic potency, related to log Q (regression line in Fig. 3B). Ether-linked drugs were relatively more potent than amide-linked drugs in this context as well as in those previously described (Fig. 2) with regard to both negative chronotropic and negative inotropic actions of these same local anesthetics.

Finally, the combined effects of blocking of excitability during action potentials and unblocking between action potentials should be considered. The arrows in Fig. 3B are coded by size with those drugs showing the slowest recovery kinetics having the largest arrow. The drugs that have slower unblocking rates will accumulate greater losses of excitability at any given heart rate when these drugs are used at their standard concentrations (kappa = 0.2). Thus, bupivacaine and etidocaine will, by virtue of their 1–2 sec recovery half-times, accumulate significant depression of cardiac excitability at normal heart rates, whereas lidocaine, with its 0.1–0.2 sec half-time, should not (see Hondeghem and Clarkson, this volume, Chapter 32). These kinetically slower drugs are, in effect, more potent at normal heart rates than their standard kappa concentrations in Fig. 3B indicate; this may contribute to their reported cardiotoxicity.[1]

DISCUSSION

The effects of local anesthetics on cardiac excitability are significantly potentiated when cardiac tissues are infarcted, acidotic, or hyperkalemic. These actions are effectively explained by the modulated receptor hypothesis for such drug action, which is described in greater detail by Hondeghem and Clarkson (this volume, Chapter 32).

Previous studies regarding the effects of local anesthetics on cardiac contractility and automaticity have stressed the correlation between their potency for producing nerve block and their potentially cardiotoxic actions.[8] The observations reported here are similar to those previously published.[8] However, an important feature needs to be emphasized. Both etidocaine and bupivacaine are some 10–20 times as potent as lidocaine in depressing cardiac (atrial) contractility. These same two local anesthetics are reported to have 4 times lidocaine's potency as local anesthetics[2]; thus they may be relatively more cardiotoxic when used in equianesthetic doses. It should be noted that these inotropic actions in atrial muscle may not be the same as those that will occur in ventricular muscle (see Lynch, this volume, Chapter 34).

Etidocaine and bupivacaine depress automaticity of sinoatrial pacemaker tissue with potencies more in line with their potency as local anesthetics. They are 2 and 3 times as potent as lidocaine, respectively. The slopes of the regression lines in Figs. 2B and 3B, which describe the increases in negative inotropic and sodium channel blocking effect, respectively, with increases in log Q, are very similar (3.3-fold/log$Q$ unit). The slope in Fig. 2A is substantially lower (2.1-fold/log Q unit—note that the y-axis spans only two orders of magnitude in Fig. 2A). Thus, as one increases the log Q characteristic of an amide-linked drug, concurrent increases in negative inotropic and sodium channel blocking capabilities occur, along with smaller increases in negative chronotropic potency.

A major finding of this study is that amide-linked local anesthetics have a number of actions on cardiac tissues which increase as drug lipid distribution capability increases. This, in itself, is not surprising in view of the long history of success of Meyer–Overton correlations in anesthesia research.[9,13] However, ether-linked drugs produce actions that are relatively stronger than those produced by amide-linked drugs having similar log Q characteristics. This holds for negative inotropic, negative chronotropic, and sodium channel blocking actions in cardiac tissues. Very similar results have been published regarding the sodium channel blocking potencies (half blocking concentrations) of these same drugs in myelinated nerve and skeletal muscle. These results were summarized by Courtney and Etter,[7] who showed that ether-linked drugs appeared to be the most potent, linear amides intermediate, and cyclic amides (many anticonvulsants) least potent in this regard. The hypothesis of steric hindrance at the anesthetic binding site was developed from these studies since ethers would provide the least, and cyclic amides the most, hindrance for close apposition of drug molecules to hypothesized binding sites on or near sodium channels.

It should also be pointed out that the differences in observed potencies for the different structural classes, as illustrated in Fig. 2, could be explained by an alternative hypothesis which does not involve steric hindrance at the binding site. Octanol could be an inappropriate model solvent for cell membranes; a better solvent choice, hence log Q scale, might realign the drug potencies for these different structural classes.[9]

Acknowledgments. The author would like to thank Elaine Etter for her help during portions of this study. This work was supported by NIH grant HL24156 (to KRC) and Biomedical Research Support RR05513 (to PAMF).

REFERENCES

1. Albright, GA: Cardiac arrest following regional anesthesia with etidocaine or bupivacaine. *Anesthesiology* 1979;51:285–287.
2. Covino BG, Vassallo HG: *Local Anesthetics: Mechanisms of Action and Clinical Use.* New York, Grune & Stratton, 1976.
3. Courtney KR: Interval-dependent effects of small antiarrhythmic drugs on excitability of guinea-pig myocardium. *J Mol Cell Cardiol* 1980;12:1273–1286.
4. Courtney KR: Tests of the size/solubility hypothesis. *Circulation* 1983a;68:296a.
5. Courtney KR: Quantifying antiarrhythmic drug blocking during action potentials in guinea-pig papillary muscle. *J Mol Cell Cardiol* 1983b;15:749–757.
6. Courtney KR: Relationship between excitability block and negative inotropic actions of antiarrhythmic drugs. *Proc West Pharmacol Soc* 1984;27:181–184.
7. Courtney KR, Etter EF: Modulated anticonvulsant block of sodium channels in nerve and muscle. *Eur J Pharmacol* 1983;88:1–9.
8. Feldman HS, Covino BM, Sage DJ: Direct chronotropic and inotropic effects of local anesthetic agents in isolated guinea pig atria. *Reg Anesth* 1982;7:149–156.

9. Franks NP, Lieb WR: Where do general anesthetics act? *Nature* 1978;274:339–342.
10. Hansch C, Leo A: *Substituent Constants for Correlation Analysis in Chemistry and Biology*. New York, Wiley-Interscience, 1979.
11. Hirada S, Ban T, Fujita T, Koshiro A: Negative inotropic effects and the hydrophobicity of beta-adrenergic blocking agents. *Arch Int Pharmacol Ther* 1981;252:262–271.
12. Hondeghem LM, Katzung BG: Time- and voltage-dependent interactions of antiarrhythmic drugs with cardiac sodium channels. *Biochim Biophys Acta* 1977;472:373–398.
13. Seeman P: The membrane actions of anesthetics and tranquilizers. *Pharmacol Rev* 1972;24:583–655.
14. Vassalle M, Bhattacharyya M: Local anesthetics and the role of sodium in force development by canine ventricular muscle and Purkinje fibers. *Circ Res* 1980;47:666–674.

32

Modulated Receptor Theory and Cardiac Toxicity of Local Anesthetics

Luc M. Hondeghem and Craig W. Clarkson

ABSTRACT. It is well known that not all local anesthetic drugs have equal cardiac toxicity. Recent reports indicate that bupivacaine may be relatively more cardiotoxic than lidocaine.[1,27] To determine a cellular electrophysiological basis for this difference in cardiotoxicity, we investigated the effects of both bupivacaine and lidocaine on the maximum upstroke velocity ($\dot{V}_{max}$) of the action potential in guinea pig papillary muscles. $\dot{V}_{max}$ was used as an indicator of peak sodium current. Papillary muscles were mounted in a single sucrose gap tissue bath, and membrane potential was controlled under voltage clamp at all times except during the action potential upstroke. Both bupivacaine (1 μg/ml, 3.5 μM) and lidocaine (10 μg/ml, 43 μM) blocked few sodium channels after a long rest at the normal diastolic potential (−85 mV), but rapidly blocked channels during the cardiac action potential. However, these two drugs differed markedly in the rate at which block dissipated between consecutive beats, dissipation of bupivacaine block being much slower (time constant ≅ 1 sec) than that of lidocaine (time constant ≅ 180 msec). Thus, if channel block is visualized as drug movement into and out of the sodium channel,[20] bupivacaine may be considered to block channels in a fast-in–slow-out fashion, while lidocaine blocks channels in a fast-in–fast-out fashion. The imbalance of drug association and dissociation rates for bupivacaine results in substantial cumulative block at even very slow heart rates (<60 beats/min). In contrast, cumulative block does not occur in the presence of lidocaine. This may explain why bupivacaine has a greater potential for producing serious disturbances of cardiac conduction and arrhythmias when compared to lidocaine.[1,27]

INTRODUCTION

Sodium channels exist in at least three primary states (Fig. 1), each of which may have several substates.[2,21] At normal diastolic membrane potentials (−80 to −90 mV in heart), the channels are rested (R): they are closed but can be opened by depolarization. At less negative and positive potentials (above −50 mV), the channels are inactivated (I): they are closed and cannot be opened, no matter how strong the depolarization. Upon transition from R to I state (by depolarization), the channels may briefly become open (O).

The affinity of local anesthetic drugs for sodium channels has been shown to be dependent on the state of the channel, in both nerve[20] and heart.[22] This state-dependent blockade of sodium channels can be well described by the modulated receptor theory (see this volume,

Luc M. Hondeghem • Department of Medicine and Pharmacology, Stahlman Cardiovascular Research Program, Vanderbilt University School of Medicine, Nashville, Tennessee 37232. *Craig W. Clarkson* • Department of Pharmacology, Tulane Medical School, New Orleans, Louisiana 70112.

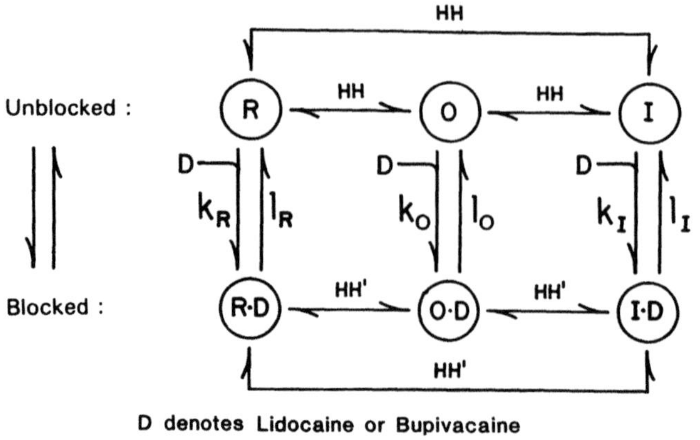

Figure 1. A schematic representation of the modulated receptor hypothesis. Drug-free (unblocked) states are shown at the top: R, rested; O, open; I, inactivated. Transitions between all three states are assumed to be governed by Hodgkin–Huxley rate constants (HH). Drug may bind to all three states, resulting in drug-associated (blocked) states (R·D, O·D, and I·D). Different association (k_R, k_O, k_I) and dissociation (l_R, $l_{O'}$, l_I) rate constants define drug interaction with each state. The Hodgkin–Huxley rate constants defining transitions between drug-associated channels (HH′) are similar to those of drug-free channels, except that they have had their voltage dependence shifted to more negative potentials by an amount ΔV (voltage shift). For further discussion see refs. 20 and 22.

Section III). In general, rested channels have a relatively low affinity for sodium channel blockers, whereas open or inactivated channels have a much higher affinity for these blocker.[23] As a result, blockade of sodium channels increases during each action potential[29] and declines during the diastolic interval between action potentials.[9] Whenever the diastolic interval is too short for complete recovery from block to occur, the sodium current during the action potential upstroke is reduced, and conduction is depressed. When conduction in the heart is substantially depressed, arrhythmias are likely to occur.[18]

Local anesthetic agents block pain by reversibly blocking impulse conduction of nerve fibers in specific areas of the body.[14] Unfortunately, local anesthetic drugs and their effects do not always remain local.[6,27] Occasionally, fast absorption from the site of administration or accidental systemic injection can lead to substantial drug levels in the blood.[14] If blood levels become high enough, then toxic effects can be observed in several organ systems, especially the central nervous and cardiovascular systems (see Feldman *et al.*, this volume, Chapter 33). Although we realize that cardiac toxicity of local anesthetics may result from several different mechanisms, i.e., from indirect effects on autonomic nerves, from effects on contractility,[6] from biochemical actions,[30] and from blockade of numerous transmembrane channels,[7] we shall limit this chapter to the direct effects of local anesthetics on the cardiac sodium current. There is ample evidence which indicates that excessive block of cardiac sodium channels may be a primary mechanism underlying clinical toxicity of local anesthetics.[26]

Blockade of cardiac sodium channels *per se* is, however, not necessarily cardiotoxic. In fact, many antiarrhythmic drugs exert their therapeutic effects primarily by blocking cardiac sodium channels.[23] Moreover, one of the most commonly used local anesthetics, lidocaine,[25] happens to also be the most commonly used intravenous antiarrhythmic drug.[24] Therapeutic plasma concentrations of lidocaine (antiarrhythmic range 2–6 μg/ml or 9–26 μM)[17] hardly reduce the sodium current of cardiac tissue paced at physiological heart rates (60–100 beats/min) and having normal (−80 to −90 mV) resting potentials.[4] In fact, approximately 50 μg/ml (213 μM) of lidocaine is required to reduce the maximum upstroke velocity of the cardiac action potential by 10% at 95 beats/min in well-polarized cardiac

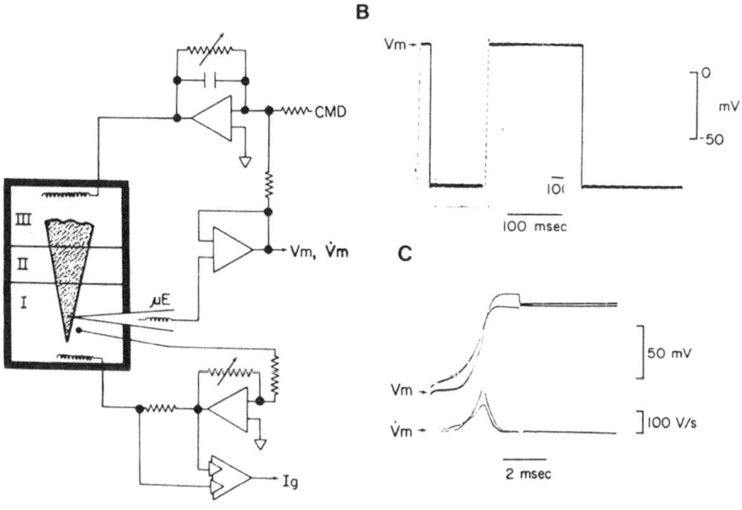

Figure 2. Methods used for voltage clamp and measurement of $\dot{V}_{max}$. (A) Schematic diagram of the clamp circuit and the three-chambered tissue bath used to produce voltage clamp employing the single sucrose gap method. Chamber III was perfused with isotonic KCl solution, chamber II with calcium-enriched isotonic sucrose, and chamber I with either drug-free or drug-enriched HEPES-buffered solution. I_g, transgap current; μE, KCl-filled glass microelectrode, electrode; CMD, command voltage; $\dot{V}_m$, dV_m/dt. (B) An oscilloscope record of transmembrane voltage obtained at a slow sweep speed. A broken vertical line was added to indicate the presence of a 5-msec free-running action potential upstroke evoked by a 1-msec current stimulus. The membrane was voltage-clamped to +20 or −85 mV at all other times than during the first 5 msec of an action potential upstroke. (C) Two superimposed oscilloscope records of transmembrane voltage and their differentiated signals, recorded at a rapid sweep speed. The amplitude of the differentiated signal ($\dot{V}_m$) indicates $\dot{V}_{max}$. The membrane was clamped to +20 MV following the initial 5 msec of the free-running action potential.

cells.[15] Consequently, lidocaine does not markedly depress conduction in healthy cardiac tissue, even in concentrations up to 10-fold above the maximum therapeutic level.

Not all local anesthetic agents share this great degree of safety against cardiac toxicity with lidocaine. Another very popular local anesthetic, bupivacaine, has recently been reported to posess marked cardiac toxicity when the drug achieves excessive levels in the blood.[1] The symptoms of cardiac toxicity range from depression of conduction to serious arrhythmias, which sometimes are fatal. Concentrations as low as 1 μg/ml (3.5 μM) of bupivacaine produce a considerable (32%) reduction of the maximum upstroke velocity of the cardiac action potential at 60 beats/min.[10] Hence, the question arises as to why some sodium channel blockers are well tolerated by the heart, whereas others readily precipitate cardiac toxicity. In an attempt to answer this question, we compared the actions of lidocaine and bupivacaine on the maximum stroke velocity of the cardiac action potential.

METHODS

The experiments were performed on guinea pig papillary muscles that were voltage-clamped using the single sucrose gap technique (Fig. 2). The maximum upstroke velocity of the free running test action potential ($\dot{V}_{max}$) was used as an indicator of the peak sodium current. The experimental chamber was perfused with a salt solution of the following composition: NaCl 143, KCL 4, $CaCl_2$ 1.8, $MgCl_2$ 1.1, *d*-glucose 5, and HEPES 5 mM. The pH was adjusted to 7.4 with approximately 3mM NaOH, and the temperature was kept at 37°C. Once control measurements were obtained, drugs were added to superfusate, and an equilibration period of at least 30 min in lidocaine or bupivacaine was allowed before the measurements were repeated. Use-dependent block was studied by applying 180-msec pulses to +20 mV at various interstimulus intervals (200–8000 msec). Diastolic recovery from block was evaluated by applying a test pulse at variable times after a conditioning train (19 pulses at 200 pulses/min). Development of block during the action potential was characterized by applying pulses of variable duration (5–2000 msec) followed by a test action potential after a fixed recovery interval (100 msec).

RESULTS

Bupivacaine (1 μg/ml, 3.5 μM) does not significantly decrease $\dot{V}_{max}$ of the first action potential after a long (20-sec) rest period (Fig. 3A). However, it reduces $\dot{V}_{max}$ in a use-

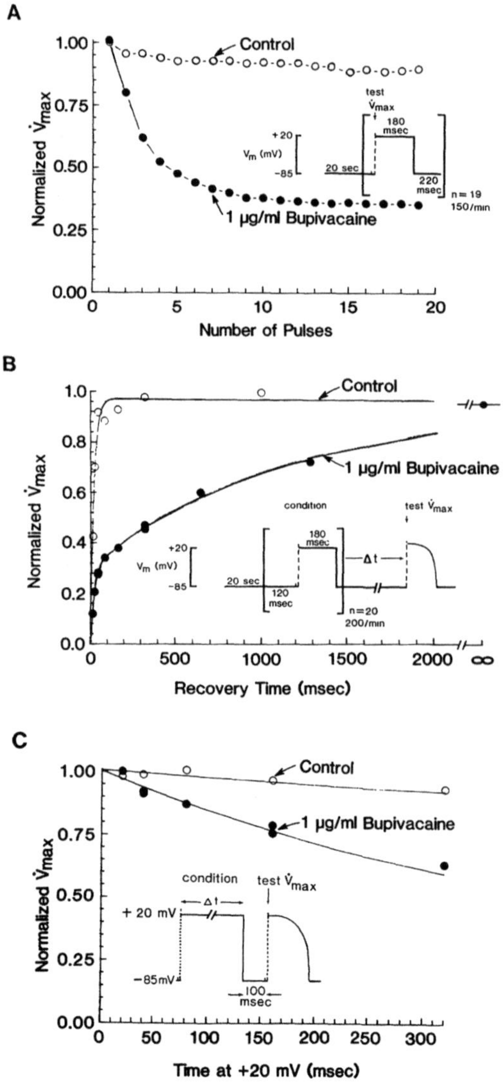

Figure 3. Characteristics of bupivacaine block of cardiac sodium channels. In all three panels, pulse protocols are shown as insets. Vertical broken lines indicate free-running action potential upstroke of 5 msec duration from which $\dot{V}_{max}$ was measured. $\dot{V}_{max}$ has been normalized to its value after a long rest in the absence of drug. (A) $\dot{V}_{max}$ during a train of 19 beats (ISI = 400 msec) evoked after a 20-sec rest at −85 mV. The membrane was voltage-clamped during the action potential plateau and diastole. There was little change in $\dot{V}_{max}$ under control conditions, but a substantial use-dependent decline in the presence of bupivacaine. (B) The time course of $\dot{V}_{max}$ recovery at −85 mV under control conditions and in bupivacaine. The curves represent least-square error fits of the data. Under control conditions $\dot{V}_{max}$ recovered exponentially (time constant = 21 msec). In bupivacaine, recovery was well described by the sum of two exponentials ($\dot{V}_{max}$ = −0.66 exp[−t/1138 msec] − 0.39 exp[−t/20 msec] + 0.96). (C) The time course of bupivacaine block development during an action potential plateau clamped to +20 mV. There was little decline in $\dot{V}_{max}$ under control conditions. However, in bupivacaine $\dot{V}_{max}$ declined exponentially ($\dot{V}_{max}$ = 0.80 exp[−t/461 msec] + 0.20) as the plateau clamp was prolonged.

dependent fashion,[10,11] so that with each successive action potential in a train, $\dot{V}_{max}$ declines until a steady state is reached. At this time the amount of the block development during each action potential equals the amount of recovery from block during diastole. In order to characterize the diastolic recovery process, we applied a test action potential at various times after a conditioning train (Fig. 3B). At a diastolic potential of −85 mV, recovery from bupivacaine block could be well approximated by the sum of two exponentials. The fast phase had a time constant of 10–20 msec and reflects recovery from inactivation of drug-free channels. The slower phase had a time constant of approximately 1 sec and reflects the time course of drug dissociation. Durg dissociation could be slowed by depolarization and made faster by hyperpolarization (not shown). In order to characterize the development of block during the action potential, we voltage-clamped the membrane during the plateau of the action potential for variable durations. The amount of block that developed during the conditioning pulse was then assayed with a test $\dot{V}_{max}$ evoked after a 100-msec recovery interval. The recovery interval was long enough to allow all drug-free channels to recover from inactivation, and short enough so that only a small (< 6%) fraction of bupivacaine-blocked channels unblock. As shown in Fig. 3C, bupivacaine causes virtually no block during the first 5 msec of the plateau clamp, but a substantial amount of block develops during the subsequent 315 msec. Thus, bupivacaine mainly blocks inactivated sodium channels and blocks open channels only to a negligible extent. Also, the amount of block development per action potential is rather large (about 40%) and recovery from block during diastole is rather slow. Therefore, bupivacaine can best be characterized as a fast-in–slow-out blocker (FISO).

Lidocaine (10 μg/ml, 43 μM) also has relatively little effect on the $\dot{V}_{max}$ of the first action potential after a long rest period and reduces $\dot{V}_{max}$ in a use-dependent fashion (Fig. 4A). However, steady-state reduction of $\dot{V}_{max}$ is achieved much more rapidly than in bupivacaine; i.e., most of the block develops within two beats. Diastolic recovery from block by lidocaine occurs with a time constant of approximately 180 msec (Fig. 4B) and hence proceeds approximately 6 times faster than for bupivacaine. In further contrast to bupivacaine, a substantial amount of block by lidocaine develops during the first 5 msec of the plateau clamp (Fig. 4C). Thereafter, as for bupivacaine, block develops progressively throughout the plateau until steady-state block is reached. However, since lidocaine recovers so quickly, during the 100-msec time period preceding the test pulse, as much as 40% of the channels may recover from lidocaine block. Therefore, the 65% reduction of $\dot{V}_{max}$ in Fig. 4C probably represents saturation of the channels with lidocaine. Over 40% of the channels block during the upstroke of the action potential (open channel block), and the remainder block during the first 100 msec of the plateau clamp (inactivated channel block). Thus, during a typical cardiac action potential, lidocaine blocks sodium channels in both the open and inactivated states. Since there is a substantial amount of block development per action potential, and recovery from block occurs rather quickly, block by lidocaine is best described as fast-in–fast-out (FIFO).

Bupivacaine is generally considered to be approximately four times more potent than lidocaine in producing local anesthesia.[14] However, according to the modulated receptor theory, potencies of drugs are difficult to compare. Comparative "potencies" vary with drug concentration, resting membrane potential, firing rate, action potential duration, and other experimental conditions such as pH.[19] To illustrate this, in Fig. 5 we computed the ratio of bupivacaine to lidocaine block using the modulated receptor theory for various stimulation rates and action potential durations. For this simulation we used 2.5 μM bupivacaine and 10 μM lidocaine. At very slow heart rates, both drugs had virtually no effect on $\dot{V}_{max}$, and consequently the ratio of block is close to unity. However, as the heart rate is increased,

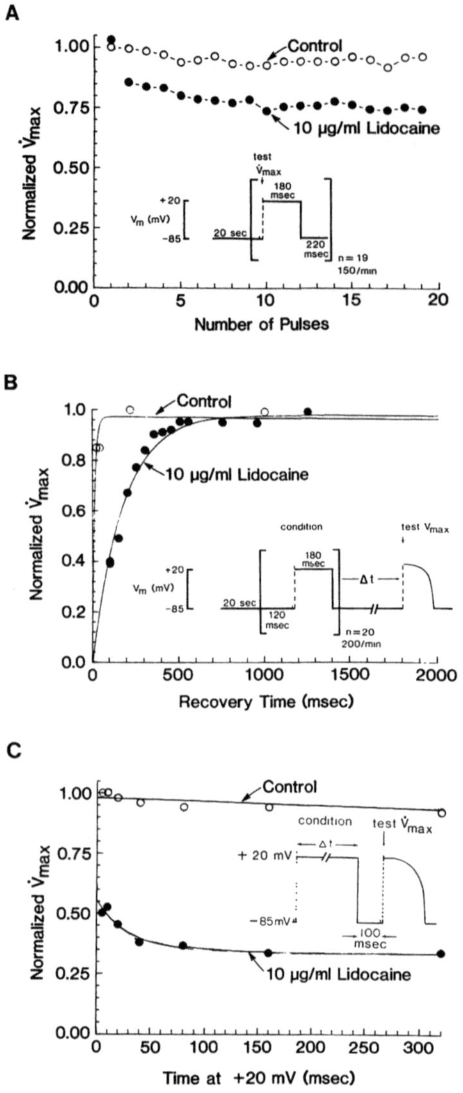

Figure 4. Characteristics of lidocaine block of cardiac sodium channels. Clamp pulse protocols are shown as insets. Vertical broken lines indicate free-running action potential upstrokes of 5 msec duration from which $\dot{V}_{max}$ was measured. $\dot{V}_{max}$ has been normalized to its value after a long rest, in the absence of drug. (A) $\dot{V}_{max}$ during a train of 19 beats (ISI = 400 msec) following a 20-sec rest at −85 mV. The membrane was clamped during the action potential plateau and diastole. There was essentially no change in $\dot{V}_{max}$ under control conditions, but a use-dependent decline in lidocaine. (B) The time course of $\dot{V}_{max}$ recovery at −85 mV under control conditions and in lidocaine. The curves represent least-square exponential fits of the data. $\dot{V}_{max}$ recovered with a time constant of 15 msec under control conditions, and with a time constant of 179 msec in lidocaine. (C) The time course of lidocaine block development during an action potential with a plateau clamped to +20 mV. There was little change of $\dot{V}_{max}$ under control conditions. However, in lidocaine there was a substantial "instantaneous" (open channel) block, as well as a rapid block of inactivated channels (time constant = 30 msec). Note that the fraction of $\dot{V}_{max}$ that does not appear to become blocked (35%) may largely result from rapid recovery of lidocaine-blocked channels during the 100-msec interpulse interval (see text).

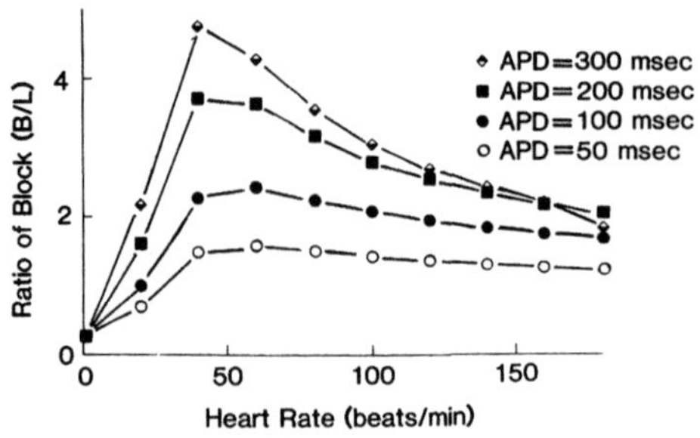

Figure 5. Effect of action potential duration (APD) and heart rate on the ratio of block produced by equianesthetic concentrations of bupivacaine (2.5 μM, 0.8 μg/ml) and lidocaine (10 μM, 2.3 μg/ml). Ratio of block indicates block produced by bupivacaine/lidocaine. Levels of block were calculated using a mathematical model of the modulated receptor hypothesis.[22] Voltage shifts and drug dissociation constants were estimated from fitting experimental data.

block by bupivacaine will accumulate at much lower (physiological) heart rates than block by lidocaine. This results mainly from the large difference in the rate of recovery from block between beats for the two drugs. Consequently, the ratio of block (bupivacaine/lidocaine) will become larger than unity at physiological heart rates. However, as heart rate is further increased, there is progressively less recovery time available, and the difference of block between the two drugs declines again. The change in ratio of block as a function of action potential duration can be explained by the same mechanism. These modulated receptor computations were confirmed experimentally. Actually, we found that at lower concentrations of bupivacaine, the differences in ratio of block could be even larger. For example, bupivacaine (0.2 μg/ml, 0.6 μM) and lidocaine (10 μg/ml, 43 μM) reduce $\dot{V}_{max}$ to similar extents at heart rates between 60 and 120 beats/min (action potential duration 180 msec, diastolic potential −85 mV).[10] Thus, under these conditions one needs 70 times less bupivacaine than lidocaine to obtain similar reductions of $\dot{V}_{max}$.

DISCUSSION

Our results demonstrate that both drugs have a relatively low affinity for sodium channels that are rested. Indeed, drug concentrations that result in marked use-dependent reduction of $\dot{V}_{max}$ will reduce the $\dot{V}_{max}$ of the first action potential after a long rest period by only a minimal extent. Moreover, the small reduction that may be observed does not necessarily reflect block of rested channels, but rather may result either from rapid block of open channels before the time of $\dot{V}_{max}$, or from "trapping" of channels in the ID state (Fig. 1) owing to the presence of a substantial voltage shift and high affinity for the I state.[23] In contrast to the low affinity for the rested state, lidocaine has a relatively high affinity for the open state, and both lidocaine and bupivacaine have high affinities for the inactivated state. Other authors have also observed that lidocaine has a low affinity for the rested state and a much higher affinity for the inactivated state.[3,28] These authors did not observe much open channel block, possibly because they usually clamped to negative membrane potentials (about −30 mV). Since open state blockade appears to be strongly voltage-dependent,[8,31] significant open channel block may only be observed for stronger depolarizations. In the present experiments the upstroke was free-running and usually reached +20 to +30 mV.

The rate of diastolic recovery from block varies greatly between various sodium channel blockers. Since the normal cardiac cycle has a diastolic interval of approximately 500 msec, for a drug to achieve better than 90% recovery during a diastolic interval, it must have a time constant for recovery of less than 250 msec. Consequently, we refer to such drugs in cardiac tissue as "fast-out," and those drugs having time constants of recovery much greater than 250 msec as "slow-out." For example, during a 500-msec diastolic recovery from lidocaine, block will be 94% complete (fast-out), whereas for bupivacaine, it will be only 28% complete (slow-out).

Rate of block development per action potential also varies greatly between local anesthetics.[12,13] Moreover, this rate is strongly dose-dependent, so that a drug that is "slow-in" at one dose may become "fast-in" at a higher dose, and vice versa. For example, therapeutic concentrations of lidocaine (2 μg/ml) can block 40% of the drug-free channels in a single action potential (fast-in). Toxic concentrations of bupivacaine (1 μg/ml) also block about 45% of the drug-free sodium channels, i.e., behave as "fast-in." (Note that since bupivacaine binds to plasma proteins, approximately 5 μg/ml bupivacaine in plasma is required to produce 1 μg/ml free bupivacaine.[16])

Although in terms of the above description lidocaine can be described as a FIFO drug and bupivacaine as a FISO drug, for several reasons we do not wish to propose this as a rigid classification for all sodium channel blockers: (1) Although the drugs discussed here are easily classified as fast-out or slow-out, there must exist numerous drugs that would be difficult to classify unambiguously. (2) The fast-in or slow-in classification could similarly be criticized. Nevertheless, the concept of comparing the rate of block development to rate of recovery from block is an important one. Indeed, if for a given drug and under particular conditions (drug concentration, membrane potential, action potential, driving rate), the rate of block development is much slower than the rate of recovery from block (SIFO), then the drug will have insignificant effects on conduction at normal rates. Conversely, if the rate of block development far exceeds the rate of recovery from block (FISO), then the conduction will be substantially depressed and toxicity will result. Only when both rates are matched (FIFO or SISO) can an effective, but not excessive, drug effect be expected. It is important to stress that an effective and nontoxic drug concentration under a particular set of conditions (normal heart rate and normal membrane potential) could be substantially more depressant on conduction under arrhythmogenic conditions (tachycardias or depolarized tissues). In this way antiarrhythmic drugs can act by selectively interfering with arrhythmogenic activity while leaving tissue with normal electrophysiological properties relatively unaffected[5] (for a review see ref. 23). Our results clearly show that the relative effects of sodium channel blockers are strongly dependent on the experimental conditions. Thus, the potencies observed in nerve tissue may be substantially different from those obtained in heart owing to the differences in cellular electrophysiology, e.g., action potential duration, firing rate, and resting potential.

It has been shown that bupivacaine produces relatively more cardiac toxicity than lidocaine at equieffective local anesthetic concentrations.[27] However, this property is not unique to bupivacaine: drugs that have slow-out properties (e.g., quinidine) are also prone to precipitate cardiac toxicity unless their plasma concentrations are meticulously controlled. Consequently, other drugs that have slow recovery from block combined with relatively fast development of block might also have the potential for cardiac toxicity (e.g., etidocaine[12]).

ACKNOWLEDGMENT. We are greatly indebted to Dr. J. Moyer for developing the software for data analysis and graphics. This work was supported in part by National Institutes of Health Grant HL-21672. Dr. Hondeghem is an Established Investigator of the American Heart Association.

REFERENCES

1. Albright GA: Cardiac arrest following regional anesthesia with etidocaine or bupivacaine. *Anesthesiology* 1979;51:285–287.
2. Armstrong CM, Benzinilla F: Inactivation of the sodium channel. II. Gating current experiments. *J Gen Physiol* 1977;70:567–590.
3. Bean BP, Cohen CJ, Tsien RW: Lidocaine block of cardiac sodium channels. *J Gen Physiol* 1983;81:613–642.
4. Bigger JT, Mandel WJ: Effects of lidocaine on the electrophysiological properties of ventricular and Purkinje fibers. *J Clin Invest* 1970;49:63–77.
5. Brennan FJ, Cranefield PF, Wit AL: Effects of lidocaine on slow response and depressed fast response action potentials of canine cardiac Purkinje fibers. *J Pharmacol Exp Ther* 1978;204:312–324.
6. Block A, Covino BG: Effects of local anesthetic agents on cardiac conduction and contractility. *Regional Anesth* 1981;6:55–61.
7. Boyden PA, Wit AL: Pharmacology of the antiarrhythmic drugs, in Rosen MR, Hofman BF (eds): *Cardiac Therapy*. New York, Martinus Nijhoff, 1983, pp 171–234.
8. Cahalan MD: Local anesthetic block of sodium channel in normal and pronase-treated squid giant axons. *Biophys J* 1978;23:285–311.
9. Chen CM, Gettes LS, Katzung BG: Effects of lidocaine and quinidine on steady state characteristics and recovery kinetics of (dV/dt)max in guinea pig ventricular myocardium. *Circ Res* 1975;37:20–29.

10. Clarkson CW, Hondeghem LM: Mechanism for bupivacaine depression of cardiac conduction: fast block of sodium channels during the action potential with slow recovery from block during diastole. *Anesthesiology* 1985a;62:396–405.
11. Clarkson CW, Hondeghem LM: Evidence for a specific receptor site for lidocaine, quinidine, and bupivacaine associated with cardiac sodium channels in Guinea pig ventricular myocardium. *Circ Res* 1985b;56:496–506.
12. Courtney KR: Quantifying antiarrhythmic drug blocking during action potentials in guinea-pig papillary muscle. *J Mol Cell Cardiol* 1983;15:749:757.
13. Courtney KR, Kendig JJ, Cohen EN: The rates of interaction of local anesthetics with sodium channels in nerve. *J Pharmacol Exp Ther* 1978;207:594–604.
14. Covino BG, Vassallo HG: *Local Anesthetics, Mechanisms of Action and Clinical Use*. New York, Grune & Stratton, 1976.
15. Davis LD, Temte JV: Electrophysiological actions of lidocaine on canine ventricular muscle and Purkinje fibers. *Circ Res* 1969;24:639–655.
16. Denson D, Coyle D, Thompson G, Meyers JS: Alpha$_1$-acid glyco-protein and albumin in human serum bupivacaine binding. *Clin Pharmacol Ther* 1984;35:409–415.
17. Gianelly R, Von der Groben JO, Spivack AP, Harrison DC: Effect of lidocaine on ventricular arrhythmia in patients with coronary artery disease. *N Engl J Med* 1967;277:1215–1219.
18. Gold, H: *Quinidine in Disorders of the Heart*. New York, Hoeber Medical Division, Harper & Row, 1950.
19. Grant AO, Strauss LJ, Wallace AG, Strauss HC: The influence of pH on the electrophysiological effects of lidocaine in guinea pig ventricular myocardium. *Circ Res* 1980;47:542–550.
20. Hille B: Local anesthetics: Hydrophilic and hydrophobic pathways for the drug–receptor reaction. *J Gen Physiol* 1977;69:497–515.
21. Hodgkin AL, Huxley AF: A quantitative description of membrane current and its application to conduction and excitation in nerve. *J Physiol (London)* 1952;117:500–544.
22. Hondeghem LM, Katzung BG: Time- and voltage-dependent interactions of antiarrhythmic drugs with cardiac sodium channels. *Biochim Biophys Acta* 1977;472:373–398.
23. Hondeghem LM, Katzung BG: Antiarrhythmic agents: Modulated receptor mechanism of action of sodium and calcium channel blocking drugs. *Annu Rev Pharmacol Toxicol* 1984;24:387–423.
24. Hondeghem LM, Mason JW: Agents used in cardiac arrhythmias, in Katzung GB (ed): *Basic and Clinical Pharmacology*. Los Altos, CA, Lange, 1982, pp 138–154.
25. Hondeghem LM, Miller RD: Local anesthetics, in Katzung GB (ed): *Basic and Clinical Pharmacology*. Los Altos, CA, Lange, 1982, pp 264–269.
26. Katzung BG: Myocardial toxicity as a result of altered membrane channel function, in Van Stee EW (ed): *Cardiovascular Toxicology*. New York, Raven Press, 1982, pp 135–179.
27. Kotelko DM, Shnider SM, Dailey PA, Brizgys RV, Levinson G, Shapiro WA, Koike M, Rose MA: Bupivacaine cardiac arrhythmias in sheep. *Anesthesiology* 1984;60:10–18.
28. Sanchez-Chapula J, Tsuda Y, Josephson IR: Voltage- and use-dependent effects of lidocaine on sodium current in rat single ventricular cells. *Circ Res* 1983;52:557–565.
29. Strichartz GR: The inhibition of sodium currents in myelinated nerve by quaternary derivatives of lidocaine. *J Gen Physiol* 1973;62:37–57.
30. Szekeres L, Papp GJ: Analysis of the mode of action antiarrhythmic drugs on the basis of their effect on myocardial metabolism, in *Experimental Cardiac Arrhythmias and Antiarrhythmic Drugs*. Budapest, Al-ademiai Kido, 1971, pp 314–324.
31. Yeh JZ, Narahashi T: Kinetic analysis of pancuronium interaction with sodium channels in squid axon membranes. *J Gen Physiol* 1977;69:2293–2323.

33

Toxicity of Intravenously Administered Local Anesthetic Agents in the Dog

Cardiovascular and Central Nervous System Effects

Hal S. Feldman, G. Richard Arthur, and Benjamin G. Covino

ABSTRACT. The accidental intravenous injection of large doses of local anesthetic agents can produce severe central nervous system and cardiovascular system toxicity. A series of experiments was performed to evaluate the cardiovascular/central nervous system toxicity of several local anesthetic agents in the dog. In the first portion of these studies the mean intravenous convulsive doses for lidocaine, etidocaine, bupivacaine, mepivacaine, and tetracaine were determined in awake dogs by increasing serial injections. In the next series of experiments, animals received similar intravenous serial bolus injections while under pentobarbital anesthesia. These experiments allowed the determination of the acute cardiovascular toxicity of the various agents without interference of central nervous system toxicity. In the final series of experiments, the mean convulsive dose of lidocaine, bupivacaine, etidocaine, and mepivacaine was administered as an intravenous bolus to unanesthetized dogs. The cardiovascular/central nervous system toxicity of these agents in awake animals was determined, and blood drug concentrations are reported.

INTRODUCTION

It is well known that the local anesthetic agents can cause severe central nervous system (CNS) and cardiovascular toxicity. In animals and man, the relative anesthetic potency of various ester- and amide-type local anesthetics correlates with the intravenous dose necessary to produce CNS toxicity.[2] In general, the dose and blood concentration of local anesthetics associated with cardiovascular toxicity are significantly greater than those which produce CNS toxicity.[17]

In recent years, several clinical reports have appeared describing cardiovascular collapse following the accidental intravenous injection of the more potent local anesthetics bupivacaine and etidocaine.[5,14] Albright postulated that the more potent lipid-soluble and protein-bound local anesthetics such as bupivacaine may be relatively more cardiotoxic than less potent agents such as lidocaine.[1] Several animal studies have been conducted to determine the relative cardiotoxicity of bupivacaine and lidocaine.[3,4,9–11,13] One of the car-

Hal S. Feldman, G. Richard Arthur, and Benjamin G. Covino • Department of Anesthesia, Harvard Medical School and Brigham and Women's Hospital, Boston, Massachusetts 02115.

diotoxic manifestations associated with rapid intravenous administration of bupivacaine appears to be the production of lethal ventricular arrhythmias. Investigations in conscious but paralyzed cats[4] and conscious spontaneously breathing sheep have demonstrated the occurrence of ventricular arrhythmias following intravenous injection of bupivacaine but not lidocaine. Studies in the anesthetized dogs with various local anesthetic agents failed to demonstrate these electrocardiographic abnormalities.[6,7]

Studies in our laboratory were initiated to investigate the CNS toxicity of various amide local anesthetic agents as well as the cardiovascular effects of these agents. Our experimental studies were conducted in three phases.

Phase I studies were designed to determine the relative convulsive dose of the various local anesthetic agents. In Phase II, the intravenous dose to produce cardiovascular toxicity was determined in anesthetized dogs. The anesthetized dog model was chosen so as to suppress overt CNS toxicity and to allow a clearer evaluation of cardiovascular toxicity. Phase III consisted of studies in which the cardiovascular effects of convulsant or supraconvulsant doses of various local anesthetics were evaluated in nonanesthetized dogs.

METHODS

"Animals used in these studies were maintained in accordance with the guidelines of the Committee on Animals of the Harvard Medical School and those prepared by the Committee on Care and Use of Laboratory Animals of the Institute of Laboratory Animal Resources, National Research Council (DHEW (now DHHS) publicaton no. (NIH) 78-23, revised 1978)."

Other published data of similar pharmacologic studies express dosage and blood concentration of drug in terms of mg/kg and μg/ml, respectively. Therefore, we present our data in a similar manner. For purposes of comparison to other studies using SI units, the following equivalents are provided: 1 mg/kg ≡ 3.69 μM/kg lidocaine HCl, 3.08 μM/kg bupivacaine HCl, 3.53 μM/kg mepivacaine HCl, 3.19 μM/kg etidocaine HCl. 1 μg/ml ≡ 3.69 μM/L lidocaine HCl, 3.08 μM/L bupivacaine HCl, 3.53 μM/L mepivacaine HCl, 3.19 μM/L etidocaine HCl.

Phase I

The determination of the relative convulsive dose was carried out in mongrel dogs weighing an average of 20 kg. The animals were secured in the standing position in a canvas sling. An indwelling venous catheter was inserted in the forelimb vein percutaneously. Intravenous injections of various local anesthetic agents were made at 30-min intervals. Volumes of injectate were kept at a minimum based on drug concentration, animal weight, and dose. Injections were made over 20- to 30-sec periods. Commercially available solutions containing no preservatives were used. Animals were observed for overt signs of CNS irritation and seizure activity. It was planned to inject doses of 1, 3, 5, 10, and 15 mg/kg. However, serial dosing was terminated when seizures occurred. These studies have been described previously (ref. 8 and H. S. Feldman, unpublished data, 1984).

Phase II

Twenty-five adult mongrel dogs having an average weight of approximately 16 kg were used. Animals were anesthetized with 30 mg/kg of intravenous sodium pentobarbital. Polyethylene catheters were surgically implanted in the abdominal aorta and inferior vena cava via the femoral vessels. A similar catheter was inserted into a forelimb vein. A catheter for measurement of cardiac output (CO) was inserted into the external jugular vein and advanced under direct pressure monitoring until the tip of the catheter lay in the pulmonary artery. Animals were intubated and ventilated with oxygen-enriched room air to maintain Po_2, Pco_2, and pH within normal limits. Drug administration was begun after a 45- to 60-min equilibration period. Again, commercially available preservative-free solutions were used. Each animal received 0.3, 1.0, 3.0, and 10.0 mg/kg of the various agents at 30-min intervals. After the 10 mg/kg dose, the dosing schedule varied at 30-min intervals as follows: 20, 30, and 40 mg/kg for lidocaine and mepivacaine, 20 and 30 mg/kg for etidocaine, and 15 mg/kg for bupivacaine.

Injections continued at 30-min intervals until cardiovascular collapse and death occurred. These studies have been reported previously (ref. 6 and H. S. Feldman, unpublished data, 1984).

Phase III

The third phase of our studies consisted of two similarly conducted double-blind crossover studies in awake dogs. The first of these experiments compared lidocaine and bupivacaine. The second set was a comparison of etidocaine and mepivacaine. Twelve adult mongrel dogs were used. Animals were anesthetized with intravenous thiamylal, 20 mg/kg. Tygon catheters were implanted in the abdominal aorta and inferior vena cava via the femoral vessels and exteriorized in the hind leg. A similar catheter was placed in the right cephalic vein. A modified pulmonary artery catheter introducer was placed in the right external jugular vein and exteriorized in the neck. A broad-spectrum antibiotic was administered, and the animals were allowed to recover for 48 hr.

On the day of the experiment, the awake unsedated animal was secured in a canvas sling in the standing position allowing attachments of EKG leads, rectal temperature probe, chest plethysmograph for measuring respiratory rate, and catheters for direct recording of blood pressure.

Previously implanted catheters were flushed intermittently with heparinized saline. A custom-made cardiac output catheter was advanced under pressure monitoring until the tip lay in a pulmonary artery. Iced normal saline, 5 ml, was injected through the right atrial port of the PA catheter, and cardiac output was calculated. Arterial, venous, and pulmonary artery pressures and respiratory rate were transduced and recorded simultaneously with the EKG. Blood samples were drawn from the aortic catheter and analyzed for pH, pO_2, pCO_2, Na, and K concentrations. The mean convulsive dose from our phase I studies were employed: 11 mg/kg for lidocaine, 3.4 mg/kg for bupivacaine, 4.6 mg/kg for etidocaine, and 13 mg/kg for mepivacaine. Animals initially received a mean convulsive dose of one of the agents in a randomized blinded fashion. Commercial preparations without preservatives of lidocaine hydrochloride 2%, bupivacaine hydrochloride 0.5%, etidocaine hydrochloride 1%, and mepivacaine hydrochloride 2% were used. Drug solutions were injected over 30 sec via the cephalic vein catheter in a volume of 15 ml. Animals that did not convulse were given a second dose 1 hr after the first dose. The second injection was 50% greater than the mean convulsive dose and was given over 40 sec in a volume of 20 ml. During the hour prior to the second injection, all control measurements were repeated. Two days following the initial exposure to the first anesthetic solution, the same dog received the opposite solution. Animals remained in the sling for 90 min after end seizure at which time they were removed and allowed freedom of movement within the room. Blood samples for drug concentration analysis continued to be withdrawn for 300 min post end seizure. All blood samples were

Table I. Convulsive Dose in the Awake Dog[a]

	Cumulative convulsive dose (mg/kg) ($\bar{x} \pm$ SEM)	Single dose to cause convulsions (mg/kg) ($\bar{x} \pm$ SEM)
Mepivacaine	28.0 ± 3.7	13.0 ± 1.23
Lidocaine	22.0 ± 6.7	11.0 ± 1.0
Etidocaine	8.0 ± 2.2	4.6 ± 0.4
Bupivacaine	5.0 ± 2.2	3.4 ± 0.4
Tetracaine	4.0 ± 0	3.0 ± 0

[a] n = 5 in all cases.

stored at −20°C in heparinized tubes until analysis for local anesthetic concentration by a gas chromatographic technique similar to that previously described by Tucker.[17]

At least 48 hr after exposure to the second drug in the crossover study, the animal was returned to the sling. At this time, two times the convulsive dose of one of the agents was administered as a bolus injection over 30–60 sec. Drugs were assigned randomly and administered in a blinded fashion. If this dose failed to cause death, the animal was observed for 30–60 min before being euthanized.

RESULTS

Phase I

Table I presents the mean cumulative dose and the single dose to cause seizures in the awake dog. A dose of 3–5 mg/kg of bupivacaine or etidocaine was required to induce seizure, compared to a dose of 10.0–15.0 mg/kg of lidocaine or mepivacaine. The amount of drug required to induce seizure activity with lidocaine and mepivacaine was significantly higher than that required for etidocaine and bupivacaine. No significant differences existed between bupivacaine and etidocaine, or between lidocaine and mepivacaine.

Phase II

In general, there were no significant changes in the cardiovascular parameters measured at doses up to 3.0 mg/kg with the various agents. Mean arterial blood pressure (MABP) did show a significant decrease in animals treated with etidocaine and bupivacaine. When 10 mg/kg of bupivacaine was administered to five dogs, four developed severe hypotension, bradycardia, and subsequent cardiovascular collapse and death. At doses higher than 10 mg/kg, all agents showed progressive depression of the various parameters. Pulmonary artery blood pressure (PAP) was consistently elevated for all agents at doses of 0.3–10.0 mg/kg. At doses of 30 mg/kg, all agents caused profound decreases in PAP.

Table II presents the doses required to cause cardiovascular collapse and death in these animals. The average single dose to cause death ranged from 11 ± 1.0 mg/kg for bupivacaine to 33.3 ± 2.1 mg/kg for mepivacaine. Cumulative lethal doses of lidocaine and mepivacaine occurred in the 80-mg/kg range and for bupivacaine and etidocaine at 20 and 40 mg/kg, respectively.

The mode of death in these animals was similar for all drugs tested. Death occurred following progressive hypotension, severe bradycardia, and ultimately complete cardiovascular collapse. No ventricular arrhythmias were noted before the terminal cardiac

Table II. Lethal Doses in Anesthetized Ventilated Dogs of Various Local Anesthetics

Drug[a]	Single dose causing death (mg/kg)	Cumulative dose causing death (mg/kg)
Bupivacaine	11 ± 1.0	20.4 ± 2.4
Etidocaine	22 ± 2.0	40.4 ± 6.0
Lidocaine	28 ± 1.22	76.2 ± 15.1
Mepivacaine	33.3 ± 2.1	80.4 ± 9.7

[a]Values are means ± S.E.; n = 5 throughout.

event. Electrical activity continued even after blood pressure and cardiac output had dropped to zero, suggesting that pump failure was the primary cause of death.

Phase III

The first group of animals received lidocaine and bupivacaine. The second group received etidocaine and mepivacaine. Two animals receiving 3.4 and 5.1 mg/kg bupivacaine died and were excluded from statistical analysis. It should be noted that a comparison of the control values for these two animals as compared to the others in the group show no statistically significant differences. For the most part, little difference was seen between drugs with regard to changes in the various parameters. Table III presents the mean percent change for the measured parameters during the seizure period. As one would expect, profound changes occurred during the seizure period. Onset and duration of seizures are presented in Table IV.

Most maximum changes occurred within 3–5 min of the onset of seizures. By the end of seizure activity heart rate and CO had returned toward predrug control values. MABP remained elevated slightly longer after the cessation of seizure activity.

Oxygenation was well maintained throughout the study. Respiratory rates varied greatly, with hyper- and hypoventilation being observed at different times. In those cases where respiratory rates could be measured during seizure, there appeared to be no change from predrug values. There were no instances of acidosis, hypoxia, hyper-, or hypokalemia observed.

Table III. Intravenous Acute Systemic Toxicity in the Unanesthetized Dog (Mean Maximum % Change during Seizure)[a]

	Bupivacaine	Lidocaine	Etidocaine	Mepivacaine	Units
Heart rate	129 ± 40.4	108 ± 37.3	128 ± 21.7	150 ± 31.5	Beats/min
MABP	61 ± 5.3	52 ± 5.6	56 ± 6.7	52 ± 9.0	mm Hg
CO	131 ± 69.9	138 ± 46.1	77 ± 30.2[4]	103 ± 55.6[3]	l/min
SV	13.1 ± 23.7	36.9 ± 28.9	−32.4 ± 6.5[4]	−31.8 ± 2.9[3]	ml
MPAP	263 ± 77.5[4]	188 ± 67.4	110 ± 10.2	160 ± 27.3[4]	mm Hg
TPR	6.3 ± 36.9	−2.3 ± 27.1	−14.9 ± 20.2[4]	−0.03 ± 26.7[3]	Arbitrary units
PVR	203 ± 53.9[3]	117.0 ± 103.2	11.5 ± 28.8[4]	60.2 ± 18.0[3]	Arbitrary units

[a]All values are $\bar{x}$ ± S.E., n = 5 unless noted in superscript. MABP, mean arterial blood pressure; CO, cardiac output; MPAP, mean pulmonary arterial pressure; TPR, total peripheral resistance; PVR, pulmonary vascular resistance.

Table IV. Seizure Activity after Intravenous Local Anesthetic ($\bar{x}$ ± S.E.M.)

	Mean convulsive dose (mg/kg)	Onset of seizures (sec./P.E.I.)[a]	Duration of seizures (min)
Bupivacaine	3.74 ± 3.4	6.2 ± 2.1	6.1 ± 2.5
Etidocaine	4.6 ± 0.0	12 ± 3.1	4.0 ± 0.8
Lidocaine	13.2 ± 1.4	6.2 ± 2.8	1.9 ± 0.5
Mepivacaine	15.6 ± 1.6	17 ± 1.5	4.9 ± 1.7

[a]Post end injection.

As noted previously, two animals receiving 3.5 and 5.1 mg/kg bupivacaine rapidly developed ventricular fibrillation and died. Fibrillation occurred at 37 and 60 sec, respectively, after ending the injection of drug. No arrhythmias were noted with either lidocaine, mepivacaine, or etidocaine.

The peak arterial and venous concentrations obtained after i.v. bolus injection are presented in Table V. Data from pilot studies which have not been described elsewhere have been included in the calculation of drug concentrations. Peak arterial concentrations were much greater than venous concentrations in all cases. However, as redistribution of drug occurred, arterial and venous concentrations became similar. As the dose of each drug was increased, peak concentrations showed a tendency to increase also. Exceptions to this were partially due to the small number of experiments at higher doses. Also, at the lower doses, the time of the peak sample was generally at the end of drug injection when convulsive activity started, whereas at the higher doses, convulsive activity occurred prior to the end of the injection of the total dose, and it was at this time that blood samples were drawn. Peak venous concentrations occurred somewhat later than peak arterial concentrations, up to 5 min after the end of the injection.

Blood samples were drawn immediately upon the termination of seizure activity, and

Table VI. Peak Arterial and Venous Drug Blood Concentrations after Intravenous Injection

Drug	Dose (mg/kg)	n	Drug concentration (μg/ml) Arterial ($\bar{x}$ ± S.D.)	Venous ($\bar{x}$ ± S.D.)
Bupivacaine	3.4	7	28.4 ± 7.0	3.20 ± 1.16
	5.1	2	27.1	5.67
	6.8	2	42.2	8.55
	10.2	1	50.1	13.9
Lidocaine	11	6	80.0 ± 27.9	12.3 ± 3.6
	16.5	2	52.5	14.3
	33	2	154	28.0
Mepivacaine	13	4	123 ± 43	10.8 ± 2.3
	19.5	2	180	27.1
	26	2	133	25.1 (n = 1)
Etidocaine	4.6	5	31.9 ± 4.3	7.06 ± 2.92
	9.2	2	37.9	8.64

Table VI. Mean Arterial Blood Concentrations at End Seizure (μg/ml)[a]

	Mean	S.D.	Range
Bupivacaine HCl (n = 7)	2.67	± 1.14	1.77–4.89
Lidocaine HCl (n = 9)	8.15	± 3.10	4.54–13.3
Mepivacaine HCl (n = 8)	12.1	± 3.7	8.74–20.1
Etidocaine HCl (n = 7)	4.42	± 1.39	3.23–7.34

[a]Concentrations have been included from all doses of drugs administered.

the resultant arterial drug concentrations are presented in Table VI. These concentrations match the apparent relative toxicity of these drugs in the dog in terms of both mean concentrations and the lowest and highest values over the range of concentrations.

Analysis of arterial blood samples, drawn at the time of death, for bupivacaine and lidocaine concentrations gave the data shown in Table VII. In the case of the bupivacaine concentrations, circulatory collapse occurred so rapidly after injection in some cases that there had been little time for drug to circulate and redistribute prior to the sample being taken. In the dog that died after injection of lidocaine, 13 min had elapsed prior to death, so there had been sufficient opportunity for redistribution of drug to occur.

DISCUSSION

These data clearly demonstrate that local anesthetic agents are capable of producing severe CNS and cardiovascular system toxicity. Based on a comparison of convulsive dose data and dose to cause fatal cardiovascular collapse, it appears that the CNS is the primary target organ for local anesthetic toxicity. In general, the ratio of dose to cause cardiovascular collapse was at least 3.5 times greater than the dose required to cause convulsions. These results support results in sheep with lidocaine[13] and the clinical impression that the CNS is the primary target organ of local anesthetic toxicity.

In the anesthetized dog studies, there was no evidence of ventricular arrhythmia being the cause of death. All animals developed hypotension, bradycardia, and finally total cardiovascular collapse. Electrical activity of the heart continued even after the blood pressure and cardiac output were at zero. These results are in agreement with previous studies in monkeys[13] but are in contradiction to reports in cats[4] and sheep.[14] This may be due to species difference or methodological differences.

Table VII. Arterial Concentrations at Death (μg/ml)[a]

Dose	Bupivacaine HCl				Lidocaine HCl
(mg/kg)	3.4	5.1	6.8	10.2	33
	(2.54)	6.36	4.35	23.8	13.9
			13.4		
			11.5		

[a]Value in parentheses was a postmortem sample from the left ventricle.

Table VIII. Summary of Overall Incidence and Cause of Death[a]

Drug	Dose (mg/kg)	*n*	Number of deaths	Mode of death
Bupivacaine	3.4	13	2	2, VF
	5.1	2	1	1, VF
	6.8	5	4	3, VF; 1, resp. arrest
	10.2	1	1	1, CV collapse/resp. arrest
	13.6	1	1	Resp. arrest
Lidocaine	11.0	7	0	
	16.5	2	0	
	33.0	1	1	Resp. arrest
Etidocaine	4.6	5	0	
	9.2	2	0	
Mepivacaine	13.0	6	0	
	19.5	2	0	
	26.0	2	0[b]	

[a]VF, ventricular fibrillation; CV, cardiovascular.
[b]After 30 min of seizure; one animal euthanized.

In our studies in awake dogs, given i.v. bolus injections of convulsant doses of the various local anesthetics, little differences in the measured parameters were observed between drugs. However, major differences existed with regard to lethality and the electrocardiogram. In two animals receiving 3.4 and 5.1 mg/kg of bupivacaine, ventricular fibrillation developed rapidly. In other animals receiving two times the convulsive dose, fatal ventricular fibrillation was also observed. Table VIII presents a summary of dogs receiving the various local anesthetic and the fatal events. Bupivacaine clearly has a higher incidence of ventricular fibrillation than any other agent. In fact, ventricular fibrillation was observed only in the bupivacaine-treated animals. These results are in agreement with several reports in cats, sheep, and man. It is interesting to note also that these animals showed no sign of acidosis, hypoxia, or hypercarbia, even associated with seizures. Therefore, it appears that the development of ventricular fibrillation in association with bupivacaine does not require acidosis and hypoxia although these conditions may exacerbate bupivacaine's toxic effects.[6,15]

Peak drug concentrations reported here (Table V) differ somewhat from those reported in other animal species. Using a cat model, de Jong and co-workers found that plasma concentrations necessary to produce seizure activity were 19.6 μg/ml lidocaine, 6.6 μg/ml etidocaine, and 3.6 μg/ml bupivacaine, which are in close agreement with data for monkeys[12] of 18.2 μg/ml lidocaine and 4.5 μg/ml bupivacaine. However, in both these studies, drug was administered as a slow intravenous infusion.

There appeared to be little relationship between peak bupivacaine concentration and death after bolus drug administration. The same was true for the two animals receiving 33 mg/kg lidocaine. The surviving animal tolerating a peak arterial concentration of 248 μg/ml and terminating seizure activity at 14.2 μg/ml. This is effectively the lidocaine concentration at death in the other animal.

The mechanism by which bupivacaine and other local anesthetics can produce fatal cardiovascular collapse may be related to their intrinsic negative inotropic action and their ability to block sodium channels. However, it is still unclear why bupivacaine has a propensity to produce fatal ventricular fibrillation.

Table IX illustrates some physicochemical properties of the amide-type local anesthetic agents. As can be seen, the ventricular arrhythmias associated with the use of bupivacaine

Table IX. Various Properties of Local Anesthetic Agents

Drug	MW (base)	Dog serum protein binding (%) (1 μg/ml)	(40 μg/ml)	Lipid solubility	pKa	Whole blood: plasma conc. ratio
Lidocaine	234	80	35	2.9	7.7	0.84
Mepivacaine	246	70	30	0.8	7.6	0.92
Etidocaine	276	95	65	141	7.7	0.58
Bupivacaine	288	98	62	28	8.1	0.73

are apparently not related to its high lipid solubility since etiodcaine, which is the most lipid-soluble agent, did not cause arrhythmias in our study. Although ventricular arrhythmias were observed by de Jong and co-workers in etidocaine-treated cats,[4] the frequency of these cardiac irregularities was far less than was seen with bupivacaine-treated animals.

With the exception of bupivacaine, the pKa values for these drugs are very similar such that at physiologic pH the proportion of free drug existing in the ionized and un-ionized forms will be the same. In the case of bupivacaine, less of the free drug will be present in the un-ionized form which diffuses most readily through cell membranes. Although this would suggest that less drug would be available for uptake by the heart, studies in sheep have shown that the heart tissue to blood concentration ratio for bupivacaine (3.3) is greater than that for lidocaine (1.8) implying that the pKa of these agents is not related to cardiac toxicity.

The free form of drug is most readily available for absorption by the myocardium and other tissues. Marked differences exist in the serum protein binding of these agents at varying concentrations and between different drugs. At the low concentrations of bupivacaine achieved after epidural blockade, less than 5% of the circulating drug is in the unbound form. Accidental intravenous injection of local anesthetic agent results in very high circulating drug concentrations which markedly increase the amount of unbound bupivacaine in blood (40–50%). This is illustrated by the free molar concentrations of drug at peak and end seizure concentrations shown in Table X. If the marked increase in free drug that occurs when agents of high protein binding are administered as an intravenous bolus is responsible for the cardiotoxicity associated with bupivacaine, then etidocaine would be expected to produce the same effects as bupivacaine, which was not the case in this study.

Looking at molecular structure, all these agents have similar molecular weights, and

Table X. Approximate Ranges of Circulating Free Drug Concentrations in Arterial Blood at Peak and End Seizure Concentrations[a]

	Peak (μM/ml)	End seizure (μM/ml)
Bupivacaine	35–80	0.2–1.3
Lidocaine	150–450	9–17
Mepivacaine	350–510	15–50
Etidocaine	40–50	0.8–4.9

[a]See Tables V and VI.

both bupivacaine and mepivacaine have piperidine rings. If the piperidine ring were the causal factor of ventricular arrhythmias after administration of bupivacaine, then mepivacaine too would be expected to cause similar cardiac effects. In this study, mepivacaine was found to have the lowest potential for central nervous system toxicity, and no cardiac effects were seen with this agent. It is possible that the toxic effects of the amide-type local anesthetic agents are due to a combination of the several factors discussed, although their relative importance remains unclear, and the reason why bupivacaine is more cardiotoxic than other drugs of the same type has yet to be elucidated.

ACKNOWLEDGMENTS. We gratefully acknowledge the expert technical assistance of Sally Norway, Ann Marie Doucette, Cheryl Latka, Cynthia Allanson, and Colette Lavoie. This work was supported in part by a grant from Breon Laboratories, New York, New York.

REFERENCES

1. Albright GA: Cardiac arrest following regional anesthesia with etidocaine or bupivacaine (editorial). *Anesthesiology* 1979;51:285–287.
2. Covino BG, Vassallo HG: *Local Anaesthetics—Mechanisms of Action and Clinical Use*. New York, Grune & Stratton, 1976, pp 125–134.
3. deJong RH, Bonin JD: Deaths from local anaesthetic-induced convulsions in mice. *Anesth Analg* 1980;59:401–405.
4. de Jong RH, Ronfeld RA, DeRosa RA: Cardiovascular effects of convulsant and supraconvulsant doses of amide local anaesthetics. *Anesth Analg* 1982;61:3–9.
5. Edde RR, Deutsch S: Cardiac arrest after interscalene brachial plexus block. *Anesth Analg* 1977;56:446–447.
6. Liu PL, Feldman HS, Covino BM, Giasi R, Covino BG: Acute cardiovascular toxicity of intravenous amide local anaesthetics in anaesthetized ventilated dogs. *Anesth Analg* 1982a;61:317–322.
7. Liu P, Feldman HS, Covino BM, Giasi R, Covino BG: Acute cardiovascular toxicity of procaine, chloroprocaine and tetracaine in anaesthetized ventilated dogs. *Reg Anesth* 1982b;7:14–19.
8. Liu PL, Feldman HS, Giasi R, Patterson MK, Covino BG: Comparative CNS toxicity of lidocaine, etidocaine, bupivacaine, and tetracaine in awake dogs following rapid intravenous administration. *Anesth Analg* 1983;62:375–379.
9. Morishima HO, Pedersen H, Finster M, Sakuma K, Bruce SL, Gutsche BB, Stark RI, Covino BG: Toxicity of lidocaine in adult, newborn, and fetal sheep. *Anesthesiology* 1981;55:57–61.
10. Morishima HO, Pedersen H, Finster M, Feldman HS, Covino BG: Is etidocaine more cardiotoxic than lidocaine? *Anesthesiology* 1982;57:A401.
11. Morishima HO, Pedersen H, Finster M, Tsuji A, Hiraoka H, Feldman HS, Arthur GR, Covino BG: Is bupivacaine more cardiotoxic than lidocaine? *Anesthesiology* 1983;59:A409.
12. Munson ES, Martucci RW, Wagman IH: Bupivacaine and lidocaine induced seizures in rhesus monkeys. *Brit J Anaesth* 1972;44:1025–1029.
13. Munson ES, Tucker WK, Ausinisch B, Malagodi M: Etidocaine, bupivacaine, and lidocaine seizure thresholds in monkeys. *Anesthesiology* 1975;42:471–478.
14. Prentiss JE: Cardiac arrest following caudal anaesthesia. *Anesthesiology* 1979;50:51–53.
15. Sage DJ, Feldman HS, Arthur GR, Datta S, Ferretti AM, Norway SB, Covino BG: Influence of lidocaine and bupivacaine on isolated guinea pig atria in the presence of acidosis and hypoxia. *Anesth Analg* 1984;63:1–7.
16. Scott DB, Cousins MJ: Clinical pharmacology of local anaesthetic agents, in Cousins MJ, Bridenbaugh PO (eds): *Neural Blockage in Clinical Anaesthesia and Management of Pain*. Philadelphia, Lippincott, 1980, p 113.
17. Tucker GT: Determination of bupivacaine (Marcaine) and other anilide-type local anesthetics in human blood and plasma by gas chromatography. *Anesthesiology* 1970;33:255–260.

34

Local Anesthetic Effects on Myocardial Excitation–Contraction Coupling

Carl Lynch III

ABSTRACT. Isolated guinea pig papillary muscles studied in normal Tyrode's solution demonstrated a frequency-dependent depression of twitch tension by 1–10 μM bupivacaine, which caused greater depression at low frequencies of stimulation (rested state to 0.5 Hz). In muscles partially depolarized in 26 mM K Tyrode's and stimulated with 0.1 μM isoproterenol, the effects of 4 and 10 μM bupivacaine and 40 and 100 μM lidocaine upon slow action potentials (SAP) and accompanying contractions were studied. No depression of the SAP rate of depolarization was observed. The late-peaking tension response in the rested state and up to 0.5 Hz stimulation was significantly more depressed by bupivacaine than the rapidly developing tension at 2–3 Hz. Lidocaine at 10-fold higher concentrations demonstrated a similar effect. Depression of the late-peaking tension response, attributed to calcium release by the sarcoplasmic reticulum, suggests a selective depression of this component of myocardial excitation–contraction coupling, with little effect on calcium entry.

INTRODUCTION

The comparative cardiovascular toxicity of the various local anesthetics has been the subject of recent investigation, since the highly lipid-soluble and protein-bound local anesthetics, specifically bupivacaine and etidocaine, have been noted to cause severe cardiovascular collapse following inadvertent intravascular injection, an effect not observed with lidocaine.[1] Resuscitation to a spontaneous cardiac rhythm with adequate blood pressure has been difficult.

In anesthetized dogs, Liu *et al.*[20] found no difference in toxicity between lidocaine and bupivacaine. However, subsequently this group noted an incidence of lethal ventricular fibrillation in awake dogs receiving a convulsant dose of bupivacaine, not observed with such a convulsant dose of lidocaine.[28] A significant incidence of cardiac arrest or cardiac arrhythmias was noted in cyclopropane-anesthetized rats[25] and in adult sheep[24,33] with bupivacaine, but not with lidocaine. More recently, Tanz *et al.*[32] found 10 μM bupivacaine (3 μg/ml) to be more cardiotoxic than a 10-fold greater concentration of 129 μM lidocaine (30 μg/ml) in a guinea pig Langendorff heart preparation. However, the authors did not control heart rate, which markedly influences contractility.

Carl Lynch III • Department of Anesthesiology, University of Virginia Medical Center, Charlottesville, Virginia 22908.

Quantitative comparison of local anesthetic effects on both cardiac and neural tissue may be made by comparing the study of Gissen *et al.*[15] on isolated rabbit vagus nerve and the study of Block and Covino[4] on isolated rabbit heart. Roughly equivalent concentrations of lidocaine (70–110 μM) depress the vagal nerve A fiber compound action potentials and also depress cardiac contractility (by 25%) and conduction (by 50%). In contrast, at least 10 times less bupivacaine is required to depress the heart as compared to nerve.

The highly lipid-soluble ester local anesthetic tetracaine is also more potent than lidocaine and is about 5 times more potent than bupivacaine or etidocaine in depressing neural conduction[15] and myocardial contractility and conduction.[4] However, tetracaine has also been shown to have effects on excitation–contraction coupling in skeletal and cardiac muscle, which are in addition to its effects on excitability. Tetracaine blocks contractions in the skeletal muscle from the frog stimulated by either caffeine[14] or depolarization.[3] Since tetracaine and other local anesthetics also depress Ca^{2+} release from isolated sarcoplasmic reticulum (SR) and the accompanying SR depolarization,[7,16] their contractile depressant effects have been attributed to this mechanism. In mammalian Purkinje fibers[13] and in frog atrial muscle,[8] tetracaine also appears to block the slow inward (Ca) current. Bupivacaine and etidocaine have a similar highly lipophilic character, and their depression of myocardial contractility may also be due in part to effects on Ca^{2+} release by the SR and/or effects on Ca^{2+} entry through slow channels. Komai and Rusy[19] have demonstrated prominent depression by bupivacaine of atrioventricular conduction, a process mediated in part by Ca^{2+}-dependent action potentials in the cardiac conduction system. Lidocaine does not appear to have any slow channel blocking effects,[5] although at very high concentration (100 μM) some preparations may show depression of presumed calcium-mediated slow action potentials.[17]

This study compared the effects of bupivacaine and lidocaine on the contractility of isolated guinea pig papillary muscle, examining contractions accompanying normal action potentials and slow action potentials. Since myocardial contractility is extremely dependent on the rate of stimulation,[2,12] contractility was examined with rested state contractions and with increasing stimulation frequencies up to 3 Hz.

METHODS

The heart was removed from guinea pigs (300–400 g) under diethyl ether anesthesia and quickly transferred to oxygenated Tyrode's solution at room temperature. Right ventricular papillary muscles were dissected out and pinned by the cut end into the base of the chamber (25 mm long, 5 mm wide, 3 mm deep), which was maintained at 37°C and superfused with Tyrode's solution at a rate of 8 ml/min. The tendonous end was attached by a strut to a Grass FT03 force transducer mounted on a micromanipulator, and the muscle length was adjusted to the smallest resting tension at which maximum twitch tension developed. Preparations were equilibrated in the chamber during constant stimulation at a rate of 0.5–1 Hz for 1 hr prior to study, and typically weighed 0.7–3 mg.

The muscles were field-stimulated by stainless steel electrodes placed along the wall of the chamber, with stimuli of 0.5 msec duration. For studies in which intracellular recordings were not employed, stimulus intensity was approximately twice threshhold. For many experiments, muscles were impaled and intracellular potential monitored by conventional 3 M KCl-filled microelectrodes attached to a WPI VF-1 preamplifier. The bath was grounded through an agar bridge–Ag/AgCl electrode. Bath potential was monitored through a similar agar bridge–Ag/AgCl electrode, and subtracted from the intracellular potential to give the membrane potential (V_m). Stimulation intensity was adjusted to provide a latency between the stimulus and action potential (AP) of 5–10 msec and adjusted during experiments to maintain the same latency duration. The membrane potential, tension, and first derivative of each signal were monitored on a Tetronix D11 storage oscilloscope and were recorded on a Gould 2400 recorder and/or a Hewlett–Packard 3960 FM tape recorder for later playback and analysis. The normal action potential first derivative was recorded from the oscilloscope.

The standard Tyrode's solution had the following composition (in mM): Na, 143; K, 4.7; Cl, 128; Ca, 2.5; Mg, 2.0; SO_4, 2.0; HCO_3, 25; glucose, 11; EDTA, 1 (added to prevent inactivation of isoproterenol). Solution circulated through the chamber from heated reservoir containers through which 95% O_2/5% CO_2 was bubbled maintaining pH at 7.45 ± 0.5. Contractions were also studied in muscles that were partially depolarized to −40 to −45 mV in 26 mM K Tyrode's (K substitution for an equivalent concentration of Na to maintain the calculated osmolality). Isoproterenol (0.1 μM) was added to elicit slow AP's, mediated by current through the slow (calcium) channel,[27] with typical maximum rates of rise ($\dot{V}_{max}$) of 15–25 *V*/sec. Slow AP $\dot{V}_{max}$ was employed as a measure of the peak slow inward current.[18]

A protocol of stimulation was employed in which after 15 min rest, a rested-state contraction (RSC) was elicited, followed by stimulation rates of 0.1, 0.25, 0.5, 1.0, 2.0, and 3 Hz. Each stimulation rate was maintained until a steady state and unchanging contractile response was achieved before proceeding to the following higher frequency. Following control studies, muscles were exposed to anesthetics for at least 15 min, which was found to be sufficient time for a stable constant drug effect, prior to the recording of responses. Recovery responses were recorded at least 25 min after washout of anesthetic from the chamber. Because of the difficulty of maintaining impalements during the vigorous contractions at higher frequencies of stimulation, some experiments were performed only at 0.25–0.5 Hz. Action potential characteristics were analyzed only in those experiments in which impalements were maintained throughout the control, experimental, and recovery periods.

Measurements during local anesthetic exposure were compared to only the control value, since complete recovery of control tension was not observed following exposure to high concentrations of local anesthetics. Changes due to aging of such preparations are usually less than 10% during equivalent periods of study, and complete recovery to control levels is observed following exposure to volatile anesthetics.[22,23] Testing for significant differences ($p < 0.05$) among the frequencies and between the anesthetics studied was carried out by one-way analysis of variance, while significance of anesthetic effect versus control was tested by Student's *t* test.

RESULTS

Frequency-Dependent Contractile Depression

The effects of bupivacaine on papillary muscle twitch responses are shown in Fig. 1A for frequencies of 0.25, 1, 2, and 3 Hz. At 0.25 Hz, there was marked depression of the late peak of contraction, while the initial rate of tension development was relatively unaffected. With increasing frequency, the amount of depression decreased. Although there was some depression of the initial rate of tension development, the later components of tension appeared to be more depressed, especially evident at 0.25 Hz. At 1 Hz, the peak tension was depressed by bupivacaine, but occurred slightly earlier, also suggesting an effect on later components of tension development. Figure 1B shows the average depression of RSC and twitch responses at 0.1–3Hz stimulation caused by 1, 4, and 10 μM bupivacaine. At each concentration, bupivacaine depressed the RSC significantly more than at 2–3 Hz. In two experiments in which intracellular impalements were maintained, the normal AP V_{max} was

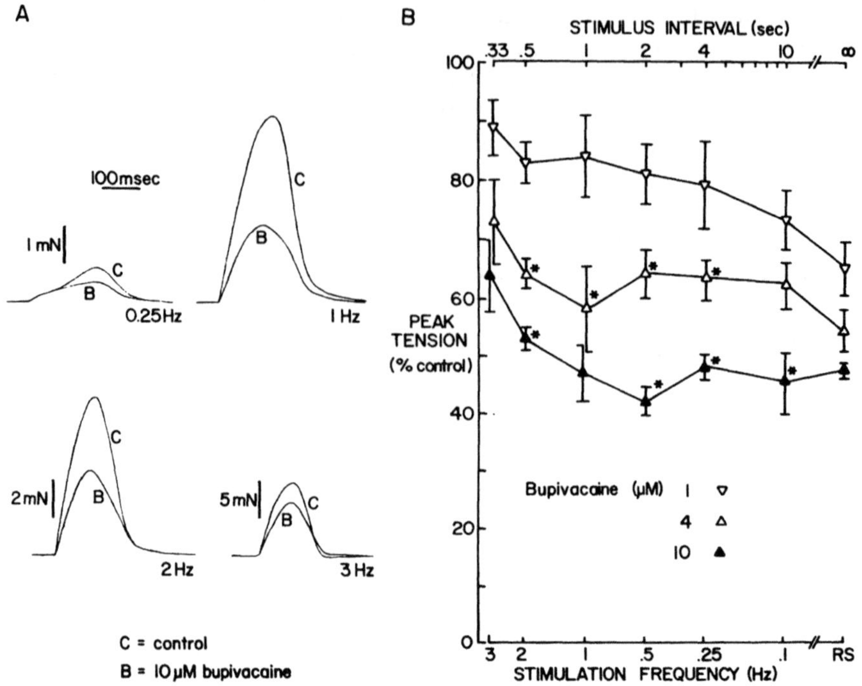

Figure 1. (A) Steady-state tension responses accompanying normal AP's at rates of stimulation of 0.25, 1, 2, and 3 Hz, in the control setting (C) and following exposure to 10 μM bupivacaine (B). Note the change in tension scale. (B) Peak tension developed (expressed as percent of control) at rested state and up to stimulation frequencies of 3 Hz in the presence of bupivacaine (1, 4, and 10 μM). Points represent average (± S.E.M., shown by error bars) of four muscles studied at each concentration. Significant ($p < 0.05$) difference between 1 and 4 μM bupivacaine and between 4 and 10 μM bupivacaine was observed at the frequencies indicated (*) on the higher concentration curve. All points were less than control values, and 10 μM was significantly more depressant than 1 μM bupivacaine at all frequencies.

depressed in a frequency-dependent fashion above 0.5 Hz, as previously described in atrial muscle[9] and ventricular muscle (Hondeghem and Clarkson, this volume, Chapter 32).

Effects on Slow Action Potentials and Tension Development

In order to better define the mechanism of the frequency-dependent depression, local anesthetic effects were studied in the presence of β-adrenergic stimulation with 0.1 μM isoproterenol. β-Adrenergic stimulation increases the Ca^{2+} influx as slow inward current,[26] while uptake of Ca^{2+} by the sarcoplasmic reticulum (SR) is also greatly enhanced.[31] The RSC is greatly increased in this setting with a distinctive late-peaking character.[30] As stimulation frequency is increased above 1 Hz in the presence of β-adrenergic stimulation, automaticity develops in the preparation and prevents practical study of the frequency dependence of the depression. However, partial depolarization in 26 mM K depolarizes the cells to −40 to −45 mV, which inactivates the sodium current and prevents automaticity. Stimulation produces propagated slow AP's with a typical late-peaking contractile response at frequencies up to 0.5 Hz,[29,30] identical to the RSC prior to partial depolarization. Figure 2 compares the tension responses in the same muscle at the frequencies indicated for fast and slow AP's. This late-peaking twitch appears to result largely from Ca^{2+} influx,[12,26] which

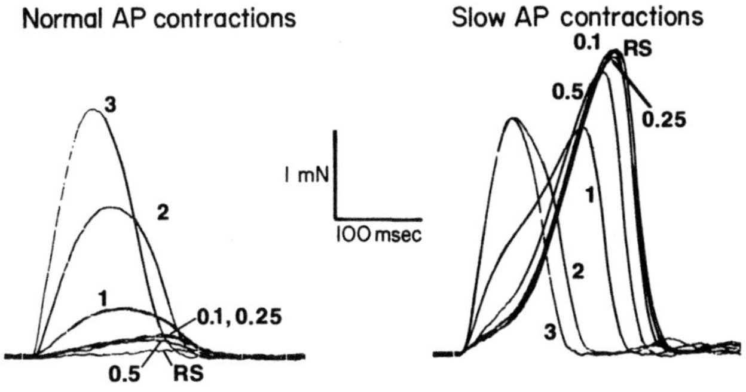

Figure 2. Contractions accompanying normal AP's (left) or slow AP's (right), induced with 0.1 μM isoproterenol in 26 mM K Tyrode. Numbers indicate the frequency at which the steady-state twitch response was observed. Note the marked enhancement of the late-peaking tension development for the rested state (RS) and contractions at lower frequencies (<0.5 Hz) which accompany slow AP's.

does not directly activate the tension development. Rather, the Ca^{2+} accumulates in the myocardial SR and is then released from the SR after a delay, near the end of the action potential.[26,30] While the late peak observed with β-adrenergic stimulation is attributable to release of Ca^{2+} from the cardiac SR, the rapid tension development at higher frequency may be due to release from a more superficial store.[30]

The upper panels in Fig. 3 show the effects of 4 and 10 μM bupivacaine on contractile response. The late-peaking tension at low frequency was markedly more depressed than the twitch at high frequency with both 4 and 10 μM bupivacaine. This is also seen in Fig. 4, which shows the average depression ($n = 4$) of peak tension, the maximum initial rate of tension development ($\dot{T}i_{max}$), and the maximum rate of tension development of the late peak ($\dot{T}l_{max}$). The late-peaking tension at frequencies less than 2 Hz was depressed by bupivacaine in a dose-dependent manner, 10 μM decreasing tension to one-half the control values. $\dot{T}l_{max}$

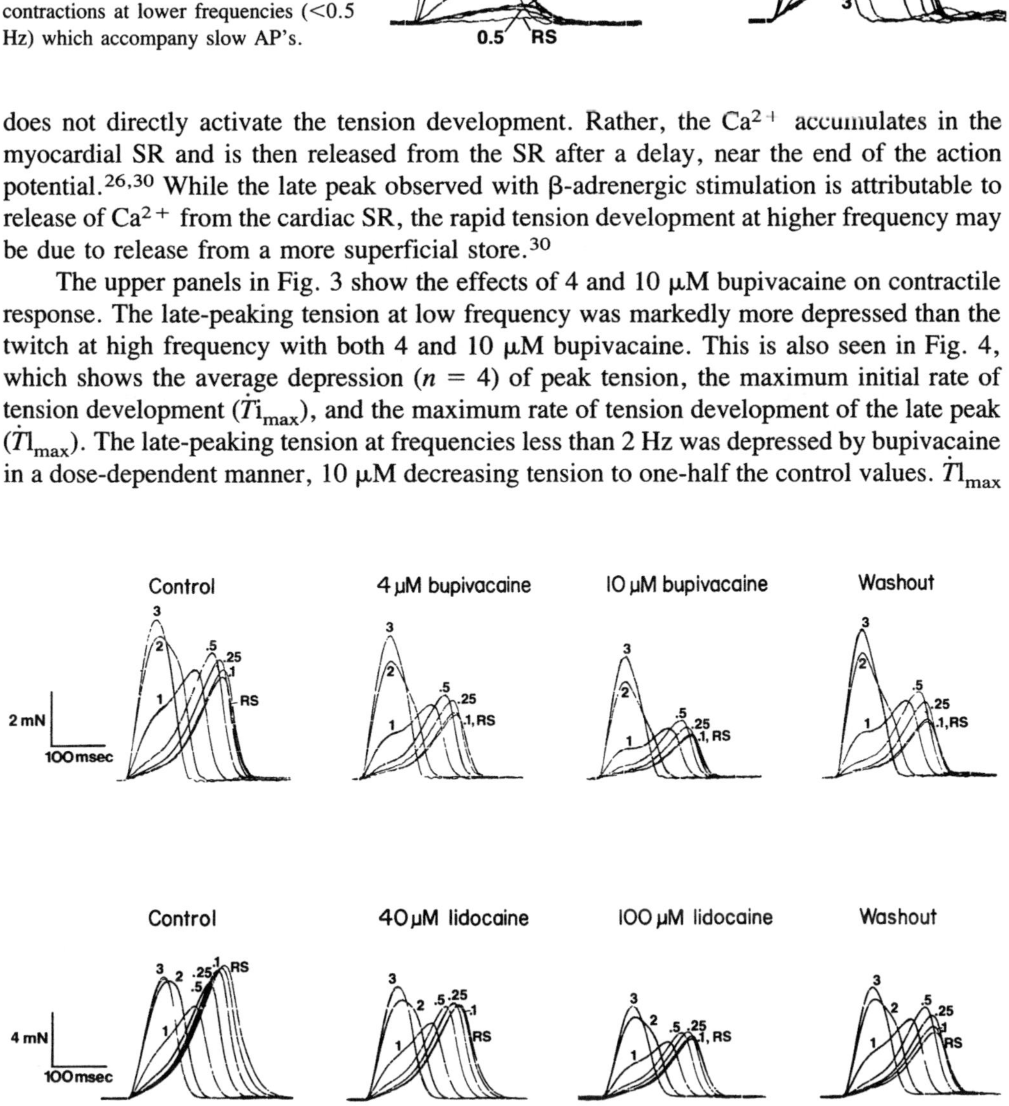

Figure 3. (Upper panels) Superimposed slow AP contractions (steady-state responses) at the frequencies indicated from a papillary muscle in 26 mM Tyrode and 0.1 μM isoproterenol. Control responses and subsequent depressed responses in 4 μM, followed by 10 μM bupivacaine. Responses are also shown after 45 min of washout. (Lower panels) Superimposed slow AP contractions (steady-state responses) at the frequencies indicated, from a different papillary muscle than shown in upper panels but under identical conditions. Tension responses in the control setting and with 40 μM, followed by 100 μM lidocaine. Washout responses are also shown.

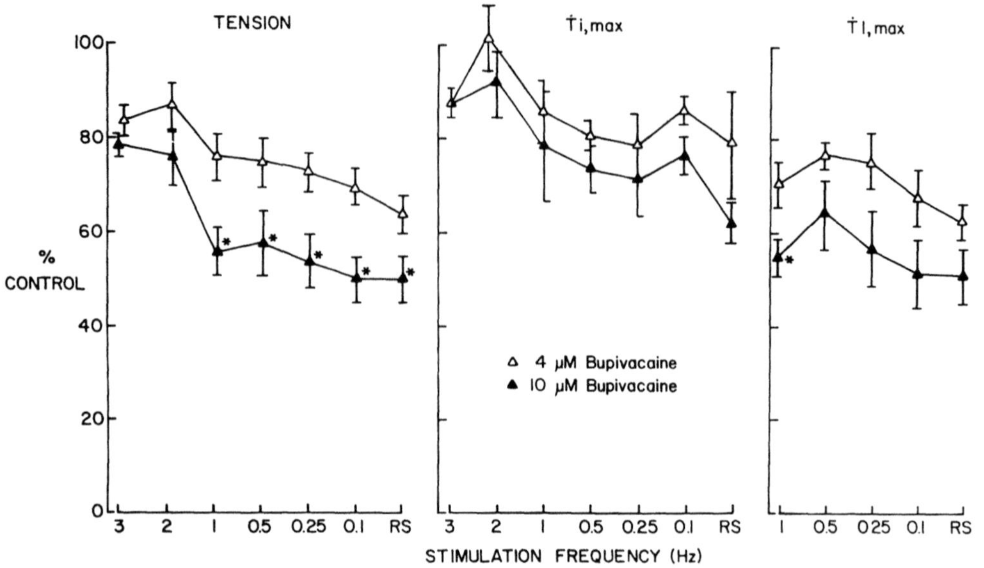

Figure 4. Effects of 4 and 10 μM bupivacaine on peak tension, the maximum rate of initial tension development ($\dot{T}i_{max}$), and the maximum rate of late-peaking tension development ($\dot{T}l_{max}$). The prominent late peak is lost at frequencies above 1 Hz (see Figs. 2 and 3), and no values for $\dot{T}l_{max}$ are plotted. All values are significantly depressed from control except those of $\dot{T}i_{max}$ at 2 Hz. Differences ($p < 0.05$) between the depressant effect at the concentrations employed are indicated (*).

was likewise depressed, although a significant difference between the effects of 4 and 10 μM bupivacaine was evident only at 1 Hz. In contrast, the early-peaking tension at 2–3 Hz was depressed by bupivacaine significantly less than at lower frequencies, and $\dot{T}i_{max}$ was depressed less than $\dot{T}l_{max}$ (at 0.1 and 1 Hz) and was not significantly depressed from control at 2 Hz. Washout of 10 μM bupivacaine caused some reversal of the depressant effects, with tension development restored to the level observed in 4 μM bupivacaine.

The depressant effects of 10-fold higher concentrations of lidocaine (40, 100 μM) were similar to or less prominent than those for bupivacaine. There was a dose-dependent depression of tension and $\dot{T}l_{max}$ at frequencies of 0.25–0.50 Hz (see Fig. 5). In contrast, there was significantly less depression of tension (early-peaking) at 3 Hz, with no depression of $\dot{T}i_{max}$ at this frequency.

Bupivacaine, 4 and 10 μM, resulted in slow AP $\dot{V}_{max}$ values (average ± S.E.M., $n = 4$) at 0.25–0.3 Hz, which were 96 ± 5 ($p > 0.2$) and 94 ± 2 ($0.10 > p > 0.05$) percent of control, respectively. There was no increased depression of $\dot{V}_{max}$ with increased frequency of stimulation (1, 2 Hz) as seen with normal AP $\dot{V}_{max}$, and due to sodium channel blockade. AP duration and amplitude were likewise unaffected. No effect of lidocaine on slow AP characteristics was observed.

DISCUSSION

Bupivacaine (4, 10 μM) and lidocaine (40, 100 μM) depress myocardial contractility in a frequency-dependent manner and independently of effects on the sodium channel. The drug concentrations employed in this study represent free drug concentrations and are similar to free drug levels that might be expected in humans following inadvertent intravascular injection during regional anesthesia (see Table I). Both bupivacaine and lidocaine depressed

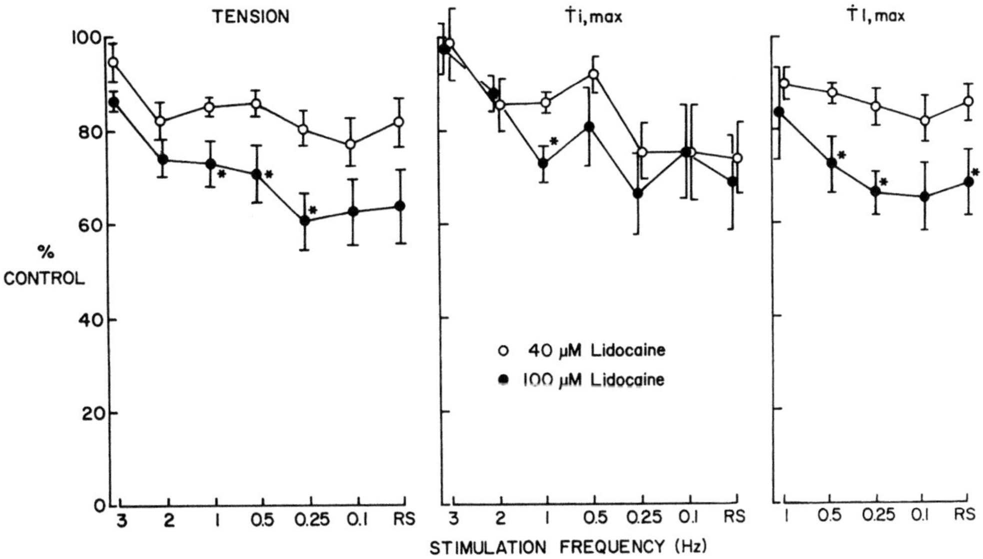

Figure 5. Effects of 40 and 100 μM lidocaine on peak tension, the maximum rate of initial tension development ($\dot{T}i_{max}$), and the maximum rate of late-peaking tension development ($\dot{T}l_{max}$). All values in the presence of lidocaine are significantly depressed compared to control responses except $\dot{T}i_{max}$ at 3 Hz, and peak tension developed in 40 μM lidocaine at 3 Hz. Differences ($p < 0.05$) between the depressant effects at the concentrations employed are indicated (*).

myocardial tension development at the doses employed, although 4 μM bupivacaine appeared to cause significantly greater depression than 40 μM lidocaine, and at most frequencies caused as much as or more depression than 100 μM lidocaine. Such results are similar to those of Tanz *et al.*[32] on Langendorff perfused guinea pig hearts. Those workers found that 10 μM free bupivacaine was more depressant than over 125 μM free lidocaine. However, the heart rate was significantly decreased by bupivacaine, which would accentuate contractile depression owing to the frequency-dependent nature of myocardial contractility.

The changing pattern and amount of tension development with increasing frequency

Table I. Free and Total Equivalent Serum Concentrations for Bupivacaine and Lidocaine[a]

Drug	Free drug concentration (μM)	Free drug concentration (μg/ml)	Approximate fraction bound to serum protein	Equivalent total serum concentration (μg/ml)
Bupivacaine	1	0.29	0.90	2.9
	4	1.15	0.70	3.8
Lidocaine	10	2.33	0.60	5.8
	40	9.32	0.40	15.5

[a]The equivalent total serum concentrations of drug (free plus protein bound) were calculated from the estimated serum protein binding of the local anesthetics, which is concentration dependent[(34)] (see also ref. 11). For the drug concentrations in this study, the table summarizes the pertinent concentrations, which are converted to μg/ml for ease of comparison with the literature. The higher equivalent total serum concentration value of each drug is approximately a preseizure or convulsant dose.

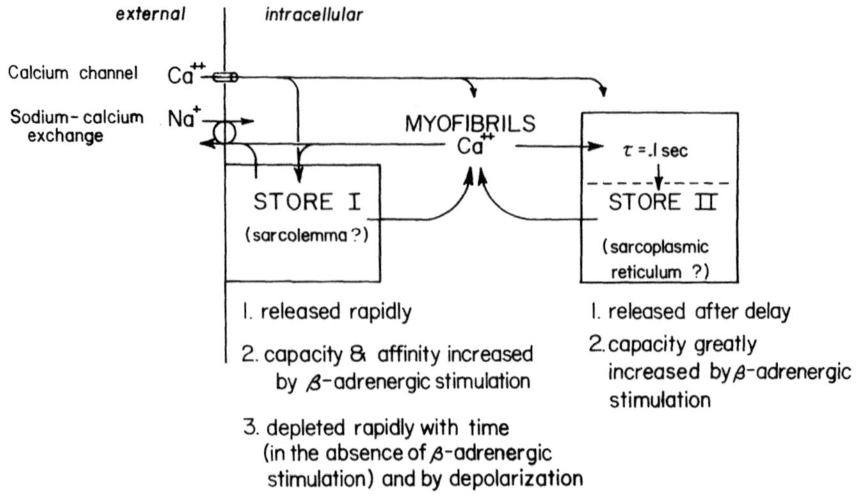

Figure 6. Schematic of cellular handling of contractile activator calcium by the myocardium. Store I is responsible for calcium release which causes the rapid initial tension development. Store II is responsible for the late calcium release which causes the late peak of tension at frequencies of <1 Hz. Bupivacaine and lidocaine appear to decrease the amount of calcium released from Store II more than from Store I, which results in greater depression of the late-peaking tension at low frequencies.

reflect release of activator Ca^{2+} from functionally distinct stores.[2,12,26,29,30] Figure 6 shows the model of Ca^{2+} control within the myocardial cell. Both Stores I and II are enhanced by β-adrenergic stimulation but can be temporally distinguished by the pattern of tension development at different frequency. Store II is responsible for the late tension peak and at high frequency appears to be functionally depleted, or requires a longer action potential depolarization (present below 1 Hz stimulation) for Ca^{2+} release.[26,30] In contrast, Store I is released rapidly and is markedly depleted at frequencies below 1–2 Hz, either in the normal state or with β-adrenergic stimulation and partial depolarization (β-adrenergic stimulation alone prevents depletion at frequencies as low as 0.05 Hz). Although Store II cannot be identified with certainty, its physiological behavior correlates with the myocardial SR. The structural identity of Store I is less well characterized, and it may represent a functionally or structurally distinct SR[2,26] or a sarcolemmal store of activator Ca^{2+}.[30]

The local anesthetics' preferential depression of the late-peaking tension suggests a distinct effect on SR store of Ca^{2+}, which could arise by a variety of mechanisms: (1) decreased Ca^{2+} entering the myocyte, which decreases the amount taken up and released by the SR; (2) decreased Ca^{2+} uptake by the SR with less present for subsequent release; or (3) impaired SR release of Ca^{2+}. Blockade of Ca^{2+} by calcium channel blockers depresses the late peak tension.[26] As assessed by slow AP rate of depolarization, however, local anesthetics do not appear to significantly depress slow (Ca^{2+}) channel current, suggesting that mechanism 1 does not contribute. Although a decrease in the SR uptake of entering Ca^{2+} (mechanism 2) cannot be ruled out, decreased SR uptake of entering Ca^{2+} might be anticipated to cause an early increase in myoplasmic Ca^{2+} and enhanced early tension development, which was not observed. Local anesthetic depression of Ca^{2+} release from the SR would explain experimental observations. Tetracaine has been well shown to interfere with SR release in skeletal muscle.[7,16] Dantrolene sodium, which interferes with skeletal muscle contractions and release of calcium from the heavy fraction of sarcoplasmic reticulum,[6] causes a similar pattern of frequency-dependent depression with predominant depression of late tension development in identical papillary muscle preparations.[21]

The local anesthetics' pattern of decreasing contractile depression with increasing frequency contrasts sharply with the depolarization-dependent[10] and frequency-dependent blockade of sodium channels, in which blockade is enhanced with increasing frequency (see Hondeghem and Clarkson, this volume, Chapter 32). The frequency-dependent block of sodium channels extends to lower frequencies with bupivacaine than with lidocaine. Consequently, bupivacaine will depress cardiac contractility increasingly at lesser frequencies (<1 Hz), while sodium-dependent fast conduction will be blocked more as frequencies of stimulation (>1 Hz) are increased. Such opposite frequency dependence of contractile and sodium channel inhibition by bupivacaine will result in a relatively limited range of heart rate in which contractility and conduction are maintained. Because the frequency-dependent blockade of sodium channels by lidocaine does not extend to frequencies less than 3 Hz, and because it has intrinsically less myocardial depression for an equivalent degree of neural blockade, it should cause less cardiac depression.

ACKNOWLEDGMENTS. This work was supported in part by a grant from the American Society of Regional Anesthesia. The author would like to thank Dr. Cosmo DiFazio for encouragement, helpful discussion, and review of the manuscript.

REFERENCES

1. Albright GA: Cardiac arrest following regional anesthesia with etidocaine or bupivacaine. *Anesthesiology* 1979;51:285–287.
2. Allen DG, Jewell BR, Wood EH: Studies of the contractility of mammalian myocardium at low rates of stimulation. *J Physiol (London)* 1976;254:1–17.
3. Almers W, Best PM: Effects of tetracaine on displacement currents and contraction of frog skeletal muscle. *J Physiol* 1976;262:583–611.
4. Block A, Covino B: Effect of local anesthetic agents on cardiac conduction and contractility. *Regional Anesth* 1981;6:55–61.
5. Brennan FJ, Cranefield PF, Wit AL: Effects of lidocaine on slow response and depressed fast response action potentials of canine cardiac Purkinje fibres. *J Pharmacol Exp Ther* 1978;204:312–324.
6. Campbell KP, Shamoo AE: Chloride-induced release of actively loaded calcium from light and heavy sarcoplasmic reticulum vesicles. *J Membr Biol* 1980;54:73–80.
7. Caputo C, Vergara J, Bezanilla F: Local anesthetics inhibit tension development and Nile blue fluorescence signals in frog muscle fibres. *Nature* 1979;277:400–402.
8. Chapman RA, Leoty C: The effects of tetracaine on the membrane currents and contraction of frog atrial muscle. *J Physiol* 1981;317:475–486.
9. Courtney KR: Fast frequency-dependent block of action potential upstroke in rabbit atrium by small local anaesthetics. *Life Sci* 1979;24:1581–1588.
10. Courtney KR, Kendig JJ, Cohen EN: The rates of interaction of local anesthetics with sodium channels in nerve. *J Pharmacol Exp Ther* 1978;207:594–604.
11. Coyle DE, Denson DD: Plasma protein binding of bupivaine and lidocaine in the dog. *Anesthesiology* 1982;57:A209.
12. Edman KAP, Johannsson M: The contractile state of rabbit papillary muscle in relation to stimulation frequency. *J Physiol* 1976;254:565–581.
13 Eisner DA, Lederer WJ, Noble D: Caffeine and tetracaine abolish the slow inward current in sheep Purkinje fibres. *J Physiol* 1979;293:76P.
14. Feinstein MB: Inhibition of caffeine rigor and radiocalcium movements by local anesthetics in frog sartorias muscle. *J Gen Physiol* 1963;47:151–172.
15. Gissen AJ, Covino BG, Gregus J: Differential sensitivities of mammalian nerve fibers to local anesthetic agents. *Anesthesiology* 1980;53:467–474.
16. Johnson PN, Inesi G: The effects of methylxanthines and local anesthetics on fragmented sarcoplasmic reticulum. *J Pharmacol Exp Ther* 1969;169:308–314.
17. Josephson I, Sperelakis N: Local anesthetic blockade of Ca mediated action potentials in cardiac muscle. *Eur J Pharmacol* 1976;40:201–208.
18. Kass RS, Seigelbaum SA, Tsien RW: Three-microelectrode voltage clamp experiments in calf cardiac Purkinje fibres: Is slow inward current adequately measured? *J Physiol* 1979;290:201–225.

19. Komai H, Rusy B: Effects of bupivacaine and lidocaine on AV conduction in the isolated rat heart: Modification by hyperkalemia. *Anesthesiology* 1981;55:281–285.
20. Liu P, Feldman H, Covino BG: Acute cardiovascular toxicity of lidocaine, bupivacaine and etidocaine in anesthetized, ventilated dogs. *Anesthesiology* 1981;53:S231.
21. Lynch C: Differential effects of local anesthetics, calcium entry blockers, and dantrolene on myocardial tension development (submitted for publication).
22. Lynch C, Vogel S, Sperelakis N: Halothane depression of myocardial slow action potentials. *Anesthesiology* 1981;55:360–369.
23. Lynch C, Vogel S, Pratila MG, Sperelakis N: Halothane depression of myocardial slow action potentials. *J Pharmacol Exp Ther* 1982;222:405–409.
24. Morishima HO, Pederson H, Finster M, Tsuji A, Hiraoka H, Feldman HS, Arthur GR, Covino BG: Is bupivacaine more cardiotoxic than lidocaine? *Anesthesiology* 1983;59:A409.
25. Niederlehner J, DiFazio C: Comparison of bupivacaine and lidocaine: General anesthetic potency and therapeutic index. *Regional Anesth* 1981;6:95–97.
26. Reiter M, Vierling W, Seibel K: Excitation–contraction coupling in rested-state contractions of guinea pig ventricular myocardium. *Naunyn-Schmiedeberg's Arch Pharmacol* 1984;325:159–169.
27. Reuter H, Scholz H: The regulation of the calcium conductance of cardiac muscle by adrenaline. *J Physiol (London)* 1977;264:49–62.
28. Sage D, Feldman H, Arthur GR, Covino BG: Cardiovascular effects of lidocaine and bupivacaine in the awake dog. *Anesthesiology* 1983;59:A210.
29. Seibel K, Reiter M: Potassium induced loss of the early component of myocardial contraction in the presence of noradrenaline. *Naunyn-Schmiedeberg's Arch Pharmacol* 1978;302:R31.
30. Seibel K, Karema E, Takeya K, Reiter M: Effects of noradrenaline on an early and a late component of the myocardial contraction. *Naunyn-Schmiedeberg's Arch Pharmacol* 1978;305:65–74.
31. Tada M, Kirchberger MA, Repke DI, Katz AH: The stimulation of calcium transport in cardiac sarcoplasmic reticulum by adenosine 3′:5′-monophosphate-dependent protein kinase. *J Biol Chem* 1974;249:6174–6180.
32. Tanz RD, Meskett T, Loehning RW, Fairfax CA: Comparative cardiotoxicity of bupivacaine and lidocaine in the isolated perfused mammalian heart. *Anesth Analg* 1984;63:549–556.
33. Thigpen JW, Kotelko DM, Shnider SM, Foutz SE, Levinson G, Kolke M, Rosen MA: Bupivacaine cardiotoxicity in hypoxic-acidotic sheep. *Anesthesiology* 1983;59:A204.
34. Tucker GT, Boyes RN, Bridenbaugh PO, Moore DC: Binding of anilide-type local anesthetics in human plasma: I. Relationships between binding, physiochemical properties, and anesthetic activity. *Anesthesiology* 1970;33:287–303.

VII

TOXICITY OF GENERAL ANESTHETICS

35

Anesthetic-Induced Hepatotoxicity in the Hyperthyroid Rat

M. Lawrence Berman, Duncan A. Holaday, Lee Kuhnert, Roger England, and James M. Phythyon

ABSTRACT. Evidence is reviewed which contradicts the hypothesis that the hepatic necrosis that occurs in hyperthyroid rats, 4–6 hr after a 2-hr exposure to halothane is a product of hypoxia and a hepatotoxic metabolite of halothane. Two experiments are reported which direct attention to the extravagant metabolism of hyperthyroidism and an effect on energy metabolism common to all general anesthetics. Body temperature falls 4–5°C during anesthesia in euthyroid animals, whereas it is maintained at a mildly hyperthermic level by hyperthyroid rats. Euthyroid rats made hyperthermic during anesthesia with halothane, isoflurane, or enflurane failed to develop a significant incidence of hepatic necrosis. Sevoflurane, an experimental anesthetic ether without demonstrable hepatotoxicity in euthyroid subjects, produced hepatic necrosis and acute death in hyperthyroid rats with an incidence similar to that produced following enflurane and isoflurane anesthesia. These observations direct attention to reported *in vitro* studies demonstrating interference by general anesthetics with oxygen utilization in normal mitochondria. This suggests that the mechanism of hepatotoxicity in the anesthetized hyperthyroid rat may be a block of carbohydrate oxidation by NAD which causes a failure of adequate energy metabolism.

INTRODUCTION

Male rats rendered hyperthyroid by pretreatment with triiodothyronine (T_3) develop hepatic centrilobular necrosis 4–6 hr after a 2 hr exposure to 1% halothane in 21% oxygen.[19] Initially it was thought that the hepatotoxicity observed in this model was due to an increase in oxygen consumption induced by the hyperthyroid state coupled with a decrease in oxygen delivery to the liver consequent to anesthetic-induced reduction of splanchnic blood flow resulting in liver hypoxia. It was further thought that this hypoxia caused an increase in the reductive biotransformation of halothane to metabolite(s) that produced hepatic necrosis.

Lack of Evidence for Anesthetic Hepatotoxic Metabolite(s)

The reductive and oxidative metabolism of halothane was compared in the T_3 model, and biochemical changes that might be related to biotransformation activity were examined.

M. Lawrence Berman, Duncan A. Holaday, Lee Kuhnert, Roger England, and James M. Phythyon • Departments of Anesthesiology and Pharmacology, Vanderbilt University, Nashville, Tennessee 37232.

Reductive metabolism of halothane as determined by tissue levels of chlorotrifluoroethane, and chlorodifluoroethylene and plasma levels of inorganic fluoride, were observed to be lower in T_3-treated rats than in euthythroid rats.[15] Urinary bromide and trifluroacetic acid, indicators of oxidative metabolism of halothane, were decreased in hyperthyroid rats compared to euthyroid rats.[15] Furthermore, the incidence and severity of the liver lesion was not changed in hyperthyroid rats exposed to deuterated halothane[17] suggesting that, in the T_3 model, the products of oxidation of halothane were not responsible for the observed hepatotoxicity. Thus, the available data indicate that the hyperthyroid state inhibited both reductive and oxidative metabolism of halothane. However, biochemical changes related to the liver microsomal mixed function oxidase system were observed in T_3-treated rats.[15] Maximum depression of cytochrome P-450 and glutathione-S-transferase and maximum increase in aniline hydroxylase activity occurred after 3 days of T_3 pretreatment. However, hepatic lesions after halothane anesthesia did not appear until after the fourth day of pretreatment. It remains to be established whether these biochemical changes and the appearance of the hepatic lesions are causally related.

The Role of Glutathione

The observation that glutathione levels fall after 1 day of pretreatment with T_3 and remain low throughout the course of pretreatment and anesthetic exposure[15] suggested that anesthetic-induced liver lesions in the hyperthyroid rats may be due to a decrease in a normally protective detoxification pathway. However, the administration of free sulfhydryl agonists *N*-acetylcysteine and cysteamine consistently exacerbated halothane-induced hepatotoxicity, whereas the administration of diethylmaleate, which depletes glutathione, reduced the incidence of hepatic lesions suggesting that glutathione may be involved in activation rather than detoxification of the anesthetic in the T_3 model.[16]

Fasting, But Not Hyperoxia, Is Protective

Other factors have been observed to affect anesthetic-induced hepatotoxicity in the T_3 model. Fasting hyperthyroid rats for 14 hr prior to anesthetic exposure reduced the incidence and severity of the liver lesions.[17] Hypoxia (14% oxygen) during anesthesia worsens the lesion, and hyperoxia (99% oxygen) decreases the severity, but not the incidence, of the hepatic lesion.[19]

Effects of Other General Anesthetics

It has been reported that both isoflurance and enflurane induce noticeable hepatic necrosis in male hyperthyroid rats.[2] Since isoflurane and enflurane are metabolized to a limited extent in oxygen and not at all in the presence of low oxygen tensions,[5] it seems unlikely that formation of reactive intermediates of these anesthetic agents may be involved in producing liver toxicity.

It is possible that the hepatic necrosis which follows isoflurane and enflurane as well as halothane anesthesia is not dependent on the drug or a drug metabolite, but on the imposition of general anesthesia *per se* on hyperthyroidism. Demonstration that sevoflurane, an experimental inhalational anesthetic which in euthyroid animals and in human volunteers causes no detectable disturbances in liver function (Fig. 1),[11,18] produces hepatic necrosis in hyperthyroid rats would be further support that the anesthetic state interacting with the biochemical derangements caused by hyperthyroidism may be responsible for the liver lesions, rather than a drug-specific effect.

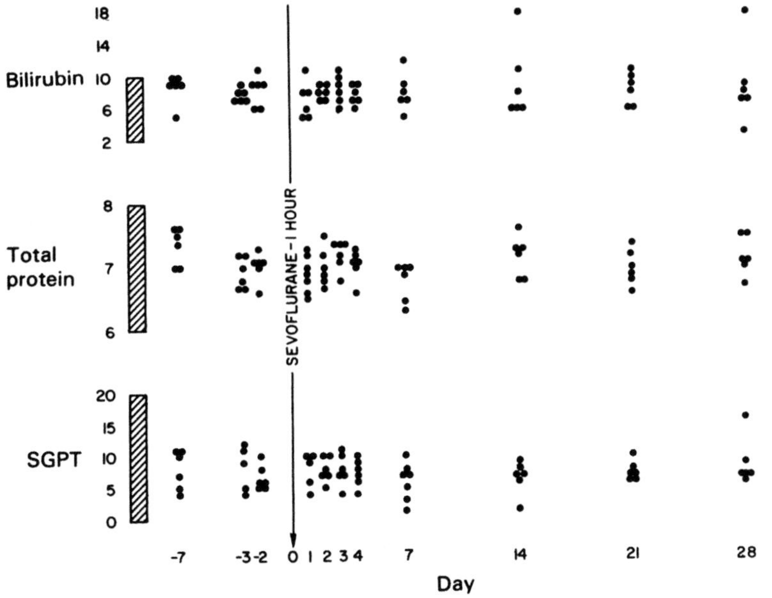

Figure 1. Liver function test values obtained from 7 days before to 28 days after 1 hr of anesthesia produced by sevoflurane in oxygen in six normal, male volunteers (after Holaday and Smith[11]). Only two values for bilirubin deviate from the normal range for the laboratory. SGPT, serum glutamate pyruvate-transaminase.

Role of Gender

Recent studies[3] have demonstrated that the incidence of halothane-induced hepatic necrosis in hyperthyroid female rats was similar to that in hyperthyroid male rats demonstrating a lack of a gender effect in the T_3 model.

Active versus Inactive Thyroid Hormones

Analogues of T_3 have been used to treat rats prior to halothane anesthesia.[15] Pretreatment with diiodothyronine, reverse T_3 or (+) thyroxin, analogues that are devoid of or have little thyromimetic activity, failed to sensitize the rat to halothane-induced hepatotoxicity. However, pretreatment with (−) thyroxin, which has hormonal activity, resulted in hepatic necrosis after halothane exposure similar to that observed in rats treated with T_3.

Is Hyperthermia a Factor?

The studies with analogues of T_3 indicate that the hyperthyroid state is essential for the development of the hepatic lesion in the T_3 model. Hyperthyroidism induces a multiplicity of physiological and biochemical derangements. Among the derangements caused by hyperthyroidism is heat production, which is reflected in a slightly elevated basal body temperature. Hyperthyroid rats after a 2-hr exposure to anesthesia with halothane, enflurane, or isoflurane were noted to be extremely warm, whereas euthyroid rats following anesthetic exposure were extremely cool. This observation suggested that the hyperthermia induced by the hyperthyroid state may contribute to the development of anesthetic-induced hepatotoxicity.

The purposes of the present studies were to determine whether hyperthermia during anesthesia is a determinant in the production of centrilobular hepatic necrosis in anesthetized hyperthyroid rats and to test the potential of sevoflurane anaesthesia to cause hepatic necrosis in hyperthyroid rats.

METHODS

Male Sprague–Dawley rats weighing 100–125 g were given T_3, 5 mg/kg intraperitoneally (i.p.) daily for 6 days. An equal number of littermates were given the vehicle for T_3 (3.0 mM sodium hydroxide in physiological saline solution) i.p. for 6 days. On day 7, the rats were anesthetized with 1% halothane, 1.8% enflurane, 1.3% isoflurane, or 2.5% sevoflurane in USP compressed air at 10 liters/min for 2 hr. Vehicle-pretreated rats were anesthetized on separate days, but at the same time of day. The anesthetic chamber, the administration of the volatile agents, and the monitoring of the concentration of anesthetic vapors and oxygen in the chamber were the same as previously described.[2] To record the body temperature, rats were anesthetized with 3% of one of the anesthetics for 10 min. The chamber was opened, and calibrated flexible vinyl thermistor probes (Yellow Springs Instrument Company, Model 44TD, Yellow Springs, Ohio) previously threaded through a sealable hole in the top of the chamber were inserted into the rat's rectum. The chamber was closed and the anesthetic concentration reduced. Temperature was recorded every 10 min using a multichannel telethermometer (Yellow Springs Instrument Company, Model 402). To induce hyperthermia in euthythroid rats during anesthesia, a hydrothermal mat (Gorman-Rupp, Bellville, Ohio) was placed on the floor of the chamber and infrared heat was focused on the top of the chamber. Six hours after the inhalational exposure the rats were killed by decapitation. The livers were removed and prepared for histological examination and graded as previously described.[2]

RESULTS

The body temperature of hyperthyroid rats at the beginning of the anesthetic exposure was slightly, but significantly, elevated compared to that of euthyroid rats (Table I). In one group of 10 hyperthyroid rats anesthetized with halothane, the body temperatures fell an average of 0.6° and in another group of eight hyperthyroid rats anesthetized with halothane, the temperature fell 1.1°. During anesthesia with enflurane, isoflurane, and sevoflurane, the body temperatures of hyperthyroid rats fell 0.7°, 1.2°, and 0.2°, respectively. However, there was a marked and significant decrease in body temperature of euthyroid rats during anesthetic exposure (Table I). The incidence of hepatic necrosis in hyperthyroid rats exposed to halothane, isoflurane, or enflurane (Table I) was similar to that reported previously.[2]

Effects of Hyperthermia on Euthyroid Rats

The mean body temperature of euthyroid rats subjected to hyperthermia during anesthesia ranged from 39° to 39.8° (Table II). This temperature elevation averaged 1.2° ± 0.2° higher than that of hyperthyroid rats during their exposure to anaesthesia, but the difference was not statistically different ($p > 0.1$). The hyperthermia induced in the euthyroid rats occurred 10–15 min after induction of anesthesia and was sustained to the end of the anesthetic exposure.

Hyperthermia during anesthetic exposure of euthyroid rats resulted in a 6.6% incidence of hepatic necrosis after halothane, 3.3% after enflurane, and no liver lesions after isoflurane (Table II).

Table I. Mean Temperature (°C ± S.E.M.) of Triiodothyronine-Pretreated Rats (Hyperthyroid) and Vehicle-Pretreated (Euthyroid) Rats at the Beginning and End of Anesthesia and Incidence of Hepatic Centrilobular Necrosis (10 Rats/Group)

Anesthetic	Beginning	End	Difference	Incidence of necrosis
Halothane				
Hyperthyroid[a]	38.9 ± 0.2	38.3 ± 0.2	−0.6	90%
Euthyroid	37.3 ± 0.1	32.3 ± 0.2	−5.0	0%
			$p < 0.001$	
Isoflurane				
Hyperthyroid[a]	39.1 ± 0.2	37.9 ± 0.3	−1.2	40%
Euthyroid	36.9 ± 0.1	32.9 ± 0.2	−4.0	0%
			$p < 0.001$	
Enflurane				
Hyperthyroid[a]	39.3 ± 0.1	38.6 ± 0.5	−0.7	30%
Euthyroid	37.1 ± 0.1	32.6 ± 0.2	−4.5	0%
			$p < 0.001$	

[a]Hyperthyroid versus euthyroid $p < 0.01$ at the beginning of anesthesia.

Hepatic Necrosis Follows Sevoflurane

Of 49 rats exposed to sevoflurane, five died during exposure (Table III). Of the 44 survivors, 21 developed hepatic necrosis. Seven rats exhibited 1+ necrosis, 10 exhibited 2+, and four exhibited 3+. The average grade, including those which did not develop hepatic necrosis, was 0.89 (S.E.M. 0.16). Exposure of 39 animals to halothane resulted in death of 13, and hepatic necrosis of all the remainder, average grade 2.19 (S.E.M. 0.14).

The body temperature of the rats exposed to sevoflurane did not change significantly during 2 hr of anaesthesia and remained above 39.6°C. The body temperature of those exposed to halothane fell significantly from 39.1°C to 38.0°C (Fig. 2), but not significantly different from those of the hyperthyroid animals used in the hyperthermia experiment.

DISCUSSION

These studies continue a search for the cause of hepatic necrosis in the hyperthyroid rat following anesthesia. The metabolic state of the animal appears to be an important factor

Table II. Mean Temperature (°C ± S.E.M.) of Vehicle-Pretreated Rats Subjected to Hyperthermia at the Beginning and End of Anesthesia and Incidence of Hepatic Centrilobular Necrosis (30 Rats/Group)

Anesthetic	Beginning	End	Difference	Incidence of necrosis
Halothane	37.0 ± 0.1	39.8 ± 0.2	+2.8	6.6%
Isoflurane	37.2 ± 0.1	39.0 ± 0.2	+2.2	0%
Enflurane	38.0 ± 0.2	39.5 ± 0.1	+1.5	3.3%

Table III. Incidence of Acute Death and Hepatic Necrosis in Anesthetized Hyperthyroid Rats

	n	Mortality	Lesions	Grade (S.E.M.)
1% Halothane	39	13 (33%)	26/26 (100%)	2.19 (0.14)
2.5% Sevoflurane	49	5 (10%)	21/44 (48%)	0.89 (0.16)

since access to standard rat chow up to the time of anesthesia is followed by a high incidence of hepatic necrosis, whereas a 14-hr fast prior to exposure is protective.[17] Making euthyroid rats hyperthermic during anesthesia with halothane, enflurane, or isoflurane was not effective in producing hepatic lesions. This suggests that the hyperthermia which occurs during anesthesia in hyperthyroid rats is not a factor in sensitizing the rat to anesthesia-induced hepatotoxicity.

Exposure of hyperthyroid rats to sevoflurane resulted in approximately the same incidence of hepatic necrosis and acute death as exposure to enflurane or isoflurane as shown in Table IV.[2] Sevoflurane, a heavily fluorinated isopropylmethyl ether, has several advantageous properties as an anesthetic.[10] It is not flammable in air and nonexplosive in oxygen and nitrous oxide. It is chemically stable without preservative. It has a low solubility in blood and does not stimulate upper airway reflexes; hence it induces anesthesia rapidly and pleasantly.[11] It does not sensitize the heart to epinephrine and causes less depression of respiration and cardiovascular functions than halothane.[18] Also, sevoflurane has been found to cause no detectable disturbance of liver functions in experimental animals or in human volunteers.[11,18] Furthermore, it is known to have a single pathway of metabolism which results in no hepatotoxin[11,13] (Fig. 3). It appears unlikely, therefore, that to continue the search for a necrotizing hepatotoxin derived from anesthetic drug metabolism will yield a clue to the cause of postanesthetic necrosis in this hyperthyroid rat model.

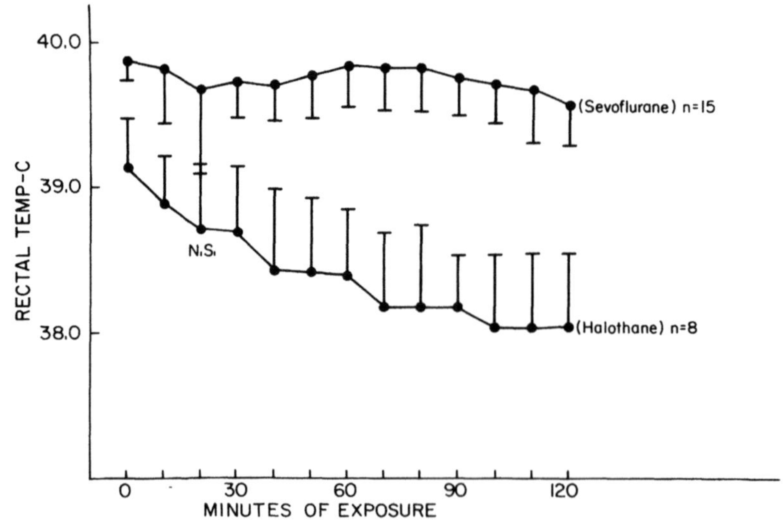

Figure 2. Body temperature remains elevated above normal euthyroid body temperatures during both sevoflurane anesthesia and halothane anesthesia. The bars are S.E.M.'s.

Table IV. Incidences of Acute Death and Hepatic Necrosis in Anesthetized Hyperthyroid Rats

	Treatment	*n*	Mortality	Lesions	Grade
1% Halothane	T_3	132	33%	92%	1.56
	Vehicle	36	0	0	
1.3% Isoflurane	T_3	45	11%	28%	0.28
	Vehicle	18	0	0	
1.8% Enflurane	T_3	40	5%	24%	0.21
	Vehicle	18	0	0	

Anesthesia Interferes with Mitochondrial Metabolism

The literature contains clues of other mechanisms possibly responsible for this phenomenon. Halothane and a number of other anesthetics block NADH dehydrogenase, the enzyme responsible for utilization of NAD-dependent substrates, including pyruvate, maleate, isocitrate, glutamate, and 3-hydroxyacyl CoA in the electron transfer chain and regeneration of ATP from ADP.[4,14] This results in a reduction of oxygen consumption in phase 3 mitochondrial metabolism.[7] It is conceivable that the explanation for the protection afforded by a period of fasting is that the ketogenic state initiated by starvation[6] has resulted in mobilization of fat that is not dependent on NAD for oxidation.[4] Rats that have subsisted on a carbohydrate diet up to the time of block of NADH dehydrogenase lack the necessary time to convert to other substrates for energy transduction. Current studies are pursuing this hypothesis.[12]

Evidence for Liver Hypoxia

Another factor differentiates the fed from the starved animals. The gut utilizes oxygen to absorb food; portal vein oxygen saturation is reduced in the fed rat.[9] In the starved rat, portal vein oxygen saturation approaches arterial levels. It is possible that portal vein hypoxia in the fed rat contributes to hepatic necrosis.

Further support for a hypoxic factor is derived from the fact that halothane decreases hepatic blood flow in the euthyroid subject by reducing splanchnic blood flow more than isoflurane[8] and also by causing hepatic artery spasm during anesthesia.[1] Although the effects of halothane and other anesthetics on hepatic blood flow in the hyperthyroid rat have not been measured, we have observed that arterial blood pressure rises progressively in the T_3-pretreated rat during halothane anesthesia, in contrast to the response of the euthyroid rat

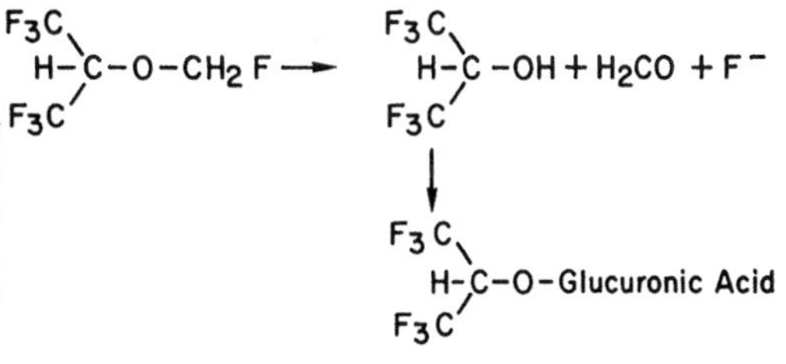

Figure 3. Sevoflurane undergoes cleavage of the ether linkage and prompt glucuronidation in the rat, dog, and man. Formaldehyde is incorporated into the one-carbon chain. Half of the fluoride ion is incorporated into bone. The remaining fluoride and the glucuronide are rapidly excreted in the urine. Adapted from Martis *et al.*[13]

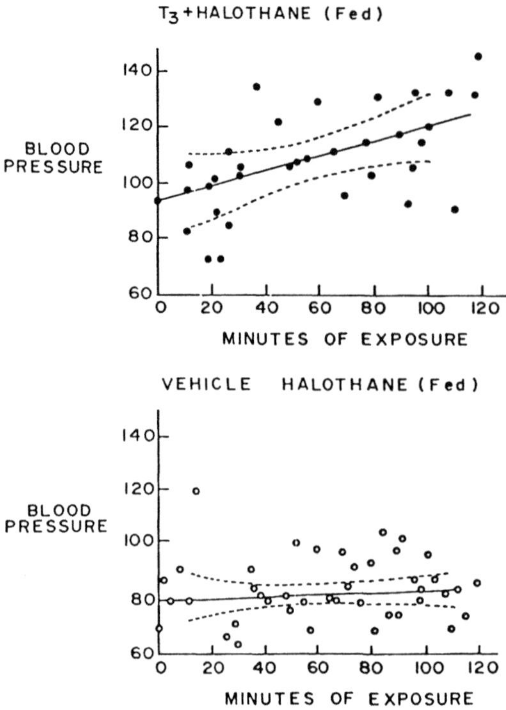

Figure 4. Systolic blood pressure values obtained from hyperthyroid (T_3) and euthyroid (vehicle), fed rats during halothane anesthesia. Measurements were made by an indirect, oncometric method using a Pneumatic Pulse Transducer, model 320-0112B, and tail cuff (Narco Biosystems, Inc., Houston, Texas). The solid lines are least-squares fits and the dotted lines are the 95% confidence limits.

(Fig. 4) (unpublished observations). The latter response may be a compensatory adjustment to the physiological changes produced by hyperthyroidism and the anesthetic state of the animal that favor the needs of oxidative metabolism.

It appears that the mechanism(s) whereby halogenated anesthetics produced hepatic necrosis in the hyperthyroid rat cannot be explained by the biotransformation of anesthetic molecules to hepatotoxic metabolites. However, a discrepancy between energy requirements and energy production in this model may have an important role. Hyperthyroidism stimulates oxygen consumption; the anesthetics depress spontaneous ventilation, reduce splanchnic blood flow, and inhibit oxygen uptake by mitochondria.[7] Thus, an imbalance of oxygen demand and/or substrate availability causing depression of mitochondrial function may be the key factors that result in liver necrosis in the hyperthyroid rat following anesthetic exposure to halothane, isoflurane, enflurane, and sevoflurane.

REFERENCES

1. Benumof NL, Bookstein JJ, Saidman LJ, Harris R: Diminished hepatic arterial flow during halothane administration. *Anesthesiology* 1976;45:545–551.
2. Berman ML, Kuhnert L, Phythyon JM, Holaday DA: Isoflurane and enflurane-induced hepatic necrosis in triiodothyronine-pretreated rats. *Anesthesiology* 1983;58:1–5.
3. Berman ML, Kuhnert L, Phythyon JM, Holaday DA: Lack of gender effect of triiodothyronine enhancment of halothane-induced toxicity (abst). *Anesthesiology* 1984;61:A275.
4. Biebuyck JF: Effects of anesthetic agents on metabolic pathways. *Br J Anaesth* 1973;45:263–268.
5. Burke JR Jr, Branchflower RV, Lees DE, Pohl LR: Mechanism of defluorination of enflurane: Identification of an organic metabolite in rat and man. *Drug Metab Disposition* 1980;9:19–24.
6. Cahill GF Jr: Starvation in man. *N Engl J Med* 1970;282:668–675.
7. Cohen PJ, Marshall BE: Effects of halothane on respiratory control and oxygen consumption of rat liver mitochondria. *Anesthesiology* 1968;29:181.

8. Gelman SI, Fowler KC, Smith LR: Regional blood flow during isoflurane and halothane anesthesia. *Anesth Analg* 1984;63:557–565.
9. Greenway CV, Stark RD: Hepatic vascular bed. *Physiol Rev* 1971;51:23–65.
10. Holaday DA: Sevoflurane: An experimental anesthetic, in Brown BR Jr (ed): *New Pharmacological Vistas in Anesthesia.* Philadelphia, F A Davis Co, 1983, chap 4.
11. Holaday DA, Smith FR: Clinical characteristics and biotransformation of sevoflurane in healthy human volunteers. *Anesthesiology* 1981;54:100–106.
12. Holaday DA, Berman ML, England R, Abumrad N: The effect of halothane on metabolic pathways in the hyperthyroid rat. (abstr). *Anesthesiology* 1984;61:A266.
13. Martis L, Lynch B, Napoli MD, Woods EF: Biotransformation of sevoflurane in dogs and rats. *Anesth Analg* 1981;60:186–191.
14. Nahrwold ML, Cohen PJ: Anesthetics and mitochondrial respiration, in Cohen PJ (ed): *Clinical Anesthesia* Vol 11. Philadelphia, F A Davis Co, 1975, No 1, chap 2.
15. Smith AC, Berman ML, James RC, Harbison RD: Characterization of hyperthyroidism enhancement of halothane-induced hepatotoxicity. *Biochem Pharmacol* 1983;32:3531–3539.
16. Smith AC, James RC, Berman ML, Harbison RD: Paradoxical effects of pertubation of intracellular levels of glutathione on halothane-induced hepatotoxicity in hyperthyroid rats. *Fundamental Appl Toxicol* 1984;4:221–230.
17. Uetrecht J, Wood AJJ, Phythyon JM, Wood M: Contrasting effects on halothane hepatotoxicity in the phenobarbital-hypoxia and triiodothyronine model: Mechanistic implications. *Anesthesiology* 1983;59:196–201.
18. Wallin RF, Regan BM, Napoli MD, Stern, IJ: Sevoflurane: A new Inhalational Anesthetic. *Anesth Analg* 1975;54:758–765.
19. Wood M, Berman ML, Harbison RD, Hoyle P, Phythyon JM, Wood AJJ: Halothane-induced hepatic necrosis in triiodothyronine pretreated rats. *Anesthesiology* 1980;52:470–476.

36

Role of Oxygen Availability to the Liver in Anesthesia-Induced Hepatotoxicity

Simon Gelman

ABSTRACT. The mechanism of anesthesia-induced hepatotoxicity remains obscure. Hepatic centrolobular necrosis was described in rats pretreated with phenobarbital to produce enzyme induction and then anesthetized with halothane under hypoxic conditions. It has been suggested that the intermediate products accumulating during reductive metabolism of halothane which takes place under hypoxic conditions are responsible for hepatic injury. Another hypothesis assumes that hypoxia of the liver *per se*, regardless of halothane metabolic intermediates, produces liver damage. According to the hypoxic hypothesis, hepatic injury occurs when liver oxygen supply/demand is severely out of balance. Many anesthetics, inhalational and intravenous, which do not produce any toxic intermediates can also produce liver injury in the rat model. Barbiturate pretreatment increases liver oxygen requirements, while various anesthetic drugs create different degrees of respiratory and circulatory (systemic and splanchnic) depression, resulting in a reduction of oxygen delivered to the liver. Available data suggest that liver damage in the anesthesia-induced hepatotoxicity rat model is strongly associated with a decrease in oxygen delivery to the liver, fitting both the metabolic and hypoxic hypotheses of the hepatotoxicity. Further studies are required to elucidate the cause of the hepatotoxicity.

An anesthesia-induced hepatotoxicity model was introduced in 1979 when centrolobular necrosis in the liver was demonstrated in rats pretreated with pentobarbital (to produce enzyme induction) and anesthetized with halothane under hypoxic conditions.[11,17] The mechanisms of damage to the liver has not been established. One hypothesis suggests that liver damage is related to the reductive metabolism of halothane, which takes place in hypoxic conditions and results in the accumulation of products that may be hepatotoxic.[3,11,13,14] Cytochrome P-450 is involved in the reductive metabolism of halothane. It appears, however, that damage to the liver occurs during exposure to anesthesia and hypoxia, at a time prior to maximal metabolism, and binding of reactive metabolites to the intracellular structures.[20] Neither observation fits the metabolic hypothesis. Both halothane and hypoxia *per se* reduce the cytochrome P-450 level, thereby decreasing the metabolism of halothane. These observations do not support the metabolic hypothesis either.[20] Another hypothesis assumes that the liver ischemia *per se*, regardless of halothane metabolic inter-

Simon Gelman • Department of Anesthesiology, The University of Alabama at Birmingham, Birmingham, Alabama 35294.

Table I. Metabolic and Hypoxic Hypotheses for Anesthesia-Induced Hepatotoxicity Model in Rats

Events observed	Metabolic hypothesis	Hypoxic hypothesis
Barbiturate pretreatment, halothane, and hypoxia produce liver necrosis[1,2]	Reductive halothane metabolism produces hepatotoxic intermediates	Barbiturate pretreatment increases oxygen demand, anesthesia decreases oxygen supply
Centrolobular (Zone III) location of the necrosis[1,2,7]	Edema in Zone I resulting in ischemia in Zone III	Ischemia in Zone III > ischemia in Zone I
Liver damage is reproduced only in male rats[5,11]	Specific enzyme profile	High oxygen demand and specific difference in hypoxic brain/heart and hypoxic hepatic injury in male rats
Triiodothyronine pretreatment and halothane anesthesia without hypoxia produce liver damage[6,15,17]		Decrease in cytochrome P-450 and increase in oxygen demand
Other inhalational and intravenous anesthetics can produce liver necrosis[6,8,14,16,17]		The drugs decrease oxygen delivery to the liver
Food deprivation enhances liver necrosis[6,8]	Specific substrate protecting the liver is not delivered with food	Liver blood flow and, therefore, liver oxygen delivery is lower when animals are deprived from food
Hypothermia reduces liver necrosis[9,10]	Decrease in enzyme activity results in a decrease in halothane metabolism	Decrease in oxygen demand results in an improvement of oxygen supply/demand ratio
The degree of liver necrosis correlates with amount of cytochrome P-450[10]	Halothane metabolism is increased	Oxygen demand is increased resulting in reduction in oxygen supply/demand ratio
The degree of liver necrosis correlates with halothane concentration used[10]	Increase in halothane	Decrease in oxygen supply
Metyrapone, cimetidine decrease liver necrosis[12,13]	A decrease in cytochrome P-450 reduces halothane	A decrease in oxygen demand increases oxygen supply/demand ratio

mediates, produces liver injury. In other words, according to the hypoxic hypothesis, barbiturates induce liver enzymes (cytochrome P-450), thereby increasing liver oxygen demand; anesthesia leads to different degrees of respiratory and/or circulatory depression, resulting in a reduction of liver oxygen supply and, consequently, a decrease in oxygen supply/demand ratio for the liver. The majority of available data in this field can be interpreted from both standpoints, metabolic and hypoxic (Table I).

Anesthesia-induced liver necrosis has been mainly described in the centrolobular areas, zone III according to Rappaport,[15] which are located at the periphery of the liver unit, remote from hepatic arterioles and portal venulae more than other parts of the hepatic acinus (zones II and III). These observations support the hypoxic theory and contradict, to a certain extent, the metabolic hypothesis: toxic intermediates would be delivered more to the areas around hepatic arterioles and portal venulae (zone I), which are better supplied with oxygen, while the centrolobular areas (zone III) would be less exposed to toxic products and would have a relatively lower blood and oxygen supply than other parts of liver acinus.[15]

It has been demonstrated that fasting rats develop much greater liver necrosis than fed animals under otherwise similar conditions.[19] The authors say: ". . . there is something in food which protects the liver from damage." Liver blood flow might very well be this "something": a meal increases splanchnic blood flow, which apparently decreases hypoxia of the liver. Hypothermia also decreases liver damage in the rat model.[2,10] This effect can be attributed to a reduction in metabolism of halothane as well as a decrease in oxygen demand by the liver, resulting in an improved oxygen supply/demand ratio.

Certain genetic differences in reductive metabolism and hepatotoxicity of halothane in different rat strains[7] and differences in degrees of liver damage between male and female rats[14] can be explained from both hypoxic and metabolic standpoints: male rats, as well as some particular rat strains, might have a higher oxygen demand than other animals and, therefore, a lower supply/demand ratio. A similar hypothetical explanation can be assumed for the fact that liver injury can be reproduced only in rats: the brain and heart of other species may be more susceptible to hypoxia than the liver. Consequently, the proper degree of hypoxia that would lead to liver injury without early deaths from brain and/or heart damage cannot be found. In other words, animals die from heart and brain damage prior to the development of liver damage.

The degree of hepatotoxicity in the rat model correlates well with the concentrations of hepatic cytochrome P-450 and concentrations of inspired halothane.[10] These observations also fit both metabolic and hypoxic theories: the concentration of cytochrome P-450 influences the oxygen demand, while the halothane concentration influences the oxygen supply to the liver, altering the liver oxygen supply/demand ratio.

Metyrapone[12] and cimetidine[22] protect rats against halothane-induced liver damage. These drugs decrease the cytochrome P-450 activity and, therefore, decrease the metabolism of halothane (which fits the metabolic theory) or, on the other hand, both drugs may reduce oxygen demand, benefiting the liver oxygen supply/demand ratio.

It has been demonstrated that higher doses with short exposures to halothane are accompanied by a more pronounced liver injury than lower concentrations and long exposures to halothane.[18] These data fit the hypoxic theory better than the metabolic theory since a lengthy exposure would increase the amount of halothane metabolic intermediates.

Triiodothyronine (T_3) pretreatment of rats exposed to halothane, enflurane, and isoflurane without hypoxia is also accompanied by the development of liver necrosis[1,21] T_3 treatment reduces cytochrome P-450 content (which contradicts the metabolic theory) and increases oxygen demand, deteriorating the oxygen supply/demand ratio. Therefore, the T_3 model supports the hypoxic hypothesis.[20]

Other inhalational agents such as enflurane and isoflurane, and even intravenous anesthetics and narcotics such as thiopental and fentanyl, which do not produce any toxic products, can also produce liver necrosis in the rat model.[1,8,18–20] However, halothane appears to be more destructive to the liver than other anesthetics. The difference, again, can be explained by the reductive metabolism of halothane, or by more pronounced liver hypoxia (resulting from respiratory and/or circulatory depression) during halothane compared to other anesthetics. It has been demonstrated in dogs, when used in equipotent doses, that halothane decreases oxygen delivery to the liver to a much larger etent than either enflurane[9] or isoflurane.[5] The difference in the effect of these anesthetics was mainly related to the difference in a reduction of hepatic arterial blood flow rather than to a decrease in portal blood flow.[5,9] In one study on rats hepatic arterial blood flow was found to be almost twice as high in hypoxic awake rats than in hypoxic animals anesthetized with halothane. The difference in values of the flow did not reach a statistically significant level, and the authors concluded that ". . . decrement in total hepatic blood flow . . . is unlikely to be sufficient to account for the liver necrosis. . . ."[16] However, absence of significance should not be

Table II. Postanesthetic Liver Injury

1. Postoperative liver insufficiency
   a. Liver hypoxia
   b. Intermediate products (?)
2. Viral hepatitis, acute (1:10,000)
3. Aggravated chronic hepatitis
   a. Immunity depression by anesthesia and/or surgery
   b. Respiratory/circulatory depression resulting in liver hypoxia
   c. Deteriorated hepatic artery blood flow autoregulation
   d. Relative overdose (altered pharmacokinetics)
4. Fulminant hepatitis with specific halothane-related antibodies
5. Specific drug therapy

interpreted as a significance of absence. Recently, we performed a study evaluating the role of respiratory and circulatory depression in anesthesia-induced hepatotoxicity in rats. The animals were pretreated with pentobarbital and exposed to hypoxia ($FiO_2 = 0.14$) for 2 hr. The animals were divided into four groups: Group 1 received 1% inspired halothane (1 MAC) in the hypoxic gas mixture; Group 2 received 1.4% inspired isoflurane (1 MAC) and hypoxia; Group 3 had 25–30% of their blood volume removed 2 hr prior to exposure to hypoxia; and Group 4 served as a control with no treatment other than hypoxia. Hepatic blood flow was studied using microspheres. Oxygen availability to the liver was calculated by using values of hepatic blood flow and oxygen content of arterial and portal venous blood; liver injury was quantitatively evaluated. The lowest portal blood flow was observed in Groups 1 and 3, while hepatic arterial blood flow was the lowest in Group 1 and highest in Group 3. There was an inverse association between hepatic oxygen availability and histological lesions. The most severe hepatic lesions and the lowest oxygen availability were in Group 1 (halothane group), followed by Group 3 (hemorrhage), then Group 2 (isoflurane), and, finally, the control group, which had normal liver and the highest oxygen availability.[6]

It seems that in the clinical arena, we are dealing with different types of liver damage (Table II) and that the anesthesia-induced hepatotoxicity rat model reflects possibly only one type, namely, hypoxic injury to the liver.[4] It appears that liver damage in the anesthesia-induced hepatotoxicity rat model is associated with a decrease in oxygen delivery to the liver, which fits both the metabolic and hypoxic hypotheses of hepatotoxicity. The hypoxic theory seems to be better substantiated than the metabolic hypothesis. However, it is obvious that further studies are required to elucidate the cause of hepatotoxicity.

REFERENCES

1. Berman ML, Kuhnert L, Phythyon JM, Holaday DA: Isoflurane and enflurane-induced hepatic necrosis in triiodothryonine-pretreated rats. *Anesthesiology* 1983;58:192–193.
2. Ford D, Coyle DE: Effect of body temperature on the halothane–hypoxia-induction model of halothane hepatitis. *Anesthesiology* 1983;59:A222.
3. Gandolfi AJ, Brown BR: Hypoxia and halothane hepatotoxicity. *Anesth Analg* 1983;62:859–861.
4. Gelman S: Anesthesia induced hepatotoxicity. *Anesthesiology* 1983;59:487–488.
5. Gelman S, Fowler K, Smith L: Liver circulation and function during isoflurane and halothane anesthesia and hypoxia. *Anesthesiology* 1984a;61:726–730.
6. Gelman S, Rimerman V, Fowler K, Bishop S, Bradley EL: The effect of halothane, isoflurane, and blood loss on hepatotoxicity and hepatic oxygen availability in phenobarbital-pretreated hypoxic rats. *Anesth Analg* 1984b;63:965–972.
7. Gourlay DN, Adams JF, Cousins MJ, Hall P: Genetic differences in reductive metabolism and hepatotoxicity of halothane in three rat strains. *Anesthesiology* 1981;55:96–103.

8. Harper MH, Collins P, Johnson BH, Eger EI, Biava CG: Postanesthetic hepatic injury in rats: Influence of lateratios in hepatic blood flow, surgery, and anesthesia time. *Anesth Analg* 1982;61:79–82.
9. Hughes RL, Campbell D, Fitch W: Effects of enflurane and halothane on liver blood flow and oxygen consumption in the greyhound. *Br J Anesth* 1980;52:1079–1086.
10. Jee RC, Sipes IG, Gandolfi AJ, Brown BR: Factors influencing halothane hepatotoxicity in the rat hypoxic model. *Toxicol Appl Pharmacol* 1980;52:267–277.
11. McLain GE, Sipes IG, Brown BR: An animal model of halothane hepatotoxicity: Roles of enzyme induction and hypoxia. *Anesthesiology* 1979;51:321–326.
12. Nastainczyk W, Rendic S, Ullrich V, Kruger R: The prevention of halothane-induced liver cell necrosis in rats with metyrapone. *Anaesthesist* 1982;31:181–184.
13. Plummer JL, Hall de la M, Bastin FN, Iisley AH: Hepatic injury in rats due to sub-acute halothane exposure. *Anesthesiology* 1982;57:A222.
14. Plummer JL, Cousins MJ, Hall P: Hypoxia and halothane hepatotoxicity. *Anesth Analg* 1983;62:859.
15. Rappaport AM: Hepatic blood flow: Morphologic aspects and physiologic regulation. *Liver Biliary Tract Physiol I.* 1980;21:1–63.
16. Ross WT Jr, Daggy BP: Hepatic blood flow in phenobarbital-pretreated rats during halothane anesthesia and hypoxia. *Anesth Analg* 1981;60:306–309.
17. Ross WT Jr, Daggy BP, Cardell RR: Hepatic necrosis caused by halothane and hypoxia in phenobarbital-treated rats. *Anesthesiology* 1979;51:327–333.
18. Shingu K, Eger EI, Johnson BH, VanDyke RA, Lurz FW, Harper MH, Cheng A: Hepatic injury induced by anesthetic agents in rats. *Anesth Analg* 1983;62:140–145.
19. Van Dyke RA: Effect of fasting on anesthetic-associated liver toxicity. *Anesthesiology* 1981;55:A181.
20. Van Dyke RA: Halogenative anesthetic hepatotoxicity—Is the answer close at hand? *Clin Anaesthesiol* 1983;1:485–506.
21. Wood M, Berman ML, Harbison RD, Hoyle R, Phythyon JM, Wood AJJ: Halothane-induced hepatic necrosis in triiodothryonine-treated rats. *Anesthesiology* 1980;52:47–476.
22. Wood M, Uetrecht J, Phythyon JM, Wood AJJ: Cimetidine protects against halothane-induced hepatotoxicity. *Anesthesiology* 1982;57:A221.

37

Pathogenic Mechanisms for Halothane Hepatotoxicity in Animal Models and Man

Pauline de la M. Hall, Michael J. Cousins, John L. Plummer, and Christine A. Lunam

ABSTRACT. Two animal models have been used in this laboratory to study halothane hepatotoxicity: the mildly hypoxic (14% O_2) phenobarbital-induced Fischer 344 rat and the guinea pig model. The hypoxic rat model has been most useful in the study of metabolite-mediated halothane hepatotoxicity and was used to determine the time course of hepatic injury, metabolism of halothane, effects of inhibitors such as cimetidine on halothane metabolism, and identification of free radicals of halothane *in vivo*.

The guinea pig is a unique model because exogenous "conditioning factors" such as enzyme induction and hypoxia are not required, and thus the pathogenesis of halothane hepatotoxicity can be studied *per se*. This model was used to study the hemodynamic changes that occurred during anesthesia, the role of halothane metabolism, and the ultrastructural changes of hepatocyte membranes following exposure to halothane. Hepatic injury occurred only in a proportion of guinea pigs, which is in keeping with the heterogeneity of the human response. The extent of halothane metabolism in the guinea pig is also closer to man than the rat model.

The evidence provided indicates that the etiology of halothane hepatotoxicity is multifactorial with halothane metabolism playing a central role. It is proposed that individual differences in metabolism and/or susceptibility of the liver to toxic intermediates, are important factors in the development of liver damage, which may explain the rarity of halothane hepatotoxicity in man.

INTRODUCTION

Progress in the understanding of halothane hepatotoxicity in man has been aided by the development of animal models with similar metabolism of halothane and similar patterns of hepatic damage. Rat models for halothane hepatotoxicity include the mildly hypoxic, phenobarbitone-induced rat,[1] the triiodothyronine-pretreated rat,[15] and the grossly hypoxic (8–10% O_2) rat.[13]

In our laboratory we use the mildly hypoxic phenobarbitone-induced rat model[2] and have developed a guinea pig model for halothane hepatotoxicity.[9] The guinea pig has unique advantages over other animal models in that (1) exogenous "conditioning factors" such as

Pauline de la M. Hall and Christine A. Lunam • Department of Pathology, Flinders Medical Centre, Adelaide, S.A. 5042, Australia *Michael J. Cousins and John L. Plummer* • Department of Anaesthesia and Intensive Care, Flinders Medical Centre, Adelaide, S.A. 5042, Australia.

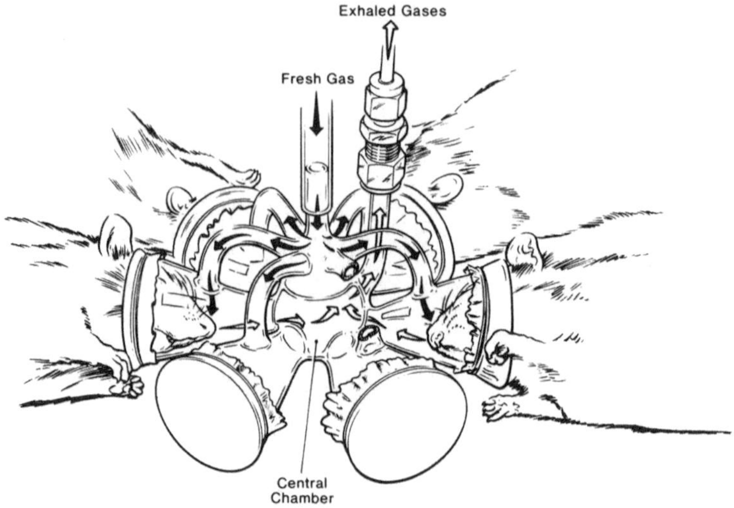

Figure 1. Glass mask for anesthesia and collection of pooled volatile metabolites in breath. Solid arrows show fresh gas flow; open arrows show exhaled breath. Reproduced with the permission of the American Society of Anesthesiologists from Cousins.[3]

enzyme induction and hypoxia are not required, allowing the study of the pathogenesis of halothane hepatotoxicity *per se,* (2) hepatic injury occurs in only a proportion of animals, which is more in keeping with heterogeneity of the response in man, and (3) the extent of halothane metabolism in the guinea pig is closer to that in humans than is the case with the rat model.

RAT MODEL

Methods

When phenobarbitone-induced male rats were anesthetized with 1% halothane in 14% oxygen/85% nitrogen for 2 hr and killed 24 hr after anesthesia, centrilobular hepatic necrosis was seen in all animals.[2] We found that the Fischer 344 rat developed more severe liver injury than Sprague–Dawley and Wistar rats[4] and have used the Fischer 344 rat for all subsequent studies.

An important aspect of our studies has been the precise control and documentation of the conditions of anesthesia and the physiological status of the animals. Inspired and end-expired concentrations of halothane, O_2, and CO_2 were measured intermittently by placing the probe of a respiratory mass spectrometer in the rats' nostrils. A six-port glass mask (Fig. 1) was designed to allow the collection of pooled exhaled breath samples for the measurement of volatile metabolites.

Time Course of Hepatic Injury

By 12 hr after anesthesia all rats exhibited confluent centrilobular liver cell necrosis (zone 3 necrosis). By 10–15 days the livers of most animals were indistinguishable from livers of control animals (Fig. 2).[5]

Studies of hepatic microsomal function following halothane anesthesia showed marked reductions in microsomal cytochrome P-450 and aminopyrine demethylase activity. These

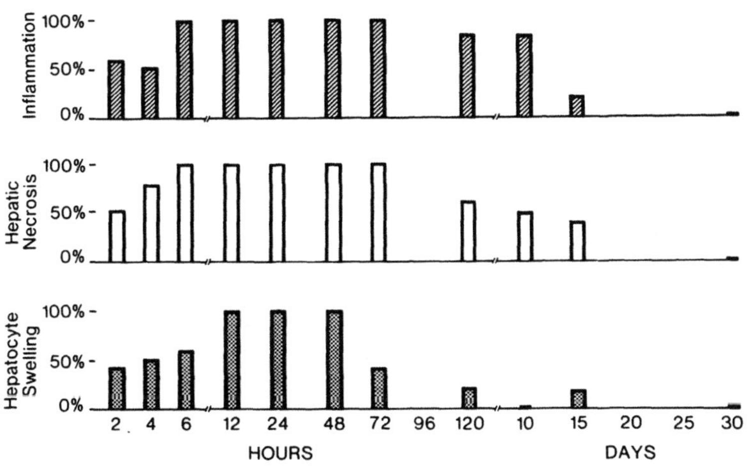

Figure 2. Time course of liver injury. The incidence of liver cell degeneration and necrosis following halothane anesthesia was assessed over the course of 30 days.

changes were maximal at 1–2 hr after commencement of anesthesia and persisted for more than 24 hr.[6]

Metabolism of Halothane in the Rat Model

As in man, the major pathway of halothane metabolism is O_2 dependent (oxidative) and yields water-soluble metabolites trifluoroacetic acid and bromide. A minor non-oxygen-dependent pathway (reductive) results in the formation of the volatile metabolites 2-chloro-1,1,1-trifluoroethane (CTF) and 2-chloro-1,1-difluorethylene, which can be detected in the breath. Both pathways lead to the production of potentially damaging, short-lived intermediates.

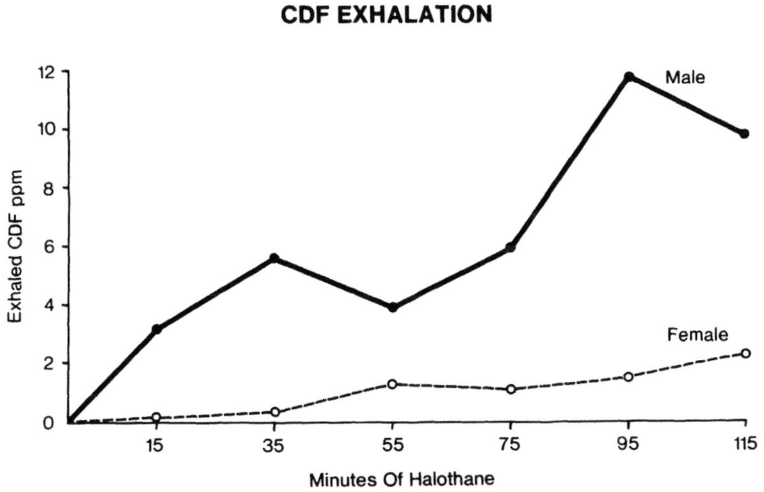

Figure 3. Sex differences in the amount of reductive metabolism of halothane. The amount of the volatile metabolite CTF produced by male and female rats during a typical 2-hr halothane anesthesia is plotted against time. Reproduced with the permission of the American Society of Anesthesiologists from Cousins.[3]

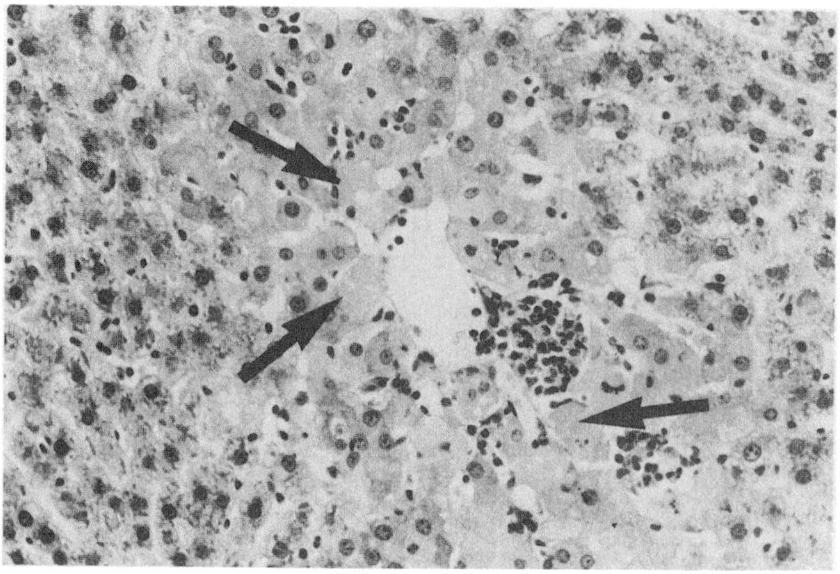

Figure 4. Light micrograph of a liver section, from a phenobarbitone-induced female rat killed 24 hr after 1% halothane in 14% O_2 for 2 hr. Liver cell necrosis extends two to three cells away from a central vein (arrows). Foci of lymphocytes are apparent in the region of necrosis. H&E stain × 264.

Evidence Linking Reductive Metabolism with Hepatic Injury

Three factors are necessary for the reproducible development of hepatic necrosis in the rat: male sex, phenobarbitone induction, and mild hypoxia. Each of these factors is associated with increased levels of reductive metabolism. A comparison of sex-related differences in reductive metabolism of halothane in the rat model[12] showed that the male rats produced

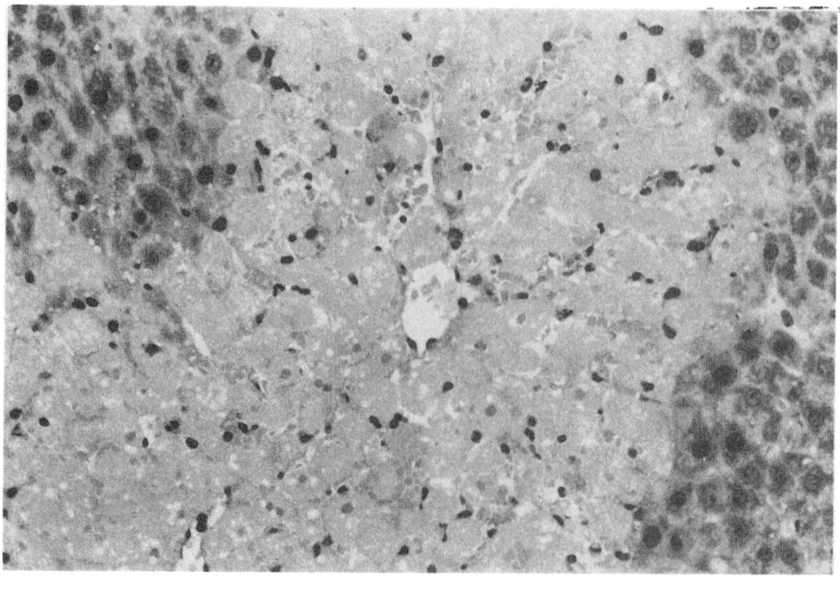

Figure 5. Light micrograph of a liver section, from a phenobarbitone-induced male rat killed 24 hr after 1% halothane in 14% O_2 for 2 hr. An extensive confluent zone of liver cell necrosis surrounds a central vein. H&E × 264.

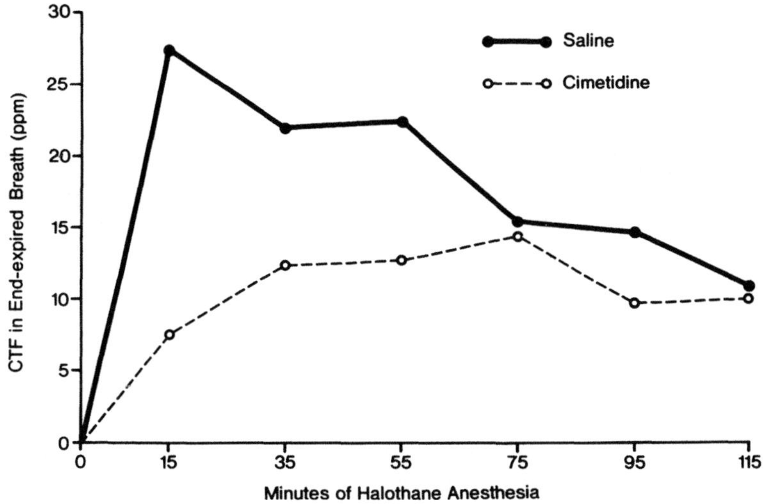

Figure 6. The effect of cimetidine pretreatment on the amount of reductive metabolism of halothane. The amount of the reductive metabolite CTF, plotted against time, is markedly decreased by cimetidine in comparison to saline pretreatment.

2–4 times as much reductive metabolites, e.g., volatile metabolite (CTF) as shown in Fig. 3, as do female rats. Female rats had no damage, mild focal necrosis of liver cells, or at the most confluent necrosis around the central veins extending only two or three cells from the vein wall (Fig. 4); this was in marked contrast to male rats, all of whom had confluent necrosis of zone 3 (Fig. 5).

Direct confirmation of the role of reductive metabolism is provided by pretreatment with cimetidine, which selectively decreases the amount of reductive metabolism of halothane (Fig. 6) and gives partial protection against halothane hepatotoxicity.[11]

We have shown that free radicals (short-lived intermediates capable of damaging macromolecules) are formed during the reductive metabolism of halothane.[10] Trudell *et al.*[14] have shown that human cytochrome P-450 *in vitro* can form a free radical from halothane and that this free radical can bind to phospholipids. Damage to proteins and lipids in organelle

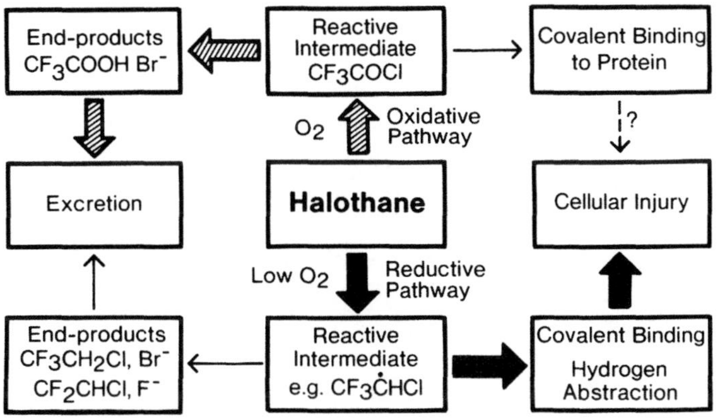

Figure 7. Relationship of halothane metabolism to hepatotoxicity in the hypoxic rat model. Reductive metabolism of halothane (dark arrows) results in the formation of potentially toxic reactive intermediates which may initiate cell damage. Reproduced with the permission of the American Society of Anesthesiologists from Cousins.[3]

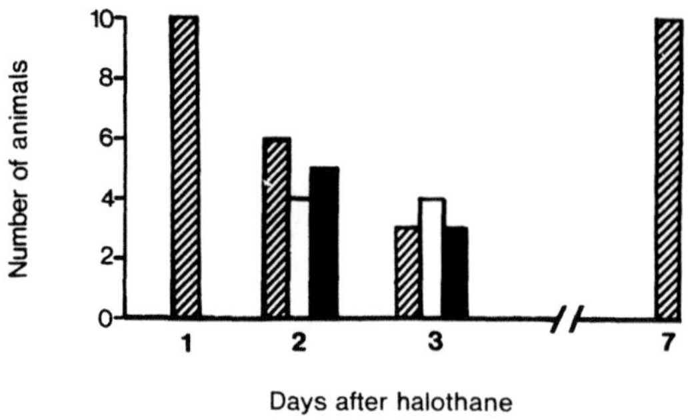

Figure 8. The time course, in guinea pigs, of the development of and recovery from halothane-induced liver injury. The hatched bars indicate the number of animals with normal livers, the open bars indicate animals with mild focal necrosis of liver cells, and the solid bars indicate the number of animals with marked liver injury, i.e., zone 3 degeneration and necrosis.

membranes, by free radicals or other intermediates of halothane metabolism, may be the initial event leading to liver cell necrosis.

This rat model has been most useful in the study of metabolite-mediated halothane hepatotoxicity. The proposed steps in the pathogenesis of halothane hepatotoxicity in this rat model are shown in Fig. 7.

GUINEA PIG MODEL

Methods

Outbred male guinea pigs developed liver damage after a 4-hr halothane anesthesia in 21% O_2 (Fig. 8). Approximately two-thirds of the animals were susceptible to halothane hepatotoxicity. Light microscopic examination of liver sections revealed two distinct distributions of liver necrosis: marked damage—centrilobular necrosis extending with a stellate distribution to the portal tracts, i.e., necrosis of zone 3 (Fig. 9); and mild damage—several

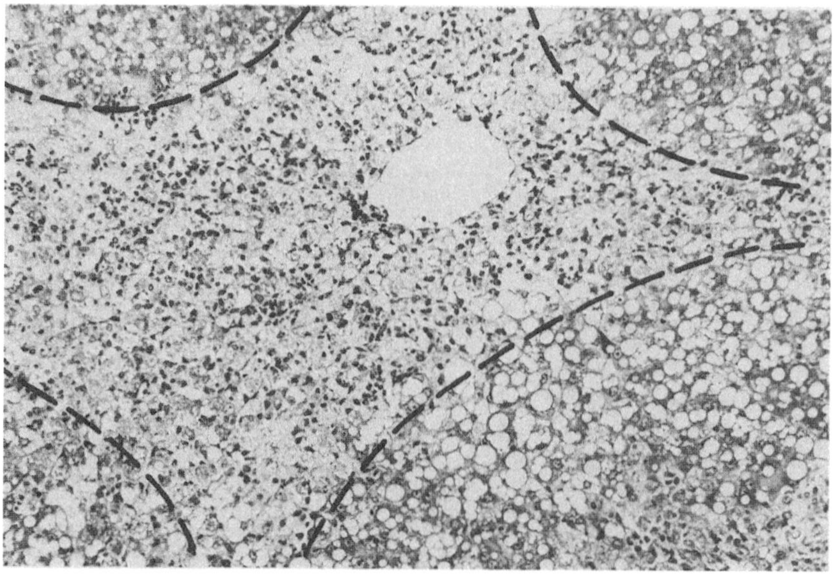

Figure 9. Light micrograph of a liver section, from a guinea pig killed 2 days after receiving 1% halothane in 21% O_2 for 4 hr. The liver injury is confluent around the central vein and extends into the adjacent liver parenchyma (outlined by broken black lines). The surviving hepatocytes show fatty change. H&E × 125.

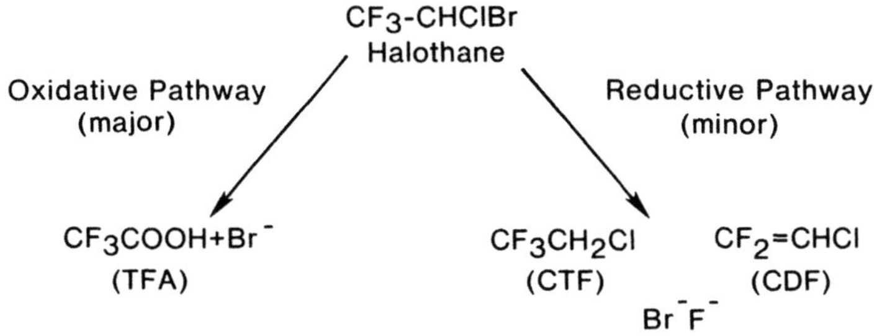

Figure 10. Oxidative and reductive pathways of halothane metabolism, in the guinea pig.

foci of necrosis within the liver loubules, in a nonzonal distribution. Necrosis occurred within 48 hr posthalothane and by day 7 after anesthesia was no longer apparent. The metabolism of halothane by the guinea pig (Fig. 10) and the distribution of marked liver injury are both similar to those observed in man. Animals that developed liver injury following halothane anesthesia all developed a similar degree of liver damage when a second halothane anesthetic was given after a recovery period. In contrast, animals that did not develop liver damage with the first anesthetic did not develop liver injury on reexposure.[2]

Hemodynamic Changes during Anesthesia

Halothane and isoflurane at equianesthetic concentrations caused similar falls in mean arterial pressure. Blood gases were similar during anesthesia with the two agents. However, in contrast to halothane, isoflurane was not hepatotoxic.[8] It seems unlikely that halothane-associated liver injury is due to anesthesia-induced hypotension; however, the possibility that halothane may reduce total blood flow to the liver and/or cause redistribution of blood flow within the liver has yet to be investigated.

The Role of Halothane Metabolism in Liver Injury

Inhibition of hepatic mixed-function oxidases by agents such as SKF-525A significantly decreases halothane metabolism. Table I shows a decrease in the extent of reductive metabolism as indicated by the decreased amounts of the volatile metabolite CTF. Pretreatment with SKF-525A also reduces the incidence and severity of liver damage (Table II), indicating that metabolism plays a role in the pathogenesis of halothane hepatotoxicity in the guinea pig. However, as only some animals develop liver injury, other factors, including individual

Table I. The Effect of SKF-525A on the Reductive Metabolism of Halothane

	CTF (ppm)
1% halothane, 21% O_2–78% N_2, for 4 hr	2.6 ± 0.1[a]
SKF-525A, 125 mg/kg, i.p. 1 hr before 1% halothane for 4 hr	1.3 ± 0.2

[a] $p < 0.001$ versus SKF-525A pretreatment (Student's unpaired t-test). Values are means and S.E.M., for 20 guinea pigs.

Table II. Effect of Pretreatment with SKF-525A on Incidence and Severity of Halothane Hepatotoxicity[a]

	Histologic grade of damage		
	None	Mild	Marked
1% halothane 4 hr, in 21% O_2–78% N_2	$n = 10$	$n = 4$	$n = 6$
SKF-525A, 125 mg/kg, i.p. 1 hr before 1% halothane for 4 hr	$n = 16$	$n = 4$	$n = 0$

[a] *n* is the number of guinea pigs.

susceptibility, which is possibly genetically determined, and an immunological response may contribute to liver injury.

Ultrastructural Membrane Changes associated with Halothane-Induced Liver Damage

In freeze-fracture replicas, the tight junctions between hepatocytes of guinea pigs anesthetized with isoflurane were similar to those of untreated controls. These tight junctions consisted of a narrow network of three to five interconnecting strands, oriented parallel to the lumen of the bile canaliculus. Tight junctions of livers of some guinea pigs that had received halothane were modified in several regions, showing disarray with strands which were loosened and fragmented. These tight junctions invariably contained regions in which only one or two strands were seen.[7] The fragmentation and decreased number of junctional strands suggests that the altered tight junctions may become "leaky" allowing blood and bile constituents to mix, thereby contributing to liver damage.

CONCLUSION

Evidence from our studies indicates that the etiology of halothane hepatotoxicity is multifactorial, with halothane metabolism playing a central role. We propose that individual differences in the amount of reductive metabolism of halothane and/or in susceptibility of the liver to toxic intermediates, following either a single or repeated halothane anesthesias, are important factors in the development of liver damage. Such individual differences may also help explain the rarity of halothane hepatotoxicity in humans.

ACKNOWLEDGMENTS. This work was supported by National Health and Medical Research Council Grants.

REFERENCES

1. Brown BR, Sipes IG: Biotransformation and hepatotoxicity of halothane. *Biochem Pharmacol* 1977;26:2091–2094.
2. Cousins MJ, Sharp JH, Gourlay GK, Adams JF, Haynes WDG, Whitehead R: Hepatotoxicity and halothane metabolism in an animal model with application for human toxicity. *Anaesth Intens Care* 1979;7:9–24.
3. Cousins MJ, Plummer JL, Hall P de la M: Mechanisms and evaluation of hepatotoxicity, in, Barash PG (ed): *Refresher Courses in Anaesthesiology,* Vol 13. Philadelphia, J B Lippincott C, 1985, in press.
4. Gourlay GK, Adams JF, Cousins MJ, Hall P: Genetic differences in reductive metabolism and hepatotoxicity of halothane in three rat strains. *Anaesthesiology* 1981;55:96–103.

5. Hall P, Cousins MJ, Knights KM, Gourlay G: Halothane hepatitis in an animal model: Time course of hepatic injury. *Hepatology* 1982;2:131.
6. Knights KM, Gourlay GK, Cousins MJ: Changes in rat hepatic microsomal mixed-function oxidase activity following exposure to halothane. *Clin Exp Pharmacol Physiol* 1983;10:655.
7. Lunam CA, Hall P, Cousins MJ, Dwarte D: Reorganization of tight junctions in the guinea pig liver following halothane anaesthesia. *Micron* 1982;13:361–362.
8. Lunam CA, Hall P, Cousins MJ: Cardiovascular and hepatic effects of halothane and isoflurane in the guinea pig model of halothane hepatitis. *Clin Exp Pharmacol Physiol* 1983;10:726.
9. Lunam CA, Cousins MJ, Hall P de la M: A guinea pig model of halothane-associated hepatotoxicity. *J Pharmacol Exp Ther* 1985;232:802–809.
10. Plummer JL, Beckwith ALJ, Bastin FN, Adams JF, Cousins MJ, Hall P de la M: Free radical formation *in vivo* and hepatotoxicity due to anesthesia with halothane. Anaesthesiology 1982;57:160–166.
11. Plummer JL, Wanwimolruk S, Jenner MA, Hall P de la M, Cousins MJ: Effects of cimetidine and ranitidine on halothane metabolism and hepatotoxicity in an animal model. *Drug Metab Disposit* 1984;12:106–110.
12. Plummer JL, Jenner MA, Cousins MJ, Hall P de la M: Sex differences in halothane metabolism and heptatotoxicity in a rat model. *Anesthesia and Analgesia* 1985;64:563–569.
13. Shingu K, Eger EI, Johnson BH: Hypoxia *per se* can produce hepatic damage without death in rats. *Anesth Analg* 1982;61:820–823.
14. Trudell JR, Bosterling B, Trevor A: 1-Chloro-2,2,2-trifluorethyl radical: Formation from halothane by human cytochrome p 450 in reconstituted vesicles and binding to phospholipids. *Biochem Biophys Res Commun* 1981;102:372–377.
15. Wood M, Berman ML, Harbison RD, *et al.*: Halothane-induced hepatic necrosis in triiodothyronine-pretreated rats. *Anaesthesiology* 1982;52:470–476.

38

Evidence for a Role of the Immune System in the Pathogenesis of Halothane Hepatitis

Andrea H. Callis, Stephen D. Brooks, Stephen J. Waters, David O. Lucas, A. Jay Gandolfi, I. Glenn Sipes, Hiroko Satoh, and Lance R. Pohl

ABSTRACT. The hepatotoxicity of halothane in animal models is via its bioactivation to a reactive intermediate. Clinically, however, the hepatic injury mimics an immunological hypersensitivity response. This study determines whether the host develops an immune sensitivity to a self-component through its conjugation with a reactive intermediate of halothane (trifluoroacetyl chloride, TFA-). TFA- conjugated with rabbit serum albumin (RSA) was used as a tool to detect antibody that was specific or cross-reacted with this complex. Any antibody directed against this antigen was quantified by an enzyme-linked immunosorbent assay. Blood samples were obtained from rabbits that were repeatedly induced with phenobarbital alone or with phenobarbital and beta-naphthoflavone and exposed to halothane under hypoxic conditions. Blood samples were also obtained from halothane hepatitis patients. Rabbits that had been exposed to halothane three or more times had antibodies to TFA–RSA at levels three- to fourfold above control animals. Elevated serum glutamate pyruvate transaminase levels were also found, but they did not correlate with the number of exposures, the type of induction, or the TFA–RSA antibody titer. High levels of antibodies to TFA–RSA were found in the halothane hepatitis patients up to 4 years postexposure. Rabbits immunized with the TFA–RSA developed high antibody titers but were not found to be more prone to halothane-induced hepatotoxicity. This study demonstrates for the first time that halothane exposure can produce in a laboratory animal and in humans an antibody against a covalently bound oxidative intermediate of halothane that has altered a host component into a non-self-immunogen.

INTRODUCTION

There is considerable evidence for the direct hepatotoxic action of halothane in laboratory animals. This toxicity is mediated by the metabolism of halothane to reactive intermediates. These reactive intermediates covalently bind to subcellular macromolecules and initiate the sequence of events leading to the hepatocellular necrosis.[14] However, the hepatic damage produced by halothane in these animal models is only a mild, transient injury.[1] This is

Andrea H. Callis, Stephen D. Brooks, Stephen J. Waters, David O. Lucas, A. Jay Gandolfi, and I. Glenn Sipes • Department of Anesthesiology, University of Arizona, Tucson, Arizona 85724. *Hiroko Satoh and Lance R. Pohl* • Laboratory of Chemical Pharmacology, National Institutes of Health, Bethesda, Maryland 20205.

contrary to the hepatic complications observed in humans, where a delayed progressive injury is observed, which may culminate in fulminant hepatic necrosis and death. Epidemiological data indicate that the hepatic injury in humans does not occur until 4–10 days after anesthesia, that patients with multiple exposures have a higher incidence of hepatitis, and that the time between exposure and evidence of hepatic injury is decreased with increasing exposures. This evidence suggests that the hepatotoxicity of halothane results from an immunological hypersensitivity response.[2]

Convincing evidence of sensitization to halothane has proven difficult to obtain. In early studies it was demonstrated that in animals exposed to halothane, lymphocytes became sensitized and underwent blast transformation[9] or produced a lymphokine that inhibited macrophage migration.[10] Similar results were demonstrated in an *in vitro* system where halothane was being metabolized. Liver homogenates from rabbits exposed to halothane stimulated sensitized lymphocytes to release macrophage inhibition factor.[15] Researchers rationalize that it was probably the trifluoroacetyl chloride intermediate of the oxidative metabolism of halothane that was reacting with host protein to form a hapten–carrier complex.[1]

A number of studies have been performed with synthetic trifluoroacetyl-carrier protein serving as an immunogen. The trifluoroacetyl moiety conjugated with albumin has been reported to be a weak immunogen.[7] Animals innoculated with this immunogen demonstrated an increased skin sensitivity to the trifluoroacetyl-carrier protein[5] and evidenced a humoral immune response as well.[7] When animals were challenged with this immunogen and then exposed to halothane, there was no observed increase in liver toxicity as determined by serum enzyme levels and histology.[3,6,11,12] These results do not rule out the role of an immune response in mediating the halothane-associated liver injury and do add to the confusion.

The strongest evidence supporting a role for the immune system in the expression of the halothane-induced liver injury is demonstrated in research using the sera from patients who have had liver complications following halothane anesthesia.[16] These studies detected a circulating antibody from patients with severe "halothane hepatitis" which reacted to the surface of halothane-altered rabbit hepatocytes. In an antibody-dependent cellular cytotoxicity assay, as well as in an indirect immunofluoresence test, only sera from the "halothane hepatitis" patients were positive with the hepatocytes from halothane-exposed rabbits.

These investigators also performed some preliminary studies in which they reported that the maximum antibody-dependent cellular cytotoxicity response was observed when the hepatocytes were obtained from rabbits induced with beta-naphthoflavone and exposed to halothane under a 99% oxygen atmosphere.[8] These conditions for the metabolism of halothane to toxic reactive intermediates are in contrast to conditions used by many researchers in which reductive metabolism of halothane produces hepatotoxicity in phenobarbital-induced animals.

In recent studies with the synthetic antigen, trifluoroacetyl–rabbit serum albumin (TFA–RSA), it was found that antibodies could be elicited against this antigen in rabbits.[13] These antibodies were specific for the trifluoroacetyl moiety since neither trifluoroacetic acid nor halothane could block the antibody binding. The antibody, however, could be blocked by trifluoroacetyl lysine but not by acetyllysine nor by rabbit serum albumin alone. Studies using these antibodies demonstrated the presence of the trifluoroacetyl moiety attached to cellular macromolecules in the centrilobular region of the liver from rats exposed to halothane. This antibody reaction was not present in unexposed rats and almost absent in rats exposed to deuterated halothane. Deuteration of halothane greatly decreased its oxidative metabolism suggesting that it is the oxidative intermediate of halothane that is binding to subcellular tissue in the centrilobular region. Additional studies with hepatocytes detected the presence of this trifluoroacetylated entity on the surface of hepatocytes exposed to halothane. These studies indicated the presence of a reactive intermediate of halothane attached to cellular macromolecules and that the entity has the potential of being recognized by the immune system.

Based on previous work demonstrating that patients with liver injury following halothane anesthesia have circulating antibodies that react with liver tissue which has metabolized halothane, and based on the fact that antibodies to a synthetic antigen, trifluoroacetyl–RSA, can detect this moiety in the livers of rats exposed to halothane, we proposed the following studies: (1) to determine whether antibodies to TFA–RSA are produced in rabbits following exposure to halothane, (2) to determine the specificity of these antibodies, (3) to determine whether similar antibodies to TFA–RSA are present in patients who have had liver injury following halothane anesthesia, and (4) to determine whether rabbits prechallenged with TFA–RSA are more susceptible to halothane-induced liver injury.

METHODS

Synthesis of TFA–RSA

Ethyl thioltrifluoroacetate was reacted with rabbit serum albumin to produce trifluoroacetyl rabbit serum albumin. As previously reported, the conjugation is through the e-amino amino group on the lysine amino acids in the albumin.[4] The conjugate is purified by isoelectric precipitation, ultrafiltration, and dialysis. Although the resulting product is not completely soluble in aqueous solutions, this does not appear to affect its immunogenicity or reactivity.

Inoculation of Rabbits with TFA–RSA

TFA–RSA (1 mg) was emulsified in 0.5 ml water and 1.5 ml of Freund adjuvant. The suspension was injected in 0.1 ml injections in the legs (i.m.) and back (s.c.) of 1.5-kg male New Zealand white rabbits. A booster immunization was repeated 4 weeks later with the same amount of TFA–RSA in adjuvant. Antibody titers were monitored by an enzyme-linked immunosorbent assay (ELISA) on weekly blood samples. These animals were also used in the prechallenge studies described later.

Rabbit Halothane Exposures

Male New Zealand white rabbits (1.5–2 kg) underwent one of two induction protocols: daily s.c. injections of phenobarbital (90 mg/kg) for 5 days or daily s.c. injections of phenobarbital (90 mg/kg; 5 days) combined with two s.c. injections of beta-naphthoflavone (80 mg/kg) on days 1 and 5. Twenty-four hours postinduction, the rabbits were exposed to halothane (1%) under hypoxic conditions (14%) for 2 hr. At 24, 48, or 144 hr postexposure blood samples were drawn from the central auricular artery. Hepatic toxicity was assessed by elevations in serum glutamate pyruvate transaminase (SGPT) levels. Plasma samples were stored at −70°C for subsequent analysis of TFA–RSA antibodies by ELISA. The rabbits were allowed an 11-day recovery period at which time the induction/exposure regimen was repeated.

To determine whether rabbits that were prechallenged with TFA–RSA had increased hepatotoxicity to halothane, prechallenged rabbits and untreated rabbits were induced with phenobarbital as described and exposed to halothane (1%) for 2 hr in either a 14% or 21% oxygen atmosphere. Blood samples were taken at 72, 96, or 168 hr post exposure and assayed for SGPT, serum fluoride, and antibodies to TFA–RSA.

Patient Plasma Samples

Blood samples were obtained from two advised and consenting patients who had had liver complications following halothane anesthesia. One patient was a woman (aged 45 years) with typical elevation of temperature and serum enzymes 4 days after halothane anesthesia. Blood samples were obtained at 11 days, 1 month, and 3 months after the episodes. The other patient was a man (aged 50 years) with halothane administration almost 5 years previously. Volunteers from the operating room (anesthesiologists) and laboratory served as donors of control blood samples.

Enzyme Linked Immunosorbent Assay (ELISA)

Antibody titers were measured by ELISA with TFA–RSA serving as the antigen. ELISA plates were coated with TFA–RSA (5.0 μg/well) and incubated overnight at 4°C. The following day, the plates were washed, and plasma from rabbits or humans was added. Each plasma sample was serially diluted and each dilution tested in triplicate in the ELISA to determine its titer. Following a 2-hr incubation, either horseradish peroxidase-conjugated goat antirabbit (IgG, IgM, IgA) or HRP goat antihuman (G, M, A) was added for a 1-hr incubation. The plates were again washed and the substrate for the peroxidase added (ABTS). The reaction was terminated with 5% SDS. The presence of antibody was determined spectrally with a Dynatech Microelisa Reader (410 nm).

In blocking studies, the blocking agent was added to an antiserum one dilution from its endpoint. The blocking agent and the diluted antiserum were incubated for 2 hr at 37°C and then overnight at 4°C. They were used the following day in an ELISA. A range of concentrations of the blocking agents was tested to show any dose–response relationship. ELISA readings were reported as absorbance at 410 nm at a specified dilution.

RESULTS

Antibodies in TFA–RSA Immunized Rabbits

TFA–RSA immunized rabbits developed an antibody titer to TFA–RSA as quantified by ELISA (Table I). Even though each rabbit was treated identically, there was a wide

Table I. Serum Titers in Rabbits Receiving TFA–RSA Innoculations

Animal no.	ELISA (410 nm)[a]
1	0.242 ± .022
2	0.978 ± .040
3	1.776 ± .033
4	0.207 ± .057
5	0.290 ± .016
6	0.792 ± .014

[a] 1:2000 dilution of serum. Control rabbits had no titer to TFA–RSA.

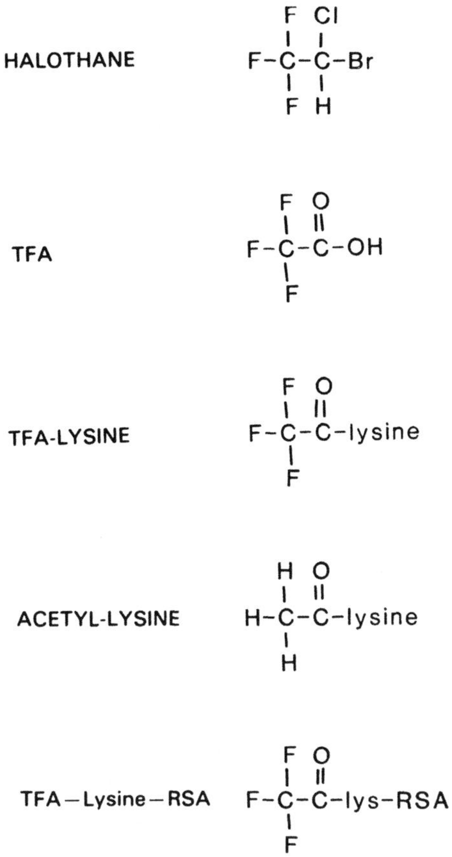

Figure 1. Structures of the TFA–RSA, halothane, and blocking agents.

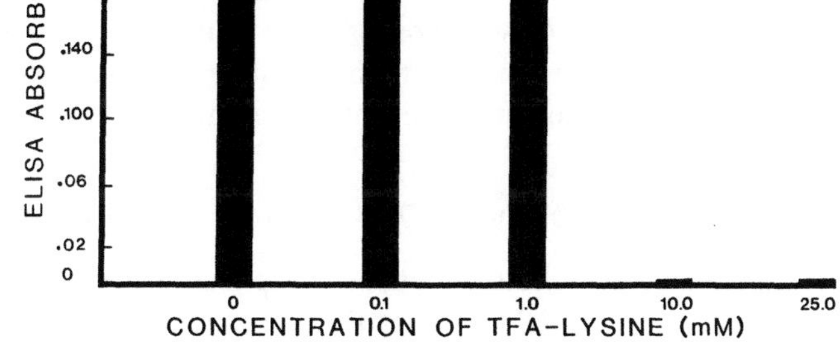

Figure 2. Inhibition of binding to TFA–RSA by TFA–lysine: inoculated rabbits. Plasma from rabbits inoculated 4 weeks prior with TFA–RSA was incubated with varying concentrations of TFA–lysine. This plasma was tested in an ELISA for its ability to bind to TFA–RSA coated wells.

Table II. Presence of Antibody to TFA–RSA in Multiple Exposed Rabbits and "Halothane Hepatitis" Patients

Plasma source	TFA–RSA Antibody[a]
Rabbit (PB/BNF)	1.7 (2)
Rabbit (PB/BNF)	3.0 (3)
Rabbit (PB)	2.3 (+4)
Halothane patient A	4–5
Halothane patient B	3

[a]Fold increase above normal control plasma. ( ), number of halothane exposures.

variation in this humoral response. The antibody titer was maximal 1–2 weeks after the booster immunization and remained elevated for several weeks with only a small decrease in the titer.

This antibody was determined to be specific for the trifluoroacetyl moiety (Fig. 1) since it could not be blocked by halothane or trifluoroacetic acid. Thus, the antibody recognized the intermediary product between halothane and trifluoroacetic acid. Trifluoroacetyl lysine was used to block the antibody-antigen reaction (Fig. 2), since the conjugation of the trifluoroacetyl group was through the lysine constituents on the albumin. Whereas trifluoroacetyl lysine was capable of blocking the binding of antibody to TFA–RSA, lysine, acetyl lysine, and rabbit serum albumin were ineffective. Thus, the antibody directed against TFA–RSA specifically recognizes the trifluoroacetyl entity as the hapten. Trifluoroacetyl lysine was used throughout the study to demonstrate the specificity of any antibody–antigen reaction.

Presence of Antibody to TFA–RSA in Multiple Halothane-Exposed Rabbits

Antibody to TFA–RSA was present in rabbits exposed repeatedly to halothane (Table II). The antibody titer increased twofold between two and three exposures. Although the titer was highest in rabbits induced with both phenobarbital and beta-naphthoflavone, phenobarbital-induced rabbits also produced antibody to TFA–RSA. After every exposure, the rabbits were monitored for increases in SGPT as a marker of liver injury (data not shown). All the rabbits had elevated SGPT, but there was no relationship between the increase in SGPT and either the number of exposures or the antibody titer to TFA–RSA.

Although the antisera from the multiple halothane-exposed rabbits did indeed react with TFA–RSA in the ELISA assay, the reaction could not be blocked by TFA–lysine (Fig. 3). This may indicate a heterogeneity in antigen specificity in antibodies directed against TFA–RSA in the animals. The affinity of certain populations of antibodies may vary for TFA–lysine allowing the dissociation of these complexes in the presence of TFA–RSA. Alternatively, the antigen binding site may not be fully saturated with this small hapten moiety, TFA–lysine, thus not fully preventing binding to TFA–RSA.

Presence of Antibody to TFA–RSA in "Halothane Hepatitis" Patients

Both patients who had hepatic complications following halothane anesthesia had antibody titers to TFA–RSA (Table II). The patient with the most recent anesthesia had a high

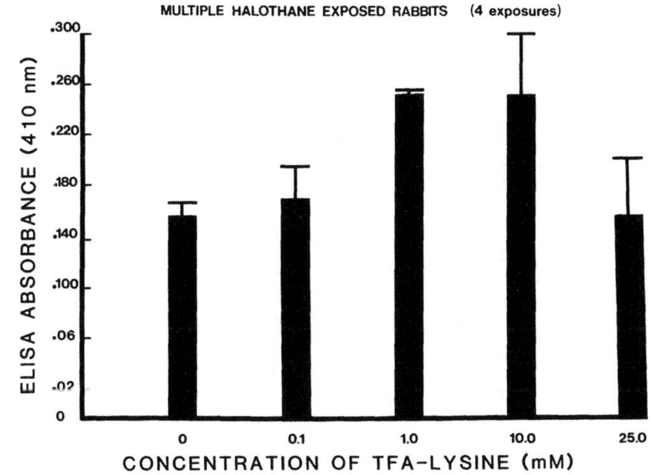

Figure 3. Inhibition of binding to TFA–RSA by TFA–lysine: multiple exposed rabbits. Plasma from rabbits exposed to halothane 4 times at 16-day intervals was incubated with varying concentrations of TFA–lysine. This plasma was tested in an ELISA for its ability to bind to TFA–RSA coated wells.

titer by 11 days after anesthesia, which increased over the next few weeks. Although the antibody level in the other patient was not as high, this lower titer may reflect the length of time (4 years) since the last exposure.

Studies were conducted to ascertain that the patient's antibody was reacting with the TFA entity and not just the RSA. When RSA and TFA–RSA were tested in the ELISA assay with serial dilutions of the patient's antiserum, response to RSA was readily eliminated by dilution, whereas maximal response to TFA–RSA was still present (Fig. 4). In fact, in the patient's antisera collected within 2 weeks to 3 months after halothane exposure, TFA–lysine caused a dose-related inhibition of the antibody binding to TFA–RSA (Fig. 5) comparable to the inhibition seen with the rabbits immunized with TFA–RSA (Fig. 2).

Effect of TFA–RSA Immunization on Halothane Hepatotoxicity

Both unimmunized rabbits and TFA–RSA immunized rabbits had increases in SGPT following exposure to halothane in either a 14% or a 21% oxygen atmosphere (Table III).

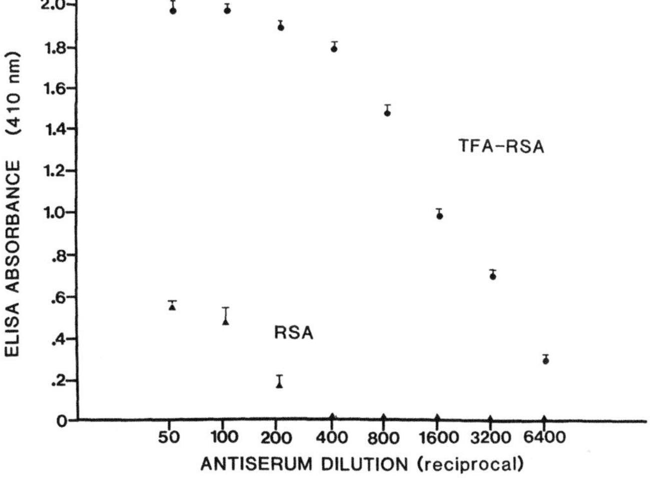

Figure 4. Presence and specificity of halothane hepatitis patient's antiserum. Serum from a halothane hepatitis patient was tested in an ELISA for binding to either TFA–RSA or RSA-coated wells. Binding to RSA was negated at a 1/400 dilution, whereas significant binding to TFA–RSA remained at a 1/6400 dilution.

Table III. Effect of TFA–RSA Innoculation on Halothane Hepatotoxicity

Rabbit no.		TFA–RSA Ab[a]	Exposure 1[b] (SGPT[d])	Exposure 2[c] F$^-$(μM)[e]	Exposure 2[c] SGPT[f]
Control	1	0	71	40	100
	2	0	140	48	700
	4	0	62	36	115
TFA–RSA	1[g]	0.560	86	24	75
	2	1.690	107	30	45
	4	0.420	112	65	345
	6	1.130	48	34	105

[a]Antibody titer to TFA–RSA; ELISA (410 nm), 1:1000 dilution.
[b]1% Halothane, 14% O_2, fasted 24 hr, 2 hr exposure.
[c]1% Halothane, 21% O_2, unfasted, 2 hr exposure.
[d]4 days postexposure.
[e]Immediately postexposure.
[f]Three days postexposure; units/ml.
[g]TFA–RSA: rabbits innoculated with TFA–RSA.

Both groups also produced equivalently high concentrations of fluoride in their serum at the end of exposure. However, there was no difference noted between the two groups in relation to these hepatotoxicity profiles. The SGPT scores were elevated for a longer period of time than in previous halothane studies with rats, but the lesion was completely resolved by 7 days postexposure. Histological examination of the livers following these exposures has not yet been performed.

DISCUSSION

This study demonstrates that halothane exposure can result in the production of an antibody directed against the biotransformation intermediate that produces the oxidative

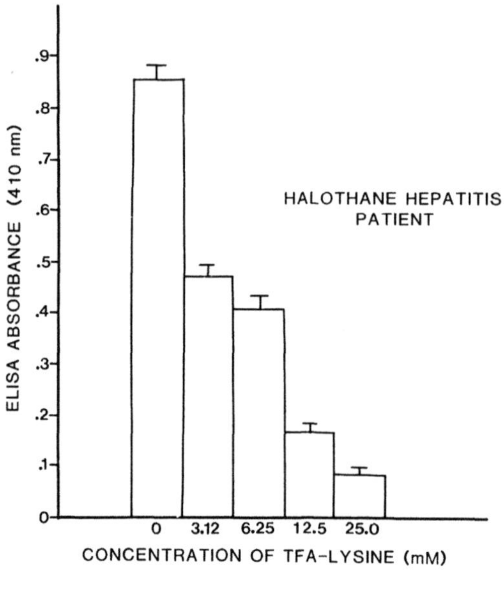

Figure 5. Inhibition of binding to TFA/RSA by TFA–lysine: halothane patient's serum. Serum collected 11 days after the halothane hepatitis episode was incubated with varying concentrations of TFA–lysine. This serum was then tested in an ELISA for binding to TFA–RSA coated wells.

metabolite, TFA. This antibody is specific for the TFA moiety attached to a macromolecule. The antibody production occurs both in laboratory animals exposed to halothane and in human patients who have had halothane-induced liver complications. This is the first such demonstration that halothane exposure can produce an immunological response against a specific, synthetic antigen.

The results presented here complement previous studies using sera from "halothane hepatitis" patients.[15] Sera from these patients was found to produce an antibody-dependent cellular cytotoxicity when reacted with hepatocytes from animals exposed to halothane. The antibodies in the patients' antisera were recognizing the halothane-altered hepatocyte membranes. Those investigators also noted that patients' antisera had a strong reaction with hepatocytes from animals exposed to halothane under oxidative conditions,[8] corresponding with the TFA conjugate used in these studies.

The recent report by Satoh *et al.*[13] demonstrated that TFA–RSA was a strong immunogen. This information can be integrated with the previous reports using antisera from human halothane hepatitis patients[16] and the studies presented here. Satoh *et al.*[13] found that antibody directed against TFA–RSA would react with the centrilobular region of the liver of rats exposed to halothane. If the rats were exposed to deuterated halothane to inhibit the oxidative metabolism of halothane, there was little antibody reaction seen. In addition, the antibody to TFA–RSA was found to bind to the membrane surface of hepatocytes from the halothane-exposed rats. These results indicate that the antibody to TFA–RSA is recognizing a component in the altered hepatocyte membrane which presumably results from the covalent binding of the TFA intermediate metabolite to a membrane protein. In the previous reports with sera from halothane hepatitis patients, this same process was proposed to explain the results.

An altered hepatocyte membrane resulting from the bioactivation of halothane to TFA and its covalent binding to cellular protein may elicit a humoral immune response. Our studies demonstrated that in both laboratory animals and halothane hepatitis patients, an antibody to this covalently bound reactive intermediate was present. In fact, in the majority of the cases the antibody binding was readily blocked only by trifluoroacetyl–lysine. The lack of specific blocking in the rabbits that were exposed repeatedly to halothane might be explained by the exposure inducing subpopulations of antibodies against TFA–conjugated to a variety of hepatocellular proteins. The TFA–RSA is readily recognized by all the antibodies, but TFA–lysine may have neither the affinity nor the size to block all the binding sites on the antibodies.

The lack of potentiation of the hepatotoxicity of halothane by prechallenging the rabbits with TFA–RSA confirms earlier studies.[3,6,11,12] In those studies other laboratory animals were used (rat and guinea pig) and either the animals were subjected to multiple halothane exposures or the animals were induced with phenobarbital and exposed to halothane in a 14% oxygen atmosphere. Thus, at this stage it appears that prechallenging with TFA conjugates does not enhance the subsequent hepatotoxicity of halothane. This does not rule out that an immunological hypersensitivity reaction is an integral component in the halothane hepatotoxicity observed in patients. In fact, our preliminary halothane hepatotoxicity studies with these prechallenged rabbits indicate that they may even be protected from the hepatotoxicity of halothane by the prechallenge with TFA–RSA.

There are numerous avenues of research to pursue in ultimately determining the role of the immune system in halothane hepatotoxicity. Although we have a defined hapten-carrier protein immunogen, do we have the correct immunogen? Conjugates with other reactive intermediates of halothane should also be synthesized to determine whether only TFA- is an immunological hapten. How long after immunization should the animal be exposed to halothane? In other words, how much of an existing titer to a metabolite is required for potentiation? At present, the halothane exposures have been designed to produce the mild transient liver injury observed in past studies. It is unclear whether the liver should be insulted to a maximum level of cellular damage or whether a milder injury would be more appropriate to sensitize the immune system and cause a more progressive injury. In addition, immunostimulators may well be needed to augment the immune system and enhance its reactivity. The scenario observed in the clinical setting following halothane-associated liver injury may well be unique to man and may not be readily mimicked in laboratory animals.

Finally, the real question as to whether there is any role for the immune system in halothane-associated liver injury should be addressed. All the existing animal models of halothane liver injury produce only a mild transient necrosis. The chronology for the injury is completely different from that seen in "halothane hepatitis" patients. Although these models are convenient and reproducible, they are not addressing the true problem. Care, too, must be taken in interpreting the immunological hypersensitivity data. The principle question that arises is whether the formation of antibodies in response to the halothane exposures is a "result" of the liver injury or in some way a "cause" of the liver injury. Answering this "chicken or the egg" question appears to be the charge for anesthetic hepatotoxicity researchers in the 1980s.

ACKNOWLEDGMENT. The research was supported by NIH Grant No. AM 16715 and GM 34788. The authors express their gratitude to Stephen T. Weldon for his help in preparing the manuscript.

REFERENCES

1. Davis M, Vergani D, Eddleston A, Williams R: Halothane hepatitis—Toxicity and immunity, in Eddleston A, Weber W, Williams R (eds): *Immune Reactions in Liver Disease*. London, Pitman Medical Limited, 1979, pp 235–246.
2. Dienstag JL: Halothane hepatitis, allergy or idiosyncracy? *N Engl J Med* 1980;303:102–104.
3. Ford D, Coyle D, Harrington FJ: Effects of hypersensitivity to a halothane metabolite on halothane-induced liver damage. *Anesthesiology* 1984;60:141–143.
4. Goldberger R, Anfinsen C: The reversible masking of amino groups in ribonuclease and its possible usefulness in the synthesis of the protein. *Biochemistry* 1962;1:401–405.
5. Mathieu A, DiPadua D, Mills J, Kahan B: Experimental immunity to a metabolite of halothane and fluroxene. Cutaneous delayed-type hypersensitivity. *Anesthesiology* 1974;40:385–390.
6. Mathieu A, DiPadua D, Kahan BD, Galdabini JJ, Mills J: Correlation between specific immunity to a metabolite of halothane and hepatic lesions after multiple exposures. *Anesth Analg* 1975a;54:332–339.
7. Matheiu A, DiPadua D, Kahan BD, Mills J: Humoral immunity to a metabolite of halothane, fluroxene, and enflurane. *Anesthesiology* 1975b;42:612–616.
8. Neuberger J, Mieli-Vergani G, Tredger J, Davis M, Williams R: Oxidative metabolism of halothane in the production of altered hepatocyte membrane antigens in acute halothane-induced hepatic necrosis. *Gut* 1981;22:669–672.
9. Paronetto F, Popper H: Lymphocyte stimulation induced by halothane in patients with hepatitis following exposure to halothane. *N Engl J Med* 1970;283:277–280.
10. Price C, Gibbs A, Williams W: Halothane macrophage migration inhibition factor test in halothane-associated hepatitis. *J Clin Pathol* 1977;30:312–316.
11. Reves J, McCracken L: Failure to induce hepatic pathology in animals sensitized to a halothane metabolite and subsequently challenged with halothane. *Anesth Analg* 1976a;55:235–242.
12. Reves J, McCracken L: Hepatic pathology and skin test reactions to trifluoro-acetylated autologous protein after repeated halothane anesthesia in the guinea pig. *Br J Anaesth* 1976b;48:419–424.
13. Satoh H, Pohl L, Anderson D, Fukuda Y, Ferrans V, Gillette J: Immunological studies on the mechanism of halothane-induced hepatotoxicity: Immunochemical evidence of trifluoroacetylated hepatocytes. *Fed Proc* 1984;43:934.
14. Sipes IG, Gandolfi AJ, Brown B: Halothane-associated hepatitis: Evidence for direct hepatotoxicity, in Davis

M, Tredger JM, Williams R (eds): *Drug Reactions and the Liver*. London, Pitman Medical Limited, 1981, pp 219–231.

15. Vergani D, Eddleston A, Tsantoulas D, Davis M, Williams R: Sensitization to halothane-altered liver components in severe hepatic necrosis after halothane anesthesia. *Lancet* 1978;1:801–803.
16. Vergani D, Mieli-Vergani G, Alberti A, Neuberger J, Adrian L, Eddleston A, Davis M, Williams R: Antibodies to the surface of halothane-altered rabbit hepatocytes in patients with severe halothane associated hepatitis. *N Engl J Med* 1980;303:66–71.

Appendix

Tables of Physical and Pharmacological Properties of Anesthetics

Leonard L. Firestone, Janet C. Miller, and Keith W. Miller

INTRODUCTION

These tables bring together physical and pharmacological properties of 47 general anesthetics. Under each anesthetic will be found the physical properties of the agent itself, followed by those physical properties which involve interactions with other substances, and finally the pharmacological properties are listed.

We have made use of secondary sources of data in many cases. This reduces the work of compilation and provides the reader with references to useful sources with which he may be unfamiliar. In many cases such sources contain additional information not listed here, but they also carry the risk of propagating errors. In other cases not all the data have been compiled previously, and original articles are cited.

Agents have been chosen for their theoretical importance, their clinical usefulness, or because they are commonly used in research. In some cases we have omitted agents for which little information is available.

The tables are not intended to be encyclopedic. We have chosen to collect properties that are useful in a practical or theoretical sense. Often we have passed over data because only a few agents have been examined. To keep the tables compact, we have necessarily made some arbitrary choices, but we are willing to entertain suggestions for future editions of these tables.

NOTES ON USE OF TABLES

Most of the entries in the tables should be self-explanatory, but the notes below contain important definitions and qualifications.

Temperature is always given in degrees centrigrade unless otherwise noted. In many cases properties depend on temperature, and the value given is then followed by the temperature it was obtained at.

In the specific case of vapor pressure, the constants of the Antoine equation[62] are given. This equation predicts the vapor pressure (in units of mm Hg or atmospheres) over a limited range of temperature only; check the source before extrapolating wildly[62];

$$\text{Log (vapor pressure)} = A - [B/(t + C)] \tag{1}$$

Leonard L. Firestone and Keith W. Miller • Departments of Pharmacology and Anaesthesia, Harvard Medical School and Massachusetts General Hospital, Boston, Massachusetts 02114. *Janet C. Miller* • Department of Anesthesia, Massachusetts General Hospital, Boston, Massachusetts 02114.

where t is temperature (centrigrade), and A, B, and C are constants which are given in the tables.

Second virial coefficients (B) are given in ml/mole. They express the deviation from perfect gas behavior according to

$$PV/RT = 1 + B/V \tag{2}$$

where T is temperature in degrees Kelvin and the other symbols have their usual meaning. Such deviations are around 5% for many volatile liquid anesthetics.

For gaseous and volatile anesthetics, partition coefficients are given relative to the gas phase. The partition coefficient quoted is the Bunsen partition coefficient, which is defined as the liters of gas (expressed at 1 atm and 0°C) dissolved in 1 liter of solvent when the partial pressure of the gas above it is 1 atm. The Ostwald partition coefficient is similarly defined except that the volume of gas is expressed at the temperature of the experiment.

The limiting solubility in water of a volatile liquid can be calculated from the Bunsen coefficient since the vapor pressure at the given temperature is the maximum partial pressure obtainable. It is usually reasonably accurate to assume Henry's law applies provided concentrations are in the millimolar range. We have also quoted some saturated solubilities obtained empirically in *Torpedo* Ringer (NaCl 250 mM, KCl 5 mM, $CaCl_2$ 3 mM, Na_2PO_4/$NaHPO_4$ 5 mM, at pH 7.0) solutions in our laboratory. In general, the solubility of anesthetics in saline solution is less than that in water by 10–20%.[6,64]

All partition coefficients contained in the tables are values determined experimentally and are not the result of extrapolations from measurements obtained in a few members of chemically related series. Partition coefficients between two liquid phases are given in units of (moles per liter in solvent A)/(moles per liter in solvent B), except that in biological membranes most authors adopt (moles per kilogram membrane). The density of the red cell membrane is 1.2 g/ml,[31] while that of lipid bilayers is about 1.0 g/ml. Only in the former case is the difference appreciable. Phosphotidylcholine/cholesterol bilayers are usually made in a 2 : 1 molar ratio.

For barbiturates, the charged and uncharged moieties have very different partition coefficients, and the pH of the experiment relative to the pK of the compound must be considered.[4,59] In the following tables, the octanol/water (buffer) partition coefficients for barbiturates are given uniformly for the acid form unless otherwise stated, but if in doubt, check the original papers carefully. Partition coefficients which were measured at room temperature are indicated by R.T.

The formulae of the barbiturates are given by defining the radicals X, R_1, R_2, and R_3.

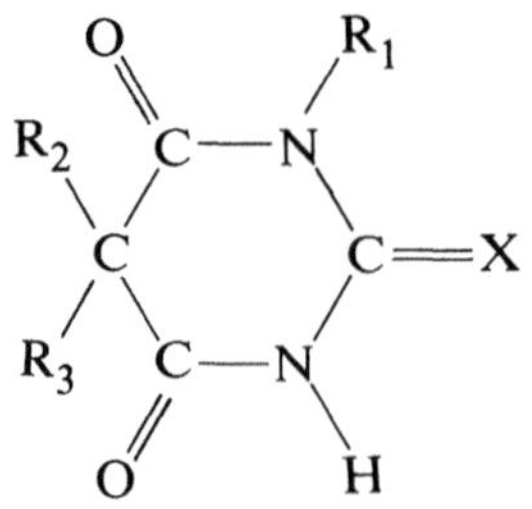

The pKs of barbiturates depend on the ionic strength (μ) of the buffer and the temperature.[27] We have quoted a value at low or zero ionic strength and a value corresponding approximately to physiological saline where possible.

Lastly, the ED50s reported are those that cause loss of righting reflex (LRR), unless specifically noted otherwise [i.e., loss of response to painful stimuli (MAC)]. For a discussion of the relationships between LRR and other anesthetic endpoints the reader is referred to Reference 78.

REFERENCES

1. Secher O: *Acta Anaes Scand* 1971;42(Suppl):1–93
2. Miller KW, Smith EB, in Featherstone RM (ed): *A Guide to Molecular Pharmacology and Toxicology*, Pt. II. New York, Marcel Dekker, 1973,pp 427–476.
3. Miller KW, Hammond L, Porter EG: *Chem Phys Lipids* 1977;20:229–241.
4. Miller KW, Yu S-C: *Br. J. Pharmacol* 1977;61:57–63.
5. Rogers RC, Hill GE: *Br J Anaesth* 1978;50:415–423.
6. Smith RA, Porter EG, Miller KW: *Biochim Biophys Acta* 1981;645:327–338.
7. *Handbook of Chemistry and Physics*, Ed. 62. Cleveland, OH, Chemical Rubber Company Press, 1981.
8. Bell GH: *Chem. Phys. Lipids* 1973;10:1–10.
9. Dymond JH, Smith EB, *The Virial Coefficients of Pure Gases and Mixtures*. Oxford, Oxford University Press, 1980.
10. Haydon DA, Urban, BW: *J. Physiol.* 1985; in press.
11. Pringle MJ, Brown KB, Miller KW: *Mol Pharmacol* 1981;19:49–55.
12. Pang K-Y, Braswell LM, Chang L, Sommer TJ, Miller KW: *Mol Pharmacol* 1980;18:84–90.
13. Saidman LJ, Eger EI, Munson ES, Babad AA, Muallem M: *Anesthesiology* 1967;28:994–1002.
14. Eger EI, Lundgren C, Miller SL, Stevens WC: *Anesthesiology* 1969;30:129–135.
15. Gion H, Saidman LJ: *Anesthesiology* 1971;35:361–364.
16. Stevens WC, Dolan WM, Gibbons RT, White A, Eger EI, Miller RD, de Jong RH, Elashoff RM: *Anesthesiology* 1975;42:197–200.
17. Smith RA: Investigations into the mechanism of anaesthesia: The relationship between anaesthesia and pressure. D. Phil. Thesis, Oxford, 1974.
18. Drummond JC: *Anesthesiology* 1985;62:336–338.
19. Mazze RI, Rice SA, Baden JM: *Anesthesiology* 1985;62:339–341.
20. Drummond JC, Todd MM, Shapiro HM: *J Am Vet Med Assoc* 1983;182:1099–1101.
21. Munson ES, Saidman LJ, Eger EI: *Anesthesiology* 1965;26:134–139.
22. Eger EI, Brandstater B, Saidman LJ, Regan MJ, Severinghaus JW, Munson ES: *Anesthesiology* 1965;26:771–777.
23. Dodson BA, Furmaniak ZW, Miller KW: *J. Physiol* 1985;362:233–244.
24. Miller KW, Wilson MW, Smith RA: *Mol Pharmacol* 1978;14:950–959.
25. Epstein RM, Ngai SH, Brody DC, Rittenberg DM: *Anesthesiology* 1963;24:130–131.
26. Steffey EP: *Am J Vet Res* 1977;38:1833–1836.
27. Krahl ME: *J Physiol Chem* 1940;44:449–463.
28. Eger EI: Cited in Dobkin AB, *et al.*: *Anesthesiology* 1968;29:275–287.
29. Bush MT: Sedatives and hypnotics. 1. Absorption, fate and excretion, in Root WS, Hofman FS (eds): *Physiological Pharmacology*, Vol I. New York, Academic Press, 1963.
30. Kita Y, Bennett LJ, Miller KW: *Biochim Biophys Acta* 1981;647:130–139.
31. Kita Y, Miller KW: *Biochemistry* 1982;21:2840–2846.
32. Korten K, Sommer TJ, Miller KW: *Biochim Biophys Acta* 1980;599:271–279.
33. Korten K, Miller KW: *Can J Physiol Pharmacol* 1979;57:325–328.
34. Dawson R, Elliott D, Elliot W, Jones K (eds): *Data for Biochemical Research* Oxford, Oxford Press, 1969.
35. Koblin D, Eger EI: in Miller RD (ed): *Anesthesia* ed II. New York, Churchill-Livingstone (in press).
36. Tanifuji Y, Eger EI, Terrell RC: *Anesth Analg* 1977;56:387–390.
37. Lee-Son S, Waud BE, Waud DR: *J Pharm Exp Ther* 1975;195:251–256.
38. Waud BE, Waud DR: *Anesthesiology* 1975;43:540–547.
39. Miller KW, Paton WDM, Smith EB: *Nature* 1965;206:574–577.
40. Seeman P: *Pharm Rev* 1972;24:583–634.
41. Halsey MJ: in Gray TC, Utting JE, Nunn JF (eds): *General Anaesthesia*. London, Butterworths, London, 1980, pp 45–65.
42. Meyer K: *Trans Faraday Soc* 1937;33:1062.
43. Miller KW, Paton WDM, Smith EB, Smith RA: *Anesthesiology* 1972;36:339–351.

44. Lever MJ, Miller KW, Paton WDM, Smith EB: *Nature* 1971;231:368–371.
45. Kent DW, Halsey MJ, Eger EI: in Lambertson CJ (ed): *Underwater Physiology V*. Bethesda, FASEB, 1976, pp 581–586.
46. Miller KW, Paton WDM, Smith RA, Smith EB: *Mol Pharm* 1973;9:131–143.
47. Horiuti J: *Sci Papers Instit Phys Chem Res (Tokyo)* 1931;17:125–256.
48. Kritchevsky J, Ilinskaya A: *Acta Physicochem USSR* 1945;20:327.
49. Masterton WL: *J Chem Phys* 1954;22:1830.
50 Ng K, Walkley J: *J Phys Chem* 1969;73:2274.
51. Linford RG, Hildebrand JH: *Trans Farad Soc* 1969;65:1470–1472.
52. Rottenberg H, Waring A, Rubin E: *Science* 1981;213:583–585.
53. Battino R: *Chem Rev* 1971;71:5–45.
54. Franks NP, Lieb WR: *Nature* 1981;292:248–250.
55. Lieb WR, Kovalycsik M, Mendelsohn R: *Biochim Biophys Acta* 1982;688:388–398.
56. Johnson SM, Miller KW, Bangham AD: *Biochim Biophys Acta* 1973;307:42–57.
57. Colley CM, Metcalfe SM, Turner B, Metcalfe JC: *Biochim Biophys Acta* 1971;233:720–729.
58. Colley CM, Metcalfe JC: *FEBS Letts* 1972;24:241–246.
59. Korten K, Miller KW: *Can J Physiol Pharm* 1978;57:325–328.
60. Power GG, Stegall H: *J Appl Physiol* 1970;29:145–149.
61. *Matheson Gas Data Book,* ed 5. Milwaukee, Matheson Gas Products, 1971.
62. Riddick JA, Bunger WB: in Weissberger A (ed): *Organic Solvents, Physical Properties and Methods of Purification,* ed 3. *Techniques of Chemistry,* Vol II. New York, Wiley–Interscience, 1970.
63. Hildebrand JH, Prausnitz JM, Scott RL: *Regular and Related Solutions.* New York, Van Nostrand, 1970.
64. Lerman J, Willis M, Gregory G, Eger E: *Anesthesiology* 1983;59:554–558.
65. Miller SL: *Proc Natl Acad Sci USA* 1961;47:1515–1524.
66. Wilhelm E, Battino R, Wilcock R: *Chem Rev* 1977;77:219–262.
67. Wilcock R, Battino R, Danforth W, Wilhelm E: *J Chem Thermodynamics* 1978;10:817–822.
68. Hansch C, Vittoria A, Silipo C, Jow P: *J Medicin Chem* 1975;18:546–548.
69. Yih T, Rossum J: *Biochem Pharmacol* 1977;26:2117–2120.
70. Leo A, Hansch C, Elkins D: *Chem Rev* 1971;71:525–615.
71. Kakemi K, Arita T, Hori R, Konishi R: *Chem Pharm Bull* 1967;15:1705–1712.
72. Backes W, Means M, Canady W: *J Biol Chem* 1984;259:10092–10099.
73. Seevers M, Waters R: *Physiol Rev* 1938;18:447–479.
74. Roth SH, Smith RA, Paton WDM: *Brit J Anaesth* 1976;48:621–628.
75. Smith RA, Winter PM, Smith M, Eger EI: *Anesthesiology* 1979;50:496–500.
76. Smith RA, Winter PM, Smith M, Eger EI: *Anesthesiology* 1979;50:505–509.
77. Kent DW, Halsey MJ, Eger EI: *Anesth Analg* 1977;56:97–101.
78. Deady JE, Koblin DD, Eger EI et al. *Anesth Analg* 1981;60:380–384.
79. Merck Index, 16th ed., Rahway, NJ, Merck and Co., 1983.

NAME: **Helium**
FORMULA: He
MOL. WT., g/mole: 4.0
M.P.; B.P., °C: −272.0[61]; −268.9[1]
HEAT OF VAP. kcal/mole: 0.0194[61]
VIRIAL COEFF. B, ml: 11.66, 25°C[9]
CRITICAL TEMP., °C: −267.95[1]
MOLAR VOL AT B.P., ml: 32[2]
PARTITION COEFFICIENT:
Water/gas: 0.0085, 25°C; 0.0085, 35°C[2]
Oil/gas: 0.016, 25°C; 0.016, 37°C[2]
Octanol/gas: 0.017, 25°C[67]
Octanol/water: 1.91, 24°C[68]
PARTIAL MOLAR VOLUME, ml, 25°C: (In benzene): 36 estimated[46]
HYDRATE DISSOCIATION PRESSURE: None formed
ED_{50}: No anesthesia observed in tolerable pressure range.[2,24]

NAME: **Neon**
FORMULA: Ne
MOL. WT., g/mole: 20.2
M.P.; B.P., °C: −248.6; −246.1[61]
HEAT OF VAP, kcal/mole: 0.429[61]
VIRIAL COEFF B, ml: 11.42, 25°C[9]
CRITICAL TEMP., °C: −228.76[1]
MOLAR VOL AT BP, ml: 17[2]
PARTITION COEFFICIENT:
Water/gas: 0.010, 25°C; 0.0098, 35°C[2]
Oil/gas: 0.019, 25°C; 0.019, 37°C[2]
Octanol/gas: 0.024, 25°C[67]
Erythrocyte membrane/gas: 0.012, 37°C[60]
Octanol/water: 1.91, 24°C[68]
PARTIAL MOLAR VOLUME, ml, 25°C: (In benzene): 33 estimated[46]
HYDRATE DISSOCIATION PRESSURE: None formed
ED_{50}: No anesthesia observed in tolerable pressure range.[44]

NAME: **Argon**
FORMULA: Ar
MOL. WT., g/mole: 40.0
M.P.; B.P., °C: −189.2; −185.9[61]
HEAT OF VAP, kcal/mole: 1.56[61]
VIRIAL COEFF B, ml: −15.5, 27°C[9]
CRITICAL TEMP., °C: −122.5[61]
MOLAR VOL AT BP, ml: 29[2]
PARTITION COEFFICIENT:
Water/gas: 0.031, 25°C; 0.026, 35°C[2]
Oil/gas: 0.14, 25°C; 0.13, 37°C[2]
Octanol/gas: 0.18, 25°C; 0.17, 40°C[67]
Octanol/water: 5.5, 0°C[68]
PARTIAL MOLAR VOLUME, ml, 25°C: (In benzene): 45*
HYDRATE DISSOCIATION PRESSURE, 95 atm, 0°C[65]
ED_{50}: Mouse: 18.1 atm[24]; 15.2 atm[43]
Newt: 16.8 atm[17]
Tadpole: 24.4 atm[23]
C_{50}: Nerve block: 325 atm (rat, 37°C)[40]

NAME: **Krypton**
FORMULA: Kr
MOL. WT., g/mole: 83.8
M.P.; B.P., °C: −156.6[7]; −153.6[61]
HEAT OF VAP, kcal/mole: 2.15[61]
VIRIAL COEFF B, ml: −50.5, 25°C[9]
CRITICAL TEMP., °C: −63.6[61]
MOLAR VOL AT BP, ml: 35[2]
PARTITION COEFFICIENT:
Water/gas: 0.053, 25°C; 0.045, 35°C[2]
Oil/gas: 0.44, 25°C; 0.4, 37°C[2]
Octanol/gas: 0.54, 25°C[67]
Octanol/water: 7.8, 24°C[68]
HYDRATE DISSOCIATION PRESSURE: 14.5 atm, 0°C[65]
ED_{50}: Mouse: 4.5 atm[2]

*J. Hildebrand, personal communication.

NAME: **Xenon**

FORMULA: Xe

MOL. WT., g/mole: 131.3

M.P.; B.P., °C: −111.9[7]; −108.1[61]

HEAT OF VAP, kcal/mole: 3.0[61]

VIRIAL COEFF B, ml: −133, 27°C[9]

CRITICAL TEMP., °C: 16.6[5]

MOLAR VOL AT BP, ml: 43[2]

ANTOINE EQUATION DATA: A: 4.14959 B: 705.9352 C: 278.342 atm[5]

PARTITION COEFFICIENT:
Water/gas: 0.095, 25°C; 0.075, 35°C[2]
Oil/gas: 1.9, 25°C; 1.8, 37°C[2]
PC bilayer/gas: 1.36, 25°C[6]
PC+chol bilayer/gas: 1.23, 25°C[6]
Octanol/water: 19.1, 24°C[68]

PARTIAL MOLAR VOLUME, ml, 25°C: (In benzene): 60 estimated*

HYDRATE DISSOCIATION PRESSURE: 1.2 atm, 0°C[65]

ED_{50}: Man: 0.71 atm[13] (MAC)
Dog: 1.19 atm[22] (MAC)
Mouse: 0.95 atm[2]
Newt: Approximately 0.5 atm[17]

NAME: **Hydrogen**

FORMULA: H_2

MOL. WT., g/mole: 2.0

M.P.; B.P., °C: −259.1[7]; −252.9[61]

HEAT OF VAP, kcal/mole: 0.215[61]

VIRIAL COEFF B, ml: 14.8, 27°C[9]

CRITICAL TEMP., °C: −240.2[61]

MOLAR VOL AT BP, ml: 28[2]

PARTITION COEFFICIENT:
Water/gas: 0.017, 25°C; 0.017, 35°C[2]
Oil/gas: 0.05, 37°C[2]
Erythrocyte membrane/gas: 0.024, 37°C[60]
Octanol/water: 2.8, 0°C[68]

PARTIAL MOLAR VOLUME, ml, 25°C: (In benzene): 35[47] (In water): 26[48]

ED_{50}: Mouse: 130–180 atm (mixtures with N_2O)[2,45]
Newt: > 204 atm[17]
Tadpole: 198 atm[23]

*J. Hildebrand, personal communication.

NAME: **Nitrogen**

FORMULA: N_2

MOL. WT., g/mole: 28.0

M.P.; B.P., °C: −209.9; −195.8[1]

HEAT OF VAP, kcal/mole: 1.3[61]

VIRIAL COEFF B, ml: −4.2, 27°C[9]

CRITICAL TEMP., °C: −147[1]

MOLAR VOL AT BP, ml: 35[2]

PARTITION COEFFICIENT:
Water/gas: 0.015, 25°C; 0.013, 35°C[2]
Oil/gas: 0.07, 25°C; 0.66, 37°C[2]
Octanol/gas: 0.088, 25°C; 0.089, 40°C[67]
Erythrocyte membrane/gas: 0.096, 37°C[60]
Octanol/water: 4.68, 24°C[68]

PARTIAL MOLAR VOLUME, ml, 25°C: (In benzene): 53[46] (In water): 40[48]

HYDRATE DISSOCIATION PRESSURE: 160 atm, 0°C[65]

ED_{50}: Dog: > 43.5 atm[35] (MAC)
Mouse: 38.9 atm[24]
Newt: 22 atm[46]
Tadpole: 41.5 atm[23]

NAME: **Nitrous oxide**

FORMULA: N_2O

MOL. WT., g/mole: 44.0

M.P.; B.P., °C: −90.8°C, −88.5°C[1]

HEAT OF VAP, kcal/mole: 4.34[1]

VIRIAL COEFF B, ml: −128, 27°C[9]

CRITICAL TEMP., °C: 36.5°C[1]

MOLAR VOL AT BP, ml: 35[61]

ANTOINE EQUATION DATA: A: 4.69718 B: 912.8988 C: 285.309 atm[5]

PARTITION COEFFICIENT:
Water/Gas: 0.55, 25°C; 0.39, 37°C[2]
Oil/gas: 1.3, 37°C[2]
Octanol/water: 2.7, 0°C[68]

PARTIAL MOLAR VOLUME, ml, 25°C: (In benzene): 47[47]

HYDRATE DISSOCIATION PRESSURE: 10 atm, 0°C[65]

ED_{50}: Man: 1.01 atm[13,16] (MAC)
Dog: 1.9 atm[2] (MAC)
Mouse: 1.2–1.5 atm[24,45]
Frog: 0.84 atm[17]
Newt: 0.7 atm[46]

C_{50}: Nerve Block: 11.5 atm (rat, 37°C)[40]; 4.9 atm (frog, 20°C)[74]

NAME: **Methane**

FORMULA: CH_4

MOL. WT., g/mole: 16.0

M.P.; B.P., °C: −182.6; −161.5[61]

HEAT OF VAP, kcal/mole: 1.9[61]

VIRIAL COEFF B, ml: −42, 27°C[9]

CRITICAL TEMP., °C: −82[61]

MOLAR VOL AT BP, ml: 38[2]

PARTITION COEFFICIENT:
Water/gas: 0.031, 25°C; 0.026, 37°C[66]
Oil/gas: 0.25, 37°C[2]
Octanol/gas: 0.38, 25°C; 0.36, 40°C[67]
PC bilayer/gas: 0.2, 25°C[3]
PC+chol bilayer/gas: 0.18, 25°C[3]
Octanol/water: 12.3, 24°C[68]

PARTIAL MOLAR VOLUME, ml, 25°C: (In benzene): 52[47] (In water): 37[48]

HYDRATE DISSOCIATION PRESSURE: 26 atm, 0°C[65]

ED_{50}: Mouse: 4.6 atm[39]; 5.4 atm*

C_{50}: Nerve block: 101 atm (rat, 37°C)[40]

NAME: **Ethane**

FORMULA: C_2H_6

MOL. WT., g/mole: 30.1

M.P.; B.P., °C: −183.3[7]; −88.2[61]

HEAT OF VAP, kcal/mole: 3.5[61]

VIRIAL COEFF B, ml: −182, 27°C[9]

CRITICAL TEMP., °C: 32.3[61]

MOLAR VOL AT BP, ml: 55†

PARTITION COEFFICIENT:
Water/gas: 0.042, 25°C; 0.031, 37°C[66]
PC bilayer/gas: 0.97, 25°C[3]
PC+chol bilayer/gas: 1.12, 25°C[3]
Octanol/water: 64.6, 24°C[68]

PARTIAL MOLAR VOLUME, ml, 25°C: (In benzene): 66 estimated[47] (In water): 51[49]

HYDRATE DISSOCIATION PRESSURE: 5.2 atm, 0°C[65]

ED_{50}: Mouse: 0.89 atm*

NAME: **Ethylene**

FORMULA: C_2H_4

MOL. WT., g/mole: 28.1

M.P.; B.P., °C: −171; −103.7[61]

HEAT OF VAP, kcal/mole: 3.24[61]

VIRIAL COEFF B, ml: −138, 27°C[9]

CRITICAL TEMP., °C: 9.5°C[1]

MOLAR VOL AT BP, ml: 49[61]

ANTOINE EQUATION DATA: A: 4.07057 B: 640.1445 C: 260.868, atm[5]

PARTITION COEFFICIENT:
Water/gas: 0.107, 25°C; 0.085, 37°C[66]; 0.085, 37°C[1]
Oil/gas: 1.26, 37°C[35]; 1.14, 37°C[2]
Octanol/water: 13.5, 24°C[68]

PARTIAL MOLAR VOLUME, ml, 25°C: (In benzene): 61 estimated[47]

HYDRATE DISSOCIATION PRESSURE: 5.5 atm, 0°C[65]

ED_{50}: Man: 0.67 atm[35] (MAC)
Mouse: 1.36 atm[25]; 1.31 atm with balance of 1 atm O_2 + 13.6 atm He[75]
Newt: 0.60 atm[17]

C_{50}: Nerve block: 11.2 atm (rat, 37°C)[40]; 6.6 atm (frog, 20°C)[74]

NAME: **Propane**

FORMULA: C_3H_8

MOL. WT., g/mole: 44.1

M.P.; B.P., °C: −188; −42[61]

HEAT OF VAP, kcal/mole: 4.5[61]

VIRIAL COEFF B, ml: −382, 27°C[9]

CRITICAL TEMP., °C: 96.8[61]

MOLAR VOL AT BP, ml: 75[63]

PARTITION COEFFICIENT:
Water/gas: 0.034, 25°C; 0.024, 37°C[66]
PC bilayer/gas: 3.6, 25°C[3]
PC+chol bilayer/gas: 2.6, 25°C[3]
Octanol/water: 229, 24°C[68]

PARTIAL MOLAR VOLUME, ml, 25°C: (In benzene): 91[50] (In water): 67[49]

HYDRATE DISSOCIATION PRESSURE: 1.74 atm, 0°C[65]

ED_{50}: Man: approximately 0.9 atm[73]

*M. Wilson and K. W. Miller, unpublished observations.
†J. Hildebrand, personal communication.

NAME: **Cyclopropane**

FORMULA: c-C_3H_6

MOL. WT., g/mole: 42.1

M.P.; B.P., °C: −126.6, −32.9°C[1]

HEAT OF VAP, kcal/mole: 4.8[61]

VIRIAL COEFF B, ml: −346.3, 30°C[9]

CRITICAL TEMP., °C: 124.7°C[1,3]

MOLAR VOL AT BP, ml: 63[2]

ANTOINE EQUATION DATA: A: 3.76400 B: 723.3158 C: 225.699, atm[5]

PARTITION COEFFICIENT:
Water/gas: 0.26, 25°C; 0.20, 37°C[66]; 0.203, 37°C[1]
Oil/gas: 14.1, 25°C[2]; 9.7, 37°C[2]
PC bilayer/gas: 11, 25°C[3]
PC+chol bilayer/gas: 10, 25°C[3]
Octanol/water: 52.5, 24°C[68]

HYDRATE DISSOCIATION PRESSURE: 0.63 atm, 0°C[65]

ED_{50}: Man: 0.092 atm[13] (MAC)
Dog: 0.175 atm[22] (MAC)
Mouse: 0.16 atm[2]; 0.13 atm[75]
Newt: 0.049 atm[17]

C_{50}: Nerve block: 1.8 atm (rat, 37°C)[40]; 0.69 atm (frog, 20°C)[74]
Na channel: 8.4 mM (6°C)*

NAME: **Butane**

FORMULA: C_4H_{10}

MOL. WT., g/mole: 58.1

M.P.; B.P., °C: −138; −0.5[61]

HEAT OF VAP, kcal/mole: 5.3[61]

VIRIAL COEFF B, ml: −722, 27°C[9]

CRITICAL TEMP., °C: 152[61]

MOLAR VOL AT BP, ml: 97†

PARTITION COEFFICIENT:
Water/gas: 0.027, 25°C[6]; 0.019, 37°C[66]
PC bilayer/gas: 16.6, 25°C[6]
PC+chol bilayer/gas: 14.4, 25°C[6]

PARTIAL MOLAR VOLUME, ml, 25°C: (In benzene): 105[50]

ED_{50}: Man: approximately 0.2 atm[73]

C_{50}: Na channel: 0.7 mM (6°C)[10]

NAME: **Sulfur hexafluoride**

FORMULA: SF_6

MOL. WT., g/mole: 146.1

M.P.; B.P., °C: −50.5; sublimes at −64[7]

VIRIAL COEFF B, ml: −277, 27°C[9]

CRITICAL TEMP., °C: 45.6[61]

MOLAR VOL AT BP, ml: 76 (estimated)[2]

PARTITION COEFFICIENT:
Water/gas: 0.0054, 25°C[6]; 0.0042, 37°C[66]
Olive oil/gas: 0.30, 25°C; 0.26, 37°C[2]
Octanol/gas: 0.46,[67] 0.26,[68] 25°C; 0.398, 40°C[67]
PC bilayer/gas: 0.238, 25°C[6]
PC+chol bilayer/gas: 0.218, 25°C[6]
Erythrocyte membrane/gas: 0.151, 37°C[60]

PARTIAL MOLAR VOLUME, mi, 25°C: (In benzene): 97[46]

HYDRATE DISSOCIATION PRESSURE: 1.0 atm, 0°C[65]

ED_{50}: Dog: 4.9 atm[14] (MAC)
Mouse: 5.58 atm, with balance of 1 atm 0_2, 5.5 atm He[24]; 6.1 atm[2]
Newt: 1.8 atm[46]

C_{50}: Nerve block: 50 atm (rat, 37°C)[40]

NAME: **Carbon tetrafluoride (Perfluoromethane)**

FORMULA: CF_4

MOL. WT., g/mole: 88.0

M.P.; B.P., °C: −184; −128[61]

HEAT OF VAP, kcal/mole: 2.9[61]

VIRIAL COEFF B, ml: −87, 27°C[9]

CRITICAL TEMP., °C: −45.5[61]

MOLAR VOL AT BP, ml: 54[2]

PARTITION COEFFICIENT:
Water/gas: 0.0047, 25°C[6]; 0.0039, 37°C[66]
Oil/gas: 0.06, 37°C[2]
Octanol/gas: 0.10,[67] 0.07,[68] 25°C; 0.102, 40°C[67]
PC bilayer/gas: 0.056, 25°C[6]
PC+chol bilayer/gas: 0.044, 25°C[6]

PARTIAL MOLAR VOLUME, ml, 25°C: (in benzene): 82[46]

HYDRATE DISSOCIATION PRESSURE: 41 atm[43]

ED_{50}: Dog: 26 atm[14] (MAC)
Mouse: 19 atm[2]; 26 atm[24]
Newt: 11 atm[46]

*Chapter 22, this volume.

†J. Hildebrand, personal communication.

NAME: **Perfluoroethane (Hexafluoroethane)**

FORMULA: C_2F_6

MOL. WT., g/mole: 138.0

M.P.; B.P., °C: −100.6; −78.2[61]

HEAT OF VAP, kcal/mole: 3.9[61]

VIRIAL COEFF B, ml: −210, 25°C[9]

CRITICAL TEMP., °C: 24.3[61]

MOLAR VOL AT BP, ml: 86[63]

PARTITION COEFFICIENT:
Water/gas: 0.008 estimated, 25°C[6]
Oil/gas: 0.12, 30°C[2]; 0.126, 37°C[35]
PC bilayer/gas: 0.34, 25°C[6]
PC+chol bilayer/gas: 0.33, 25°C[6]
Octanol/water: 100, 24°C[68]

PARTIAL MOLAR VOLUME, ml, 25°C: (in cyclohexane): 111[51]

HYDRATE DISSOCIATION PRESSURE: No hydrate formed[43]

ED_{50}: Mouse: 17.1 atm[43]; 18 atm[24]

NAME: **Perfluoropropane**

FORMULA: C_3F_8

MOL. WT., g/mole: 188.0

M.P.; B.P., °C: −183; −37[61]

HEAT OF VAP, kcal/mole: 4.7[61]

VIRIAL COEFF B, ml: −525, 25°C[9]

CRITICAL TEMP., °C: 71.9[61]

MOLAR VOL AT BP, ml: 117[63]

PARTITION COEFFICIENT:
Oil/gas: 0.23, 30°C[43]

PARTIAL MOLAR VOLUME, ml, 25°C: (In cyclohexane): 140[51]

HYDRATE DISSOCIATION PRESSURE: No hydrate formed[43]

ED_{50}: Mouse: 18 atm (extrapolated from mixtures with N_2O)[2]

NAME: **Ethanol**

FORMULA: C_2H_5OH

MOL. WT., g/mole: 46.1

DENSITY g/ml: 0.78937, 20°C[62]

MOLAR VOLUME, ml (20°C): 58.4

M.P.; B.P., °C: −114.1; 78.3[62]

VAPOR PRESSURE, mm Hg: 59.77, 25°C[62]

HEAT OF VAP, kcal/mole: 10.11, 25°C[62]

VIRIAL COEFF B, ml: −2000, 40°C[9]

CRITICAL TEMP., °C: 243.1 C[5]

ANTOINE EQUATION DATA: A: 5.50154 B: 1739.914 C: 238.131, atm[5]

DIELECTRIC CONST.: 24.55, 25°C[62]

DIPOLE MOMENT, Debye: 1.66, 20°C[62]

PARTITION COEFFICIENT:
Octanol/water: 0.48, R.T.[70]
PC bilayer/water: 0.48, 30°C[34]
Synaptosomal membrane/buffer: 1.0, (?R.T.)[52]

PARTIAL MOLAR VOLUME, ml, 25°C: (In benzene): 61[53] (In water): 55[53]

ED_{50}: Newt: 0.000175 atm[17]
Tadpole: 190 mM*

C_{50}: Nerve block: 500 mM (rat, 37°C)[40]
Firefly luciferase: 600 mM†
Antihemolysis: 600 mM[40]

NAME: **Butanol**

FORMULA: C_4H_9OH

MOL. WT., g/mole: 74.1

DENSITY, g/ml: 0.8097, 20°C[62]

MOLAR VOLUME, ml (20°C): 91.5

M.P.; B.P., °C: −88.6; 117.7[62]

VAPOR PRESS., mm Hg: 6.18, 25°C[62]

HEAT OF VAP, kcal/mole, 12.54, 25°C[62]

VIRIAL COEFF B, ml: −1700, 77°C[9]

CRITICAL TEMP., °C: 289.8[62]

ANTOINE EQUATION DATA: A: 7.47680 B: 1362.39 C: 178.77, mm Hg[62]

DIELECTRIC CONST.: 17.8, 20°C[7]; 17.5, 25°C[62]

DIPOLE MOMENT, Debye: 1.75, 25°C[62]

PARTITION COEFFICIENT:
PC bilayer/water: 3.2, 25°C[12]

PARTIAL MOLAR VOLUME, ml, 25°C: (In benzene): 94[53] (In water): 86[53]

ED_{50}: Newt: 76 mM[17]
Tadpole: 12 mM[11]

C_{50}: Nerve block: 68 mM (frog, R.T.)[40]
Firefly luciferase: 9.7 mM†
Antihemolysis: 42 mM[40]

*L. Firestone, J. K. Alifimoff, L. M. Braswell, and K. W. Miller, unpublished observations.

†N. P. Franks and W. R. Lieb, personal communication.

NAME: **Hexanol**

FORMULA: $C_6H_{13}OH$

MOL. WT., g/mole: 102.2

DENSITY, g/ml: 0.8196, 20°C[62]

MOLAR VOLUME, ml (20°C): 124.7

M.P.; B.P., °C: −44.6; 157.0[62]

VAPOR PRESS., mm Hg: 0.82, 25°C[62]

HEAT OF VAP, kcal/mole: 14.8, 25°C[62]

CRITICAL TEMP., °C: 337[62]

ANTOINE EQUATION DATA: A: 7.8604 B: 1761.26 C: 196.66, mm Hg[62]

DIELECTRIC CONST.: 13.3, 25°C[7]

DIPOLE MOMENT, Debye: 1.55, 20°C[62]

SOLUBILITY (mM): Water: 61[11]

PARTITION COEFFICIENT:
Octanol/water: 107, R. T.[70]
Egg PC/buffer: 50, 24°C[11]
Erythrocyte/buffer: 13, 22°C[11]

PARTIAL MOLAR VOLUME, ml, 25°C: (In benzene) 128[53]

ED_{50}: Tadpole: 0.7 mM[11]

C_{50}: Nerve block: 6 mM (rat, 37°C)[40]
Na channel: 3.5 mM (6°C)[10]
K channel: 9.0 mM (6°C)[10]
Firefly luciferase: 0.99 mM*
Antihemolysis: 3 mM[40]

NAME: **Benzyl alcohol**

FORMULA: $C_6H_5CH_2OH$

MOL. WT., g/mole: 108.1

DENSITY, g/ml: 1.04127, 25°C[62]

MOLAR VOLUME, ml (20°C): —; 103.8, 25°C[62]

M.P.; B.P., °C: −15.3; 205.5[62]

VAPOR PRESS., mm Hg: 0.11, 25°C[62]

HEAT OF VAP, kcal/mole: 12.1, bp[62]

DIELECTRIC CONST.: 13.1, 20°C[62]

DIPOLE MOMENT, Debye: 1.66[62]

SOLUBILITY (mM): Water (25°C): 339[57]

PARTITION COEFFICIENT:
Octanol/water: 12.6, R.T.[70]
Egg PC/buffer: 18, 25°C[58]
Egg PC+chol bilayer/buffer: 11, 25°C[58]
Erythrocyte/buffer: 3.5, 25°C[57]

PARTIAL MOLAR VOLUME, ml, 25°C: (In water): 101[30] (In egg PC+ chol): 105[30]

ED_{50}: Tadpole: 2.0 mM[30]

C_{50}: Nerve block: 20 mM (rat, 37°C)[40]
Na channel: 8 mM (6°C)[10]
Firefly luciferase: 1.7 mM*
Antihemolysis: 22 mM[40]

NAME: **Decanol**

FORMULA: $C_{10}H_{21}OH$

MOL. WT., g/mole: 158.3

DENSITY, g/ml: 0.8297, 20°C[7]

MOLAR VOLUME, ml (20°C): 190.8

M.P.; B.P., °C: 7; 229[7]

DIELECTRIC CONST.: 8.1, 20°C[7]

SOLUBILITY (mM): Water: 0.316[11]

PARTITION COEFFICIENT:
Erythrocyte/buffer: 1222[11]
Brush border membrane/buffer: 1975[11]

PARTIAL MOLAR VOLUME, ml, 25°C: (in benzene): 195[53]

ED_{50}: Tadpole: 0.013mM[11]

C_{50}: Na channel: 0.022 mM (6°C)[10]
K channel: 0.21 mM (6°C)[10]
Firefly luciferase: 0.0056 mM*
Antihemolysis: 0.010 mM[40]

NAME: **Octanol**

FORMULA: $C_8H_{17}OH$

MOL. WT., g/mole: 130.2

DENSITY, g/ml: 0.8256, 20°C[62]

MOLAR VOLUME, ml (20°C): 157.7

M.P.; B.P., °C: −15.0; 195.2[62]

VAPOR PRESS., mm Hg: 0.08, 25°C[62]

HEAT OF VAP, kcal/mole: 17.2, 25°C[62]

CRITICAL TEMP., °C: 390[62]

ANTOINE EQUATION DATA: A: 7.08451 B: 1457.76 C: 151.58, mm Hg[62]

DIELECTRIC CONST.: 10.3, 20°C[7,62]

DIPOLE MOMENT, Debye: 1.76, 25°C[62]

SOLUBILITY (mM): Water: 4.5[8]

PARTITION COEFFICIENT:
Octanol/water: 1413, R.T.[70]
Egg PC/buffer: 387, 24°C[11]
Erythrocyte/buffer: 152, 22°C[11]

PARTIAL MOLAR VOLUME, ml, 25°C: (In benzene); 162[53] (In water): 149[31] (In egg PC): 159[31] (In erythrocyte): 159[31]

ED_{50}: Tadpole: 0.06 mM[11]

C_{50}: Na channel: 0.29 mM (6°C)[10]
K channel: 1.0 mM (6°C)[10]
Firefly luciferase: 0.28 mM*
Antihemolysis: 0.24 mM[40]

*N. Franks and W. R. Lieb, personal communication.

NAME: **Halothane**

FORMULA: $CF_3CHClBr$

MOL. WT., g/mole: 197.4

DENSITY, g/ml: 1.87, 20°C[1]

MOLAR VOLUME, ml (20°C): 106 ml[1]

M.P.; B.P., °C: −118.3; 50[1]

VAPOR PRESS., mm Hg: 300, 25°C[5]

HEAT OF VAP, kcal/mole: 7.30[1]

VIRIAL COEFF B, ml: −1371, 25°C[9]

MOLAR VOL AT BP, ml: 114[2]

ANTOINE EQUATION DATA: A: 3.88718 B: 1043.697 C: 218.262 atm[5]

SOLUBILITY (mM): *Torpedo* Ringer, 20°C: 17 mM*

PARTITION COEFFICIENT:
- Water/gas: 1.20, 25°C; 0.63, 37°C[6]
- Oil/gas: 330, 25°C; 197, 37°C[2]
- PC bilayer/gas: 148, 25°C; 90.4, 37°C[6]
- PC+chol bilayer/gas: 93.0, 25°C; 54.4, 37°C[6]
- Octanol/water: 199, 24°C[68]
- PC+chol bilayer/buffer: 50, 20°C[12]
- Synaptosomal membrane/buffer: 27.5, (?R.T.)[52]

PARTIAL MOLAR VOLUME, ml, 25°C: (In water): 89.7[30] (In olive oil): 102.6[30] (In PC bilayer): 103.2[30] (In PC+ chol bilayer): 106.7[30] (In erythrocyte): 134[54]

HYDRATE DISSOCIATION PRESSURE: None formed[14]

ED_{50}:
- Man: 0.0077 atm[13] (MAC)
- Dog: 0.0087 atm[22] (MAC)
- Cat: 0.012 atm[20] (MAC)
- Rabbit: 0.014 atm[18] (MAC)
- Rat: 0.0103 atm[19] (MAC)
- Mouse: 0.0095 atm[19] (MAC); 0.0077 atm[2]; 0.0054 atm[6]
- Newt: 0.0038 atm[17]
- Tadpole: 0.23 mM[30]

C_{50}:
- Nerve block: 5 mM (frog, 22°C)[40]
- Antihemolysis: 8 mM[40]
- Na channel: 20 mM (6°C)[10]
- K channel: 6.4 mM (6°C)[10]
- Firefly luciferase: 0.39 mM†
- Postsynaptic *n*-AChR blockade: 2.4 %(v/v)[38]

NAME: **Trichloroethylene**

FORMULA: C_2HCl_3

MOL. WT., g/mole: 131.4

DENSITY, g/ml: 1.465, 20°C[1]

MOLAR VOLUME, ml (20°C): 89.7[1]

M.P.; B.P., °C: −86.4; 87.2[62]

VAPOR PRESS., mm Hg: 64.5, 20°C[1]; 47.31, 25°C[62]

HEAT OF VAP., kcal/mole: 8.19, 25°C[62]

VIRIAL COEFF B, ml: −1900, 50°C[9]

CRITICAL TEMP., °C: 298[62]

ANTOINE EQUATION DATA: A: 3.95614 B: 1198.47 C: 216.436, atm[5]

DIELECTRIC CONST.: 3.42, ca. 16°C[62]

DIPOLE MOMENT, Debye: 0.8[62]

PARTITION COEFFICIENT:
- Water/gas: 1.55, 37°C[1]
- Oil/water: 400, 20°C[1]

ED_{50}: Rat: 0.6% (v/v)[41] (MAC)

NAME: **Dichlorodifluoromethane**

FORMULA: CCl_2F_2

MOL. WT., g/mole: 120.9

M.P.; B.P.,°C: −158; −29.8[61]

HEAT OF VAP, kcal/mole: 4.8[61]

VIRIAL COEFF B, ml: −460, 25°C[9]

CRITICAL TEMP., °C: 112[61]

MOLAR VOL AT BP, ml: 81[2]

PARTITION COEFFICIENT:
- Water/gas: 0.05, 25°C; 0.044, 37°C[2]
- Oil/gas: 5.1, 37°C[2]
- Octanol/water: 145, 24°C[68]

HYDRATE DISSOCIATION PRESSURE: 0.9 atm, 0°C[65]

ED_{50}:
- Mouse: 0.4 atm[2]
- Newt: 0.19 atm[17]

C_{50}: Nerve block (frog, 20°C): 1.07 atm[74]

*K. W. Miller and L. M. Braswell, unpublished observations.

†N. P. Franks and W. R. Lieb, personal communication.

NAME: **Chlorodifluouromethane**

FORMULA: $CHClF_2$

MOL. WT., g/mole: 86.5

M.P.; B.P., °C: −160; −40.8[61]

HEAT OF VAP, kcal/mole: 4.8[61]

VIRIAL COEFF B, ml: −356, 25°C[9]

CRITICAL TEMP., °C: 96[61]

MOLAR VOL AT BP, ml: 61[2]

PARTITION COEFFICIENT:
Water/gas: 0.77, 25°C; 0.51, 37°C[66]
Oil/gas: 11.0, 37°C[2]

ED_{50}: Mouse: 0.16 atm[2]

NAME: **Ethylchloride**

FORMULA: C_2H_5Cl

MOL. WT., g/mole: 64.5

DENSITY, g/ml: 0.9028, 15°C[62]

MOLAR VOLUME, ml (20°C):—; 71.4, 15°C

M.P.; B.P., °C: −136.4; 12.3[62]

HEAT OF VAP, kcal/mole: 5.892, bp[62]

VIRIAL COEFF B, ml: −750, 25°C[9]

CRITICAL TEMP., °C: 187.2[62]

MOLAR VOL AT BP, ml: 71[62]

ANTOINE EQUATION DATA: A: 4.50965
B: 1269.62 C: 269.157, atm[5]

DIELECTRIC CONST.: 9.45, 20°C[62]

DIPOLE MOMENT, Debye: 1.96, 20°C[62]

PARTITION COEFFICIENT:
Water/gas: 2.5, 37°C[1]
Octanol/water: 27, 24°C[68]

HYDRATE DISSOCIATION PRESSURE: 0.27 atm, 0°C[65]

ED_{50}: Mouse: 0.04 atm[39]

NAME: **Chloroform**

FORMULA: $CHCl_3$

MOL. WT., g/mole: 119.4

DENSITY, g/ml: 1.47, 20°C[62]

MOLAR VOLUME, ml (20°C): 80.7[1]

M.P.; B.P., °C: −63.5; 61.2[1]

VAPOR PRESS., mm Hg: 160, 20°C; 194.8, 25°C[62]

HEAT OF VAP, kcal/mole: 7.97, 20°C[62]

VIRIAL COEFF B, ml: −1000, 40°C[9]

CRITICAL TEMP., °C: 263.4[62]

MOLAR VOL AT BP, ml: 85[2]

ANTOINE EQUATION DATA: A: 3.97345
B: 1125.046 C: 222.014, atm[5]

DIELECTRIC CONST.: 4.806, 20°C[62]

DIPOLE MOMENT, Debye: 1.15, 25°C[62]

SOLUBILITY (mM): Water: 66.6[1]

PARTITION COEFFICIENT:
Water/gas: 3.8, 37°C[1]
Oil/gas: 330, 25°C; 233, 37°C[2]
Octanol/water: 93, 24°C[68]
PC+chol/buffer: 5.8, 25°C[55]

HYDRATE DISSOCIATION PRESSURE: 0.062 atm, 0°C[65]

ED_{50}: Dog: 0.0077 atm[14] (MAC)
Mouse: 0.0084 atm[2], 0.0036 atm[35] (MAC)
Newt: 0.0048 atm[17]; 0.89 mM[46]

C_{50}: Nerve block: 5 mM (frog, 22°C)[40]
Na channel: 3.0 mM (6°C)[10]
Firefly luciferase: 1.4 mM*
Antihemolysis: 9 mM[40]

*N. P. Franks and W. R. Lieb, personal communication.

NAME: **Diethylether**

FORMULA: $(C_2H_5)_2O$

MOL. WT., g/mole: 74.1

DENSITY, g/ml: 0.718, 20°C[1]

MOLAR VOLUME, ml (20°C): 103.2[1]

M.P.; B.P., °C: −116.2°C, 34.6°C[1]

VAPOR PRESS., mm Hg: 534.2, 25°C[62]

HEAT OF VAP, kcal/mole: 6.51, 25°C[62]

VIRIAL COEFF B, ml: −1200, 27°C[9]

CRITICAL TEMP., °C: 193.55[62]

MOLAR VOL AT BP, ml: 106[2]

ANTOINE EQUATION DATA: A: 4.14602 B: 1109.577 C: 233.155, atm[5]

DIELECTRIC CONST.: 4.335, 20°C[62]

DIPOLE MOMENT, Debye: 1.15, 20°C[62]

PARTITION COEFFICIENT:
- Water/gas: 24, 25°C; 11, 37°C[2]
- Oil/gas: 156, 25°C; 57, 37°C[2]
- Octanol/water: 7.8, 24°C[68]

ED_{50}:
- Man: 0.019 atm[13] (MAC)
- Dog: 0.030 atm[22] (MAC)
- Mouse: 0.030 atm[2]
- Newt: 0.009 atm[17]
- Tadpole: 24 mM[42]

C_{50}:
- Nerve block: 50 mM (frog, 22°C)[40]
- Na channel: 75 mM (6°C)[10]
- K channel: 150 mM (6°C)[10]
- Antihemolysis: 9 mM[40]
- Firefly luciferase: 12 mM*
- Postsynaptic *n*-AChR blockade: 4.1 %(v/v)[38]

NAME: **Divinylether**

FORMULA: $(CH_2CH)_2O$

MOL. WT., g/mole: 70.1

DENSITY, g/ml: 0.77, 20°C[1]

MOLAR VOLUME, ml (20°C): 91[1]

M.P.; B.P., °C:—; 28.3°C[1]

VAPOR PRESS., mm Hg: 670, 25°C[5]

HEAT OF VAP, kcal/mole: 6.4[1]

ANTOINE EQUATION DATA: A: 4.08898 B: 1044.143 C: 226.982, atm[5]

DIELECTRIC CONST.: 3.94, 20°C[7]

PARTITION COEFFICIENT:
- Water/gas: 1.32, 37°C[1]
- Oil/gas: 54.6, 37°C[1]

NAME: **Fluroxene**

FORMULA: $CF_3CH_2OCHCH_2$

MOL. WT., g/mole: 126.0

DENSITY, g/ml: 1.13, 20°C[1]

MOLAR VOLUME, ml (20°C): 111.5

M.P.; B.P., °C:—; 43.2°C[1]

VAPOR PRESS., mm Hg: 369, 25[5]

ANTOINE EQUATION DATA: A: 1.37193 B: 136.8463 C: 56.1910, atm[5]

SOLUBILITY (mM): *Torpedo* Ringer, 20°C: 30†

PARTITION COEFFICIENT:
- Water/gas: 1.24, 25°C; 0.71, 37°C[6]
- PC bilayer/gas: 51.5, 25°C; 29.3, 37°C[6]
- PC+chol bilayer/gas: 29.8, 25°C; 18.8, 37°C[6]
- Oil/water: 94, 37°C[1]

ED_{50}:
- Man: 0.034 atm[21] (MAC)
- Dog: 0.06 atm[22] (MAC)
- Mouse: 0.0345 atm[35] (MAC)
- Tadpole: 2.3 mM†

C_{50}:
- Postsynaptic *n*-AChR blockade: 7.2 %(v/v)[38]

NAME: **Enflurane**

FORMULA: CHF_2OCF_2CHFCl

MOL. WT., g/mole: 184.5

DENSITY, g/ml: 1.52[5] (temperature not specified, probably 20–25°C)

M.P.; B.P., °C:—; 56.5°C[5]

VAPOR PRESS, mm Hg: 216, 25°C[5]

ANTOINE EQUATION DATA: A: 4.10759 B: 1107.839 C: 213.063, atm[5]

SOLUBILITY (mM): *Torpedo* Ringer, 20°C; 15†

PARTITION COEFFICIENT:
- Water/gas: 0.78, 37°C[28]
- Oil/gas: 96.5, 37°C[35]
- Octanol/water: 126, 24°C[68]

ED_{50}:
- Man: 0.0168 atm[15] (MAC)
- Dog: 0.0267 atm[35] (MAC)
- Cat: 0.024 atm[20] (MAC)
- Rabbit: 0.029 atm[18] (MAC)
- Rat: 0.022 atm[19] (MAC)
- Mouse: 0.020 atm[19] (MAC)

C_{50}:
- Na channel: 3.9 mM (6°C)[10]
- Postsynaptic *n*-AChR blockade: 2.4 %(v/v)[38]

*N. P. Franks and W. R. Lieb, personal communication.

†E. Ambalavanar, L. M. Braswell, and K. W. Miller, unpublished observations.

NAME: **Isoflurane**

FORMULA: CF_3 $CHClOCHF_2$

MOL. WT., g/mole: 184.5

DENSITY, g/ml: 1.49[5] (temperature not specified, probably 20–25°C)

M.P.; B.P., °C: —, 48.5°C[5]

VAPOR PRESS, mm Hg: 292, 25[5]

ANTOINE EQUATION DATA: A: 2.81697 B: 536.4589 C: 140.991, atm[5]

SOLUBILITY (mM): *Torpedo* Ringer, 20°C: 15*

PARTITION COEFFICIENT:
Water/gas: 1.08, 25°C; 0.54, 37°C[6]
Oil/gas: 90.8, 37°C[35]
PC bilayer/gas: 140, 25°C; 78.2 37°C[6]
PC+chol bilayer/gas: 65.3, 25°C; 42.0, 37°C[6]

ED_{50}: Man: 0.0115 atm[35] (MAC)
Dog: 0.0141 atm[35] (MAC); 0.0128 atm[26] (MAC)
Cat: 0.016 atm[20,26] (MAC)
Rabbit: 0.021 atm[18] (MAC)
Rat: 0.015 atm[19] (MAC)
Mouse: 0.013 atm[19] (MAC); 0.0057 atm[76,77]
Tadpole: 0.29 mM†

C_{50}: Na channel: 3.7 mM (6°C)†
Postsynaptic *n*-AChR blockade: 3.4 %(v/v)[38]

NAME: **Thiomethoxyflurane**

FORMULA: $CHCl_2CF_2SCH_3$

MOL. WT., g/mole: 180.0

VAPOR PRESS, mm Hg: 6, 25°C‡

SOLUBILITY (mM): *Torpedo* Ringer, 20°C: 5†

PARTITION COEFFICIENT:
Water/gas: 5.4, 37°C[36]
Oil/gas: 7230, 37°C[36]

ED_{50}: Dog: 0.00035 atm[36] (MAC)
Tadpole: 0.063 mM†

NAME: **Methoxyflurane**

FORMULA: $CHCl_2CF_2OCH_3$

MOL. WT., g/mole: 165.0

DENSITY, g/ml: 1.43, 20°C[1]

MOLAR VOLUME, ml (20°C): 115

M.P.; B.P., °C: −35; 104.8[1]

VAPOR PRESS, mm Hg: 30.0, 25°C[5]

HEAT OF VAP, kcal/mole: 9.7[1]

ANTOINE EQUATION DATA: A: 4.20138 B: 1336.580 C: 213.48, atm[5]

SOLUBILITY (mM): *Torpedo* Ringer, 20°C: 8.7†

PARTITION COEFFICIENT:
Water/gas: 7.0, 20°C; 4.2, 37°C[2]
Oil/gas: 1560, 25°C; 850, 37°C[2]
PC bilayer/gas: 1050, 20°C; 493, 37°C[6]
PC+chol bilayer/gas: 635, 20°C; 306, 37°C[6]
Octanol/water: 162, 24°C[68]

PARTIAL MOLAR VOLUME, 25°C, ml: (In water): 107.4[30] (In PC bilayer): 113[30] (In PC+chol bilayer): 119.5[30]

ED_{50}: Man: 0.0016 atm[13] (MAC)
Dog: 0.0023 atm[22] (MAC)
Mouse: 0.0022 atm[2]
Tadpole: 0.21 mM[30]

C_{50}: Na channel: 1.5 mM (6°C)[10]
K channel: 2.1 mM (6°C)[10]
Firefly luciferase: 0.29 mM**
Postsynaptic *n*-AChR blockade: 0.4 %(v/v)[38]

NAME: **Amobarbital**

FORMULA: X=O; R_1=H; $R_2=C_2H_5$; $R_3=CH_2CH_2CH(CH_3)_2$

MOL. WT., g/mole: 226

M.P.; B.P., °C: 156-8[34]

SOLUBILITY (mM): *Torpedo* Ringer, 20°C (pH 7): 5†

PARTITION COEFFICIENT:
Octanol/water: 129, (?R.T.)[70]
Octanol/phosphate buffer (0.2 M): 168 (acid form, 25°C)[72]

pK: 7.92, μ 0.002; 7.75, μ 0.16, 25°C[27]

ED_{50}: Tadpole: 0.44 mM at pH 7.4[37]

C_{50}: Postsynaptic *n*-AChR blockade: 0.63 mM at pH 7.4[37]

*B. A. Dodson, E. Ambalavanar, L. M. Braswell, and K. W. Miller, unpublished observations.
†Chapter 22, this volume.
‡R. C. Terrell, personal communication.
**N. P. Franks and W. R. Lieb, personal communication.

NAME: **Barbital**

FORMULA: X=O; R_1=H; $R_2=C_2H_5$; $R_3=C_2H_5$

MOL. WT., g/mole: 184

M.P.; B.P., °C: 189–92[34]

SOLUBILITY (mM): *Torpedo* Ringer, 20°C (pH 7): 35*

PARTITION COEFFICIENT:
Octanol/buffer: 4.5 (acid form, 25°C)[33]
Octanol/phosphate buffer (0.2 M): 5.3 (acid form, 25°C)[72]
Erythrocyte/buffer: 0.68 (pH 7, R.T.)[40]

pK: 7.9, μ 0.002; 7.74, μ 0.16[27]

ED_{50}: Tadpole: 14.6 mM at pH 7.4[37]

C_{50}: Nerve block: 28 mM (frog, 22°C)[40]
Postsynaptic *n*-AChR blockade: effect not reversible[37]

NAME: **Butabarbital**

FORMULA: X=O; R_1=H; $R_2=C_2H_5$; $R_3=CH(CH_3)C_2H_5$

MOL. WT., g/mole: 212

SOLUBILITY (mM): *Torpedo* Ringer, 20°C (pH 7): 12*

PARTITION COEFFICIENT:
Octanol/phosphate buffer (0.2 M): 39.2 (at pH 7.4 25°C)[72]

ED_{50}: Tadpole: 0.56 mM at pH 7.4[37]

C_{50}: Postsynaptic *n*-AChR blockade: 2.2 mM at pH 7.4[37]

NAME: **Dimethylbutylethylbarbital**

FORMULA: X=O; R_1=H; $R_2=C_2H_5$; $R_3=CH_2CH(CH_3)_2$

MOL. WT., g/mole: 212

SOLUBILITY (mM): *Torpedo* Ringer, 20°C (pH 7): 1.5*

pK: 8.14, μ 0.002; 7.96, μ 0.16[27]

EC_{50}: Tadpole: 0.05 mM at pH 7.4[37]

C_{50}: Postsynaptic *n*-AChR blockade: 0.17 mM at pH 7.4[37]

*B. A. Dodson, L. Firestone, L. M. Braswell, and K. W. Miller, unpublished observations.

NAME: **Hexobarbital**

FORMULA: X=O; $R_1=CH_3$; $R_2=CH_3$; $R_3=cC_6H_9$

MOL. WT., g/mole: 236

PARTITION COEFFICIENT:
Octanol/water: 110, (?R.T.)[69]
Octanol/phosphate buffer (0.2 M): 446 (acid form, 25°C)[72]

pK: 8.32, μ 0.002; 8.18, μ 0.16[27]

ED_{50}: Tadpole: 0.16 mM at pH 7.4[37]

C_{50}: Postsynaptic *n*-AChR blockade: 0.92 mM at pH 7.4[37]

NAME: **Methohexital**

FORMULA: X=O; $R_1=CH_3$; $R_2=CH_2CH{=}CH_2$; $R_3=CH(CH_3)C{\equiv}CC_2H_5$

MOL. WT.: 262

SOLUBILITY (mM): *Torpedo* Ringer, 20°C (pH 7): 0.6*

PARTITION COEFFICIENT:
Octanol/phosphate buffer (0.2 M): 3333 (acid form, 25°C)[72]

pK: 8.4[29]

ED_{50}: Tadpole: 0.02 mM at pH 7.4[37]

C_{50}: Postsynaptic *n*-AChR blockade: 0.10 mM at pH 7.4[37]

NAME: **Pentobarbital**

FORMULA: X=O; R_1=H; $R_2=C_2H_5$; $R_3=CH(CH_3)C_3H_7$

MOL. WT., g/mole: 226

M. P.; B.P., °C: 128.5–130 (free acid)[34]

SOLUBILITY (mM): *Torpedo* Ringer, 20°C (pH 7): 7*

PARTITION COEFFICIENT:
Octanol/buffer: 106, 25°C[33]
Octanol/water: 135, (?R.T.)[69]
Octanol/phosphate buffer (0.2 M): 168 (acid form, 25°C)[72]
PC/buffer: 200 (acid form, 25°C)[4]
PC+chol bilayer/water: 56, pH 7.0[12]
Erythrocyte ghost/buffer: 12.7 (acid form, 25°C)[33]; 9.6 (pH 7, 23°C)[80]

PARTIAL MOLAR VOLUME, ml, 25°C: (In ethanol): 185

pK: 8.07, μ 0.002; 7.96, μ 0.16[27]

ED_{50}: Newt: 0.32 mM at pH 7.1[17]
Tadpole: 0.16 mM at pH 7.4[37]

C_{50}: Nerve block: 1.7 mM (frog, 22°C)[40]
Postsynaptic *n*-AChR blockade: 0.44 mM at pH 7.4[37]

NAME: **Phenobarbital**

FORMULA: X=O; R_1=H; R_2=C_2H_5; R_3=cC_6H_5

MOL. WT., g/mole: 232

M.P., B.P., °C: 174–178[79]; —

SOLUBILITY (mM): *Torpedo* Ringer, 20°C (pH 7): 6*

PARTITION COEFFICIENT:
- Octanol/buffer: 25, 25°C[71]
- PC/buffer: 89 (pH 7.0, 25°C)[12]
- PC+chol/buffer: 59 (pH 7.0, 25°C)[12]
- Erythrocyte ghost/buffer: 9.5 (acid form, 25°C)[33]; 6.4 (pH 7, R.T.)[40]

pK: 7.4, μ 0.002; 7.24, μ 0.16[27]

ED_{50}: Tadpole: 3.3 mM at pH 7.4[37]

C_{50}: Nerve block: 5.7 mM (frog, 22°C)[40]
Postsynaptic *n*-AChR blockade: 4.6 mM at pH 7.4[37]

NAME: **Secobarbital**

FORMULA: X=O; R_1=H; R_2=CH_2CH=CH_2; R_3=$CH(CH_3)C_3H_7$

MOL. WT., g/mole: 238

SOLUBILITY (mM): *Torpedo* Ringer, 20°C (pH 7): 10*

PARTITION COEFFICIENT:
- Octanol/water: 389, (?R.T.)[69]
- Octanol/phospate buffer (0.2 M): 236 (acid form, 25°C)[72]

pK: 8.06, μ 0.002; 7.90, μ 0.16[27]

ED_{50}: Tadpole: 0.08 mM at pH 7.4[37]

C_{50}: Postsynaptic *n*-AChR blockade: 0.20 mM at pH 7.4[37]

NAME: **Thiopental**

FORMULA: X=S; R_1=H; R_2=C_2H_5; R_3=$CH(CH_3)C_3H_7$

MOL. WT.: 242

SOLUBILITY (mM): *Torpedo* Ringer, 20°C (pH 7): 0.6 mM*

PARTITION COEFFICIENT:
- Octanol/buffer: 390, (acid form, 25°C)[33]
- PC/buffer: 437 (pH 6.6, 25°C)[32]
- PC+chol/buffer: 215 (pH 6.6, 25°C)[32]
- Synaptosomal membrane/buffer: 32 (pH 6.6, 25°C)[32]
- Erythrocyte ghost/buffer: 24 (pH 6.6, 25°C)[32]

pK: 7.4[29]

ED_{50}: Tadpole: 0.03 mM at pH 7.4[37]

C_{50}: Postsynaptic *n*-AChR blockade: 0.24 mM at pH 7.4[37]

*B. A. Dodson, L. M. Braswell, and K. W. Miller, unpublished observations.

Contributors

G. Richard Arthur • Department of Anesthesia, Harvard Medical School and Brigham and Women's Hospital, Boston, Massachusetts 02115

Jeffery L. Barker • Laboratory of Neurophysiology, NINCDS, NIH, Bethesda, Maryland 20205

Bruce P. Bean • Department of Physiology and Biophysics, University of Iowa, Iowa City, Iowa 52242; *present address:* Department of Neurobiology, Harvard Medical School, Boston, Massachusctts 02115

Elliott Bedows • Departments of Anesthesiology and Epidemiology, University of Michigan, Ann Arbor, Michigan 48109

M. Lawrence Berman • Departments of Anesthesiology and Pharmacology, Vanderbilt University, Nashville, Tennessee 37232

Timothy J. Blaxter • Addiction Research Foundation Clinical Institute and Playfair Neuroscience Unit, Toronto Western Hospital, Departments of Medicine and Physiology, University of Toronto, Toronto, Ontario M5S 2S1, Canada

Frank Bowser-Riley • Oxford Hyperbaric Group, Physical Chemistry Laboratory, OX1 3QZ, England

Leon M. Braswell • Department of Anesthesia, Massachusetts General Hospital, Boston, Massachusetts 02114

Ralph W. Brauer • Institute of Marine Biomedical Research, University of North Carolina at Wilmington, Wilmington, North Carolina 28403

Stephen D. Brooks • Department of Anesthesiology, University of Arizona, Tucson, Arizona 85724

Robert Dale Brown • Division of Pharmacology, University of California, San Diego, La Jolla, California 92093

Keith W. Butler • Division of Biological Sciences, National Research Council of Canada, Ottawa K1A OR6, Canada

Andrea H. Callis • Department of Anesthesiology, University of Arizona, Tucson, Arizona 85724

Peter L. Carlen • Addiction Research Foundation Clinical Institute and Playfair Neuroscience Unit, Toronto Western Hospital, Departments of Medicine and Physiology, University of Toronto, Toronto, Ontario M5S 2S1, Canada

William A. Catterall • Department of Pharmacology, University of Washington, Seattle, Washington 98195

Craig W. Clarkson • Department of Pharmacology, Tulane Medical School, New Orleans, Louisiana 70112.

Charles J. Cohen • Miles Institute, New Haven, Connecticut 06509

Jonathan B. Cohen • Department of Anatomy and Neurobiology, Washington University School of Medicine, St. Louis, Missouri 63110

Peter J. Cohen • Departments of Anesthesiology and Epidemiology, University of Michigan, Ann Arbor, Michigan 48109; *present address:* Department of Anesthesia, University of Pennsylvania, Philadelphia, Pennsylvania 19104.

Leslay A. Correll • Department of Anatomy and Neurobiology, Washington University School of Medicine, St. Louis, Missouri 63110

Kenneth R. Courtney • Palo Alto Medical Foundation, Palo Alto, California 94301

Michael J. Cousins • Department of Anaesthesia and Intensive Care, Flinders Medical Centre, Adelaide, S.A. 5042, Australia

Benjamin G. Covino • Department of Anesthesia, Harvard Medical School and Brigham and Women's Hospital, Boston, Massachusetts 02115

Paul Culver • Division of Pharmacology, University of California, San Diego, La Jolla, California 92093

Stephen Daniels • Oxford Hyperbaric Group, Physical Chemistry Laboratory, Oxford OX1 3QZ, England

Pauline de la M. Hall • Department of Pathology, Flinders Medical Centre, Adelaide, S.A. 5042, Australia

Barbara A. Dodson • Department of Anesthesia, Massachusetts General Hospital, Boston, Massachusetts 02114

Evan B. Dreyer • Department of Anatomy and Neurobiology, Washington University School of Medicine, St. Louis, Missouri 63110

Thomas V. Dunwiddie • Department of Pharmacology, University of Colorado Health Sciences Center, Denver, Colorado 80262

Dominque Durand • Applied Neural Control Laboratory, Department of Biomedical Engineering, Case Western Reserve University, Cleveland, Ohio 44106

Julie Dutcher • Institute of Marine Biomedical Research, University of North Carolina at Wilmington, Wilmington, North Carolina 28403

James R. Elliott • Physiological Laboratory, University of Cambridge, Cambridge CB2 3EG, England, and the Laboratory of the Marine Biological Association, Plymouth PL1 2PB, England

Roger England • Departments of Anesthesiology and Pharmacology, Vanderbilt University, Nashville, Tennessee 37232

Hal S. Feldman • Department of Anesthesia, Harvard Medical School and Brigham and Women's Hospital, Boston, Massachusetts 02115

B. Raymond Fink • Department of Anesthesia, University of Washington, Seattle, Washington 98195

Leonard L. Firestone • Departments of Pharmacology and Anaesthesia, Harvard Medical School and Massachusetts General Hospital, Boston, Massachusetts 02114

James B. Fischer • Department of Pharmacology, School of Medicine, and Brain Research Institute, University of California, Los Angeles, California 90024

Stuart A. Forman • Department of Biophysics, Harvard University and Harvard Medical School, Boston, Massachusetts 02115

N. P. Franks • Biophysics Section, Blackett Laboratory, Imperial College of Science and Technology, London SW7 2BZ, England

Evan B. Friedman • Addiction Research Foundation Clinical Institute and Playfair Neuroscience Unit, Toronto Western Hospital, Departments of Medicine and Physiology, University of Toronto, Toronto, Ontario M5S 2S1, Canada

Peter W. Gage • Department of Physiology, John Curtin School of Medical Research, Australian National University, Canberra City A.C.T. 2601, Australia

A. Jay Gandolfi • Department of Anesthesiology, University of Arizona, Tucson, Arizona 85724

Simon Gelman • Department of Anesthesiology, The University of Alabama at Birmingham, Birmingham, Alabama 35294

Tohru Gonoi • Department of Pharmacology, University of Washington, Seattle, Washington 98195

Yoram Grossman • Department of Physiology, Faculty of Health Sciences, Ben-Gurion University of the Negev, Beer-Sheva 84 105, Israel

Denis A. Haydon • Physiological Laboratory, University of Cambridge, Cambridge CB2 3EG, England, and the Laboratory of the Marine Biological Association, Plymouth PL1 2PB, England

Bruce M. Hendry • Physiological Laboratory, University of Cambridge, Cambridge CB2 3EG, England, and the Laboratory of the Marine Biological Association, Plymouth PL1 2PB, England

Jeffrey Herz • Division of Pharmacology, University of California, San Diego, La Jolla, California 92093

Michael L. Hines • Department of Physiology, Duke University Medical Center, Durham, North Carolina 27710

William M. Hinson • Institute of Marine Biomedical Research, University of North Carolina at Wilmington, Wilmington, North Carolina 28403

Duncan A. Holaday • Departments of Anesthesiology and Pharmacology, Vanderbilt University, Nashville, Tennessee 37232

Luc M. Hondeghem • Department of Medicine and Pharmacology, Stahlman Cardiovascular Research Program, Vanderbilt University School of Medicine, Nashville, Tennessee 37232

Russell E. Jacobs • Department of Physiology and Biophysics, University of California at Irvine, Irvine, California 92717

Jørgen Johansen • Department of Neurology, Yale University School of Medicine, New Haven, Connecticut 06510

David A. Johnson • Division of Pharmacology, University of California, San Diego, La Jolla, California 92093; *present address:* Division of Biomedical Sciences, University of California, Riverside, California 92521

Joan J. Kendig • Department of Anesthesia, Stanford University School of Medicine, Stanford, California 94305

Glen I. King • Department of Physiology and Biophysics, University of California at Irvine, Irvine, California 92717

Anna L. Kleinhaus • Department of Neurology, Yale University School of Medicine, New Haven, Connecticut 06510

Paul R. Knight • Departments of Anesthesiology and Epidemiology, University of Michigan, Ann Arbor, Michigan 48109

Krešimir Krnjević • Departments of Anaesthesia Research and Physiology, McGill University, Montréal, Québec H3G 1Y6, Canada

Lee Kuhnert • Departments of Anesthesiology and Pharmacology, Vanderbilt University, Nashville, Tennessee 37232

Ilmar R. Kuisk • Department of Anatomy and Neurobiology, Washington University School of Medicine, St. Louis, Missouri 63110

W. R. Lieb • Biophysics Section, Blackett Laboratory, Imperial College of Science and Technology, London SW7 2BZ, England

David O. Lucas • Department of Anesthesiology, University of Arizona, Tucson, Arizona 85724

Christine A. Lunam • Department of Pathology, Flinders Medical Centre, Adelaide, S.A. 5042, Australia

Carl Lynch III • Department of Anesthesiology, University of Virginia Medical Center, Charlottesville, Virginia 22908

Robert L. Macdonald • Department of Neurology, University of Michigan, Ann Arbor, Michigan 48104

M. Bruce MacIver • Department of Pharmacology and Therapeutics, Faculty of Medicine, The University of Calgary, Calgary, Alberta T2N 4N1, Canada

William A. McCarthy, Jr. • Department of Pharmacology, Northwestern University Medical School, Chicago, Illinois 60611

Thomas J. McIntosh • Department of Anatomy, Duke University Medical Center, Durham, North Carolina 27710

David McKinnon • Department of Physiology, John Curtin School of Medical Research, Australian National University, Canberra City A.C.T. 2601, Australia

James G. McLarnon • Department of Pharmacology and Therapeutics, Faculty of Medicine, The University of British Columbia, Vancouver, British Columbia V6T 1W5, Canada

Daniel C. Medynski • Department of Anatomy and Neurobiology, Washington University School of Medicine, St. Louis, Missouri 63110

Janet C. Miller • Department of Anesthesia, Massachusetts General Hospital, Boston, Massachusetts 02114

Keith W. Miller • Departments of Pharmacology and Anaesthesia, Harvard Medical School and Massachusetts General Hospital, Boston, Massachusetts 02114

Mary E. Morris • Departments of Anesthesia and Pharmacology and the Playfair Neuroscience Unit, University of Toronto, Toronto, Ontario M5S 1A1, Canada

Richard W. Olsen • Department of Pharmacology, School of Medicine, and Brain Research Institute, University of California, Los Angeles, California 90024

David G. Owen • Laboratory of Neurophysiology, NINCDS, NIH, Bethesda, Maryland 20205

Peter Pennefather • Department of Neurobiology and Behavior, State University of New York, Stony Brook, New York 11794

James M. Phythyon • Departments of Anesthesiology and Pharmacology, Vanderbilt University, Nashville, Tennessee 37232

John L. Plummer • Department of Anaesthesia and Intensive Care, Flinders Medical Centre, Adelaide, S.A. 5042, Australia

Lance R. Pohl • Laboratory of Chemical Pharmacology, National Institutes of Health, Bethesda, Maryland 20205

Sidney W. Postma • Departments of Pharmacology and Anesthesiology, University of Washington, Seattle, Washington, 98195

E. Puil • Department of Anesthesia and Department of Pharmacology and Therapeutics, Faculty of Medicine, The University of British Columbia, Vancouver, British Columbia V6T 1W5, Canada

Fred N. Quandt • Department of Physiology, University of Calgary/Medical School, Calgary, Alberta T2N 4N1, Canada

David M. J. Quastel • Department of Pharmacology and Therapeutics, Faculty of Medicine, The University of British Columbia, Vancouver, British Columbia V6T 1W5, Canada

Santosh K. Rastogi • Department of Pharmacology, The University of Texas Health Science Center, San Antonio, Texas 78284

J. Murdoch Ritchie • Department of Pharmacology, Yale University School of Medicine, New Haven, Connecticut 06510

Brian Robertson • Department of Physiology, John Curtin School of Medical Research, Australian National University, Canberra City A.C.T. 2601, Australia

Sheldon H. Roth • Department of Pharmacology and Therapeutics, Faculty of Medicine, The University of Calgary, Calgary, Alberta T2N 4N1, Canada

Hiroko Satoh • Laboratory of Chemical Pharmacology, National Institutes of Health, Bethesda, Maryland 20205

Menachem Segal • Weizmann Institute, Rehovot, Israel

Sidney A. Simon • Departments of Physiology and Anesthesiology, Duke University Medical Center, Durham, North Carolina 27710

I. Glenn Sipes • Department of Anesthesiology, University of Arizona, Tucson, Arizona 85724

John H. Skerritt • Department of Neurology, University of Michigan, Ann Arbor, Michigan 48104

E. Brian Smith • Oxford Hyperbaric Group, Physical Chemistry Laboratory, Oxford OX1 3QZ, England

Ian C. P. Smith • Division of Biological Sciences, National Research Council of Canada, Ottawa, K1A OR6, Canada

Gary Strichartz • Anesthesia Research Laboratory, Brigham and Women's Hospital, and Department of Pharmacology, Harvard Medical School, Boston, Massachusetts 02115

Nina P. Strnad • Department of Anatomy and Neurobiology, Washington University School of Medicine, St. Louis, Missouri 63110

Robert E. Study • Laboratory of Neurophysiology, NINCDS, NIH, Bethesda, Maryland 20205

Kok-Swang Tan • Department of Pharmacology and Therapeutics, Faculty of Medicine, The University of Calgary, Calgary, Alberta T2N 4N1, Canada

Rosemarie C. Tan • Department of Physiology and Biophysics, University of Iowa, Iowa City, Iowa 52242

Palmer Taylor • Division of Pharmacology, University of California, San Diego, La Jolla, California 92093

Maharaj K. Ticku • Department of Pharmacology, The University of Texas Health Science Center, San Antonio, Texas 78284

Richard W. Tsien • Department of Physiology, Yale University, New Haven, Connecticut 06510

Bernd W. Urban • Departments of Anesthesiology and Physiology, Cornell University Medical College, New York, New York 10021

Ging Kuo Wang • Anesthesia Research Laboratory, Brigham and Women's Hospital, and Department of Pharmacology, Harvard Medical School, Boston, Massachusetts 02115

Stephen J. Waters • Department of Anesthesiology, University of Arizona, Tucson, Arizona 85724

Mary Ann Werz • Department of Neurology, University of Michigan, Ann Arbor, Michigan 48104

Stephen H. White • Department of Physiology and Biophysics, University of California at Irvine, Irvine, California 92717

Max Willow • Department of Pharmacology, University of Washington, Seattle, Washington 98195; *present address:* School of Medicine, McGill University, Montreal, Quebec H3G 1Y6, Canada

Daisuke Yamamoto • Mitsubishie-Kasei Institute of Life Sciences, Tokyo, Japan

Jay Yang • Department of Neurology, Yale University School of Medicine, New Haven, Connecticut 06510; *present address:* Department of Anatomy and Neurobiology, Washington University School of Medicine, St. Louis, Missouri 63110

Jay Z. Yeh • Department of Pharmacology, Northwestern University Medical School, Chicago, Illinois 60611

Index

GPSR Compliance

The European Union's (EU) General Product Safety Regulation (GPSR) is a set of rules that requires consumer products to be safe and our obligations to ensure this.

If you have any concerns about our products, you can contact us on ProductSafety@springernature.com

In case Publisher is established outside the EU, the EU authorized representative is:

Springer Nature Customer Service Center GmbH
Europaplatz 3
69115 Heidelberg, Germany

Batch number: 10692442

Printed by Printforce, the Netherlands